Universitext

Universitext

Universitext is a series of textbooks that presents material from a wide variety of mathematical disciplines at master's level and beyond. The books, often well class-tested by their author, may have an informal, personal even experimental approach to their subject matter. Some of the most successful and established books in the series have evolved through several editions, always following the evolution of teaching curricula, to very polished texts.

Thus as research topics trickle down into graduate-level teaching, first textbooks written for new, cutting-edge courses may make their way into *Universitext*.

More information about this series at http://www.springer.com/series/223

Gilles Pagès

Numerical Probability

An Introduction with Applications to Finance

Springer

Gilles Pagès
Laboratoire de Probabilités,
Statistique et Modélisation
Sorbonne Université
Paris
France

ISSN 0172-5939 ISSN 2191-6675 (electronic)
Universitext
ISBN 978-3-319-90274-6 ISBN 978-3-319-90276-0 (eBook)
https://doi.org/10.1007/978-3-319-90276-0

Library of Congress Control Number: 2018939129

Mathematics Subject Classification (2010): 65C05, 91G60, 65C30, 68U20, 60H35, 62L15, 60G40, 62L20, 91B25, 91G20

Printed on acid-free paper

This Springer imprint is published by the registered company Springer International Publishing AG part of Springer Nature
The registered company address is: Gewerbestrasse 11, 6330 Cham, Switzerland

To Julie, Romain and Nicolas

Preface

This book is an extended written version of the Master 2 course "Probabilités Numériques" (*i.e.,* Numerical Probability or Numerical Methods in Probability) which has been part of the Master program "Probability and Finance" since 2007. This Master program has been jointly run in Paris by Université Pierre et Marie Curie and École Polytechnique since its creation by Nicole El Karoui in 1990.

Our aim is to present different aspects of the Monte Carlo method and its various avatars (the quasi-Monte Carlo method, stochastic approximation, quantization-based cubature formulas, etc.), both on a theoretical level – most of the stated results are rigorously proved – as well as on a practical level through numerous examples arising in the pricing and hedging of derivatives.

This book is divided into two parts, one devoted to exact simulation, hence unbiased, and the other to approximate, hence biased, simulation. The first part is subsequently mostly devoted to general simulation methods (Chap. 1) and basic principles of Monte Carlo simulation (confidence interval in Chap. 2), sensitivity computation (still in Chap. 2), and variance reduction (in Chap. 3). The second part deals with approximate – and subsequently biased – simulation through the presentation and analysis of discretization schemes of Brownian diffusions (defined as solutions to Stochastic Differential Equations driven by Brownian motions), especially Euler–Maruyama and Milstein schemes (in Chap. 7). The analysis of their convergence in both the strong sense (L^p) and the weak sense (higher-order expansions of bias error in the scale of the scale of the discretization step) is presented in detail with their consequences on the Monte Carlo simulation of "vanilla" payoffs in local volatility models, *i.e.,* of functionals only depending on the diffusion at a fixed time T. As a second step (Chap. 8), this analysis is partially extended to path-dependent functionals relying on the diffusion bridge (or Brownian bridge) with their applications to the pricing of various classes of exotic options. The computation of sensitivities – known as Greeks in finance – is revisited in this more general biased setting in Chap. 10 (shock method, tangent flow, the log-likelihood method for Euler scheme, Malliavin–Monte Carlo).

Once these biased methods have been designed, it remains to devise a "bias chase." This is the objective of Chap. 9, entirely devoted to recent developments in the multilevel paradigm with or without weights, where we successively consider a "smooth" and "non-smooth" frameworks.

Beyond these questions directly connected to Monte Carlo simulation, three chapters deal with less classical aspects: Chap. 6 investigates recursive stochastic approximation "à la Robbins–Monro", viewed as a tool for implicitation or calibration but also for risk measure computation (quantiles, VaR). The quasi-Monte Carlo method, its theoretical foundation, and its limitations, are developed and analyzed in Chap. 4, supported by plenty of practical advice to practitioners (not only for the celebrated Sobol sequences). Finally, we dedicate a chapter (Chap. 5) to optimal vector quantization-based cubature formulas for the fast computation of expectations in medium dimensions (say up to 10) with possible applications to a universal stratified sampling variance reduction method.

In the – last – Chap. 12, we present a nonlinear problem (compared to expectation computation), optimal stopping theory, mostly in a Markovian framework. Time discretization schemes of the so-called Snell envelope are briefly presented, corresponding in finance of derivative products to the approximation of American style options by Bermudan options which can only be exercised at discrete time instants up to the maturity. Due to the nonlinear Backward Dynamic Programming Principle, such a discretization is not enough to devise implementable algorithms. A spatial discretization is necessary: To this end, we present a detailed analysis of regression methods ("à la Longstaff–Schwartz") and of the quantization tree approach, both developed during the decade 1990–2000.

Each chapter has been written in such a way that, as far as possible, an autonomous reading independent of former chapters is possible. However, some remain connected (in particular Chaps. 7 and 8) and the reading of Chap. 1, devoted to basic simulation, or the beginning of Chap. 2, which presents Monte Carlo simulation and the crucial notion of confidence interval, remains mandatory to benefit from the whole book. Within each chapter, we have purposefully scattered various exercises, in direct connection with their environment, either of a theoretical or applied nature (simulations).

We have added a large bibliography (more than 250 articles or books) with two objectives: first to refer to seminal, or at least important papers, on a topic developed in the book and, secondly, propose further readings which complete partially tackled topics in the book: For instance, in Chap. 12, most efforts are focused on numerical methods for pricing of American options. For the adaptation to the computation of δ-hedge or the extension to swing options (such as take-or-pay or swing option contracts), we only provide references in order to keep the size of the book within a reasonable limit.

The mathematical prerequisites to successfully approach the reading or the consultation of this book are, on the one hand, a solid foundation in the theory of Lebesgue integration (including σ-fields and measurability), Probability and

Measure Theory (see among others [44, 52, 154]). The reader is also assumed to be familiar with discrete- and continuous-time martingale theory with a focus on stochastic calculus, at least in a Brownian framework (see [183] for an introduction or [251], [162, 163] for more advanced textbooks). A basic background on discrete- and continuous-time Markov processes is also requested, but we will not go far beyond their very definitions in our proofs. As for programming ability, we leave to the reader the choice of his/her favorite language. Algorithms are always described in a meta-language. Of course, experts in high-performance computing and parallelized programming (multi-core, CUDA, etc.) are invited to take advantage of their expertise to convince themselves of the high compatibility of Monte Carlo simulation with these modern technologies.

Of course, we would like to mention that, though most of our examples are drawn or inspired by the pricing and hedging of derivatives products, the field of application of the presented methods is far larger. Thus, stratified sampling or importance sampling to the best of our knowledge are not often used by quants, structurers, and traders, but given their global importance in Monte Carlo simulation, we decided to present them in detail (with some clues on how to apply them in the finance of derivative products). By contrast, to limit the size of the manuscript, we renounced to present more advanced and important notions such as the discretization and simulation of Stochastic Differential Equations with jumps, typically driven by Lévy processes. However we hope that, having assimilated the Brownian case, our readers will be well prepared to successfully immerse themselves in the huge literature, from which we selected the most significant references for further reading.

The book contains more than 150 exercises, some theoretical, others devoted to simulation of numerical experiments. The exercises are distributed over the chapters, usually close to the theoretical results necessary to solve them. Several of them are accompanied by hints.

To conclude, I wish to thank those who preceded me in teaching this course at Université Pierre and Marie Curie: Bruno Bouchard, Benjamin Jourdain, and Bernard Lapeyre but also Damien Lamberton and Annie Millet who taught similar courses in other universities in Paris. Their approaches inspired me on several occasions. I would like to thank the students of the Master 2 "Probabilités and Finance" course since I began to teach it in the academic year 2007–08 who had access to the successive versions of the manuscript in its evolutionary phases: Many of their questions, criticisms, and suggestions helped improving it. I would like to express my special gratitude to one of them, Emilio Saïd, for his enthusiastic involvement in this task. I am grateful to my colleagues Fabienne Comte, Sylvain Corlay, Noufel Frikha, Daphné Giorgi, Damien Lamberton, Yating Liu, Sophie Laruelle, Vincent Lemaire, Harald Luschgy, Thibaut Montes, Fabien Panloup, Victor Reutenauer, and Abass Sagna, who volunteered to read carefully some chapters at different stages of drafting. Of course, all remaining errors are mine. I also have widely benefited from many illustrations and simulations provided by

Daphné Giorgi, Jean-Claude Fort, Vincent Lemaire, Jacques Printems, and Benedikt Wilbertz. Many of them are also long-term collaborators with whom I have shared exciting mathematical adventures during so many years. I thank them for that.

Paris, France
May 2017

Gilles Pagès

Contents

Notation

▷ General Notation

- $\mathbb{N} = \{0, 1, 2, \ldots\}$ and $\mathbb{N}^* = \{1, 2, \ldots\} = \mathbb{N}\backslash\{0\}$. $\mathbb{R}$ denotes the set of real numbers, $\mathbb{R}_+$ denotes the set of nonnegative real numbers, etc.
- The cardinality of a set A is denoted by $|A|$ or card(A) (in case of possible ambiguity with a norm).
- $\lfloor x \rfloor$ denotes the integer part of the real number x, *i.e.*, the largest integer not greater than x. $\{x\} = x - \lfloor x \rfloor$ denotes the fractional part of the real number x. $x_\pm = \max(\pm x, 0)$.
- The base of imaginary numbers in $\mathbb{C}$ is denoted by $\tilde{\imath}$ (with $\tilde{\imath}^2 = -1$).
- The notation $u \in \mathbb{K}^d$ denotes the *column* vector u of the vector space $\mathbb{K}^d$, $\mathbb{K} = \mathbb{R}$ or $\mathbb{C}$. A *row* vector will be denoted by u^* or u^t.
- $(u|v) = u^* v = \sum_{1 \le i \le d} u^i v^i$ denotes the canonical inner product of vectors $u = (u^1, \ldots, u^d)$ and $v = (v^1, \ldots, v^d)$ of $\mathbb{R}^d$. $|u| = \sqrt{(u|u)}$ denotes the canonical Euclidean norm on $\mathbb{R}^d$ derived from this inner product (unless stated otherwise). Sometimes, it will be specified as $|\cdot|_d$.
- A^* or A^t denotes the transpose of a matrix A, depending on the notational environment.
- $S(d, \mathbb{R})$ denotes the set of symmetric $d \times d$ matrices and $S^+(d, \mathbb{R})$ the set of positive semidefinite (symmetric) matrices. For $A \in S(d, \mathbb{R})$ and $u \in \mathbb{R}^d$, we write $Au^{\otimes 2} = u^* A u$.
- $\mathcal{M}(d, q, \mathbb{K})$ denotes a vector space of matrices with d rows and q columns with $\mathbb{K}$-valued entries. For $A \in \mathcal{M}(d, q, \mathbb{K})$, the Fröbenius norm of A, denoted by $\|A\|$, is defined by $\|A\| = \sqrt{A\bar{A}^*} = \sqrt{\sum_{i,j} |a_{ij}|^2}$ and the operator norm is denoted by $|||A|||$.
- $C_b(S, \mathbb{R}^d) := \big\{f : (S, \delta) \to \mathbb{R}^d, \text{ continuous and bounded}\big\}$, where (S, δ) denotes a metric space.

- $\mathbb{D}([0,T],\mathbb{R}^d) = \{f : [0,T] \to \mathbb{R}^d,\ \text{càdlàg}\}$ (càdlàg: a commonly used French acronym for "continu à droite, limité à gauche" which means "right continuous left limited").
- For a function $f : (\mathbb{R}^d, |\cdot|_d) \to (\mathbb{R}^p, |\cdot|_p)$, $[f]_{\text{Lip}} = \sup_{x \neq y} \frac{|f(x)-f(y)|_p}{|x-y|_d}$ and f is Lipschitz continuous with coefficient $[f]_{\text{Lip}}$ if $[f]_{\text{Lip}} < +\infty$.
- An assertion $\mathcal{P}(x)$ depending on a generic element x of a measure space (E, ε, μ) is true *μ-almost everywhere* (denoted *μ-a.e.*) if it is true outside a μ-negligible set of E.
- For a function $f : X \to (\mathbb{R}^d, |\cdot|)$, $\|f\|_{\text{sup}} = \sup_{x \in X} |f(x)|$ denotes the sup-norm.
- $C^n(X,F)$ denotes the set of n times continuously differentiable functions $f : X \to F$ ($X \subset E$, open subset of a Banach space E, and F also a Banach space). When $E = \mathbb{R}$, X may be any non-trivial interval of $\mathbb{R}$ with the usual conventions at the endpoints if contained in X.
- For any sequence $(a_n)_{n \geq 0}$ having values in vector space (or any additive structure), we define $\Delta a_n = a_n - a_{n-1}$, $n \geq 1$.
- $\overline{\lim}_n\, a_n$ and $\underline{\lim}_n\, a_n$ will denote the limsup and the liminf of the sequence $(a_n)_n$ respectively.
- For $(a_n)_{n \geq 1}$ and $(b_n)_{n \geq 1}$ two sequences of real numbers, $a_n \lesssim b_n$ if $\overline{\lim}_n \frac{a_n}{b_n} \leq 1$.

▷ Probability (and Integration) Theory Notation

- λ_d denotes the Lebesgue measure on $(\mathbb{R}^d, \mathcal{B}or(\mathbb{R}^d))$ where $\mathcal{B}or(\mathbb{R}^d)$ denotes the σ-field of Borel subsets of $\mathbb{R}^d$.
- Let $p \in (0, +\infty)$. $\mathcal{L}^p_{\mathbb{R}^d}(X, \mathcal{A}, \mu) = \{f : (X, \mathcal{A}) \to \mathbb{R}^d \ s.t. \int_X |f|^p d\mu < +\infty\}$, where μ is nonnegative measure (changing the canonical Euclidean norm $|\cdot|$ on $\mathbb{R}^d$ for another norm has no impact on this space but does have an impact on the value of the integral itself). $L^p_{\mathbb{R}^d}(X, \mathcal{A}, \mu)$ denotes the set of equivalence classes of $\mathcal{L}^p_{\mathbb{R}^d}(X, \mathcal{A}, \mu)$ with respect to the μ-*a.e.* equality. We write $\|f\|_p = \left[\int_X |f|^p d\mu\right]^{\frac{1}{p}}$.
- $\mathbb{P}_X$ or $\mathbb{P} \circ X^{-1}$ (or $\mathcal{L}(X)$) denotes the distribution (or law) on $(\mathbb{R}^d, \mathcal{B}or(\mathbb{R}^d))$ of a random vector $X : (\Omega, \mathcal{A}, \mathbb{P}) \to \mathbb{R}^d$ but also of an $\mathbb{R}^d$-valued stochastic process $X = (X_t)_{t \in [0,T]}$.
- Let Ω be a set. If $\mathcal{C} \subset \mathcal{P}(\Omega)$, $\sigma(\mathcal{C})$ denotes the σ-field spanned by $\mathcal{C}$ or, equivalently, the smallest σ-field on Ω which contains $\mathcal{C}$.
- The *cumulative distribution function* (cdf) of a random variable X, usually denoted F_X or Φ_X, is defined for every $x \in \mathbb{R}$, $F_X(x) = \mathbb{P}(X \leq x)$.

- χ_X will denote the characteristic function of an $\mathbb{R}^d$-valued random vector X. It is defined by $\chi_X(u) = \mathbb{E}\, e^{\tilde{\imath}(u\,|\,X)}$ (where $\tilde{\imath}^2 = -1$).
- $X \stackrel{d}{=} Y$ stands for equality in distribution of the random vectors (or processes) X and Y.
- $\mu_n \stackrel{(S)}{\Rightarrow} \mu$ denotes the weak convergence of μ_n toward μ, where μ_n and μ are probability measures on the metric space (S, δ) equipped with its Borel σ-field $\mathcal{B}or(S))$.
- $\stackrel{\mathcal{L}}{\to}$ denotes the convergence in distribution of random vectors (*i.e.*, the weak convergence of their distributions).
- The acronym i.i.d. stands for *independent identically distributed.*
- $\mathcal{N}(m; \sigma^2)$ denotes the Gaussian distribution with mean m and variance σ^2 on the real line. Φ_0 always denotes the cdf of the normal distribution $\mathcal{N}(0; 1)$. In higher dimensions, $\mathcal{N}(m; \Sigma)$ denotes the distribution of the Gaussian vector of mean m and covariance matrix Σ.

Chapter 1
Simulation of Random Variables

1.1 Pseudo-random Numbers

From a mathematical point of view, the definition of a sequence of (uniformly distributed) random numbers (over the unit interval $[0, 1]$) should be:

Definition. *A sequence* $(x_n)_{n \geq 1}$ *of* $[0, 1]$*-valued real numbers is a sequence of random numbers if there exists a probability space* $(\Omega, \mathcal{A}, \mathbb{P})$*, a sequence* U_n*,* $n \geq 1$*, of i.i.d. random variables with uniform distribution* $\mathcal{U}([0, 1])$ *and* $\omega \in \Omega$ *such that* $x_n = U_n(\omega)$ *for every* $n \geq 1$*.*

However, this naive and abstract definition is not satisfactory because the "scenario" $\omega \in \Omega$ may be not a "good" one, *i.e.* not a "generic"...? Many probabilistic properties (like the law of large numbers, to quote the most basic one) are only satisfied $\mathbb{P}$-$a.s$. Thus, if ω lies in the negligible set that does not satisfy one of them, the induced sequence will not be "admissible".

In any case, one usually cannot have access to an i.i.d. sequence of random variables (U_n) with distribution $\mathcal{U}([0, 1])$! Any physical device producing such a sequence would be too slow and unreliable. The works of by logicians like Martin-Löf lead to the conclusion that, roughly speaking, a sequence (x_n) that can be generated by an algorithm cannot be considered as "random" one. Thus the digits of the real number π are not random in that sense. This is quite embarrassing since an essential requested feature for such sequences is to be generated almost instantly on a computer!

The approach coming from the computer and algorithmic sciences is not much more tractable since their definition of a sequence of random numbers is that the complexity of the algorithm to generate the first n terms behaves like $O(n)$. The rapidly growing need for good (pseudo-)random sequences that came with the explosion of Monte Carlo simulation in many fields of Science and Technology after World War II (we refer not only to neutronics) led to the adoption of a more pragmatic approac – say heuristic – based on statistical tests. The idea is to submit some sequences to statistical tests (uniform distribution, block non-correlation, rank tests, etc).

G. Pagès, *Numerical Probability*, Universitext,
https://doi.org/10.1007/978-3-319-90276-0_1

For practical implementation, such sequences are finite, as the accuracy of computers is. One considers some sequences (x_n) of so-called *pseudo-random* numbers defined by

$$x_n = \frac{y_n}{N}, \quad y_n \in \{0, \dots, N-1\}.$$

One classical process is to generate the integers y_n by a congruential induction:

$$y_{n+1} \equiv a y_n + b \mod N$$

where $\gcd(a, N) = 1$, so that $\bar{a}$ (class of a modulo N) is invertible with respect to the multiplication (modulo N). Let $(\mathbb{Z}/N\mathbb{Z})^*$ denote the set of such invertible classes (modulo N). The multiplication of classes (*modulo* N) is an internal law on $(\mathbb{Z}/N\mathbb{Z})^*$ and $\big((\mathbb{Z}/N\mathbb{Z})^*, \times\big)$ is a commutative group for this operation. By the very definition of the Euler indicator function $\varphi(N)$ as the number of integers a in $\{0, \dots, N-1\}$ such that $\gcd(a, N) = 1$, the cardinality of $(\mathbb{Z}/N\mathbb{Z})^*$ is equal to $\varphi(N)$. Let us recall that the Euler function is multiplicative and given by the following closed formula

$$\varphi(N) = N \prod_{p|N,\ p \text{ prime}} \left(1 - \frac{1}{p}\right).$$

Thus $\varphi(p) = p - 1$ for every prime integer $p \in \mathbb{N}$ and $\varphi(p^k) = p^k - p^{k-1}$ (primary numbers), etc. In particular, if p is prime, then $(\mathbb{Z}/N\mathbb{Z})^* = (\mathbb{Z}/N\mathbb{Z}) \setminus \{0\}$.
If $b = 0$ (the most common case), one speaks of a *homogeneous generator*. We will focus on this type of generator in what follows.
▷ *Homogeneous congruent generators.* When $b = 0$, the period of the sequence (y_n) is given by the multiplicative order of a in $((\mathbb{Z}/N\mathbb{Z})^*, \times)$, *i.e.*

$$\tau_a := \min\big\{k \geq 1 : a^k \equiv 1 \bmod N\big\} = \min\big\{k \geq 1 : \bar{a}^k = 1\big\}.$$

Moreover, we know by Lagrange's Theorem that τ_a divides the cardinality $\varphi(N)$ of $(\mathbb{Z}/N\mathbb{Z})^*$.

For pseudo-random number simulation purposes, we search for pairs (N, a) such that the period τ_a of a in $\big((\mathbb{Z}/N\mathbb{Z})^*, \times\big)$ is very large. This needs an in-depth study of the multiplicative groups $\big((\mathbb{Z}/N\mathbb{Z})^*, \times\big)$, bearing in mind that N itself should be large to allow the element a to have a large period. This suggests to focus on prime or primary Numbers since, as seen above, in these cases $\varphi(N)$ is itself large.

In fact, the structure of these groups has long been understood. We summarize these results below.

Theorem 1.1 *Let $N = p^\alpha$, p prime, $\alpha \in \mathbb{N}^*$ be a primary number.*
(a) If $\alpha = 1$ (i.e. $N = p$ prime), then $\big((\mathbb{Z}/N\,\mathbb{Z})^*, \times\big)$ (whose cardinality is $p - 1$) is a cyclic group. This means that there exists an $a \in \{1, \dots, p-1\}$ s.t. $(\mathbb{Z}/p\,\mathbb{Z})^* = \langle \overline{a} \rangle$. Hence the maximal period is $\tau = p - 1$.

(b) If $p = 2$, $\alpha \geq 3$, $(\mathbb{Z}/N\mathbb{Z})^*$ (whose cardinality is $2^{\alpha-1} = N/2$) is not cyclic. The maximal period is then $\tau = 2^{\alpha-2}$ with $a \equiv \pm 3$ mod 8. (If $\alpha = 2$ ($N = 4$), the group of size 2 is trivially cyclic!)
(c) If $p \neq 2$, then $(\mathbb{Z}/N\mathbb{Z})^*$ (whose cardinality is $p^{\alpha-1}(p-1)$) is cyclic, hence $\tau = p^{\alpha-1}(p-1)$. It is generated by any element a whose class $\widetilde{a}$ in $(\mathbb{Z}/p\,\mathbb{Z})$ spans the cyclic group $\big((\mathbb{Z}/p\mathbb{Z})^*, \times\big)$.

What does this theorem say in connection with our pseudo-random number generation problem?

First some very good news: *when N is a prime number*, the group $\big((\mathbb{Z}/N\mathbb{Z})^*, \times\big)$ is *cyclic*, *i.e.* there exists an $a \in \{1, \ldots, N\}$ such that $(\mathbb{Z}/N\mathbb{Z})^* = \big\{\bar{a}^n,\ 1 \leq n \leq N-1\big\}$. The bad news is that *we do not know which a satisfies this property*, not all do and, even worse, we do not know how to find any. Thus, if $p = 7$, $\varphi(7) = 7 - 1 = 6$ and $o(3) = o(5) = 6$ but $o(2) = o(4) = 3$ (which divides 6) and $o(6) = 2$ (which again divides 6).

The second bad news is that the length of a sequence, though a necessary property of a sequence $(y_n)_n$, provides no guarantee or even clue that $(x_n)_n$ is a *good* sequence of pseudo-random numbers! Thus, the (homogeneous) generator of the FORTRAN IMSL library does not fit into the formerly described setting: one sets $N := 2^{31} - 1 = 2\,147\,483\,647$ (which is a Mersenne prime, see below) discovered by Leonhard Euler in 1772), $a := 7^5$, and $b := 0$ ($a \not\equiv 0$ mod 8), the point being that the period of 7^5 is not maximal.

Another approach to random number generation is based on a shift register and relies upon the theory of finite fields.

At this stage, a sequence must pass successfully through various statistical tests, keeping in mind that such a sequence is finite by construction and consequently cannot satisfy asymptotically such common properties as the Law of Large Numbers, the Central Limit Theorem or the Law of the Iterated Logarithm (see the next chapter). Dedicated statistical toolboxes like *DIEHARD* (Marsaglia, 1998) have been devised to test and "certify" sequences of pseudo-random numbers.

The aim of this introductory section is just to give the reader the flavor of pseudo-random number generation, but in no case we do recommend the specific use of any of the above generators or discuss the virtues of such or such a generator.

For more recent developments on random numbers generator (like a shift register, for example, we refer *e.g.* to [77, 219]). Nevertheless, let us mention the Mersenne twister generators. This family of generator has been introduced in 1997 by Makoto Matsumata and Takuji Nishimura in [211]. The first level of Mersenne Twister Generators (denoted by MT-p) are congruential generators whose period N_p is a Mersenne prime, *i.e.* an prime number of the form $N_p = 2^p - 1$ where p is itself prime. The most popular and now worldwide implemented generator is the MT-19937, owing to its unrivaled period $2^{19937} - 1 \simeq 10^{6000}$ since $2^{10} \simeq 10^3$). It can simulate a uniform distribution in 623 dimensions (*i.e.* on $[0, 1]^{623}$). A second "shuffling" device improves it further. For recent improvements and their implementations in C++, see

`www.math.sci.hiroshima-u.ac.jp/~m-mat/MT/emt.html`

Recently, new developments in massively parallel computing have drawn the attention back to pseudo-random number simulation. In particular, consider the GPU-based intensive computations which use the graphics device of a computer as a computing unit which may run hundreds of parallel computations. One can imagine that each pixel is a (virtual) small computing unit achieving a small chain of elementary computations (a thread). What is really new is that access to such intensive parallel computing becomes cheap, although it requires some specific programming language (like CUDA on Nvidia GPU). As concerns its use for intensively parallel Monte Carlo simulation, some new questions arise: in particular, the ability to generate independently (in parallel!) many "independent" sequences of pseudo-random numbers, since the computing units of a GPU never "speak" to each other or to anybody else while running: each pixel is a separate (virtual) thread. The Wichmann–Hill pseudo-random number generator (which is in fact a family of 273 different methods) seems to be a good candidate for Monte Carlo simulation on a GPU. For more insight on this topic, we refer to [268] and the references therein.

1.2 The Fundamental Principle of Simulation

Theorem 1.2 (Fundamental theorem of simulation) *Let (E, d_E) be a Polish space (complete and separable) and let $X : (\Omega, \mathcal{A}, \mathbb{P}) \to (E, \mathcal{B}or_{d_E}(E))$ be a random variable with distribution $\mathbb{P}_X$. Then there exists a Borel function $\varphi : ([0,1], \mathcal{B}([0,1]), \lambda_{[0,1]}) \to (E, \mathcal{B}or_{d_E}(E), \mathbb{P}_X)$ such that*

$$\mathbb{P}_X = \lambda_{[0,1]} \circ \varphi^{-1},$$

where $\lambda_{[0,1]} \circ \varphi^{-1}$ denotes the image measure by φ of the Lebesgue measure $\lambda_{[0,1]}$ on the unit interval.

We will admit this theorem. For a proof, we refer to [48] (Theorem A.3.1, p. 38). It also appears as a "building block" in the proof of the Skorokhod representation theorem for weakly converging sequences of random variables having values in a Polish space.

As a consequence this means that, if U denotes any random variable with uniform distribution on $(0, 1)$ defined on a probability space $(\Omega, \mathcal{A}, \mathbb{P})$, then

$$X \stackrel{d}{=} \varphi(U).$$

The interpretation is that any E-valued random variable can be simulated from a uniform distribution…provided the function φ can be computed (at a reasonable computational cost, *i.e.* in algorithmic terms with reasonable complexity). If this is the case, the *yield* of the simulation is 1 since every (pseudo-)random number $u \in [0, 1]$ produces a $\mathbb{P}_X$-distributed random number. Except in very special situations (see below), this result turns out to be of purely theoretical interest and is of little help for practical simulation. However, the fundamental theorem of simulation is important from a theoretical point view in Probability Theory since, as mentioned above, it provides a fundamental step in the proof of the Skorokhod representation theorem (see *e.g* [45], Chap. 5).

In the three sections below we provide a short background on the most classical simulation methods (inversion of the distribution function, the acceptance-rejection method, Box–Muller for Gaussian vectors). This is, of course, far from being exhaustive. For an overview of the different aspects of simulation of non-uniform random variables or vectors, we refer to [77]. In fact, many results from Probability Theory can give rise to simulation methods.

1.3 The (Inverse) Distribution Function Method

Let μ be a probability distribution on $(\mathbb{R}, \mathcal{B}or(\mathbb{R}))$ with distribution function F defined for every $x \in \mathbb{R}$ by

$$F(x) := \mu\big((-\infty, x]\big).$$

The function F is always non-decreasing, "càdlàg"(a French acronym for "right continuous with left limits" which we shall adopt in what follows) and $\lim_{x\to+\infty} F(x) = 1$, $\lim_{x\to-\infty} F(x) = 0$.

One can always associate to F its canonical left inverse F_l^{-1} defined on the open unit interval $(0, 1)$ by

$$\forall u \in (0,1),\quad F_l^{-1}(u) = \inf\big\{x \in \mathbb{R} : F(x) \geq u\big\}.$$

One easily checks that F_l^{-1} is non-decreasing, left-continuous and satisfies

$$\forall\, u \in (0,1), \qquad F_l^{-1}(u) \leq x \Longleftrightarrow u \leq F(x).$$

Proposition 1.1 *If* $U \stackrel{d}{=} U((0,1))$, *then* $X := F_l^{-1}(U) \stackrel{d}{=} \mu$.

Proof. Let $x \in \mathbb{R}$. It follows from the preceding that

$$\{X \leq x\} = \{F_l^{-1}(U) \leq x\} = \{U \leq F(x)\}$$

so that $\mathbb{P}(X \leq x) = \mathbb{P}(F_l^{-1}(U) \leq x) = \mathbb{P}(U \leq F(X)) = F(x)$. ◇

Remarks. • If F is increasing and continuous on the real line, then F has an inverse function denoted by F^{-1} defined $(0, 1)$ (also increasing and continuous) such that $F \circ F^{-1} = \mathrm{Id}_{|(0,1)}$ and $F^{-1} \circ F = \mathrm{Id}_{\mathbb{R}}$. Clearly $F^{-1} = F_l^{-1}$ by the very definition of F_l^{-1}. The above proof can be made even more straightforward since the event $\{F^{-1}(U) \leq x\} = \{U \leq F(x)\}$ by simple left composition of F^{-1} by the increasing function F.

• If μ has a probability density f such that $\{f = 0\}$ has an empty interior, then $F(x) = \int_{-\infty}^{x} f(u)du$ is continuous and increasing.
• One can replace $\mathbb{R}$ by any interval $[a, b] \subset \mathbb{R}$ or $\overline{\mathbb{R}}$ (with obvious conventions).
• One could also have considered the *right continuous canonical inverse* F_r^{-1} by:

$$\forall u \in (0,1), \quad F_r^{-1}(u) = \inf\big\{x : F(x) > u\big\}.$$

One shows that F_r^{-1} is non-decreasing, right continuous and that

$$F_r^{-1}(u) \le x \Longrightarrow u \le F(x) \quad \text{and} \quad u < F(x) \Longrightarrow F_r^{-1}(u) \le x.$$

Hence $F_r^{-1}(U) \stackrel{d}{=} X$ since

$$\mathbb{P}(F_r^{-1}(U) \le x) \left\{ \begin{array}{l} \le \mathbb{P}(F(x) \ge U) = F(x) \\ \ge \mathbb{P}(F(x) > U) = F(x) \end{array} \right\} = \mathbb{P}(X \le x) = F(x).$$

When X takes finitely many values in $\mathbb{R}$, we will see in Example 4 below that this simulation method corresponds to the standard simulation method of such random variables.

▶ **Exercise.** (a) Show that, for every $u \in (0,1)$,

$$F\big(F_l^{-1}(u)_-\big) \le u \le F \circ F_l^{-1}(u),$$

so that if F is continuous (or equivalently μ has no atom: $\mu(\{x\}) = 0$ for every $x \in \mathbb{R}$), then $F \circ F_l^{-1} = Id_{(0,1)}$

(b) Show that if F is continuous, then $F(X) \stackrel{d}{=} \mathcal{U}([0,1])$.
(c) Show that if F is (strictly) increasing, then F_l^{-1} is continuous and $F_l^{-1} \circ F = \mathrm{Id}_{\mathbb{R}}$.
(d) One defines the survival function of μ by $\bar{F}(x) = 1 - F(x) = \mu\big((x, +\infty)\big)$, $x \in \mathbb{R}$. One defines the canonical *right* inverse of $\bar{F}$ by

$$\forall u \in (0,1), \quad \bar{F}_r^{-1}(u) = \inf\big\{x : \bar{F}(x) \le u\big\}.$$

Show that $\bar{F}_r^{-1}(u) = F_l^{-1}(1-u)$. Deduce that $\bar{F}_r^{-1}$ is right continuous on $(0,1)$ and that $\bar{F}_r^{-1}(U)$ has distribution μ. Define $\bar{F}_l^{-1}$ and show that $\bar{F}_r^{-1}(U)$ has distribution μ. Finally, establish for $\bar{F}_r^{-1}$ similar properties to (a)–(b)–(c).

(Informal) Definition. *The yield (often denoted by r) of a simulation procedure is defined as the inverse of the number of pseudo-random numbers used to generate one $\mathbb{P}_X$-distributed random number.*

One must keep in mind that the yield is attached to a simulation method, not to a probability distribution (the fundamental theorem of simulation always provides a simulation method with yield 1, except that it is usually not tractable).

Typically, if $X = \varphi(U_1, \dots, U_m)$, where φ is a Borel function defined on $[0, 1]^m$ and $U_1, \dots, U_m$ are independent and uniformly distributed over $[0, 1]$, the yield of this φ-based procedure to simulate the distribution of X is $r = 1/m$.

Thus, the yield r of the (inverse) distribution function is consequently always equal to $r = 1$.

$\triangleright$ **Examples. 1.** *Simulation of an exponential distribution. Let* $X \stackrel{d}{=} \mathcal{E}(\lambda),\ \lambda > 0$. Then

$$\forall\, x \in (0, +\infty), \qquad F_X(x) = \lambda \int_0^x e^{-\lambda\xi} d\xi = 1 - e^{-\lambda x}.$$

Consequently, for every $u \in (0, 1)$, $F_X^{-1}(u) = -\log(1 - u)/\lambda$. Now, using that $U \stackrel{d}{=} 1 - U$ if $U \stackrel{d}{=} U((0, 1))$ yields

$$X = -\frac{\log(U)}{\lambda} \stackrel{d}{=} \mathcal{E}(\lambda).$$

2. *Simulation of a Cauchy(c), $c > 0$, distribution.* We know that $\mathbb{P}_X(dx) = \dfrac{c}{\pi(x^2 + c^2)} dx$.

$$\forall x \in \mathbb{R}, \qquad F_X(x) = \int_{-\infty}^x \frac{c}{u^2 + c^2} \frac{du}{\pi} = \frac{1}{\pi}\left(\text{Arctan}\left(\frac{x}{c}\right) + \frac{\pi}{2}\right).$$

Hence $F_X^{-1}(x) = c \tan\big(\pi(u - 1/2)\big)$. It follows that

$$X = c \tan\big(\pi(U - 1/2)\big) \stackrel{d}{=} \text{Cauchy}(c).$$

3. *Simulation of a Pareto* (θ), $\theta > 0$, *distribution.* This distribution reads $\mathbb{P}_X(dx) = \dfrac{\theta}{x^{1+\theta}} \mathbf{1}_{\{x \ge 1\}} dx$. Its distribution function $F_X(x) = 1 - x^{-\theta}$ so that, still using $U \stackrel{d}{=} 1 - U$ if $U \stackrel{d}{=} U\big((0, 1)\big)$,

$$X = U^{-\frac{1}{\theta}} \stackrel{d}{=} \text{Pareto}(\theta).$$

4. *Simulation of a distribution supported by a finite set E.* Let $E := \{x_1, \dots, x_N\}$ be a subset of $\mathbb{R}$ indexed so that $i \mapsto x_i$ is increasing. Let $X : (\Omega, \mathcal{A}, \mathbb{P}) \to E$ be an E-valued random variable with distribution $\mathbb{P}(X = x_k) = p_k$, $1 \le k \le N$, where $p_k \in [0, 1]$, $p_1 + \dots + p_N = 1$. Then, one checks that its distribution function F_X reads

$$\forall\, x \in \mathbb{R}, \quad F_X(x) = p_1 + \dots + p_i \ \text{ if } x_i \le x < x_{i+1}$$

with the convention $x_0 = -\infty$ and $x_{N+1} = +\infty$. As a consequence, its left continuous canonical inverse is given by

$$\forall u \in (0,1), \qquad F_{X,l}^{-1}(u) = \sum_{k=1}^{N} x_k \mathbf{1}_{\{p_1+\cdots+p_{k-1} < u \le p_1+\cdots+p_k\}}$$

so that

$$X \stackrel{d}{=} \sum_{k=1}^{N} x_k \mathbf{1}_{\{p_1+\cdots+p_{k-1} < U \le p_1+\cdots+p_k\}}.$$

The yield of the procedure is still $r = 1$. However, when implemented naively, its complexity – which corresponds to (at most) N comparisons for every simulation – may be quite high. See [77] for some considerations (in the spirit of quick sort algorithms) which lead to a $O(\log N)$ complexity. Furthermore, this procedure underlines that one has access to the probability weights p_k with an arbitrary accuracy. This is not always the case, even in *a priori simple* situations, as emphasized in Example 6 below.

It should be observed of course that the above simulation formula is still appropriate for a random variable taking values in *any* set E, not only for subsets of $\mathbb{R}$!

5. *Simulation of a Bernoulli random variable* $B(p)$, $p \in (0,1)$. This is the simplest application of the previous method since

$$X = \mathbf{1}_{\{U \le p\}} \stackrel{d}{=} B(p),$$

6. *Simulation of a Binomial random variable* $B(n,p)$, $p \in (0,1)$, $n \ge 1$. One relies on the very definition of the binomial distribution as the law of the sum of n independent $B(p)$-distributed random variables, *i.e.*

$$X = \sum_{k=1}^{n} \mathbf{1}_{\{U_k \le p\}} \stackrel{d}{=} B(n,p).$$

where $U_1, \ldots, U_n$ are i.i.d. random variables, uniformly distributed over $[0,1]$.

Note that this procedure has a very bad yield, namely $r = \frac{1}{n}$. Moreover, it needs n comparisons like the standard method (without any shortcut).

Why not use the above standard method for a random variable taking finitely many values? Because the cost of the computation of the probability weights p_k is much too high as n grows.

7. *Simulation of geometric random variables* $G(p)$ *and* $G^*(p)$, $p \in (0,1)$. This is the distribution of the first success instant when independently repeating the same Bernoulli experiment with parameter p. Conventionally, $G(p)$ starts at time 0 whereas $G^*(p)$ starts at time 1.

To be precise, if $(X_k)_{k\geq 0}$ denotes an i.i.d. sequence of random variables with distribution $B(p)$, $p \in (0,1)$, then

$$\tau^* := \min\{k \geq 1 : X_k = 1\} \stackrel{d}{=} G^*(p)$$

and

$$\tau := \min\{k \geq 0 \,:\, X_k = 1\} \stackrel{d}{=} G(p).$$

Hence

$$\mathbb{P}(\tau^* = k) = p(1-p)^{k-1},\;\; k \in \mathbb{N}^* := \{1, 2, \ldots, n, \ldots\}$$

and

$$\mathbb{P}(\tau = k) = p(1-p)^{k},\;\; k \in \mathbb{N} := \{0, 1, 2, \ldots, n, \ldots\}$$

(so that both random variables are $\mathbb{P}$-$a.s.$ finite since one has $\sum_{k\geq 0}\mathbb{P}(\tau = k) = \sum_{k\geq 1}\mathbb{P}(\tau^* = k) = 1$). Note that $\tau + 1$ has the same $G^*(p)$-distribution as τ^*.

The (random) yields of the above two procedures are $r^* = \frac{1}{\tau^*}$ and $r = \frac{1}{\tau+1}$, respectively. Their common mean (*average yield*) $\bar{r} = \bar{r}^*$ is given by

$$\begin{aligned}
\mathbb{E}\left(\frac{1}{\tau+1}\right) = \mathbb{E}\left(\frac{1}{\tau^*}\right) &= \sum_{k\geq 1}\frac{1}{k}p(1-p)^{k-1} \\
&= \frac{p}{1-p}\sum_{k\geq 1}\frac{1}{k}(1-p)^k \\
&= -\frac{p}{1-p}\log(1-x)_{|x=1-p} \\
&= -\frac{p}{1-p}\log(p).
\end{aligned}$$

► **Exercises. 1.** (*Conditional distributions*). Let $X : (\Omega, \mathcal{A}, \mathbb{P}) \to (\mathbb{R}, \mathcal{B}or(\mathbb{R}))$ be a real-valued random variable with distribution function F and left continuous canonical inverse F_l^{-1}. Let $I = [a, b]$, $-\infty \leq a < b \leq +\infty$, be a nontrivial interval of $\mathbb{R}$. Show that, if $U \stackrel{d}{=} \mathcal{U}([0, 1])$, then

$$F_l^{-1}\big(F(a) + (F(b) - F(a))U\big) \stackrel{d}{=} \mathcal{L}(X \mid X \in I).$$

2. *Negative binomial distributions*. The negative binomial distribution with parameter (n, p) is the law $\mu_{n,p}$ of the n-th success in an infinite sequence of independent Bernoulli trials, namely, using the above notations or the geometric distributions, the distribution of

$$\tau^{(n)} = \min\big\{k \geq 1 : \mathrm{card}\{1 \leq \ell \leq k : X_\ell = 1\} = n\big\}.$$

Show that

$$\mu_{n,p}(k) = 0,\ k \leq n-1, \quad \mu_{n,p}(k) = C_{k-1}^{n-1} p^n (1-p)^{k-n},\ k \geq n.$$

Compute the mean yield of its (natural and straightforward) simulation method.

1.4 The Acceptance-Rejection Method

This method is due to Von Neumann (1951). It is contemporary with the development of computers and of the Monte Carlo method. The original motivation was to devise a simulation method for a probability distribution ν on a measurable space $(E, \mathcal{E})$, absolutely continuous with respect to a σ-finite non-negative measure μ with a density given, *up to a constant*, by $f : (E, \mathcal{E}) \to \mathbb{R}_+$ when we know that f is *dominated* by a probability distribution $g \cdot \mu$ which can be simulated at "low computational cost". (Note that $\nu = \frac{f}{\int_E f d\mu} \cdot \mu$.)

In most elementary applications (see below), E is either a Borel set of $\mathbb{R}^d$ equipped with its Borel σ-field and μ is the trace of the Lebesgue measure on E or a subset $E \subset \mathbb{Z}^d$ equipped with the counting measure.

Let us be more precise. So, let μ be a non-negative σ-finite measure on $(E, \mathcal{E})$ and let $f, g : (E, \mathcal{E}) \longrightarrow \mathbb{R}_+$ be two Borel functions. Assume that $f \in L^1_{\mathbb{R}_+}(\mu)$ with $\int_E f\, d\mu > 0$ and that g is a *probability density* with respect to μ satisfying furthermore $g > 0$ μ-$a.s.$ and there exists a positive real constant $c > 0$ such that

$$f(x) \leq c\, g(x) \quad \mu(dx)\text{-}a.e.$$

Note that this implies $c \geq \displaystyle\int_E f d\mu$. As mentioned above, the aim of this section is to show how to simulate some random numbers distributed according to the probability distribution

$$\nu = \frac{f}{\int_{\mathbb{R}^d} f\, d\mu} \cdot \mu$$

using some $g \cdot \mu$-distributed (pseudo-)random numbers. In particular, to make the problem consistent, we will assume that $\nu \neq \mu$, which in turn implies that

$$c > \int_{\mathbb{R}^d} f d\mu.$$

The underlying requirements on f, g and μ to undertake a practical implementation of the method described below are the following:

- the numerical value of the real constant c is known;

- we know how to simulate (at a reasonable cost) on a computer a sequence of i.i.d. random vectors $(Y_k)_{k\geq 1}$ with the distribution $g\cdot\mu$
- we can compute on a computer the ratio $\frac{f}{g}(x)$ at every $x\in\mathbb{R}^d$ (again at a reasonable cost).

As a first (not so) preliminary step, we will explore a natural connection (in distribution) between an E-valued random variable X with distribution ν and Y an E-valued random variable with distribution $g\cdot\mu$. We will see that the key idea is completely elementary. Let $h:(E,\mathcal{E})\to\mathbb{R}$ be a test function (measurable and bounded or non-negative). On the one hand

$$\begin{aligned}
\mathbb{E}\,h(X) &= \frac{1}{\int_E f d\mu}\int_E h(x)f(x)\mu(dx)\\
&= \frac{1}{\int_E f d\mu}\int_E h(y)\frac{f}{g}(y)g(y)\mu(dy) \text{ since } g>0\ \mu\text{–a.e.}\\
&= \frac{1}{\int_E f d\mu}\mathbb{E}\left(h(Y)\frac{f}{g}(Y)\right).
\end{aligned}$$

We can also forget about the last line, stay on the state space E and note in a somewhat artificial way that

$$\begin{aligned}
\mathbb{E}\,h(X) &= \frac{c}{\int_E f d\mu}\int_E h(y)\left(\int_0^1 \mathbf{1}_{\{u\leq\frac{1}{c}\frac{f}{g}(y)\}}du\right)g(y)\mu(dy)\\
&= \frac{c}{\int_E f d\mu}\int_E\int_0^1 h(y)\mathbf{1}_{\{u\leq\frac{1}{c}\frac{f}{g}(y)\}}g(y)\mu(dy)du\\
&= \frac{c}{\int_E f d\mu}\mathbb{E}\left(h(Y)\mathbf{1}_{\{cU\leq\frac{f}{g}(Y)\}}\right)
\end{aligned}$$

where U is uniformly distributed over $[0,1]$ and *independent* of Y.

By considering $h\equiv 1$, we derive from the above identity that $\frac{c}{\int_E f d\mu} = \frac{1}{\mathbb{P}\left(cU\leq\frac{f}{g}(Y)\right)}$ so that finally

$$\mathbb{E}\,h(X) = \mathbb{E}\left(h(Y)\,\big|\,\{cU\leq\frac{f}{g}(Y))\}\right).$$

The proposition below takes advantage of this identity in distribution to propose a simulation procedure. In fact, it is simply a reverse way to make (and interpret) the above computations.

Proposition 1.2 (Acceptance-rejection simulation method) *Let* $(U_n, Y_n)_{n\geq 1}$ *be a sequence of i.i.d. random variables with distribution* $\mathcal{U}([0,1]) \otimes \mathbb{P}_Y$ *(independent marginals) defined on* $(\Omega, \mathcal{A}, \mathbb{P})$ *where* $\mathbb{P}_Y(dy) = g(y)\mu(dy)$ *is the distribution of* Y *on* $(E, \mathcal{E})$. *Set*

$$\tau := \min\big\{k \geq 1 : c\, U_k g(Y_k) \leq f(Y_k)\big\}.$$

Then, τ *has a geometric distribution* $G^*(p)$ *with parameter given by* $p := \mathbb{P}\big(c\, U_1 g(Y_1) \leq f(Y_1)\big) = \dfrac{\int_E f\, d\mu}{c}$ *and*

$$X := Y_\tau \stackrel{d}{=} \nu.$$

Remark. The (random) *yield* of the method is $\frac{1}{\tau}$. Hence, we know that its mean yield is given by

$$\mathbb{E}\Big(\frac{1}{\tau}\Big) = -\frac{p\log p}{1-p} = \frac{\int_E f\, d\mu}{c - \int_E f\, d\mu}\log\left(\frac{c}{\int_E f\, d\mu}\right).$$

Since $\lim\limits_{p\to 1} -\dfrac{p\log p}{1-p} = 1$, the closer to $\displaystyle\int_E f\, d\mu$ the constant c is, the higher the yield of the simulation.

Proof. *Step 1*. Let (U, Y) be a pair of random variables with distribution $\mathcal{U}([0,1]) \otimes \mathbb{P}_Y$. Let $h : \mathbb{R}^d \to \mathbb{R}$ be a bounded Borel test function. We have

$$\begin{aligned}
\mathbb{E}\left(h(Y)\mathbf{1}_{\{c\, U g(Y) \leq f(Y)\}}\right) &= \int_{E\times[0,1]} h(y)\mathbf{1}_{\{c\, u g(y) \leq f(y)\}} g(y)\mu(dy)\otimes du \\
&= \int_E \left[\int_{[0,1]} \mathbf{1}_{\{c\, u g(y) \leq f(y)\}\cap\{g(y)>0\}} du\right] h(y)g(y)\mu(dy) \\
&= \int_E h(y)\left[\int_0^1 \mathbf{1}_{\{u\leq \frac{f(y)}{cg(y)}\}\cap\{g(y)>0\}} du\right] g(y)\mu(dy) \\
&= \int_{\{g(y)>0\}} h(y)\frac{f(y)}{cg(y)} g(y)\mu(dy) \\
&= \frac{1}{c}\int_E h(y)f(y)\mu(dy),
\end{aligned}$$

where we used successively Fubini's Theorem, $g(y) > 0$ $\mu(dy)$-*a.e.*, Fubini's Theorem again and $\frac{f}{cg}(y) \leq 1$ $\mu(dy)$-*a.e.* Note that we can apply Fubini's Theorem since the reference measure μ is σ-finite.

Putting $h \equiv 1$ yields

$$c = \frac{\int_E f d\mu}{\mathbb{P}(c\, U g(Y) \leq f(Y))}.$$

Then, elementary conditioning yields

$$\mathbb{E}\left(h(Y)|\{c\,Ug(Y) \leq f(Y)\}\right) = \int_E h(y)\frac{f(y)}{\int_E f\,d\mu}\mu(dy) = \int_E h(y)\nu(dy),$$

i.e.

$$\mathcal{L}\left(Y|\{c\,Ug(Y) \leq f(Y)\}\right) = \nu.$$

Step 2. Then (using that τ is $\mathbb{P}$-$a.s.$ finite)

$$\begin{aligned}
\mathbb{E}\left(h(Y_\tau)\right) &= \sum_{n\geq 1}\mathbb{E}\left(\mathbf{1}_{\{\tau=n\}}h(Y_n)\right)\\
&= \sum_{n\geq 1}\mathbb{P}\big(\{c\,U_1 g(Y_1) > f(Y_1)\}\big)^{n-1}\mathbb{E}\left(h(Y_1)\mathbf{1}_{\{c\,U_1 g(Y_1)\leq f(Y_1)\}}\right)\\
&= \sum_{n\geq 1}(1-p)^{n-1}\mathbb{E}\left(h(Y_1)\mathbf{1}_{\{c\,U_1 g(Y_1)\leq f(Y_1)\}}\right)\\
&= \sum_{n\geq 1}p(1-p)^{n-1}\mathbb{E}\left(h(Y_1)\,|\,\{c\,U_1 g(Y_1)\leq f(Y_1)\}\right)\\
&= 1\times\int_E h(y)\nu(dy)\\
&= \int_E h(y)\nu(dy).
\end{aligned}$$

◇

Remark. An important point to be noticed is that we do not need to know the numerical value of $\int_E f\,d\mu$ to implement the above acceptance-rejection procedure.

Corollary 1.1 *Set by induction for every* $n \geq 1$

$$\tau_1 := \min\left\{k \geq 1 : c\,U_k g(Y_k) \leq f(Y_k)\right\}$$

and

$$\tau_{n+1} := \min\left\{k \geq \tau_n + 1 : c\,U_k g(Y_k) \leq f(Y_k)\right\}.$$

(a) The sequence $(\tau_n - \tau_{n-1})_{n\geq 1}$ *(with the convention* $\tau_0 = 0$*) is i.i.d. with distribution* $G^*(p)$ *and the sequence*

$$X_n := Y_{\tau_n}$$

is an i.i.d. $\mathbb{P}_X$*-distributed sequence of random variables.*
(b) Furthermore, the random yield of the simulation of the first n $\mathbb{P}_X$*-distributed random variables* Y_{τ_k}*,* $k = 1, \ldots, n$*, is*

$$\rho_n = \frac{n}{\tau_n} \xrightarrow{a.s.} p \quad as \quad n \to +\infty.$$

Proof. (a) Left as an exercise (see below).
(b) The fact that $\rho_n = \frac{n}{\tau_n}$ is obvious. The announced $a.s.$ convergence follows from the Strong Law of Large Numbers since $(\tau_n - \tau_{n-1})_{n\geq 1}$ is i.i.d. and $\mathbb{E}\,\tau_1 = \frac{1}{p}$, which implies that $\frac{\tau_n}{n} \stackrel{a.s.}{\longrightarrow} \frac{1}{p}$. ◇

▶ **Exercise.** Prove item (a) of the above corollary.

Before proposing some first applications, let us briefly present a more applied point of view which is closer to what is really implemented in practice when performing a Monte Carlo simulation based on the acceptance-rejection method.

ℵ Practitioner's corner (the Practitioner's point of view)

The practical implementation of the acceptance-rejection method is rather simple. Let $h : E \to \mathbb{R}$ be a $\mathbb{P}_X$-integrable Borel function. How to compute $\mathbb{E}\,h(X)$ using Von Neumann's acceptance-rejection method? It amounts to the simulation of an n-sample $(U_k, Y_k)_{1\leq k\leq n}$ on a computer and to the computation of

$$\frac{\displaystyle\sum_{k=1}^{n} \mathbf{1}_{\{cU_k g(Y_k)\leq f(Y_k)\}} h(Y_k)}{\displaystyle\sum_{k=1}^{n} \mathbf{1}_{\{cU_k g(Y_k)\leq f(Y_k)\}}}.$$

Note that

$$\frac{\displaystyle\sum_{k=1}^{n} \mathbf{1}_{\{cU_k g(Y_k)\leq f(Y_k)\}} h(Y_k)}{\displaystyle\sum_{k=1}^{n} \mathbf{1}_{\{cU_k g(Y_k)\leq f(Y_k)\}}} = \frac{\displaystyle\frac{1}{n}\sum_{k=1}^{n} \mathbf{1}_{\{cU_k g(Y_k)\leq f(Y_k)\}} h(Y_k)}{\displaystyle\frac{1}{n}\sum_{k=1}^{n} \mathbf{1}_{\{cU_k g(Y_k)\leq f(Y_k)\}}}, \quad n \geq 1.$$

Hence, owing to the strong Law of Large Numbers (see the next chapter if necessary) this quantity $a.s.$ converges as n goes to infinity toward

$$\begin{aligned}
\frac{\displaystyle\int_0^1 du \int \mu(dy)\mathbf{1}_{\{cug(y)\leq f(y)\}} h(y)g(y)}{\displaystyle\int_0^1 du \int \mu(dy)\mathbf{1}_{\{cug(y)\leq f(y)\}} g(y)} &= \frac{\displaystyle\int_E \frac{f(y)}{cg(y)} h(y)g(y)\mu(dy)}{\displaystyle\int_E \frac{f(y)}{cg(y)} g(y)\mu(dy)} \\
&= \frac{\displaystyle\int_E \frac{f(y)}{c} h(y)\mu(dy)}{\displaystyle\int_E \frac{f(y)}{c} \mu(dy)} \\
&= \int_E h(y) \frac{f(y)}{\int_{\mathbb{R}^d} f\, d\mu} \mu(dy) \\
&= \int_E h(y)\nu(dy).
\end{aligned}$$

This third way to present the same computations shows that in terms of practical implementation, this method is in fact very elementary.

Classical applications

$\triangleright$ *Uniform distributions on a bounded domain* $D \subset \mathbb{R}^d$.
Let $D \subset a + [-M, M]^d$, $\lambda_d(D) > 0$ (where $a \in \mathbb{R}^d$ and $M > 0$) and let $Y \stackrel{d}{=} \mathcal{U}(a + [-M, M]^d)$, let $\tau_{_D} := \min\{n : Y_n \in D\}$, where $(Y_n)_{n\geq 1}$ is an i.i.d. sequence defined on a probability space $(\Omega, \mathcal{A}, \mathbb{P})$ with the same distribution as Y. Then,

$$Y_{\tau_D} \stackrel{d}{=} U(D),$$

where $U(D)$ denotes the uniform distribution over D.

This follows from the above Proposition 1.2 with $E = a + [-M, M]^d$, $\mu := \lambda_{d|(a+[-M,M]^d)}$ (Lebesgue measure on $a + [-M, M]^d$),

$$g(u) := (2M)^{-d}\mathbf{1}_{a+[-M,M]^d}(u)$$

and

$$f(x) = \mathbf{1}_D(x) \leq \underbrace{(2M)^d}_{=:c} g(x)$$

so that $\dfrac{f}{\int_{\mathbb{R}^d} f d\mu} \cdot \mu = \mathcal{U}(D)$.

As a matter of fact, with the notations of the above proposition,

$$\tau = \min\left\{k \geq 1 : c\, U_k \leq \frac{f}{g}(Y_k)\right\}.$$

However, $\frac{f}{g}(y) > 0$ if and only if $y \in D$ and if so, $\frac{f}{g}(y) = 1$. Consequently $\tau = \tau_{_D}$.

A standard application is to consider the unit ball of $\mathbb{R}^d$, $D := B_d(0; 1)$. When $d = 2$, this is a step of the so-called *polar method*, see below, for the simulation of $\mathcal{N}(0; I_2)$ random vectors.

$\triangleright$ *The* $\gamma(\alpha)$*-distribution.*

Let $\alpha > 0$ and $\mathbb{P}_{_X}(dx) = f_\alpha(x)\frac{dx}{\Gamma(\alpha)}$ where

$$f_\alpha(x) = x^{\alpha-1}e^{-x}\mathbf{1}_{\{x>0\}}(x).$$

(Keep in mind that $\Gamma(a) = \int_0^{+\infty} u^{a-1}e^{-u}du$, $a > 0$). Note that when $\alpha = 1$ the gamma distribution is simply the exponential distribution. We will consider $E = (0, +\infty)$ and the reference σ-finite measure $\mu = \lambda_{1|(0,+\infty)}$.

– *Case* $0 < \alpha < 1$. We use the rejection method, based on the probability density

$$g_\alpha(x) = \frac{\alpha e}{\alpha + e}\Big(x^{\alpha-1}\mathbf{1}_{\{0<x<1\}} + e^{-x}\mathbf{1}_{\{x\geq 1\}}\Big).$$

The fact that g_α is a probability density function follows from an elementary computation. First, one checks that $f_\alpha(x) \le c_\alpha g_\alpha(x)$ for every $x \in \mathbb{R}_+$, where

$$c_\alpha = \frac{\alpha + e}{\alpha e}.$$

Note that this choice of c_α is optimal since $f_\alpha(1) = c_\alpha g_\alpha(1)$. Then, one uses the inverse distribution function to simulate the random variable with distribution $\mathbb{P}_Y(dy) = g_\alpha(y)\lambda(dy)$. Namely, if G_α denotes the distribution function of Y, one checks that, for every $x \in \mathbb{R}$,

$$G_\alpha(x) = \frac{e}{\alpha + e} x^\alpha \mathbf{1}_{\{0<x<1\}} + \frac{\alpha e}{\alpha + e}\left(\frac{1}{e} + \frac{1}{\alpha} - e^{-x}\right)\mathbf{1}_{\{x>1\}}$$

so that for every $u \in (0, 1)$,

$$G_\alpha^{-1}(u) = \left(\frac{\alpha + e}{e} u\right)^{\frac{1}{\alpha}} \mathbf{1}_{\left\{u < \frac{e}{\alpha+e}\right\}} - \log\left((1-u)\frac{\alpha + e}{\alpha e}\right)\mathbf{1}_{\left\{u \ge \frac{e}{\alpha+e}\right\}}.$$

Note that the computation of the Γ function is never required to implement this method.

– *Case* $\alpha \ge 1$. We rely on the following classical lemma about the γ distribution that we leave as an exercise to the reader.

Lemma 1.1 *Let X' and X'' two independent random variables with distributions $\gamma(\alpha')$ and $\gamma(\alpha'')$, respectively. Then $X = X' + X''$ has distribution $\gamma(\alpha' + \alpha'')$.*

Consequently, if $\alpha = n \in \mathbb{N}$, an induction based on the lemma shows that

$$X = \xi_1 + \cdots + \xi_n$$

where the random variables ξ_k are i.i.d. with exponential distribution since $\gamma(1) = \mathcal{E}(1)$. Consequently, if $U_1, \ldots, U_n$ are i.i.d. uniformly distributed random variables

$$X \stackrel{d}{=} -\log\left(\prod_{k=1}^{n} U_k\right).$$

To simulate a random variable with general distribution $\gamma(\alpha)$, one writes $\alpha = \lfloor\alpha\rfloor + \{\alpha\}$, where $\lfloor\alpha\rfloor := \max\{k \le \alpha,\ k \in \mathbb{N}\}$ denotes the integer part of α and $\{\alpha\} \in [0, 1)$ its fractional part.

► **Exercises. 1.** Prove the above lemma.

2. Show that considering the normalized probability density function of the $\gamma(\alpha)$-distribution (which involves the computation of $\Gamma(\alpha)$) as a function f_α will not improve the yield of the simulation.

3. Let $\alpha = \alpha' + n$, $\alpha' = \lfloor \alpha \rfloor \in [0, 1)$. Show that the yield of the simulation is given by $r = \frac{1}{n+\tau_\alpha}$, where τ_α has a $G^*(p_\alpha)$ distribution with p_α related to the simulation of the $\gamma(\lfloor \alpha \rfloor)$-distribution. Show that

$$\bar{r} := \mathbb{E}\, r = -\frac{p}{(1-p)^{n+1}}\Big(\log p + \sum_{k=1}^{n} \frac{(1-p)^k}{k}\Big).$$

▷ *Acceptance-rejection method for a bounded density with compact support.*

Let $f : \mathbb{R}^d \to \mathbb{R}_+$ be a bounded Borel function with compact support (hence integral with respect to the Lebesgue measure). If f can be computed at a reasonable cost, one may simulate the distribution $\nu = \frac{f}{\int_{\mathbb{R}^d} f.d\lambda_d}\lambda_d$ by simply considering a uniform distribution on a hypercube dominating f. To be more precise, let $a = (a_1, \dots, a_d)$, $b = (b_1, \dots, b_d)$, $\kappa \in \mathbb{R}^d$ such that

$$\mathrm{supp}(f) = \overline{\{f \neq 0\}} \subset K = \kappa + \prod_{i=1}^{d} [a_i, b_i].$$

Let $E = K$, let $\mu = \lambda_{d|E}$ be the reference measure and $g = \frac{1}{\lambda_d(K)}\mathbf{1}_K$ the density of the uniform distribution over K (this distribution is very easy to simulate as emphasized in a former example). Then

$$f(x) \le c\, g(x),\ x \in K \quad \text{with} \quad c = \|f\|_{\sup}\lambda_d(K) = \|f\|_{\sup}\prod_{i=1}^{d}(b_i - a_i).$$

Then, if $(Y_n)_{n\ge 1}$ denotes an i.i.d. sequence defined on a probability space $(\Omega, \mathcal{A}, \mathbb{P})$ with the uniform distribution over K, the stopping strategy τ of Von Neumann's acceptance-rejection method reads

$$\tau = \min\big\{k : \|f\|_{\sup} U_k \le f(Y_k)\big\}.$$

Equivalently this can be rewritten in a more intuitive way as follows: let $V_n = (V_n^1, V_n^2)_{n\ge 1}$ be an i.i.d. sequence of random vectors defined on a probability space $(\Omega, \mathcal{A}, \mathbb{P})$ having *a uniform distribution over* $K \times [0, \|f\|_{\sup}]$. Then

$$V_\tau^1 \stackrel{d}{=} \nu \quad \text{where} \quad \tau = \min\big\{k \ge 1 : V_k^2 \le f(V_k^1)\big\}.$$

▷ *Simulation of a discrete distribution supported by the non-negative integers.*

Let $Y \stackrel{d}{=} G^*(p)$, $p \in (0, 1)$., *i.e.* such that distribution satisfies $\mathbb{P}_Y = g \cdot m$, where m is the counting measure on $\mathbb{N}^*$ ($m(k) = 1$, $k \in \mathbb{N}^*$) and $g_k = p(1-p)^{k-1}$, $k \ge 1$.

Let $f = (f_k)_{k\geq 1}$ be a function from $\mathbb{N} \to \mathbb{R}_+$ defined by $f_k = a_k(1-p)^{k-1}$ and satisfying $\kappa := \sup_n a_n < +\infty$ (so that $\sum_n f_n < +\infty$). Then

$$f_k \leq c\, g_k \quad \text{with} \quad c = \frac{\kappa}{p}.$$

Consequently, if $\tau := \min\left\{k \geq 1 : U_k \leq \frac{p a_k}{\kappa}\right\}$, then $Y_\tau \stackrel{d}{=} \nu$ where $\nu_k := \dfrac{a_k(1-p)^{k-1}}{\sum_n a_n(1-p)^{n-1}}$, $k \geq 1$.

There are many other applications of Von Neumann's acceptance-rejection method, *e.g.* in Physics, to take advantage of the fact the density to be simulated is only known up to a constant. Several methods have been devised to speed it up, *i.e.* to increase its yield. Among them let us cite the *Ziggurat method* for which we refer to [212]. It was developed by Marsaglia and Tsang in the early 2000s.

1.5 Simulation of Poisson Distributions (and Poisson Processes)

The Poisson distribution with parameter $\lambda > 0$, denoted by $\mathcal{P}(\lambda)$, is an integer-valued probability measure analytically defined by

$$\forall\, k \in \mathbb{N}, \quad \mathcal{P}(\lambda)\big(\{k\}\big) = e^{-\lambda}\frac{\lambda^k}{k!}.$$

To simulate this distribution in an exact way, one relies on its close connection with the Poisson counting process. The (normalized) Poisson counting process is the counting process induced by the Exponential random walk (with parameter 1). It is defined by

$$\forall\, t \geq 0, \quad N_t = \sum_{n\geq 1} \mathbf{1}_{\{S_n \leq t\}} = \min\left\{n : S_{n+1} > t\right\}$$

where $S_n = X_1 + \cdots + X_n$, $n \geq 1$, $S_0 = 0$ and $(X_n)_{n\geq 1}$ is an i.i.d. sequence of random variables with distribution $\mathcal{E}(1)$ defined on a probability space $(\Omega, \mathcal{A}, \mathbb{P})$.

Proposition 1.3 *The process $(N_t)_{t\geq 0}$ has càdlàg* ([1]) *paths and independent stationary increments. In particular, for every $s,\, t \geq 0$, $s \leq t$, $N_t - N_s$ is independent of N_s and has the same distribution as N_{t-s}. Furthermore, for every $t \geq 0$, N_t has a Poisson distribution with parameter t.*

---

[1] A French acronym for right continuous with left limits (*continu à droite limitée à gauche*).

Proof. Let $(X_k)_{k\geq 1}$ be a sequence of i.i.d. random variables with an exponential distribution $\mathcal{E}(1)$. Set, for every $n \geq 1$,

$$S_n = X_1 + \cdots + X_n.$$

Let $t_1, t_2 \in \mathbb{R}_+$, $t_1 < t_2$ and let $k_1, k_2 \in \mathbb{N}$. Assume temporarily that $k_2 \geq 1$. Then

$$\begin{aligned}\mathbb{P}(N_{t_2} - N_{t_1} = k_2,\ N_{t_1} = k_1) &= \mathbb{P}(N_{t_1} = k_1,\ N_{t_2} = k_1 + k_2)\\ &= \mathbb{P}(S_{k_1} \leq t_1 < S_{k_1+1} \leq S_{k_1+k_2} \leq t_2 < S_{k_1+k_2+1}).\end{aligned}$$

Now, if we set $A = \mathbb{P}(S_{k_1} \leq t_1 < S_{k_1+1} \leq S_{k_1+k_2} \leq t_2 < S_{k_1+k_2+1})$ for convenience, we get

$$A = \int_{\mathbb{R}_+^{k_1+k_2+1}} \mathbf{1}_{\{x_1+\cdots+x_{k_1} \leq t_1 \leq x_1+\cdots+x_{k_1+1},\, x_1+\cdots+x_{k_1+k_2} \leq t_2 < x_1+\cdots+x_{k_1+k_2+1}\}} e^{-(x_1+\cdots+x_{k_1+k_2+1})} dx_1 \cdots dx_{k_1+k_2+1}$$

The usual change of variable $x_1 = u_1$ and $x_i = u_i - u_{i-1}$, $i = 2, \ldots, k_1 + k_2 + 1$, yields

$$A = \int_{\{0 \leq u_1 \leq \cdots \leq u_{k_1} \leq t_1 \leq u_{k_1+1} \leq \cdots \leq u_{k_1+k_2} \leq t_2 < u_{k_1+k_2+1}\}} e^{-u_{k_1+k_2+1}} du_1 \cdots du_{k_1+k_2+1}.$$

Integrating downward from $u_{k_1+k_2+1}$ to u_1 we get, owing to Fubini's Theorem,

$$\begin{aligned}A &= \int_{\{0 \leq u_1 \leq \cdots \leq u_{k_1} \leq t_1 \leq u_{k_1+1} \leq \cdots \leq u_{k_1+k_2} \leq t_2 < u_{k_1+k_2+1}\}} du_1 \cdots du_{k_1+k_2}\, e^{-t_2}\\ &= \int_{\{0 \leq u_1 \leq \cdots \leq u_{k_1} \leq t_1\}} du_1 \cdots du_{k_1} \frac{(t_2-t_1)^{k_2}}{k_2!} e^{-t_2}\\ &= \frac{t_1^{k_1}}{k_1!} \frac{(t_2-t_1)^{k_2}}{k_2!} e^{-t_2}\\ &= e^{-t_1} \frac{t_1^{k_1}}{k_1!} \times e^{-(t_2-t_1)} \frac{(t_2-t_1)^{k_2}}{k_2!}.\end{aligned}$$

When $k_2 = 0$, one computes likewise

$$\mathbb{P}(S_{k_1} \leq t_1 < t_2 < S_{k_1+1}) = \frac{t_1^{k_1}}{k_1!} e^{-t_2} = e^{-t_1} \frac{t_1^{k_1}}{k_1!} e^{-(t_2-t_1)}.$$

Summing over $k_2 \in \mathbb{N}$ shows that, for every $k_1 \in \mathbb{N}$,

$$\mathbb{P}(N_{t_1} = k_1) = e^{-t_1} \frac{t_1^{k_1}}{k_1!}$$

i.e. $N_{t_1} \stackrel{d}{=} \mathcal{P}(t_1)$. Summing over $k_1 \in \mathbb{N}$ shows that, for every $k_2 \in \mathbb{N}$,

$$N_{t_2} - N_{t_1} \stackrel{d}{=} N_{t_2 - t_1} \stackrel{d}{=} \mathcal{P}(t_2 - t_1).$$

Finally, this yields for every $k_1, k_2 \in \mathbb{N}$,

$$\mathbb{P}(N_{t_2} - N_{t_1} = k_2,\ N_{t_1} = k_1) = \mathbb{P}(N_{t_2} - N_{t_1} = k_2) \times \mathbb{P}(N_{t_1} = k_1),$$

i.e. the increments $N_{t_2} - N_{t_1}$ and N_{t_1} are independent.

One shows likewise, with a few more technicalities, that in fact, if $0 < t_1 < t_2 < \cdots < t_n, n \geq 1$, then the increments $(N_{t_i} - N_{t_{i-1}})_{i=1,\ldots,n}$ are independent and stationary in the sense that $N_{t_i} - N_{t_{i-1}} \stackrel{d}{=} \mathcal{P}(t_i - t_{i-1})$. ◇

Corollary 1.2 (Simulation of a Poisson distribution) *Let $(U_n)_{n\geq 1}$ be an i.i.d. sequence of uniformly distributed random variables on the unit interval. The process null at zero and defined for every $t > 0$ by*

$$N_t = \min\left\{k \geq 0 : U_1 \cdots U_{k+1} < e^{-t}\right\} \stackrel{d}{=} \mathcal{P}(t)$$

is a normalized Poisson process.

Proof. It follows from Example 1 in the former Sect. 1.3 that the exponentially distributed i.i.d. sequence $(X_k)_{k\geq 1}$ can be written in the following form

$$X_k = -\log U_k, \quad k \geq 1.$$

Using that the random walk $(S_n)_{n\geq 1}$ is non-decreasing it follows from the definition of a Poisson process that for every $t \geq 0$,

$$\begin{aligned} N_t &= \min\{k \geq 0 \text{ such that } S_{k+1} > t\}, \\ &= \min\left\{k \geq 0 \text{ such that } -\log\left(U_1 \cdots U_{k+1}\right) > t\right\}, \\ &= \min\left\{k \geq 0 \text{ such that } U_1 \cdots U_{k+1} < e^{-t}\right\}. \end{aligned}$$ ◇

1.6 Simulation of Gaussian Random Vectors

1.6.1 d-dimensional Standard Normal Vectors

A first method to simulate a bi-variate normal distribution $\mathcal{N}(0; I_2)$ is the so-called Box–Muller method, described below.

Proposition 1.4 *Let R^2 and $\Theta : (\Omega, \mathcal{A}, \mathbb{P}) \to \mathbb{R}$ be two independent r.v. with distributions $\mathcal{L}(R^2) = \mathcal{E}(\frac{1}{2})$ and $\mathcal{L}(\Theta) = U([0, 2\pi])$, respectively. Then*

$$X := (R\cos\Theta,\, R\sin\Theta) \stackrel{d}{=} \mathcal{N}(0; I_2),$$

where $R := \sqrt{R^2}$.

Proof. Let f be a bounded Borel function.

$$\iint_{\mathbb{R}^2} f(x_1,x_2)\exp\left(-\frac{x_1^2+x_2^2}{2}\right)\frac{dx_1dx_2}{2\pi} = \iint f(\rho\cos\theta,\rho\sin\theta)e^{-\frac{\rho^2}{2}}\mathbf{1}_{\mathbb{R}_+^*}(\rho)\mathbf{1}_{(0,2\pi)}(\theta)\rho\frac{d\rho d\theta}{2\pi}$$

using the standard change of variable: $x_1 = \rho\cos\theta$, $x_2 = \rho\sin\theta$. We use the facts that $(\rho,\theta) \mapsto (\rho\cos\theta, \rho\sin\theta)$ is a $\mathcal{C}^1$-diffeomorphism from $(0, 2\pi) \times (0, +\infty) \to \mathbb{R}^2 \setminus (\mathbb{R}_+ \times \{0\})$ and $\lambda_2(\mathbb{R}_+ \times \{0\}) = 0$. Setting now $\rho = \sqrt{r}$, one has:

$$\begin{aligned}\iint_{\mathbb{R}^2} f(x_1,x_2)\exp\left(-\frac{x_1^2+x_2^2}{2}\right)\frac{dx_1dx_2}{2\pi} &= \iint f(\sqrt{r}\cos\theta,\sqrt{r}\sin\theta)\frac{e^{-\frac{r}{2}}}{2}\mathbf{1}_{\mathbb{R}_+^*}(\rho)\mathbf{1}_{(0,2\pi)}(\theta)\frac{drd\theta}{2\pi}\\ &= \mathbb{E}\left(f\left(\sqrt{R^2}\cos\Theta,\sqrt{R^2}\sin\Theta\right)\right)\\ &= \mathbb{E}(f(X)).\end{aligned}$$

◇

Corollary 1.3 (Box–Muller method) *One can simulate a distribution $\mathcal{N}(0; I_2)$ from a pair (U_1, U_2) of independent random variables with distribution $\mathcal{U}([0, 1])$ by setting*

$$X := \left(\sqrt{-2\log(U_1)}\cos(2\pi U_2),\, \sqrt{-2\log(U_1)}\sin(2\pi U_2)\right).$$

The yield of the simulation is $r = 1/2$ with respect to the $\mathcal{N}(0; 1)$ distribution and $r = 1$ when the aim is to simulate an $\mathcal{N}(0; I_2)$-distributed (pseudo-)random vector or, equivalently, two $\mathcal{N}(0; 1)$-distributed (pseudo-)random numbers.

Proof. Simulate the exponential distribution using the inverse distribution function using $U_1 \stackrel{d}{=} \mathcal{U}([0, 1])$ and note that if $U_2 \stackrel{d}{=} \mathcal{U}([0, 1])$, then $2\pi U_2 \stackrel{d}{=} \mathcal{U}([0, 2\pi])$ (where U_2 is taken independent of U_1). ◇

Application to the simulation of the multivariate normal distribution

To simulate a d-dimensional vector $X = (X^1, \ldots, X^d)$ with $\mathcal{N}(0; I_d)$ distribution, one may assume that d is even and "concatenate" the above process. We consider a d-tuple $(U_1, \ldots, U_d) \stackrel{d}{=} \mathcal{U}([0, 1]^d)$ random vector (so that $U_1, \ldots, U_d$ are i.i.d. with distribution $\mathcal{U}([0, 1])$) and we set

$$\left(X^{2i-1}, X^{2i}\right) = \left(\sqrt{-2\log(U_{2i-1})}\cos(2\pi U_{2i}),\, \sqrt{-2\log(U_{2i-1})}\sin(2\pi U_{2i})\right), \tag{1.1}$$

$$i = 1, \ldots, d/2.$$

A second popular method for simulating bi-variate normal distributions is the Polar method due to Marsaglia. It relies on the simulation of uniformly distributed random variables on the 2-dimensional Euclidean unit ball by an acceptance-rejection method.

► **Exercise** (*Marsaglia's Polar method*). See [210]. Let $(V_1, V_2) \stackrel{d}{=} \mathcal{U}(B(0;1))$, where $B(0,1)$ denotes the canonical Euclidean unit ball in $\mathbb{R}^2$. Set $R^2 = V_1^2 + V_2^2$ and

$$X := \sqrt{-2\log(R^2)/R^2}\Big(V_1, V_2\Big).$$

(a) Show that $R^2 \stackrel{d}{=} \mathcal{U}([0,1])$, $\left(\frac{V_1}{R}, \frac{V_2}{R}\right) \sim \big(\cos(\theta), \sin(\theta)\big)$, R^2 and $\left(\frac{V_1}{R}, \frac{V_2}{R}\right)$ are independent. Deduce that $X \stackrel{d}{=} \mathcal{N}(0; I_2)$.
(b) Let $(U_1, U_2) \stackrel{d}{=} \mathcal{U}([-1,1]^2)$. Show that $\mathcal{L}\big((U_1, U_2)\,|\,U_1^2 + U_2^2 \le 1\big) = \mathcal{U}\big(B(0;1)\big)$. Derive a simulation method for $\mathcal{N}(0; I_2)$ combining the above identity and an appropriate acceptance-rejection algorithm to simulate the $\mathcal{N}(0; I_2)$ distribution. What is the yield of the resulting procedure?
(c) Compare the performances of Marsaglia's polar method with those of the Box–Muller algorithm (*i.e.* the acceptance-rejection rule *versus* the computation of trigonometric functions). Conclude.

1.6.2 Correlated d-dimensional Gaussian Vectors, Gaussian Processes

Let $X = (X^1, \ldots, X^d)$ be a centered $\mathbb{R}^d$-valued Gaussian vector with covariance matrix $\Sigma = \big[\mathrm{Cov}(X^i, X^j)\big]_{1\le i,j\le d}$. The symmetric matrix Σ is positive semidefinite (but possibly non-invertible). It follows from the very definition of Gaussian vectors that any linear combination $(u|X) = \sum_i u^i X^i$ has a (centered) Gaussian distribution with variance $u^*\Sigma u$ so that the characteristic function of X is given by

$$\chi_X(u) := \mathbb{E}\left(e^{i(u|X)}\right) = e^{-\frac{1}{2}u^*\Sigma u}, \quad u \in \mathbb{R}^d.$$

As a well-known consequence the covariance matrix Σ characterizes the distribution of X, which allows us to denote $\mathcal{N}(0; \Sigma)$ the distribution of X.

The key to simulating such a random vector is the following general lemma (which has nothing to do with Gaussian vectors). It describes how the covariance is modified by a linear transform.

Lemma 1.2 *Let Y be an $\mathbb{R}^d$-valued square integrable random vector and let $A \in \mathcal{M}(q, d)$ be a $q \times d$ matrix. Then the covariance matrix C_{AY} of the random vector AY is given by*

$$C_{AY} = AC_Y A^*$$

where A^* stands for the transpose of A.

This result can be used in two different ways.

– *Square root of* Σ. It is a classical result that there exists a unique positive semidefinite matrix $\sqrt{\Sigma}$ commuting with Σ such that $\Sigma = \sqrt{\Sigma}^2 = \sqrt{\Sigma}\sqrt{\Sigma}^*$. Consequently, owing to the above lemma,

$$\text{If} \quad Z \stackrel{d}{=} \mathcal{N}(0; I_d) \quad \text{then} \quad \sqrt{\Sigma} Z \stackrel{d}{=} \mathcal{N}(0; \Sigma).$$

One can compute $\sqrt{\Sigma}$ by diagonalizing Σ in the orthogonal group: if $\Sigma = P\,\mathrm{Diag}(\lambda_1, \dots, \lambda_d)P^*$ with $PP^* = I_d$ and $\lambda_1, \dots, \lambda_d \geq 0$, then, by uniqueness of the square root of Σ as defined above, it is clear that $\sqrt{\Sigma} = P\mathrm{Diag}(\sqrt{\lambda_1}, \dots, \sqrt{\lambda_d})P^*$.

– *Cholesky decomposition of* Σ. When the covariance matrix Σ is invertible (*i.e.* definite), it is much more efficient to rely on the Cholesky decomposition (see *e.g.* Numerical Recipes [220]) by decomposing Σ as

$$\Sigma = TT^*$$

where T is a lower triangular matrix (*i.e.* such that $T_{ij} = 0$ if $i < j$). Then, owing to Lemma 1.2,

$$TZ \stackrel{d}{=} \mathcal{N}(0; \Sigma).$$

In fact, the Cholesky decomposition is the matrix formulation of the Hilbert–Schmidt orthonormalization procedure. In particular, there exists *a unique such lower triangular matrix* T *with positive diagonal terms* ($T_{ii} > 0, i = 1, \dots, d$) called the Cholesky matrix of Σ.

The Cholesky-based approach performs better since it approximately divides the complexity of this phase of the simulation by a factor almost 2.

► **Exercises. 1.** Let $Z = (Z_1, Z_2)$ be a Gaussian vector such that $Z_1 \stackrel{d}{=} Z_2 \stackrel{d}{=} \mathcal{N}(0; 1)$ and $\mathrm{Cov}(Z_1, Z_2) = \rho \in [-1, 1]$.
(a) Compute for every $u \in \mathbb{R}^2$ the Laplace transform $L(u) = \mathbb{E}\, e^{(u|Z)}$ of Z.
(b) Compute for every $\sigma_1, \sigma_2 > 0$ the correlation ([2]) ρ_{X_1,X_2} between the random variables $X_1 = e^{\sigma_1 Z_1}$ and $X_2 = e^{\sigma_2 Z_2}$.
(c) Show that $\inf_{\rho\in[-1,1]} \rho_{X_1,X_2} \in (-1, 0)$ and that, when $\sigma_i = \sigma > 0$, $\inf_{\rho\in[-1,1]} \rho_{X_1,X_2} = -e^{-\sigma^2}$.
2. Let Σ be a positive definite matrix. Show the existence of a unique lower triangular matrix T and a diagonal matrix D such that both T and D have positive

[2]The correlation ρ_{X_1,X_2} between two square integrable non-*a.s.* constant random variables defined on the same probability space is defined by

$$\rho_{X_1,X_2} = \frac{\mathrm{Cov}(X_1, X_2)}{\sigma(X_1)\sigma(X_2)} = \frac{\mathrm{Cov}(X_1, X_2)}{\sqrt{\mathrm{Var}(X_1)\mathrm{Var}(X_2)}}.$$

diagonal entries and $\sum_{i,j=1} T_{ij}^2 = 1$ for every $i = 1, \dots, d$. [Hint: change the reference Euclidean norm to perform the Hilbert–Schmidt decomposition] ([3]).

Application to the simulation of a standard Brownian motion at fixed times

Let $W = (W_t)_{t \geq 0}$ be a standard Brownian motion defined on a probability space $(\Omega, \mathcal{A}, \mathbb{P})$. Let $(t_1, \dots, t_n)$ be a non-decreasing n-tuple $(0 \leq t_1 < t_2 < \dots, t_{n-1} < t_n)$ of instants. One elementary definition of a standard Brownian motion is that it is a centered Gaussian process with covariance given by $C^W(s, t) = \mathbb{E}(W_s W_t) = s \wedge t$ ([4]). The resulting simulation method relying on the Cholesky decomposition of the covariance structure of the Gaussian vector $(W_{t_1}, \dots, W_{t_n})$ given by

$$\Sigma^W_{(t_1, \dots, t_n)} = \big[t_i \wedge t_j\big]_{1 \leq i,j \leq n}$$

is a first possibility.

However, it seems more natural to use the independence and the stationarity of the increments, *i.e.* that

$$\Big(W_{t_1}, W_{t_2} - W_{t_1}, \dots, W_{t_n} - W_{t_{n-1}}\Big) \stackrel{d}{=} \mathcal{N}\Big(0; \text{Diag}(t_1, t_2 - t_1, \dots, t_n - t_{n-1})\Big)$$

so that

$$\begin{bmatrix} W_{t_1} \\ W_{t_2} - W_{t_1} \\ \vdots \\ W_{t_n} - W_{t_{n-1}} \end{bmatrix} \stackrel{d}{=} \text{Diag}\big(\sqrt{t_1}, \sqrt{t_2 - t_1}, \dots, \sqrt{t_n - t_{n-1}}\big) \begin{bmatrix} Z_1 \\ \vdots \\ Z_n \end{bmatrix}$$

where $(Z_1, \dots, Z_n) \stackrel{d}{=} \mathcal{N}\big(0; I_n\big)$. The simulation of $(W_{t_1}, \dots, W_{t_n})$ follows by summing the increments.

Remark. To be more precise, one derives from the above result that

$$\begin{bmatrix} W_{t_1} \\ W_{t_2} \\ \vdots \\ W_{t_n} \end{bmatrix} = L \begin{bmatrix} W_{t_1} \\ W_{t_2} - W_{t_1} \\ \vdots \\ W_{t_n} - W_{t_{n-1}} \end{bmatrix} \quad \text{where} \quad L = \big[\mathbf{1}_{\{i \geq j\}}\big]_{1 \leq i,j \leq n}.$$

[3]This modified Cholesky decomposition is faster than the standard one (see *e.g.* [275]) since it avoids square root computations even if, in practice, the cost of the decomposition itself remains negligible compared to that of a large-scale Monte Carlo simulation.

[4]This definition does not include the fact that W has continuous paths, however, it can be derived using the celebrated Kolmogorov criterion, that it has a modification with continuous paths (see *e.g.* [251]).

Hence, if we set $T = L\,\mathrm{Diag}\big(\sqrt{t_1}, \sqrt{t_2 - t_1}, \dots, \sqrt{t_n - t_{n-1}}\big)$, we checks on the one hand that $T = \big[\sqrt{t_j - t_{j-1}}\mathbf{1}_{\{i \ge j\}}\big]_{1 \le i,j \le n}$ and on the other hand that

$$\begin{bmatrix} W_{t_1} \\ W_{t_2} \\ \vdots \\ W_{t_n} \end{bmatrix} = T \begin{bmatrix} Z_1 \\ \vdots \\ Z_n \end{bmatrix}.$$

We derive, owing to Lemma 1.2, that $TT^* = TI_nT^* = \Sigma^W_{t_1,\dots,t_n}$. The matrix T being lower triangular, it provides the Cholesky decomposition of the covariance matrix $\Sigma^W_{t_1,\dots,t_n}$.

ℵ Practitioner's corner (Warning!)

In quantitative finance, especially when modeling the dynamics of several risky assets, say d, the correlation between the Brownian sources of randomness $B = (B^1, \dots, B^d)$ attached to the log-return is often misleading in terms of notations: since it is usually written as

$$\forall\, i \in \{1, \dots, d\},\ B^i_t = \sum_{j=1}^{q} \sigma_{ij} W^j_t,$$

where $W = (W^1, \dots, W^q)$ is a standard q-dimensional Brownian motion (*i.e.* each component W^j, $j = 1, \dots, q$, is a standard Brownian motion and all these components are independent). The normalized covariance matrix of B (at time 1) is given by

$$\mathrm{Cov}\big(B^i_1, B^j_1,\big) = \sum_{\ell=1}^{q} \sigma_{i\ell}\sigma_{j\ell} = \big(\sigma_{i.}|\sigma_{j.}\big) = (\sigma\sigma^*)_{ij},$$

where $\sigma_{i.}$ is the column vector $[\sigma_{ij}]_{1\le j\le q}$, $\sigma = [\sigma_{ij}]_{1\le i\le d, 1\le j\le q}$ and $(\cdot|\cdot)$ denotes the canonical inner product on $\mathbb{R}^q$. So one should process the Cholesky decomposition on the symmetric non-negative matrix $\Sigma_{B_1} = \sigma\sigma^*$. Also keep in mind that, if $q < d$, then σ_{B_1} has rank at most q and cannot be invertible.

Application to the simulation of a Fractional Brownian motion at fixed times

A fractional Brownian motion (fBm) $W^H = (W^H_t)_{t\ge 0}$ with Hurst coefficient $H \in (0, 1)$ is defined as a centered Gaussian process with a correlation function given for every $s, t \in \mathbb{R}_+$ by

$$C^{W^H}(s, t) = \frac{1}{2}\Big(t^{2H} + s^{2H} - |t - s|^{2H}\Big).$$

When $H = \frac{1}{2}$, W^H is simply a standard Brownian motion. When $H \ne \frac{1}{2}$, W^H has none of the usual properties of a Brownian motion (except the stationarity of its increments and some self-similarity properties): it has dependent increments, it is not a martingale (or even a semi-martingale). Neither is it a Markov process.

A natural approach to simulating the fBm W^H at times $t_1, \ldots, t_n$ is to rely on the Cholesky decomposition of its auto-covariance matrix $[C^{W^H}(t_i, t_j)]_{1 \leq i,j \leq n}$. On the one hand, this matrix is ill-conditioned, which induces instability in the computation of the Cholesky decomposition. This should balanced out with the fact that such a computation can be made only once and stored off-line (at least for usual values of the t_i like $t_i = \frac{iT}{n}, i = 1, \ldots, n$).

Other methods have been introduced in which the auto-covariance matrix is embedded in a circuit matrix. It relies on a (fast) Fourier transform procedure. This method, originally introduced in [73], has recently been improved in [277], where a precise algorithm is described.

Chapter 2
The Monte Carlo Method and Applications to Option Pricing

2.1 The Monte Carlo Method

The basic principle of the *Monte Carlo* method is to implement on a computer the Strong Law of Large Numbers (*SLLN*): if $(X_k)_{k\geq 1}$ denotes a sequence, defined on a probability space $(\Omega, \mathcal{A}, \mathbb{P})$, of independent integrable random variables with the same distribution as X, then

$$\mathbb{P}(d\omega)\text{-}a.s. \qquad \overline{X}_{M}(\omega) := \frac{X_1(\omega) + \cdots + X_M(\omega)}{M} \stackrel{M\to+\infty}{\longrightarrow} m_{X} := \mathbb{E}\,X.$$

The sequence $(X_k)_{k\geq 1}$ is also called an i.i.d. sequence of random variables with distribution $\mu = \mathbb{P}_{X}$ (that of X) or a(n infinite) *sample* of the distribution μ. Two conditions are requested to undertake the *Monte Carlo simulation* of (the distribution μ of) X, *i.e.* to *implement the above SLLN on a computer* for the distribution μ or for the random vector X.

$\triangleright$ One can generate on a computer some as perfect as possible (pseudo-)random numbers which can be regarded as a "generic" realization $(U_k(\omega))_{k\geq 1}$ of a sample $(U_k)_{k\geq 1}$ of the uniform distribution over the unit interval $[0, 1]$. Note that, as a consequence, for any integer $d \geq 1$, $\big((U_{(k-1)d+\ell}(\omega))_{1\leq\ell\leq d}\big)_{k\geq 1}$ is also generic for the sample $((U_{(k-1)d+\ell})_{1\leq\ell\leq d})_{k\geq 1}$ of the uniform distribution over $[0, 1]^d$ (at least theoretically) since $\mathcal{U}([0, 1]^d) = \big(\mathcal{U}([0, 1])\big)^{\otimes d}$. This problem has already been briefly discussed in the introductory chapter.

$\triangleright$ In practice, X reads or is represented either as

– $X \stackrel{d}{=} \varphi(U)$, where U is a uniformly distributed random vector on $[0, 1]^d$ where the Borel function $\varphi : u \mapsto \varphi(u)$ can be computed at any $u \in [0, 1]^d$

or

– $X \stackrel{d}{=} \varphi_\tau(U_1, \ldots, U_\tau)$, where $(U_n)_{n\geq 1}$ is a sequence of i.i.d. random variables, uniformly distributed over $[0, 1]$, τ is a simulable *finite stopping rule* (or stopping time) for the sequence $(U_k)_{k\geq 1}$, taking values in $\mathbb{N}^*$. By "stopping rule" we mean that the event $\{\tau = k\} \in \sigma(U_1, \ldots, U_k)$ for every $k \geq 1$ and by "simulable" we

G. Pagès, *Numerical Probability*, Universitext,
https://doi.org/10.1007/978-3-319-90276-0_2

mean that $\mathbf{1}_{\{\tau=k\}} = \psi_k(U_1, \ldots, U_k)$ where ψ_k has an explicit form for every $k \geq 1$. Of course, we also assume that, for every $k \geq 1$, the function φ_k defined on the set of finite $[0, 1]^k$-valued sequences is a computable function as well.

This procedure is at the core of Monte Carlo simulation. We provided several examples of such representations in the previous chapter. For further developments on this wide topic, we refer to [77] and the references therein, but in some way, one may consider that a significant part of scientific activity in Probability Theory is motivated by or can be applied to simulation.

Once these conditions are fulfilled, a Monte Carlo simulation can be performed. But this leads to two important issues:

– what about the rate of convergence of the method?

and

– how can the resulting error be controlled?

The answers to these questions relies on fundamental results in Probability Theory and Statistics.

2.1.1 *Rate(s) of Convergence*

The (weak) rate of convergence in the *SLLN* is ruled by the Central Limit Theorem (*CLT*), which says that if X is square integrable ($X \in L^2(\mathbb{P})$), then

$$\sqrt{M}\left(\overline{X}_M - m_X\right) \xrightarrow{\mathcal{L}} \mathcal{N}(0; \sigma_X^2) \quad \text{as} \quad M \to +\infty,$$

where $\sigma_X^2 = \mathrm{Var}(X) := \mathbb{E}\left((X - \mathbb{E}\,X)^2\right) = \mathbb{E}\,X^2 - (\mathbb{E}\,X)^2$ is the variance (its square root σ_X is called *standard deviation*) ([1]). Also note that the mean quadratic error of convergence (*i.e.* the convergence rate in $L^2(\mathbb{P})$) is exactly

$$\left\|\overline{X}_M - m_X\right\|_2 = \frac{\sigma_X}{\sqrt{M}}.$$

This error is also known as the *RMSE* (for Root Mean Square Error).

---

[1] The symbol $\xrightarrow{\mathcal{L}}$ stands for convergence in distribution (or "in law" whence the notation "$\mathcal{L}$"): a sequence of random variables $(Y_n)_{n\geq 1}$ converges in distribution toward Y_∞ if

$$\forall\, f \in \mathcal{C}_b(\mathbb{R}, \mathbb{R}), \quad \mathbb{E}\, f(Y_n) \longrightarrow \mathbb{E}\, f(Y_\infty) \quad \text{as } n \to +\infty.$$

It can be defined equivalently as the weak convergence of the distributions $\mathbb{P}_{Y_n}$ toward the distribution $\mathbb{P}_{Y_\infty}$. Convergence in distribution is also characterized by the following property

$$\forall\, A \in \mathcal{B}(\mathbb{R}),\ \mathbb{P}(Y_\infty \in \partial A) = 0 \Longrightarrow \mathbb{P}(Y_n \in A) \longrightarrow \mathbb{P}(Y_\infty \in A) \quad \text{as } n \to +\infty.$$

The extension to $\mathbb{R}^d$-valued random vectors is straightforward. See [45] for a presentation of weak convergence of probability measures in a general framework.

If $\sigma_X = 0$, then $\overline{X}_M = X = m_X$ $\mathbb{P}$-$a.s$. So, throughout this chapter, we may assume without loss of generality that $\sigma_X > 0$.

The Law of the Iterated Logarithm (*LIL*) provides an $a.s.$ rate of convergence, namely

$$\overline{\lim_M} \sqrt{\frac{M}{2\log(\log M)}}\left(\overline{X}_M - m_X\right) = \sigma_X \quad \text{and} \quad \underline{\lim_M} \sqrt{\frac{M}{2\log(\log M)}}\left(\overline{X}_M - m_X\right) = -\sigma_X.$$

A proof of this (difficult) result can be found *e.g.* in [263]. All these rates stress the main drawback of the Monte Carlo method: it is a slow method since dividing the error by 2 requires an increase of the size of the simulation by 4 (even a bit more viewed from the *LIL*).

2.1.2 Data Driven Control of the Error: Confidence Level and Confidence Interval

Assume $\sigma_X > 0$. As concerns the control of the error, one relies on the *CLT*. It is obvious from the definition of convergence in distribution that the *CLT* also reads

$$\sqrt{M}\frac{\overline{X}_M - m_X}{\sigma_X} \stackrel{\mathcal{L}}{\longrightarrow} \mathcal{N}(0;1) \quad \text{as} \quad M \to +\infty$$

since $\sigma_X Z \stackrel{d}{=} \mathcal{N}(0;\sigma_X^2)$ if $Z \stackrel{d}{=} \mathcal{N}(0;1)$ and $f\big(\,\cdot\,/\sigma_X\big) \in \mathcal{C}_b(\mathbb{R},\mathbb{R})$ if $f \in \mathcal{C}_b(\mathbb{R},\mathbb{R})$.

Moreover, the normal distribution having a density, it has no atom. Consequently, this convergence implies (in fact it is equivalent, see [45]) that for all real numbers $a, b, a > b$,

$$\begin{aligned}\lim_M \mathbb{P}\left(\sqrt{M}\,\frac{\overline{X}_M - m_X}{\sigma_X} \in [b,a]\right) &= \mathbb{P}\big(\mathcal{N}(0;1) \in [b,a]\big) \quad \text{as} \quad M \to +\infty.\\ &= \Phi_0(a) - \Phi_0(b),\end{aligned}$$

where Φ_0 denotes the cumulative distribution function (c.d.f.) of the standard normal distribution $\mathcal{N}(0;1)$ defined by

$$\Phi_0(x) = \int_{-\infty}^{x} e^{-\frac{\xi^2}{2}} \frac{d\xi}{\sqrt{2\pi}}$$

(see Sect. 12.1.3 for a tabulation of Φ_0).

► **Exercise.** Show that, like any distribution function of a symmetric random variable without atom, the distribution function of the centered normal distribution on the real line satisfies

$$\forall x \in \mathbb{R}, \qquad \Phi_0(x) + \Phi_0(-x) = 1.$$

Deduce that $\mathbb{P}\big(|\mathcal{N}(0;1)| \le a\big) = 2\Phi_0(a) - 1$ for every $a > 0$.

In turn, if $X_1 \in L^3(\mathbb{P})$, the convergence rate in the Central Limit Theorem is ruled by the following Berry–Esseen Theorem (see [263], p 344).

Theorem 2.1 (Berry–Esseen Theorem) *Let $X_1 \in L^3(\mathbb{P})$ with $\sigma_X > 0$ and let $(X_k)_{k\ge1}$ be an i.i.d. sequence of real-valued random variables defined on $(\Omega, \mathcal{A}, \mathbb{P})$. Then*

$$\forall n \ge 1, \quad \forall x \in \mathbb{R}, \quad \Big|\mathbb{P}\big(\sqrt{M}(\bar{X}_M - m_X) \le x\,\sigma_X\big) - \Phi_0(x)\Big| \le \frac{C\,\mathbb{E}\,|X_1 - \mathbb{E}\,X_1|^3}{\sigma_X^3\sqrt{M}}\,\frac{1}{1+|x|^3}.$$

Hence, the rate of convergence in the *CLT* is again $1/\sqrt{M}$, which is rather slow, at least from a statistical viewpoint. However, this is not a real problem within the usual range of Monte Carlo simulations (at least many thousands, usually one hundred thousand or one million paths). Consequently, one can assume that $\sqrt{M}\,\dfrac{\overline{X}_M - m_X}{\sigma_X}$ has a standard normal distribution. This means in particular that one can design a *probabilistic control of the error* directly derived from statistical concepts: let $\alpha \in (0, 1)$ denote a *confidence level* (close to 1 in practice) and let q_α be the *two-sided α-quantile* defined as the unique solution to the equation

$$\mathbb{P}\big(|\mathcal{N}(0;1)| \le q_\alpha\big) = \alpha \qquad i.e. \qquad 2\,\Phi_0(q_\alpha) - 1 = \alpha.$$

Then, setting $a = q_\alpha$ and $b = -q_\alpha$, one defines a *theoretical* random confidence interval at level α by

$$J_M^\alpha := \left[\overline{X}_M - q_\alpha\,\frac{\sigma_X}{\sqrt{M}},\ \overline{X}_M + q_\alpha\,\frac{\sigma_X}{\sqrt{M}}\right]$$

which satisfies

$$\begin{aligned}\mathbb{P}\big(m_X \in J_M^\alpha\big) &= \mathbb{P}\left(\sqrt{M}\frac{|\overline{X}_M - m_X|}{\sigma_X} \le q_\alpha\right)\\ &\xrightarrow{M\to+\infty} \mathbb{P}\big(|\mathcal{N}(0;1)| \le q_\alpha\big) = \alpha.\end{aligned}$$

However, at this stage this procedure remains purely theoretical since the confidence interval J_M^α involves the standard deviation σ_X of X, which is usually unknown.

Here comes the "magic trick" which led to the tremendous success of the Monte Carlo method: the variance of X can in turn be evaluated on-line, without any additional assumption, by simply adding a *companion Monte Carlo simulation* to estimate the variance σ_X^2, namely

$$\overline{V}_M = \frac{1}{M-1}\sum_{k=1}^{M}(X_k - \overline{X}_M)^2 \tag{2.1}$$

$$= \frac{1}{M-1}\sum_{k=1}^{M} X_k^2 - \frac{M}{M-1}\overline{X}_M^2 \longrightarrow \mathrm{Var}(X) = \sigma_X^2 \quad \text{as} \quad M \to +\infty.$$

The above convergence ([2]) follows from the *SLLN* applied to the i.i.d. sequence of integrable random variables $(X_k^2)_{k\geq 1}$ (and the convergence of $\overline{X}_M$, which also follows from the *SLLN*). It is an easy exercise to show that moreover $\mathbb{E}\,\overline{V}_M = \sigma_X^2$, *i.e.* using the terminology of Statistics, $\overline{V}_M$ is an *unbiased* estimator of σ_X^2. This remark is of little importance in practice due to the common large ranges of Monte Carlo simulations. Note that the above *a.s.* convergence is again ruled by a *CLT* if $X \in L^4(\mathbb{P})$ (so that $X^2 \in L^2(\mathbb{P})$).

► **Exercises. 1.** Show that $\overline{V}_M$ is unbiased, that is $\mathbb{E}\,\overline{V}_M = \sigma_X^2$.

2. Show that the sequence $(\overline{X}_M, \overline{V}_M)_{M\geq 1}$ satisfies the following recursive equation

$$\overline{X}_M = \overline{X}_{M-1} + \frac{1}{M}\big(X_M - \overline{X}_{M-1}\big)$$
$$= \frac{M-1}{M}\overline{X}_{M-1} + \frac{X_M}{M}, \quad M \geq 1$$

(with the convention $\overline{X}_0 = 0$) and

$$\overline{V}_M = \overline{V}_{M-1} - \frac{1}{M-1}\Big(\overline{V}_{M-1} - (X_M - \overline{X}_{M-1})(X_M - \overline{X}_M)\Big)$$
$$= \frac{M-2}{M-1}\overline{V}_{M-1} + \frac{(X_M - \overline{X}_{M-1})^2}{M}, \quad M \geq 2.$$

As a consequence, one derives from Slutsky's Theorem ([3]) that

$$\sqrt{M}\,\frac{\overline{X}_M - m_X}{\sqrt{\overline{V}_M}} = \sqrt{M}\,\frac{\overline{X}_M - m_X}{\sigma_X} \times \frac{\sigma_X}{\sqrt{\overline{V}_M}} \xrightarrow{\mathcal{L}} \mathcal{N}(0;1) \quad \text{as} \quad M \to +\infty.$$

Of course, within the usual range of Monte Carlo simulations, one can always consider that, for large M,

---

[2] When X "already" has an $\mathcal{N}(0;1)$ distribution, then $\overline{V}_M$ has a $\chi^2(M-1)$-distribution. The $\chi^2(\nu)$-distribution, known as the χ^2-distribution with $\nu \in \mathbb{N}^*$ degrees of freedom, is the distribution of the sum $Z_1^2 + \cdots + Z_\nu^2$, where $Z_1, \ldots, Z_\nu$ are i.i.d. with $\mathcal{N}(0;1)$ distribution. The loss of one degree of freedom for $\overline{V}_M$ comes from the fact that $X_1, \ldots, X_M$ and $\overline{X}_M$ satisfies a linear equation which induces a linear constraint. This result is known as Cochran's Theorem.

[3] If a sequence of $\mathbb{R}^d$-valued random vectors satisfy $Z_n \xrightarrow{\mathcal{L}} Z_\infty$ and a sequence of random variables $C_n \xrightarrow{\mathbb{P}} c_\infty \in \mathbb{R}$, then $C_n Z_n \xrightarrow{\mathcal{L}} c_\infty Z_\infty$, see *e.g.* [155].

$$\sqrt{M}\,\frac{\overline{X}_M - m_{_X}}{\sqrt{\overline{V}_{_M}}} \simeq \mathcal{N}(0;1).$$

Note that when X is itself normally distributed, one shows that the empirical mean $\overline{X}_{_M}$ and the empirical variance $\overline{V}_{_M}$ are independent, so that the true distribution of the left-hand side of the above (approximate) equation is a *Student distribution with* $M-1$ *degrees of freedom* ([4]), denoted by $T(M-1)$.

Finally, one defines the *confidence interval* at level α of the Monte Carlo simulation by

$$I^{\alpha}_{_M} = \left[\overline{X}_M - q_\alpha\sqrt{\frac{\overline{V}_{_M}}{M}},\ \overline{X}_M + q_\alpha\sqrt{\frac{\overline{V}_{_M}}{M}}\right]$$

which will still satisfy (for large M)

$$\mathbb{P}\big(m_{_X}\in I^{\alpha}_{_M}\big) \simeq \mathbb{P}\big(|\mathcal{N}(0;1)| \le q_\alpha\big) = \alpha.$$

For numerical implementation one often considers $q_\alpha = 2$, which corresponds to the confidence level $\alpha = 95.44\,\% \simeq 95\,\%$. With the constant increase in performance of computing devices, higher confidence levels α became common, typically $\alpha = 99\%$ or 99.5% (like for the computation of Value-at-Risk, following Basel II recommendations).

The "academic" birth date of the Monte Carlo method is 1949, with the publication of a seemingly seminal paper by Metropolis and Ulam "The Monte Carlo method" in *J. of American Statistics Association* (**44**, 335–341). However, the method had already been extensively used for several years as a secret project of the U.S. Defense Department.

One can also consider that, in fact, the Monte Carlo method goes back to the celebrated Buffon's needle experiment and should subsequently be credited to the French naturalist Georges Louis Leclerc, Comte de Buffon (1707–1788).

As concerns quantitative Finance and more precisely the pricing of derivatives, it seems difficult to trace the origin of the implementation of the Monte Carlo method for the pricing and hedging of options. In the academic literature, the first academic paper dealing in a systematic manner with a Monte Carlo approach seems to go back to Boyle in [50].

[4]The Student distribution with ν degrees of freedom, denoted by $T(\nu)$, is the law of the random variable $\dfrac{\sqrt{\nu}Z_0}{\sqrt{Z_1^2+\cdots+Z_\nu^2}}$, where $Z_0, Z_1, \ldots, Z_\nu$ are i.i.d. with $\mathcal{N}(0;1)$ distribution. As expected for the coherence of the preceding, one shows that $T(\nu)$ converges in distribution to the normal distribution $\mathcal{N}(0;1)$ as $\nu \to +\infty$.

2.1.3 *Vanilla Option Pricing in a Black–Scholes Model: The Premium*

For the sake of simplicity, we consider a 2-dimensional risk-neutral correlated Black–Scholes model for two risky assets X^1 and X^2 under its unique risk neutral probability (but a general d-dimensional model can be defined likewise):

$$\begin{aligned} &dX_t^0 = rX_t^0 dt,\ X_0^0 = 1, \\ &dX_t^1 = X_t^1\big(rdt + \sigma_1 dW_t^1\big),\ X_0^1 = x_0^1,\ dX_t^2 = X_t^2\big(rdt + \sigma_2 dW_t^2\big),\ X_0^2 = x_0^2, \end{aligned} \tag{2.2}$$

with the usual notations (r interest rate, $\sigma_i > 0$ volatility of X^i). In particular, $W = (W^1, W^2)$ denotes a *correlated* bi-dimensional Brownian motion such that

$$\langle W^1, W^2\rangle_t = \rho t,\ \rho \in [-1, 1].$$

This implies that W^2 can be decomposed as $W_t^2 = \rho W_t^1 + \sqrt{1-\rho^2}\,\widetilde{W}_t^2$, where $(W^1, \widetilde{W}^2)$ is a standard 2-dimensional Brownian motion. The filtration $(\mathcal{F}_t)_{t\in[0,T]}$ of this market is the augmented filtration of W, *i.e.* $\mathcal{F}_t = \mathcal{F}_t^W := \sigma(W_s,\ 0 \le s \le t,\ \mathcal{N}_{\mathbb{P}})$ where $\mathcal{N}_{\mathbb{P}}$ denotes the family of $\mathbb{P}$-negligible sets of $\mathcal{A}$ ([5]). By "filtration of the market", we mean that $(\mathcal{F}_t)_{t\in[0,T]}$ is the smallest filtration satisfying the usual conditions to which the process $(X_t)_{t\in[0,T]}$ is adapted. By "risk-neutral", we mean that $e^{-rt}X_t$ is a $\big(\mathbb{P}, (\mathcal{F}_t)_t\big)$-martingale. We will not go further into financial modeling at this stage, for which we refer *e.g.* to [185] or [163], but focus instead on numerical aspects.

For every $t \in [0, T]$, we have

$$X_t^0 = e^{rt},\quad X_t^i = x_0^i e^{(r-\frac{\sigma_i^2}{2})t+\sigma_i W_t^i},\ i = 1, 2.$$

(One easily verifies using Itô's Lemma, see Sect. 12.8, that X_t thus defined satisfies (2.2); formally finding the solution by applying Itô's Lemma to $\log X_t^i$, $i = 1, 2$, assuming *a priori* that the solutions of (2.2) are positive).

When $r = 0$, X^i is called a geometric Brownian motion associated to W^i with volatility $\sigma_i > 0$.

A European *vanilla* option with maturity $T > 0$ is an option related to a European payoff

$$h_T := h(X_T)$$

which only depends on X at time T. In such a complete market the option premium at time 0 is given by

$$V_0 = e^{-rt}\mathbb{E}\, h(X_T)$$

[5] One shows that, owing to the 0-1 Kolmogorov law, this filtration is right continuous, *i.e.* $\mathcal{F}_t = \bigcap_{s>t}\mathcal{F}_s$. A right continuous filtration which contains the $\mathbb{P}$-negligible sets satisfies the so-called "usual conditions".

and more generally, at any time $t \in [0, T]$,

$$V_t = e^{-r(T-t)}\mathbb{E}\left(h(X_T) \mid \mathcal{F}_t\right).$$

The fact that W has independent stationary increments implies that X^1 and X^2 have independent stationary ratios, *i.e.*

$$\left(\frac{X^i_T}{X^i_t}\right)_{i=1,2} \stackrel{d}{=} \left(\frac{X^i_{T-t}}{x^i_0}\right)_{i=1,2} \quad \text{is independent of } \mathcal{F}_t.$$

As a consequence, if we define for every $T > 0$, $x_0 = (x^1_0, x^2_0) \in (0, +\infty)^2$,

$$v(x_0, T) = e^{-rt}\mathbb{E}\, h(X_T),$$

then

$$\begin{aligned}
V_t &= e^{-r(T-t)}\mathbb{E}\left(h(X_T) \mid \mathcal{F}_t\right)\\
&= e^{-r(T-t)}\mathbb{E}\left(h\left(X^i_t \times \left(\frac{X^i_T}{X^i_t}\right)_{i=1,2}\right) \Big| \mathcal{F}_t\right)\\
&= e^{-r(T-t)}\left[\mathbb{E}\, h\left(\left(x^i \frac{X^i_{T-t}}{x^i_0}\right)_{i=1,2}\right)\right]_{|x^i = X^i_t,\, i=1,2} \qquad \text{by independence}\\
&= v(X_t, T-t).
\end{aligned}$$

⊳ **Examples. 1.** *Vanilla call* with strike price K:

$$h(x^1, x^2) = \left(x^1 - K\right)_+.$$

There is a closed form for such a *call* option – the celebrated *Black–Scholes formula* for option on stock (without dividend) – given by

$$\text{Call}^{BS}_0 = C(x_0, K, r, \sigma_1, T) = x_0\Phi_0(d_1) - e^{-rt}K\,\Phi_0(d_2) \tag{2.3}$$

$$\text{with } d_1 = \frac{\log(x_0/K) + (r + \frac{\sigma_1^2}{2})T}{\sigma_1\sqrt{T}}, \quad d_2 = d_1 - \sigma_1\sqrt{T}, \tag{2.4}$$

where Φ_0 denotes the c.d.f. of the $\mathcal{N}(0; 1)$-distribution.

2. *Best-of-call* with strike price K:

$$h_T = \left(\max(X^1_T, X^2_T) - K\right)_+.$$

A quasi-closed form is available involving the distribution function of the bi-variate (correlated) normal distribution. Laziness may lead to price it by Monte Carlo simulation (a *PDE* approach is also appropriate but needs more care) as detailed below.

3. *Exchange Call Spread* with strike price K:

$$h_T = \big((X_T^1 - X_T^2) - K\big)_+.$$

For this payoff no closed form is available. One has a choice between a *PDE* approach (quite appropriate in this 2-dimensional setting but requiring some specific developments) and a Monte Carlo simulation.

We will illustrate below the regular Monte Carlo procedure on the example of a *Best-of-Call* which is traded on an organized market, unlike its cousin the *Exchange Call Spread*.

Pricing a *best-of-call* by a monte carlo simulation

To implement a (crude) Monte Carlo simulation we need to write the payoff as a function of independent uniformly distributed random variables, or, equivalently, as a tractable function of such random variables. In our case, we write it as a function of two standard normal variables, *i.e.* a bi-variate standard normal distribution (Z^1, Z^2), namely

$$\begin{aligned} e^{-rt} h_T &\stackrel{d}{=} \varphi(Z^1, Z^2) \\ &:= \left(\max\left(x_0^1 \exp\left(-\frac{\sigma_1^2}{2}T + \sigma_1\sqrt{T}Z^1\right), x_0^2 \exp\left(-\frac{\sigma_2^2}{2}T + \sigma_2\sqrt{T}\big(\rho Z^1 + \sqrt{1-\rho^2}Z^2\big)\right)\right) - Ke^{-rt}\right)_+ \end{aligned}$$

where $Z = (Z^1, Z^2) \stackrel{d}{=} \mathcal{N}(0; I_2)$ (the dependence of φ in x_0^i, etc, is dropped). Then, simulating a M-sample $(Z_m)_{1 \le m \le M}$ of the $\mathcal{N}(0; I_2)$ distribution using *e.g.* the Box–Muller method yields the estimate

$$\begin{aligned} Best\text{-}of\text{-}Call_0 &= e^{-rt}\mathbb{E}\Big(\big(\max(X_T^1, X_T^2) - K\big)_+\Big) \\ &= \mathbb{E}\,\varphi(Z^1, Z^2) \\ &\simeq \overline{\varphi}_M := \frac{1}{M}\sum_{m=1}^{M} \varphi(Z_m). \end{aligned}$$

One computes an estimate for the variance using the same sample

$$\overline{V}_M(\varphi) = \frac{1}{M-1}\sum_{m=1}^{M} \varphi(Z_m)^2 - \frac{M}{M-1}\overline{\varphi}_M^2 \simeq \mathrm{Var}(\varphi(Z))$$

since M is large enough. Then one designs a confidence interval for $\mathbb{E}\,\varphi(Z)$ at level $\alpha \in (0, 1)$ by setting

$$I_M^\alpha = \left[\overline{\varphi}_M - q_\alpha \sqrt{\frac{\overline{V}_M(\varphi)}{M}}, \overline{\varphi}_M + q_\alpha \sqrt{\frac{\overline{V}_M(\varphi)}{M}}\right]$$

where q_α is defined by $\mathbb{P}\big(|\mathcal{N}(0;1)| \le q_\alpha\big) = \alpha$ (or equivalently by $2\Phi_0(q_\alpha) - 1 = \alpha$).

▷ *Numerical Application.* We consider a European "Best-of-Call" option with the following parameters

$$r = 0.1,\ \sigma_i = 0.2 = 20\,\%,\ \rho = 0.5,\ X_0^i = 100,\ T = 1,\ K = 100.$$

The confidence level is set at $\alpha = 0.95$.

The Monte Carlo simulation parameters are $M = 2^m$, $m = 10, \dots, 20$ (keep in mind that $2^{10} = 1024$). The (typical) results of a numerical simulation are reported in Table 2.1 below, see also Fig. 2.1.

▶ **Exercise.** Proceed likewise with an *Exchange Call Spread* option.

Remark. Once the script is written for one option, *i.e.* one *payoff* function, it is almost instantaneous to modify it to price another option based on a new payoff function: the Monte Carlo method is very flexible, much more than a *PDE* approach.

Table 2.1 BLACK–SCHOLES BEST - OF- CALL. *Pointwise estimate and confidence intervals as a function of the size M of the Monte Carlo simulation, $M = 2^k$, $k = 10, \dots, 20, 2^{30}$*

M	$\overline{\varphi}_M$	$I_{\alpha,M}$
$2^{10} = 1\,024$	18.5133	[17.4093; 19.6173]
$2^{11} = 2\,048$	18.7398	[17.9425; 19.5371]
$2^{12} = 4\,096$	18.8370	[18.2798; 19.3942]
$2^{13} = 8\,192$	19.3155	[18.9196; 19.7114]
$2^{14} = 16\,384$	19.1575	[18.8789; 19.4361]
$2^{15} = 32\,768$	19.0911	[18.8936; 19.2886]
$2^{16} = 65\,536$	19.0475	[18.9079; 19.1872]
$2^{17} = 131\,072$	19.0566	[18.9576; 19.1556]
$2^{18} = 262\,144$	19.0373	[18.9673; 19.1073]
$2^{19} = 524\,288$	19.0719	[19.0223; 19.1214]
$2^{20} = 1\,048\,576$	19.0542	[19.0191 19.0892]
...	...	...
$2^{30} = 1.0737.10^9$	19.0766	[19.0756; 19.0777]

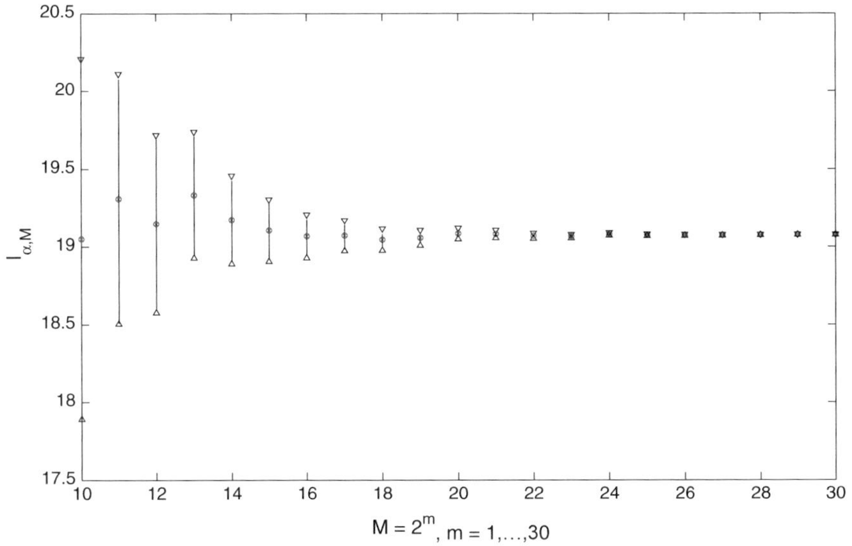

Fig. 2.1 BLACK–SCHOLES BEST- OF- CALL. The Monte Carlo estimator (⊗) and its confidence interval at level $\alpha = 95\%$ for sizes $M \in \{2^k,\ k = 10, \ldots, 30\}$

2.1.4 ℵ *Practitioner's Corner*

The practitioner should never forget that performing a **Monte Carlo simulation** to compute $m_{_X} = \mathbb{E}\, X$, consists of **three mandatory steps**:

1. Specification of a confidence level $\alpha \in (0, 1)$ ($\alpha \simeq 1$, typically, $\alpha = 95\%, 99\%, 99.5\%$, etc).
2. Simulation of an M-sample $X_1, X_2, \ldots, X_M$ of i.i.d. random vectors having the same distribution as X and (possibly recursive) computation of both its empirical mean $\bar{X}_{_M}$ and its empirical variance $\bar{V}_{_M}$.
3. Computation of the resulting *confidence interval I_M at confidence level α, which will be the only trustable result of the Monte Carlo simulation.*

We will see further on in Chap. 3 how to specify the size M of the simulation to comply with an *a priori* accuracy level. The case of *biased* simulation is deeply investigated in Chap. 9 (see Sect. 9.3 for the analysis of a crude Monte Carlo simulation in a *biased* framework).

2.2 Greeks (Sensitivity to the Option Parameters): A First Approach

2.2.1 *Background on Differentiation of Functions Defined by an Integral*

The *greeks* or sensitivities denote the set of parameters obtained as derivatives of the premium of an option with respect to some of its parameters: the starting value, the volatility, etc. It applies more generally to any function defined by an expectation. In elementary situations, one simply needs to apply some more or less standard theorems like the one reproduced below (see [52], Chap. 8 for a proof). A typical example of such a "elementary situation" is the case of a possibly multi-dimensional risk neutral Black–Scholes model.

Theorem 2.2 (Interchanging differentiation and expectation) *Let* $(\Omega, \mathcal{A}, \mathbb{P})$ *be a probability space, let* I *be a nontrivial interval of* $\mathbb{R}$. *Let* $\varphi : I \times \Omega \to \mathbb{R}$ *be a* $\mathcal{B}or(I) \otimes \mathcal{A}$*-measurable function.*

(a) LOCAL VERSION. *Let* $x_0 \in I$. *If the function* φ *satisfies:*

(i) *for every* $x \in I$, *the random variable* $\varphi(x, \,.\,) \in L^1_{\mathbb{R}}(\Omega, \mathcal{A}, \mathbb{P})$,

(ii) $\mathbb{P}(d\omega)$*-a.s.* $\dfrac{\partial \varphi}{\partial x}(x_0, \omega)$ *exists,*

(iii) *there exists a* $Y \in L^1_{\mathbb{R}_+}(\mathbb{P})$ *such that, for every* $x \in I$,

$$\mathbb{P}(d\omega)\text{-}a.s. \;\; \big|\varphi(x, \omega) - \varphi(x_0, \omega)\big| \le Y(\omega)|x - x_0|,$$

then, the function $\Phi(x) := \mathbb{E}\,\varphi(x, \,.\,) = \displaystyle\int_\Omega \varphi(x, \omega)\mathbb{P}(d\omega)$ *is defined at every* $x \in I$, *differentiable at* x_0 *with derivative*

$$\Phi'(x_0) = \mathbb{E}\left(\frac{\partial \varphi}{\partial x}(x_0, \cdot)\right).$$

(b) GLOBAL VERSION. *If* φ *satisfies* (i) *and*

$(ii)_{glob}$ $\mathbb{P}(d\omega)$*-a.s.,* $\dfrac{\partial \varphi}{\partial x}(x, \omega)$ *exists at every* $x \in I$,

$(iii)_{glob}$ *There exists a* $Y \in L^1_{\mathbb{R}_+}(\Omega, \mathcal{A}, \mathbb{P})$ *such that, for every* $x \in I$,

$$\mathbb{P}(d\omega)\text{-}a.s. \;\; \left|\frac{\partial \varphi(x, \omega)}{\partial x}\right| \le Y(\omega),$$

then, the function $\Phi(x) := \mathbb{E}\,\varphi(x, \,.\,)$ *is defined and differentiable at every* $x \in I$, *with derivative*

$$\Phi'(x) = \mathbb{E}\left(\frac{\partial \varphi}{\partial x}(x, \,.\,)\right).$$

Remarks. • The local version of the above theorem may be necessary to prove the differentiability of a function defined by an expectation over the whole real line (see the exercise that follows).
• All of the preceding remains true if one replaces the probability space $(\Omega, \mathcal{A}, \mathbb{P})$ by any measurable space $(E, \mathcal{E}, \mu)$ where μ is a non-negative measure (see again Chap. 8 in [52]). However, this extension is no longer true as seen when dealing with the uniform integrability assumption mentioned in the exercises hereafter.
• Some variants of the result can be established to obtain a theorem for right or left differentiability of functions $\Phi = \mathbb{E}_\omega \varphi(x, \omega)$ defined on the real line, (partially) differentiable functions defined on $\mathbb{R}^d$, for holomorphic functions on $\mathbb{C}$, etc. The proofs follow the same lines.
• There exists a local continuity result for such functions φ defined as an expectation which is quite similar to Claim (a). The domination property by an integrable non-negative random variable Y is requested on $\varphi(x, \omega)$ itself. A precise statement is provided in Chap. 12 (with the same notations). For a proof we still refer to [52], Chap. 8. This result is often useful to establish the (continuous) differentiability of a multi-variable function by combining the existence and the continuity of its partial derivatives.

► **Exercise.** Let $Z \stackrel{d}{=} \mathcal{N}(0; 1)$ be defined on a probability space $(\Omega, \mathcal{A}, \mathbb{P})$, $\varphi(x, \omega) = \big(x - Z(\omega)\big)_+$ and $\Phi(x) = \mathbb{E}\,\varphi(x, Z) = \mathbb{E}\,(x - Z)_+$, $x \in \mathbb{R}$.
(a) Show that Φ is differentiable on the real line by applying the local version of Theorem 2.2 and compute its derivative.
(b) Let I denote a non-trivial interval of $\mathbb{R}$. Show that if $\omega \in \{Z \in I\}$ (*i.e.* $Z(\omega) \in I$), the function $x \mapsto \big(x - Z(\omega)\big)_+$ is never differentiable on the whole interval I.

► **Exercises (*Extension to uniform integrability*).** One can replace the domination property (iii) in Claim (a) (local version) of the above Theorem 2.2 by the less stringent *uniform integrability* assumption

$$(iii)_{ui} \quad \left(\frac{\varphi(x, .) - \varphi(x_0, .)}{x - x_0}\right)_{x \in I \setminus \{x_0\}} \quad \text{is } \mathbb{P}\text{-uniformly integrable on } (\Omega, \mathcal{A}, \mathbb{P}).$$

For the definition and some background on uniform integrability, see Chap. 12, Sect. 12.4.

1. Show that $(iii)_{ui}$ implies (iii).
2. Show that (i)–(ii)–$(iii)_{ui}$ implies the conclusion of Claim (a) (local version) in the above Theorem 2.2.
3. State a "uniform integrable" counterpart of $(iii)_{glob}$ to extend Claim (b) (global version) of Theorem 2.2.
4. Show that uniform integrability of the above family of random variables follows from its L^p-boundedness for (any) $p > 1$.

2.2.2 *Working on the Scenarii Space (Black–Scholes Model)*

To illustrate the different methods to compute the sensitivities, we will consider the one dimensional risk-neutral Black–Scholes model with constant interest rate r and volatility $\sigma > 0$:

$$dX_t^x = X_t^x\big(r\,dt + \sigma dW_t\big), \quad X_0^x = x > 0,$$

so that $X_t^x = x\exp\big((r - \frac{\sigma^2}{2})t + \sigma W_t\big)$. Then, we consider for every $x \in (0, +\infty)$,

$$\Phi(x, r, \sigma, T) := \mathbb{E}\,\varphi(X_T^x), \tag{2.5}$$

where $\varphi : (0, +\infty) \to \mathbb{R}$ lies in $L^1(\mathbb{P}_{X_T^x})$ for every $x \in (0, +\infty)$ and $T > 0$. This corresponds (when φ is non-negative), to *vanilla payoffs* with maturity T. However, we skip on purpose the discounting factor in what follows to alleviate notation: one can always imagine it is included as a constant in the function φ since we will work at a fixed time T. The updating of formulas is obvious. Note that, in many cases, new parameters directly attached to the function φ itself are of interest, typically the strike price K when $\varphi(x) = (x - K)_+$ (call option), $(K - x)_+$ (Put option), $|x - K|$ (butterfly option), etc.

First, we are interested in computing the first two derivatives $\Phi'(x)$ and $\Phi''(x)$ of the function Φ which correspond (up to the discounting factor) to the δ-hedge of the option and its γ parameter, respectively. The second parameter γ is involved in the so-called "tracking error". But other sensitivities are of interest to the practitioners like the *vega*, *i.e.* the derivative of the (discounted) function Φ with respect to the volatility parameter, the ρ (derivative with respect to the interest rate r), etc. The aim is to derive representations of these sensitivities as expectations in order to compute them using a Monte Carlo simulation, in parallel with the computation of the premium.

The Black–Scholes model is here clearly a toy model to illustrate our approach since, for such a model, closed forms s exist for standard payoffs (Call, Put and their linear combinations) and more efficient methods can be successfully implemented like solving the *PDE* derived from the Feynman–Kac formula (see Theorem 7.11). At least in a one – like here – or a low-dimensional model (say $d \le 3$) using a finite difference (elements, volumes) scheme after an appropriate change of variable. For *PDE* methods we refer to [2].

We will first work on the scenarii space $(\Omega, \mathcal{A}, \mathbb{P})$, because this approach contains the "seed" of methods that can be developed in much more general settings in which the *SDE* no longer has an explicit solution. On the other hand, as soon as an explicit expression is available for the density $p_T(x, y)$ of X_T^x, it is more efficient to use the method described in the next Sect. 2.2.3.

Proposition 2.1 *(a) If $\varphi : (0, +\infty) \to \mathbb{R}$ is differentiable and φ' has polynomial growth, then the function Φ defined by (2.5) is differentiable and*

$$\forall x > 0, \quad \Phi'(x) = \mathbb{E}\left(\varphi'(X_T^x)\frac{X_T^x}{x}\right). \tag{2.6}$$

(b) If $\varphi : (0, +\infty) \to \mathbb{R}$ is differentiable outside a countable set and is locally Lipschitz continuous with polynomial growth *in the following sense*

$$\exists\, m \geq 0,\ \exists\, C > 0,\ \forall\, u,\ v \in \mathbb{R}_+, \quad |\varphi(u) - \varphi(v)| \leq C|u - v|(1 + |u|^m + |v|^m),$$

then Φ is differentiable everywhere on $(0, +\infty)$ and Φ' is given by (2.6).
(c) If $\varphi : (0, +\infty) \to \mathbb{R}$ is simply a Borel function with polynomial growth, then Φ is still differentiable and

$$\forall x > 0, \quad \Phi'(x) = \mathbb{E}\left(\varphi(X_T^x)\frac{W_T}{x\,\sigma\,T}\right). \tag{2.7}$$

Remark. The above formula (2.7) can be seen as a first example of *Malliavin weight* to compute a greek – here the δ-hedge – and the first method that we will use to establish it, based on an integration by parts, as the embryo of the so-called Malliavin Monte Carlo approach to Greek computation. See, among other references, [47] for more developments in this direction when the underlying diffusion process is no longer an elementary geometric Brownian motion.

Proof. (a) This straightforwardly follows from the explicit expression for X_T^x and the above differentiation Theorem 2.2 (global version): for every $x \in (0, +\infty)$,

$$\frac{\partial}{\partial x}\varphi(X_T^x) = \varphi'(X_T^x)\frac{\partial X_T^x}{\partial x} = \varphi'(X_T^x)\frac{X_T^x}{x}.$$

Now $|\varphi'(u)| \leq C\big(1 + |u|^m\big)$, where $m \in \mathbb{N}$ and $C \in (0, +\infty)$, so that, if $0 < x \leq L < +\infty$,

$$\left|\frac{\partial}{\partial x}\varphi(X_T^x)\right| \leq C'_{r,\sigma,T}\Big(1 + L^m \exp\big((m+1)\sigma W_T\big)\Big) \in L^1(\mathbb{P}),$$

where $C'_{r,\sigma,T}$ is another positive real constant. This yields the domination condition of the derivative.
(b) This claim follows from Theorem 2.2(a) (local version) and the fact that, for every $T > 0$, $\mathbb{P}(X_T^x = \xi) = 0$ for every $\xi \geq 0$.
(c) First, we still assume that the assumption (a) is in force. Then

$$\begin{aligned}
\Phi'(x) &= \int_{\mathbb{R}} \varphi'\big(x \exp\big(\mu T + \sigma\sqrt{T}u\big)\big) \exp\big(\mu T + \sigma\sqrt{T}u\big) e^{-u^2/2} \frac{du}{\sqrt{2\pi}} \\
&= \frac{1}{x\,\sigma\sqrt{T}} \int_{\mathbb{R}} \frac{\partial \varphi\big(x \exp\big(\mu T + \sigma\sqrt{T}u\big)\big)}{\partial u} e^{-u^2/2} \frac{du}{\sqrt{2\pi}} \\
&= -\frac{1}{x\,\sigma\sqrt{T}} \int_{\mathbb{R}} \varphi\big(x \exp\big(\mu T + \sigma\sqrt{T}u\big)\big) \frac{\partial e^{-u^2/2}}{\partial u} \frac{du}{\sqrt{2\pi}} \\
&= \frac{1}{x\,\sigma\sqrt{T}} \int_{\mathbb{R}} \varphi\big(x \exp\big(\mu T + \sigma\sqrt{T}u\big)\big) u\, e^{-u^2/2} \frac{du}{\sqrt{2\pi}} \\
&= \frac{1}{x\,\sigma T} \int_{\mathbb{R}} \varphi\big(x \exp\big(\mu T + \sigma\sqrt{T}u\big)\big) \sqrt{T}\, u\, e^{-u^2/2} \frac{du}{\sqrt{2\pi}},
\end{aligned}$$

where we used an integration by parts to obtain the third equality, taking advantage of the fact that, owing to the polynomial growth assumptions on φ,

$$\lim_{|u|\to+\infty} \varphi\big(x \exp\big(\mu T + \sigma\sqrt{T}u\big)\big) e^{-u^2/2} = 0.$$

Finally, returning to Ω,

$$\Phi'(x) = \frac{1}{x\,\sigma T}\, \mathbb{E}\big(\varphi(X_T^x) W_T\big). \tag{2.8}$$

When φ is not differentiable, let us first sketch the extension of the formula by a density argument. When φ is continuous and has compact support in $\mathbb{R}_+$, one may assume without loss of generality that φ is defined on the whole real line as a continuous function with compact support. Then φ can be uniformly approximated by differentiable functions φ_n with compact support (use a convolution by *mollifiers*, see [52], Chap. 8). Then, with obvious notations, $\Phi_n'(x) := \frac{1}{x\,\sigma T} \mathbb{E}\left(\varphi_n(X_T^x) W_T\right)$ converges uniformly on compact sets of $(0, +\infty)$ to $\Phi'(x)$ defined by (2.8) since

$$|\Phi_n'(x) - \Phi'(x)| \le \|\varphi_n - \varphi\|_{\sup} \frac{\mathbb{E}\,|W_T|}{x\,\sigma\, T}.$$

Furthermore, $\Phi_n(x)$ converges (uniformly) toward $f(x)$ on $(0, +\infty)$. Consequently, Φ is differentiable with derivative f'. ◇

Remark. We will see in the next section a much quicker way to establish claim (c). The above method of proof, based on an integration by parts, can be seen as a toy-introduction to a systematic way to produce random weights like $\frac{W_T}{x\sigma T}$ in the differentiation procedure of Φ, especially when the differential of φ does not exist. The most general extension of this approach, developed on the Wiener space ([6]) for functionals of the Brownian motion is known as the Malliavin-Monte Carlo method.

[6]The Wiener space $\mathcal{C}(\mathbb{R}_+, \mathbb{R}^d)$ and its Borel σ-field for the topology of uniform convergence on compact sets, namely $\sigma\big(\omega \mapsto \omega(t),\ t \in \mathbb{R}_+\big)$, endowed with the Wiener measure, *i.e.* the distribution of a standard d-dimensional Brownian motion $(W^1, \ldots, W^d)$.

► **Exercise** (*Extension to Borel functions with polynomial growth*). (a) Show that as soon as φ is a Borel function with polynomial growth, the function f defined by (2.5) is continuous. [Hint: use that the distribution X_T^x has a probability density $p_T(x, y)$ on the positive real line which continuously depends on x and apply the continuity theorem for functions defined by an integral, see Theorem 12.5(a) in the Miscellany Chap. 12.]
(b) Show that (2.8) holds true as soon as φ is a bounded Borel function. [Hint: Apply the Functional Monotone Class Theorem (see Theorem 12.2 in the Miscellany Chap. 12) to an appropriate vector subspace of functions φ and use the Baire σ-field Theorem.]
(c) Extend the result to Borel functions φ with polynomial growth. [Hint: use that $\varphi(X_T^x) \in L^1(\mathbb{P})$ and $\varphi = \lim_n \varphi_n$ with $\varphi_n = (n \wedge \varphi \vee (-n))$.]
(d) Derive from the preceding a simple expression for $\Phi(x)$ when $\varphi = \mathbf{1}_I$ is the indicator function of a nontrivial interval I.

COMMENTS: The extension to Borel functions φ always needs at some place an argument based on the regularizing effect of the diffusion induced by the Brownian motion. As a matter of fact, if X_t^x were the solution to a regular *ODE* this extension to non-continuous functions φ would fail. We propose in the next section an approach – the log-likelihood method – directly based on this regularizing effect through the direct differentiation of the probability density $p_T(x, y)$ of X_T^x.

► **Exercise.** Prove claim (b) in detail.

Note that the assumptions of claim (b) are satisfied by usual payoff functions like $\varphi_{Call}(x) = (x - K)_+$ or $\varphi_{Put} := (K - x)_+$ (when X_T^x has a continuous distribution). In particular, this shows that

$$\frac{\partial \mathbb{E}\left(\varphi_{Call}(X_T^x)\right)}{\partial x} = \mathbb{E}\left(\mathbf{1}_{\{X_T^x \geq K\}} \frac{X_T^x}{x}\right).$$

The computation of this quantity – which is part of that of the Black–Scholes formula – finally yields as expected:

$$\frac{\partial \mathbb{E}\left(\varphi_{Call}(X_T^x)\right)}{\partial x} = e^{rt}\,\Phi_0(d_1),$$

where d_1 is given by (2.4) (keep in mind that the discounting factor is missing).

► **Exercises. 0.** *A comparison.* Try a direct differentiation of the Black–Scholes formula (2.3) and compare with a (formal) differentiation based on Theorem 2.2. You should find by both methods

$$\frac{\partial Call_0^{BS}}{\partial x}(x) = \Phi_0(d_1).$$

But the true question is: "how long did it take you to proceed?"

1. *Application to the computation of the* γ *(i.e.* $\Phi''(x)$*).* Show that, if φ is differentiable with a derivative having polynomial growth,

$$\Phi''(x) := \frac{1}{x^2\sigma T}\mathbb{E}\Big(\big(\varphi'(X_T^x)X_T^x - \varphi(X_T^x)\big)W_T\Big)$$

and that, if φ is continuous with compact support,

$$\Phi''(x) := \frac{1}{x^2\sigma T}\mathbb{E}\left(\varphi(X_T^x)\left(\frac{W_T^2}{\sigma T} - W_T - \frac{1}{\sigma}\right)\right).$$

Extend this identity to the case where φ is simply Borel with polynomial growth. Note that a (somewhat simpler) formula also exists when the function φ is itself twice differentiable, but such a smoothness assumption is not realistic, at least for financial applications.

2. *Variance reduction for the* δ ([7]). The above formulas are clearly not the unique representations of the δ as an expectation: using that $\mathbb{E}\, W_T = 0$ and $\mathbb{E}\, X_T^x = xe^{rt}$, one derives immediately that

$$\Phi'(x) = \varphi'(xe^{rt})e^{rt} + \mathbb{E}\left(\big(\varphi'(X_T^x) - \varphi'(xe^{rt})\big)\frac{X_T^x}{x}\right)$$

as soon as φ is differentiable at xe^{rt}. When φ is simply Borel

$$\Phi'(x) = \frac{1}{x\,\sigma\,T}\mathbb{E}\Big(\big(\varphi(X_T^x) - \varphi(xe^{rt})\big)W_T\Big).$$

3. *Variance reduction for the* γ. Show that

$$\Phi''(x) = \frac{1}{x^2\sigma T}\mathbb{E}\Big(\big(\varphi'(X_T^x)X_T^x - \varphi(X_T^x) - xe^{rt}\varphi'(xe^{rt}) + \varphi(xe^{rt})\big)W_T\Big).$$

4. *Testing the variance reduction, if any.* Although the former two exercises are entitled "variance reduction" the above formulas do not guarantee a variance reduction at a fixed time T. It seems intuitive that they do only when the maturity T is small. Perform some numerical experiments to test whether or not the above formulas induce some variance reduction.

As the maturity increases, test whether or not the regression method introduced in Sect. 3.2 works with these "control variates".

5. *Computation of the vega* ([8]). Show likewise that $\mathbb{E}\,\varphi(X_T^x)$ is differentiable with respect to the volatility parameter σ under the same assumptions on φ, namely

[7] In this exercise we slightly anticipate the next chapter, which is entirely devoted to variance reduction.

[8] Which is not a greek letter...

$$\frac{\partial}{\partial\sigma}\mathbb{E}\,\varphi(X_T^x) = \mathbb{E}\Big(\varphi'(X_T^x)X_T^x\big(W_T - \sigma T\big)\Big)$$

if φ is differentiable with a derivative having polynomial growth. Derive without any further computations – but with the help of the previous exercises – that

$$\frac{\partial}{\partial\sigma}\mathbb{E}\,\varphi(X_T^x) = \mathbb{E}\left(\varphi(X_T^x)\left(\frac{W_T^2}{\sigma T} - W_T - \frac{1}{\sigma}\right)\right)$$

if φ is simply Borel with polynomial growth. [Hint: use the former exercises.]

This derivative is known (up to an appropriate discounting) as the *vega* of the option related to the payoff $\varphi(X_T^x)$. Note that the γ and the *vega* of a Call satisfy (after discounting by e^{-rt})

$$vega(x, K, r, \sigma, T) = x^2\sigma T\gamma(x, K, r, \sigma, T),$$

which is the key of the tracking error formula.

In fact, the beginning of this section can be seen as an introduction to the so-called tangent process method (see Sect. 2.2.4 at the end of this chapter and Sect. 10.2.2).

2.2.3 Direct Differentiation on the State Space: The Log-Likelihood Method

In fact, the formulas established in the former section for the Black–Scholes model can be obtained by working directly on the state space $(0, +\infty)$, taking advantage of the fact that X_T^x has a smooth and explicit probability density $p_T(x, y)$ with respect to the Lebesgue measure on $(0, +\infty)$, *which is known explicitly* since it is a log-normal distribution.

This probability density also depends on the other parameters of the model like the volatility, σ, the interest rate r and the maturity T. Let us denote by θ one of these parameters which is assumed to lie in a parameter set Θ. More generally, we could imagine that $X_T^x(\theta)$ is an $\mathbb{R}^d$-valued solution at time T to a stochastic differential equation whose coefficients $b(\theta, x)$ and $\sigma(\theta, x)$ depend on a parameter $\theta \in \Theta \subset \mathbb{R}$. An important result of stochastic analysis for Brownian diffusions is that, under uniform ellipticity assumptions (or the less stringent "parabolic Hörmander ellipticity assumptions", see [24, 139]), combined with smoothness assumptions on both the drift and the diffusion coefficient, such a solution of an *SDE* does have a smooth density $p_T(\theta, x, y)$ – at least in (x, y) – with respect to the Lebesgue measure on $\mathbb{R}^d$. For more details, we refer to [25] or [11, 98]. Formally, we then get

$$\Phi(\theta) = \mathbb{E}\,\varphi(X_T^x(\theta)) = \int_{\mathbb{R}^d}\varphi(y)p_T(\theta, x, y)\mu(dy)$$

so that, formally,

$$\begin{aligned}\Phi'(\theta) &= \int_{\mathbb{R}^d} \varphi(y)\frac{\partial p_T}{\partial \theta}(\theta, x, y)\mu(dy)\\ &= \int_{\mathbb{R}^d} \varphi(y)\frac{\frac{\partial p_T}{\partial \theta}(\theta, x, y)}{p_T(\theta, x, y)} p_T(\theta, x, y)\mu(dy)\\ &= \mathbb{E}\left(\varphi(X_T^x)\frac{\partial \log p_T}{\partial \theta}(\theta, x, X_T^x)\right). \end{aligned} \tag{2.9}$$

Of course, to be valid, the above computations need appropriate assumptions (domination, uniform integrability, etc) to justify interchange of integration and differentiation (see Exercises below and also Sect. 10.3.1).

Using this approach in the simple case of a Black–Scholes model, one immediately retrieves the formulas established in Proposition 2.1(c) of Sect. 2.2.2 for the δ-hedge (in particular, when the function φ defining the payoff is only Borel). One can also retrieve by the same method the formulas for all the greeks.

However, this straightforward and simple approach to "greek" computation remains limited beyond the Black–Scholes world by the fact that it is mandatory to have access not only to the regularity of the probability density $p_T(\theta, x, y)$ of the asset at time T but also to *its explicit expression* as well as that of the partial derivative of its logarithm $\frac{\partial \log p_T}{\partial \theta}(\theta, x, X_T^x)$ in order to include it in a simulation process.

A solution in practice is to replace the true diffusion process $\big(X_t^x(\theta)\big)_{t\in[0,T]}$ (starting at x) by an approximation, typically its discrete time Euler scheme with step $\frac{T}{n}$, denoted by $(\bar{X}_{t_k^n}^x(\theta))_{0\le k\le n}$, where $\bar{X}_0^x(\theta) = x$ and $t_k^n := \frac{kT}{n}$ (see Sect. 7.1 in Chap. 7 for its definition and first properties). Then, under a light ellipticity assumption (non-degeneracy of the volatility), the whole scheme $(\bar{X}_{t_k^n}^x)_{k=1,\dots,n}$ starting at x (with step $\frac{T}{n}$) does have an explicit density $\bar{p}_{t_1^n,\dots,t_n^n}(\theta, x, y_1, \dots, y_n)$ (see Proposition 10.8). On the other hand the Euler scheme is of course simulable so that an approximation of Eq. (2.9) where $X_T^x(\theta)$ is replaced by $\bar{X}_T^x(\theta)$ can be computed by a Monte Carlo simulation. More details are given in Sect. 10.3.1.

An alternative is to "go back" to the "scenarii" space Ω. Then, some extensions of the first two approaches are possible when the risky asset prices follow a Brownian diffusion: if the payoff function and the diffusion coefficients are both smooth, one may rely on the tangent process (derivative of the process with respect to its starting value, or more generally with respect to one of its structure parameters, see below).

When the payoff function does not have enough regularity to map possible path wise differentiation, a more sophisticated method is to call upon Malliavin calculus or stochastic variational analysis. In (very) short, it provides a differentiation theory with respect to the generic Brownian paths. In particular, an integration by parts formula can be established which plays the role on the Wiener space of the elementary integration by parts used in Sect. 2.2.2. This second topic is beyond the scope of the present monograph. However, a flavor of Malliavin calculus is proposed in Sect. 10.4 through the Bismut and Clark-Occone formulas. These formulas, which can be viewed as ancestors of Malliavin calculus, provide the δ-hedge for vanilla options in local volatility models (see Theorem 10.2 and the application that follows).

▶ **Exercises. 1.** Provide simple assumptions to justify the above formal computations in (2.9), at some point θ_0 or for all θ running over a non-empty open interval Θ of $\mathbb{R}$ (or domain of $\mathbb{R}^d$ if θ is vector valued). [Hint: use the remark directly below Theorem 2.2.]

2. Compute the probability density $p_{_T}(\sigma, x, y)$ of $X_{_T}^{x,\sigma}$ in a Black–Scholes model ($\sigma > 0$ stands for the volatility parameter).

3. Re-establish all the sensitivity formulas established in the former Sect. 2.2.2 (including the exercises at the end of the section) using this approach.

4. Apply these formulas to the case $\varphi(x) := e^{-rt}(x - K)_+$ and retrieve the classical expressions for the greeks in a Black–Scholes model: the δ, the γ and the *vega*.

In this section we focused on the case of the marginal $X_{_T}^x(\theta)$ at time T of a Brownian diffusion as encountered in local volatility models viewed as a generalization of the Black–Scholes models investigated in the former section. In fact, this method, known as the log-likelihood method, has a much wider range of application since it works for any family $\big(X(\theta)\big)_{\theta\in\Theta}$ of $\mathbb{R}^d$-valued vectors, ($\Theta \subset \mathbb{R}^q$) such that, for every $\theta \in \Theta$, the distribution of $X(\theta)$ has a probability density $p(\theta, y)$ with respect to a reference measure μ on $\mathbb{R}^d$, usually the Lebesgue measure.

2.2.4 The Tangent Process Method

In fact, when both the payoff function/functional and the coefficients of the *SDE* are regular enough, one can differentiate the function/functional of the process directly with respect to a given parameter. The former Sect. 2.2.2 was a special case of this method for vanilla payoffs in a Black–Scholes model. We refer to Sect. 10.2.2 for more detailed developments.

Chapter 3
Variance Reduction

3.1 The Monte Carlo Method Revisited: Static Control Variate

Let $X \in L^2_{\mathbb{R}}(\Omega, \mathcal{A}, \mathbb{P})$ be a random variable, assumed to be easy to simulate at a reasonable computational cost. We wish to compute

$$m_{_X} = \mathbb{E}\, X \in \mathbb{R}$$

as the result of a Monte Carlo simulation.

Confidence interval revisited from the simulation viewpoint

The parameter m is to be computed by a Monte Carlo simulation. Let X_k, $k \geq 1$, be a sequence of i.i.d. copies of X. Then the Strong Law of Large Numbers (*SLLN*) yields

$$m_{_X} = \lim_{M \to +\infty} \overline{X}_{_M} \quad \mathbb{P}\text{-}a.s. \quad \text{with} \quad \overline{X}_{_M} := \frac{1}{M} \sum_{k=1}^{M} X_k.$$

This convergence is ruled by the Central Limit Theorem (*CLT*)

$$\sqrt{M}\,\big(\overline{X}_{_M} - m_{_X}\big) \xrightarrow{\mathcal{L}} \mathcal{N}\big(0; \mathrm{Var}(X)\big) \quad \text{as} \quad M \to +\infty.$$

Hence, for every $q \in (0, +\infty)$, and for large enough M

$$\mathbb{P}\left(m \in \left[\overline{X}_{_M} - q\sqrt{\frac{\overline{V}_{_M}}{M}}, \overline{X}_M + q\sqrt{\frac{\overline{V}_{_M}}{M}}\right]\right) \simeq 2\Phi_0(q) - 1,$$

G. Pagès, *Numerical Probability*, Universitext,
https://doi.org/10.1007/978-3-319-90276-0_3

where $\overline{V}_M \simeq \sqrt{\text{Var}(X)}$ is an estimator of the variance (see (2.1)) and $\Phi_0(x) = \int_{-\infty}^{x} e^{-\frac{\xi^2}{2}} \frac{d\xi}{\sqrt{2\pi}}$ is the c.d.f. of the normal distribution.

In numerical probability, we adopt the following *reverse point of view* based on the *target* or *prescribed accuracy* $\varepsilon > 0$: to make $\overline{X}_M$ enter a confidence interval $[m - \varepsilon, m + \varepsilon]$ with a confidence level $\alpha := 2\Phi_0(q_\alpha) - 1$, one needs to perform a Monte Carlo simulation of size

$$M \geq M^X(\varepsilon, \alpha) = \frac{q_\alpha^2 \text{Var}(X)}{\varepsilon^2}. \tag{3.1}$$

In practice, of course, the estimator $\overline{V}_M$ is computed on-line to estimate the variance $\text{Var}(X)$ as presented in the previous chapter. This estimate need not to be as sharp as the estimation of m, so it can be processed at the beginning of the simulation on a smaller sample size.

As a first conclusion, this shows that, a confidence level being fixed, the size of a Monte Carlo simulation grows linearly with the variance of X for a given accuracy and quadratically as the inverse of the prescribed accuracy for a given variance.

Variance reduction: (not so) naive approach

Assume now that we know *two* random variables $X,\ X' \in L^2_{\mathbb{R}}(\Omega, \mathcal{A}, \mathbb{P})$ satisfying

$$m = \mathbb{E}\, X = \mathbb{E}\, X' \in \mathbb{R}, \quad \text{Var}(X),\ \text{Var}(X'),\ \text{Var}(X - X') = \mathbb{E}\,(X - X')^2 > 0$$

(the last condition only says that X and X' are not *a.s.* equal).
Question: Which random vector (distribution…) is more appropriate?

Several examples of such a situation have already been pointed out in the previous chapter: usually many formulas are available to compute a greek parameter, even more if one takes into account the (potential) control variates introduced in the exercises.

A natural answer is: if both X and X' *can be simulated with an equivalent cost* (complexity), then the one with the lowest variance is the best choice, *i.e.*

$$X' \quad \text{if} \quad \text{Var}(X') < \text{Var}(X), \qquad X \text{ otherwise},$$

provided this fact is known *a priori*.

ℵ Practitioner's corner

Usually, the problem appears as follows: there exists a random variable $\Xi \in L^2_{\mathbb{R}}(\Omega, \mathcal{A}, \mathbb{P})$ such that:

(i) $\mathbb{E}\,\Xi$ can be computed at a very low cost by a deterministic method (closed form, numerical analysis method),

(ii) the random variable $X - \Xi$ can be simulated with the same cost (complexity) as X,

(iii) the variance $\mathrm{Var}(X - \Xi) < \mathrm{Var}(X)$.

Then, the random variable

$$X' = X - \Xi + \mathbb{E}\,\Xi$$

can be simulated at the same cost as X,

$$\mathbb{E}\,X' = \mathbb{E}\,X = m \quad \text{and} \quad \mathrm{Var}(X') = \mathrm{Var}(X - \Xi) < \mathrm{Var}(X).$$

Definition 3.1 *A random variable* Ξ *satisfying* (i)–(ii)–(iii) *is called a* control variate *for* X.

▶ **Exercise.** Show that if the simulation process of X and $X - \Xi$ have complexity κ and κ' respectively, then (iii) becomes

$$(iii)' \quad \kappa' \,\mathrm{Var}(X - \Xi) < \kappa\,\mathrm{Var}(X).$$

The product of the variance of a random variable by its simulation complexity is called the *effort*. It will be a central notion when we will introduce and analyze Multilevel methods in Chap. 9.

Toy-example. In the previous chapter, we showed in Proposition 2.1 that, in a risk-neutral Black–Scholes model $X_t^x = x\,e^{(r-\frac{\sigma^2}{2})t+\sigma W_t}$ $(x > 0, \sigma > 0, t \in [0, T])$, if the payoff function is differentiable outside a countable set and locally Lipschitz continuous with polynomial growth at infinity then the function $\Phi(x) = \mathbb{E}\,\varphi(X_T^x)$ is differentiable on $(0, +\infty)$ and

$$\Phi'(x) = \mathbb{E}\left(\varphi'(X_T^x)\frac{X_T^x}{x}\right) = \mathbb{E}\left(\varphi(X_T^x)\frac{W_T}{x\sigma T}\right). \tag{3.2}$$

So we have at hand two formulas for $\Phi'(x)$ that can be implemented in a Monte Carlo simulation. Which one should we choose to compute $\Phi'(x)$ (*i.e.* the δ-hedge up to a factor of e^{-rt})? Since they have the same expectations, the two random variables should be discriminated through (the square of) their L^2-norm, namely

$$\mathbb{E}\left[\left(\varphi'(X_T^x)\frac{X_T^x}{x}\right)^2\right] \quad \text{and} \quad \mathbb{E}\left[\left(\varphi(X_T^x)\frac{W_T}{x\sigma T}\right)^2\right].$$

▷ *Short maturities.* It is clear that, if $\varphi(x) \neq 0$,

$$\lim_{T\to 0} \mathbb{E}\left[\left(\varphi(X_T^x)\frac{W_T}{x\sigma T}\right)^2\right] \sim \frac{\varphi(x)^2}{Tx^2\sigma^2} \to +\infty \quad \text{as } T \to 0,$$

whereas, at least if φ' has polynomial growth,

$$\lim_{T\to 0}\mathbb{E}\left[\left(\varphi'(X^x_T)\frac{X^x_T}{x}\right)^2\right]=\left(\varphi'(x)\right)^2.$$

Since $\Phi'(x)\to\varphi'(x)$ as $T\to 0$, it follows that, if φ has polynomial growth and $\varphi(x)\neq 0$,

$$\mathrm{Var}\left(\varphi'(X^x_T)\frac{X^x_T}{x}\right)\to 0 \quad\text{and}\quad \mathrm{Var}\left(\varphi(X^x_T)\frac{W_T}{x\sigma T}\right)\to+\infty \quad\text{as}\quad T\to 0.$$

Such is the case for an in-the-money Call option with payoff $\varphi(\xi)=(\xi-K)_+$ if $X^x_0=x>K$.

$\triangleright$ *Long maturities.* On the other hand, at least if φ is bounded,

$$\mathbb{E}\left[\left(\varphi(X^x_T)\frac{W_T}{x\sigma T}\right)^2\right]=O\left(\frac{1}{T}\right)\to 0 \quad\text{as}\quad T\to+\infty,$$

whereas, if φ' is bounded away from zero at infinity, easy computations show that

$$\lim_{T\to+\infty}\mathbb{E}\left[\left(\varphi'(X^x_T)\frac{X^x_T}{x}\right)^2\right]=+\infty.$$

However, these last two conditions on the function φ conflict with each other.

In practice, one observes on the greeks of usual payoffs that *the pathwise differentiated estimator has a significantly lower variance* (even when this variance goes to 0 like for Puts in long maturities).

Numerical Experiment *(Short maturities)*. We consider the Call payoff $\varphi(\xi)=(\xi-K)_+$, with $K=95$ and $x=100$, still $r=0$ and a volatility $\sigma=0.5$ in the Black–Scholes model. Fig. 3.1 (left) depicts the variance of the pathwise differentiated and the weighted estimators of the δ-hedge for maturities running from $T=0.001$ up to $T=1$.

These estimators can be improved (at least for short maturities) by introducing control variates as follows (see Exercise **2.**, Sect. 2.2.2 of the former chapter):

$$\Phi'(x)=\varphi'(x)+\mathbb{E}\left(\left(\varphi'(X^x_T)-\varphi'(x)\right)\frac{X^x_T}{x}\right)=\mathbb{E}\left(\left(\varphi(X^x_T)-\varphi(x)\right)\frac{W_T}{x\sigma T}\right). \quad (3.3)$$

However the comparisons carried out with these new estimators tend to confirm the above heuristics, as illustrated by Fig. 3.1 (right).

A variant (pseudo-control variate)

In option pricing, when the random variable X is a payoff it is usually non-negative. In that case, any random variable Ξ satisfying (i)–(ii) and

$$(iii)'' \qquad 0\le\Xi\le X$$

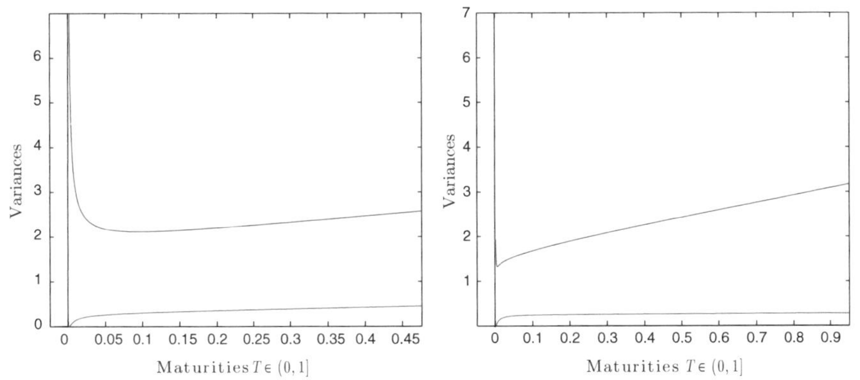

Fig. 3.1 BLACK–SCHOLES CALLS: *Left*: $T \mapsto \mathrm{Var}\big(\varphi'(X_T^x)\frac{X_T^x}{x}\big)$ *(blue line) and* $T \mapsto \mathrm{Var}\big(\varphi(X_T^x)\frac{W_T}{x\sigma T}\big)$ *(red line)*, $T \in (0, 1]$, $x = 100$, $K = 95$, $r = 0$, $\sigma = 0.5$. *Right*: $T \mapsto \mathrm{Var}\Big(\big(\varphi'(X_T^x) - \varphi'(x)\big)\frac{X_T^x}{x}\Big)$ *(blue line) and* $T \mapsto \mathrm{Var}\Big(\big(\varphi(X_T^x) - \varphi(x)\big)\frac{W_T}{x\sigma T}\Big)$ (red line), $T \in (0, 1]$

can be considered as a good *candidate* to reduce the variance, especially if Ξ is close to X so that $X - \Xi$ is "small".

However, note that it does not imply (iii). Here is a trivial counter-example: let $X \equiv 1$, then $\mathrm{Var}(X) = 0$ whereas a uniformly distributed random variable Ξ on the interval $[1 - \eta, 1]$, $0 < \eta < 1$, will (almost) satisfy (i)–(ii) but $\mathrm{Var}(X - \Xi) > 0$. Consequently, this variant is only a heuristic method to reduce the variance which often works, but with no *a priori* guarantee.

3.1.1 Jensen's Inequality and Variance Reduction

This section is in fact an illustration of the notion of pseudo-control variate described above.

Proposition 3.1 (Jensen's Inequality) *Let* $X : (\Omega, \mathcal{A}, \mathbb{P}) \to \mathbb{R}$ *be a random variable and let* $g : \mathbb{R} \to \mathbb{R}$ *be a* convex *function. Suppose* X *and* $g(X)$ *are integrable. Then, for any sub-*σ*-field* $\mathcal{B}$ *of* $\mathcal{A}$,

$$g\big(\mathbb{E}\,(X \mid \mathcal{B})\big) \le \mathbb{E}\,\big(g(X) \mid \mathcal{B}\big) \qquad \mathbb{P}\text{-}a.s.$$

In particular, considering $\mathcal{B} = \{\varnothing, \Omega\}$ *yields the above inequality for regular expectation, i.e.*

$$g\big(\mathbb{E}\,X\big) \le \mathbb{E}\,g(X).$$

Proof. The inequality is a straightforward consequence of the linearity of conditional expectation and the following classical characterization of a convex function

$$g \text{ is convex if and only if } \forall\, x \in \mathbb{R},\ g(x) = \sup\Big\{ax + b,\ a,\ b \in \mathbb{Q},\ ay + b \le g(y),\ \forall\, y \in \mathbb{R}\Big\}. \quad \diamond$$

Jensen's inequality is an efficient tool for designing control variates when dealing with path-dependent or multi-asset options as emphasized by the following examples.

⊳ **Examples. 1. Basket or index option**. We consider a payoff on a basket of d (positive) risky assets (this basket can be an index). For the sake of simplicity we suppose it is a Call with strike K, *i.e.*

$$h_T = \left(\sum_{i=1}^{d} \alpha_i X_T^{i,x_i} - K\right)_+$$

where $(X^{1,x_1}, \ldots, X^{d,x_d})$ models the price of d traded risky assets on a market and the α_i are some positive ($\alpha_i > 0$) weights satisfying $\sum_{1\le i\le d} \alpha_i = 1$. Then, the convexity of the exponential implies that

$$0 \le e^{\sum_{1\le i\le d} \alpha_i \log(X_T^{i,x_i})} \le \sum_{i=1}^{d} \alpha_i X_T^{i,x_i}$$

so that

$$h_T \ge k_T := \left(e^{\sum_{1\le i\le d} \alpha_i \log(X_T^{i,x_i})} - K\right)_+ \ge 0.$$

The motivation for this example is that in a (possibly correlated) d-dimensional Black–Scholes model (see below), $\sum_{1\le i\le d} \alpha_i \log(X_T^{i,x_i})$ still has a normal distribution so that the European Call option written on the payoff

$$k_T := \left(e^{\sum_{1\le i\le d} \alpha_i \log(X_T^{i,x_i})} - K\right)_+$$

has a closed form.

Let us be more specific on the model and the variance reduction procedure.

The *correlated d-dimensional Black–Scholes model* (under the risk-neutral probability measure with $r > 0$ denoting the interest rate) can be defined by the following system of *SDE* s which governs the price of d risky assets denoted by $i = 1, \ldots, d$:

$$dX_t^{i,x_i} = X_t^{i,x_i}\left(r\,dt + \sum_{j=1}^{q} \sigma_{ij} dW_t^j\right), \quad t \in [0, T], \quad x_i > 0, \ i = 1, \ldots, d,$$

where $W = (W^1, \ldots, W^q)$ is a standard q-dimensional Brownian motion and $\sigma = [\sigma_{ij}]_{1\le i\le d, 1\le j\le q} \in \mathcal{M}(d, q, \mathbb{R})$ (matrix with d rows and q columns with real entries). Its solution is

$$X_t^{i,x_i} = x_i \exp\left(\left(r - \frac{\sigma_{i.}^2}{2}\right)t + \sum_{j=1}^{q} \sigma_{ij} W_t^j\right), \quad t \in [0, T], \quad x_i > 0, \ i = 1, \ldots, d,$$

where

$$\sigma_{i.}^2 = \sum_{j=1}^{q} \sigma_{ij}^2, \quad i = 1, \ldots, d.$$

► **Exercise.** Show that if the matrix $\sigma\sigma^*$ is positive definite (then $q \geq d$) and one may assume without modifying the model that X^{i,x_i} only depends on the first i components of a d-dimensional standard Brownian motion. [Hint: consider the Cholesky decomposition in Sect. 1.6.2.]

Now, let us describe the two phases of the variance reduction procedure:

– PHASE I: $\Xi = e^{-rt} k_T$ *as a pseudo-control variate and computation of its expectation* $\mathbb{E}\,\Xi$.

The vanilla Call option has a closed form in a Black–Scholes model and elementary computations show that

$$\sum_{1 \leq i \leq d} \alpha_i \log(X_T^{i,x_i}/x_i) \stackrel{d}{=} \mathcal{N}\left(\left(r - \frac{1}{2} \sum_{1 \leq i \leq d} \alpha_i \sigma_{i.}^2\right)T;\, \alpha^* \sigma\sigma^* \alpha\, T\right)$$

where α is the column vector with components α_i, $i = 1, \ldots, d$.

Consequently, the premium at the origin $e^{-rt}\mathbb{E}\, k_T$ admits a closed form (see Sect. 12.2 in the Miscellany Chapter) given by

$$e^{-rt}\mathbb{E}\, k_T = \mathrm{Call}_{BS}\left(\prod_{i=1}^{d} x_i^{\alpha_i} e^{-\frac{1}{2}\left(\sum_{1 \leq i \leq d} \alpha_i \sigma_{i.}^2 - \alpha^* \sigma\sigma^* \alpha\right)T}, K, r, \sqrt{\alpha^* \sigma\sigma^* \alpha}, T\right).$$

– PHASE II: *Joint simulation of the pair* (h_T, k_T).

We need to simulate M independent copies of the pair (h_T, k_T) or, to be more precise of the quantity

$$e^{-rt}(h_T - k_T) = e^{-rt}\left(\left(\sum_{i=1}^{d} \alpha_i X_T^{i,x_i} - K\right)_+ - \left(e^{\sum_{1 \leq i \leq d} \alpha_i \log(X_T^{i,x_i})} - K\right)_+\right).$$

This task clearly amounts to simulating M independent copies, of the q-dimensional standard Brownian motion W at time T, namely

$$W_T^{(m)} = (W_T^{1,(m)}, \ldots, W_T^{q,(m)}),\ m = 1, \ldots, M,$$

i.e. M independent copies $Z^{(m)} = (Z_1^{(m)}, \dots, Z_q^{(m)})$ of the $\mathcal{N}(0; I_q)$ distribution in order to set $W_T^{(m)} \stackrel{d}{=} \sqrt{T} Z^{(m)}$, $m = 1, \dots, M$.

The resulting pointwise estimator of the premium is given, with obvious notations, by

$$\frac{e^{-rt}}{M}\sum_{m=1}^{M}\big(h_T^{(m)} - k_T^{(m)}\big) + \mathrm{Call}_{BS}\left(\prod_{i=1}^{d} x_i^{\alpha_i} e^{-\frac{1}{2}\left(\sum_{1\le i\le d}\alpha_i\sigma_{i.}^2 - \alpha^*\sigma\sigma^*\alpha\right)T}, K, r, \sqrt{\alpha^*\sigma\sigma^*\alpha}, T\right).$$

Remark. The extension to more general payoffs of the form $\varphi\Big(\sum_{1\le i\le d}\alpha_i X_T^{i,x_i}\Big)$ is straightforward provided φ is non-decreasing and a closed form exists for the vanilla option with payoff $\varphi\Big(e^{\sum_{1\le i\le d}\alpha_i \log(X_T^{i,x_i})}\Big)$.

► **Exercise.** Other ways to take advantage of the convexity of the exponential function can be explored: thus one can start from

$$\sum_{1\le i\le d}\alpha_i X_T^{i,x_i} = \Big(\sum_{1\le i\le d}\alpha_i x_i\Big)\sum_{1\le i\le d}\widetilde{\alpha}_i\frac{X_T^{i,x_i}}{x_i},$$

where $\widetilde{\alpha}_i = \dfrac{\alpha_i x_i}{\sum_{1\le k\le d}\alpha_k x_k}$, $i = 1, \dots, d$. Compare on simulations the respective performances of these different approaches.

2. Asian options and the Kemna–Vorst control variate in a Black–Scholes model (see [166]). Let

$$h_T = \varphi\left(\frac{1}{T}\int_0^T X_t^x\,dt\right)$$

be a generic *Asian* payoff where φ is non-negative, non-decreasing function defined on $\mathbb{R}_+$ and let

$$X_t^x = x\exp\left(\left(r - \frac{\sigma^2}{2}\right)t + \sigma W_t\right),\quad x > 0,\ t \in [0, T],$$

be a regular Black–Scholes dynamics with volatility $\sigma > 0$ and interest rate r. Then, the (standard) Jensen inequality applied to the probability measure $\frac{1}{T}\mathbf{1}_{[0,T]}(t)dt$ implies

$$\begin{aligned}\frac{1}{T}\int_0^T X_t^x\,dt &\ge x\exp\left(\frac{1}{T}\int_0^T\Big((r - \sigma^2/2)t + \sigma W_t\Big)dt\right)\\ &= x\exp\left((r - \sigma^2/2)\frac{T}{2} + \frac{\sigma}{T}\int_0^T W_t dt\right).\end{aligned}$$

Now

$$\int_0^T W_t dt = T\, W_T - \int_0^T s dW_s = \int_0^T (T-s)dW_s$$

so that

$$\frac{1}{T}\int_0^T W_t dt \stackrel{d}{=} \mathcal{N}\left(0; \frac{1}{T^2}\int_0^T s^2 ds\right) = \mathcal{N}\left(0; \frac{T}{3}\right).$$

This suggests to rewrite the right-hand side of the above inequality in a "Black–Scholes asset" style, namely:

$$\frac{1}{T}\int_0^T X_t^x\, dt \geq x\, e^{-(\frac{r}{2}+\frac{\sigma^2}{12})T} \exp\left((r-(\sigma^2/3)/2)T + \frac{\sigma}{T}\int_0^T W_t dt\right).$$

This naturally leads us to introduce the so-called *Kemna–Vorst* (pseudo-)control variate

$$k_T^{KV} := \varphi\left(x\, e^{-(\frac{r}{2}+\frac{\sigma^2}{12})T} \exp\left(\left(r - \frac{1}{2}\frac{\sigma^2}{3}\right)T + \sigma\frac{1}{T}\int_0^T W_t\, dt\right)\right)$$

which is clearly of Black–Scholes type and moreover satisfies

$$h_T \geq k_T^{KV}.$$

– PHASE I: The random variable k_T^{KV} is an admissible control variate *as soon as the vanilla option related to the payoff* $\varphi(X_T^x)$ *has a closed form*. Indeed, if a vanilla option related to the payoff $\varphi(X_T^x)$ has a closed form

$$e^{-rt}\mathbb{E}\,\varphi(X_T^x) = \text{Premium}_{BS}^{\varphi}(x, r, \sigma, T),$$

then, one has

$$e^{-rt}\mathbb{E}\, k_T^{KV} = \text{Premium}_{BS}^{\varphi}\left(x\, e^{-(\frac{r}{2}+\frac{\sigma^2}{12})T}, r, \frac{\sigma}{\sqrt{3}}, T\right).$$

– PHASE II: One has to simulate independent copies of $h_T - k_T^{KV}$, *i.e.* in practice, independent copies of the pair (h_T, k_T^{KV}). Theoretically speaking, this requires us to know how to simulate paths of the standard Brownian motion $(W_t)_{t\in[0,T]}$ exactly and, moreover, to compute with an infinite accuracy integrals of the form $\frac{1}{T}\int_0^T f(t)dt$.

In practice these two tasks are clearly impossible (one cannot even compute a real-valued function $f(t)$ *at every* $t \in [0, T]$ with a computer). In fact, one relies on quadrature formulas to approximate the time integrals in both payoffs which makes this simulation possible since only finitely many random marginals of the Brownian motion, say $W_{t_1}, \ldots, W_{t_n}$, are necessary, which is then quite realistic. Typically, one uses a mid-point quadrature formula

$$\frac{1}{T}\int_0^T f(t)dt \simeq \frac{1}{n}\sum_{k=1}^{n} f\left(\frac{2k-1}{2n}T\right)$$

or any other numerical integration method, keeping in mind nevertheless that the (continuous) functions f of interest are here given for the first and the second payoff functions by

$$f(t) = \varphi\left(x\exp\left(\left(r-\frac{\sigma^2}{2}\right)t+\sigma W_t(\omega)\right)\right) \quad \text{and} \quad f(t) = W_t(\omega),$$

respectively. Hence, their regularity is $\frac{1}{2}^-$-Hölder (*i.e.* α-Hölder on $[0, T]$, for every $\alpha < \frac{1}{2}$, like for the payoff h_T). Finally, in practice, it will amount to simulating independent copies of the n-tuple

$$\left(W_{\frac{T}{2n}}, \ldots, W_{\frac{(2k-1)T}{2n}}, \ldots, W_{\frac{(2n-1)T}{2n}}\right)$$

from which one can reconstruct both a mid-point approximation of both integrals appearing in h_T and k_T.

In fact, one can improve this first approach by taking advantage of the fact that W is a Gaussian process as detailed in the exercise below.

Further developments to reduce the time discretization error are proposed in Sect. 8.2.5 (see [187] where an in-depth study of the Asian option pricing in a Black–Scholes model is carried out).

► **Exercises. 1.** (a) Show that if $f : [0, T] \to \mathbb{R}$ is continuous then

$$\lim_n \frac{1}{n}\sum_{k=1}^{n} f\left(\frac{kT}{n}\right) = \frac{1}{T}\int_0^T f(t)\,dt.$$

Show that $t \mapsto x\exp\left(\left(r-\frac{\sigma^2}{2}\right)t+\sigma W_t(\omega)\right)$ is α-Hölder for every $\alpha \in (0, \frac{1}{2})$ (with a random Hölder ratio, of course).

(b) Show that

$$\mathcal{L}\left(\int_0^T W_t\,dt \,\middle|\, W_{\frac{kT}{n}} - W_{\frac{(k-1)T}{n}} = \delta w_k,\ 1 \le k \le n\right) = \mathcal{N}\left(\sum_{k=1}^{n} a_k\,\delta w_k;\ \frac{T^3}{12n^2}\right)$$

with $a_k = \frac{2(n-k)+1}{2n}T$, $k = 1, \ldots, n$. [Hint: for Gaussian vectors, conditional expectation and affine regression coincide.]

(c) Propose a variance reduction method in which the pseudo-control variate $e^{-rt}k_T$ will be simulated exactly.

2. Check that the preceding can be applied to payoffs of the form

$$\varphi\left(\frac{1}{T}\int_0^T X_t^x dt - X_T^x\right)$$

where φ is still a non-negative, non-decreasing function defined on $\mathbb{R}$.

3. *Best-of-Call option.* We consider the *Best-of-Call* payoff given by

$$h_T = \left(\max(X_T^1, X_T^2) - K\right)_+.$$

(a) Using the convexity inequality (that can still be seen as an application of Jensen's inequality)

$$\sqrt{ab} \leq \max(a, b), \quad a,\, b > 0,$$

show that

$$k_T := \left(\sqrt{X_T^1 X_T^2} - K\right)_+$$

is a natural (pseudo-)control variate for h_T.

(b) Show that, in a 2-dimensional Black–Scholes (possibly correlated) model (see example in Sect. 2.1.3), the premium of the option with payoff k_T (known as the *geometric mean option*) has a closed form. Show that this closed form can be written as a Black–Scholes formula with appropriate parameters (and maturity T). [Hint: see Sect. 12.2 in the Miscellany Chapter.]

(c) Check on (at least one) simulation(s) that this procedure does reduce the variance (use the parameters of the model specified in Sect. 2.1.3).

(d) When σ_1 and σ_2 are not equal, improve the above variance reduction protocol by considering a parametrized family of (pseudo-)control variate, obtained from the more general inequality $a^\theta b^{1-\theta} \leq \max(a, b)$ when $\theta \in (0, 1)$.

3.1.2 Negatively Correlated Variables, Antithetic Method

In this section we assume that X and X' have not only the same expectation m_X but also the same variance, *i.e.* $\mathrm{Var}(X) = \mathrm{Var}(X')$, can be simulated with the same complexity $\kappa = \kappa_X = \kappa_{X'}$. We also assume that $\mathbb{E}(X - X')^2 > 0$ so that X and X' are not $a.s.$ equal. In such a situation, choosing between X or X' may seem *a priori* a question of little interest. However, it is possible to take advantage of this situation to reduce the variance of a simulation when X and X' are *negatively correlated.*

Set

$$\chi = \frac{X + X'}{2}.$$

This corresponds to $\Xi = \frac{X - X'}{2}$ with our formalism. It is reasonable (when no further information on (X, X') is available) to assume that the simulation complexity of χ is twice that of X and X', *i.e.* $\kappa_\chi = 2\kappa$. On the other hand

$$\begin{aligned}\text{Var}\,(\chi) &= \frac{1}{4}\,\text{Var}\left(X + X'\right)\\ &= \frac{1}{4}\left(\text{Var}(X) + \text{Var}(X') + 2\text{Cov}(X, X')\right)\\ &= \frac{\text{Var}(X) + \text{Cov}(X, X')}{2}.\end{aligned}$$

The size $M^X(\varepsilon, \alpha)$ and $M^\chi(\varepsilon, \alpha)$ of the simulation using X and χ respectively to enter a given interval $[m - \varepsilon, m + \varepsilon]$ with the same confidence level α is given, following (3.1), by

$$M^X = \left(\frac{q_\alpha}{\varepsilon}\right)^2 \text{Var}(X) \quad \text{for } X \quad \text{and} \quad M^\chi = \left(\frac{q_\alpha}{\varepsilon}\right)^2 \text{Var}(\chi) \quad \text{for } \chi.$$

Taking into account the complexity as in the exercise that follows Definition 3.1, this means in terms of CPU computation time that one should better use χ if and only if

$$\kappa_\chi M^\chi < \kappa_X M^X \iff 2\kappa\, M^\chi < \kappa\, M^X,$$

i.e.

$$2\text{Var}(\chi) < \text{Var}(X).$$

Given the above inequality, this reads

$$\text{Cov}(X, X') < 0.$$

To take advantage of this remark in practice, one usually relies on the following result.

Proposition 3.2 (co-monotony) *(a) Let $Z : (\Omega, \mathcal{A}, \mathbb{P}) \to \mathbb{R}$ be a random variable and let $\varphi, \psi : \mathbb{R} \to \mathbb{R}$ be two monotonic (hence Borel) functions with the same monotonicity. Assume that $\varphi(Z), \psi(Z) \in L^2_{\mathbb{R}}(\Omega, \mathcal{A}, \mathbb{P})$. Then*

$$\text{Cov}\big(\varphi(Z), \psi(Z)\big) = \mathbb{E}\big(\varphi(Z)\psi(Z)\big) - \mathbb{E}\,\varphi(Z)\mathbb{E}\,\psi(Z) \geq 0. \tag{3.4}$$

If, mutatis mutandis, *φ and ψ have opposite monotonicity, then*

$$\text{Cov}\big(\varphi(Z), \psi(Z)\big) \leq 0.$$

Furthermore, the inequality holds as an equality if and only if $\varphi(Z) = \mathbb{E}\,\varphi(Z)$ $\mathbb{P}$-a.s. or $\psi(Z) = \mathbb{E}\,\psi(Z)$ $\mathbb{P}$-a.s.
(b) Assume there exists a non-increasing (hence Borel) function $T : \mathbb{R} \to \mathbb{R}$ such that $Z \stackrel{d}{=} T(Z)$. Then $X = \varphi(Z)$ and $X' = \varphi(T(Z))$ are identically distributed and satisfy

$$\text{Cov}(X, X') \leq 0.$$

In that case, the random variables X and X' are called antithetic.

Proof. (*a*) *Inequality.* Let Z, Z' be two independent random variables defined on the same probability space $(\Omega, \mathcal{A}, \mathbb{P})$ with distribution $\mathbb{P}_Z$ ([1]). Then, using that φ and ψ are monotonic with the same monotonicity, we have

$$\big(\varphi(Z)-\varphi(Z')\big)\big(\psi(Z)-\psi(Z')\big)\geq 0$$

so that the expectation of this product is non-negative (and finite since all random variables are square integrable). Consequently

$$\mathbb{E}\,\big(\varphi(Z)\psi(Z)\big)+\mathbb{E}\,(\varphi(Z')\psi(Z'))-\mathbb{E}\,\big(\varphi(Z)\psi(Z')\big)-\mathbb{E}\,\big(\varphi(Z')\psi(Z)\big)\geq 0.$$

Using that $Z' \stackrel{d}{=} Z$ and that Z, Z' are independent, we get

$$2\,\mathbb{E}\,\big(\varphi(Z)\psi(Z)\big)\geq \mathbb{E}\,\varphi(Z)\mathbb{E}\,\psi(Z')+\mathbb{E}\,\varphi(Z')\,\mathbb{E}\,\psi(Z)=2\,\mathbb{E}\,\varphi(Z)\,\mathbb{E}\,\psi(Z),$$

that is

$$\operatorname{Cov}\big(\varphi(Z),\psi(Z)\big)=\mathbb{E}\,\big(\varphi(Z)\psi(Z)\big)-\mathbb{E}\,\varphi(Z)\,\mathbb{E}\,\psi(Z)\geq 0.$$

If the functions φ and ψ have opposite monotonicity, then

$$\big(\varphi(Z)-\varphi(Z')\big)\big(\psi(Z)-\psi(Z')\big)\leq 0$$

and one concludes as above up to sign changes.
Equality case. As for the equality case under the co-monotony assumption, we may assume without loss of generality that φ and ψ are non-decreasing. Moreover, we make the following convention: if a is not an atom of the distribution $\mathbb{P}_Z$ of Z, then set $\varphi(a)=\varphi(a_+)$, $\psi(a)=\psi(a_+)$, idem for b.

Now, if $\varphi(Z)$ or $\psi(Z)$ are $\mathbb{P}$-$a.s.$ constant (hence equal to their expectation) then equality clearly holds.

Conversely, it follows by reading the above proof backwards that if equality holds, then $\mathbb{E}\,\big(\varphi(Z)-\varphi(Z')\big)\big(\psi(Z)-\psi(Z')\big)=0$ so that $\big(\varphi(Z)-\varphi(Z')\big)\big(\psi(Z)-\psi(Z')\big)$ $\mathbb{P}$-$a.s.$ Now, let I be the (closed) convex hull of the support of the distribution $\mu=\mathbb{P}_Z$ of Z on the real line. Assume *e.g.* that $I=[a,b]\subset\mathbb{R}$, $a,b\in\mathbb{R}$, $a<b$ (other cases can be adapted easily from this one).

By construction a and b are in the support of μ so that for every ε, $\varepsilon\in(0,b-a)$, both $\mathbb{P}(a\leq Z<a+\varepsilon)$ and $\mathbb{P}(b-\varepsilon<Z\leq b)$ are (strictly) positive. If a is an atom of $\mathbb{P}_Z$, one may choose $\varepsilon_a=0$, idem for b. Hence the event $C_\varepsilon=\{a\leq Z<a+\varepsilon\}\cap\{b-\varepsilon<Z'\leq b\}$ has a positive probability since Z and Z' are independent.

Now assume that $\varphi(Z)$ is not $\mathbb{P}$-$a.s.$ constant. Then, φ cannot be constant on I and $\varphi(a)<\varphi(b)$ (with the above convention on atoms). Consequently, on C_ε,

---

[1] This is always possible owing to Fubini's Theorem for product measures by considering the probability space $(\Omega^2, \mathcal{A}^{\otimes 2}, \mathbb{P}^{\otimes 2})$: extend Z by $Z(\omega,\omega')=Z(\omega)$ and define Z' by $Z'(\omega,\omega')=Z(\omega')$.

$\varphi(Z) - \varphi(Z') > 0\, a.s.$ so that $\psi(Z) - \Psi(Z') = 0\ \mathbb{P}\text{-}a.s.$; which in turn implies that $\psi(a+\varepsilon) = \psi(b-\varepsilon)$. Then, letting ε go to 0, one derives that $\psi(a) = \psi(b)$ (still keeping in mind the convention on atoms). Finally, this shows that $\psi(Z)$ is $\mathbb{P}$-$a.s.$ constant.

(b) Set $\psi = \varphi \circ T$ so that φ and ψ have opposite monotonicity. Noting that X and X' have the same distribution and applying claim (a) completes the proof. ◇

This leads to the well-known "*antithetic random variables method*".

The antithetic random variables method

This terminology is shared by two classical situations in which the above approach applies:

– the *symmetric random variable* Z: $Z \stackrel{d}{=} -Z$ (*i.e.* $T(z) = -z$);

– the $[0, L]$-valued random variable Z such that $Z \stackrel{d}{=} L - Z$ (*i.e.* $T(z) = L - z$). This is satisfied by $U \stackrel{d}{=} \mathcal{U}([0,1])$ with $L = 1$.

⊳ **Examples. 1.** *European option pricing in a BS model.* Let $h_T = \varphi(X_T^x)$ with φ monotonic (like for Calls, Puts, spreads, etc). Then $h_T = \varphi\Big(xe^{(r-\frac{\sigma^2}{2})T+\sigma\sqrt{T}Z}\Big)$, $Z = \frac{W_T}{\sqrt{T}} \stackrel{d}{=} \mathcal{N}(0;1)$. The function $z \mapsto \varphi\Big(xe^{(r-\frac{\sigma^2}{2})T+\sigma\sqrt{T}z}\Big)$ is monotonic as the composition of two monotonic functions and Z is symmetric.

2. *Uniform distribution on the unit interval.* If φ is monotonic on $[0, 1]$ and $U \stackrel{d}{=} \mathcal{U}([0,1])$ then

$$\operatorname{Var}\left(\frac{\varphi(U) + \varphi(1-U)}{2}\right) \leq \frac{1}{2}\operatorname{Var}(\varphi(U)).$$

The above one-dimensional Proposition 3.2 admits a multi-dimensional extension that reads as follows.

Theorem 3.1 *Let $d \in \mathbb{N}^*$ and let $\varphi, \psi : \mathbb{R}^d \to \mathbb{R}$ be two functions satisfying the following joint marginal monotonicity assumption:*

for every $i \in \{1, \ldots, d\}$ and every $(z_{i+1}, \ldots, z_d) \in \mathbb{R}^{d-i}$, the functions

$z_i \mapsto \varphi(z_1, \ldots, z_i, \ldots, z_d)$ *and* $z_i \mapsto \psi(z_1, \ldots, z_i, \ldots, z_d)$ *have the same monotonicity*

not depending on $(z_1, \ldots, z_{i-1}) \in \mathbb{R}^{i-1}$ (but possibly on i and on $(z_{i+1}, \ldots, z_d)$). Let $Z_1, \ldots, Z_d$ be independent real-valued random variables defined on a probability space $(\Omega, \mathcal{A}, \mathbb{P})$.

Then, if $\varphi(Z_1, \ldots, Z_d), \psi(Z_1, \ldots, Z_d) \in L^2(\Omega, \mathcal{A}, \mathbb{P})$,

$$\operatorname{Cov}\Big(\varphi(Z_1, \ldots, Z_d), \psi\big(Z_1, \ldots, Z_d\big)\Big) \geq 0.$$

Corollary 3.1 *If $T_i : \mathbb{R} \to \mathbb{R}$, $i = 1, \ldots, d$, are non-increasing functions such that $T_i(Z_i) \stackrel{d}{=} Z_i$, then if φ and ψ are as in the above theorem*

$$\mathrm{Cov}\Big(\varphi(Z_1,\ldots,Z_d),\psi\big(T_1(Z_1),\ldots,T_d(Z_d)\big)\Big)\le 0.$$

Proof of Theorem 3.1. We proceed by induction on the dimension d using the notation $z_{d:d'}$ to denote $(z_d,\ldots,z_{d'})$. When $d=1$ the result follows from Proposition 3.2(a). Then, assume it holds true on $\mathbb{R}^d$. By Fubini's Theorem

$$\begin{aligned}\mathbb{E}\,\varphi(Z_{1:d+1})\psi(Z_{1:d+1}) &= \int_{\mathbb{R}} \mathbb{E}\big[\varphi(Z_{1:d},z_{d+1})\psi(Z_{1:d},z_{d+1})\big]\,\mathbb{P}_{Z_{d+1}}(dz_{d+1})\\ &\ge \int_{\mathbb{R}} \mathbb{E}\,\varphi(Z_{1:d},z_{d+1})\,\mathbb{E}\,\psi(Z_{1:d},z_{d+1})\mathbb{P}_{Z_{d+1}}(dz_{d+1})\end{aligned}$$

since, z_{d+1} being fixed, the functions $z_{1:d}\mapsto\varphi(z_{1:d},z_{d+1})$ and $z_{1:d}\mapsto\psi(z_{1:d},z_{d+1})$ satisfy the above joint marginal co-monotonicity assumption on $\mathbb{R}^d$.

Now, the functions $z_{d+1}\mapsto\varphi(z_{1:d},z_{d+1})$ and $z_{d+1}\mapsto\psi(z_{1:d},z_{d+1})$ have the same monotonicity, not depending on $z_{1:d}$ so that $\Phi: z_{d+1}\mapsto\mathbb{E}\,\varphi(Z_{1:d},z_{d+1})$ and $\Psi: z_{d+1}\mapsto\mathbb{E}\,\psi(Z_{1:d},z_{d+1})$ have the same monotonicity. Hence

$$\begin{aligned}\int_{\mathbb{R}}\mathbb{E}\,\varphi(Z_{d+1})\mathbb{E}\,\psi(Z_{1:d},z_{d+1})\mathbb{P}_{Z_{d+1}}(dz_{d+1}) &= \mathbb{E}\,\Phi(Z_{d+1})\Psi(Z_{d+1})\\ &\ge \mathbb{E}\,\Phi(Z_{d+1})\mathbb{E}\,\Psi(Z_{d+1})\\ &= \mathbb{E}\,\varphi(Z_{1:d+1})\,\mathbb{E}\,\psi(Z_{1:d+1}),\end{aligned}$$

where we used Fubini's Theorem twice (in a reverse way) in the last line. This completes the proof. ◇

The proof of the corollary is obvious.

► **Exercise.** (a) Show that if there is a permutation $\sigma:\{1,\ldots,d\}\to\{1,\ldots,d\}$ such that the functions φ_σ and ψ_σ respectively defined by $\varphi_\sigma(z_1,\ldots,z_d)=\varphi(z_{\sigma(1)},\ldots,z_{\sigma(d)})$ and $\psi_\sigma(z_1,\ldots,z_d)=\psi(z_{\sigma(1)},\ldots,z_{\sigma(d)})$ satisfy the above joint marginal monotonicity assumption and if $Z_1,\ldots,Z_d$ are i.i.d. then the conclusion of Theorem 3.1 remains valid.
(b) Show that if $\sigma(i)=d+1-i$, the conclusion remains true when $Z_1,\ldots,Z_d$ are simply independent.

Remarks. • This result may be successfully applied to functions of the form $f\big(\bar{X}_{\frac{T}{n}},\ldots,\bar{X}_{k\frac{T}{n}},\ldots,\bar{X}_{n\frac{T}{n}}\big)$ of the Euler scheme with step $\frac{T}{n}$ of a one-dimensional Brownian diffusion with a non-decreasing drift and a deterministic strictly positive diffusion coefficient, provided f is "marginally monotonic", *i.e.* monotonic in each of its variable with the same monotonicity. (We refer to Chap. 7 for an introduction to the Euler scheme of a diffusion.) The idea is to rewrite the functional as a "marginally monotonic" function of the n (independent) Brownian increments which play the role of the random variables Z_i. Furthermore, passing to the limit as the step size goes to zero yields some correlation results for a class of monotonic continuous functionals defined on the canonical space $\mathcal{C}([0,T],\mathbb{R})$ of the diffusion itself (the monotonicity should be understood with respect to the naive pointwise partial order: $f \le g$ if $f(t) \le g(t)$, $t \in [0,T]$).

• The co-monotony inequality (3.4) is one of the most powerful tool to establish lower bounds for expectations. For more insight about these kinds of *co-monotony* properties and their consequences for the pricing of derivatives, we refer to [227].

▶ **Exercises. 1.** *A toy simulation*. Let f and g be two functions defined on the real line by $f(u) = \frac{u}{\sqrt{u^2+1}}$ and $g(u) = \tanh u$, $u \in \mathbb{R}$. Set

$$\varphi(z_1, z_2) = f\Big(z_1 e^{a z_2}\Big) \quad \text{and} \quad \psi(z_1, z_2) = g\Big(z_1 e^{b z_2}\Big) \quad \text{with} \quad a,\, b > 0.$$

Show that, if Z_1, Z_2 are two independent random variables, then

$$\mathbb{E}\,\varphi(Z_1, Z_2)\psi(Z_1, Z_2) \ge \mathbb{E}\,\varphi(Z_1, Z_2)\,\mathbb{E}\,\psi(Z_1, Z_2).$$

2. Show that if φ and ψ are *non-negative*, Borel functions defined on $\mathbb{R}$, monotonic with *opposite monotonicity*, then

$$\mathbb{E}\,\big(\varphi(Z)\psi(Z)\big) \le \mathbb{E}\,\varphi(Z)\,\mathbb{E}\,\psi(Z) \le +\infty$$

so that, if $\varphi(Z), \psi(Z) \in L^1(\mathbb{P})$, then $\varphi(Z)\psi(Z) \in L^1(\mathbb{P})$.

3. Use Propositions 3.2(a) and 2.1(b) to derive directly from its representation as an expectation that, in the Black–Scholes model, the δ-hedge of a European option whose payoff function is convex is non-negative.

3.2 Regression-Based Control Variates

3.2.1 Optimal Mean Square Control Variates

We return to the original situation of two *square integrable* random variables X and X', having the same expectation

$$\mathbb{E}\,X = \mathbb{E}\,X' = m$$

with non-zero variances, *i.e.*

$$\text{Var}(X),\ \text{Var}(X') > 0.$$

We assume again that X and X' are not identical in the sense that $\mathbb{P}(X \ne X') > 0$, which turns out to be equivalent to

$$\mathrm{Var}(X - X') > 0.$$

We saw that if $\mathrm{Var}(X') \ll \mathrm{Var}(X)$, one will naturally choose X' to implement the Monte Carlo simulation and we provided several classical examples in that direction. However, we will see that with a little more effort it is possible to improve this naive strategy.

This time we simply (and temporarily) set

$$\Xi := X - X' \qquad \text{with} \quad \mathbb{E}\,\Xi = 0 \qquad \text{and} \qquad \mathrm{Var}(\Xi) > 0.$$

The idea is simply to parametrize the impact of the control variate Ξ by a factor λ, *i.e.* we set for every $\lambda \in \mathbb{R}$,

$$X^\lambda := X - \lambda\,\Xi = (1-\lambda)X + \lambda X'.$$

Then the strictly convex parabolic function Φ defined by

$$\Phi(\lambda) := \mathrm{Var}(X^\lambda) = \lambda^2 \mathrm{Var}(\Xi) - 2\,\lambda\,\mathrm{Cov}(X, \Xi) + \mathrm{Var}(X)$$

attains its minimum value at $\lambda_{\min}$ defined by

$$\begin{aligned}\lambda_{\min} &:= \frac{\mathrm{Cov}(X, \Xi)}{\mathrm{Var}(\Xi)} = \frac{\mathbb{E}\,(X\,\Xi)}{\mathbb{E}\,\Xi^2}\\ &= 1 + \frac{\mathrm{Cov}(X', \Xi)}{\mathrm{Var}(\Xi)} = 1 + \frac{\mathbb{E}\,(X'\,\Xi)}{\mathbb{E}\,\Xi^2}.\end{aligned}$$

Consequently

$$\sigma^2_{\min} := \mathrm{Var}(X^{\lambda_{\min}}) = \mathrm{Var}(X) - \frac{(\mathrm{Cov}(X, \Xi))^2}{\mathrm{Var}(\Xi)} = \mathrm{Var}(X') - \frac{(\mathrm{Cov}(X', \Xi))^2}{\mathrm{Var}(\Xi)}.$$

so that

$$\sigma^2_{\min} \le \min\big(\mathrm{Var}(X), \mathrm{Var}(X')\big)$$

and $\sigma^2_{\min} = \mathrm{Var}(X)$ if and only if $\mathrm{Cov}(X, \Xi) = 0$.

Remark. Note that $\mathrm{Cov}(X, \Xi) = 0$ if and only if $\lambda_{\min} = 0$, *i.e.* $\mathrm{Var}(X) = \min_{\lambda \in \mathbb{R}} \Phi(\lambda)$.

If we denote by $\rho_{X,\Xi}$ the correlation coefficient between X and Ξ, one gets

$$\sigma^2_{\min} = \mathrm{Var}(X)(1 - \rho^2_{X,\Xi}) = \mathrm{Var}(X')(1 - \rho^2_{X',\Xi}).$$

A more symmetric expression for $\mathrm{Var}(X^{\lambda_{\min}})$ is

$$\sigma^2_{\min} = \frac{\mathrm{Var}(X)\mathrm{Var}(X')(1-\rho^2_{X,X'})}{\left(\sqrt{\mathrm{Var}(X)}-\sqrt{\mathrm{Var}(X')}\right)^2 + 2\sqrt{\mathrm{Var}(X)\mathrm{Var}(X')}(1-\rho_{X,X'})}$$
$$\leq \sigma_X \sigma_{X'} \frac{1+\rho_{X,X'}}{2}$$

where σ_X and $\sigma_{X'}$ denote the standard deviations of X and X', respectively, and $\rho_{X,X'}$ is the correlation coefficient between X and X'.

► **Exercise.** We go back to the Toy-example from Sect. 3.1 with $\varphi(\xi) = (\xi - K)_+$ (with $x > K$) and $\Phi(x) = \mathbb{E}\,\varphi(X^x_T)$ (see Eq. (3.2)). In order to reduce the variance for short maturities of the estimators of the δ-hedge, we note that, for every λ, $\mu \in \mathbb{R}$,

$$\Phi'(x) = \lambda\varphi'(x) + \mathbb{E}\Big(\big(\varphi'(X^x_T) - \lambda\varphi'(x)\big)\frac{X^x_T}{x}\Big) = \mathbb{E}\Big(\big(\varphi(X^x_T) - \varphi(x)\big)\frac{W_T}{x\sigma T}\Big).$$

Apply the preceding to this example and implement it with $x = 100$, $K = 95$, $\sigma = 0.5$ and $T \in (0, 1]$ (after reading the next section). Compare the numerical results with the "naive" variance reduction obtained by (3.3).

3.2.2 *Implementation of the Variance Reduction: Batch versus Adaptive*

Let $(X_k, X'_k)_{k\geq 1}$ be an i.i.d. sequence of random vectors with the same distribution as (X, X') and let $\lambda \in \mathbb{R}$. Set, for every $k \geq 1$,

$$\Xi_k = X_k - X'_k, \quad X^\lambda_k = X_k - \lambda\,\Xi_k.$$

Now, set for every size $M \geq 1$ of the simulation:

$$V_M := \frac{1}{M}\sum_{k=1}^{M} \Xi_k^2, \qquad C_M := \frac{1}{M}\sum_{k=1}^{M} X_k\,\Xi_k$$

and

$$\lambda_M := \frac{C_M}{V_M} \qquad \text{(convention: } \lambda_0 = 0\text{)}. \tag{3.5}$$

The "batch" approach

▷ *Definition of the batch estimator.* The Strong Law of Large Numbers implies that both

$$V_M \longrightarrow \mathrm{Var}(X - X') \quad \text{and} \quad C_M \longrightarrow \mathrm{Cov}(X, X - X') \quad \mathbb{P}\text{-}a.s. \quad as \quad M \to +\infty$$

so that

$$\lambda_M \to \lambda_{\min} \quad \mathbb{P}\text{-}a.s. \quad \text{as} \quad M \to +\infty.$$

This suggests to introduce the *batch* estimator of m, defined for every size $M \geq 1$ of the simulation by

$$\overline{X}^{\lambda_M}_M = \frac{1}{M}\sum_{k=1}^{M} X_k^{\lambda_M}.$$

One checks that, for every $M \geq 1$,

$$\begin{aligned}\overline{X}^{\lambda_M}_M &= \frac{1}{M}\sum_{k=1}^{M} X_k - \lambda_M \frac{1}{M}\sum_{k=1}^{M} \Xi_k \\ &= \overline{X}_M - \lambda_M \overline{\Xi}_M \end{aligned} \tag{3.6}$$

with standard notations for empirical means.

▷ *Convergence of the batch estimator.* The asymptotic behavior of the batch estimator is summed up in the proposition below.

Proposition 3.3 *The* batch *estimator* $\mathbb{P}$-$a.s.$ *converges to m (consistency), i.e.*

$$\overline{X}^{\lambda_M}_M = \frac{1}{M}\sum_{k=1}^{M} X_k^{\lambda_M} \xrightarrow{a.s.} \mathbb{E}\, X = m$$

and satisfies a CLT (asymptotic normality) with an optimal asymptotic variance $\sigma^2_{\min}$, *i.e.*

$$\sqrt{M}\Big(\overline{X}^{\lambda_M}_M - m\Big) \xrightarrow{\mathcal{L}} \mathcal{N}\big(0; \sigma^2_{\min}\big).$$

Remark. However, note that the batch estimator is a *biased* estimator of m since $\mathbb{E}\,\lambda_M \overline{\Xi}_M \neq 0$.

Proof. First, one checks from (3.6) that

$$\frac{1}{M}\sum_{k=1}^{M} X_k^{\lambda_M} \xrightarrow{a.s.} m - \lambda_{\min} \times 0 = m.$$

Now, it follows from the regular *CLT* that

$$\sqrt{M}\left(\frac{1}{M}\sum_{k=1}^{M} X_k^{\lambda_{\min}} - m\right) \xrightarrow{\mathcal{L}} \mathcal{N}\big(0; \sigma^2_{\min}\big)$$

since $\mathrm{Var}(X - \lambda_{\min}\,\Xi) = \sigma^2_{\min}$. On the other hand

$$\sqrt{M}\left(\frac{1}{M}\sum_{k=1}^{M}X_k^{\lambda_M}-X_k^{\lambda_{\min}}\right)=\left(\lambda_{M}-\lambda_{\min}\right)\times\frac{1}{\sqrt{M}}\sum_{k=1}^{M}\Xi_k\xrightarrow{\mathbb{P}}0$$

owing to Slutsky's Lemma ([2]) since $\lambda_{M}-\lambda_{\min}\to 0\ a.s.$ as $M\to+\infty$ and

$$\frac{1}{\sqrt{M}}\sum_{k=1}^{M}\Xi_k\xrightarrow{\mathcal{L}}\mathcal{N}\big(0;\mathbb{E}\,\Xi^2\big)$$

by the regular *CLT* applied to the centered square integrable i.i.d. random variables $\Xi_k, k\ge 1$. Combining these two convergence results yields the announced *CLT*. $\diamond$

► **Exercise.** Let $\overline{X}_m$ and $\overline{\Xi}_m$ denote the empirical mean processes of the sequences $(X_k)_{k\ge1}$ and $(\Xi_k)_{k\ge1}$, respectively. Show that the quadruplet $(\overline{X}_{M},\overline{\Xi}_{M},C_{M},V_{M})$ can be computed in a recursive way from the sequence $(X_k,X'_k)_{k\ge1}$. Derive a recursive way to compute the batch estimator.

ℵ Practitioner's corner

One may proceed as follows:

– *Recursive implementation*: Use the recursion satisfied by the sequence $(\overline{X}_k,\overline{\Xi}_k,C_k,V_k)_{k\ge1}$ to compute λ_{M} and the resulting batch estimator for each size M.

– *True batch implementation*: A first phase of the simulation of size M', $M'\ll M$ (say $M'\simeq 5\%$ or 10% of the total budget M of the simulation) devoted to a rough estimate $\lambda_{M'}$ of $\lambda_{\min}$, based on the Monte Carlo estimator (3.5).

A second phase of the simulation to compute the estimator of m defined by

$$\frac{1}{M-M'}\sum_{k=M'+1}^{M}X_k^{\lambda_{M'}}$$

whose asymptotic variance – given the first phase of the simulation – is $\dfrac{\mathrm{Var}(X^\lambda)_{|\lambda=\lambda_{M'}}}{M-M'}$. This approach is not recursive at all. On the other hand, the above resulting estimator satisfies a *CLT* with asymptotic variance $\Phi(\lambda_{M'})=\mathrm{Var}(X^\lambda)_{|\lambda=\lambda_{M'}}$. In particular, we will most likely observe a significant – although not optimal – variance reduction. So, from this point of view, you can stop reading this section at this point.

The adaptive unbiased approach

Another approach is to design an adaptive estimator of m by considering at each step k the (predictable) estimator λ_{k-1} of $\lambda_{\min}$. This adaptive estimator is defined and analyzed below.

---

[2]If $Y_n\to c$ in probability and $Z_n\to Z_\infty$ in distribution, then $Y_nZ_n\to c\,Z_\infty$ in distribution. In particular, if $c=0$ the last convergence holds in probability.

Theorem 3.2 *Assume $X,\ X' \in L^{2+\delta}(\mathbb{P})$ for some $\delta > 0$. Let $(X_k, X'_k)_{k\ge 1}$ be an i.i.d. sequence with the same distribution as (X, X'). We set, for every $k \ge 1$,*

$$\widetilde{X}_k = X_k - \widetilde{\lambda}_{k-1}\,\Xi_k = (1-\widetilde{\lambda}_{k-1})X_k + \widetilde{\lambda}_{k-1}X'_k \quad \textit{where} \quad \widetilde{\lambda}_k = (-k)\vee(\lambda_k \wedge k)$$

and λ_k is defined by (3.5). Then, the adaptive estimator of m defined by

$$\overline{\overline{X}}^{\widetilde{\lambda}}_{M} = \frac{1}{M}\sum_{k=1}^{M}\widetilde{X}_k$$

is unbiased $\left(\mathbb{E}\,\overline{X}^{\widetilde{\lambda}}_{M} = m\right)$, convergent, i.e.

$$\overline{\overline{X}}^{\widetilde{\lambda}}_{M} \xrightarrow{a.s.} m \quad as \quad M \to +\infty,$$

and asymptotically normal with minimal variance, i.e.

$$\sqrt{M}\left(\overline{\overline{X}}^{\widetilde{\lambda}}_{M} - m\right) \xrightarrow{\mathcal{L}} \mathcal{N}\big(0;\sigma^2_{\min}\big) \quad as \quad M \to +\infty.$$

What follows in this section can be omitted at the occasion of a first reading although the method of proof exposed below is quite standard when dealing with the efficiency of an estimator by martingale methods the preceding

Proof. Step 1 (*a.s. convergence*). Let $\mathcal{F}_0 = \{\varnothing, \Omega\}$ and, for every $k \ge 1$, let $\mathcal{F}_k := \sigma(X_1, X'_1, \ldots, X_k, X'_k)$, be the filtration of the simulation.

First, we show that $(\widetilde{X}_k - m)_{k\ge1}$ is a sequence of square integrable $(\mathcal{F}_k, \mathbb{P})$-martingale increments. It is clear by construction that $\widetilde{X}_k$ is $\mathcal{F}_k$-measurable. Moreover,

$$\mathbb{E}\,(\widetilde{X}_k)^2 \le 2\big(\mathbb{E}\,X_k^2 + \mathbb{E}\,(\widetilde{\lambda}_{k-1}\Xi_k)^2\big)$$

$$= 2\big(\mathbb{E}\,X_k^2 + \mathbb{E}\,\widetilde{\lambda}^2_{k-1}\mathbb{E}\,\Xi_k^2\big) < +\infty,$$

where we used that Ξ_k and $\widetilde{\lambda}_{k-1}$ are independent and $\widetilde{\lambda}_{k-1} \le k-1$. Finally, for every $k \ge 1$,

$$\mathbb{E}\,(\widetilde{X}_k \mid \mathcal{F}_{k-1}) = \mathbb{E}\,(X_k \mid \mathcal{F}_{k-1}) - \widetilde{\lambda}_{k-1}\mathbb{E}\,(\Xi_k \mid \mathcal{F}_{k-1}) = m.$$

This shows that the adaptive estimator is unbiased since $\mathbb{E}\widetilde{X}_k = m$ for every $k \ge 1$. In fact, we can also compute the conditional variance increment process:

$$\mathbb{E}\left((\widetilde{X}_k - m)^2 \mid \mathcal{F}_{k-1}\right) = \mathrm{Var}(X^{\lambda})_{|\lambda=\widetilde{\lambda}_{k-1}} = \Phi(\widetilde{\lambda}_{k-1}).$$

Now, we set for every $k \ge 1$,

$$N_k := \sum_{\ell=1}^{k} \frac{\widetilde{X}_\ell - m}{\ell}.$$

It follows from the preceding that the sequence $(N_k)_{k\geq 1}$ is a square integrable $\big((\mathcal{F}_k)_k, \mathbb{P}\big)$-martingale since $(\widetilde{X}_k - m)_{k\geq 1}$, is a sequence of square integrable $\big((\mathcal{F}_k)_k, \mathbb{P}\big)$-martingale increments. Its conditional variance increment process (also known as "bracket process") $\langle N\rangle_k$, $k \geq 1$, is given by

$$\begin{aligned}\langle N\rangle_k &= \sum_{\ell=1}^{k} \frac{\mathbb{E}\,((\widetilde{X}_\ell - m)^2 \,|\, \mathcal{F}_{\ell-1})}{\ell^2}\\ &= \sum_{\ell=1}^{k} \frac{\Phi(\widetilde{\lambda}_{\ell-1})}{\ell^2}.\end{aligned}$$

Now, the above series is $a.s.$ convergent since we know that $\Phi(\widetilde{\lambda}_k)$ $a.s.$ converges towards $\Phi(\lambda_{\min})$ as $k \to +\infty$ since $\widetilde{\lambda}_k$ $a.s.$ converges toward $\lambda_{\min}$ and Φ is continuous. Consequently,

$$\langle N\rangle_\infty = a.s. \lim_{M\to+\infty} \langle N\rangle_M < +\infty \quad a.s.$$

Hence, it follows from Theorem 12.7 in the Miscellany Chapter that $N_M \to N_\infty$ $\mathbb{P}$-$a.s.$ as $M \to +\infty$ where N_∞ is an $a.s.$ finite random variable. In turn, the Kronecker Lemma (see Lemma 12.1, Sect. 12.7 of the Miscellany Chapter) implies,

$$\frac{1}{M}\sum_{k=1}^{M} \widetilde{X}_k - m \overset{a.s.}{\longrightarrow} 0 \quad \text{as} \quad M \to +\infty,$$

i.e.

$$\overline{\widetilde{X}}_M := \frac{1}{M}\sum_{k=1}^{M} \widetilde{X}_k \overset{a.s.}{\longrightarrow} m \quad \text{as} \quad M \to +\infty.$$

STEP 2 (*CLT, weak rate of convergence*). This is a consequence of the Lindeberg Central Limit Theorem for (square integrable) martingale increments (see Theorem 12.8 in the Miscellany Chapter or Theorem 3.2 and its Corollary 3.1, p. 58 in [142] referred to as Lindeberg's *CLT* in what follows). In our case, the array of martingale increments is defined by

$$X_{M,k} := \frac{\widetilde{X}_k - m}{\sqrt{M}}, \quad 1 \leq k \leq M.$$

There are essentially two assumptions to be checked. First the convergence of the conditional variance increment process toward $\sigma^2_{\min}$:

$$\begin{aligned}\sum_{k=1}^{M}\mathbb{E}\left(X_{M,k}^2\,|\,\mathcal{F}_{k-1}\right) &= \frac{1}{M}\sum_{k=1}^{M}\mathbb{E}\left((\widetilde{X}_k-m)^2\,|\,\mathcal{F}_{k-1}\right)\\ &= \frac{1}{M}\sum_{k=1}^{M}\Phi(\widetilde{\lambda}_{k-1})\\ &\longrightarrow \sigma^2_{\min} := \min_{\lambda}\Phi(\lambda).\end{aligned}$$

The second one is the so-called Lindeberg condition (see again Theorem 12.8 or [142], p. 58) which reads in our framework:

$$\forall\,\varepsilon>0,\quad \sum_{\ell=1}^{M}\mathbb{E}\left(X_{M,\ell}^2\mathbf{1}_{\{|X_{M,\ell}|>\varepsilon\}}\,|\,\mathcal{F}_{\ell-1}\right)\stackrel{\mathbb{P}}{\longrightarrow}0.$$

In turn, owing to the conditional Markov inequality and the definition of $X_{M,\ell}$, this condition classically follows from the slightly stronger: there exists a real number $\delta>0$ such that

$$\sup_{\ell\geq 1}\mathbb{E}\left(|\widetilde{X}_\ell-m|^{2+\delta}\,|\,\mathcal{F}_{\ell-1}\right)<+\infty\quad \mathbb{P}\text{-}a.s.$$

since

$$\sum_{\ell=1}^{M}\mathbb{E}\left(X_{M,\ell}^2\mathbf{1}_{\{|X_{M,\ell}|>\varepsilon\}}\,|\,\mathcal{F}_{\ell-1}\right)\leq\frac{1}{\varepsilon^\delta M^{1+\frac{\delta}{2}}}\sum_{\ell=1}^{M}\mathbb{E}\left(|\widetilde{X}_\ell-m|^{2+\delta}\,|\,\mathcal{F}_{\ell-1}\right).$$

Now, using that $(u+v)^{2+\delta}\leq 2^{1+\delta}(u^{2+\delta}+v^{2+\delta})$, $u,v\geq 0$, and the fact that $X, X'\in L^{2+\delta}(\mathbb{P})$, one gets

$$\mathbb{E}\,(|\widetilde{X}_\ell-m|^{2+\delta}\,|\,\mathcal{F}_{\ell-1})\leq 2^{1+\delta}\big(\mathbb{E}\,|X-m|^{2+\delta}+|\widetilde{\lambda}_{\ell-1}|^{2+\delta}\mathbb{E}\,|\Xi|^{2+\delta}\big).$$

Finally, the Lindeberg Central Limit Theorem implies

$$\sqrt{M}\left(\frac{1}{M}\sum_{k=1}^{M}\widetilde{X}_k-m\right)\stackrel{\mathcal{L}}{\longrightarrow}\mathcal{N}\big(0;\sigma^2_{\min}\big).$$

This means that the expected variance reduction does occur if one implements the recursive approach described above. ◇

3.3 Application to Option Pricing: Using Parity Equations to Produce Control Variates

The variance reduction by regression introduced in the former section still relies on the fact that $\kappa_X \simeq \kappa_{X-\lambda\Xi}$ or, equivalently, that the additional complexity induced by the simulation of Ξ given that of X is negligible. This condition may look demanding but we will see that in the framework of derivative pricing this requirement is always fulfilled as soon as the payoff of interest satisfies a so-called *parity equation*, *i.e.* that the original payoff can be *duplicated* by a "synthetic" version.

Furthermore, these *parity equations* are *model free* so they can be applied for various specifications of the dynamics of the underlying asset.

In this section, we denote by $(S_t)_{t\geq 0}$ the risky asset (with $S_0 = s_0 > 0$) and set $S_t^0 = e^{rt}$, the riskless asset. We work under the risk-neutral *risk-neutral* probability $\mathbb{P}$ (supposed to exist), which means that

$$\left(e^{-rt}S_t\right)_{t\in[0,T]} \quad \text{is a martingale on the scenarii space } (\Omega, \mathcal{A}, \mathbb{P})$$

(with respect to the augmented filtration of $(S_t)_{t\in[0,T]}$). Furthermore, to comply with usual assumptions of AOA theory, we will assume that this risk neutral probability is unique (complete market) to justify that we may price any derivative under this probability. However this has no real impact on what follows.

Vanilla Call-Put parity ($d = 1$)

We consider a Call and a Put with common maturity T and strike K. We denote by

$$\text{Call}_0(K,T) = e^{-rt}\mathbb{E}\left((S_T - K)_+\right) \qquad \text{and} \qquad \text{Put}_0(K,T) = e^{-rt}\mathbb{E}\left((K - S_T)_+\right)$$

the premium of this Call and this Put option, respectively. Since

$$(S_T - K)_+ - (K - S_T)_+ = S_T - K$$

and $\left(e^{-rt}S_t\right)_{t\in[0,T]}$ is a martingale, one derives the classical Call-Put parity equation:

$$\text{Call}_0(K,T) - \text{Put}_0(K,T) = s_0 - e^{-rt}K$$

so that $\text{Call}_0(K,T) = \mathbb{E}(X) = \mathbb{E}(X')$ with

$$X := e^{-rt}(S_T - K)_+ \quad \text{and} \quad X' := e^{-rt}(K - S_T)_+ + s_0 - e^{-rt}K.$$

As a result one sets

$$\Xi = X - X' = e^{-rt}S_T - s_0,$$

which turns out to be the terminal value of a martingale null at time 0 (this is in fact the generic situation of application of this parity method).

Note that the simulation of X involves that of S_T so that the additional cost of the simulation of Ξ is definitely negligible.

Asian Call-Put parity

We consider an Asian Call and an Asian Put with common maturity T, strike K and averaging period $[T_0, T]$, $0 \le T_0 < T$.

$$\text{Call}_0^{As} = e^{-rt}\mathbb{E}\left[\left(\tfrac{1}{T-T_0}\int_{T_0}^{T} S_t\,dt - K\right)_+\right]$$

and

$$\text{Put}_0^{As} = e^{-rt}\mathbb{E}\left[\left(K - \tfrac{1}{T-T_0}\int_{T_0}^{T} S_t\,dt\right)_+\right].$$

Still using that $\widetilde{S}_t = e^{-rt}S_t$ is a $\mathbb{P}$-martingale and, this time, the Fubini–Tonelli Theorem yield

$$\text{Call}_0^{As} - \text{Put}_0^{As} = s_0\frac{1-e^{-r(T-T_0)}}{r(T-T_0)} - e^{-rt}K$$

so that

$$\text{Call}_0^{As} = \mathbb{E}\,(X) = \mathbb{E}\,(X')$$

with

$$X := e^{-rt}\left(\frac{1}{T-T_0}\int_{T_0}^{T} S_t\,dt - K\right)_+$$
$$X' := s_0\frac{1-e^{-r(T-T_0)}}{r(T-T_0)} - e^{-rt}K + e^{-rt}\left(K - \frac{1}{T-T_0}\int_{T_0}^{T} S_t\,dt\right)_+.$$

This leads to

$$\Xi = e^{-rt}\frac{1}{T-T_0}\int_{T_0}^{T} S_t dt - s_0\frac{1-e^{-r(T-T_0)}}{r(T-T_0)}.$$

Remark. In both cases, the parity equation directly follows from the $\mathbb{P}$-*martingale property* of $\widetilde{S}_t = e^{-rt}S_t$.

3.3.1 Complexity Aspects in the General Case

In practical implementations, one often neglects the cost of the computation of $\lambda_{\min}$ since only a rough estimate is computed: this leads us to stop its computation after the first 5% or 10% of the simulation.

– However, one must be aware that the case of the existence of parity equations is quite specific since the random variable Ξ is involved in the simulation of X, so the complexity of the simulation process *is not* increased: thus in the recursive approach the updating of $\lambda_{_M}$ and of (the empirical mean) $\overline{\widetilde{X}}_{_M}$ is (almost) costless. Similar observations can be made to some extent on batch approaches. As a consequence, in that specific setting, the complexity of the adaptive linear regression procedure and the original one are (almost) the same!

– **Warning!** This is no longer true in general…and in a general setting the complexity of the simulation of X and X' is double that of X itself. Then the regression method is efficient if and only if

$$\sigma^2_{\min} < \frac{1}{2}\min\big(\mathrm{Var}(X),\,\mathrm{Var}(X')\big)$$

(provided one neglects the cost of the estimation of the coefficient $\lambda_{\min}$).

The exercise below shows the connection with antithetic variables which then appears as a special case of regression methods.

► **Exercise** (*Connection with the antithetic variable method*). Let $X,\ X' \in L^2(\mathbb{P})$ such that $\mathbb{E}\,X = \mathbb{E}\,X' = m$ and $\mathrm{Var}(X) = \mathrm{Var}(X')$.

(a) Show that $\lambda_{\min} = \dfrac{1}{2}$.

(b) Show that $X^{\lambda_{\min}} = \dfrac{X+X'}{2}$ and $\mathrm{Var}\left(\dfrac{X+X'}{2}\right) = \dfrac{1}{2}\big(\mathrm{Var}(X) + \mathrm{Cov}(X,\,X')\big)$.

Characterize the pairs $(X,\,X')$ for which the regression method does reduce the variance. Make the connection with the antithetic method.

3.3.2 *Examples of Numerical Simulations*

Vanilla B-S **Calls** (See Figs. 3.2, 3.3 and 3.4)

The model parameters are specified as follows

$$T = 1,\ x_0 = 100,\ r = 5,\ \sigma = 20,\ K = 90, \ldots, 120.$$

The simulation size is set at $M = 10^6$.

Asian Calls in a Heston model (See Figs. 3.5, 3.6 **and** 3.7)

The dynamics of the risky asset is this time a stochastic volatility model, namely the Heston model, defined as follows. Let ϑ, k, a such that $\vartheta^2/(2ak) \leq 1$ (so that v_t remains $a.s.$ positive, see [183], Proposition 6.2.4, p. 130).

$$\begin{aligned} dS_t &= S_t\big(r\,dt + \sqrt{v_t}\,dW_t^1\big), \qquad s_0 = x_0 > 0,\ t \in [0, T], \quad \text{(risky asset)} \\ dv_t &= k(a - v_t)dt + \vartheta\sqrt{v_t}\,dW_t^2,\ v_0 > 0 \\ &\text{with } \langle W^1, W^2\rangle_t = \rho\, t,\ \rho \in [-1, 1],\ t \in [0, T]. \end{aligned}$$

The payoff is an *Asian call* with strike price K

$$\text{AsCall}^{Hest} = e^{-rt}\mathbb{E}\left[\left(\frac{1}{T}\int_0^T S_s ds - K\right)_+\right].$$

Usually, no closed forms are available for Asian payoffs, even in the Black–Scholes model, and this is also the case in the Heston model. Note however that (quasi-)closed forms do exist for vanilla European options in this model (see [150]), which is the origin of its success. The simulation has been carried out by replacing the above diffusion by an Euler scheme (see Chap. 7 for an introduction to the Euler time discretization scheme). In fact, the dynamics of the stochastic volatility process does not fulfill the standard Lipschitz continuous assumptions required to make the Euler scheme converge, at least at its usual rate. In the present case it is even difficult to define this scheme because of the term $\sqrt{v_t}$. Since our purpose here is to illustrate

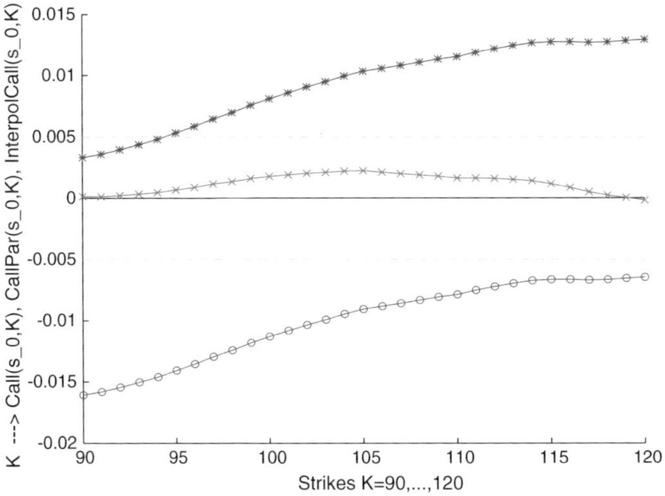

Fig. 3.2 Black–Scholes Calls: Error = Reference BS-(MC Premium). $K = 90, \ldots, 120$. –o–o–o– Crude Call. –∗–∗–∗– Synthetic Parity Call. –×–×–×– Interpolated Synthetic Call

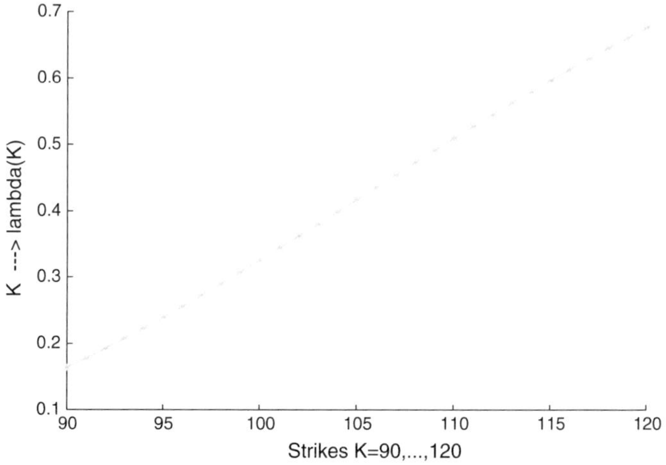

Fig. 3.3 BLACK–SCHOLES CALLS: $K \mapsto 1 - \lambda_{\min}(K)$, $K = 90, \ldots, 120$, for the Interpolated Synthetic Call

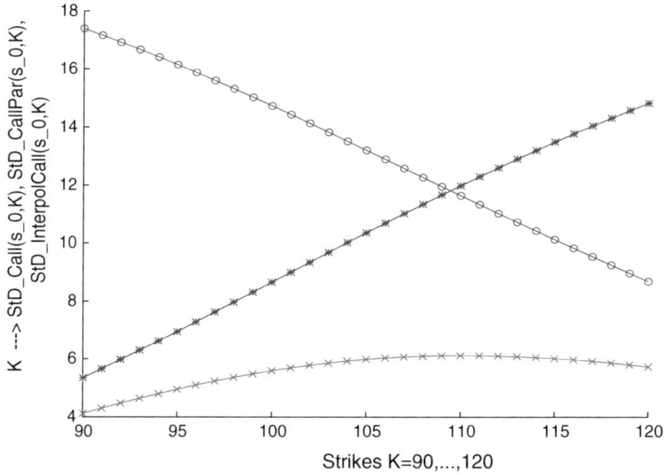

Fig. 3.4 BLACK–SCHOLES CALLS. Standard Deviation (MC Premium). $K = 90, \ldots, 120$. –o–o–o– Crude Call. –*–*–*– Parity Synthetic Call. –×–×–×– Interpolated Synthetic Call

the efficiency of parity relations to reduce variance, we adopted a rather "basic" scheme, namely

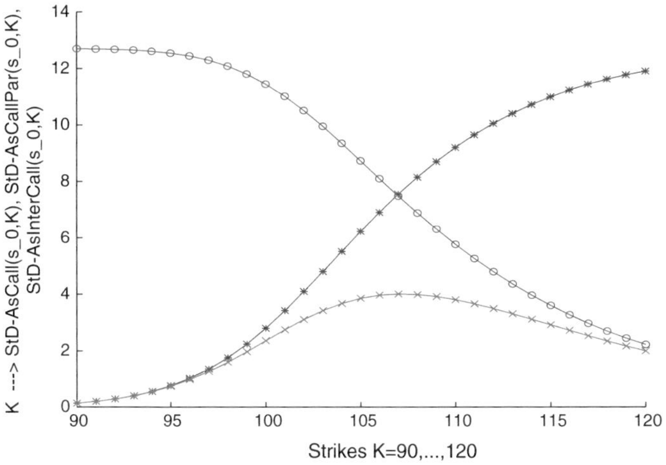

Fig. 3.5 HESTON ASIAN CALLS. *Standard Deviation (MC Premium).* $K = 90, \dots, 120$. *-o-o-o- Crude Call. -*-*-*- Synthetic Parity Call. -×××- Interpolated Synthetic Call*

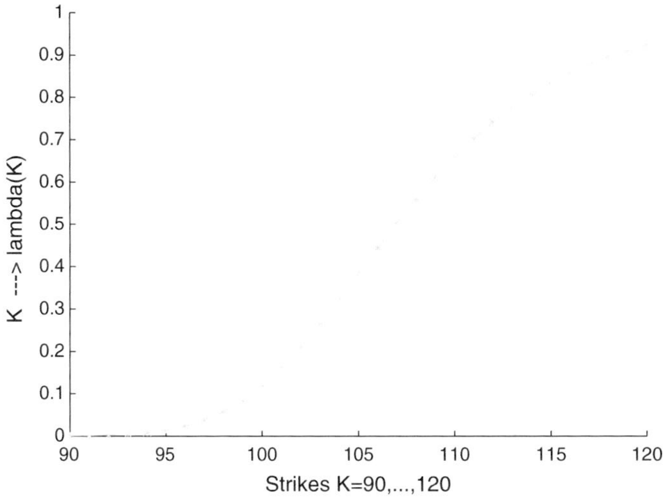

Fig. 3.6 HESTON ASIAN CALLS. $K \mapsto 1 - \lambda_{\min}(K)$, $K = 90, \dots, 120$, *for the Interpolated Synthetic Asian Call*

$$
\begin{aligned}
\bar{S}_{\frac{kT}{n}} &= \bar{S}_{\frac{(k-1)T}{n}}\Big(1 + \frac{rT}{n} + \sqrt{|\bar{v}_{\frac{(k-1)T}{n}}|}\sqrt{\frac{T}{n}}\big(\rho Z_k^2 + \sqrt{1-\rho^2}\,Z_k^1\big)\Big),\\
\bar{S}_0 &= s_0 > 0,\\
\bar{v}_{\frac{kT}{n}} &= k\big(a - \bar{v}_{\frac{(k-1)T}{n}}\big)\frac{T}{n} + \vartheta\sqrt{|\bar{v}_{\frac{(k-1)T}{n}}|}\,Z_k^2, \quad \bar{v}_0 = v_0 > 0,
\end{aligned}
$$

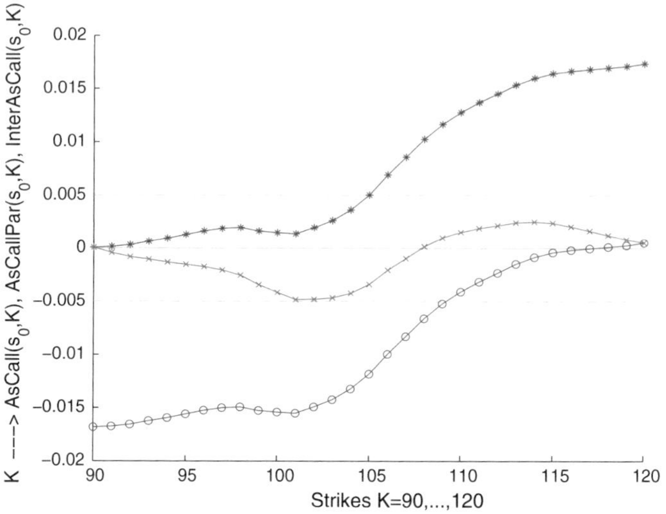

Fig. 3.7 HESTON ASIAN CALLS. $M = 10^6$ *(Reference*: *MC with* $M = 10^8$). $K = 90, \ldots, 120$. *-o-o-o- Crude Call. -*-*-*- Parity Synthetic Call. -×××- Interpolated Synthetic Call*

where $Z_k = (Z_k^1, Z_k^2)_{k\geq 1}$ is an i.i.d. sequence of $\mathcal{N}(0; I_2)$-distributed random vectors.

This scheme is consistent but its rate of convergence is not optimal. The simulation of the Heston model has given rise to an extensive literature. See *e.g.* [3, 7, 41, 115] and more recently [4] devoted to general affine diffusion processes.

– Parameters of the model:

$$s_0 = 100, k = 2, a = 0.01, \rho = 0.5, v_0 = 10\%, \vartheta = 20\%.$$

– Parameters of the option portfolio:

$$T = 1, K = 90, \ldots, 120 \qquad \text{(31 strikes)}.$$

▶ **Exercises.** We consider a one dimensional Black–Scholes model with market parameters

$$r = 0, \quad \sigma = 0.3, \quad x_0 = 100, \quad T = 1.$$

1. Consider a vanilla Call with strike $K = 80$. The random variable Ξ is defined as above. Estimate the $\lambda_{\min}$ (this should be not too far from 0.825). Then, compute a confidence interval for the Monte Carlo pricing of the Call with and without the linear variance reduction for the following sizes of the simulation: $M = 5\,000,\ 10\,000, 100\,000,\ 500\,000$.

2. Proceed as above but with $K = 150$ (true price 1.49). What do you observe? Provide an interpretation.

3.3.3 *The Multi-dimensional Case*

Let $X := (X^1, \ldots, X^d) : (\Omega, \mathcal{A}, \mathbb{P}) \longrightarrow \mathbb{R}^d$, $\Xi := (\Xi^1, \ldots, \Xi^q) : (\Omega, \mathcal{A}, \mathbb{P}) \longrightarrow \mathbb{R}^q$, be *square integrable* random vectors and assume

$$\mathbb{E}\, X = m \in \mathbb{R}^d, \qquad \mathbb{E}\, \Xi = 0 \in \mathbb{R}^q.$$

Let $D(X) := \big[\mathrm{Cov}(X^i, X^j)\big]_{1\le i,j\le d}$ and let $D(\Xi)$ denote the covariance (dispersion) matrices of X and Ξ, respectively. Assume

$$D(X) \text{ and } D(\Xi) > 0$$

as positive definite symmetric matrices.
The problem is to find a matrix solution $\Lambda \in \mathcal{M}(d, q, \mathbb{R})$ to the optimization problem

$$\mathrm{Var}(X - \Lambda\, \Xi) = \min\big\{\mathrm{Var}\big(X - L\, \Xi\big),\ L \in \mathcal{M}(d, q, \mathbb{R})\big\}$$

where $\mathrm{Var}(Y)$ is defined by $\mathrm{Var}(Y) := \mathbb{E}\,|Y - \mathbb{E}\,Y|^2 = \mathbb{E}\,|Y|^2 - |\mathbb{E}\,Y|^2$ for any $\mathbb{R}^d$-valued random vector.

The solution is given by

$$\Lambda = D(\Xi)^{-1} C(X, \Xi),$$

where

$$C(X, \Xi) = [\mathrm{Cov}(X^i, \Xi^j)]_{1\le i\le d, 1\le j\le q}.$$

⊳ **Examples-exercises.** Let $X_t = (X_t^1, \ldots, X_t^d)$, $t \in [0, T]$, be the price process of d risky traded assets (be careful about the notations that collide at this point: here X is used to denote the traded assets and the aim is to reduce the variance of the discounted payoff, usually denoted by the letter h).

1. Options on various baskets:

$$h_T^i = \left(\sum_{j=1}^{d} \theta_j^i X_T^j - K\right)_+, \quad i = 1, \ldots, d,$$

where θ_j^i are positive real coefficients.

Remark. This approach also produces an *optimal asset selection* (since it is essentially a PCA), which helps for hedging.

2. Portfolio of *forward start options*

$$h^{i,j} = \left(X_{T_{i+1}}^j - X_{T_i}^j\right)_+, \quad i = 1, \ldots, d-1,$$

where T_i, $i = 1, \ldots, d$ is an increasing sequence of maturities.

3.4 Pre-conditioning

The principle of the pre-conditioning method – also known as the Blackwell–Rao method – is based on the very definition of conditional expectation: let $(\Omega, \mathcal{A}, \mathbb{P})$ be probability space and let $X : (\Omega, \mathcal{A}, \mathbb{P}) \to \mathbb{R}$ be a square integrable random variable. The practical constraint for implementation is the ability to simulate $\mathbb{E}\,(X \mid \mathcal{B})$ at a competitive computational cost. Such is the case in the typical examples hereafter.

For every sub-σ-field $\mathcal{B} \subset \mathcal{A}$

$$\mathbb{E}\,X = \mathbb{E}\,\big(\mathbb{E}\,(X \mid \mathcal{B})\big)$$

and

$$\operatorname{Var}\big(\mathbb{E}\,(X \mid \mathcal{B})\big) = \mathbb{E}\,\big(\mathbb{E}\,(X \mid \mathcal{B})^2\big) - (\mathbb{E}\,X)^2 \leq \mathbb{E}\,X^2 - (\mathbb{E}\,X)^2 = \operatorname{Var}(X) \qquad (3.7)$$

since conditional expectation is a contraction in $L^2(\mathbb{P})$ as an orthogonal projection. In fact, one easily checks the following more precise result

$$\operatorname{Var}\big(\mathbb{E}\,(X \mid \mathcal{B})\big) = \operatorname{Var}(X) - \big\|X - \mathbb{E}\,(X \mid \mathcal{B})\big\|_2^2.$$

This shows that the above inequality (3.7) is strict, except if X is $\mathcal{B}$-measurable, *i.e.* is strict in any case of interest.

The archetypal situation is the following. Assume that

$$X = g(Z_1, Z_2), \quad g \in L^2\big(\mathbb{R}^2, \mathcal{B}or(\mathbb{R}^2), \mathbb{P}_{(Z_1,Z_2)}\big),$$

where Z_1, Z_2 are independent random vectors. Set $\mathcal{B} := \sigma(Z_2)$. Then standard results on conditional expectations show that

$$\mathbb{E}\,X = \mathbb{E}\,G(Z_2) \quad \text{where} \quad G(z_2) = \mathbb{E}\,\big(g(Z_1, Z_2) \mid Z_2 = z_2\big) = \mathbb{E}\,g(Z_1, z_2)$$

is a version of the conditional expectation of $g(Z_1, Z_2)$ given $\sigma(Z_2)$. At this stage, the pre-conditioning method can be implemented as soon as the following conditions are satisfied:

– a closed form is available for the function G and

– Z_2 can be simulated with the same complexity as X.

▷ **Examples. 1.** *Exchange spread options.* Let $X_T^i = x_i e^{(r-\frac{\sigma_i^2}{2})T+\sigma_i W_T^i}$, $x_i, \sigma_i > 0$, $i = 1, 2$, be two "Black–Scholes" assets at time T related to two Brownian motions W_T^i, $i = 1, 2$, with correlation $\rho \in [-1, 1]$. One considers an exchange spread option with strike K, *i.e.* related to the payoff

$$h_T = (X_T^1 - X_T^2 - K)_+.$$

Then one can write

$$(W_T^1, W_T^2) = \sqrt{T}\big(\sqrt{1-\rho^2}Z_1 + \rho\, Z_2, Z_2\big),$$

where $Z = (Z_1, Z_2)$ is an $\mathcal{N}(0; I_2)$-distributed random vector. Then, see *e.g.* Sect. 12.2 in the Miscellany Chapter),

$$\begin{aligned} e^{-rt}\mathbb{E}\left(h_T \mid Z_2\right) &= e^{-rt}\left[\mathbb{E}\left(\left(x_1 e^{(r-\frac{\sigma_1^2}{2})T+\sigma_1\sqrt{T}(\sqrt{1-\rho^2}\,Z_1+\rho z_2)}\right.\right.\right. \\ &\qquad \left.\left.\left. -x_2 e^{(r-\frac{\sigma_2^2}{2})T+\sigma_2\sqrt{T}z_2} - K\right)_+\right)\right]_{|z_2=Z_2} \\ &= \mathrm{Call}_{BS}\left(x_1 e^{-\frac{\rho^2\sigma_1^2 T}{2}+\sigma_1\rho\sqrt{T}Z_2},\, x_2 e^{(r-\frac{\sigma_2^2}{2})T+\sigma_2\sqrt{T}Z_2}\right. \\ &\qquad \left. +K, r, \sigma_1\sqrt{1-\rho^2}, T\right). \end{aligned}$$

Finally, one takes advantage of the closed form available for vanilla Call options in a Black–Scholes model to compute

$$\mathrm{Premium}_{BS}(x_1, x_2, K, \sigma_1, \sigma_2, r, T) = \mathbb{E}\Big(\mathbb{E}\,(e^{-rt}h_T \mid Z_2)\Big)$$

with a smaller variance than with the original payoff.

2. *Barrier options.* This example will be detailed in Sect. 8.2.3 devoted to the pricing (of some classes) of barrier options in a general model using the simulation of a continuous Euler scheme (using the so-called Brownian bridge method).

3.5 Stratified Sampling

The starting idea of stratification is to localize the Monte Carlo method on the elements of a measurable partition of the state space E of a random variable $X : (\Omega, \mathcal{A}, \mathbb{P}) \to (E, \mathcal{E})$.

Let $(A_i)_{i\in I}$ be a finite $\mathcal{E}$-measurable partition of the state space E. The A_i's are called *strata* and $(A_i)_{i\in I}$ a *stratification* of E. Assume that the weights

$$p_i = \mathbb{P}(X \in A_i),\ \ i \in I,$$

are *known*, (strictly) *positive* and that, still for every $i \in I$,

$$\mathcal{L}(X \mid X \in A_i) \stackrel{d}{=} \varphi_i(U),$$

where U is uniformly distributed over $[0,1]^{r_i}$ (with $r_i \in \mathbb{N} \cup \{\infty\}$, the case $r_i = +\infty$ corresponding to the acceptance-rejection method) and $\varphi_i : [0,1]^{r_i} \to E$ is an (easily) computable function. This second condition simply means that the conditional distribution $\mathcal{L}(X \mid X \in A_i)$ is easy to simulate on a computer. To be more precise we implicitly assume in what follows that the simulation of X and of the conditional distributions $\mathcal{L}(X \mid X \in A_i)$, $i \in I$, or, equivalently, the random vectors $\varphi_i(U_i)$, have approximately the same complexity. One must always keep this in mind since *it is a major constraint for practical implementations of stratification methods*.

This simulability condition usually has a strong impact on the possible design of the strata. For convenience, we will assume in what follows that $r_i = r$.

Let $F : (E, \mathcal{E}) \to (\mathbb{R}, \mathcal{B}or(\mathbb{R}))$ such that $\mathbb{E}\, F^2(X) < +\infty$. By elementary conditioning, we get

$$\mathbb{E}\, F(X) = \sum_{i \in I} \mathbb{E}\Big(\mathbf{1}_{\{X \in A_i\}} F(X)\Big) = \sum_{i \in I} p_i\, \mathbb{E}\big(F(X) \mid X \in A_i\big) = \sum_{i \in I} p_i\, \mathbb{E}\big(F(\varphi_i(U_i))\big),$$

where the random variables U_i, $i \in I$, are i.i.d. with uniform distribution over $[0,1]^r$. This is where the stratification idea is introduced. Let M be the global "budget" allocated to the simulation of $\mathbb{E}\, F(X)$. We split this budget into $|I|$ groups by setting

$$M_i = q_i M, \quad i \in I,$$

to be the allocated budget to compute $\mathbb{E}\, F(\varphi_i(U))$ in each stratum "A_i". This leads us to define the following (unbiased) estimator

$$\widehat{F(X)}^{I}_{M} := \sum_{i \in I} p_i \frac{1}{M_i} \sum_{k=1}^{M_i} F\big(\varphi_i(U_i^k)\big)$$

where $(U_i^k)_{1 \le k \le M_i, i \in I}$ are uniformly distributed on $[0,1]^r$, i.i.d. random variables (with an abuse of notation since the estimator actually depends on all the M_i). Then, elementary computations show that

$$\operatorname{Var}\Big(\widehat{F(X)}^{I}_{M}\Big) = \frac{1}{M} \sum_{i \in I} \frac{p_i^2}{q_i} \sigma^2_{F,i},$$

where, for every $i \in I$, the *local inertia*, reads

$$\begin{aligned}
\sigma^2_{F,i} &= \operatorname{Var}\Big(F(\varphi_i(U))\Big) = \operatorname{Var}\big(F(X) \mid X \in A_i\big)\\
&= \mathbb{E}\Big(\big(F(X) - \mathbb{E}\,(F(X) \mid X \in A_i)\big)^2 \mid X \in A_i\Big)\\
&= \frac{\mathbb{E}\Big(\big(F(X) - \mathbb{E}\,(F(X) \mid X \in A_i)\big)^2 \mathbf{1}_{\{X \in A_i\}}\Big)}{p_i}.
\end{aligned}$$

Optimizing the simulation budget allocation to each stratum amounts to solving the following minimization problem

$$\min_{(q_i)\in\mathcal{S}_I} \sum_{i\in I} \frac{p_i^2}{q_i}\sigma_{F,i}^2 \quad \text{where} \quad \mathcal{S}_I := \Big\{(q_i)_{i\in I} \in (0,1)^{|I|} \,\Big|\, \sum_{i\in I} q_i = 1\Big\}.$$

▷ A *sub-optimal choice*. It is natural and simple to set

$$q_i = p_i, \quad i \in I.$$

Such a choice is first motivated by the fact that the weights p_i are known and of course because it does reduce the variance since

$$\begin{aligned}
\sum_{i\in I} \frac{p_i^2}{q_i}\sigma_{F,i}^2 &= \sum_{i\in I} p_i\,\sigma_{F,i}^2 \\
&= \sum_{i\in I} \mathbb{E}\Big(\mathbf{1}_{\{X\in A_i\}}\big(F(X) - \mathbb{E}\,(F(X)\,|\,X \in A_i)\big)^2\Big) \\
&= \big\|F(X) - \mathbb{E}\,\big(F(X)\big|\,\mathcal{A}_I^X\big)\big\|_2^2,
\end{aligned} \tag{3.8}$$

where $\mathcal{A}_I^X = \sigma\big(\{X \in A_i\}, i \in I\big)$ denotes the σ-field spanned by the measurable partition $\{X \in A_i\}, i \in I$, of Ω and

$$\mathbb{E}\,\big(F(X)\big|\,\mathcal{A}_I^X\big) = \sum_{i\in I} \mathbb{E}\,\big(F(X)\,|\,X \in A_i\big)\mathbf{1}_{\{X\in A_i\}}.$$

Consequently

$$\sum_{i\in I} \frac{p_i^2}{q_i}\sigma_{F,i}^2 \le \big\|F(X) - \mathbb{E}\,F(X)\big\|_2^2 = \mathrm{Var}\big(F(X)\big) \tag{3.9}$$

with equality if and only if $\mathbb{E}\,\big(F(X)\,|\,\mathcal{A}_I^X\big) = \mathbb{E}\,F(X)$ $\mathbb{P}$-$a.s$. Or, equivalently, equality holds in this inequality if and only if

$$\mathbb{E}\,\big(F(X)\,|\,X \in A_i\big) = \mathbb{E}\,F(X), \;\; i \in I.$$

So this choice always reduces the variance of the estimator since we assumed that the stratification is not trivial. It corresponds in the *opinion poll* world to the so-called *quota method*.

▷ *Optimal choice*. The optimal choice is a solution to the above constrained minimization problem. It follows from a simple application of the Schwarz Inequality (and its equality case) that

$$\sum_{i\in I} p_i\sigma_{F,i} = \sum_{i\in I} \frac{p_i\sigma_{F,i}}{\sqrt{q_i}}\sqrt{q_i} \le \left(\sum_{i\in I}\frac{p_i^2\sigma_{F,i}^2}{q_i}\right)^{\frac{1}{2}}\left(\sum_{i\in I} q_i\right)^{\frac{1}{2}} = \left(\sum_{i\in I}\frac{p_i^2\sigma_{F,i}^2}{q_i}\right)^{\frac{1}{2}} \times 1^{\frac{1}{2}}$$
$$= \left(\sum_{i\in I}\frac{p_i^2\sigma_{F,i}^2}{q_i}\right)^{\frac{1}{2}}.$$

Consequently, the optimal choice for the allocation parameters q_i's, *i.e.* the solution to the above constrained minimization problem, is given by

$$q^*_{F,i} = \frac{p_i\sigma_{F,i}}{\sum_j p_j\sigma_{F,j}}, \quad i \in I, \tag{3.10}$$

with a resulting minimal variance

$$\left(\sum_{i\in I} p_i\sigma_{F,i}\right)^2.$$

At this stage the problem is that *unlike the weights p_i, the local inertia $\sigma^2_{F,i}$ are not known*, which makes the implementation less straightforward and sometimes questionable.

Some attempts have been made to circumvent this problem, see *e.g.* [86] for a recent reference based on an adaptive procedure for the computation of the local F-inertia $\sigma^2_{F,i}$.

However, using that the L^p-norms with respect to a probability measure are non-decreasing in p, one derives that

$$\begin{aligned}
\sigma_{F,i} &= \Big[\mathbb{E}\Big(\big|F(X) - \mathbb{E}\big(F(X)\,|\,\{X\in A_i\}\big)\big|^2\;|\,\{X\in A_i\}\Big)\Big]^{\frac{1}{2}}\\
&\ge \mathbb{E}\Big(\big|F(X) - \mathbb{E}\big(F(X)\,|\,\{X\in A_i\}\big)\big|\;|\,\{X\in A_i\}\Big)\\
&= \frac{\mathbb{E}\Big(\big|F(X) - \mathbb{E}\left(F(X)|X\in A_i\right)\big|\;|\,\{X\in A_i\}\Big)}{p_i}
\end{aligned}$$

so that, owing to Minkowski's Inequality,

$$\left(\sum_{i\in I} p_i\sigma_{F,i}\right)^2 \ge \big\|F(X) - \mathbb{E}\big(F(X)\,|\,\mathcal{A}_I\big)\big\|_1^2.$$

When compared to the resulting variance in (3.9) obtained with the suboptimal choice $q_i = p_i$, this illustrates the magnitude of the gain that can be expected from the optimal choice $q_i = q_i^*$: it lies in between $\big\|F(X) - \mathbb{E}\big(F(X)\,|\,\mathcal{A}_I\big)\big\|_1^2$ and $\big\|F(X) - \mathbb{E}\big(F(X)\,|\,\mathcal{A}_I\big)\big\|_2^2$.

⊳ **Examples.** Stratifications for the computation of $\mathbb{E}\,F(X)$, $X \stackrel{d}{=} \mathcal{N}(0; I_d)$, $d \geq 1$.

(a) *Stripes.* Let v be a fixed unitary vector (a simple and natural choice for v is $v = e_1 = (1, 0, 0, \ldots, 0)$: it is natural to define the strata as hyper-stripes perpendicular to the main axis $\mathbb{R}e_1$ of X). So, we set, for a given size N of the stratification $(I = \{1, \ldots, N\})$,

$$A_i := \Big\{x \in \mathbb{R}^d \;\text{ s.t. }\; (v|x) \in [y_{i-1}, y_i]\Big\}, \quad i = 1, \ldots, N,$$

where $y_i \in \overline{\mathbb{R}}$ is defined by $\Phi_0(y_i) = \frac{i}{N}$, $i = 0, \ldots, N$ (the N-quantiles of the $\mathcal{N}(0; 1)$ distribution). In particular, $y_0 = -\infty$ and $y_N = +\infty$. Then, if Z denotes an $\mathcal{N}(0; 1)$-distributed random variable,

$$p_i = \mathbb{P}(X \in A_i) = \mathbb{P}(Z \in [y_{i-1}, y_i]) = \Phi_0(y_i) - \Phi_0(y_{i-1}) = \frac{1}{N},$$

where Φ_0 denotes the c.d.f. of the $\mathcal{N}(0; 1)$-distribution. Other choices are possible for the y_i, leading to a non-uniform distribution of the p_i's. The simulation of the conditional distributions follows from the fact that

$$\mathcal{L}\big(X \,|\, (v|X) \in [a, b]\big) = \mathcal{L}\big(\xi_1 v + \pi_{v^\perp}^\perp(\xi_2)\big),$$

where $\xi_1 \stackrel{d}{=} \mathcal{L}(Z \,|\, Z \in [a, b])$ is independent of $\xi_2 \stackrel{d}{=} \mathcal{N}(0; I_{d-1})$,

$$\mathcal{L}(Z \,|\, Z \in [a, b]) \stackrel{d}{=} \Phi_0^{-1}\Big((\Phi_0(b) - \Phi_0(a))U + \Phi_0(a)\Big), \quad U \stackrel{d}{=} \mathcal{U}([0, 1])$$

and $\pi_{v^\perp}^\perp$ denotes the orthogonal projection on $v^\perp$. When $v = e_1$, this reads simply as

$$\mathcal{L}(X \,|\, (v|X) \in [a, b]) = \mathcal{L}(Z \,|\, Z \in [a, b]) \otimes \mathcal{N}(0; I_{d-1}).$$

(b) *Hyper-rectangles.* We still consider $X = (X^1, \ldots, X^d) \stackrel{d}{=} \mathcal{N}(0; I_d)$, $d \geq 2$. Let $(e_1, \ldots, e_d)$ denote the canonical basis of $\mathbb{R}^d$. We define the strata as hyper-rectangles. Let $N_1, \ldots, N_d \geq 1$.

$$A_{\underline{i}} := \prod_{\ell=1}^{d} \Big\{x \in \mathbb{R}^d \;\text{ s.t. }\; (e_\ell|x) \in [y^\ell_{i_\ell - 1}, y^\ell_{i_\ell}]\Big\}, \quad \underline{i} \in \prod_{\ell=1}^{d}\{1, \ldots, N_\ell\},$$

where the $y_i^\ell \in \overline{\mathbb{R}}$ are defined by $\Phi_0(y_i^\ell) = \frac{i}{N_\ell}$, $i = 0, \ldots, N_\ell$. Then, for every multi-index $\underline{i} = (i_1, \ldots, i_d) \in \prod_{\ell=1}^d \{1, \ldots, N_\ell\}$,

$$\mathcal{L}(X \mid X \in A_{\underline{i}}) = \bigotimes_{\ell=1}^{d} \mathcal{L}(Z \mid Z \in [y^{\ell}_{i_\ell - 1}, y^{\ell}_{i_\ell}]). \tag{3.11}$$

Optimizing the allocation to each stratum in the simulation for a given function F in order to reduce the variance is of course interesting and can be highly efficient but with the drawback of being strongly F-dependent, especially when this allocation needs an extra procedure like in [86]. An alternative and somewhat dual approach is to try optimizing the strata themselves uniformly with respect to a class of functions F (namely Lipschitz continuous functions) prior to the allocation across the strata.

This approach emphasizes the connections between stratification and optimal quantization and provides bounds on the best possible variance reduction factor that can be expected from a stratification. Some elements are provided in Chap. 5, see also [70] for further developments in infinite dimensions.

3.6 Importance Sampling

3.6.1 The Abstract Paradigm of Important Sampling

The basic principle of importance sampling is the following: let $X : (\Omega, \mathcal{A}, \mathbb{P}) \to (E, \mathcal{E})$ be an E-valued random variable. Let μ be a σ-finite *reference* measure on $(E, \mathcal{E})$ such that $\mathbb{P}_X \ll \mu$, *i.e.* there exists a probability density $f : (E, \mathcal{E}) \to (\mathbb{R}_+, \mathcal{B}(\mathbb{R}_+))$ such that

$$\mathbb{P}_X = f \cdot \mu.$$

In practice, we will have to simulate several r.v.s, whose distributions are all absolutely continuous with respect to this *reference* measure μ. For a first reading, one may assume that $E = \mathbb{R}^d$ and μ is the Lebesgue measure λ_d, but what follows can also be applied to more general measure spaces like the Wiener space (equipped with the Wiener measure), or countable sets (with the counting measure), etc. Let $h \in L^1(\mathbb{P}_X)$. Then,

$$\mathbb{E}\, h(X) = \int_E h(x)\mathbb{P}_X(dx) = \int_E h(x) f(x)\mu(dx).$$

Now, for any μ-$a.s.$ positive probability density function g defined on $(E, \mathcal{E})$ (with respect to μ), one has

$$\mathbb{E}\, h(X) = \int_E h(x) f(x)\mu(dx) = \int_E \frac{h(x) f(x)}{g(x)} g(x)\mu(dx).$$

One can always enlarge (if necessary) the original probability space $(\Omega, \mathcal{A}, \mathbb{P})$ to design a random variable $Y : (\Omega, \mathcal{A}, \mathbb{P}) \to (E, \mathcal{E})$ having g as a probability density with respect to μ. Then, going back to the probability space yields for every non-negative or $\mathbb{P}_X$-integrable function $h : E \to \mathbb{R}$,

$$\mathbb{E}\, h(X) = \mathbb{E}\left(\frac{h(Y)f(Y)}{g(Y)}\right). \tag{3.12}$$

So, in order to compute $\mathbb{E}\, h(X)$, one may also implement a Monte Carlo simulation based on the simulation of independent copies of the random variable Y, *i.e.*

$$\mathbb{E}\, h(X) = \mathbb{E}\left(\frac{h(Y)f(Y)}{g(Y)}\right) = \lim_{M\to+\infty} \frac{1}{M}\sum_{k=1}^{M} h(Y_k)\frac{f(Y_k)}{g(Y_k)} \quad a.s.$$

ℵ Practitioner's corner.

▷ *Practical requirements (to undertake the simulation).* To proceed, it is necessary to simulate independent copies of Y and to compute the ratio of density functions f/g at a reasonable cost. Note that only the ratio is needed which makes useless the computation of some "structural" constants like $(2\pi)^{d/2}$, *e.g.* when both f and g are Gaussian densities with different means (see below). By "reasonable cost" for the simulation of Y, we mean *at the same cost as* that of X (in terms of complexity). As concerns the ratio f/g, this means that its computational cost remains negligible with respect to that of h or, which may be the case in some slightly different situations, that the computational cost of $h \times \frac{f}{g}$ is equivalent to that of h alone.

▷ *Sufficient conditions (to undertake the simulation).* Once the above conditions are fulfilled, the question is: is it profitable to proceed like this? This is the case if the complexity of the simulation for a given accuracy (in terms of confidence interval) is lower with the second method. If one assumes as above that simulating X and Y on the one hand, and computing $h(x)$ and $(hf/g)(x)$ on the other hand are both comparable in terms of complexity, *the question amounts to comparing the variances* or, equivalently, the squared quadratic norm of the estimators since they have the same expectation $\mathbb{E}\, h(X)$.

Now

$$\begin{aligned}
\mathbb{E}\left[\left(\frac{h(Y)f(Y)}{g(Y)}\right)^2\right] = \mathbb{E}\left[\left(\frac{hf}{g}(Y)\right)^2\right] &= \int_E \left(\frac{h(x)f(x)}{g(x)}\right)^2 g(x)\,\mu(dx)\\
&= \int_E h(x)^2\frac{f(x)}{g(x)} f(x)\,\mu(dx)\\
&= \mathbb{E}\left(h(X)^2\frac{f}{g}(X)\right).
\end{aligned}$$

As a consequence, simulating $\frac{hf}{g}(Y)$ rather than $h(X)$ will reduce the variance if and only if

$$\mathbb{E}\left(h(X)^2\frac{f}{g}(X)\right) < \mathbb{E}\left(h(X)^2\right). \tag{3.13}$$

Remark. In fact, theoretically, as soon as h is non-negative and $\mathbb{E}\,h(X) \neq 0$, one may reduce the variance of the new simulation to…0. As a matter of fact, using the Schwarz Inequality one gets, as if trying to "reprove" that $\mathrm{Var}\left(\frac{h(Y)f(Y)}{g(Y)}\right) \geq 0$,

$$\begin{aligned}\left(\mathbb{E}\,h(X)\right)^2 = \left(\int_E h(x)f(x)\mu(dx)\right)^2 &= \left(\int_E \frac{h(x)f(x)}{\sqrt{g(x)}}\sqrt{g(x)}\mu(dx)\right)^2\\ &\leq \int_E \frac{(h(x)f(x))^2}{g(x)}\mu(dx) \times \int_E g\,d\mu\\ &= \int_E \frac{(h(x)f(x))^2}{g(x)}\mu(dx)\end{aligned}$$

since g is a probability density function. Now the equality case in the Schwarz inequality says that the variance is 0 if and only if $\sqrt{g(x)}$ and $\frac{h(x)f(x)}{\sqrt{g(x)}}$ are proportional $\mu(dx)$-*a.s.*, *i.e.* $h(x)f(x) = c\,g(x)\ \mu(dx)$-*a.s.* for a (non-negative) real constant c. Finally, *when h has a constant sign and* $\mathbb{E}\,h(X) \neq 0$ this leads to

$$g(x) = f(x)\frac{h(x)}{\mathbb{E}\,h(X)}\ \mu(dx)\quad \mathbb{P}\text{-}a.s.$$

This choice is clearly impossible to make since it would mean that $\mathbb{E}\,h(X)$ is known since it is involved in the formula…and would then be of no use. *A contrario* this may suggest a direction for designing the (distribution) of Y.

3.6.2 How to Design and Implement Importance Sampling

The intuition that must guide practitioners when designing an importance sampling method is to replace a random variable X by a random variable Y so that $\frac{hf}{g}(Y)$ is in some way often "closer" than $h(X)$ to their common mean. Let us be more specific.

We consider a Call on the risky asset $(X_t)_{t\in[0,T]}$ with strike price K and maturity $T > 0$ (with interest rate $r \equiv 0$ for simplicity). If $X_0 = x \ll K$, *i.e.* the option is deep out-of-the money at the origin of time so that most of the scenarii $X_T(\omega)$ will satisfy $X_T(\omega) \leq K$ or equivalently $(X_T(\omega) - K)_+ = 0$. In such a setting, the event $\{(X_T - K)_+ > 0\}$ – the payoff is positive – is a *rare event* so that the number of scenarii that will produce a non-zero value for $(X_T - K)_+$ will be small, inducing a too rough estimate of the quantity of interest $\mathbb{E}\,(X_T - K)_+$. Put in a more quantitative way, it means that, even if both the expectation and the standard deviation of the payoff are both small in absolute value, their *ratio* (standard deviation over expectation) will be very large.

By contrast, if we switch from $(X_t)_{t\in[0,T]}$ to $(Y_t)_{t\in[0,T]}$ so that:

– we can compute the ratio $\frac{f_{X_T}}{g_{Y_T}}(y)$ where f_{X_T} and g_{Y_T} are the probability densities of X_T and Y_T, respectively,

– Y_T takes most, or at least a significant part, of its values in $[K, +\infty)$.

Then

$$\mathbb{E}\,(X_T - K)_+ = \mathbb{E}\left((Y_T - K)_+ \frac{f_{X_T}}{g_{Y_T}}(Y_T)\right)$$

and we can reasonably hope that we will simulate more significant scenarii for $(Y_T - K)_+ \frac{f_{X_T}}{g_{Y_T}}(Y_T)$ than for $(X_T - K)_+$. This effect will be measured by the variance reduction.

This interpretation in terms of "rare events" is in fact the core of importance sampling, more than the plain "variance reduction" feature. In particular, this is what a practitioner must have in mind when searching for a "good" probability distribution g: importance sampling is more a matter of "focusing light where it is needed" than reducing variance.

When dealing with vanilla options in simple models (typically local volatility), one usually works on the state space $E = \mathbb{R}_+$ and importance sampling amounts to a change of variable in one-dimensional integrals as emphasized above. However, in more involved frameworks, one considers the scenarii space as a state space, typically $E = \Omega = \mathcal{C}(\mathbb{R}_+, \mathbb{R}^d)$ and uses Girsanov's Theorem instead of the usual change of variable with respect to the Lebesgue measure.

3.6.3 Parametric Importance Sampling

In practice, the starting idea is to introduce a parametric family of random variables $(Y_\theta)_{\theta\in\Theta}$ (often defined on the same probability space $(\Omega, \mathcal{A}, \mathbb{P})$ as X) such that

– for every $\theta \in \Theta$, Y_θ has a probability density $g_\theta > 0$ μ-$a.e.$ with respect to a reference measure μ and Y_θ is as easy to simulate as X in terms of complexity,

– the ratio $\frac{f}{g_\theta}$ has a small computational cost where f is the probability density of the distribution of X with respect to μ.

Furthermore, we can always assume, by adding a value to Θ if necessary, that for a value $\theta_0 \in \Theta$ of the parameter, $Y_{\theta_0} = X$ (at least in distribution).

The problem becomes a parametric optimization problem, typically solving the minimization problem

$$\min_{\theta\in\Theta}\left\{\mathbb{E}\left[\left(h(Y_\theta)\frac{f}{g_\theta}(Y_\theta)\right)^2\right] = \mathbb{E}\left(h(X)^2\frac{f}{g_\theta}(X)\right)\right\}.$$

Of course there is no reason why the solution to the above problem should be θ_0 (if so, such a parametric model is inappropriate). At this stage one can follow two strategies:

– Try to solve by numerical means the above minimization problem.

– Use one's intuition to select *a priori* a good (though sub-optimal) $\theta \in \Theta$ by applying the heuristic principle: "focus light where needed".

$\rhd$ **Example** *(The Cameron–Martin formula and Importance Sampling by mean translation)*. This example takes place in a Gaussian framework. We consider (as a starting motivation) a one dimensional Black–Scholes model defined by

$$X_T^x = xe^{\mu T+\sigma W_T} = xe^{\mu T+\sigma\sqrt{T}Z}, \quad Z \stackrel{d}{=} \mathcal{N}(0;1),$$

with $x > 0$, $\sigma > 0$ and $\mu = r - \frac{\sigma^2}{2}$. Then, the premium of an option with payoff $h : (0, +\infty) \to (0, +\infty)$ reads

$$e^{-rt}\mathbb{E}\, h(X_T^x) = \mathbb{E}\,\varphi(Z) = \int_{\mathbb{R}} \varphi(z)e^{-\frac{z^2}{2}}\frac{dz}{\sqrt{2\pi}},$$

where $\varphi(z) = e^{-rt}h\Big(xe^{\mu T+\sigma\sqrt{T}z}\Big)$, $z \in \mathbb{R}$.

From now on, we forget about the financial framework and deal with

$$\mathbb{E}\,\varphi(Z) = \int_{\mathbb{R}} \varphi(z)g_0(z)dz \quad \text{where} \quad g_0(z) = \frac{e^{-\frac{z^2}{2}}}{\sqrt{2\pi}}$$

and the random variable Z plays the role of X in the above theoretical part. The idea is to introduce the parametric family

$$Y_\theta = Z + \theta, \quad \theta \in \Theta := \mathbb{R}.$$

We consider the Lebesgue measure on the real line λ_1 as a reference measure, so that

$$g_\theta(y) = \frac{e^{-\frac{(y-\theta)^2}{2}}}{\sqrt{2\pi}}, \quad y \in \mathbb{R},\ \theta \in \Theta := \mathbb{R}.$$

Elementary computations show that

$$\frac{g_0}{g_\theta}(y) = e^{-\theta y+\frac{\theta^2}{2}}, \quad y \in \mathbb{R},\ \theta \in \Theta := \mathbb{R}.$$

Hence, we derive the *Cameron–Martin formula*

$$\mathbb{E}\,\varphi(Z) = e^{\frac{\theta^2}{2}}\mathbb{E}\Big(\varphi(Y_\theta)e^{-\theta Y_\theta}\Big)$$
$$= e^{\frac{\theta^2}{2}}\mathbb{E}\Big(\varphi(Z+\theta)e^{-\theta(Z+\theta)}\Big) = e^{-\frac{\theta^2}{2}}\mathbb{E}\Big(\varphi(Z+\theta)e^{-\theta Z}\Big).$$

Remark. In fact, a *standard change of variable* based on the invariance of the Lebesgue measure by translation yields the same result in a much more straightforward way: setting $z = u + \theta$ shows that

$$\mathbb{E}\,\varphi(Z) = \int_{\mathbb{R}} \varphi(u+\theta)e^{-\frac{\theta^2}{2}-\theta u-\frac{u^2}{2}}\frac{du}{\sqrt{2\pi}} = e^{-\frac{\theta^2}{2}}\,\mathbb{E}\left(e^{-\theta Z}\varphi(Z+\theta)\right)$$
$$= e^{\frac{\theta^2}{2}}\mathbb{E}\Big(\varphi(Z+\theta)e^{-\theta(Z+\theta)}\Big).$$

It is to be noticed again that there is no need to account for the normalization constants to compute the ratio $\dfrac{g_0}{g_\theta}$.

The next step is to choose a "good" θ which significantly reduces the variance, *i.e.* following Condition (3.13) (using the formulation involving "$Y_\theta = Z + \theta$"), such that

$$\mathbb{E}\left(e^{\frac{\theta^2}{2}}\varphi(Z+\theta)e^{-\theta(Z+\theta)}\right)^2 < \mathbb{E}\,\varphi^2(Z),$$

i.e.

$$e^{-\theta^2}\mathbb{E}\Big(\varphi^2(Z+\theta)e^{-2\theta Z}\Big) < \mathbb{E}\,\varphi^2(Z)$$

or, equivalently, if one uses the formulation of (3.13) based on the original random variable (here Z),

$$\mathbb{E}\Big(\varphi^2(Z)e^{\frac{\theta^2}{2}-\theta Z}\Big) < \mathbb{E}\,\varphi^2(Z).$$

Consequently the variance minimization amounts to the following problem

$$\min_{\theta\in\mathbb{R}}\left[e^{\frac{\theta^2}{2}}\mathbb{E}\left(\varphi^2(Z)e^{-\theta Z}\right) = e^{-\theta^2}\mathbb{E}\left(\varphi^2(Z+\theta)e^{-2\theta Z}\right)\right].$$

It is clear that the solution of this optimization problem and the resulting choice of θ highly depends on the function h.

– *Optimization approach*: When h is smooth enough, an approach based on large deviation estimates has been proposed by Glasserman et al. (see [115]). We propose a simple recursive/adaptive approach in Sect. 6.3.1 of Chap. 6 based on Stochastic Approximation which does not depend upon the regularity of the function h (see also [12] for a pioneering work in that direction).

– *Heuristic suboptimal approach*: Let us temporarily return to our pricing problem involving the specified function $\varphi(z) = e^{-rt}\big(x\exp(\mu T + \sigma\sqrt{T}z) - K\big)_+$, $z \in \mathbb{R}$. When $x \ll K$ (deep-out-of-the-money option), most simulations of $\varphi(Z)$ will produce 0 as a result. A first simple idea – if one does not wish to carry out the above optimization – can be to "re-center the simulation" of X^x_T around K by replacing Z by $Z+\theta$, where θ satisfies

$$\mathbb{E}\left(x\exp\left(\mu T+\sigma\sqrt{T}(Z+\theta)\right)\right)=K$$

which yields, since $\mathbb{E}\,X^x_T = xe^{rt}$,

$$\theta := -\frac{\log(x/K)+r\,T}{\sigma\sqrt{T}}. \tag{3.14}$$

Solving the similar, though slightly different equation,

$$\mathbb{E}\left(x\exp\left(\mu T+\sigma\sqrt{T}(Z+\theta)\right)\right)=e^{rt}K$$

would lead to

$$\theta := -\frac{\log(x/K)}{\sigma\sqrt{T}}. \tag{3.15}$$

A third simple, intuitive idea is to search for θ such that

$$\mathbb{P}\Big(x\exp\left(\mu T+\sigma\sqrt{T}(Z+\theta)\right)<K\Big)=\frac{1}{2},$$

which yields

$$\theta := -\frac{\log(x/K)+\mu\,T}{\sigma\sqrt{T}}. \tag{3.16}$$

This choice is also the solution to the equation $x\,e^{\mu T+\sigma\sqrt{T}\theta}=K$, etc.

All these choices are suboptimal but reasonable when $x \ll K$. However, if we need to price a whole portfolio including many options with various strikes, maturities (and underlyings...), the above approach is no longer possible and a data driven optimization method like the one developed in Chap. 6 becomes mandatory.

Other parametric methods can be introduced, especially in non-Gaussian frameworks, like for example the so-called "exponential tilting" (or Esscher transform) for distributions having a Laplace transform on the whole real line (see *e.g.* [199]). Thus, when dealing with the NIG distribution (for Normal Inverse Gaussian) this transform has an impact on the thickness of the tail of the distribution. Of course, there is no *a priori* limit to what can be designed on a specific problem. When dealing with path-dependent options, one usually relies on the Girsanov theorem to modify likewise the drift of the risky asset dynamics (see [199]). All the preceding can be adapted to multi-dimensional models.

▶ **Exercises. 1.** (a) Show that under appropriate integrability assumptions on h to be specified, the function

$$\theta \mapsto \mathbb{E}\Big(\varphi^2(Z)e^{\frac{\theta^2}{2}-\theta Z}\Big)$$

is strictly convex and differentiable on the whole real line with a derivative given by

$$\theta \mapsto \mathbb{E}\Big(\varphi^2(Z)(\theta - Z)e^{\frac{\theta^2}{2}-\theta Z}\Big).$$

(b) Show that if φ is an even function, then this parametric importance sampling procedure by mean translation is useless. Give a necessary and sufficient condition (involving φ and Z) that makes it always useful.

2. Set $r = 0$, $\sigma = 0.2$, $X_0 = x = 70$, $T = 1$. One wishes to price a Call option with strike price $K = 100$ (*i.e.* deep out-of-the-money). The true Black–Scholes price is 0.248 (see Sect. 12.2).

Compare the performances of

(i) a "crude" Monte Carlo simulation,

(ii) the above "intuitively guided" heuristic choices for θ.

Assume now that $x = K = 100$. What do you think of the heuristic suboptimal choice?

3. Write all the preceding when $Z \stackrel{d}{=} \mathcal{N}(0; I_d)$.

4. *Randomization of an integral.* Let $h \in L^1(\mathbb{R}^d, \mathcal{B}or(\mathbb{R}^d), \lambda_d)$.

(a) Show that for any $\mathbb{R}^d$-valued random vector Y having an absolutely continuous distribution $\mathbb{P}_Y = g.\lambda_d$ with $g > 0$, λ_d $a.s.$ on $\{h > 0\}$, one has

$$\int_{\mathbb{R}^d} h d\lambda_d = \mathbb{E}\left(\frac{h}{g}(X)\right).$$

Derive a probabilistic method to compute $\displaystyle\int_{\mathbb{R}^d} h\, d\lambda_d$.

(b) Propose an importance sampling approach to this problem inspired by the above examples.

3.6.4 *Computing the Value-At-Risk by Monte Carlo Simulation: First Approach*

Let X be a real-valued random variable defined on a probability space, representative of a *loss*. For the sake of simplicity we suppose here that X has a continuous distribution, *i.e.* that its distribution function defined for every $x \in \mathbb{R}$ by $F(x) := \mathbb{P}(X \le x)$ is continuous. For a given confidence level $\alpha \in (0, 1)$ (usually closed to 1), the Value-at-Risk at level α (denoted by VaR_α or $\mathrm{VaR}_{\alpha,X}$ following [92]) is any real number satisfying the equation

$$\mathbb{P}\big(X \le \mathrm{VaR}_{\alpha,X}\big) = \alpha \in (0,1). \tag{3.17}$$

Equation (3.17) has at least one solution since F is continuous which may not be unique in general. For convenience one often assumes that the lowest solution of Equation (3.17) is the $\mathrm{VaR}_{\alpha,X}$. In fact, Value-at-Risk (of X) is not consistent as a measure of risk (as emphasized in [92]), but nowadays it is still widely used to measure financial risk.

One naive way to compute $\mathrm{VaR}_{\alpha,X}$ is to estimate the empirical distribution function of a (large enough) Monte Carlo simulation at some points ξ lying in a grid $\Gamma := \{\xi_i,\ i \in I\}$, namely

$$\widehat{F(\xi)}_M := \frac{1}{M}\sum_{k=1}^{M} \mathbf{1}_{\{X_k \le \xi\}},\ \ \xi \in \Gamma,$$

where $(X_k)_{k\ge 1}$ is an i.i.d. sequence of X-distributed random variables. Then one solves the equation $\widehat{F}(\xi)_M = \alpha$ (using an interpolation step of course).

Such an approach based on the empirical distribution of X needs to simulate extreme values of X since α is usually close to 1. So implementing a Monte Carlo simulation directly on the above equation is usually a slightly meaningless exercise. Importance sampling becomes the natural way to "re-center" the equation.

Assume, for example, that

$$X := \varphi(Z), \qquad Z \stackrel{d}{=} \mathcal{N}(0;1).$$

Then, for every $\xi \in \mathbb{R}$,

$$\mathbb{P}(X \le \xi) = e^{-\frac{\theta^2}{2}}\,\mathbb{E}\left(\mathbf{1}_{\{\varphi(Z+\theta)\le \xi\}} e^{-\theta Z}\right)$$

so that the above Eq. (3.17) now reads (θ being fixed)

$$\mathbb{E}\left(\mathbf{1}_{\{\varphi(Z+\theta)\le \mathrm{VaR}_{\alpha,X}\}} e^{-\theta Z}\right) = e^{\frac{\theta^2}{2}}\alpha.$$

It remains to find good variance reducers θ. This choice depends of course on ξ but in practice it should be fitted to reduce the variance in the neighborhood of $\mathrm{VaR}_{\alpha,X}$. We will see in Chap. 6 that more efficient methods based on Stochastic Approximation can be devised. But they also need variance reduction to be implemented. Furthermore, similar ideas can be used to compute a consistent measure of risk called the Conditional Value-at-Risk (or Averaged Value-at-Risk).

Chapter 4
The Quasi-Monte Carlo Method

In this chapter we present the so-called *Quasi-Monte Carlo* (*QMC*) method, which can be seen as a deterministic alternative to the standard Monte Carlo method: the pseudo-random numbers are replaced by deterministic computable sequences of $[0, 1]^d$-valued vectors which, once substituted *mutatis mutandis* in place of pseudo-random numbers in the Monte Carlo method, may significantly speed up its rate of convergence, making it *almost* independent of the structural dimension d of the simulation.

4.1 Motivation and Definitions

Computing an expectation $\mathbb{E}\,\varphi(X)$ using a Monte Carlo simulation ultimately amounts to computing either a finite-dimensional integral

$$\int_{[0,1]^d} f(u^1, \dots, u^d)du^1 \cdots du^d$$

or an infinite-dimensional integral

$$\int_{[0,1]^{(\mathbb{N}^*)}} f(u)\lambda_\infty(du),$$

where $[0, 1]^{(\mathbb{N}^*)}$ denotes the set of finite $[0, 1]$-valued sequences (or, equivalently, sequences vanishing at a finite range) and $\lambda_\infty = \lambda^{\otimes \mathbb{N}}$ is the Lebesgue measure on $\big([0, 1]^{\mathbb{N}^*}, \mathcal{B}or([0, 1]^{\mathbb{N}^*})\big)$. Integrals of the first type show up when X can be simulated by standard methods like the inverse distribution function, Box–Muller simulation method of Gaussian distributions, etc, so that $X = g(U)$, $U = (U^1, \cdots, U^d) \stackrel{d}{=}$

G. Pagès, *Numerical Probability*, Universitext,
https://doi.org/10.1007/978-3-319-90276-0_4

$\mathcal{U}([0,1]^d)$ whereas the second type is typical of a simulation using an acceptance-rejection method (like the polar method for Gaussian distributions).

As concerns finite-dimensional integrals, we saw that, if $(U_n)_{n\geq 1}$ denotes an i.i.d. sequence of uniformly distributed random vectors on $[0,1]^d$, then, for every function $f \in L^1_{\mathbb{R}}([0,1]^d, \lambda_d)$,

$$\mathbb{P}(d\omega)\text{-}a.s.\quad \frac{1}{n}\sum_{k=1}^{n} f\big(U_k(\omega)\big) \longrightarrow \mathbb{E}\, f(U_1) = \int_{[0,1]^d} f(u^1,\dots,u^d)du^1\cdots du^d, \tag{4.1}$$

where the subset Ω_f of $\mathbb{P}$-probability 1 on which this convergence holds true depends on the function f. In particular, the above $a.s.$-convergence holds for any continuous function on $[0,1]^d$. But in fact, taking advantage of the separability of the space of continuous functions, we will show below that this convergence *simultaneously* holds for all continuous functions on $[0,1]^d$ and even on the larger class of *Riemann integrable functions* on $[0,1]^d$.

First we briefly recall the basic definition of weak convergence of probability measures on metric spaces (see [45] Chap. 1 for a general introduction to weak convergence of probability measures on metric spaces).

Definition 4.1 (Weak convergence) *Let (S,δ) be a metric space and let $\mathcal{S} := \mathcal{B}or_\delta(S)$ be its Borel σ-field. Let $(\mu_n)_{n\geq 1}$ be a sequence of probability measures on $(S,\mathcal{S})$ and μ a probability measure on the same space. The sequence $(\mu_n)_{n\geq 1}$* weakly converges *to μ (denoted by $\mu_n \overset{(S)}{\Longrightarrow} \mu$) if, for every function $f \in \mathcal{C}_b(S,\mathbb{R})$,*

$$\int_S f\,d\mu_n \longrightarrow \int_S f\,d\mu \quad as \quad n\to+\infty. \tag{4.2}$$

In view of applications, the first important result on weak convergence of probability measures of this chapter is the following.

Proposition 4.1 *(See Theorem 5.1 in [45]) If $\mu_n \overset{(S)}{\Longrightarrow} \mu$, then the above convergence (4.2) holds for every bounded Borel function $f : (S,\mathcal{S}) \to \mathbb{R}$ such that*

$$\mu\big(\mathrm{Disc}(f)\big) = 0 \quad where \quad \mathrm{Disc}(f) = \big\{x \in S,\ f \text{ is discontinuous at } x\big\} \in \mathcal{S}.$$

Functions f such that $\mu(\mathrm{Disc}(f)) = 0$ are called *μ-$a.s.$ continuous functions.*

Theorem 4.1 (Glivenko–Cantelli Theorem) *If $(U_n)_{n\geq 1}$ is an i.i.d. sequence of uniformly distributed random variables on the unit hypercube $[0,1]^d$, then*

$$\mathbb{P}(d\omega)\text{-}a.s.\qquad \frac{1}{n}\sum_{k=1}^{n} \delta_{U_k(\omega)} \overset{([0,1]^d)}{\Longrightarrow} \lambda_{d|[0,1]^d} = \mathcal{U}([0,1]^d) \quad as \quad n\to+\infty,$$

i.e.

$$\mathbb{P}(d\omega)\text{-}a.s. \quad \forall f \in \mathcal{C}([0,1]^d, \mathbb{R}),$$
$$\frac{1}{n}\sum_{k=1}^{n} f(U_k(\omega)) \overset{n\to+\infty}{\longrightarrow} \int_{[0,1]^d} f(x)\lambda_d(dx). \tag{4.3}$$

Proof. The vector space $\mathcal{C}([0,1]^d, \mathbb{R})$ endowed with the sup-norm $\|f\|_{\sup} := \sup_{x\in[0,1]^d} |f(x)|$ is separable in the sense that there exists a sequence $(f_m)_{m\geq 1}$ of continuous functions on $[0,1]^d$ which is everywhere dense in $\mathcal{C}([0,1]^d, \mathbb{R})$ with respect to the sup-norm ([1]).

Now, for every $m \geq 1$, the convergence (4.1) holds with $f = f_m$ for every $\omega \notin \Omega_m$, where $\Omega_m \in \mathcal{A}$ and $\mathbb{P}(\Omega_m) = 1$. Set $\Omega_0 = \bigcap_{m\in\mathbb{N}^*} \Omega_m$. One has $\mathbb{P}(\Omega_0^c) \leq \sum_{m\geq 1} \mathbb{P}(\Omega_m^c) = 0$ by σ-sub-additivity of probability measures so that $\mathbb{P}(\Omega_0) = 1$. Then,

$$\forall\, \omega \in \Omega_0,\ \forall\, m \geq 1,$$
$$\frac{1}{n}\sum_{k=1}^{n} f_m(U_k(\omega)) \overset{n\to+\infty}{\longrightarrow} \mathbb{E}\, f_m(U_1) = \int_{[0,1]^d} f_m(u)\lambda_d(du).$$

On the other hand, it is straightforward that, for every $f \in \mathcal{C}([0,1]^d, \mathbb{R})$, for every $n \geq 1$ and every $\omega \in \Omega$,

$$\left|\frac{1}{n}\sum_{k=1}^{n} f_m(U_k(\omega)) - \frac{1}{n}\sum_{k=1}^{n} f(U_k(\omega))\right| \leq \|f - f_m\|_{\sup}$$

and $\left|\mathbb{E}\,(f_m(U_1)) - \mathbb{E}\,(f(U_1))\right| \leq \|f - f_m\|_{\sup}$. As a consequence, for every $\omega \in \Omega_0$ and every $m \geq 1$,

$$\overline{\lim_n} \left|\frac{1}{n}\sum_{k=1}^{n} f(U_k(\omega)) - \mathbb{E}\, f(U_1)\right| \leq 2\|f - f_m\|_{\sup}.$$

Now, the fact that the sequence $(f_m)_{m\geq 1}$ is everywhere dense in $\mathcal{C}([0,1]^d, \mathbb{R})$ with respect to the sup-norm means precisely that

$$\underline{\lim_m} \|f - f_m\|_{\sup} = 0.$$

[1] When $d = 1$, an easy way to construct this sequence is to consider the countable family of continuous piecewise affine functions with monotonicity breaks at rational points of the unit interval and taking rational values at these break points (and at 0 and 1). The density follows from that of the set $\mathbb{Q}$ of rational numbers. When $d \geq 2$, one proceeds likewise by considering continuous functions which are affine on hyper-rectangles with rational vertices which tile the unit hypercube $[0,1]^d$. We refer to [45] for more details.

Consequently, for every $f \in \mathcal{C}([0,1]^d, \mathbb{R})$,

$$\frac{1}{n}\sum_{k=1}^{n} f(U_k(\omega)) - \mathbb{E}\, f(U_1) \longrightarrow 0 \quad \text{as} \quad n \to +\infty.$$

This completes the proof since it shows that, for every $\omega \in \Omega_0$, the expected convergence holds for every continuous function on $[0,1]^d$. ◇

Corollary 4.1 *Owing to Proposition 4.1, one may replace in (4.3) the set of continuous functions on* $[0,1]^d$ *by that of all bounded Borel* λ_d*-a.s. continuous functions on* $[0,1]^d$.

Definition 4.2 *A bounded* λ_d*-a.s. continuous Borel function* $f : [0,1]^d \to \mathbb{R}$ *is called* Riemann integrable.

Remark. In fact, in the above definition one may even replace the Borel measurability by "Lebesgue"-measurability. This means, for a function $f : [0,1]^d \to \mathbb{R}$, replacing $\big(\mathcal{B}or([0,1]^d), \mathcal{B}or(\mathbb{R})\big)$-measurability by $\big(\mathcal{L}([0,1]^d), \mathcal{B}or(\mathbb{R})\big)$-measurability, where $\mathcal{L}([0,1]^d)$ denotes the completion of the Borel σ-field on $[0,1]^d$ by the λ_d-negligible sets (see [52], Chap. 13). Such functions are known as *Riemann integrable functions* on $[0,1]^d$ (see again [52], Chap. 13).

The preceding suggests that, as long as one wishes to compute some quantities $\mathbb{E}\, f(U)$ for (reasonably) smooth functions f, we only need to have access to a sequence that satisfies the above convergence property for its empirical distribution. Furthermore, we know from the fundamental theorem of simulation (see Theorem 1.2) that this situation is generic since all distributions can be simulated from a uniformly distributed random variable over $[0,1]$, at least theoretically. This leads us to formulate the following definition of a *uniformly distributed* sequence.

Definition 4.3 *A* $[0,1]^d$*-valued sequence* $(\xi_n)_{n\geq 1}$ *is uniformly distributed on* $[0,1]^d$ *(or simply uniformly distributed in what follows) if*

$$\frac{1}{n}\sum_{k=1}^{n} \delta_{\xi_k} \stackrel{([0,1]^d)}{\Longrightarrow} \mathcal{U}([0,1]^d) \quad \textit{as} \quad n \to +\infty.$$

We need some characterizations of uniform distribution which can be established more easily on examples. These are provided by the next proposition. To this end we need to introduce further definitions and notations.

Definition 4.4 *(a) We define a componentwise partial order on* $[0,1]^d$, *simply denoted by "$\leq$", by: for every* $x = (x^1, \dots, x^d)$, $y = (y^1, \dots, y^d) \in [0,1]^d$

$$x \leq y \ \text{ if } \ x^i \leq y^i, \ 1 \leq i \leq d.$$

(b) The "box" $[\![x, y]\!]$ *is defined for every* $x = (x^1, \dots, x^d)$, $y = (y^1, \dots, y^d) \in [0,1]^d$ *by*

$$[\![x, y]\!] := \{\xi \in [0,1]^d, \ x \leq \xi \leq y\}.$$

Note that $[\![x, y]\!] \neq \varnothing$ *if and only if* $x \leq y$ *and, if this is the case,* $[\![x, y]\!] = \Pi_{i=1}^d [x^i, y^i]$.

Notation. In particular, the unit hypercube $[0,1]^d$ can be denoted by $[\![0, \mathbf{1}]\!]$, where $\mathbf{1} = (1, \dots, 1) \in \mathbb{R}^d$.

Proposition 4.2 (Portmanteau Theorem) *(See among other references [175] or [48] ([2])) Let* $(\xi_n)_{n\geq 1}$ *be a* $[0,1]^d$*-valued sequence. The following assertions are equivalent.*

(i) $(\xi_n)_{n\geq 1}$ *is uniformly distributed on* $[0,1]^d$.

(ii) Convergence of distribution functions: for every $x \in [0,1]^d$,

$$\frac{1}{n}\sum_{k=1}^{n} \mathbf{1}_{[\![0,x]\!]}(\xi_k) \longrightarrow \lambda_d([\![0, x]\!]) = \prod_{i=1}^{d} x^i \quad \text{as} \quad n \to +\infty.$$

(iii) "Discrepancy at the origin" or "star discrepancy":

$$D_n^*(\xi) := \sup_{x \in [0,1]^d} \left| \frac{1}{n}\sum_{k=1}^{n} \mathbf{1}_{[\![0,x]\!]}(\xi_k) - \prod_{i=1}^{d} x^i \right| \longrightarrow 0 \quad \text{as} \quad n \to +\infty. \tag{4.4}$$

(iv) "Extreme discrepancy":

$$D_n^\infty(\xi) := \sup_{x,y \in [0,1]^d} \left| \frac{1}{n}\sum_{k=1}^{n} \mathbf{1}_{[\![x,y]\!]}(\xi_k) - \prod_{i=1}^{d} (y^i - x^i) \right| \longrightarrow 0 \quad \text{as} \quad n \to +\infty. \tag{4.5}$$

[2]The name of this theorem looks mysterious. Intuitively, it can be simply justified by the multiple properties established as equivalent to the weak convergence of a sequence of probability measures. However, it is sometimes credited to Jean-Pierre Portmanteau in the paper: Espoir pour l'ensemble vide, *Annales de l'Université de Felletin* (1915), 322–325. In fact, one can easily check that no mathematician called Jean-Pierre Portmanteau ever existed and that there is no university in the very small French town of Felletin. This reference is just a joke hidden in the bibliography of the second edition of [45]. The empty set is definitely hopeless…

(v) *Weyl's criterion: for every integer* $p \in \mathbb{N}^d \setminus \{0\}$

$$\frac{1}{n}\sum_{k=1}^{n} e^{2\tilde{\imath}\pi(p|\xi_k)} \longrightarrow 0 \qquad as \;\; n \to +\infty \;\; (with\; \tilde{\imath}^2 = -1).$$

(vi) *Bounded Riemann integrable function: for every bounded* λ_d*-a.s. continuous Lebesgue-measurable function* $f : [0, 1]^d \to \mathbb{R}$

$$\frac{1}{n}\sum_{k=1}^{n} f(\xi_k) \longrightarrow \int_{[0,1]^d} f(x)\lambda_d(dx) \;\; as \quad n \to +\infty.$$

Definition 4.5 *The two moduli introduced in items* (iii) *by* (4.4) *and* (iv) *by* (4.5) *define the* discrepancy at the origin *and the* extreme discrepancy, *respectively.*

Remark. By construction these two discrepancies take their values in the unit interval $[0, 1]$.

Sketch of proof. The ingredients of the proof come from theory of weak convergence of probability measures. For more details in the multi-dimensional setting we refer to [45] (Chap. 1 devoted to the general theory of weak convergence of probability measures on a Polish space) or [175] (an old but great book devoted to *uniformly distributed* sequences). We provide hereafter some elements of proof in the one dimensional case.

The equivalence $(i) \Longleftrightarrow (ii)$ is simply the characterization of weak convergence of probability measures by the convergence of their distributions functions ([3]) since the distribution function F_{μ_n} of $\mu_n = \frac{1}{n}\sum_{1\le k\le n}\delta_{\xi_k}$ is given by $F_{\mu_n}(x) = \frac{1}{n}\sum_{1\le k\le n}\mathbf{1}_{\{0\le \xi_k\le x\}}$.

Owing to Dini's second Lemma, this convergence of non-decreasing (distribution) functions is uniform as soon as it holds pointwise since its pointwise limiting function, $F_{\mathcal{U}([0,1])}(x) = x$, is continuous. This remark yields the equivalence $(ii) \Longleftrightarrow (iii)$. Although more technical, the d-dimensional extension remains elementary and relies on a similar principle.

The equivalence $(iii) \Longleftrightarrow (iv)$ is trivial since $D_n^*(\xi) \le D^\infty(\xi) \le 2D_n^*(\xi)$ in one dimension. Note that in d dimensions the inequality reads $D_n^*(\xi) \le D^\infty(\xi) \le 2^d D_n^*(\xi)$.

Item (v) is based on the fact that weak convergence of finite measures on $[0, 1]^d$ is characterized by that of the sequences of their *Fourier coefficients*. The Fourier coefficients of a finite measure μ on $\big([0, 1]^d, \mathcal{B}or([0, 1]^d)\big)$ are defined by

[3] The distribution function F_μ of a probability measure μ on $[0, 1]$ is defined by $F_\mu(x) = \mu([0, x])$. One shows that a sequence of probability measures μ_n converges toward a probability measure μ if and only if their distribution functions F_{μ_n} and F_μ satisfy that $F_{\mu_n}(x)$ converges to $F_\mu(x)$ at every continuity point $x \in \mathbb{R}$ of F_μ (see [45], Chap. 1).

$c_p(\mu) := \int_{[0,1]^d} e^{2\tilde{\imath}\pi(p|u)}\mu(du)$, $p \in \mathbb{Z}^d$, $\tilde{\imath}^2 = -1$ (see *e.g.* [156]). One checks that the Fourier coefficients $\big(c_p(\lambda_{d|[0,1]^d})\big)_{p\in\mathbb{Z}^d}$ are simply $c_p(\lambda_{d|[0,1]^d}) = 0$ if $p \neq 0$ and 1 if $p = 0$.

Item (vi) follows from (i) and Proposition 4.1 since for every $x \in [\![0, 1]\!]$, $f_c(\xi) := \mathbf{1}_{\{\xi \le x\}}$ is continuous outside $\{x\}$, which is clearly Lebesgue negligible. Conversely, (vi) implies the pointwise convergence of the distribution functions F_{μ_n} as defined above toward $F_{\mathcal{U}([0,1])}$. ◇

The discrepancy at the origin $D_n^*(\xi)$ plays a central rôle in theory of uniformly distributed sequences: it does not only provide a criterion for uniform distribution, it also appears as an upper error modulus for numerical integration when the function f has the appropriate regularity (see Koksma–Hlawka Inequality below).

Definition 4.6 (Discrepancy of an n-tuple) *One defines the star discrepancy* $D_n^*(\xi_1, \ldots, \xi_n)$ *at the origin of a n-tuple* $(\xi_1, \ldots, \xi_n) \in ([0, 1]^d)^n$ *by* (4.4) *of the above Proposition and by* (4.5) *for the extreme discrepancy.*

The geometric interpretation of the star discrepancy is the following: if $x = (x^1, \ldots, x^d) \in [\![0, \mathbf{1}]\!]$, the hyper-volume of $[\![0, x]\!]$ is equal to the product $x^1 \cdots x^d$ and

$$\frac{1}{n}\sum_{k=1}^{n} \mathbf{1}_{[\![0,x]\!]}(\xi_k) = \frac{\operatorname{card}\{k \in \{1, \ldots, n\},\ \xi_k \in [\![0, x]\!]\}}{n}$$

is simply the frequency with which the first n points ξ_k's of the sequence fall into $[\![0, x]\!]$. The star discrepancy measures the maximal resulting error when x runs over $[\![0, \mathbf{1}]\!]$.

Exercise **1.** below provides a first example of a uniformly distributed sequence.

► **Exercises.** **1.** (*Rotations of the torus* $([0, 1]^d, +)$). Let $(\alpha^1, \ldots, \alpha^d) \in (\mathbb{R} \setminus \mathbb{Q})^d$ be (irrational numbers) such that $(1, \alpha^1, \ldots, \alpha^d)$ are linearly independent over $\mathbb{Q}$ ([4]). Let $x = (x^1, \ldots, x^d) \in \mathbb{R}^d$. For every $n \ge 1$, set

$$\xi_n := \big(\{x^i + n\,\alpha^i\}\big)_{1\le i\le d},$$

where $\{x\}$ denotes the fractional part of a real number x. Show that the sequence $(\xi_n)_{n\ge1}$ is uniformly distributed on $[0, 1]^d$ (and can be recursively generated). [Hint: use Weyl's criterion.]

2. *More on the one dimensional case.* (a) Assume $d = 1$. Show that, for every n-tuple $(\xi_1, \ldots, \xi_n) \in [0, 1]^n$

[4] This means that if the rational scalars $\lambda^i \in \mathbb{Q}$, $i = 0, \ldots, d$ satisfy $\lambda^0 + \lambda^1\alpha^1 + \cdots + \lambda^d\alpha^d = 0$ then $\lambda^0 = \lambda^1 = \cdots = \lambda^d = 0$. Thus $\alpha \in \mathbb{R}$ is irrational if and only if $(1, \alpha)$ are linearly independent on $\mathbb{Q}$.

$$D_n^*(\xi_1,\ldots,\xi_n) = \max_{1\le k\le n}\left(\left|\frac{k-1}{n} - \xi_k^{(n)}\right|, \left|\frac{k}{n} - \xi_k^{(n)}\right|\right)$$

where $(\xi_k^{(n)})_{1\le k\le n}$ is the/a reordering of the n-tuple $(\xi_1,\ldots,\xi_n)$ defined by $k\mapsto\xi_k^{(n)}$ is non-decreasing and $\{\xi_1^{(n)},\ldots,\xi_n^{(n)}\} = \{\xi_1,\ldots,\xi_n\}$. [Hint: Where does the càdlàg function $x\mapsto \frac{1}{n}\sum_{k=1}^n \mathbf{1}_{\{\xi_k\le x\}} - x$ attain its infimum and supremum?]
(b) Deduce that

$$D_n^*(\xi_1,\ldots,\xi_n) = \frac{1}{2n} + \max_{1\le k\le n}\left|\xi_k^{(n)} - \frac{2k-1}{2n}\right|.$$

(c) *Minimal discrepancy at the origin.* Show that the n-tuple with the lowest discrepancy (at the origin) is $\left(\frac{2k-1}{2n}\right)_{1\le k\le n}$ (the "mid-point" uniform n-tuple) with discrepancy $\frac{1}{2n}$.

4.2 Application to Numerical Integration: Functions with Finite Variation

Definition 4.7 *(see [48, 237]) A function $f:[0,1]^d\to\mathbb{R}$ has finite variation in the measure sense if there exists a signed measure ([5]) ν on $\big([0,1]^d, \mathcal{B}or([0,1]^d)\big)$ such that $\nu(\{0\}) = 0$ and*

$$\forall\, x\in[0,1]^d,\quad f(x) = f(\mathbf{1}) + \nu([\![0,\mathbf{1}-x]\!])$$

(or equivalently $f(x) = f(0) - \nu({}^c[\![0,\mathbf{1}-x]\!])$). The variation $V(f)$ is defined by

$$V(f) := |\nu|([0,1]^d),$$

where $|\nu|$ is the variation measure of ν.

▸ **Exercises.** **1.** Show that f has finite variation in the measure sense if and only if there exists a signed ν measure with $\nu(\{\mathbf{1}\}) = 0$ such that

$$\forall\, x\in[0,1]^d,\qquad f(x) = f(\mathbf{1}) + \nu([\![x,\mathbf{1}]\!]) = f(0) - \nu({}^c[\![x,\mathbf{1}]\!])$$

and that its variation is given by $|\nu|([0,1]^d)$. This could of course be taken as the definition of finite variation equivalently to the above one.

---

[5] A signed measure ν on a space $(X,\mathcal{X})$ is a mapping from $\mathcal{X}$ to $\mathbb{R}$ which satisfies the two axioms of a measure, namely $\nu(\varnothing) = 0$ and if A_n, $n\ge1$, are pairwise disjoint, then $\nu(\cup_n A_n) = \sum_{n\ge1}\nu(A_n)$ (the series is commutatively convergent hence absolutely convergent). Such a measure is finite and can be decomposed as $\nu = \nu_1 - \nu_2$, where ν_1, ν_2 are non-negative finite measures supported by disjoint sets, *i.e.* there exists $A\in\mathcal{X}$ such that $\nu_1(A^c) = \nu_2(A) = 0$ (see [258]).

2. Show that the function f defined on $[0,1]^2$ by

$$f(x^1, x^2) := (x^1 + x^2) \wedge 1, \quad (x^1, x^2) \in [0,1]^2$$

has finite variation in the measure sense [Hint: consider the distribution of $(U, 1-U)$, $U \stackrel{d}{=} \mathcal{U}([0,1])$].

For the class of functions with finite variation, the Koksma–Hlawka Inequality provides an error bound for $\displaystyle\frac{1}{n}\sum_{k=1}^{n} f(\xi_k) - \int_{[0,1]^d} f(x)dx$ based on the star discrepancy.

Proposition 4.3 **(Koksma–Hlawka Inequality (1943 when** *d=1*)**)** *Let* $(\xi_1, \ldots, \xi_n)$ *be an* n*-tuple of* $[0,1]^d$*-valued vectors and let* $f : [0,1]^d \to \mathbb{R}$ *be a function with finite variation (in the measure sense). Then*

$$\left| \frac{1}{n}\sum_{k=1}^{n} f(\xi_k) - \int_{[0,1]^d} f(x)\lambda_d(dx) \right| \le V(f) D_n^*\big((\xi_1, \ldots, \xi_n)\big).$$

Proof. Set $\widetilde{\mu}_n = \frac{1}{n}\sum_{k=1}^{n} \delta_{\xi_k} - \lambda_{d|[0,1]^d}$. It is a signed measure with 0-mass. Then, if f has finite variation with respect to a signed measure ν,

$$\begin{aligned}
\frac{1}{n}\sum_{k=1}^{n} f(\xi_k) - \int_{[0,1]^d} f(x)\lambda_d(dx) &= \int f(x)\widetilde{\mu}_n(dx) \\
&= f(\mathbf{1})\widetilde{\mu}_n([0,1]^d) + \int_{[0,1]^d} \nu([\![0, \mathbf{1} - x]\!])\widetilde{\mu}_n(dx) \\
&= 0 + \int_{[0,1]^d} \left(\int \mathbf{1}_{\{v \le \mathbf{1} - x\}} \nu(dv) \right) \widetilde{\mu}_n(dx) \\
&= \int_{[0,1]^d} \widetilde{\mu}_n([\![0, \mathbf{1} - v]\!])\nu(dv),
\end{aligned}$$

where we used Fubini's Theorem to interchange the integration order (which is possible since $|\nu| \otimes |\widetilde{\mu}_n|$ is a finite measure). Finally, using the extended triangle inequality for integrals with respect to signed measures,

$$\begin{aligned}\left|\frac{1}{n}\sum_{k=1}^{n} f(\xi_k) - \int_{[0,1]^d} f(x)\lambda_d(dx)\right| &= \left|\int_{[0,1]^d} \widetilde{\mu}_n([\![0, \mathbf{1} - v]\!])\nu(dv)\right| \\ &\le \int_{[0,1]^d} |\widetilde{\mu}_n([\![0, \mathbf{1} - v]\!])|\,|\nu|(dv) \\ &\le \sup_{v\in[0,1]^d} |\widetilde{\mu}_n([\![0, v]\!])||\nu|([0, 1]^d) \\ &= D_n^*(\xi)V(f). \end{aligned}$$

◇

Remarks. • The notion of finite variation in the measure sense has been introduced in [48, 237]. When $d = 1$, it coincides with the notion of *left continuous functions* with finite variation. The most general multi-dimensional extension to higher dimensions $d \ge 2$ is the notion of "finite variation" in the Hardy and Krause sense. We refer *e.g.* to [175, 219] for its definition and properties. Essentially, it relies on a geometric extension of the one-dimensional finite variation where meshes on the unit interval are replaced by tilings of $[0, 1]^d$ with hyper-rectangles with an appropriate notion of increment of the function on these hyper-rectangles. However, finite variation in the measure sense is much easier to handle, In particular, to establish the above Koksma–Hlawka Inequality. Furthermore, $V(f) \le V_{H\&K}(f)$. Conversely, one shows that a function with finite variation in the Hardy and Krause sense is λ_d-*a.s.* equal to a function $\widetilde{f}$ having finite variations in the measure sense satisfying $V(\widetilde{f}) \le V_{H\&K}(f)$. In one dimension, finite variation in the Hardy and Krause sense exactly coincides with the standard definition of finite variation.

• A classical criterion (see [48, 237]) for finite variation in the measure sense is the following: if $f : [0, 1]^d \to \mathbb{R}$ has a cross derivative $\dfrac{\partial^d f}{\partial x^1 \cdots \partial x^d}$ in the distribution sense which is an integrable function, *i.e.*

$$\int_{[0,1]^d} \left|\frac{\partial^d f}{\partial x^1 \cdots \partial x^d}(x^1, \dots, x^d)\right| dx^1 \cdots dx^d < +\infty,$$

then f has finite variation in the measure sense. This class includes the functions f defined by

$$f(x) = f(\mathbf{1}) + \int_{[\![0,\mathbf{1}-x]\!]} \varphi(u^1, \dots, u^d)du^1 \dots du^d, \quad \varphi \in L^1([0, 1]^d, \lambda_d). \quad (4.6)$$

• One derives from the above Koksma–Hlawka Inequality that

$$D_n^*(\xi_1, \dots, \xi_n) = \sup\left\{\left|\frac{1}{n}\sum_{k=1}^{n} f(\xi_k) - \int_{[0,1]^d} f d\lambda_d\right|, \; f : [0, 1]^d \to \mathbb{R},\; V(f) \le 1\right\}. \quad (4.7)$$

The inequality $\leq$ is obvious. The reverse inequality follows from the fact that the functions $f_x(u) = \mathbf{1}_{[\![0,\mathbf{1}-x]\!]}(u) = \delta_x\big([\![0, \mathbf{1} - u]\!]\big)$ have finite variation as soon as $x \neq 0$. Moreover, $\frac{1}{n}\sum_{k=1}^n f_0(\xi_k) - \int_{[0,1]^d} f_0 d\lambda_d = 0$ so that $D_n^*(\xi_1, \ldots, \xi_n) = \sup_{x\in[0,1]^d}\left\{\left|\frac{1}{n}\sum_{k=1}^n f_x(\xi_k) - \int_{[0,1]^d} f_x d\lambda_d\right|\right\}$.

▶ **Exercises.** **1.** Show that the function f on $[0, 1]^3$ defined by

$$f(x^1, x^2, x^3) := (x^1 + x^2 + x^3) \wedge 1$$

does not have finite variation in the measure sense. [Hint: its third derivative in the distribution sense is not a measure.] ([6])

2. (a) Show directly that if f satisfies (4.6), then for any n-tuple $(\xi_1, \ldots, \xi_n)$,

$$\left|\frac{1}{n}\sum_{k=1}^n f(\xi_k) - \int_{[0,1]^d} f(x)\lambda_d(dx)\right| \leq \|\varphi\|_{L^1(\lambda_{d|[0,1]^d})} D_n^*(\xi_1, \ldots, \xi_n).$$

(b) Show that if φ is also in $L^p([0, 1]^d, \lambda_d)$, for a $p \in (1, +\infty]$ with a Hölder conjugate q, then

$$\left|\frac{1}{n}\sum_{k=1}^n f(\xi_k) - \int_{[0,1]^d} f(x)\lambda_d(dx)\right| \leq \|\varphi\|_{L^p(\lambda_{d|[0,1]^d})} D_n^{(q)}(\xi_1, \ldots, \xi_n),$$

where

$$D_n^{(q)}(\xi_1, \ldots, \xi_n) = \left(\int_{[0,1]^d}\left|\frac{1}{n}\sum_{k=1}^n \mathbf{1}_{[\![0,x]\!]}(\xi_k) - \prod_{i=1}^d x^i\right|^q \lambda_d(dx)\right)^{\frac{1}{q}}.$$

This modulus is called the L^q-discrepancy of $(\xi_1, \ldots, \xi_n)$.

3. *Other forms of finite variation and the Koksma–Hlawka inequality.* Let $f : [0, 1]^d \to \mathbb{R}$ be defined by

$$f(x) = \nu([\![0, x]\!]), \quad x \in [0, 1]^d,$$

where ν is a signed measure on $[0, 1]^d$. Show that

$$\left|\frac{1}{n}\sum_{k=1}^n f(\xi_k) - \int_{[0,1]^d} f(x)\lambda_d(dx)\right| \leq |\nu|([0, 1]^d) D_n^{\infty}(\xi_1, \ldots, \xi_n).$$

---

[6] In fact, its variation in the Hardy and Krause sense is not finite either.

4. *L^1-discrepancy in one dimension.* Let $d = 1$ and $q = 1$. Show that the L^1-discrepancy satisfies

$$D_n^{(1)}(\xi_1, \dots, \xi_n) = \sum_{k=0}^{n} \int_{\xi_k^{(n)}}^{\xi_{k+1}^{(n)}} \left| \frac{k}{n} - u \right| du,$$

where $\xi_0^{(n)} = 0$, $\xi_{n+1}^{(n)} = 1$ and $\xi_1^{(n)} \le \cdots \le \xi_n^{(n)}$ is the/a reordering of $(\xi_1, \dots, \xi_n)$.

4.3 Sequences with Low Discrepancy: Definition(s) and Examples

4.3.1 Back Again to the Monte Carlo Method on $[0, 1]^d$

Let $(U_n)_{n\ge1}$ be an i.i.d. sequence of random vectors uniformly distributed over $[0, 1]^d$ defined on a probability space $(\Omega, \mathcal{A}, \mathbb{P})$. We saw that

$$\mathbb{P}(d\omega)\text{-}a.s., \quad \text{the sequence } \big(U_n(\omega)\big)_{n\ge1} \text{ is uniformly distributed}$$

so that, by the Portmanteau Theorem,

$$\mathbb{P}(d\omega)\text{-}a.s., \quad D_n^*\big(U_1(\omega), \dots, U_n(\omega)\big) \to 0 \quad \text{as} \quad n \to +\infty.$$

So, it is natural to evaluate its (random) discrepancy $D_n^*\big(U_1(\omega), \dots, U_n(\omega)\big)$ as a measure of its uniform distribution and one may wonder at which rate it goes to zero. To be more precise, is there a kind of transfer of the Central Limit Theorem (*CLT*) and the Law of the Iterated Logarithm (*LIL*) – which rule the weak and strong rate of convergence in the Strong Law of Large Numbers (*SLLN*) respectively – to this discrepancy modulus? The answer is positive since $D_n^*\big(U_1, \dots, U_n\big)$ satisfies both a *CLT* and a *LIL*. Both results are due to Chung (see *e.g.* [63]).

▷ **Chung's *CLT* for the star discrepancy.** The random sequence $D_n^*\big(U_1, \dots, U_n\big)$, $n \ge 1$, satisfies

$$\sqrt{n}\, D_n^*\big(U_1, \dots, U_n\big) \xrightarrow{\mathcal{L}} \sup_{x\in[0,1]^d} \big|Z_x^{(d)}\big| \quad \text{as } n \to +\infty,$$

and

$$\mathbb{E}\left(D_n^*\big(U_1, \dots, U_n\big)\right) \sim \frac{\mathbb{E}\left(\sup_{x\in[0,1]^d} |Z_x^{(d)}|\right)}{\sqrt{n}} \quad \text{as } n \to +\infty,$$

where $(Z_x^{(d)})_{x\in[0,1]^d}$ denotes the centered Gaussian multi-index process (or "bridged hyper-sheet") with covariance structure given by

$$\forall\, x = (x^1, \ldots, x^d),\ y = (y^1, \ldots, y^d) \in [0, 1]^d,$$
$$\mathrm{Cov}\big(Z_x^{(d)}, Z_y^{(d)}\big) = \prod_{i=1}^d x^i \wedge y^i - \Big(\prod_{i=1}^d x^i\Big)\Big(\prod_{i=1}^d y^i\Big).$$

Remarks. • A Brownian hyper-sheet is a centered Gaussian field $\big(W_x^{(d)}\big)_{x\in[0,1]^d}$ characterized by its covariance structure

$$\forall\, x,\ y \in [0, 1]^d, \quad \mathbb{E}\, W_x^{(d)} W_y^{(d)} = \prod_{i=1}^d x^i \wedge y^i.$$

Then the bridged hyper-sheet is defined by

$$\forall\, x = (x^1, \ldots, x^d) \in [0, 1]^d, \quad Z_x^{(d)} = W_x^{(d)} - \Big(\prod_{i=1}^d x^i\Big) W_{(1,\ldots,1)}^{(d)}.$$

• In particular, when $d = 1$, $Z = Z^{(1)}$ is simply the Brownian bridge over the unit interval [0, 1] defined by

$$Z_x = W_x - x W_1,\ x \in [0, 1],$$

where $(W_x)_x \in [0, 1]$ is a standard Brownian motion on [0,1].

The distribution of its sup-norm is characterized, through its tail (or survival) distribution function, by

$$\forall\, z \in \mathbb{R}_+,\quad \mathbb{P}\Big(\sup_{x\in[0,1]} |Z_x| \geq z\Big) = 2\sum_{k\geq 1}(-1)^{k-1} e^{-2k^2z^2} = 1 - \frac{\sqrt{2\pi}}{z}\sum_{k\geq 1} e^{-\frac{(2k-1)\pi^2}{8z^2}} \tag{4.8}$$

(see [72] or [45], Chap. 2). This distribution is also known as the *Kolmogorov–Smirnov* distribution distribution since it appears as the limit distribution used in the non-parametric eponymous goodness-of-fit statistical test.

• The above *CLT* points out a possible somewhat hidden dependence of the Monte Carlo method upon the dimension d. As a matter of fact, let $X^\varphi \stackrel{d}{=} \varphi(U)$, $U \stackrel{d}{=} \mathcal{U}([0, 1]^d)$ and let φ have finite variation $V(\varphi)$. Then, if $\overline{X}_n^\varphi$ denotes for every $n \geq 1$ the empirical mean of n independent copies of $X_k^\varphi = \varphi(U_k)$, one has, owing to (4.7),

$$\sqrt{n}\,\Big\| \sup_{\varphi: V(\varphi)=1} \big|\overline{X}_n^\varphi - \mathbb{E}\, X^\varphi\big| \Big\|_1 = \sqrt{n}\,\mathbb{E}\, D_n^*(U_1, \ldots, U_n)$$
$$\to V(\varphi)\mathbb{E}\Big[\sup_{x\in[0,1]^d} |Z_x^{(d)}|\Big] \quad \text{as} \quad n \to +\infty.$$

This dependence with respect to the dimension appears more clearly when dealing with Lipschitz continuous functions. For more details we refer to the third paragraph of Sect. 7.3.

$\triangleright$ **Chung's *LIL* for the star discrepancy.** The random sequence $D_n^*\big(U_1, \ldots, U_n\big)$, $n \geq 1$, satisfies the following Law of the Iterated Logarithm:

$$\overline{\lim_n} \sqrt{\frac{2n}{\log(\log n)}} D_n^*\big(U_1(\omega), \ldots, U_n(\omega)\big) = 1 \qquad \mathbb{P}(d\omega)\text{-}a.s.$$

At this stage, this suggests a temporary definition of a *sequence with low discrepancy* on $[0, 1]^d$ as a $[0, 1]^d$-valued sequence $\xi := (\xi_n)_{n\geq 1}$ such that

$$D_n^*(\xi) = o\left(\sqrt{\frac{\log(\log n)}{n}}\right) \quad \text{as} \quad n \to +\infty,$$

which means that its implementation with a function with finite variation will speed up the convergence rate of numerical integration by the empirical measure with respect to the worst rate of the Monte Carlo simulation.

$\blacktriangleright$ **Exercise.** Show, using the standard *LIL*, the easy part of Chung's *LIL*, that is

$$\begin{aligned}\overline{\lim_n} \sqrt{\frac{2n}{\log(\log n)}} D_n^*\big(U_1(\omega), \ldots, U_n(\omega)\big) &\geq 2 \sup_{x\in[0,1]^d} \sqrt{\lambda_d([\![0, x]\!])(1 - \lambda_d([\![0, x]\!]))} \\ &= 1 \;\; \mathbb{P}(d\omega)\text{-a.s.}\end{aligned}$$

4.3.2 Roth's Lower Bounds for the Star Discrepancy

Before providing examples of sequences with low discrepancy, let us first describe some results concerning the known lower bounds for the asymptotics of the (star) discrepancy of a uniformly distributed sequence.

The first results are due to Roth (see [257]): there exists a universal constant $c_d \in (0, +\infty)$ such that, for any $[0, 1]^d$-valued n-tuple $(\xi_1, \ldots, \xi_n)$,

$$D_n^*(\xi_1, \ldots, \xi_n) \geq c_d \frac{(\log n)^{\frac{d-1}{2}}}{n}. \tag{4.9}$$

Furthermore, there exists a real constant $\widetilde{c}_d \in (0, +\infty)$ such that, *for every sequence* $\xi = (\xi_n)_{n\geq 1}$,

$$D_n^*(\xi) \geq \widetilde{c}_d \frac{(\log n)^{\frac{d}{2}}}{n} \quad \text{for infinitely many } n. \tag{4.10}$$

This second lower bound can be derived from the first one, using the Hammersley procedure introduced and analyzed in the next section (see the exercise at the end of Sect. 4.3.4).

On the other hand, there exists (see Sect. 4.3.3 that follows) sequences for which

$$\forall n \geq 1, \quad D_n^*(\xi) = C(\xi)\frac{(\log n)^d}{n} \quad \text{where} \quad C(\xi) < +\infty.$$

Based on this, one can derive from the Hammersley procedure (see again Sect. 4.3.4 below) the existence of a real constant $C_d \in (0, +\infty)$ such that

$$\forall n \geq 1, \; \exists\, (\xi_1, \ldots, \xi_n) \in \big([0,1]^d\big)^n, \; D_n^*(\xi_1, \ldots, \xi_n) \leq C_d \frac{(\log n)^{d-1}}{n}.$$

In spite of more than fifty years of investigation, the gap between these asymptotic lower and upper-bounds have not been significantly reduced: it has still not been proved whether there exists a sequence for which $C(\xi) = 0$, *i.e.* for which the rate $\frac{(\log n)^d}{n}$ would not be optimal.

In fact, it is widely shared in the *QMC* community that in the above lower bounds, $\frac{d-1}{2}$ can be replaced by $d-1$ in (4.9) and $\frac{d}{2}$ by d in (4.10) so that the rate $O\left(\frac{(\log n)^d}{n}\right)$ is commonly considered as the lowest possible rate of convergence to 0 for the star discrepancy of a uniformly distributed sequence. When $d = 1$, Schmidt proved that this conjecture is true.

This leads to a more convincing definition of a *sequence with low discrepancy*.

Definition 4.8 *A* $[0,1]^d$*-valued sequence* $(\xi_n)_{n\geq 1}$ *is a sequence with low discrepancy if*

$$D_n^*(\xi) = O\left(\frac{(\log n)^d}{n}\right) \quad \textit{as} \quad n \to +\infty.$$

For more insight about the other measures of uniform distribution (L^p-discrepancy $D_n^{(p)}(\xi)$, diaphony, etc), we refer *e.g.* to [46, 219].

4.3.3 Examples of Sequences

▷ Van der Corput and Halton sequences

Let $p_1, \ldots, p_d$ be the first d prime numbers. The d-dimensional Halton sequence is defined, for every $n \geq 1$, by:

$$\xi_n = \big(\Phi_{p_1}(n), \ldots, \Phi_{p_d}(n)\big) \tag{4.11}$$

where the so-called "radical inverse functions" Φ_p defined for every integer $p \geq 2$ by

$$\Phi_p(n) = \sum_{k=0}^{r} \frac{a_k}{p^{k+1}}$$

with $n = a_0 + a_1 p + \cdots + a_r p^r,\ a_i \in \{0, \ldots, p-1\},\ a_r \neq 0$, denotes the p-adic expansion of n.

Theorem 4.2 (see [165]) *Let* $\xi = (\xi_n)_{n\geq 1}$ *be defined by* (4.11). *For every* $n \geq 1$,

$$D_n^*(\xi) \leq \frac{1}{n}\prod_{i=1}^{d}\left((p_i - 1)\left\lfloor \frac{\log(p_i n)}{\log(p_i)} \right\rfloor\right) = O\left(\frac{(\log n)^d}{n}\right) \quad \textit{as}\quad n \to +\infty. \tag{4.12}$$

The proof of this upper-bound, due to H.L. Keng and W. Yu, essentially relies on the Chinese Remainder Theorem (known as "Théorème chinois" in French and Sunzi's Theorem in China, see [165], Sect. 3.5, among others). Since the methods of proof of such bounds for sequences with low discrepancy are usually highly technical and rely on combinatorics and Number theory arguments not very familiar to specialists in Probability theory, we decided to provide a proof for this first upper-bound which turns out to be more accessible. This proof is postponed to Sect. 12.10.

In fact, the above upper-bound (4.12) remains true if the sequence $(\xi_n)_{n\geq 1}$ is defined with integers $p_1, \ldots, p_d \geq 2$ which are simply *pairwise prime*, *i.e.* $\gcd(p_i, p_j) = 1$, $i \neq j$, $1 \leq i, j \leq d$. In particular, if $d = 1$, the (one-dimensional) *Van der Corput sequence* $\mathrm{VdC}(p)$ defined by

$$\xi_n = \Phi_p(n)$$

is uniformly distributed with $D_n^*(\xi) = O\left(\frac{\log n}{n}\right)$ for every integer $p \geq 2$.

Several improvements of this classical bound have been established: some non-asymptotic and of numerical interest (see *e.g.* [222, 278]), some more theoretical. Among them let us cite the following one established by H. Faure (see [88], see also [219], p. 29)

$$D_n^*(\xi) \leq \frac{1}{n}\left(d + \prod_{i=1}^{d}\left((p_i - 1)\frac{\log n}{2\log p_i} + \frac{p_i + 2}{2}\right)\right),\quad n \geq 1,$$

which provides a lower constant in front of $\frac{(\log n)^d}{n}$ than in (4.12).

One easily checks that the first terms of the $\mathrm{VdC}(2)$ sequence are as follows

$$\xi_1 = \frac{1}{2},\ \xi_2 = \frac{1}{4},\ \xi_3 = \frac{3}{4},\ \xi_4 = \frac{1}{8},\ \xi_5 = \frac{5}{8},\ \xi_6 = \frac{3}{8},\ \xi_7 = \frac{7}{8}, \ldots$$

▶ **Exercise.** Let $\xi = (\xi_n)_{n\geq 1}$ denote the p-adic Van der Corput sequence and let $\xi_1^{(n)} \leq \cdots \leq \xi_n^{(n)}$ be the reordering of its first n terms.
(a) Show that, for every $k \in \{1, \ldots, n\}$,

$$\xi_k^{(n)} \leq \frac{k}{n+1}.$$

[Hint: Use an induction on q where $n = q\,p + r,\, 0 \leq r \leq p-1$.]
(b) Derive that, for every $n \geq 1$,

$$\frac{\xi_1 + \cdots + \xi_n}{n} \leq \frac{1}{2}.$$

(c) One considers the p-adic Van der Corput sequence $(\tilde{\xi}_n)_{n\geq 1}$ *starting at* 0, *i.e.*

$$\tilde{\xi}_1 = 0, \quad \tilde{\xi}_n = \xi_{n-1},\; n \geq 2,$$

where $(\xi_n)_{n\geq 1}$ is the regular p-adic Van der Corput sequence. Show that $\tilde{\xi}_k^{(n+1)} \leq \frac{k-1}{n+1}$, $k = 1, \ldots, n+1$. Deduce that the L^1-discrepancy of the sequence $\tilde{\xi}$ satisfies

$$D_n^{(1)}(\tilde{\xi}) = \frac{1}{2} - \frac{\xi_1 + \cdots + \xi_n}{n}.$$

▷ **Kakutani sequences**

This family of sequences was first obtained as a by-product while trying to generate the Halton sequence as the orbit of an ergodic transform (see [181, 185, 223]). This extension is based on p-adic addition on $[0, 1]$, p integer, $p \geq 2$. It is also known as the Kakutani adding machine. It is defined on *regular* p-adic expansions of real numbers of $[0, 1]$ ([7]) as the *addition from the left to the right of the regular p-adic expansions with carry-over*. The p-adic regular expansion of 1 is conventionally set to $1 = \overline{0.(p-1)(p-1)(p-1)\ldots}^{p}$ and 1 is not considered as a p-adic rational number in the rest of this section.

Let $\oplus_p$ denote this addition (or Kakutani's adding machine). Thus, as an example

$$0.123333\ldots \oplus_{10} 0.412777\ldots = 0.535011\ldots$$

[7]Every real number in $[0, 1)$ admits a p-adic expansion $x = \sum_{k\geq 1} \frac{x_k}{p^k}$, $x_k \in \{0, \ldots, p-1\}$, $k \geq 1$. If x is not a p-adic rational, this expansion is unique. If x is a p-adic rational number, *i.e.* of the form $x = \frac{N}{p^r}$ for some $r \in \mathbb{N}$ and $N \in \{0, \ldots, p^r - 1\}$, then x has two p-adic expansions, one of the form $x = \sum_{k=1}^{\ell} \frac{x_k}{p^k}$ with $x_\ell \neq 0$ and a second reading $x = \sum_{k=1}^{\ell-1} \frac{x_k}{p^k} + \frac{x_\ell - 1}{p^\ell} \sum_{k\geq \ell+1} \frac{p-1}{p^k}$. It is clear that if x is not a p-adic rational number, its p-adic "digits" x_k cannot all be equal to $p-1$ for k large enough. By definition the *regular* p-adic expansion of $x \in [0, 1)$ is the unique expansion of x whose digits x_k will be infinitely often not equal to $p-1$. The case of 1 is specific: its unique p-adic expansion $1 = \sum_{k\geq 1} \frac{p-1}{p^k}$ will be considered as regular. This regular expansion is denoted by $x = \overline{0.x_1x_2\ldots x_k\ldots}^{p}$ for every $x \in [0, 1]$.

In a more formal way if $x,\ y \in [0,1]$ with respective regular p-adic expansions $\overline{0, x_1 x_2 \cdots x_k \cdots}$ and $\overline{0, y_1 y_2 \cdots y_k \cdots}$, then $x \oplus_p y$ is defined by $x \oplus_p y = \sum_{k \geq 1} \frac{z_k}{p^k}$ where the $\{0, \ldots, p-1\}$-valued sequence $(z_k)_{k \geq 1}$ is given by

$$z_k = x_k + y_k + \varepsilon_{k-1} \mod p \quad \text{and} \quad \varepsilon_k = \mathbf{1}_{\{x_k + y_k + \varepsilon_{k-1} \geq p\}}, \quad k \geq 1,$$

with $\varepsilon_0 = 0$.

– If x or y is a p-adic rational number and $x,\ y \neq 1$, then one easily checks that $z_k = y_k$ or $z_k = y_k$ for every large enough k so that this defines a regular expansions, *i.e.* the digits $(x \oplus_p y)_k$ of $x \oplus_p y$ are $(x \oplus_p y)_k = z_k$, $k \geq 1$.

– If both x and y are not p-adic rational numbers, then it may happen that $z_k = p - 1$ for every large enough k so that $\sum_{k \geq 1} \frac{z_k}{p^k}$ *is not the regular p -adic expansion of* $x \oplus_p y$. So is the case, for example, in the following pseudo-sum:

$$0.123333... \oplus_{10} 0.412666... = 0.535999... = 0.536$$

where $(x \oplus_p y)_1 = 5$, $(x \oplus_p y)_2 = 3$, $(x \oplus_p y)_3 = 5$ and $(x \oplus_p y)_k = 9$, $k \geq 4$.

Then, for every $y \in [0,1]$, one defines the associated *p -adic rotation* with angle y by

$$T_{p,y}(x) := x \oplus_p y.$$

Proposition 4.4 (see [185, 223]) *Let $p_1, \ldots, p_d$ denote the first d prime numbers, $y_1, \ldots, y_d \in (0,1)$, where y_i is a p_i-adic rational number satisfying $y_i \geq 1/p_i$, $i = 1, \ldots, d$ and $x_1, \ldots, x_d \in [0,1]$. Then the sequence $(\xi)_{n \geq 1}$ defined by*

$$\xi_n := \left(T^{n-1}_{p_i, y_i}(x_i)\right)_{1 \leq i \leq d}, \quad n \geq 1,$$

has a discrepancy at the origin $D^*_n(\xi)$ satisfying a similar upper-bound to (4.12) as the Halton sequence, namely, for every integer $n \geq 1$,

$$D^*_n(\xi) \leq \frac{1}{n}\left(d - 1 + \prod_{i=1}^{d}\left((p_i - 1)\left\lfloor \frac{\log(p_i n)}{\log(p_i)} \right\rfloor\right)\right).$$

Remarks. • Note that if $y_i = x_i = 1/p_i = \overline{0.1}^p$, $i = 1, \ldots, d$, the sequence ξ is simply the regular Halton sequence.

• This upper-bound is obtained by adapting the proof of Theorem 4.2 (see Sect. 12.10) and we do not pretend it is optimal as a universal bound when the starting values x_i and the angles y_i vary. Its main interest is to provide a large family in which sequences with better performances than the regular Halton sequences are "hidden", at least at finite range.

ℵ **Practitioner's corner.** ▷ *Iterative generation.* One asset of this approach is to provide an easy recursive form for the computation of ξ_n since

$$\xi_n = \xi_{n-1} \oplus_p (y_1, \ldots, y_d),$$

where, with a slight abuse of notation, $\oplus_p$ denotes the componentwise pseudo-addition with $p = (p_1, \ldots, p_d)$. Once the pseudo-addition has been coded, this method is tremendously fast.

Appropriate choices of the starting vector and the "angle" can significantly reduce the discrepancy, at least in a finite range (see below).

▷ *A "super-Halton" sequence.* Heuristic arguments too lengthy to be developed here suggest that a good choice for the "angles" y_i and the starting values x_i is

$$y_i = 1/p_i, \; x_i = 1/5, \; i \neq 3, 4, \; x_i = \frac{2p_i - 1 - \sqrt{(p_i+2)^2 + 4p_i}}{3}, \; i = 3, 4.$$

This specified Kakutani – or "super-Halton" – sequence is much easier to implement than the Sobol' sequences and behaves quite well up to medium dimensions d, say $1 \le d \le 20$ (see [48, 237]).

▷ **Faure sequences.**

These sequences were introduced in [89]. Let p be the smallest *prime* integer satisfying $p \ge d$. The d-dimensional Faure sequence is defined for every $n \ge 1$ by

$$\xi_n = \Big(\Phi_p(n-1), C_p\big(\Phi_p(n-1)\big), \cdots, C_p^{d-1}\big(\Phi_p(n-1)\big)\Big)$$

where Φ_p still denotes the p-adic radical inverse function and, for every p-adic rational number u with regular p-adic expansion $u = \sum_{k\ge 0} u_k p^{-(k+1)} \in [0, 1]$ (note that $u_k = 0$ for large enough k)

$$C_p(u) = \sum_{k\ge 0} \underbrace{\left(\Big(\sum_{j\ge k} \tbinom{j}{k} u_j\Big) \bmod p\right)}_{\in\{0,\ldots,p-1\}} p^{-(k+1)}.$$

The discrepancy at the origin of these sequences satisfies (see [89])

$$D_n^*(\xi) \le \frac{1}{n}\left(\frac{1}{d!}\left(\frac{p-1}{2\log p}\right)^d (\log n)^d + O((\log n)^{d-1})\right). \tag{4.13}$$

It was later shown in [219] that they share the following $P_{p,d}$-property (see also [48], p. 79).

Proposition 4.5 *For every $m \in \mathbb{N}$ and every $\ell \in \mathbb{N}^*$, for any $r_1, \ldots, r_d \in \mathbb{N}$ such that $r_1 + \cdots + r_d = m$ and every $x_1, \ldots, x_d \in \mathbb{N}$ such that $x_k \le p^{r_k} - 1$, $k = 1, \ldots, d$, there is exactly one term in the sequence $\xi_{\ell p^m + i}$, $i = 0, \ldots, p^m - 1$, lying in the hyper-cube $\prod_{k=1}^{d} \big[x_k p^{-r_k}, (x_k + 1) p^{-r_k}\big)$.*

This property is a special case of (t, d)-sequences in base p as defined in [219] (see Definitions 4.1 and 4.2, p. 48) corresponding to $t = 0$.

The prominent feature of Faure's estimate (4.13) is that the coefficient of the leading error term (in the $(\log n)^k$-scale) satisfies

$$\lim_d \frac{1}{d!}\left(\frac{p-1}{2\log p}\right)^d = 0,$$

which seems to suggest that the rate is asymptotically better than $O\Big(\frac{(\log n)^d}{n}\Big)$ as d increases.

The above convergence result is an easy consequence of Stirling's formula and Bertrand's conjecture which says that, for every integer $d > 1$, there exists a prime number p such that $d < p < 2d$ ([8]). To be more precise, the function $x \mapsto \frac{x-1}{\log x}$ being increasing on $(0, +\infty)$,

$$\frac{1}{d!}\left(\frac{p-1}{2\log p}\right)^d \le \frac{1}{d!}\left(\frac{2d-1}{2\log 2d}\right)^d \precsim \left(\frac{e}{\log 2d}\right)^d \frac{1}{\sqrt{2\pi d}} \to 0 \quad \text{as} \quad d \to +\infty.$$

A non-asymptotic upper-bound is provided in [237] (due to Y.-J. Xiao in his PhD thesis [278]):

$$\forall\, n \ge 1, \quad D_n^*(\xi) \le \frac{1}{n}\left(\frac{1}{d!}\left(\frac{p-1}{2}\right)^d \left(\left\lfloor \frac{\log(2n)}{\log p}\right\rfloor + d + 1\right)^d\right).$$

Note that this bound has the same coefficient in its leading term as the asymptotic error bound obtained by Faure. Unfortunately, from a numerical point of view, it becomes efficient only for very large n: thus if $d = p = 5$ and $n = 1\,000$,

$$D_n^*(\xi) \le \frac{1}{n}\left(\frac{1}{d!}\left(\frac{p-1}{2}\right)^d \left(\left\lfloor \frac{\log(2n)}{\log p}\right\rfloor + d + 1\right)^d\right) \simeq 1.18,$$

which is of little interest if one keeps in mind that, by construction, the discrepancy takes its values in $[0, 1]$. This can be explained by the form of the "constant" term in the $(\log n)^k$-scale in the above upper-bound: one has

[8] Bertrand's conjecture was stated in 1845 but it is no longer a conjecture since it was proved by P. Tchebychev in 1850.

$$\lim_{d \to +\infty} \frac{(d+1)^d}{d!} \left(\frac{p-1}{2}\right)^d = +\infty.$$

A better bound is provided in Y.-J. Xiao's PhD thesis [278], provided $n \geq p^{d+2}/2$. But once again it is of little interest for applications when d increases since, for $p \geq d$, $p^{d+2}/2 \geq d^{d+2}/2$.

▷ **The Sobol' sequences**

The discovery by Ilia Sobol' of his eponymous sequences has clearly been, not only a pioneering but also a striking contribution to sequences with low discrepancy and quasi-Monte Carlo simulation. The publication of his work goes back to 1967 (see [265]). Although it was soon translated into English, it remained ignored to a large extent for many years, at least by practitioners in the western world. From a purely theoretical point of view, it is the first family of sequences ever discovered satisfying the $P_{p,d}$-property. In the modern classification of sequences with low discrepancy they appear as subfamilies of the largest family of Niederreiter's sequences (see below), discovered later on. However, now that they are widely used especially for (Quasi-)Monte Carlo simulation in Quantitative Finance, their impact, among theorists and practitioners, is unrivaled, compared to any other uniformly distributed sequence.

In terms of implementation, Antonov and Saleev proposed in [10] a new implementation based on the Gray code, which dramatically speeds up the computation of these sequences.

In practice, the construction of the Sobol' sequences relies on "direction numbers", which turn out to be crucial for the efficiency of the sequences. Not all admissible choices are equivalent and many authors proposed efficient initializations of these numbers after Sobol' himself (see [266]), who have proposed a solution up to $d = 51$ in 1976. Implementations are also available in [247]. For recent developments on this topic, see *e.g.* [276].

Even if (see below) some sequences proved to share slightly better "academic" performances, no major progress has been made since in the search for good sequences with low discrepancy. Sobol' sequences remain unrivaled among practitioners and are massively used in Quasi-Monte Carlo simulation, now becoming a synonym for *QMC* in the quantitative finance world.

The main advances come from post-processing of the sequences like *randomization* and/or *scrambling*. These points are briefly discussed in Sect. 4.4.2.

▷ **The Niederreiter sequences.**

These sequences were designed as generalizations of Faure and Sobol' sequences (see [219]).

Let $q \geq d$ be the smallest primary integer not lower than d (a primary integer is an integer of the form $q = p^r$ with p prime, $r \in \mathbb{N}^*$). The $(0, d)$-Niederreiter sequence is defined for every integer $n \geq 1$ by

$$\xi_n = \big(\Psi_{q,1}(n-1), \Psi_{q,2}(n-1), \cdots, \Psi_{q,d}(n-1)\big),$$

where

$$\Psi_{q,i}(n) := \sum_j \psi^{-1}\left(\sum_k C^{(i)}_{(j,k)}\Psi(a_k)\right)q^{-j},$$

$\Psi : \{0, \dots, q-1\} \to \mathbb{F}_q$ is a one-to-one correspondence between $\{0, \dots, q-1\}$ (to be specified) and the finite field $\mathbb{F}_q$ with cardinality q, satisfying $\Psi(0) = 0$, and

$$C^{(i)}_{(j,k)} = \binom{k}{j-1}\Psi(i-1).$$

This quite general family of sequences contains both the Faure and the Sobol' sequences. To be more precise:

- when q is the smallest prime number not less than d, one retrieves the Faure sequences,
- when $q = 2^r$, with $2^{r-1} < d \le 2^r$, the sequence coincides with the Sobol' sequences (in their original form).

The main feature of Niederreiter sequences is to be (t, d)-sequences in base q and consequently have a discrepancy satisfying an upper-bound with a structure similar to that of the Faure or Sobol' sequences (which correspond to $t = 0$). For a precise definition of (t, d)-sequences (and (t, m, d)-nets) as well as an in-depth analysis of their properties in terms of discrepancy, we refer to [219]. Note that $(0, d)$-sequences in base p reduce to the $P_{d,p}$-property mentioned above. Records in terms of low discrepancy are usually held within this family, and then beaten by other sequences from this family.

4.3.4 The Hammersley Procedure

The Hammersley procedure is a canonical method for designing a $[0,1]^d$-valued n-tuple from a $[0,1]^{d-1}$-valued one with a discrepancy at the origin ruled by the latter $(d-1)$-dimensional one.

Proposition 4.6 *Let $d \ge 2$. Let $(\zeta_1, \dots, \zeta_n)$ be a $[0,1]^{d-1}$-valued n-tuple. Then, the $[0,1]^d$-valued n-tuple defined by*

$$(\xi_k)_{1\le k\le n} = \left(\zeta_k, \frac{k}{n}\right)_{1\le k\le n}$$

satisfies

$$\frac{\max_{1\le k\le n} k D^*_k(\zeta_1, \dots, \zeta_k)}{n} \le D^*_n\left((\xi_k)_{1\le k\le n}\right) \le \frac{1 + \max_{1\le k\le n} k D^*_k(\zeta_1, \dots, \zeta_k)}{n}. \tag{4.14}$$

Proof. It follows from the very definition of the discrepancy at the origin that

$$
\begin{aligned}
D_n^*\big((\xi_k)_{1\le k\le n}\big) &= \sup_{(x,y)\in[0,1]^{d-1}\times[0,1]} \left|\frac{1}{n}\sum_{k=1}^{n}\mathbf{1}_{\{\zeta_k\in[[0,x]],\,\frac{k}{n}\le y\}} - y\prod_{k=1}^{d-1}x^i\right| \\
&= \sup_{x\in[0,1]^{d-1}}\left[\sup_{y\in[0,1]}\left|\frac{1}{n}\sum_{k=1}^{n}\mathbf{1}_{\{\zeta_k\in[[0,x]],\,\frac{k}{n}\le y\}} - y\prod_{k=1}^{d-1}x^i\right|\right] \\
&= \max_{1\le k\le n}\left[\sup_{x\in[0,1]^{d-1}}\left|\frac{1}{n}\sum_{\ell=1}^{k}\mathbf{1}_{\{\zeta_\ell\in[[0,x]]\}} - \frac{k}{n}\prod_{i=1}^{d-1}x^i\right|\right. \\
&\qquad \left.\vee \sup_{x\in[0,1]^{d-1}}\left|\frac{1}{n}\sum_{\ell=1}^{k-1}\mathbf{1}_{\{\zeta_\ell\in[[0,x]]\}} - \frac{k}{n}\prod_{i=1}^{d-1}x^i\right|\right]
\end{aligned}
$$

since one can easily check that the functions of the form $y \mapsto \left|\frac{1}{n}\sum_{k=1}^{n} a_k\mathbf{1}_{\{\frac{k}{n}\le y\}} - b\,y\right|$ $(a_k,\ b\ge 0)$ attain their supremum either at some $y=\frac{k}{n}$ or at its left limit "$y_- = \left(\frac{k}{n}\right)_-$", $k\in\{1,\dots,n\}$. Consequently,

$$
\begin{aligned}
D_n^*\big((\xi_k)_{1\le k\le n}\big) &= \frac{1}{n}\max_{1\le k\le n}\left[k\,D_k^*(\zeta_1,\dots,\zeta_k)\right. \\
&\qquad \left.\vee\left[(k-1)\sup_{x\in[0,1]^{d-1}}\left|\frac{1}{k-1}\sum_{\ell=1}^{k-1}\mathbf{1}_{\{\zeta_\ell\in[[0,x]]\}} - \frac{k}{k-1}\prod_{i=1}^{d-1}x^i\right|\right]\right] \\
&\le \frac{1}{n}\max_{1\le k\le n}\Big(k\,D_k^*(\zeta_1,\dots,\zeta_k)\vee\big((k-1)D_{k-1}^*(\zeta_1,\dots,\zeta_{k-1})+1\big)\Big) \\
&\le \frac{1+\max_{1\le k\le n} k\,D_k^*(\zeta_1,\dots,\zeta_k)}{n}.
\end{aligned}
\tag{4.15}
$$

The lower bound is obvious from (4.15). ◇

Corollary 4.2 *Let $d\ge 1$. There exists a real constant $C_d\in(0,+\infty)$ such that, for every $n\ge 1$, there exists an n-tuple $(\xi_1^n,\dots,\xi_n^n)\in([0,1]^d)^n$ satisfying*

$$
D_n^*(\xi_1^n,\dots,\xi_n^n) \le C_d\,\frac{1+(\log n)^{d-1}}{n}.
$$

Proof. If $d=1$, a solution is given for a fixed integer $n\ge 1$ by setting $\xi_k=\frac{k}{n}$, $k=1,\dots,n$ (or $\xi_k=\frac{2k-1}{2n}$, $k=1,\dots,n$, etc). If $d\ge 2$, it suffices to apply for every $n\ge 1$ the Hammersley procedure to any $(d-1)$-dimensional sequence $\zeta=(\zeta_n)_{n\ge 1}$ with low discrepancy in the sense of Definition 4.8. In this case, the constant C_d

can be taken equal to $2\,(c_{d-1}(\zeta)\vee 1)$, where $D_k^*(\zeta)\le c_{d-1}(\zeta)\big(1+(\log k)^{d-1}\big)/k$ for every $k\ge 1$. ◇

The main drawback of this procedure is that if one starts from a *sequence* with low discrepancy (often defined recursively), one loses the "telescopic" feature of such a sequence. If one wishes, for a given function f defined on $[0,1]^d$, to increase n in order to improve the accuracy of the approximation, all the terms of the sum in the empirical mean have to be re-computed.

► **Exercises.** **1.** (*Roth's lower-bound*). Derive the theoretical lower bound (4.10) for infinite sequences from the one in (4.9).

2. *Extension of Hammersley's procedure.*

(*a*) Let $(\xi_1,\dots,\xi_n)$ be a $[0,1]^d$-valued n-tuple such that $0\le\xi_1^d<\xi_2^d<\dots<\xi_n^d\le 1$. Prove that

$$D_n^*(\xi_1^{1:d-1},\dots,\xi_n^{1:d-1})\le D_n^*(\xi_1,\dots,\xi_n)\le\frac{\max_{1\le k\le n}k\,D_k^*(\xi_1^{1:d-1},\dots,\xi_k^{1:d-1})}{n}+D_n^*(\xi_1^d,\dots,\xi_n^d),$$

where $\xi_k^{1:d-1}=(\xi_k^1,\dots,\xi_n^{d-1})$. [Hint: Follow the lines of the proof of Proposition 4.6, using that the supremum of the càdlàg function $\varphi: y\mapsto\left|\frac{1}{n}\sum_{k=1}^n a_k\mathbf{1}_{\{\xi_k^d\le y\}}-b\,y\right|$ $(a_k,\ b\ge 0)$ is $\max_{1\le k\le n}|\varphi(\xi_k^d)|\vee|\varphi((\xi_k^d)_-)|\vee|\varphi(1)|$.]

(*b*) Deduce that the upper-bound in (4.14) can be slightly improved by an appropriate choice of the dth component of the terms of the n-tuple $(\xi_k)_{1\le k\le n}$ [Hint: what can be better that $(\frac{k}{n})_{1\le k\le n}$ in one dimension?]

4.3.5 *Pros and Cons of Sequences with Low Discrepancy*

The use of sequences with low discrepancy to compute integrals instead of the Monte Carlo method (based on pseudo-random numbers) is known as the *Quasi-Monte Carlo* method (*QMC*). This terminology extends to so-called *good lattice points* not described here (see [219]).

THE PROS.

▷ The main attracting feature of sequences with low discrepancy is the combination of the Koksma–Hlawka inequality with the rate of decay of the discrepancy. It suggests that the *QMC* method is almost dimension free. This fact should be tempered in practice, standard *a priori* bounds for discrepancy do not allow for the use of this inequality to provide some "100%-confidence intervals".

▷ When the sequence ξ can be obtained as the orbit $\xi_n = T^{n-1}(\xi_1)$, $n \geq 1$, of an *ergodic*, or, better, a *uniquely ergodic* transform $T : [0, 1]^d \to [0, 1]^d$ ([9]) one shows that the integration rate of a function $f : [0, 1]^d \to \mathbb{R}$ can be $O(1/n)$ if f is a *coboundary* for T, *i.e.* can be written

$$f - \int_{[0,1]^d} f(u)du = g - g \circ T,$$

where g is a bounded Borel function (see *e.g.* [223]). As a matter of fact, for such coboundaries,

$$\frac{1}{n}\sum_{k=1}^{n} f(\xi_k) - \int_{[0,1]^d} f(u)du = \frac{g(\xi_1) - g(\xi_{n+1})}{n} = O\Big(\frac{1}{n}\Big).$$

The main difficulty is to determine practical criteria.

The Kakutani transforms (rotations with respect to $\oplus_p$) and the rotations of the torus are typical examples of ergodic transforms, the former being uniquely ergodic (keep in mind that the p-adic Van der Corput sequence is an orbit of the Kakutani transform $T_{p,1/p}$). The Kakutani transforms – or, to be precise, their representation on $[0, 1]^d$ – are unfortunately not continuous. However, their (original) natural representations on the space of power series with coefficients in $\{0, \dots, p-1\}$, endowed with its natural metric compact topology, definitely are (see [223]). One can take advantage of this unique ergodicity and characterize their coboundaries. Easy-to-check criteria based on the rate of decay of the Fourier coefficients $c_p(f)$, $p = (p^1, \dots, p^d) \in \mathbb{Z}^d$, of a function f as $\|p\| := p^1 \times \cdots \times p^d$ goes to infinity have been established (see [223, 278, 279]), at least for the p-adic Van der Corput

---

[9]Let $(X, \mathcal{X}, \mu)$ be a probability space. A mapping $T : (X, \mathcal{X}) \to (X, \mathcal{X})$ is ergodic if

$$(i)\ \ \mu \circ T^{-1} = \mu \ \ i.e.\ \ \mu \text{ is invariant under } T,$$

$$(ii)\ \forall A \in \mathcal{X},\ T^{-1}(A) = A \Longrightarrow \mu(A) = 0 \text{ or } 1.$$

Then Birkhoff's pointwise ergodic Theorem (see [174]) implies that, for every $f \in L^1(\mu)$,

$$\mu(dx)\text{-}a.s. \qquad \frac{1}{n}\sum_{k=1}^{n} f(T^{k-1}(x)) \longrightarrow \int_X f\, d\mu.$$

The mapping T is *uniquely ergodic* if μ is the only measure satisfying T. If X is a topological space, $\mathcal{X} = \mathcal{B}or(X)$ and T is continuous, then, for any continuous function $f : X \to \mathbb{R}$,

$$\sup_{x\in X}\left|\frac{1}{n}\sum_{k=1}^{n} f(T^{k-1}(x)) - \int_X f d\mu\right| \longrightarrow 0 \quad \text{as} \quad n \to +\infty.$$

In particular, it shows that any orbit $(T^{n-1}(x))_{n\geq 1}$ is μ-distributed. When $X = [0, 1]^d$ and $\mu = \lambda_d$, one retrieves the notion of uniformly distributed sequence. This provides a powerful tool for devising and studying uniformly distributed sequences. This is the case *e.g.* for Kakutani sequences or rotations of the torus.

sequences in one dimension and other orbits of the Kakutani transforms. Similar results also exist for the rotations of the torus.

Extensive numerical tests on problems involving some smooth (periodic) functions on $[0,1]^d$, $d \geq 2$, have been carried out, see *e.g.* [48, 237]. They suggest that this improvement still holds in higher dimensions, at least partially.

▷ It remains that, at least for structural dimensions d up to a few tens, Quasi-Monte Carlo integration empirically usually outperforms regular Monte Carlo simulation even if the integrated function does not have finite variation. We refer to Fig. 4.1 further on (see Application to the Box–Muller method), where $\mathbb{E}\,|X^1 - X^2|$, $X = (X^1, X^2) \stackrel{d}{=} \mathcal{N}(0; I_2)$ is computed by using a simple Halton(2, 3) sequence and pseudo-random numbers. More generally we refer to [237] for extensive numerical comparisons between Quasi- and regular Monte Carlo methods.

This concludes the pros.

THE CONS.

▷ As concerns the cons, the first is that all the non-asymptotic bounds for the discrepancy at the origin are very poor from a numerical point of view. We again refer to [237] for some examples which emphasize that these bounds cannot be relied on to provide (deterministic) error intervals for numerical integration. This is a major drawback compared to the regular Monte Carlo method, which *automatically* provides, almost for free, a confidence interval at any desired level.

▷ The second significant drawback concerns the family of functions for which we know that the *QMC* numerical integration is speeded up thanks to the Koksma–Hlawka Inequality. This family – mainly the functions with finite variation over $[0,1]^d$ in some sense – somehow becomes sparser and sparser in the space of Borel functions as the dimension d increases since the requested condition becomes more and more stringent (see Exercise 2 immediately before the Koksma–Hlawka formula in Proposition 4.3 and Exercise 1 which follows).

More importantly, if one is interested in integrating functions sharing a *"standard" regularity like Lipschitz continuity*, the following theorem due to Proïnov ([248]) shows that the *curse of dimensionality* comes back into the game in a striking way, without any possible escape.

Theorem 4.3 *(Proïnov). Assume $\mathbb{R}^d$ is equipped with the ℓ^∞-norm defined by $|(x^1,\ldots,x^d)|_\infty := \max_{1\leq i\leq d} |x^i|$. Let*

$$w(f,\delta) := \sup_{x,y\in[0,1]^d,\ |x-y|_\infty\leq\delta} |f(x)-f(y)|, \quad \delta\in(0,1)$$

denote the uniform continuity modulus of f (with respect to the ℓ^∞-norm).
(a) Let $(\xi_1,\ldots,\xi_n)\in([0,1]^d)^n$. For every continuous function $f:[0,1]^d\to\mathbb{R}$,

$$\left|\int_{[0,1]^d} f(x)dx - \frac{1}{n}\sum_{k=1}^n f(\xi_k)\right| \leq C_d\, w\big(f, D_n^*(\xi_1,\ldots,\xi_n)^{\frac{1}{d}}\big),$$

where $C_d \in (0, +\infty)$ is a universal optimal real constant only depending on d.

In particular, if f is Lipschitz continuous with coefficient $[f]_{\rm Lip}^{\infty} := \sup_{x,y\in[0,1]^d} \frac{|f(x)-f(y)|}{|x-y|_\infty}$, then

$$\left|\int_{[0,1]^d} f(x)dx - \frac{1}{n}\sum_{k=1}^{n} f(\xi_k)\right| \le C_d\, [f]_{\rm Lip}^{\infty} D_n^*(\xi_1,\ldots,\xi_n)^{\frac{1}{d}}.$$

(b) If $d = 1$, $C_d = 1$ and if $d \ge 2$, $C_d \in [1, 4]$.

Claim (b) should be understood in the following sense: there exist (families of) functions f with Lipschitz coefficient $[f]_{\rm Lip} = 1$ for which the above inequality holds (at least asymptotically) as an equality for a value of the constant C_d equal to 1 or lying in $[1, 4]$ depending on the dimension d.

► **Exercise.** Show, using the Van der Corput sequences starting at 0 (see the exercise in the paragraph devoted to Van der Corput and Halton sequences in Sect. 4.3.3) and the function $f(x) = x$ on $[0, 1]$, that the above Proïnov Inequality cannot be improved for Lipschitz continuous functions even in one dimension. [Hint: Reformulate some results of the Exercise in Sect. 4.3.3.]

▷ A third drawback of using *QMC* for numerical integration is that all functions need to be defined on unit hypercubes. One way to partially get rid of that may be to consider integration on some domains $C \subset [0, 1]^d$ having a regular boundary in the Jordan sense ([10]). Then a Koksma–Hlawka-like inequality holds true:

$$\left|\int_{C} f(x)\lambda_d(dx) - \frac{1}{n}\sum_{k=1}^{n} \mathbf{1}_{\{\xi_k\in C\}} f(\xi_k)\right| \le V(f)\tilde{D}_n^{\infty}(\xi)^{\frac{1}{d}}$$

where $V(f)$ denotes the variations of f (in the Hardy and Krause or measure sense) and $\tilde{D}_n^{\infty}(\xi)$ denotes the extreme discrepancy of $(\xi_1, \ldots, \xi_n)$ (see again [219]). The simple fact to integrate over such a set annihilates the low discrepancy effect (at least from a theoretical point of view).

► **Exercise.** Prove Proïnov's Theorem when $d = 1$. [Hint: read the next chapter and compare the star discrepancy modulus and the L^1-mean quantization error.]

This suggests that the rate of numerical integration in dimension d by a sequence with low discrepancy of Lipschitz continuous functions is $O\left(\frac{\log n}{n^{\frac{1}{d}}}\right)$ as $n \to +\infty$, or $O\left(\frac{(\log n)^{\frac{d-1}{d}}}{n^{\frac{1}{d}}}\right)$ when considering, for a fixed n, an n-tuple designed by the Hammersley method. This emphasizes that sequences with low discrepancy are not spared by the curse of dimensionality when implemented on functions with standard regularity...

[10]Namely that for every $\varepsilon > 0$, $\lambda_d(\{u \in [0, 1]^d \,:\, {\rm dist}(u, \partial C) < \varepsilon\}) \le \kappa_C\,\varepsilon$.

4.3.6 ℵ *Practitioner's Corner*

▷ **WARNING!** The dimensional trap. Although it is not strictly speaking a drawback, this last con is undoubtedly the most dangerous trap, at least for beginners: *a given (one-dimensional) sequence* $(\xi_n)_{n\geq 1}$ *does not "simulate" independence*, as emphasized by the classical exercise below.

▶ **Exercise.** (a) Let $\xi = (\xi_n)_{n\geq 1}$ denote the dyadic Van der Corput sequence. Show that, for every $n \geq 0$,

$$\xi_{2n+1} = \xi_{2n} + \frac{1}{2} \qquad \text{and} \qquad \xi_{2n} = \frac{\xi_n}{2}$$

with the convention $\xi_0 = 0$. Deduce that

$$\lim_n \frac{1}{n}\sum_{k=1}^{n} \xi_{2k}\xi_{2k+1} = \frac{5}{24}.$$

Compare with $\mathbb{E}\,(UV)$, where $U,\ V$ are independent with uniform distribution over $[0, 1]$. Conclude.

(b) Show, still for the dyadic Van der Corput sequence, that $\frac{1}{n}\sum_{k=1}^{n}\delta_{(\xi_{2k},\xi_{2k+1})}$ weakly converges toward a Borel distribution μ on $[0, 1]^2$ to be specified explicitly. Extend this result to p-adic Van der Corput sequences.

In fact, this phenomenon is typical of the price to pay for "filling up the gaps" faster than random numbers do. This is the reason why *it is absolutely mandatory to use d -dimensional sequences with low discrepancy to perform QMC computations related to a random vector X of the form* $X = \Psi(U_d)$, $U_d \stackrel{d}{=} \mathcal{U}([0, 1]^d)$ (d is sometimes called the *structural dimension* of the simulation). The d components of these d-dimensional sequences do simulate independence.

This has important consequences on very standard simulation methods, as illustrated below.

Application to the " ***QMC*** **Box–Muller method"**. To adapt the Box–Muller method of simulation of a normal distribution $\mathcal{N}(0; 1)$ introduced in Corollary 1.3, we proceed as follows: let $\xi = (\xi_n^1, \xi_n^2)_{n\geq 1}$ be a uniformly distributed sequence over $[0, 1]^2$ (in practice chosen with low discrepancy). We set, for every $n \geq 1$,

$$\zeta_n = (\zeta_n^1, \zeta_n^2) := \left(\sqrt{-2\log(\xi_n^1)}\,\sin\left(2\pi\xi_n^2\right), \sqrt{-2\log(\xi_n^1)}\,\cos\left(2\pi\xi_n^2\right)\right).$$

Then, for every bounded continuous function $f_1 : \mathbb{R} \to \mathbb{R}$,

$$\lim_n \frac{1}{n}\sum_{k=1}^{n} f_1\left(\zeta_k^1\right) \longrightarrow \mathbb{E}\, f_1(Z_1), \quad Z_1 \stackrel{d}{=} \mathcal{N}(0; 1)$$

since $(\xi^1, \xi^2) \longmapsto f\left(\sqrt{-2\log\xi^1}\sin(2\pi\xi^2)\right)$ is continuous on $(0, 1]^2$ and bounded, hence Riemann integrable on $[0, 1]^2$. Likewise, for every bounded continuous function $f_2 : \mathbb{R}^2 \to \mathbb{R}$,

$$\lim_n \frac{1}{n}\sum_{k=1}^{n} f_2\big(\zeta_k\big) \longrightarrow \mathbb{E}\, f_2(Z), \quad Z \stackrel{d}{=} \mathcal{N}\big(0; I_2\big).$$

These continuity assumptions on f_1 and f_2 can be relaxed, *e.g.* for f_2 to: the function $\widetilde{f_2}$ defined on $(0, 1]^2$ by

$$(\xi^1, \xi^2) \longmapsto \widetilde{f_2}(\xi^1, \xi^2) := f_2\left(\sqrt{-2\log\xi^1}\sin\big(2\pi\xi^2\big), \sqrt{-2\log\xi^1}\cos\big(2\pi\xi^2\big)\right)$$

is Riemann integrable. To benefit from the Koksma–Hlawka inequality, we need that $\widetilde{f_2}$ has finite variation. Establishing that $\widetilde{f_2}$ does have finite variation is clearly quite demanding, even when it is true, which is clearly not the generic situation.

All these considerations admit easy extensions to functions $f(Z)$, $Z \stackrel{d}{=} \mathcal{N}\big(0; I_d\big)$. To be more precise, the extension of Box–Muller method to multivariate normal distributions (see (1.1)) should be performed following the same rule of the structural dimension (here d, assumed to be even for convenience): such a sequence $(\zeta_n)_{n\geq 1}$ can be constructed from a uniformly distributed sequence $\xi = (\xi_n)_{n\geq 1}$ over $[0, 1]^d$ by plugging the components of ξ into (1.1) in place of $(U_1, \ldots, U_d)$, that is

$$(\zeta_n^{2i-1}, \zeta_n^{2i}) = \left(\sqrt{-2\log(\xi_n^{2i-1})}\cos(2\pi\xi_n^{2i}),\ \sqrt{-2\log(\xi_n^{2i-1})}\ \sin(2\pi\xi_n^{2i})\right),$$
$$i = 1, \ldots, d/2.$$

In particular, we will see further on in Chap. 7 that simulating the Euler scheme with step $\frac{T}{m}$ of a d-dimensional diffusion over $[0, T]$ with an underlying q-dimensional Brownian motion consumes m independent $\mathcal{N}\big(0; I_q\big)$-distributed random vectors, *i.e.* $m \times q$ independent $\mathcal{N}\big(0; 1\big)$ random variables. To perform a *QMC* simulation of a function of this Euler scheme at time T, we consequently need to consider a sequence with low discrepancy over $[0, 1]^{mq}$. Existing error bounds on sequences with low discrepancy and the sparsity of functions with finite variation make essentially meaningless any use of Koksma–Hlawka's inequality to produce error bounds. Proïnov's theorem itself is difficult to use, owing to the difficult evaluation of $[f]_{\text{Lip}}$. Not to mention that in the latter case, the curse of dimensionality will lead to extremely poor theoretical bounds for Lipschitz functions (like for $\widetilde{f_2}$ in dimension 2).

$\triangleright$ **Example.** We depict below in Fig. 4.1 a "competition" between pseudo-random numbers and a basic Halton(2, 3) sequence to compute $\mathbb{E}\,|X^1 - X^2|$, $X = (X^1, X^2) \stackrel{d}{=} \mathcal{N}(0; I_2)$, on a short trial of size 1 500.

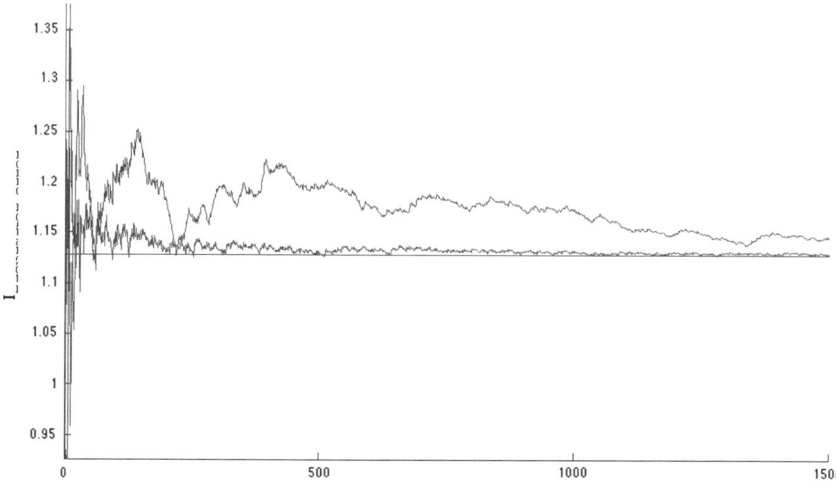

Fig. 4.1 MC VERSUS QMC. $\mathbb{E}\,|X^1 - X^2|$ computed by simulation with 1 500 trials of pseudo-random numbers (red) and a Halton(2, 3) sequence (blue) plugged into a Box–Muller formula (Reference value $\frac{2}{\sqrt{\pi}}$)

► **Exercise.** (*a*) Implement a *QMC*-adapted Box–Muller simulation method for $\mathcal{N}(0; I_2)$ (based on the sequence with low discrepancy of your choice) and organize a race *MC vs QMC* to compute various *calls*, say $\text{Call}_{BS}(K, T)$ ($T = 1$, $K \in \{95, 96, \dots, 104, 105\}$) in a Black–Scholes model (with $r = 2\%$, $\sigma = 30\%$, $x = 100$, $T = 1$). To simulate this underlying Black–Scholes risky asset, first use the closed expression

$$X_t^x = x\, e^{(r - \frac{\sigma^2}{2})t + \sigma\sqrt{T}Z}.$$

(*b*) Anticipating Chap. 7, implement the Euler scheme (7.5) of the Black–Scholes dynamics

$$dX_t^x = X_t^x(r dt + \sigma dW_t).$$

Consider steps of the form $\frac{T}{m}$ with $m = 10, 20, 50, 100$. What conclusions can be drawn?

▷ *Statistical correlation at finite range.* A final difficulty encountered by practitioners is to "reach" the effective statistical independence between the coordinates of a d-dimensional sequence with low discrepancy. This independence is only true asymptotically so that, in high dimensions and for small values of n, the coordinates of the Halton sequence remain highly correlated for a long time. As a matter of fact, the ith component of the canonical d-dimensional Halton sequence (*i.e.* designed from the first d prime numbers $p_1, \dots, p_d$) starts as follows

$$\xi_n^i = \frac{n}{p_i},\; n \in \{1, \dots, p_i - 1\},\; \xi_n^i = \frac{1}{p_i^2} + \frac{n - p_i}{p_i},\; n \in \{p_i, \dots, 2p_i - 1\}, \dots$$

so it is clear that the *ith and the* $(i+1)$*th components will remain highly correlated* if $d = 81$ and $(i, i+1) = (d-1, d)$ then $p_{d-1} = 503$ and $p_d = 509, \ldots$

To overcome this correlation observed for (not so) small values of n, the usual method is to *discard* the first values of a sequence.

► **Exercise.** Let $\xi^1 = \mathrm{VdC}(p_1)$ and $\xi^2 = \mathrm{VdC}(p_2)$ where p_1 and p_2 are two (reasonably) distinct large prime numbers satisfying $p_1 < p_2 < 2p_1$ (p_2 exists owing to Bertrand's conjecture). Show that the pseudo-estimator of the correlation between these two sequences at order $n = p_1$ satisfies

$$\frac{1}{n}\sum_{k=1}^{n}\xi_k^1\xi_k^2 - \left(\frac{1}{n}\sum_{k=1}^{n}\xi_k^1\right)\left(\frac{1}{n}\sum_{k=1}^{n}\xi_k^2\right) = \left(1+\frac{1}{p_1}\right)\frac{4p_1 - p_2 + 1}{12p_2}$$
$$> \frac{1}{12}\left(1+\frac{1}{p_1}\right)\left(1+\frac{1}{2p_1}\right) > \frac{1}{12}.$$

▷ *MC* versus *QMC : a visual point of view.* As a conclusion to this section, let us emphasize *graphically* in Fig. 4.2 the differences of *textures* between *MC* and *QMC* "sampling", *i.e.* between (pseudo-)randomly generated points (say 60 000) and the same number of terms of the Halton sequence (with $p_1 = 2$ and $p_2 = 3$).

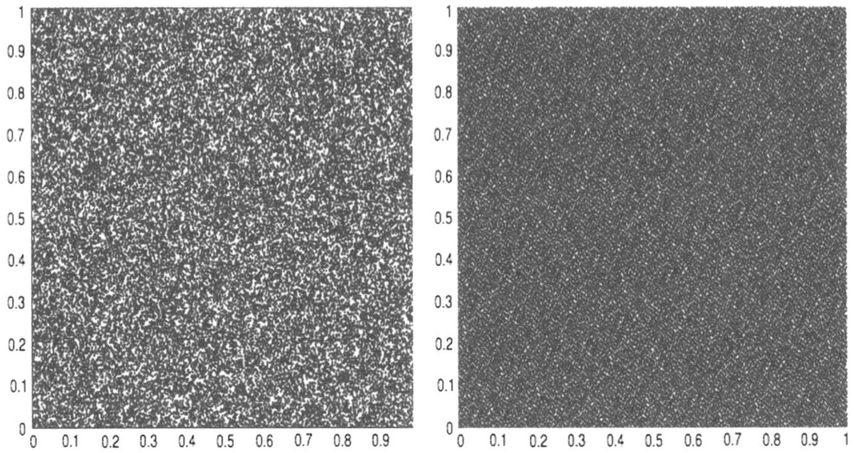

Fig. 4.2 MC VERSUS QMC Left: 6.10^4 randomly generated points; Right: 6.10^4 terms of the Halton(2, 3)-sequence

4.4 Randomized *QMC*

The principle at the origin of randomized *QMC* is to introduce some randomness in the *QMC* method in order to produce a confidence interval or, returning to a regular *MC* viewpoint, to use *QMC* sequences as a variance reducer in the Monte Carlo method.

Throughout this section, $\{x\}$ denotes the componentwise fractional part of $x = (x^1, \dots, x^d) \in \mathbb{R}^d$, *i.e.* $\{x\} = \big(\{x^1\}, \dots, g\{x^d\}\big)$.

Moreover, we denote by $(\xi_n)_{n\geq 1}$ a uniformly distributed sequence over $[0, 1]^d$.

4.4.1 Randomization by Shifting

The starting idea is to note that a shifted uniformly distributed sequence is still uniformly distributed. The second stage is to randomly shift such a sequence to combine the properties of the Quasi- and regular Monte Carlo simulation methods.

Proposition 4.7 *(a) Let U be a uniformly distributed random variable on $[0, 1]^d$. Then, for every $a \in \mathbb{R}^d$*

$$\{a + U\} \stackrel{d}{=} U.$$

(b) Let $a = (a^1, \dots, a^d) \in \mathbb{R}^d$. The sequence $(\{a + \xi_n\})_{n\geq 1}$ is uniformly distributed.

Proof. (*a*) One easily checks that the characteristic functions of U and $\{a + U\}$ coincide on $\mathbb{Z}^d$ since, for every $p \in \mathbb{Z}^d$,

$$\begin{aligned}\mathbb{E}\, e^{2\tilde{\imath}\pi(p|\{a+U\})} = \mathbb{E}\, e^{2\tilde{\imath}\pi(p|a+U)} &= e^{2\tilde{\imath}\pi(p|a)}\mathbb{E}\, e^{2\tilde{\imath}\pi(p|U)}\\ &= e^{2\tilde{\imath}\pi(p|a)}\mathbf{1}_{\{p=0\}} = \mathbf{1}_{\{p=0\}} = \mathbb{E}\, e^{2\tilde{\imath}\pi(p|U)}.\end{aligned}$$

Hence (see [155]), U and $\{a + U\}$ have the same distributions since they are $[0, 1]^d$-valued (in fact this is the static version of Weyl's criterion used to prove claim (*b*) below).

(*b*) This follows from Weyl's criterion: let $p \in \mathbb{N}^d \setminus \{0\}$,

$$\begin{aligned}\frac{1}{n}\sum_{k=1}^{n} e^{2\tilde{\imath}\pi(p|\{a+\xi_k\})} &= \frac{1}{n}\sum_{k=1}^{n} e^{2\tilde{\imath}\pi(p^1\{a^1+\xi_k^1\}+\cdots+p^d\{a^d+\xi_k^d\})}\\ &= \frac{1}{n}\sum_{k=1}^{n} e^{2\tilde{\imath}\pi(p^1(a^1+\xi_k^1)+\cdots+p^d(a^d+\xi_k^d))}\\ &= e^{2\tilde{\imath}\pi(p|a)}\frac{1}{n}\sum_{k=1}^{n} e^{2\tilde{\imath}\pi(p|\xi_k)}\\ &\longrightarrow 0 \quad \text{as} \quad n \to +\infty.\end{aligned}$$

◇

Consequently, if U is a uniformly distributed random variable on $[0,1]^d$ and $f:[0,1]^d \to \mathbb{R}$ is a Riemann integrable function, the random variable

$$\chi = \chi(f,U) := \frac{1}{n}\sum_{k=1}^{n} f(\{U+\xi_k\})$$

satisfies

$$\mathbb{E}\,\chi = \frac{1}{n} \times n\,\mathbb{E}\,f(U) = \mathbb{E}\,f(U)$$

owing to claim (b). Then, starting from an M-sample $U_1,\ldots,U_{_M}$ of the uniform distribution over $[0,1]^d$, one defines the empirical Monte Carlo estimator of size M attached to χ by

$$\widehat{I}(f,\xi)_{n,M} := \frac{1}{M}\sum_{m=1}^{M}\chi(f,U_m) = \frac{1}{nM}\sum_{m=1}^{M}\sum_{k=1}^{n} f(\{U_m+\xi_k\}).$$

This estimator has a complexity approximately equal to $\kappa \times nM$, where κ is the unitary complexity induced by the computation of one value of the function f. From the preceding, it satisfies

$$\widehat{I}(f,\xi)_{n,M} \xrightarrow{a.s.} \mathbb{E}\,f(U)$$

by the Strong Law of Large Numbers and its (weak) rate of convergence is ruled by a *CLT*

$$\sqrt{M}\big(\widehat{I}(f,\xi)_{n,M} - \mathbb{E}\,f(U)\big) \xrightarrow{\mathcal{L}} \mathcal{N}\big(0;\sigma_n^2(f,\xi)\big)$$

with

$$\sigma_n^2(f,\xi) = \mathrm{Var}\left(\frac{1}{n}\sum_{k=1}^{n} f(\{U+\xi_k\})\right).$$

Hence, the specific rate of convergence of the *QMC* is irremediably lost. In particular, this hybrid method should be compared to a regular Monte Carlo simulation of size nM through their respective variances. It is clear that we will observe a variance reduction if and only if

$$\frac{\sigma_n^2(f,\xi)}{M} < \frac{\mathrm{Var}(f(U))}{nM},$$

i.e.

$$\mathrm{Var}\left(\frac{1}{n}\sum_{k=1}^{n} f(\{U+\xi_k\})\right) \le \frac{\mathrm{Var}(f(U))}{n}.$$

The only natural upper-bound for the left-hand side of this inequality is

$$\sigma_n^2(f,\xi) = \int_{[0,1]^d}\left(\frac{1}{n}\sum_{k=1}^{n} f(\{u+\xi_k\}) - \int_{[0,1]^d} f\,d\lambda_d\right)^2 du$$
$$\leq \sup_{u\in[0,1]^d}\left|\frac{1}{n}\sum_{k=1}^{n} f(\{u+\xi_k\}) - \int_{[0,1]^d} f\,d\lambda_d\right|^2.$$

One can show that $f_u : v \mapsto f(\{u+v\})$ has finite variation on $[0,1]^d$ as soon as f has (in the same sense) and that $\sup_{u\in[0,1]^d} V(f_u) < +\infty$ (more precise results can be established). Consequently

$$\sigma_n^2(f,\xi) \leq \sup_{u\in[0,1]^d} V(f_u)^2 D_n^*(\xi_1,\ldots,\xi_n)^2$$

so that, if $\xi = (\xi_n)_{n\geq 1}$ is a sequence with low discrepancy (say Faure, Halton, Kakutani or Sobol', etc),

$$\sigma_n^2(f,\xi) \leq C_{f,\xi}^2 \frac{(\log n)^{2d}}{n^2}, \quad n \geq 1.$$

Consequently, in that case, it is clear that randomized *QMC* provides a very significant variance reduction (for the same complexity) of a magnitude proportional to $\frac{(\log n)^{2d}}{n}$ (with an impact of magnitude $\frac{(\log n)^d}{\sqrt{n}}$ on the confidence interval). But one must bear in mind once again that such functions with finite variations become dramatically sparse among Riemann integrable functions as d increases.

In fact, an even better bound of the form $\sigma_n^2(f,\xi) \leq \dfrac{C_{f,\xi}^2}{n^2}$ can be obtained for some classes of functions as emphasized in the Pros part of Sect. 4.3.5: when the sequence $(\xi_n)_{n\geq 1}$ is the orbit of a (uniquely) ergodic transform and f is a coboundary for this transform. But of course this class is even sparser. For sequences obtained by iterating rotations – of the torus or of the Kakutani adding machine – some criteria can be obtained involving the rate of decay of the Fourier coefficients $c_p(f)$, $p = (p^1,\ldots,p^d) \in \mathbb{Z}^d$, of f as $\|p\| := p^1 \times \cdots \times p^d$ goes to infinity since, in that case, one has $\sigma_n^2(f,\xi) \leq \dfrac{C_{f,\xi}^2}{n^2}$. Hence, the gain in terms of variance becomes proportional to $\frac{1}{n}$ for such functions (a global budget/complexity being prescribed for the simulation).

By contrast, if we consider Lipschitz continuous functions, things go radically differently: assume that $f : [0,1]^d \to \mathbb{R}$ is Lipschitz continuous and isotropically periodic, *i.e.* for every $x \in [0,1]^d$ and every vector $e_i = (\delta_{ij})_{1\leq j\leq d}$, $i = 1,\ldots,d$ of the canonical basis of $\mathbb{R}^d$ (δ_{ij} stands for the Kronecker symbol) $f(x+e_i) = f(x)$ as soon as $x + e_i \in [0,1]^d$, then f can be extended as a Lipschitz continuous function on the whole $\mathbb{R}^d$ with the same Lipschitz coefficient, say $[f]_{\text{Lip}}$. Furthermore, it satisfies $f(x) = f(\{x\})$ for every $x \in \mathbb{R}^d$. Then, it follows from Proïnov's Theorem 4.3 that

$$\sup_{u\in[0,1]^d}\left|\frac{1}{n}\sum_{k=1}^{n} f(\{u+\xi_k\}) - \int_{[0,1]^d} f\,d\lambda_d\right|^2 \le [f]^2_{\text{Lip}} D^*_n(\xi_1,\ldots,\xi_n)^{\frac{2}{d}}$$
$$\le C_d^2 [f]^2_{\text{Lip}} C_\xi^{\frac{2}{d}} \frac{(\log n)^2}{n^{\frac{2}{d}}}$$

(where C_d is Proïnov's constant). This time, still for a prescribed budget, the "gain" factor in terms of variance is proportional to $n^{1-\frac{2}{d}}(\log n)^2$, which is no longer a gain... but a loss as soon as $d \ge 2$!

For more results and details, we refer to the survey [271] on randomized *QMC* and the references therein.

Finally, randomized *QMC* is a specific (and not so easy to handle) variance reduction method, not a *QMC* speeding up method. It suffers from one drawback shared by all *QMC*-based simulation methods: the sparsity of the class of functions with finite variation and the difficulty for identifying them in practice when $d > 1$.

4.4.2 *Scrambled (Randomized)* QMC

If the principle of mixing randomness and the Quasi-Monte Carlo method is undoubtedly a way to improve rates of convergence of numerical integration over unit hypercubes, the approach based on randomly "shifted" sequences with low discrepancy (or nets) described in the former section turned out to be not completely satisfactory and it is no longer considered as the most efficient way to proceed by the *QMC* community.

A new idea emerged at the very end of the 20th century inspired by the pioneering work by A. Owen (see [221]): to break the undesired regularity which appears even in the most popular sequences with low discrepancy (like Sobol' sequences), he proposed to *scramble* them in an i.i.d. random way so that these regularity features disappear while preserving the quality, in terms of discrepancy, of these resulting sequences (or nets).

The underlying principle – or constraint – was to preserve their "geometric-combinatorial" properties. Typically, if a sequence shares the (s, d)-property in a given base (or the (s, m, d)-property for a net), its scrambled version should share it too. Several attempts to produce efficient deterministic scrambling procedures have been made as well, but of course the most radical way to get rid of regularity features was to consider a kind of i.i.d. scrambling as originally developed in [221]. This has been successfully applied to the "best" Sobol' sequences by various authors.

We will not go deeper into the detail and technicalities as it would lead us too far from probabilistic concepts and clearly beyond the mathematical scope of this textbook.

Nevertheless, one should keep in mind that these types of improvements are mostly devoted to highly regular functions. In the same spirit, several extensions of the Koksma–Hlawka Inequality have been established (see [78]) for differentiable functions with finite variation (in the Hardy and Krause sense) whose partial derivatives also have finite variations up to a given order.

4.5 *QMC* in Unbounded Dimension: The Acceptance-Rejection Method

If one looks at the remark "The practitioner's viewpoint" in Sect. 1.4 devoted to Von Neumann's acceptance-rejection method, it is a simple exercise to check that one can replace pseudo-random numbers by a uniformly distributed sequence in the procedure (almost) *mutatis mutandis*, except for some more stringent regularity assumptions.

We adopt the notations of this section and assume that the $\mathbb{R}^d$-valued random vectors X and Y have absolutely continuous distributions with respect to a reference σ-finite measure μ on $\big(\mathbb{R}^d, \mathcal{B}or(\mathbb{R}^d)\big)$. Assume that $\mathbb{P}_X$ has a density *proportional* to f, that $\mathbb{P}_Y = g\cdot\mu$ and that f and g satisfy

$$f \le c\,g \quad \mu\text{-}a.e. \qquad \text{and} \qquad g > 0 \quad \mu\text{-}a.e.$$

where c is a positive real constant.

Furthermore, we make the assumption that Y can be simulated at a reasonable cost like in the original rejection-acceptation method, *i.e.* that

$$Y = \Psi(U),\ \ U \stackrel{d}{=} \mathcal{U}([0,1]^r)$$

for some $r \in \mathbb{N}^*$ where $\Psi : [0,1]^r \to \mathbb{R}$.

Additional " QMC assumptions":

$\triangleright$ The first additional assumption in this *QMC* framework is that we ask Ψ to be a *Riemann integrable* function (*i.e.* Borel, bounded and λ_r-$a.s.$ continuous).

$\triangleright$ We also assume that the function

$$\mathcal{I} : (u^1, u^2) \mapsto \mathbf{1}_{\{c\,u^1 g(\Psi(u^2)) \le f(\Psi(u^2))\}} \ \ \text{is } \lambda_{r+1} - \text{a.s. continuous on}[0,1]^{r+1} \tag{4.16}$$

(which also amounts to Riemann integrability since $\mathcal{I}$ is bounded).

$\triangleright$ Our aim is to compute $\mathbb{E}\,\varphi(X)$, where $\varphi \in \mathcal{L}^1(\mathbb{P}_X)$. Since we will use $\varphi(Y) = \varphi \circ \Psi(Y)$ to perform this integration (see below), we also ask φ to be such that

$$\varphi \circ \Psi \quad \textit{is Riemann integrable.}$$

This classically holds true if φ is continuous (see *e.g.* [52], Chap. 3).

Let $\xi = (\xi_n^1, \xi_n^2)_{n\geq 1}$ be a $[0,1] \times [0,1]^r$-valued sequence, assumed to be with low discrepancy (or simply uniformly distributed) over $[0,1]^{r+1}$. Hence, $(\xi_n^1)_{n\geq 1}$ and $(\xi_n^2)_{n\geq 1}$ are In particular, uniformly distributed over $[0,1]$ and $[0,1]^r$, respectively.

If $(U, V) \stackrel{d}{=} \mathcal{U}([0,1] \times [0,1]^r)$, then $(U, \Psi(V)) \stackrel{d}{=} (U, Y)$. Consequently, the product of two Riemann integrable functions being Riemann integrable,

$$\begin{aligned}
\frac{\sum_{k=1}^n \mathbf{1}_{\{c\,\xi_k^1 g(\Psi(\xi_k^2)) \leq f(\Psi(\xi_k^2))\}}\varphi(\Psi(\xi_k^2))}{\sum_{k=1}^n \mathbf{1}_{\{c\,\xi_k^1 g(\Psi(\xi_k^2)) \leq f(\Psi(\xi_k^2))\}}} \stackrel{n\to+\infty}{\longrightarrow} & \frac{\mathbb{E}\,(\mathbf{1}_{\{c\,U g(Y) \leq f(Y)\}}\varphi(Y))}{\mathbb{P}(c\,U g(Y) \leq f(Y))} \\
&= \int_{\mathbb{R}^d} \varphi(x) f(x)dx \\
&= \mathbb{E}\,\varphi(X), \qquad (4.17)
\end{aligned}$$

where the last two lines follow from computations carried out in Sect. 1.4.

The main gap to apply the method in a *QMC* framework is the $a.s.$ continuity assumption (4.16). The following proposition yields an easy and natural criterion.

Proposition 4.8 *If the function* $\frac{f}{g} \circ \Psi$ *is* λ_r-$a.s.$ *continuous on* $[0,1]^r$, *then Assumption (4.16) is satisfied.*

Proof. First we note that

$$\begin{aligned}
\mathrm{Disc}(\mathcal{I}) \subset [0,1] \times \mathrm{Disc}\Big(\frac{f}{g} \circ \Psi\Big) \\
\cup \Big(\Big\{(\xi^1, \xi^2) \in [0,1]^{r+1} \ \text{ s.t. } c\,\xi^1 = \frac{f}{g} \circ \Psi(\xi^2)\Big\}\Big)
\end{aligned}$$

where $\mathcal{I}$ denotes the function defined in (4.16). Now, it is clear that

$$\lambda_{r+1}\Big([0,1] \times \mathrm{Disc}(\frac{f}{g} \circ \Psi)\Big) = \lambda_1([0,1]) \times \lambda_r\Big(\mathrm{Disc}(\frac{f}{g} \circ \Psi)\Big) = 1 \times 0 = 0$$

owing to the λ_r-$a.s.$ continuity of $\frac{f}{g} \circ \Psi$. Consequently

$$\lambda_{r+1}\big(\mathrm{Disc}(\mathcal{I})\big) = \lambda_{r+1}\Big(\Big\{(\xi^1, \xi^2) \in [0,1]^{r+1} \ \text{ s.t. } c\,\xi^1 = \frac{f}{g} \circ \Psi(\xi^2)\Big\}\Big).$$

In turn, this subset of $[0,1]^{r+1}$ is negligible with respect to the Lebesgue measure λ_{r+1} since, returning to the independent random variables U and Y and keeping in mind that $g(Y) > 0$ $\mathbb{P}$-$a.s.$,

$$\lambda_{r+1}\Big(\Big\{(\xi^1,\xi^2)\in[0,1]^{r+1} \text{ s.t. } c\,\xi^1 = \frac{f}{g}\circ\Psi(\xi^2)\Big\}\Big) = \mathbb{P}(c\,U g(Y) = f(Y))$$

$$= \mathbb{P}\left(U = \frac{f}{c\,g}(Y)\right) = 0$$

where we used (see exercise below) that U and Y are independent by construction and that U has a diffuse distribution (no atom). ◇

Remark. The criterion of the proposition is trivially satisfied when $\frac{f}{g}$ and Ψ are continuous on $\mathbb{R}^d$ and $[0,1]^r$, respectively.

► **Exercise.** Show that if X and Y are independent and X or Y has no atom then

$$\mathbb{P}(X = Y) = 0.$$

As a conclusion note that in this section we provide *no rate of convergence* for this acceptance-rejection method by quasi-Monte Carlo. In fact, there is no such error bound under realistic assumptions on f, g, φ and Ψ. Only empirical evidence can justify its use in practice.

4.6 Quasi-stochastic Approximation I

It is natural to try to replace regular pseudo-random numbers by quasi-random numbers in other procedures where they are commonly implemented. This is the case for stochastic Approximation which can be seen as the stochastic counterpart of recursive zero search or optimization procedures like the Newton–Raphson algorithm, etc. These aspects of *QMC* will be investigated in Chap. 6 (Sect. 6.5).

Chapter 5
Optimal Quantization Methods I: Cubatures

Optimal Vector Quantization is a method coming from Signal Processing originally devised and developed in the 1950's (see [105]) to optimally discretize a continuous (stationary) signal in view of its transmission. It was introduced as a quadrature formula for numerical integration in the early 1990's (see [224]), and for conditional expectation approximations in the early 2000's, in order to price multi-asset American style options [19–21]. In this brief chapter, we focus on the cubature formulas for numerical integration with respect to the distribution of a random vector X taking values in $\mathbb{R}^d$.

In view of applications, we will only deal in this monograph with the canonical Euclidean quadratic setting (*i.e.* the L^2 optimal vector quantization in $\mathbb{R}^d$ equipped with the canonical Euclidean norm), but a general theory of optimal vector quantization can be developed in a general framework (with any norm on $\mathbb{R}^d$ and an L^p-norm or pseudo-norm – $0 < p < +\infty$ – on the probability space $(\Omega, \mathcal{A}, \mathbb{P})$). For a more comprehensive introduction to optimal vector quantization theory, we refer to [129] and for an introduction more oriented toward applications in Numerical Probability we refer to [228]. For an extension to infinite-dimensional spaces (known as *functional quantization*) see, among other references [204, 233].

Optimal quantization is closely connected to unsupervised automatic classification since it is the natural theoretical framework for modeling and analyzing celebrated classification procedures like k-means. We will not investigate these aspects of quantization in this chapter, for which we refer to [106], for example.

We recall that the canonical Euclidean norm on the vector space $\mathbb{R}^d$ is denoted by $|\cdot|$.

G. Pagès, *Numerical Probability*, Universitext,
https://doi.org/10.1007/978-3-319-90276-0_5

5.1 Theoretical Background on Vector Quantization

Let X be an $\mathbb{R}^d$-valued random vector defined on a probability space $(\Omega, \mathcal{A}, \mathbb{P})$. Unless stated otherwise, we will assume throughout this chapter that $X \in L^2_{\mathbb{R}^d}(\Omega, \mathcal{A}, \mathbb{P})$. The purpose of vector quantization is to study the best approximation of X by random vectors taking at most N fixed values $x_1, \ldots, x_N \in \mathbb{R}^d$.

Definition 5.1 *(a) Let $\Gamma = \{x_1, \ldots, x_N\} \subset \mathbb{R}^d$ be a subset of size N, called an N-quantizer. A Borel partition $\big(C_i(\Gamma)\big)_{i=1,\ldots,N}$ of $\mathbb{R}^d$ is a* Voronoi partition *of $\mathbb{R}^d$ induced by the N-quantizer Γ if, for every $i \in \{1, \ldots, N\}$,*

$$C_i(\Gamma) \subset \Big\{\xi \in \mathbb{R}^d,\ |\xi - x_i| \le \min_{j \neq i} |\xi - x_j|\Big\}.$$

The Borel sets $C_i(\Gamma)$ are called Voronoi cells *of the partition induced by* Γ.

In vector quantization theory, an N-quantizer Γ is also called a *codebook*, as a testimony to its links with Information theory, or a *grid* in applications to Numerical Probability. Elements of Γ are also called *codewords*, again in reference to Information theory.

NOTATION AND TERMINOLOGY. Any such N-quantizer is in correspondence with the N-tuple $x = (x_1, \ldots, x_N) \in (\mathbb{R}^d)^N$ as well as with all N-tuples obtained by a permutation of the components of x. In many applications, when there is no ambiguity regarding the fact that the points x_i are pairwise distinct, we will use the notation x rather than Γ to designate such an N-quantizer. Conversely an N-tuple $x = (x_1, \ldots, x_N) \in (\mathbb{R}^d)^N$ is in correspondence with the grid (or codebook) $\Gamma_x = \{x_1, \ldots, x_N\}$ made up of its components (or codewords). However, one must keep in mind that in general Γ_x has size $|\Gamma_x| \in \{1, \ldots, N\}$, which may be lower than N when several components of x are equal. If this is the case the indexing of Γ_x is not one–to-one.

Remarks. • Although, in the above definition, $|\cdot|$ denotes the canonical Euclidean norm, many results in what follows still hold true for any norm on $\mathbb{R}^d$ (see [129]), except for some differentiability results on the quadratic distortion function (see Proposition 6.3.1 in Sect. 6.3.5).

• In our regular setting where $|\cdot|$ denotes the canonical Euclidean norm, the closure and the interior of the Voronoi cells $C_i(\Gamma)$ satisfy, for every $i \in \{1, \ldots, N\}$,

$$\overline{C}_i(\Gamma) = \Big\{\xi \in \mathbb{R}^d,\ |\xi - x_i| = \min_{1 \le j \le N} |\xi - x_j|\Big\}$$

and

$$\overset{\circ}{C}_i(\Gamma) = \Big\{\xi \in \mathbb{R}^d,\ |\xi - x_i| < \min_{i \neq j} |\xi - x_j|\Big\}.$$

Furthermore, $\overline{C}_i(\Gamma)$ and $\overset{\circ}{C}_i(\Gamma)$ are polyhedral convex sets as the intersection of the half-spaces containing x_i defined by the median hyperplanes H_{ij} of the pairs (x_i, x_j), $j \neq i$.

Let $\Gamma = \{x_1, \ldots, x_N\}$ be an N-quantizer. The nearest neighbor projection $\mathrm{Proj}_\Gamma : \mathbb{R}^d \to \{x_1, \ldots, x_N\}$ induced by a Voronoi partition $\big(C_i(x)\big)_{i=1,\ldots,N}$ is defined by

$$\forall\, \xi \in \mathbb{R}^d, \qquad \mathrm{Proj}_\Gamma(\xi) := \sum_{i=1}^{N} x_i \mathbf{1}_{\{\xi \in C_i(\Gamma)\}}.$$

Then, we define the resulting *quantization* of X by composing Proj_Γ and X, namely

$$\widehat{X}^\Gamma = \mathrm{Proj}_\Gamma(X) = \sum_{i=1}^{N} x_i \mathbf{1}_{\{X \in C_i(\Gamma)\}}. \tag{5.1}$$

There are as many quantizations of X as Voronoi partitions induced by Γ (denoting all of them by $\widehat{X}^\Gamma$ is then an abuse of notation). The *pointwise error* induced when replacing X by $\widehat{X}^\Gamma$ is given by

$$|X - \widehat{X}^\Gamma| = \mathrm{dist}\big(X, \{x_1, \ldots, x_N\}\big) = \min_{1 \le i \le N} |X - x_i|$$

and *does not depend on the selected nearest neighbour projection* on Γ. When X has a strongly continuous distribution, *i.e.* $\mathbb{P}(X \in H) = 0$ for any hyperplane H of $\mathbb{R}^d$, the boundaries of the Voronoi cells $C_i(\Gamma)$ are $\mathbb{P}$-negligible so that any two quantizations induced by Γ are $\mathbb{P}$-*a.s.* equal.

Definition 5.2 *(a) The* mean quadratic quantization error *induced by an N-quantizer $\Gamma \subset \mathbb{R}^d$ is defined as the quadratic norm of the pointwise error, i.e.*

$$\|X - \widehat{X}^\Gamma\|_2 = \Big(\mathbb{E} \min_{1 \le i \le N} |X - x_i|^2\Big)^{\frac{1}{2}} = \Big(\int_{\mathbb{R}^d} \min_{1 \le i \le N} |\xi - x_i|^2 \mathbb{P}_X(d\xi)\Big)^{\frac{1}{2}}. \tag{5.2}$$

(b) We define the quadratic distortion function *at level N, defined as the squared mean quadratic quantization error on $(\mathbb{R}^d)^N$, namely*

$$\mathcal{Q}_{2,N} : x = (x_1, \ldots, x_N) \longmapsto \mathbb{E}\Big(\min_{1 \le i \le N} |X - x_i|^2\Big) = \|X - \widehat{X}^{\Gamma_x}\|_2^2. \tag{5.3}$$

The following facts are obvious:

– if $\Gamma = \{x_1, \ldots, x_N\}$ is an N-quantizer, then $\mathcal{Q}_{2,N}(x_1, \ldots, x_N) = \|X - \widehat{X}^\Gamma\|_2^2$,

– the quadratic distortion function clearly satisfies

$$\inf_{x\in(\mathbb{R}^d)^N} \mathcal{Q}_{2,N}(x) = \inf\left\{\|X-\widehat{X}^{\Gamma}\|_2^2,\ \Gamma\subset\mathbb{R}^d,\ |\Gamma|\le N\right\}$$

since any grid Γ with cardinality at most N can be "represented" by an N-tuple by repeating some components in an appropriate way.

We briefly recall some classical facts about theoretical and numerical aspects of Optimal Quantization. For further details, we refer *e.g.* to [129, 228, 231–233].

Theorem 5.1 (Existence of optimal N-quantizers, [129, 224]) *Let $X\in L^2_{\mathbb{R}^d}(\mathbb{P})$ and let $N\in\mathbb{N}^*$.*

(a) The quadratic distortion function $\mathcal{Q}_{2,N}$ at level N attains a minimum at an N-tuple $x^{(N)}\in(\mathbb{R}^d)^N$ and $\Gamma_{x^{(N)}}=\{x_i^{(N)},\ i=1,\dots,N\}$ is an optimal quantizer at level N (though its cardinality may be lower than N, see the remark below).

(b) If the support of the distribution $\mathbb{P}_X$ of X has at least N elements, then $x^{(N)}=(x_1^N,\dots,x_N^N)$ has pairwise distinct components, $\mathbb{P}\big(X\in C_i(x^{(N)})\big)>0$, $i=1,\dots,N$ (and $\min_{x\in(\mathbb{R}^d)^{N-1}}\mathcal{Q}_{2,N-1}(x)>0$). Furthermore, the sequence $N\mapsto \inf_{x\in(\mathbb{R}^d)^N}\mathcal{Q}_{2,N}(x)$ converges to 0 and is (strictly) decreasing as long as it is positive.

Remark. If $\mathrm{supp}(\mathbb{P}_X)$ is finite, say $\mathrm{supp}(\mathbb{P}_X)=\{x_1,\dots,x_{N_0}\}\subset\mathbb{R}^d$, $N_0\ge 1$ (with pairwise distinct x_i), then $x^{(N_0)}=(x_1,\dots,x_{N_0})$ is an optimal quantizer at level N_0 and $\min_{x\in(\mathbb{R}^d)^{N_0}}\mathcal{Q}_{2,N_0}(x)=0$ and for every level $N\ge N_0$, $\min_{x\in(\mathbb{R}^d)^N}\mathcal{Q}_{2,N}(x)=0$.

Proof. (a) We will proceed by induction on the level N. First note that the L^2-mean quantization error function defined on $(\mathbb{R}^d)^N$ by

$$(x_1,\dots,x_N)\longmapsto\sqrt{\mathcal{Q}_{2,N}(x_1,\dots,x_N)}=\Big\|\min_{1\le i\le N}|X-x_i|\Big\|_2$$

is clearly 1-Lipschitz continuous with respect to the ℓ^∞-norm on $(\mathbb{R}^d)^N$ defined by $\big|(x_1,\dots,x_N)\big|_{\ell^\infty}:=\max_{1\le i\le N}|x_i|$. This is a straightforward consequence of Minkowski's inequality combined with the more elementary inequality

$$\Big|\min_{1\le i\le N}a_i-\min_{1\le i\le N}b_i\Big|\le\max_{1\le i\le N}|a_i-b_i|$$

(whose proof is left to the reader). As a consequence, it implies the continuity of its square, the quadratic distortion function $\mathcal{Q}_{2,N}$.

As a preliminary remark, note that by its very definition the sequence $N\mapsto \inf_{x\in(\mathbb{R}^d)^N}\|X-\widehat{X}^{\Gamma_x}\|_2$ is non-increasing.

$\triangleright$ $N=1$. The non-negative strictly convex function $\mathcal{Q}_{2,1}$ clearly goes to $+\infty$ as $|x_1|\to+\infty$. Hence, $\mathcal{Q}_{2,1}$ attains a unique minimum at the mean of X, *i.e.* $x_1^{(1)}=\mathbb{E}\,X$. So $\{x_1^{(1)}\}$ is an optimal quantization grid at level 1.

$\triangleright$ $N \Rightarrow N+1$. Assume there exists an $x^{(N)} \in (\mathbb{R}^d)^N$ such that $\mathcal{Q}_{2,N}(x^{(N)}) = \min_{(\mathbb{R}^d)^N} \mathcal{Q}_{2,N}$. Set $\Gamma_{x^{(N)}} = \{x_i^{(N)},\ i = 1, \dots, N\}$. Then, either $\mathrm{supp}(\mathbb{P}_X) \setminus \Gamma_{x^{(N)}} = \varnothing$ and any $(N+1)$-tuple of $(\mathbb{R}^d)^{N+1}$ which "exhausts" the grid $\Gamma_{x^{(N)}}$ makes the function $\mathcal{Q}_{2,N+1}$ equal to 0, its lowest possible value, or there exists a $\xi_{N+1} \in \mathrm{supp}(\mathbb{P}_X) \setminus \Gamma_{x^{(N)}}$.

In this second case, let $\Gamma^* = \Gamma_{x^{(N)}} \cup \{\xi_{N+1}\}$ and let $\big(C_i(\Gamma^*)\big)_{1 \le i \le N+1}$ be a Voronoi partition of $\mathbb{R}^d$ where $C_{N+1}(\Gamma^*)$ is the Voronoi cell of ξ_{N+1}. As $\xi_{N+1} \notin \Gamma_{x^{(N)}}$, it is clear that $\overset{\circ}{C}_{N+1}(\Gamma^*) \neq \varnothing$ and that $|X - \xi_{N+1}| < \min_{1 \le i \le N} |X - x_i^{(N)}|$ on the interior of this cell. Furthermore, $\mathbb{P}\big(X \in C_{N+1}(\Gamma^*)\big) \ge \mathbb{P}\big(X \in \overset{\circ}{C}_{N+1}(\Gamma^*)\big) > 0$ since $\xi_{N+1} \in \overset{\circ}{C}_{N+1}(\Gamma^*)$ and $\xi_{N+1} \in \mathrm{supp}(\mathbb{P}_X)$. Note that, everywhere on $(\mathbb{R}^d)^N$, one has $|X - \xi_{N+1}| \wedge \min\limits_{1 \le i \le N} |X - x_i^{(N)}| \le \min\limits_{1 \le i \le N} |X - x_i^{(N)}|$, so that, combining both inequalities yields

$$
\begin{aligned}
\lambda_{N+1} = \mathbb{E}\left(|X - \widehat{X}^{\Gamma^*}|^2\right) &= \mathbb{E}\left(|X - \xi_{N+1}|^2 \wedge \min_{1 \le i \le N} |X - x_i^{(N)}|^2\right) \\
&< \mathbb{E}\left(\min_{1 \le i \le N} |X - x_i^{(N)}|^2\right) \\
&= \mathcal{Q}_{2,N}(x^{(N)}) = \min_{x \in (\mathbb{R}^d)^N} \left\|X - \widehat{X}^x\right\|_2^2.
\end{aligned}
$$

In particular, $\mathrm{card}(\Gamma_{x^{(N)}}) = N$, otherwise $\mathrm{card}(\Gamma^*) \le N - 1 + 1 = N$, but this would contradict that $\mathcal{Q}_{2,N}$ is minimum at $x^{(N)}$ since any N-tuple x^* "exhausting" the values of Γ^* would satisfy $\mathcal{Q}_{2,N}(x^*) < \mathcal{Q}_{2,N}(x^{(N)})$. Hence, the set $K_{N+1} = \Big\{x \in (\mathbb{R}^d)^{N+1} : \mathcal{Q}_{2,N+1}(x) \le \lambda_{N+1}\Big\}$ is non-empty since it contains all the $(N+1)$-tuples which "exhaust" the elements of Γ^*. It is closed since $\mathcal{Q}_{2,N+1}$ is continuous. Let us show that it is also a bounded subset of $(\mathbb{R}^d)^{N+1}$. Let $x_{[k]} = \big(x_{[k],1}, \dots, x_{[k],N+1}\big)$, $k \in \mathbb{N}^*$, be a K_{N+1}-valued sequence of $(N+1)$-tuples. Up to at most $N+1$ extractions, one may assume without loss of generality that there exists a subset $I \subset \{1, \dots, N+1\}$ such that for every $i \in I$, $x_{[k],i} \to x_{[\infty],i} \in \mathbb{R}^d$ and for every $i \notin I$, $|x_{[k],i}| \to +\infty$ as $k \to +\infty$. By a straightforward application of Fatou's Lemma (where $|I|$ denotes here the cardinality of the index set I)

$$
\underline{\lim}_{k} \mathcal{Q}_{2,N+1}(x_{[k]}) \ge \left\| \min_{i \in I} |X - x_{[\infty],i}| \right\|_2^2 \ge \inf_{y \in (\mathbb{R}^d)^{|I|}} \mathcal{Q}_{2,|I|}(y).
$$

The sequence $\big(x_{[k]}\big)_{k \in \mathbb{N}^*}$ being K_{N+1}-valued, one has

$$
\inf_{y \in (\mathbb{R}^d)^{|I|}} \mathcal{Q}_{2,|I|}(y) \le \lambda_{N+1} < \inf_{x \in (\mathbb{R}^d)^N} \mathcal{Q}_{2,N}(x).
$$

In turn, this implies that $|I| = N+1$, *i.e.* the sequence of $(N+1)$-tuples $\big(x_{[k]}\big)_{k \ge 1}$ is bounded. As a consequence, the set K_{N+1} is compact and the function $\mathcal{Q}_{2,N+1}$ attains a minimum over K_{N+1} at an $x^{(N+1)}$ which is obviously its absolute minimum and has pairwise components such that, with obvious notation, $\mathrm{card}(\Gamma_{x^{(N+1)}}) = N+1$ and $\mathbb{P}\big(X \in C_i(\Gamma_{x^{(N+1)}})\big) > 0$, $i = 1, \dots, N+1$.

(b) The strict decrease of $N \mapsto \inf_{(\mathbb{R}^d)^N} \mathcal{Q}_{2,N}$ as long as it is not 0 is a straightforward consequence of the proof of the Claim (a). If $(z_N)_{N\geq 1}$ is an everywhere dense sequence in $\mathbb{R}^d$, then

$$0 \leq \inf_{(\mathbb{R}^d)^N} \mathcal{Q}_{2,N} \leq \left\|X - \widehat{X}^{(z_1,\dots,z_N)}\right\|_2 = \left\|\min_{1\leq i\leq N} |X - z_i|\right\|_2 \downarrow 0 \quad \text{as} \quad N \to +\infty$$

by the Lebesgue dominated convergence theorem ($\min_{1\leq i\leq N} |X - z_i| \leq |X - z_1| \in L^2$ ensures the domination property). The other claims follow from the proof of Claim (a). ◇

The preceding leads naturally to the following definition.

Definition 5.3 *A grid associated to any N-tuple solution to the above distortion minimization problem is called an* optimal quadratic N-quantizer *or an optimal quadratic quantizer at level N (the term "quadratic" may be dropped in the absence of ambiguity).*

NOTATION (A SLIGHT ABUSE OF). When an N-tuple $x = (x_1, \dots, x_N)$ has pairwise distinct components, we will often use the notation $\widehat{X}^x$ instead of $\widehat{X}^{\Gamma_x}$. For simplicity, we will also call it an N-quantizer. Similarly, we will also denote $C_i(x)$ instead of $C_i(\Gamma_x)$ to denote Voronoi cells.

Remarks. • When $N = 1$, the N-optimal quantizer is always unique and is equal to $\mathbb{E}\,X$.

• When $N \geq 2$, the set argmin $\mathcal{Q}_{2,N}$ is never reduced to a single N-tuple (except if X is $\mathbb{P}$-*a.s.* constant), simply because argmin $\mathcal{Q}_{2,N}$ is left stable under the action of the $N!$ permutations of $\{1, \dots, N\}$. The question of the geometric uniqueness – as a grid (non-ordered set) – is much more involved. When $d \geq 2$, uniqueness usually fails if the distribution of X is invariant under isometries. Thus, the normal distribution $\mathcal{N}(0; I_d)$ is invariant under all orthogonal transforms and so is argmin $\mathcal{Q}_{2,N}$. But there are also examples (see [129]) for which optimal grids at level N do not even make up a "connected" set.

• However, in one dimension, it has been proved (see *e.g.* [168] and Sect. 5.3.1 further on) that, as soon as μ is absolutely continuous with *a* log-*concave density*, there exists exactly one optimal quantization grid at level N. This grid has full size N and is characterized by its stationarity so that argmin $\mathcal{Q}_{2,N}$ is made of the $N!$ resulting N-tuples. Such distributions are often called *unimodal.*

• This existence result admits many extensions, in particular in infinite dimensions when $\mathbb{R}^d$ is replaced by a separable Hilbert space or, more generally, a reflexive Banach space and X is a Radon random vector (but also for L^1-spaces). In such infinite-dimensional settings vector quantization is known as *functional quantization*, see [226] for a general introduction and [131, 204] for various existence and regularity results for functional quantizers.

► **Exercises. 1.** (*Unimodality*). Show that if X has a unimodal distribution, then $|X|$ also has a unimodal distribution.

2. *L^p-optimal quantizers.* Let $p \in (0, +\infty)$ and $X \in L^p_{\mathbb{R}^d}(\mathbb{P})$. Show that the L^p-distortion function at level $N \in \mathbb{N}^*$ defined by

$$\mathcal{Q}_{p,N} : x = (x_1, \ldots, x_N) \longmapsto \mathbb{E}\Big(\min_{1 \le i \le N} |X - x_i|^p\Big) = \big\|X - \widehat{X}^{\Gamma_x}\big\|_p^p \tag{5.4}$$

attains its minimum on $(\mathbb{R}^d)^N$. (Note that when $p = N = 1$ this minimum is attained at the *median* of the distribution of X).

3. *Constrained quantization* (*at* 0). (*a*) Show that, if $X \in L^2_{\mathbb{R}^d}(\mathbb{P})$ and $N \in \mathbb{N}^*$, the function defined on $(\mathbb{R}^d)^N$ by

$$(x_1, \ldots, x_N) \mapsto \mathbb{E}\Big[\min_{1 \le i \le N} |X - x_i|^2 \wedge |X|^2\Big] = \int_{\mathbb{R}^d} \min_{1 \le i \le N} |\xi - x_i|^2 \wedge |\xi|^2 \mathbb{P}_X(d\xi) \tag{5.5}$$

attains a minimum at an N-tuple $(x_1^*, \ldots, x_N^*)$.
(*b*) How would you interpret $(x_1^*, \ldots, x_N^*)$ in terms of quadratic optimal quantization? At which level?

Proposition 5.1 *Assume that the support of $\mathbb{P}_X$ has at least N elements.*
(a) Any L^2-optimal N-quantizer $x^{(N)} \in (\mathbb{R}^d)^N$ is stationary *in the following sense (see [224, 226]): for every Voronoi quantization $\widehat{X}^{x^{(N)}}$ of X,*

$$\mathbb{E}\big(X \,|\, \widehat{X}^{x^{(N)}}\big) = \widehat{X}^{x^{(N)}}. \tag{5.6}$$

(b) An optimal N-quantization $\widehat{X}^{x^{(N)}}$ of X (i.e. $x^{(N)} \in \operatorname{argmin} \mathcal{Q}_{2,N}$) is the best quadratic approximation of X among all random variables $Y : (\Omega, \mathcal{A}, \mathbb{P}) \to \mathbb{R}^d$ having at most N values:

$$\big\|X - \widehat{X}^{x^{(N)}}\big\|_2 = \min\Big\{\|X - Y\|_2,\ Y : (\Omega, \mathcal{A}, \mathbb{P}) \to \mathbb{R}^d, \text{ measurable}, |Y(\Omega)| \le N\Big\}.$$

Proof. (*a*) Let $x^{(N)}$ be an optimal N-quantizer and let $\widehat{X}^{x^{(N)}}$ be an optimal quantization of X given by (5.1), where $\big(C_i(x^{(N)})\big)_{1 \le i \le N}$ is a Voronoi partition induced by $x^{(N)}$. By the definition of conditional expectation as an orthogonal projector on $L^2\big(\sigma(\widehat{X}^{x^{(N)}})\big) = \big\{\varphi\big(\widehat{X}^{x^{(N)}}\big),\ \varphi : \Gamma_{x^{(N)}} \to \mathbb{R},\ \text{Borel}\big\}$, we know that $X - \mathbb{E}\big(X \,|\, \widehat{X}^{x^{(N)}}\big)$ $\perp L^2\big(\sigma(\widehat{X}^{x^{(N)}})\big)$ in $L^2(\Omega, \mathcal{A}, \mathbb{P})$. As $\widehat{X}^{x^{(N)}} - \mathbb{E}\big(X \,|\, \widehat{X}^{x^{(N)}}\big)$ lies in $L^2\big(\sigma(\widehat{X}^{x^{(N)}})\big)$, it follows from Pythagoras' Theorem that

$$\big\|X - \widehat{X}^{x^{(N)}}\big\|_2^2 = \big\|X - \mathbb{E}\big(X \,|\, \widehat{X}^{x^{(N)}}\big)\big\|_2^2 + \big\|\mathbb{E}\big(X \,|\, \widehat{X}^{x^{(N)}}\big) - \widehat{X}^{x^{(N)}}\big\|_2^2.$$

On the other hand, by the definition of a $\Gamma_{x^{(N)}} = \big\{x_1^{(N)}, \ldots, x_N^{(N)}\big\}$-valued Voronoi quantizer, one has

$$\big\|X - \mathbb{E}\big(X \,|\, \widehat{X}^{x^{(N)}}\big)\big\|_2 \ge \big\|\operatorname{dist}\big(X, x^{(N)}\big)\big\|_2 = \big\|X - \widehat{X}^{x^{(N)}}\big\|_2.$$

The former equality and this inequality are compatible if and only if $\mathbb{E}\left(X \mid \widehat{X}^{x^{(N)}}\right) = \widehat{X}^{x^{(N)}}$ $\mathbb{P}$-*a.s.*
(*b*) Let $Y : (\Omega, \mathcal{A}, \mathbb{P}) \to \mathbb{R}^d$ with $|Y(\Omega)| \le N$ and let $\Gamma = \{Y(\omega),\ \omega \in \Omega\}$. It is clear that $|\Gamma| \le N$ so that

$$\|X - Y\|_2 \ge \big\|\mathrm{dist}(X, \Gamma)\big\|_2 = \|X - \widehat{X}^{\Gamma}\|_2 \ge \inf_{x\in(\mathbb{R}^d)^N} \sqrt{\mathcal{Q}_{2,N}(x)} = \big\|X - \widehat{X}^{x^{(N)}}\big\|_2. \qquad \diamond$$

Remark. • An important additional property of an optimal quantizer is shown in [129] (Theorem 4.2, p. 38): the boundaries of any of its Voronoi partition are $\mathbb{P}_X$-negligible, *i.e.* $\mathbb{P}\Big(X \in \bigcup_{i=1}^N \partial C_i(x)\Big) = 0$ (even if $\mathbb{P}_X$ has atoms).

• Let $x \in (\mathbb{R}^d)^N$ be an N-tuple with pairwise distinct components and all its Voronoi partitions have a $\mathbb{P}_X$-negligible boundary, *i.e.* $\mathbb{P}\Big(X \in \bigcup_{i=1}^N \partial C_i(x)\Big) = 0$. Hence, the Voronoi quantization $\widehat{X}^x$ of X is $\mathbb{P}$-*a.s.* uniquely defined and, if x is only a local minimum, or any other kind of critical point, of the quadratic distortion function $\mathcal{Q}_{2,N}$, then x is a stationary quantizer, still in the sense that

$$\mathbb{E}\left(X \mid \widehat{X}^x\right) = \widehat{X}^x.$$

This is a straightforward consequence of the differentiability result of the quadratic distortion $\mathcal{Q}_{2,N}$ at N-tuples with pairwise distinct components and negligible Voronoi cells boundaries established further on in Chap. 6 (see Proposition 6.3.1).

An extended definition of a stationary quantizer is as follows (note it is not intrinsic since it depends upon the choice of its Voronoi partition).

Definition 5.4 *An N-tuple $x \in (\mathbb{R}^d)^N$ with pairwise distinct components is a stationary quantizer if there exists a Voronoi partition of $\mathbb{R}^d$ induced by Γ_x whose resulting quantization $\widehat{X}^x$ satisfies $\mathbb{P}\big(X \in C_i(x)\big) > 0$, $i = 1, \ldots, N$, and the stationarity property*

$$\mathbb{E}\left(X \mid \widehat{X}^x\right) = \widehat{X}^x. \tag{5.7}$$

► **Exercise** (*Non-optimal stationary quantizer*). Let $\mathbb{P}_X = \mu = \frac{1}{4}(\delta_0 + \delta_{\frac{1}{2}}) + \frac{1}{2}\delta_{\frac{3}{4}}$ be a distribution on the real line.

(*a*) Show that $(\frac{1}{4}, \frac{3}{4})$ is a stationary quantizer for μ (in the sense of the above definition).

(*b*) Show that $(\frac{1}{8}, \frac{5}{8})$ is an optimal quadratic 2-quantizer for μ.

Figure 5.1 shows a quadratic optimal – or at least close to optimal – N-quantizer for a bivariate normal distribution $\mathcal{N}(0, I_2)$ with $N = 200$.

Figures 5.2 and 5.3 (Gaussian shaped line) illustrate on the bi-variate normal distribution the intuitive fact that optimal vector quantization does not produce quantizers whose Voronoi cells all have the same "weights". One observes that the closer a cell

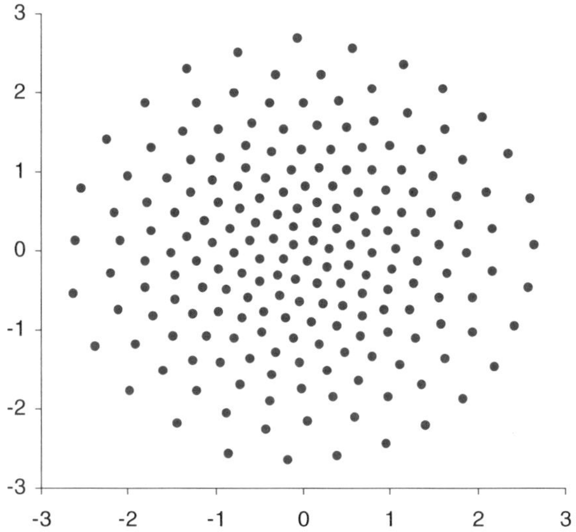

Fig. 5.1 Optimal quadratic quantization of size $N = 200$ of the bi-variate normal distribution $\mathcal{N}(0, I_2)$

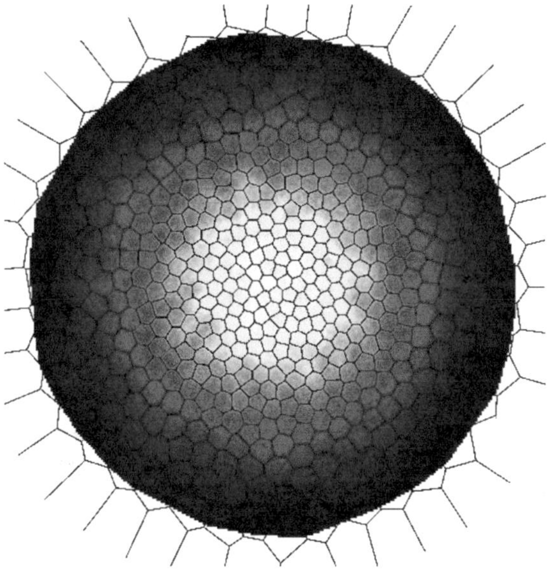

Fig. 5.2 Voronoi Tessellation of an optimal N-quantizer ($N = 500$) for the $\mathcal{N}(0; I_2)$ distribution. Color code: the heavier the cell is, the warmer (*i.e.* the lighter) the cell looks (with J. Printems)

is to the mode of the distribution, the heavier it weighs. In fact, *optimizing a quantizer tends to equalize the local inertia of the cells*, *i.e.*

$$\mathbb{E}\left(\mathbf{1}_{\{X\in C_i(x^{(N)})\}}|X - x_i^{(N)}|^2\right) \simeq \frac{\mathbb{E}\left(\min_{1\le j\le N}|X - x_j^{(N)}|^2\right)}{N},\quad i = 1, \dots, N.$$

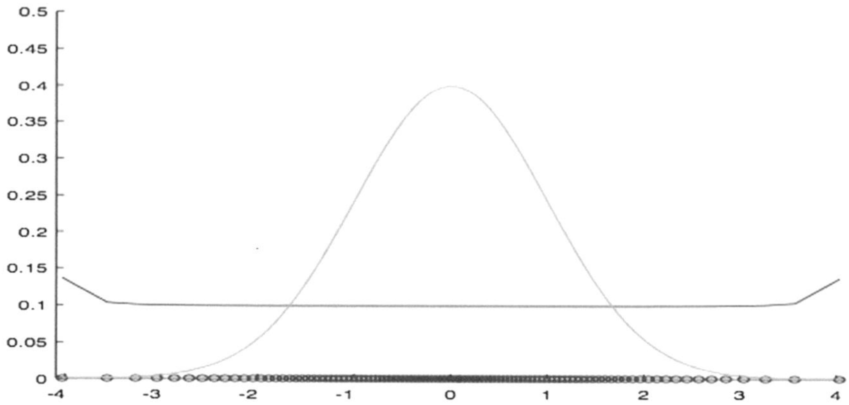

Fig. 5.3 $x_i^{(N)} \mapsto \mathbb{P}\big(X \in C_i(x^{(N)})\big)$ (flat line) and $x_i^{(N)} \mapsto \mathbb{E}\left(\mathbf{1}_{\{X \in C_i(x^{(N)})\}}|X - x_i^{(N)}|^2\right)$, $X \sim \mathcal{N}(0;1)$ (Gaussian line), $x^{(N)}$ optimal N-quantizer, $N = 50$ (with J.-C. Fort)

This fact can be easily highlighted numerically in one dimension, *e.g.* on the normal distribution, as illustrated in Fig. 5.3 ([1]).

Let us return to the study of the quadratic mean quantization error and, to be more precise, to its asymptotic behavior as the quantization level N goes to infinity. As seen in Theorem 5.1, the fact that the minimal mean quadratic quantization error goes to zero as N goes to infinity is relatively obvious since it follows from the existence of an everywhere dense sequence in $\mathbb{R}^d$. Determining the rate of this convergence is a much more involved question which is answered by the following theorem, known as Zador's Theorem, that we will essentially admit. In fact, optimal vector quantization can be defined and developed in any L^p-space and, keeping in mind some future application (in L^1), we will state the theorem in such a general framework.

Theorem 5.2 (Zador's Theorem) *Let $d \in \mathbb{N}^*$ and let $p \in (0, +\infty)$.*

(a) Sharp rate *(see [129]). Let $X \in L^{p+\delta}_{\mathbb{R}^d}(\mathbb{P})$ for some $\delta > 0$. Let $\mathbb{P}_X(d\xi) = \varphi(\xi)\lambda_d(d\xi) + \nu(d\xi)$, where $\nu \perp \lambda_d$, i.e. is singular with respect to the Lebesgue measure λ_d on $\mathbb{R}^d$ ([2]). Then, there is a constant $\widetilde{J}_{p,d} \in (0, +\infty)$ such that*

$$\lim_{N\to+\infty} N^{\frac{1}{d}} \min_{x \in (\mathbb{R}^d)^N} \|X - \widehat{X}^{\Gamma_x}\|_p = \widetilde{J}_{p,d} \left[\int_{\mathbb{R}^d} \varphi^{\frac{d}{d+p}} d\lambda_d\right]^{\frac{1}{p}+\frac{1}{d}}.$$

---

[1] The "Gaussian" solid line follows the shape of $\xi \mapsto \varphi_{\frac{1}{3}}(\xi) = \frac{e^{-\frac{\xi^2}{2\cdot 3}}}{\sqrt{3}\sqrt{2\pi}}$, *i.e.* $\mathbb{P}\big(X \in C_i(x^{(N)})\big) \simeq \varphi_{\frac{1}{3}}(x_i^{(N)})$, $i = 1, \ldots, N$, in a sense to be made precise. This is another property of optimal quantizers which is beyond the scope of this textbook, see *e.g.* [132].

[2] This means that there is a Borel set $A_\nu \in \mathcal{B}or(\mathbb{R}^d)$ such that $\lambda_d(A_\nu) = 0$ and $\nu(A_\nu) = \nu(\mathbb{R}^d)$. Such a decomposition always exists and is unique.

(*b*) NON- ASYMPTOTIC UPPER- BOUND *(see [205]). Let* $\delta > 0$. *There exists a real constant* $C_{d,p,\delta} \in (0, +\infty)$ *such that, for every* $\mathbb{R}^d$*-valued random vector* X,

$$\forall N \geq 1, \qquad \min_{x \in (\mathbb{R}^d)^N} \|X - \widehat{X}^{\Gamma_x}\|_p \leq C_{d,p,\delta}\, \sigma_{p+\delta}(X) N^{-\frac{1}{d}}$$

where, for $r \in (0, +\infty)$, $\sigma_r(X) = \min_{a \in \mathbb{R}^d} \|X - a\|_r \leq +\infty$.

Additional proof of Claim (*b*) (see also Theorem 1 in [236] and the remark that follows.) In fact, what is precisely proved in [205] (Lemma 1) is

$$\forall N \geq 1, \quad \min_{x \in (\mathbb{R}^d)^N} \|X - \widehat{X}^{\Gamma_x}\|_p \leq C_{d,p,\delta} \|X\|_{p+\delta} N^{-\frac{1}{d}}.$$

To derive the above conclusion, just note that the quantization error is invariant under translation since $\widehat{X - a}^{\Gamma_{x \ominus a}} = \widehat{X}^{\Gamma_x} - a$ (where $x \ominus a := \{x_1 - a, \ldots, x_N - a\}$), so that $X - \widehat{X}^{\Gamma_x} = (X - a) - \widehat{X - a}^{\Gamma_{x \ominus a}}$, which in turn implies

$$\begin{aligned}
\forall a \in \mathbb{R}^d, \quad \min_{x \in (\mathbb{R}^d)^N} \|X - \widehat{X}^x\|_p &= \min_{x \in (\mathbb{R}^d)^N} \|(X - a) - \widehat{X - a}^{\Gamma_{x \ominus a}}\|_p \\
&= \min_{x \in (\mathbb{R}^d)^N} \|(X - a) - \widehat{X - a}^{\Gamma_x}\|_p \\
&\leq C_{d,p,\delta} \|X - a\|_{p+\delta} N^{-\frac{1}{d}}.
\end{aligned}$$

Minimizing over a completes the proof. ◇

Remarks. • By truncating any random vector X, one easily checks that one always has

$$\underline{\lim}_{N \to +\infty} N^{\frac{1}{d}} \min_{x \in (\mathbb{R}^d)^N} \|X - \widehat{X}^{\Gamma_x}\|_p \geq \widetilde{J}_{p,d} \left[\int_{\mathbb{R}^d} \varphi^{\frac{d}{d+p}} d\lambda_d \right]^{\frac{1}{p} + \frac{1}{d}}.$$

• The first rigorous proof of claim (*a*) in a general framework is due to S. Graf and H. Luschgy in [129]. Claim (*b*) is an improved version of the so-called Pierce's Lemma, also established in [129].

• The $N^{\frac{1}{d}}$ factor is known as *the curse of dimensionality*: this is the optimal rate to "fill" a d-dimensional space by 0-dimensional objects.

• The real constant $\widetilde{J}_{p,d}$ clearly corresponds to the case of the uniform distribution $\mathcal{U}([0, 1]^d)$ over the unit hypercube $[0, 1]^d$ for which the following slightly more precise statement holds

$$\lim_N N^{\frac{1}{d}} \min_{x \in (\mathbb{R}^d)^N} \|X - \widehat{X}^x\|_p = \inf_N N^{\frac{1}{d}} \min_{x \in (\mathbb{R}^d)^N} \|X - \widehat{X}^x\|_p = \widetilde{J}_{p,d}.$$

A key part of the proof is a self-similarity argument "à la Hammersley" which establishes the theorem for the $\mathcal{U}([0,1]^d)$ distributions.

• Zador's Theorem holds true for any general – possibly non-Euclidean – norm on $\mathbb{R}^d$ and the value of $\widetilde{J}_{p,d}$ depends on the norm on $\mathbb{R}^d$ under consideration. When $d=1$, elementary computations show that $\widetilde{J}_{2,1}=\frac{1}{2\sqrt{3}}$. When $d=2$, with the canonical Euclidean norm, one shows (see [218] for a proof, see also [129]) that $\widetilde{J}_{2,d}=\sqrt{\frac{5}{18\sqrt{3}}}$. Its exact value is unknown for $d\geq 3$ but, still for the canonical Euclidean norm, one has, using some random quantization arguments (see [129]),

$$\widetilde{J}_{2,d}\sim\sqrt{\frac{d}{2\pi e}}\simeq\sqrt{\frac{d}{17,08}}\quad\text{as}\quad d\to+\infty.$$

5.2 Cubature Formulas

The random vector $\widehat{X}^x$ takes its values in the finite set (or grid) $\Gamma_x=\{x_1,\ldots,x_N\}$ (of size N), so for every continuous function $F:\mathbb{R}^d\to\mathbb{R}$ with $F(X)\in L^2(\mathbb{P})$, we have

$$\mathbb{E}\,F(\widehat{X}^x)=\sum_{i=1}^{N}F(x_i)\mathbb{P}\big(X\in C_i(x)\big),$$

which is the *quantization-based cubature formula* to approximate $\mathbb{E}\,F(X)$ (see [224, 228]). Indeed, as $\widehat{X}^x$ is close to X in $L^2(\mathbb{P})$, it is natural to estimate $\mathbb{E}\,F(X)$ by $\mathbb{E}\,F(\widehat{X}^x)$ when F is continuous. Furthermore, when F is Lipschitz continuous, one can provide an upper bound for the resulting error using the quantization errors $\|X-\widehat{X}^x\|_1$ (which comes out naturally) and $\|X-\widehat{X}^x\|_2$, or its square, when the quantizer x is stationary (see the following sections).

Likewise, one can consider *a priori* the $\sigma(\widehat{X}^x)$-measurable random variable $F(\widehat{X}^x)$ as a good approximation of the conditional expectation $\mathbb{E}\,\big(F(X)\,|\,\widehat{X}^x\big)$.

This principle also suggests to approximate the conditional expectation $\mathbb{E}\,\big(F(X)|Y\big)$ by $\mathbb{E}\,\big(F(\widehat{X}^x)|\hat{Y}^y\big)$. In that case one needs the "transition" probabilities:

$$\mathbb{P}\big(X\in C_j(x)\,|\,Y\in C_i(y)\big).$$

Numerical computation of $\mathbb{E}\,F(\widehat{X}^x)$ is possible as soon as $F(\xi)$ can be computed at any $\xi\in\mathbb{R}^d$ and the distribution $\big(\mathbb{P}(\widehat{X}=x_i)\big)_{1\leq i\leq N}$ of $\widehat{X}^x$ is known. The induced quantization error $\|X-\widehat{X}^x\|_2$ is used to control the error (see hereafter). These quantities related to the quantizer x are also called *companion parameters*.

5.2.1 Lipschitz Continuous Functions

Assume that the function $F : \mathbb{R}^d \to \mathbb{R}$ is Lipschitz continuous on $\mathbb{R}^d$ with Lipschitz coefficient $[F]_{\rm Lip}$. Then

$$\left|\mathbb{E}\,(F(X)\,|\,\widehat{X}^x) - F(\widehat{X}^x)\right| = \left|\mathbb{E}\left(F(X) - F(\widehat{X}^x)\,|\,\widehat{X}^x\right)\right| \le [F]_{\rm Lip}\mathbb{E}\left(|X - \widehat{X}^x|\,|\,\widehat{X}^x\right)$$

so that, for every real exponent $r \ge 1$,

$$\left\|\mathbb{E}\,(F(X)\,|\,\widehat{X}^x) - F(\widehat{X}^x)\right\|_r \le [F]_{\rm Lip}\|X - \widehat{X}^x\|_r$$

owing to the conditional Jensen inequality applied to the convex function $u \mapsto u^r$, see Proposition 3.1. In particular, using that $\mathbb{E}\,F(X) = \mathbb{E}\left(\mathbb{E}\,(F(X)\,|\,\widehat{X}^x)\right)$, one derives (with $r = 1$) that

$$\begin{aligned}\left|\mathbb{E}\,F(X) - \mathbb{E}\,F(\widehat{X}^x)\right| &\le \left\|\mathbb{E}\left(F(X)\,|\,\widehat{X}^x\right) - F(\widehat{X}^x)\right\|_1\\ &\le [F]_{\rm Lip}\|X - \widehat{X}^x\|_1.\end{aligned}$$

Finally, using the monotonicity of the $L^r(\mathbb{P})$-norms as a function of r yields

$$\left|\mathbb{E}\,F(X) - \mathbb{E}\,F(\widehat{X}^x)\right| \le [F]_{\rm Lip}\|X - \widehat{X}^x\|_1 \le [F]_{\rm Lip}\|X - \widehat{X}^x\|_2. \tag{5.8}$$

This universal bound is optimal in the following sense: considering the 1-Lipschitz continuous function $F(\xi) := \min_{i=1,\dots,N}|x - \xi_i| = \mathrm{dist}(\xi, \Gamma_x)$ which is equal to 0 at every component x_i of x, shows that equality may hold in (5.8) so that

$$\|X - \widehat{X}^x\|_1 = \sup_{[F]_{\rm Lip}\le 1}\left|\mathbb{E}\,F(X) - \mathbb{E}\,F(\widehat{X}^x)\right|. \tag{5.9}$$

In turn, due to the Monge–Kantorovich characterization of the L^1-Wasserstein distance ([3]) (see [272] for a definition and a characterization), (5.9) also reads

$$\|X - \widehat{X}^x\|_1 = \mathcal{W}_1\left(\mathbb{P}_X, \mathbb{P}_{\widehat{X}^x}\right).$$

---

[3]If μ and ν are distributions on $\left(\mathbb{R}^d, \mathcal{B}or(\mathbb{R}^d)\right)$ with finite pth moment ($1 \le p < +\infty$), the L^p-Wasserstein distance between μ and ν is defined by

$$\mathcal{W}_1(\mu,\nu) = \inf\left\{\left[\int_{\mathbb{R}^d\times\mathbb{R}^d}|x-y|^p\,m(dx,dy)\right]^{\frac{1}{p}},\ m \text{ Borel distribution on } \mathbb{R}^d\times\mathbb{R}^d,\ m(dx\times\mathbb{R}^d) = \mu,\ m(\mathbb{R}^d\times dy) = \nu\right\}.$$

When $p = 1$, the Monge–Kantorovich characterization of $\mathcal{W}_1$ reads as follows:

$$\mathcal{W}_1(\mu,\nu) = \sup\left\{\left|\int_{\mathbb{R}^d} f d\mu - \int_{\mathbb{R}^d} f d\nu\right|,\ f : \mathbb{R}^d \to \mathbb{R},\ [f]_{\rm Lip} \le 1\right\}.$$

Note that the absolute values can be removed in the above characterization of $\mathcal{W}_1(\mu,\nu)$ since f and $-f$ are simultaneously Lipschitz continuous with the same Lipschitz coefficient.

For an introduction to the Wasserstein distance and its main properties, we refer to [272].

Moreover, the bounded Lipschitz continuous functions making up a characterizing family for the weak convergence of probability measures on $\mathbb{R}^d$ (see Theorem 12.6(*ii*)), one derives that, for any sequence of N-quantizers $x^{(N)}$ satisfying $\|X - \widehat{X}^{x^{(N)}}\|_1 \to 0$ as $N \to +\infty$ (but *a priori* not optimal),

$$\sum_{1\le i\le N} \mathbb{P}\big(\widehat{X}^{x^{(N)}} = x_i^{(N)}\big)\, \delta_{x_i^{(N)}} \stackrel{(\mathbb{R}^d)}{\Longrightarrow} \mathbb{P}_X,$$

where $\stackrel{(\mathbb{R}^d)}{\Longrightarrow}$ denotes the weak convergence of probability measures on $\mathbb{R}^d$.

In fact, still due to the Monge–Kantorovich characterization of L^1-Wasserstein distance, we have trivially by (5.9) the more powerful result

$$\mathcal{W}_1\big(\mathbb{P}_X, \mathbb{P}_{\widehat{X}^{x^{(N)}}}\big) \longrightarrow 0 \quad \text{as} \quad n \to +\infty.$$

Variants of these cubature formulas can be found in [231] or [131] for functions F having only some local Lipschitz continuous regularity and polynomial growth.

5.2.2 Convex Functions

Let $x \in (\mathbb{R}^d)^N$ be a *stationary* quantizer, *e.g.* in the sense of Definition 5.4 (Eq. (5.7)), so that $\mathbb{E}\big(X \mid \widehat{X}^x\big) = \widehat{X}^x$ is a stationary quantization of X. Then, Jensen's inequality yields, for every convex function $F : \mathbb{R}^d \to \mathbb{R}$ such that $F(X) \in L^1(\mathbb{P})$,

$$F(\widehat{X}^x) \le \mathbb{E}\big(F(X) \,|\, \widehat{X}^x\big).$$

In particular, this implies

$$\mathbb{E}\, F(\widehat{X}^x) \le \mathbb{E}\, F(X). \tag{5.10}$$

5.2.3 Differentiable Functions With Lipschitz Continuous Gradients (C^1_{Lip})

In this chapter we will temporarily denote the canonical inner product on $\mathbb{R}^d$ by $\langle x \,|\, y\rangle$ rather than $(x \,|\, y)$ to avoid confusion with conditional expectation.

Assume now that F is differentiable on $\mathbb{R}^d$, with a Lipschitz continuous gradient ∇F. Let $x \in (\mathbb{R}^d)^N$ be a *stationary* quantizer in the sense of Definition 5.4 and let $\widehat{X}^x$ be the resulting stationary quantization. Note first that $F(X) \in L^2(\mathbb{P})$ since F has (at most) quadratic growth and $X \in L^2_{\mathbb{R}^d}(\mathbb{P})$.

We start from the first-order Taylor expansion with integral remainder of F at a point $u \in \mathbb{R}^d$: for every $v \in \mathbb{R}^d$,

$$F(v) = F(u) + \big\langle \nabla F(u) \,|\, v-u \big\rangle + \int_0^1 \big\langle \nabla F\big(tv+(1-t)u\big) - \nabla F(u) \,|\, v-u \big\rangle dt.$$

Using the Schwarz Inequality and the fact that ∇F is Lipschitz continuous, we obtain

$$\begin{aligned}\big|F(v) - F(u) - \big\langle \nabla F(u) \,|\, v-u \big\rangle\big| &\le \int_0^1 \Big|\big\langle \nabla F\big(tv+(1-t)u\big) - \nabla F(u) \,|\, v-u \big\rangle\Big| dt \\ &\le \int_0^1 \big|\nabla F\big(tv+(1-t)u\big) - \nabla F(u)\big|\, \big|v-u\big| dt \\ &\le [\nabla F]_{\text{Lip}} |v-u|^2 \int_0^1 t\,dt = \frac{[\nabla F]_{\text{Lip}}}{2} |v-u|^2.\end{aligned}$$

Then, with $v = X$ and $u = \widehat{X}^x$, the inequality reads

$$\big|F(X) - F(\widehat{X}^x) - \big\langle \nabla F(\widehat{X}^x) \,|\, X - \widehat{X}^x \big\rangle\big| \le \frac{[\nabla F]_{\text{Lip}}}{2} \big|X - \widehat{X}^x\big|^2. \tag{5.11}$$

Taking conditional expectation given $\widehat{X}^x$ yields, keeping in mind that $\big|\mathbb{E}\,\big(Z \,|\, \widehat{X}^x\big)\big| \le \mathbb{E}\,\big(|Z| \,|\, \widehat{X}^x\big)$,

$$\begin{aligned}\Big|\mathbb{E}\,(F(X) \,|\, \widehat{X}^x) - F(\widehat{X}^x) \;-\; \mathbb{E}\,\big(\langle \nabla F(\widehat{X}^x) \,|\, X - \widehat{X}^x \rangle \,\big|\, \widehat{X}^x\big)\Big| \\ \le \frac{[\nabla F]_{\text{Lip}}}{2} \mathbb{E}\,\big(|X - \widehat{X}^x|^2 \,|\, \widehat{X}^x\big).\end{aligned}$$

Now, using that the random variable $\nabla F(\widehat{X}^x)$ is $\sigma(\widehat{X}^x)$-measurable, one has

$$\mathbb{E}\,\big(\big\langle \nabla F(\widehat{X}^x) \,|\, X - \widehat{X}^x \big\rangle\big) = \mathbb{E}\,\big(\big\langle \nabla F(\widehat{X}^x) \,|\, \mathbb{E}\,(X - \widehat{X}^x \,|\, \widehat{X}^x) \big\rangle\big) = 0$$

so that

$$\big|\mathbb{E}\,(F(X) \,|\, \widehat{X}^x) - F(\widehat{X}^x)\big| \le \frac{[\nabla F]_{\text{Lip}}}{2} \mathbb{E}\,\big(|X - \widehat{X}^x|^2 \,|\, \widehat{X}^x\big).$$

Then, for every real exponent $r \ge 1$, the conditional Jensen's Inequality applied to the function $u \mapsto u^r$ yields

$$\big\|\mathbb{E}\,(F(X) \,|\, \widehat{X}^x) - F(\widehat{X}^x)\big\|_r \le \frac{[\nabla F]_{\text{Lip}}}{2} \|X - \widehat{X}^x\|_{2r}^2. \tag{5.12}$$

In particular, when $r = 1$, one derives

$$\big|\mathbb{E}\,F(X) - \mathbb{E}\,F(\widehat{X}^x)\big| \le \frac{[\nabla F]_{\text{Lip}}}{2} \|X - \widehat{X}^x\|_2^2.$$

These computations open the way to the following proposition.

Proposition 5.2 *Let $X \in L^2(\mathbb{P})$ and let x (or Γ_x) be a stationary quantizer in the sense of Definition 5.4.*

(a) If $F \in \mathcal{C}^1_{\mathrm{Lip}}(\mathbb{R}^d, \mathbb{R})$, i.e. is differentiable with a Lipschitz continuous gradient ∇F, then

$$\big|\mathbb{E}\, F(X) - \mathbb{E}\, F(\widehat{X}^x)\big| \le \frac{1}{2}[\nabla F]_{\mathrm{Lip}} \big\|X - \widehat{X}^x\big\|_2^2. \tag{5.13}$$

Moreover,

$$\sup\Big\{\big|\mathbb{E}\, F(X) - \mathbb{E}\, F(\widehat{X}^x)\big|,\ F \in \mathcal{C}^1_{\mathrm{Lip}}(\mathbb{R}^d, \mathbb{R}),\ [\nabla F]_{\mathrm{Lip}} \le 1\Big\} = \frac{1}{2}\big\|X - \widehat{X}^x\big\|_2^2. \tag{5.14}$$

In fact, this supremum holds as a maximum, attained for the function $F(\xi) = \frac{1}{2}|\xi|^2$ since $[\nabla F]_{\mathrm{Lip}} = 1$ and

$$\|X - \widehat{X}^x\|_2^2 = \mathbb{E}\,|X|^2 - \mathbb{E}\,|\widehat{X}^x|^2. \tag{5.15}$$

(b) In particular, if F is twice differentiable with a bounded Hessian D^2F, then

$$\big|\mathbb{E}\, F(X) - \mathbb{E}\, F(\widehat{X}^x)\big| \le \frac{1}{2}|||D^2F|||\, \big\|X - \widehat{X}^x\big\|_2^2,$$

where $|||D^2F||| = \sup\limits_{x\in\mathbb{R}^d}\ \sup\limits_{u:|u|=1} \big|u^*D^2F(x)u\big|$.

Proof. (a) The error bound (5.13) is proved above. Equality in (5.14) holds for the function $F(\xi) = \frac{1}{2}|\xi|^2$ since $[\nabla F]_{\mathrm{Lip}} = 1$ and

$$\begin{aligned}
\frac{1}{2}\,\mathbb{E}\,|X - \widehat{X}^x|^2 &= \frac{1}{2}\Big(\mathbb{E}\,|X|^2 - 2\,\mathbb{E}\,\langle X \,|\, \widehat{X}^x\rangle + \mathbb{E}\,|\widehat{X}^x|^2\Big)\\
&= \frac{1}{2}\Big(\mathbb{E}\,|X|^2 - 2\,\mathbb{E}\,\big\langle \mathbb{E}\,(X\,|\widehat{X}^x)\,|\,\widehat{X}^x\big\rangle + \mathbb{E}\,|\widehat{X}^x|^2\Big)\\
&= \frac{1}{2}\Big(\mathbb{E}\,|X|^2 - 2\,\mathbb{E}\,|\widehat{X}^x|^2 + \mathbb{E}\,|\widehat{X}^x|^2\Big)\\
&= \mathbb{E}\, F(X) - \mathbb{E}\, F(\widehat{X}^x)
\end{aligned}$$

where we used in the penultimate line that $\widehat{X}^x = \mathbb{E}\,(X \,|\, \widehat{X}^x)$.

(b) This error bound is straightforward by applying Taylor's formula at order 2 to F between X and $\widehat{X}^x$: this amounts to replacing $[\nabla F]_{\mathrm{Lip}}$ by $|||D^2F|||$ in (5.11). Or, equivalently, by showing that $|||D^2F||| = [\nabla F]_{\mathrm{Lip}}$ (see also footnote (3) in the proof of Step 3 of Proposition 7.4 in Chap. 7). ◇

Variants of these cubature formulas can be found in [231] or [131] for functions F whose gradient only has local Lipschitz continuous regularity and polynomial growth.

► **Exercise.** (*a*) Let $\Gamma_x \subset \mathbb{R}^d$ be a quantizer and let $\mathcal{P}_{\Gamma_x}$ denote the set of probability distributions supported by Γ_x. Show that $\|X - \widehat{X}^x\|_2 = \mathcal{W}_2(\mu_X, \mathcal{P}_{\Gamma_x})$, where μ_X denotes the distribution of X and $\mathcal{W}_2(\mu_X, \mathcal{P}_{\Gamma_x}) = \inf_{\nu \in \mathcal{P}_{\Gamma_x}} \mathcal{W}_2(\mu_X, \nu)$.
(*b*) Deduce that, *if* Γ_x *is a stationary quantizer*, then

$$\sup\left\{ \left|\mathbb{E}\, F(X) - \mathbb{E}\, F(\widehat{X}^x)\right|,\ F \in \mathcal{C}^1_{\mathrm{Lip}}(\mathbb{R}^d, \mathbb{R}),\ [\nabla F]_{\mathrm{Lip}} \le 1\right\} = \frac{1}{2}\mathcal{W}_2(\mu_X, \mathcal{P}_{\Gamma_x})^2.$$

5.2.4 Quantization-Based Cubature Formulas for $\mathbb{E}\,(F(X) \mid Y)$

Let X and Y be two $\mathbb{R}^d$-valued random vectors defined on the same probability space $(\Omega, \mathcal{A}, \mathbb{P})$ and let $F : \mathbb{R}^d \to \mathbb{R}$ be a Borel function. Assume that $F(X) \in L^2(\mathbb{P})$. The natural idea to approximate $\mathbb{E}\,(F(X) \mid Y)$ by quantization is to replace *mutatis mutandis* the random vectors X and Y by their quantizations $\widehat{X}$ and $\widehat{Y}$ (with respect to quantizers x and y that we drop in the notation for simplicity). The resulting approximation is then

$$\mathbb{E}\left(F(X) \mid Y\right) \simeq \mathbb{E}\left(F(\widehat{X}) \mid \widehat{Y}\right).$$

At this stage, a natural question is to look for *a priori* estimates for the resulting quadratic error given the quadratic mean quantization errors $\|X - \widehat{X}\|_2$ and $\|Y - \widehat{Y}\|_2$.

To this end, we need further assumptions on F. Let $\varphi_F : \mathbb{R}^d \to \mathbb{R}$ be a regular (Borel) version of the conditional expectation, *i.e.* satisfying

$$\mathbb{E}\left(F(X) \mid Y\right) = \varphi_F(Y).$$

Usually, no closed form is available for the function φ_F but some regularity property can be established, or more precisely "transmitted" or "propagated" from F to φ_F. Thus, we may assume that both F and φ_F are Lipschitz continuous with Lipschitz continuous coefficients $[F]_{\mathrm{Lip}}$ and $[\varphi_F]_{\mathrm{Lip}}$, respectively.

The main example of such a situation is the (homogeneous) Markovian framework, where $(X_n)_{n\ge 0}$ is a homogeneous Feller Markov chain, with transition $(P(y, dx))_{y\in\mathbb{R}^d}$ and when $X = X_k$ and $Y = X_{k-1}$. Then, with the above notations, $\varphi_F = PF$. If we assume that the Markov kernel P preserves and propagates Lipschitz continuity in the sense that, if $[F]_{\mathrm{Lip}} < +\infty$ then $[PF]_{\mathrm{Lip}} < +\infty$, the above assumption is clearly satisfied.

Remark. The above property is the key to quantization-based numerical schemes in Numerical probability. We refer to Chap. 11 for an application to the pricing of American options where the above principle is extensively applied to the Euler scheme which is a Markov process and propagates Lipschitz continuity.

We prove below a slightly more general proposition by considering the situation where the function F itself is replaced/approximated by a function G. This means that we approximate $\mathbb{E}\,(F(X) \mid Y)$ by $\mathbb{E}\,(G(\widehat{X}) \mid \widehat{Y})$.

Proposition 5.3 *Let* $X, Y : (\Omega, \mathcal{A}, \mathbb{P}) \to \mathbb{R}^d$ *be two random variables and let* $F, G : \mathbb{R}^d \to \mathbb{R}$ *be two Borel functions.*

(a) Quadratic case $p = 2$. *Assume* $F(X),\ G(X) \in L^2(\mathbb{P})$ *and that there exists a Lipschitz continuous function* φ_F *such that* $\mathbb{E}\big(F(X)\,|Y\big) = \varphi_F(Y) \in L^2(\mathbb{P})$. *Then*

$$\Big\|\mathbb{E}\big(F(X)\,|\,Y\big) - \mathbb{E}\big(G(\widehat{X})\,|\,\widehat{Y}\big)\Big\|_2^2 \le \Big\|F(X) - G(\widehat{X})\Big\|_2^2 + [\varphi_F]_{\mathrm{Lip}}^2 \Big\|Y - \widehat{Y}\Big\|_2^2. \tag{5.16}$$

In particular, if $G = F$ *and* F *is Lipschitz continuous*

$$\Big\|\mathbb{E}\big(F(X)\,|\,Y\big) - \mathbb{E}\big(F(\widehat{X})\,|\,\widehat{Y}\big)\Big\|_2^2 \le [F]_{\mathrm{Lip}}^2 \Big\|X - \widehat{X}\Big\|_2^2 + [\varphi_F]_{\mathrm{Lip}}^2 \Big\|Y - \widehat{Y}\Big\|_2^2. \tag{5.17}$$

(b) L^p-case $p \neq 2$. *Assume now that* $F(X),\ G(X) \in L^p(\mathbb{P}),\ p \in [1, +\infty)$. *Then* $\varphi_F(Y) \in L^p(\mathbb{P})$ *and*

$$\Big\|\mathbb{E}\big(F(X)\,|\,Y\big) - \mathbb{E}\big(G(\widehat{X})\,|\,\widehat{Y}\big)\Big\|_p \le \Big\|F(X) - G(\widehat{X})\Big\|_p + 2[\varphi_F]_{\mathrm{Lip}} \Big\|Y - \widehat{Y}\Big\|_p. \tag{5.18}$$

In particular, if $G = F$ *and* F *is Lipschitz continuous*

$$\Big\|\mathbb{E}\big(F(X)\,|\,Y\big) - \mathbb{E}\big(F(\widehat{X})\,|\,\widehat{Y}\big)\Big\|_p \le [F]_{\mathrm{Lip}} \Big\|X - \widehat{X}\Big\|_p + 2[\varphi_F]_{\mathrm{Lip}} \Big\|Y - \widehat{Y}\Big\|_p. \tag{5.19}$$

Proof. (a) We first note that

$$\begin{aligned}
\mathbb{E}\big(F(X)\,|\,Y\big) - \mathbb{E}\big(G(\widehat{X})\,|\,\widehat{Y}\big) &= \mathbb{E}\big(F(X)\,|\,Y\big) - \mathbb{E}\big(F(X)\,|\,\widehat{Y}\big) \\
&\quad + \mathbb{E}\big(F(X) - G(\widehat{X})\,|\,\widehat{Y}\big) \\
&= \mathbb{E}\big(F(X)\,|\,Y\big) - \mathbb{E}\Big(\mathbb{E}\big(F(X)\,|\,Y\big)\,|\,\widehat{Y}\Big) \\
&\quad + \mathbb{E}\big(F(X) - G(\widehat{X})\,|\,\widehat{Y}\big),
\end{aligned}$$

where we used that $\widehat{Y}$ is $\sigma(Y)$-measurable.

Now $\mathbb{E}\big(F(X)\,|\,Y\big) - \mathbb{E}\Big(\mathbb{E}\big(F(X)\,|\,Y\big)\,|\,\widehat{Y}\Big)$ and $\mathbb{E}\big(F(X) - G(\widehat{X})\,|\,\widehat{Y}\big)$ are clearly orthogonal in $L^2(\mathbb{P})$ by the very definition of the orthogonal projector $\mathbb{E}\big(\,\cdot\,|\,\widehat{Y}\big)$ on $L^2\big(\sigma(\widehat{Y}), \mathbb{P}\big)$ so that

$$\begin{aligned}
\Big\|\mathbb{E}\big(F(X)\,|\,Y\big) - \mathbb{E}\big(G(\widehat{X})\,|\,\widehat{Y}\big)\Big\|_2^2 &= \Big\|\mathbb{E}\big(F(X)\,|\,Y\big) - \mathbb{E}\Big(\mathbb{E}\big(F(X)\,|\,Y\big)\,|\,\widehat{Y}\Big)\Big\|_2^2 \\
&\quad + \Big\|\mathbb{E}\big(F(X) - G(\widehat{X})\,|\,\widehat{Y}\big)\Big\|_2^2.
\end{aligned} \tag{5.20}$$

Now using, the definition of conditional expectation given $\widehat{Y}$ as the best quadratic approximation among $\sigma(\widehat{Y})$-measurable random variables, we get

$$\left\|\mathbb{E}\left(F(X)\,|\,Y\right) - \mathbb{E}\Big(\mathbb{E}\left(F(X)\,|\,Y\right)|\,\widehat{Y}\Big)\right\|_2 = \left\|\varphi_F(Y) - \mathbb{E}\left(\varphi_F(Y)|\widehat{Y}\right)\right\|_2$$
$$\leq \left\|\varphi_F(Y) - \varphi_F(\widehat{Y})\right\|_2 \leq [\varphi_F]_{\mathrm{Lip}}\left\|Y - \widehat{Y}\right\|_2.$$

On the other hand, using that $\mathbb{E}\,(\,\cdot\,|\sigma(\widehat{Y}))$ is an L^2-contraction and that F itself is Lipschitz continuous yields

$$\left\|\mathbb{E}\,(F(X) - G(\widehat{X})\,|\,\widehat{Y})\right\|_2 \leq \left\|F(X) - G(\widehat{X})\right\|_2.$$

Finally,

$$\left\|\mathbb{E}\,(F(X)\,|\,Y) - \mathbb{E}\left(G(\widehat{X})\,|\,\widehat{Y}\right)\right\|_2^2 \leq \left\|F(X) - G(\widehat{X})\right\|_2^2 + [\varphi_F]^2_{\mathrm{Lip}}\left\|Y - \widehat{Y}\right\|_2^2. \tag{5.21}$$

The case when $G = F$ Lipschitz continuous is obvious.

(*b*) First, when $p \neq 2$, the Pythagoras-like equality (5.20) should be replaced by the standard Minkowski Inequality. Secondly, using that $\varphi_F(\widehat{Y})$ is $\sigma(\widehat{Y})$-measurable, one has

$$\left\|\varphi_F(Y) - \mathbb{E}\left(\varphi_F(Y)\,|\,\widehat{Y}\right)\right\|_p \leq \left\|\varphi_F(Y) - \varphi_F(\widehat{Y})\right\|_p + \left\|\varphi_F(\widehat{Y}) - \mathbb{E}\left(\varphi_F(Y)\,|\,\widehat{Y}\right)\right\|_p$$
$$= \left\|\varphi_F(Y) - \varphi_F(\widehat{Y})\right\|_p + \left\|\mathbb{E}\left(\varphi_F(\widehat{Y}) - \varphi_F(Y)\,|\,\widehat{Y}\right)\right\|_p$$
$$\leq 2\left\|\varphi_F(Y) - \varphi_F(\widehat{Y})\right\|_p$$

since $\mathbb{E}\,(\,\cdot\,|\,\widehat{Y})$ is an L^p-contraction when $p \in [1, +\infty)$. ◇

Remark. Markov kernels $\big(P(x, d\xi)\big)_{x\in\mathbb{R}^d}$ which propagate Lipschitz continuity in the sense that

$$[P]_{\mathrm{Lip}} = \sup\left\{[Pf]_{\mathrm{Lip}},\ f : \mathbb{R}^d \to \mathbb{R},\ [f]_{\mathrm{Lip}} \leq 1\right\} < +\infty$$

are especially well-adapted to propagate quantization errors using the above error bounds since it implies that, if F is Lipschitz continuous, then $\varphi_F = Pf$ is Lipschitz continuous too.

► **Exercises. 1.** Detail the proof of the above L^p-error bound when $p \neq 2$. [Hint: show that $\|Z - \mathbb{E}\,(\varphi(Y)|\widehat{Y})\|_p \leq \|Z - \varphi(Y)\|_p + 2\,\|\varphi(Y) - \varphi(\widehat{Y})\|_p$.]

2. Prove that the Euler scheme with step $\frac{T}{n}$ of a diffusion with Lipschitz continuous drift $b(t, x) = b(x)$ and diffusion coefficient $\sigma(t, x) = \sigma(x)$ starting at $x_0 \in \mathbb{R}^d$ at time 0 (see Chap. 7 for a definition) is an $\mathbb{R}^d$-valued homogeneous Markov chain with respect to the filtration of the Brownian increments whose transition propagates Lipschitz continuity.

5.3 How to Get Optimal Quantization?

This phase is often considered as the prominent drawback of optimal quantization-based cubature methods for expectation or conditional expectation computation, at least when compared to the Monte Carlo method. If computing optimal or optimized quantization grids and their weights is less flexible and more time consuming than simulating a random vector, one must keep in mind that such grids can be stored off line forever and made available instantly. This means that optimal quantization is mainly useful when one needs to compute many integrals (or conditional expectations) with respect to the same probability distribution such as the Gaussian distributions.

5.3.1 Dimension 1...

Though originally introduced to design weighted cubature formulas for the computation of integrals with respect to distributions in medium dimensions (from 2 to 10 or 12), the quadrature formulas derived for specific one-dimensional distributions or random variables turn out to be quite useful. This is especially the case when some commonly encountered random variables are not easy to simulate in spite of the existence of closed forms for their density, c.d.f. or cumulative first moment functions. Such is the case for, among others, the (one-sided) Lévy area, the supremum of the Brownian bridge (or Kolmogorov–Smirnov distribution) which will be investigated in exercises further on. The main assets of optimal quantizations grids and their companion weight vectors is threefold:

- Such optimal quantizers are specifically fitted to the random variable they quantize without any auxiliary "transfer" function-like for Gauss–Legendre points which are naturally adapted to uniform, Gauss–Laguerre points adapted to exponential distributions, Gauss–Hermite points adapted to Gaussian distributions.
- The error bounds in the resulting quadrature formulas for numerical integration are tailored for functions with "natural" Lipschitz or $\mathcal{C}^1_{\text{Lip}}$-regularity.
- Finally, in one dimension as will be seen below, fast deterministic algorithms, based on fixed point procedures for contracting functions or Newton–Raphson zero search algorithms, can be implemented, quickly producing optimal quantizers with high accuracy.

Although of little interest for applications (since other deterministic methods like *PDE* approach are available) we propose below for the sake of completeness a Newton–Raphson method to compute optimal quantizers of scalar *unimodal* distributions, *i.e.* absolutely continuous distributions whose density is log-concave. The starting point is the following theorem due to Kieffer [168], which gives the uniqueness of the optimal quantizer in that setting.

Theorem 5.3 **(1*D*-distributions, see [168])** *If $d = 1$ and $\mathbb{P}_X(d\xi) = \varphi(\xi)\,d\xi$ with $\log\varphi$ concave, then, for every level $N \geq 1$, there is exactly one stationary N-quantizer (up to the reordering of its components, i.e. as a grid). This unique stationary quantizer is a global (and local) minimum of the distortion function, i.e.*

$$\forall N \geq 1, \qquad \operatorname{argmin}_{\mathbb{R}^N} \mathcal{Q}_{2,N} = \{x^{(N)}\}.$$

Definition 5.5 *Absolutely continuous distributions on the real line with a* log-concave *density are called* unimodal distributions. *The support (in $\overline{\mathbb{R}}$) of such a distribution and of its density is a (closed) interval $[a, b]$, where $-\infty \leq a \leq b \leq +\infty$.*

Examples of unimodal distributions: The (non-degenerate) Gaussian distributions, the gamma distributions $\gamma(\alpha, \beta)$ with $\alpha \geq 1, \beta > 0$ (including the exponential distributions), the $B(\alpha, \beta)$-distributions with $\alpha, \beta \geq 1$ (including the uniform distribution over the unit interval), etc.

In this one-dimensional setting, a deterministic optimization approach, based on the *Newton–Raphson zero search algorithm* or on the *Lloyd fixed point algorithm*, can be developed. Both algorithms are detailed below (the procedures will be written formally for absolutely continuous distributions on the real line, not only unimodal ones).

$\triangleright$ *Specification of the Voronoi cells.* Let $x = (x_1, \ldots, x_N) \in \mathcal{S}_N^{a,b} := \{\xi \in (a, b)^N : \xi_1 < \xi_2 < \cdots < \xi_N\}$, where $[a, b]$, $a, b \in \overline{\mathbb{R}}$, denotes the convex hull of the support of $\mathbb{P}_X$ in $\overline{\mathbb{R}}$. Then we set

$$C_i(x) = \big[x_{i-\frac{1}{2}}, x_{i+\frac{1}{2}}\big),\ i = 1, \ldots, N-1,\ C_N(x) = [x_{N-\frac{1}{2}}, b]$$

with

$$x_{i+\frac{1}{2}} = \frac{x_{i+1} + x_i}{2},\ i = 1, \ldots, N-1,\ x_{\frac{1}{2}} = a,\ x_{N+\frac{1}{2}} = b$$

and the convention that, if a or b is infinite, they are "removed" from "their" Voronoi cell. Also keep in mind that if the density φ is unimodal, then $\text{supp}(\mathbb{P}_X) = \text{supp}(\varphi) = [a, b]$ (in $\overline{\mathbb{R}}$).

We will now compute the gradient and the Hessian of the quadratic distortion function $\mathcal{Q}_{2,N}$ on $\mathcal{S}_N^{a,b}$.

$\triangleright$ *Computation of the gradient* $\nabla\mathcal{Q}_{2,N}(x)$. Let $x \in \mathcal{S}_N^{a,b}$. Decomposing the quadratic distortion function $\mathcal{Q}_{2,N}(x)$ across the Voronoi cells $C_i(x)$ leads to the following expression for the quadratic distortion:

$$\mathcal{Q}_{2,N}(x) = \sum_{i=1}^{N} \int_{x_{i-\frac{1}{2}}}^{x_{i+\frac{1}{2}}} (x_i - \xi)^2 \varphi(\xi)\,d\xi.$$

If the density function is continuous, elementary differentiation with respect to every variable x_i then yields that $\mathcal{Q}_{2,N}$ is continuously differentiable at x with

$$\nabla \mathcal{Q}_{2,N}(x) := \left(\frac{\partial \mathcal{Q}_{2,N}}{\partial x_i}\right)_{1\le i\le N} = 2\left(\int_{x_{i-\frac{1}{2}}}^{x_{i+\frac{1}{2}}} (x_i - \xi)\,\varphi(\xi)\,d\xi\right)_{1\le i\le N}.$$

If $x = x^{(N)} \in \mathcal{S}_N^{a,b}$ is the optimal quantizer at level N, hence with pairwise distinct components, $\nabla \mathcal{Q}_{2,N}(x) = 0$, *i.e.* x is a zero of the gradient of the distortion function $\mathcal{Q}_{2,N}$.

If we introduce the cumulative distribution function $\Phi(u) = \int_{-\infty}^{u} \varphi(v)dv$ and the *cumulative partial first moment function* $\Psi(u) = \int_{-\infty}^{u} v\,\varphi(v)dv$, the zeros of $\nabla \mathcal{Q}_{2,N}$ are solutions to the non-linear system of equations

$$x_i\big(\Phi(x_{i+\frac{1}{2}}) - \Phi(x_{i-\frac{1}{2}})\big) = \Psi(x_{i+\frac{1}{2}}) - \Psi(x_{i-\frac{1}{2}}),\; i = 1,\dots,N. \tag{5.22}$$

Note that this formula also reads

$$x_i = \frac{\Psi(x_{i+\frac{1}{2}}) - \Psi(x_{i-\frac{1}{2}})}{\Phi(x_{i+\frac{1}{2}}) - \Phi(x_{i-\frac{1}{2}})},\quad i = 1,\dots,N,$$

which is actually a rewriting of the stationarity Eq. (5.6) since we know that $\Phi(x_{i+\frac{1}{2}}) \neq \Phi(x_{i-\frac{1}{2}})$. Hence, Eq. (5.22) and its stationarity version are of course true for any absolutely continuous distributions with density φ (keep in mind that any minimizer of the distortion function has pairwise distinct components). However, when φ is not unimodal, it may happen that $\nabla \mathcal{Q}_{2,N}$ has several zeros, *i.e.* several stationary quantizers, which do not all lie in argmin $\mathcal{Q}_{2,N}$.

These computations are special cases of a multi-dimensional result (see Sect. 6.3.5 of the next chapter devoted to stochastic approximation and optimization) which holds for any distribution $\mathbb{P}_X$ – possibly having singular component – at N-tuples with pairwise distinct components.

⊳ **Example.** Thus, if $X \stackrel{d}{=} \mathcal{N}(0;1)$, the c.d.f. $\Phi = \Phi_0$ is (tabulated and) computable (see Sect. 12.1.2) with high accuracy at a low computational cost whereas the cumulative partial first moment is simply given by

$$\Psi_0(x) = -\frac{e^{-\frac{x^2}{2}}}{\sqrt{2\pi}},\quad x \in \mathbb{R}.$$

⊳ *Computation of the Hessian* $\nabla^2 \mathcal{Q}_{2,N}(x)$. If φ is moreover continuous (at least in the neighborhood of each component x_i of x), the Hessian $\nabla^2 \mathcal{Q}_{2,N}(x)$ can in turn be computed. Note that, if φ is unimodal, φ is continuous on (a,b) but possibly discontinuous at the endpoints a, b (consider the uniform distribution). The Hessian reads at such $x = (x_1, \ldots, x_N)$,

$$\nabla^2 \mathcal{Q}_{2,N}(x) = \left[\frac{\partial^2 \mathcal{Q}_{2,N}}{\partial x_i \partial x_j}(x)\right]_{1 \le i,j \le N},$$

where, for every $i, j \in \{1, \ldots, N\}$,

$$\begin{aligned}
\frac{\partial^2 \mathcal{Q}_{2,N}}{\partial x_i^2}(x) &= 2\left[\Phi\big(x_{i+\frac{1}{2}}\big) - \Phi\big(x_{i-\frac{1}{2}}\big)\right] - \frac{x_{i+1}-x_i}{2}\,\varphi\big(x_{i+\frac{1}{2}}\big) - \frac{x_i - x_{i-1}}{2}\,\varphi\big(x_{i-\frac{1}{2}}\big),\\
\frac{\partial^2 \mathcal{Q}_{2,N}}{\partial x_i \partial x_{i+1}}(x) &= -\frac{x_{i+1} - x_i}{2}\,\varphi\big(x_{i+\frac{1}{2}}\big),\\
\frac{\partial^2 \mathcal{Q}_{2,N}}{\partial x_i \partial x_{i-1}}(x) &= -\frac{x_i - x_{i-1}}{2}\,\varphi\big(x_{i-\frac{1}{2}}\big),\\
\frac{\partial^2 \mathcal{Q}_{2,N}}{\partial x_i \partial x_j}(x) &= 0 \quad \text{otherwise},
\end{aligned}$$

when these partial derivatives make sense, with the convention (only valid in the above formulas) $\varphi\big(x_{\frac{1}{2}}\big) = \varphi\big(x_{N+\frac{1}{2}}\big) = 0$.

The Newton–Raphson zero search procedure

The Newton–Raphson can be viewed as an accelerated gradient descent. For many distributions, such as the strictly log-concave distributions, for example, it allows an almost instant search for the unique optimal N-quantizer with the requested accuracy.

Assume $\mathrm{supp}(\mathbb{P}_X) = [a, b] \cap \mathbb{R}$ (or at least that the convex hull of $\mathrm{supp}(\mathbb{P}_X)$ is equal to $[a, b] \cap \mathbb{R}$.

Let $x^{[0]} \in \mathcal{S}_N^{a,b}$. The zero search Newton–Raphson procedure starting at $x^{[0]}$ is then defined as follows:

$$x^{[n+1]} = x^{[n]} - \big(\nabla^2 \mathcal{Q}_{2,N}(x^{[n]})\big)^{-1}\big(\nabla \mathcal{Q}_{2,N}(x^{[n]})\big), \quad n \in \mathbb{N}. \tag{5.23}$$

▷ *Example of the normal distribution.* Thus, for the normal distribution $\mathcal{N}(0; 1)$, the three functions φ_0, Φ_0 and Ψ_0 are explicit and can be computed at a low computational cost. Thus, for $N = 1, \ldots, 1\,000$, tabulations with a 10^{-14} accuracy of the optimal N-quantizers

$$x^{(N)} = \big(x_1^{(N)}, \ldots, x_N^{(N)}\big)$$

have been computed and can be downloaded at the website www.quantize.maths-fi.com (package due to S. Corlay). Their companion weight parameters are computed as well, with the same accuracy, namely the weights:

$$\mathbb{P}\big(X \in C_i(x^{(N)})\big) = \Phi\big(x_{i+\frac{1}{2}}^{(N)}\big) - \Phi\big(x_{i-\frac{1}{2}}^{(N)}\big), \; i = 1, \ldots N,$$

and the resulting (squared) quadratic quantization error (through its square, the quadratic distortion) given by

$$\big\|X - \widehat{X}^{x^{(N)}}\big\|_2^2 = \mathbb{E}\,X^2 - \mathbb{E}\,\big(\widehat{X}^{x^{(N)}}\big)^2 = \mathbb{E}\,X^2 - \sum_{i=1}^{N} \big(x_i^{(N)}\big)^2 \Big(\Phi_0\big(x_{i+\frac{1}{2}}^{(N)}\big) - \Phi_0\big(x_{i-\frac{1}{2}}^{(N)}\big)\Big)$$

using (5.15). The vector of local inertia $\left(\displaystyle\int_{x_{i-\frac{1}{2}}^{(N)}}^{x_{i+\frac{1}{2}}^{(N)}} (\xi - x_i^{(N)})^2 \varphi(\xi)d\xi\right)_{i=1,\dots,N}$ is also made available.

The Lloyd method (or Lloyd's method I)

The Lloyd method (also known as Lloyd's method I) was first devised in 1957 by S.P. Lloyd but was only published in 1982 in [201]. To our knowledge, it was the first method devoted to the numerical computation of quantizers. It was rediscovered independently by Max in the early 1960s. In another seminal paper [168], Kieffer first establishes (in one dimension) the uniqueness of stationary – hence optimal – N quantizer, when X (is square integrable and) has a log-concave (unimodal) density φ. Then, he shows the convergence at an exponential rate of the Lloyd method when, furthermore, $\log \varphi$ is not piecewise affine.

Let us be more precise. Lloyd's method is essentially a fixed point algorithm based on the fact that the unique stationary quantizer $x = x^{(N)} \in \mathcal{S}_N^{a,b}$ satisfies the stationarity Eq. (5.7) (or (5.22)), which may be re-written as

$$x_i = \Lambda_i(x) := \frac{\Psi(x_{i+\frac{1}{2}}) - \Psi(x_{i-\frac{1}{2}})}{\Phi(x_{i+\frac{1}{2}}) - \Phi(x_{i-\frac{1}{2}})}, \quad i = 1, \dots, N. \tag{5.24}$$

If the density φ is log-concave and not piecewise affine, the function $\Lambda = (\Lambda_i)_{i=1,\dots,N}$ defined by the right-hand side of (5.24) from $\mathcal{S}_N^{a,b}$ onto $\mathcal{S}_N^{a,b}$ – called *Lloyd's map* – is contracting (see [168]) and the Lloyd method is defined as the iterative fixed point procedure based on Λ, namely

$$x^{[n+1]} = \Lambda\big(x^{[n]}\big), \ n \geq 0, \quad x^{[0]} \in \mathcal{S}_N^{a,b}. \tag{5.25}$$

Hence, $(x^{[n]})_{n\geq 0}$ converges exponentially fast toward $x^{(N)}$. When Λ is not contracting, no general convergence results are known, even in case of uniqueness of the optimal N-quantizer.

In terms of implementation, the Lloyd method only involves the c.d.f. Φ and the cumulative partial first moment function Ψ of the random variable X for which closed forms are required. The multi-dimensional version of this algorithm – which does not require such closed forms – is presented in Sect. 6.3.5 of Chap. 6.

When both functions Φ and Ψ admit closed forms (as for the normal distribution, for example), the procedure can be implemented. It is usually slower than the above Newton–Raphson procedure (with which is in fact an accelerated version of the deterministic "mean zero search procedure" associated with the Competitive Learning Vector Quantization algorithm presented in Sect. 6.3.5 of the next chapter). However, when the quantization level N increases the Lloyd method turns out to be more stable than the Newton–Raphson algorithm.

Like all fixed point procedures relying on a contracting function, the Lloyd method can be speeded up by storing and taking advantage of the past iterations of the procedure by an appropriate regression method, known as Anderson's acceleration method. For more details, we refer to [274].

Remark. In practice, one can successfully implement these two deterministic recursive procedures even if the distribution of X is not unimodal (see the exercises below). In particular, the procedures converge when uniqueness of the stationary quantizer holds, which is true beyond the class of unimodal distributions. Examples of uniqueness of optimal N-quantizers for non-unimodal (one-dimensional) distributions can be found in [96] (see Theorem 4): thus, uniqueness holds for the normalized Pareto distributions $\alpha x^{-(\alpha+1)}\mathbf{1}_{[1,+\infty)}$, $\alpha > 0$, or power distributions $\alpha x^{\alpha-1}\mathbf{1}_{(0,1]}$, $\alpha \in (0, 1)$, none of them being unimodal (in fact semi-closed forms are established for such distributions, from which uniqueness is derived).

ℵ Practitioner's corner: Splitting Initialization method for scalar distributions. In view of applications, one is usually interested in building a database of optimal quantizers of a random variable X for all levels between $N = 1$ and a level $N_{\max}$. In such a situation, one may take advantage of the optimal quantizer at level $N - 1$ to initialize any of the above two procedures at level N. The idea is to mimic the proof of the existence of an optimal quantizer (see Theorem 5.1).

General case. Let $N \geq 2$. Having at hand $x^{(N-1)} = \big(x_1^{(N-1)}, \dots, x_{N-1}^{(N-1)}\big)$ and $\bar{x} = \mathbb{E}\,X$. One natural and usually efficient way to initialize the procedure to compute an optimal N-quantizer is to set

$$x^{[0]} = \big(x_1^{(N-1)}, \dots, x_{i_0}^{(N-1)}, \bar{x}, x_{i_0+1}^{(N-1)}, \dots, x_{N-1}^{(N-1)}\big),$$

where $x_{i_0}^{(N-1)} < \bar{x} < x_{i_0+1}^{(N-1)}$.

Symmetric random variables. When X has a symmetric distribution (with a density or at least assigning no mass to 0), the optimal quantizers of $\mathcal{S}_N^{a,b}$ are themselves symmetric, at least under a uniqueness assumption like unimodality. Thus quantizers at odd level N are of the form $(-x^*_{1:(N-1)/2}, 0, x^*_{1:(N-1)/2})$ if N is odd and $\big(-x^*_{1:N/2}, x^*_{1:N/2}\big)$ if N is even.

Then, one checks that, if N is even, $x^*_{1:N/2}$ is obtained as the optimal $N/2$-quantizer of $|X|$ and, if N is odd, $x^*_{1:(N-1)/2}$ is obtained as the optimal $(N-1)/2$-quantizer *with constraint at* 0 solution to the minimization problem of the constrained distortion function (5.5) associated to $|X|$ instead of X.

► **Exercises. 1.** *Quantization of the non-central* $\chi^2(1)$ *distribution.* (*a*) Let $X = Z^2$, $Z \stackrel{d}{=} \mathcal{N}(0;1)$ so that $X \stackrel{d}{=} \chi^2(1)$. Let Φ_X denote the c.d.f. of X and let $\Psi_X(x) = \mathbb{E}\, X\mathbf{1}_{\{X\le x\}}$, $x \in \mathbb{R}_+$, denote its cumulated first moment. Show that, for every $x \in \mathbb{R}_+$,

$$\Phi_X(x) = 2\Phi_0(\sqrt{x}) - 1 \quad \text{and} \quad \Psi_X(x) = \Phi_X(x) - \sqrt{\frac{2x}{\pi}}\, e^{-\frac{x}{2}},$$

where as usual Φ_0 denotes the c.d.f. of the standard normal distribution $\mathcal{N}(0;1)$. Show that its density φ_X is log-convex and given by

$$\varphi_X(x) = \frac{1}{\sqrt{x}}\Phi_0'(x) = \frac{e^{-\frac{x}{2}}}{\sqrt{2\pi x}}, \quad x \in (0, +\infty)$$

so that we have no guarantee that optimal N-quantizers are unique *a priori*.

(*b*) Write and implement the Newton–Raphson zero search algorithm and the fixed point Lloyd method for the standard $\chi^2(1)$-distribution at a level $N \ge 1$. [Hint: for a practical implementation choose the starting value $\big(x_1^{(0)}, \ldots, x_N^{(0)}\big)$ carefully, keeping in mind that the density of X goes to $+\infty$ at 0.]

(*c*) Establish Φ_0-based closed formulas for Φ_X and Ψ_X when $X = (Z+m)^2$, $m \in \mathbb{R}$ (non-central $\chi^2(1)$-distribution).

(*d*) Derive and implement both the Newton–Raphson zero search algorithm and the Lloyd method to compute optimal N-quantizers for non-central $\chi^2(1)$-distributions. Compare.

This exercise is important when trying to quantize the Milstein scheme of a one dimensional diffusion (see Sect. 7.5.1 further on, in particular, the exercise after Theorem 7.5).

2. *Quantization of the* log-*normal distribution.* (*a*) Let $X = e^{\sigma Z + m}$, $\sigma > 0$, $m \in \mathbb{R}$, $Z \stackrel{d}{=} \mathcal{N}(0;1)$. Let Φ_X and Ψ_X denote the c.d.f. of X and its cumulated first moment $\Psi_X(x) = \mathbb{E}\, X\mathbf{1}_{\{X\le x\}}$, respectively. Show that, for every $x \in (0, +\infty)$,

$$\Phi_X(x) = \Phi_0\left(\frac{\log(x) - m}{\sigma}\right) \quad \text{and} \quad \Psi_X(x) = e^{m + \frac{\sigma^2}{2}}\, \Phi_0\left(\frac{\log(x) - m - \sigma^2}{\sigma}\right)$$

where Φ_0 still denotes the c.d.f. of the standard normal distribution $\mathcal{N}(0;1)$.

(*b*) Derive and implement both the Newton–Raphson zero search algorithm and Lloyd's method I for log-normal distributions. Compare.

3. *Quantization by Fourier.* Let X be a *symmetric* random variable on $(\Omega, \mathcal{A}, \mathbb{P})$ with a characteristic function $\chi(u) = \mathbb{E}\, e^{\tilde{\imath} uX}$, $u \in \mathbb{R}$, $\tilde{\imath}^2 = -1$. We assume that $\chi \in L^1(\mathbb{R}, du)$.

(*a*) Show that $\mathbb{P}_X$ is absolutely continuous with an even continuous density φ defined by $\varphi(0) = \dfrac{1}{\pi}\displaystyle\int_0^{+\infty} \chi(u)\, du$ and, for every $x \in (0, +\infty)$,

$$\varphi(x) = \frac{1}{2\pi}\int_{\mathbb{R}} e^{\tilde{\imath}xu}\chi(u)du = \frac{1}{\pi x}\int_0^{+\infty}\cos(u)\,\chi\Big(\frac{u}{x}\Big)du$$
$$= \frac{1}{\pi x}\sum_{k\geq 0}(-1)^k\int_0^{\pi}\cos(u)\,\chi\Big(\frac{u+k\pi}{x}\Big)du.$$

Furthermore, show that χ is real-valued, even and always (strictly) positive. [Hint: See an elementary course on Fourier transform and/or characteristic functions, *e.g.* [44, 52, 155, 263] among (many) others.]

(b) Show that the c.d.f. Φ_X of X is given by $\Phi_X(0) = \frac{1}{2}$ and, for every $x \in (0, +\infty)$,

$$\Phi_X(x) = \frac{1}{2} + \frac{1}{\pi}\int_0^{+\infty}\frac{\sin u}{u}\,\chi\Big(\frac{u}{x}\Big)du, \quad \Phi_X(-x) = 1 - \Phi_X(x). \tag{5.26}$$

(c) Assume furthermore that $X \in L^1(\mathbb{P})$. Prove that its cumulated partial first moment function Ψ_X is negative, even and given by $\Psi_X(0) = -C$ and, for every $x \in (0, +\infty)$,

$$\Psi_X(x) = -C + x\left(\Phi(x) - \frac{1}{2}\right) - \frac{x}{\pi}\int_0^{+\infty}\frac{1-\cos u}{u^2}\chi\Big(\frac{u}{x}\Big)du, \quad \Psi_X(-x) = \Psi_X(x), \tag{5.27}$$

where $C = \mathbb{E}\,X_+$. [Hint: Use in both cases Fubini's theorem and integration(s) by parts.]

(d) Show that $\Phi_X(x)$ can be written on $(0, +\infty)$ as an alternating series reading

$$\Phi_X(x) = \frac{1}{2} + \frac{1}{\pi}\sum_{k\geq 0}(-1)^k\int_0^{\pi}\frac{\sin u}{u+k\pi}\,\chi\Big(\frac{u+k\pi}{x}\Big)du, \quad x \in (0, +\infty).$$

Show likewise that $\Psi_X(x)$ also reads for every $x \in (0, +\infty)$

$$\Psi_X(x) = -C + x\left(\Phi_X(x) - \frac{1}{2}\right) - \frac{x}{\pi}\sum_{k\geq 0}\int_0^{\pi}\frac{1-(-1)^k\cos u}{(u+k\pi)^2}\chi\Big(\frac{u+k\pi}{x}\Big)du.$$

(e) Propose two methods to compute a (small) database of optimal quantizers $x^{(N)} = \big(x_1^{(N)}, \ldots, x_N^{(N)}\big)$ and their weight vectors $\big(p_1^{(N)}, \ldots, p_N^{(N)}\big)$ for a symmetric random variable X satisfying the above conditions, say for levels running from $N = 1$ up to $N_{\max} = 50$. Compare their respective efficiency. [Hint: Prior to the implementation have a look at the (recursive) *splitting initialization method for scalar distributions* described in the above Practitioner's corner, see also Sect. 6.3.5.]

4. *Quantized one-sided Lévy's area.* Let $W = (W^1, W^2)$ be a 2-dimensional standard Wiener process and let

$$X = \int_0^1 W_s^1 dW_s^2$$

denote the Lévy area associated to (W^1, W^2) at time 1. We admit that the characteristic function of X reads

$$\chi(u) = \mathbb{E}\, e^{\tilde{\imath} uX} = \frac{1}{\sqrt{\cosh u}}, \quad u \in \mathbb{R},$$

(see Formula (9.105) in Chap. 9 further on, applied here with $\mu = 0$, and Sect. 12.11 of the Miscellany Chapter for a proof).

(a) Show that

$$C := \mathbb{E}\, X_+ = \frac{1}{\sqrt{2\pi}}\mathbb{E}\, \|B\|_{L^2([0,1],dt)},$$

where $B = (B_t)_{t\in[0,1]}$ denotes a standard Brownian motion.

(b) Establish the elementary identity

$$\mathbb{E}\, \|B\|_{L^2([0,1],dt)} = \frac{1}{4} + \mathbb{E}\left(\|B\|_{L^2([0,1],dt)} - \frac{1}{2}\|B\|^2_{L^2([0,1],dt)}\right)$$

and justify why $\mathbb{E}\, \|B\|_{L^2([0,1],dt)}$ should be computed by a Monte Carlo simulation using this identity. [Hint: An appropriate Monte Carlo simulation should yield a result close to 0.2485, but this approximation is not accurate enough to compute optimal quantizers ([4].]

(c) Describe in detail a method (or possibly two methods) for computing a small database of N-quantizers of the Lévy area for levels running from $N = 1$ to $N_{\max} = 50$, including, for every level $N \geq 1$, both their weights and their induced quadratic mean quantization error. [Hint: Use (5.15) to compute $\|X - \widehat{X}^\Gamma\|_2^2$ when Γ is a stationary quantizer.]

5. *Clark–Cameron oscillator*. Using the identity (9.105) in its full generality, extend the quantization procedure of Exercise **4.** to the case where

$$X = \int_0^1 (W_s^1 + \mu s)\, dW_s^2,$$

with μ a fixed real constant.

6. *Supremum of the Brownian bridge*. Let

$$X = \sup_{t\in[0,1]} \left|W_t - t\, W_1\right|$$

denote the supremum of the standard Brownian bridge (see Chap. 8 for more details, see also Sect. 4.3 for the connection with uniformly distributed sequences and

---

[4]A more precise approximation is $C = 0.24852267852801818 \pm 2.033\ 10^{-7}$ obtained by implementing an *ML2R* estimator with a target *RMSE* $\varepsilon = 3.0\,10^{-7}$, see Chap. 9.

discrepancy). This distribution, also known as the Kolmogorov–Smirnov (K–S) distribution since it is the limiting distribution emerging from the non-parametric eponymous goodness-of-fit statistical test, is characterized by its survival function given (see (4.8)) by

$$\bar{\Phi}_X(x) = 2\sum_{k\geq 1}(-1)^{k-1}e^{-2k^2x^2}. \tag{5.28}$$

(a) Show that the cumulative partial first moment function Ψ_X of X is given for every $x \geq 0$ by

$$\Psi_X(x) = \mathbb{E}\, X\mathbf{1}_{\{X\leq x\}} = \int_0^x \bar{\Phi}_X(u)\,du - x\,\bar{\Phi}_X(x)$$

and deduce that

$$\forall x \geq 0,\quad \Psi_X(x) := \sqrt{2\pi}\sum_{k\geq 1}\frac{(-1)^{k-1}}{k}\left(\Phi_0(2kx) - \frac{1}{2}\right) - x\,\bar{\Phi}_X(x),$$

where Φ_0 denotes the c.d.f. of the normal distribution $\mathcal{N}(0;1)$.

(b) Compute a small database of N-quantizers, say $N = 1$ up to 50, of the K–S distribution (including the weights and the quantization error at each level), based on the fixed point Lloyd method. [Hint: Recall the *splitting method*.]

(c) Show that the K–S distribution is absolutely continuous with a density φ_X given by

$$\varphi_X(x) = 8x\sum_{k\geq 1}(-1)^{k-1}k^2e^{-2k^2x^2},\ x \geq 0$$

and implement a second method to compute the same small database. Compare their respective efficiencies.

5.3.2 *The Case of the Normal Distribution $\mathcal{N}(0; I_d)$ on $\mathbb{R}^d$, $d \geq 2$*

As soon as $d \geq 2$, most procedures to optimize the quantization error are stochastic and based on some nearest neighbor search. Let us cite:

– the randomized *Lloyd's method I* or randomized Lloyd's algorithm

and

– the Competitive Learning Vector Quantization algorithm (*CLVQ*).

The first is a randomized version of the d-dimensional version of the fixed point procedure described above in the one dimensional setting. The second is a recursive stochastic gradient approximation procedure.

These two stochastic optimization procedures are presented in more detail in Sect. 6.3.5 of Chap. 6 devoted to Stochastic Approximation.

For $\mathcal{N}(0; I_d)$, a large scale optimization has been carried out (with the support of ACI Fin'Quant) based on a mixed *CLVQ*-Lloyd's procedure. To be precise, grids have been computed for $d = 1$ up to 10 and $N = 1$ up to 5 000. Their companion parameters have also been computed (still by simulation): weight, L^1 quantization error, (squared) L^2-distortion, local L^1 and L^2-pseudo-inertia of each Voronoi cell. These enriched grids are available for downloading on the website www.quantize.maths-fi.com which also contains many papers dealing with quantization optimization.

Recent implementations of exact or approximate fast nearest neighbor search procedures has dramatically reduced the computation time in higher dimensions; not to speak of implementation on *GPU*. For further details on the theoretical aspects we refer to [224] for *CLVQ* (mainly devoted to compactly supported distributions).

As for Lloyd's method I, a huge literature is available (often under the name of *k-means algorithm*). Beyond the seminal – but purely one-dimensional – paper by J.C. Kieffer [168], let us cite [79, 80, 85, 238]. For more numerical experiments with the Gaussian distributions, we refer to [231] and for more general aspects in connection with classification and various other applications, we refer to [106].

For illustrations depicting optimal quadratic N-quantizers of the bi-variate normal distribution $\mathcal{N}(0; I_2)$, we refer to Fig. 5.1 (for $N = 200$), Fig. 6.3 (for $N = 500$, with its Voronoi diagram) and Fig. 5.2 (for the same quantizer colored with a coded representation of the weight of the cells).

5.3.3 Other Multivariate Distributions

Algorithms such as *CLVQ* and the randomized Lloyd's procedure developed to quantize the multivariate normal $\mathcal{N}(0; I_d)$ distributions in an efficient and systematic way can be successfully implemented for other multivariate distributions. Sect. 6.3.5 in the next chapter is entirely devoted to these stochastic optimization procedures, with some emphasis on their practical implementation as well as techniques to speed them up.

However, to anticipate their efficiency, we propose in Fig. 5.4 a quantization of size $N = 500$ of the joint law $(W_1, \sup_{t\in[0,1]} W_t)$ of a standard Brownian motion W.

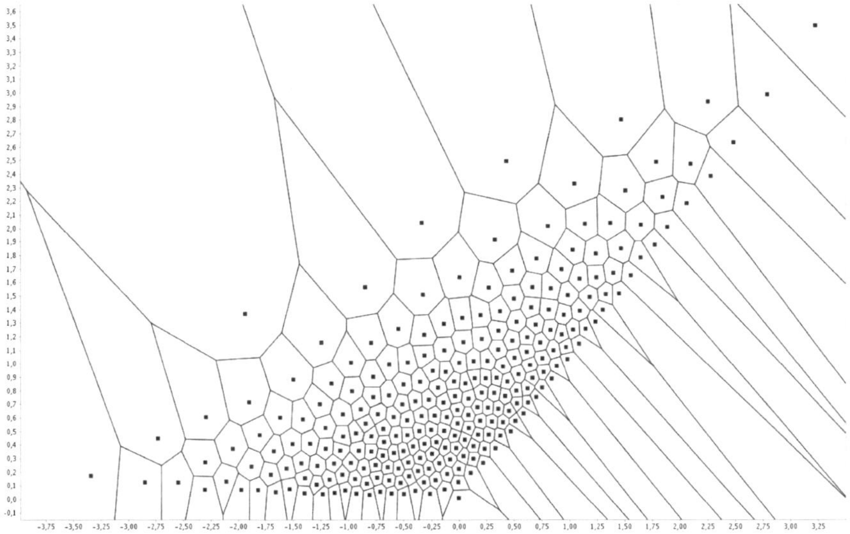

Fig. 5.4 Optimal N-quantization ($N = 500$) of $\big(W_1, \sup_{t\in[0,1]} W_t\big)$ depicted with its Voronoi tessellation, W standard Brownian motion (with B. Wilbertz)

5.4 Numerical Integration (II): Quantization-Based Richardson–Romberg Extrapolation

The challenge is to fight against the curse of dimensionality to increase the critical dimension beyond which the theoretical rate of convergence of the Monte Carlo method outperforms that of optimal quantization. Combining the above cubature formula (5.8), (5.13) and the rate of convergence of the (optimal) quantization error, it seems natural to deduce that the critical dimension to use quantization-based cubature formulas is $d = 4$ (when dealing with continuously differentiable functions), at least when compared to Monte Carlo simulation. Several tests have been carried out and reported in [229, 231] to refine this *a priori* theoretical bound. The benchmark was made of several options on a geometric index on d independent assets in a Black–Scholes model: Puts, Puts spread and the same in a smoothed version, always without any control variate. Of course, having no correlation assets is not a realistic assumption but it is clearly more challenging as far as numerical integration is concerned. Once the dimension d and the number of points N have been chosen, we compared the resulting integration error with a one standard deviation confidence interval of the corresponding Monte Carlo estimator for the same number of integration points N. The last standard deviation is computed thanks to a Monte Carlo simulation carried out using 10^4 trials.

The results turned out to be more favorable to quantization than predicted by theoretical bounds, mainly because we carried out our tests with rather small values of N, whereas the curse of dimensionality is an asymptotic bound. Up to dimension 4, the larger N is, the more quantization outperforms Monte Carlo simulation. When the dimension $d \geq 5$, quantization always outperforms Monte Carlo (in the above sense) until a critical size $N_c(d)$ which decreases as d increases.

In this section, we provide a method to push these critical sizes forward, at least for sufficiently smooth functions. Let $F : \mathbb{R}^d \to \mathbb{R}$ be a twice differentiable function with Lipschitz continuous Hessian D^2F. Let $(\widehat{X}^{(N)})_{N\geq 1}$ be a sequence of optimal quadratic quantizations. Then

$$\begin{aligned}\mathbb{E}\, F(X) = \mathbb{E}\, F(\widehat{X}^{(N)}) + \frac{1}{2}\mathbb{E}\left(D^2F(\widehat{X}^{(N)}) \cdot (X - \widehat{X}^{(N)})^{\otimes 2}\right) \\ + O\left(\mathbb{E}\,|X - \widehat{X}|^3\right).\end{aligned} \tag{5.29}$$

Under some assumptions which are satisfied by most usual distributions (including the normal distribution), it is proved in [131] as a special case of a more general result that

$$\mathbb{E}\,|X - \widehat{X}|^3 = O(N^{-\frac{3}{d}})$$

or at least (in particular, when $d = 2$) $\mathbb{E}\,|X - \widehat{X}|^3 = O(N^{-\frac{3-\varepsilon}{d}})$, $\varepsilon > 0$. If, furthermore, we conjecture the existence of a real constant $c_{F,X} \in \mathbb{R}$ such that

$$\mathbb{E}\left(D^2F(\widehat{X}^{(N)}) \cdot (X - \widehat{X}^{(N)})^{\otimes 2}\right) = \frac{c_{F,X}}{N^{\frac{2}{d}}} + o\left(N^{-\frac{3}{d}}\right), \tag{5.30}$$

one can use a Richardson–Romberg extrapolation to compute $\mathbb{E}\, F(X)$.

Quantization-based Richardson–Romberg extrapolation. We consider two sizes N_1 and N_2 (in practice one often sets $N_1 = N/2$ and $N_2 = N$ with N even). Then combining (5.29) with N_1 and N_2, we cancel the first order error term and obtain

$$\mathbb{E}\, F(X) = \frac{N_2^{\frac{2}{d}}\,\mathbb{E}\, F(\widehat{X}^{(N_2)}) - N_1^{\frac{2}{d}}\,\mathbb{E}\, F(\widehat{X}^{(N_1)})}{N_2^{\frac{2}{d}} - N_1^{\frac{2}{d}}} + O\left(\frac{1}{(N_1 \wedge N_2)^{\frac{1}{d}}(N_2^{\frac{2}{d}} - N_1^{\frac{2}{d}})}\right).$$

Numerical illustration. In order to see the effect of the extrapolation technique described above, numerical computations have been carried out with regularized versions of some Put Spread options on geometric indices in dimension $d = 4,\ 6,\ 8,\ 10$. By "regularized", we mean that the payoff at maturity T has been replaced by its price function at time $T' < T$ (with $T' \simeq T$). Numerical integration was performed using the Gaussian optimal grids of size $N = 2^k$, $k = 2, \ldots, 12$ (available at the website www.quantize.maths-fi.com).

We consider again one of the test functions implemented in [231] (p. 152). These test functions were borrowed from classical option pricing in mathematical finance, namely a *Put Spread option* (on a geometric index, which is less classical). Moreover; we will use a "regularized" version of the payoff. One considers d independent traded assets $S^1, \ldots, S^d$ following a d-dimensional Black and Scholes dynamics (under its risk neutral probability)

$$S_t^i = s_0^i \exp\left(\left(r - \frac{\sigma_i^2}{2}\right)t + \sigma_i \sqrt{t}\, Z^{i,t}\right), \quad i = 1, \dots, d,$$

where $Z^{i,t} = W_t^i$, $W = (W^1, \dots, W^d)$ is a d-dimensional standard Brownian motion. Independence is unrealistic but corresponds to the most unfavorable case for numerical experiments. We also assume that $s_0^i = s_0 > 0$, $i = 1, \dots, d$, and that the d assets share the same volatility $\sigma_i = \sigma > 0$. One considers the geometric index $I_t = \left(S_t^1 \dots S_t^d\right)^{\frac{1}{d}}$. One shows that $e^{-\frac{\sigma^2}{2}(\frac{1}{d}-1)t} I_t$ itself has a risk neutral Black–Scholes dynamics. We want to test the *regularized Put Spread option on this geometric index* with strikes $K_1 < K_2$ (at time $T/2$). Let $\psi(s_0, K_1, K_2, r, \sigma, T)$ denote the premium at time 0 of a Put Spread on any of the assets S^i. We have

$$\begin{aligned}
\psi(x, K_1, K_2, r, \sigma, T) &= \pi(x, K_2, r, \sigma, T) - \pi(x, K_1, r, \sigma, T),\\
\pi(x, K, r, \sigma, T) &= K e^{-rt} \Phi_0(-d_2) - x\, \Phi_0(-d_1),\\
d_1 &= \frac{\log(x/K) + (r + \frac{\sigma^2}{2d})T}{\sigma\sqrt{T/d}}, \quad d_2 = d_1 - \sigma\sqrt{T/d}.
\end{aligned}$$

Using the martingale property of the discounted value of the premium of a European option yields that the premium $e^{-rt}\mathbb{E}\left((K_1 - I_T)_+ - (K_2 - I_T)_+\right)$ of the Put Spread option on the index I satisfies on the one hand

$$e^{-rt}\mathbb{E}\left((K_1 - I_T)_+ - (K_2 - I_T)_+\right) = \psi\left(s_0 e^{\frac{\sigma^2}{2}(\frac{1}{d}-1)T}, K_1, K_2, r, \sigma/\sqrt{d}, T\right)$$

and, one the other hand,

$$e^{-rt}\mathbb{E}\left((K_1 - I_T)_+ - (K_2 - I_T)_+\right) = \mathbb{E}\, g(Z),$$

where

$$g(Z) = e^{-rT/2}\psi\left(e^{\frac{\sigma^2}{2}(\frac{1}{d}-1)\frac{T}{2}} I_{\frac{T}{2}}, K_1, K_2, r, \sigma, T/2\right)$$

and $Z = (Z^{1,\frac{T}{2}}, \dots, Z^{d,\frac{T}{2}}) \stackrel{d}{=} \mathcal{N}(0; I_d)$. The numerical specifications of the function g are as follows:

$$s_0 = 100, \qquad K_1 = 98, \qquad K_2 = 102, \qquad r = 5\%, \qquad \sigma = 20\%, \qquad T = 2.$$

The results are displayed in Fig. 5.5 in a log-log-scale for the dimensions $d = 4, 6, 8, 10$.

First, we recover theoretical rates (namely $-2/d$) of convergence for the error bounds. Indeed, some slopes $\beta(d)$ can be derived (using a regression) for the quantization errors and we found $\beta(4) = -0.48$, $\beta(6) = -0.33$, $\beta(8) = -0.25$ and $\beta(10) = -0.23$ for $d = 10$ (see Fig. 5.5). These rates plead for the implementation of the Richardson–Romberg extrapolation. Also note that, as already reported in [231],

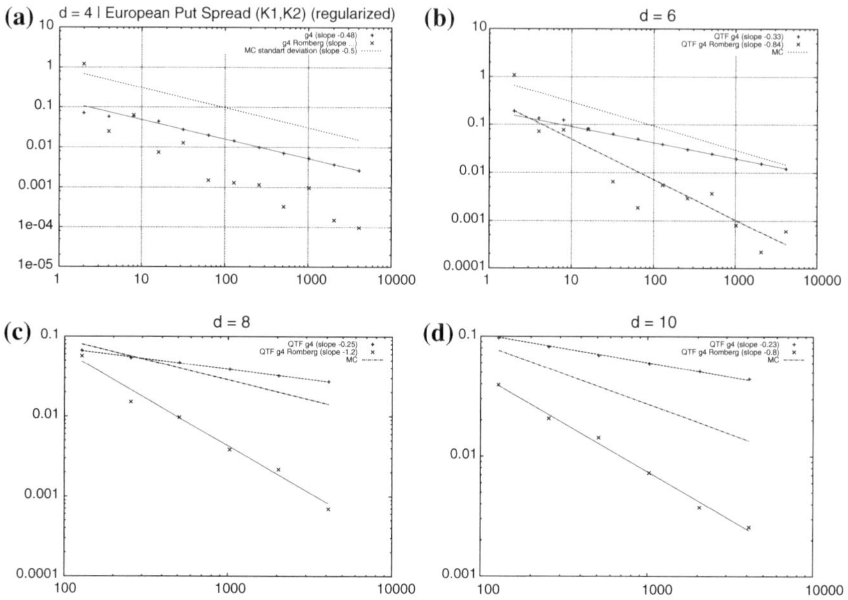

Fig. 5.5 Errors and standard deviations as functions of the number of points N in a log-log-scale. The quantization error is displayed by the cross $+$ and the Richardson–Romberg extrapolation error by the cross $\times$. The dashed line without crosses denotes the standard deviation of the Monte Carlo estimator. **a** d $= 4$, **b** d $= 6$, **c** d $= 8$, **d** d $= 10$ (with J. Printems)

when $d \geq 5$, quantization still outperforms MC simulations (in the above sense) up to a critical number $N_c(d)$ of points ($N_c(6) \sim 5000$, $N_c(7) \sim 1000$, $N_c(8) \sim 500$, etc).

As concerns the Richardson–Romberg extrapolation method itself, note first that it always gives better results than "crude" quantization. As regards now the comparison with Monte Carlo simulation, no critical number of points $N_{Romb}(d)$ comes out beyond which MC simulation outperforms Richardson–Romberg extrapolation. This means that $N_{Romb}(d)$ is greater than the range of use of quantization-based cubature formulas in our benchmark, namely 5 000.

Richardson–Romberg extrapolation techniques are sometimes considered a little unstable, and indeed, it has not always been possible to satisfactorily estimate its rate of convergence on our benchmark. But when a significant slope (in a log-log scale) can be estimated from the Richardson–Romberg errors (like for $d = 8$ and $d = 10$ in Fig. 5.5c, d), its absolute value is larger than $1/2$, and so, these *extrapolations always outperform the MC method, even for large values of* N. As a by-product, our results plead in favour of the conjecture (5.30) and lead us to think that Richardson–Romberg extrapolation is a powerful tool to accelerate numerical integration by optimal quantization, even in higher dimensions.

5.5 Hybrid Quantization-Monte Carlo Methods

In this section we explore two aspects of variance reduction by quantization. First we show how to use optimal quantization as a control variate, then we present a stratified sampling method relying on a quantization-based stratification. This second method can be seen as a guided Monte Carlo method or a hybrid Quantization-Monte Carlo method. This method was originally introduced in [231, 233] to deal with Lipschitz continuous functionals of Brownian motion. We also refer to [192] for other quantization-based variance reduction method on the Wiener space. Here, we only consider a finite-dimensional setting.

5.5.1 *Optimal Quantization as a Control Variate*

Let $X \in L^2_{\mathbb{R}^d}(\Omega, \mathcal{A}, \mathbb{P})$ take at least $N \geq 1$ values with positive probability. We assume that we have access to an N-quantizer $x := (x_1, \dots, x_N) \in (\mathbb{R}^d)^N$ with pairwise distinct components and we denote by Proj_x one of its Borel nearest neighbor projections. Let $\widehat{X}^x = \mathrm{Proj}_x(X)$. We assume that we now have the distribution of $\widehat{X}^x$ characterized by the N-tuple $(x_1, \dots, x_N)$ itself and their *weights* (probability distribution),

$$p_i := \mathbb{P}(\widehat{X}^x = x_i) = \mathbb{P}\big(X \in C_i(x)\big), \quad i = 1, \dots, N,$$

where $C_i(x) = \mathrm{Proj}_x^{-1}(\{x_i\})$, $i = 1, \dots, N$, denotes the Voronoi tessellation of the N-quantizer induced by the above nearest neighbor projection. By "know" we mean that we have access to accurate enough numerical values of the x_i and their companion weights p_i.

Let $F : \mathbb{R}^d \to \mathbb{R}^d$ be a Lipschitz continuous function such that $F(X) \in L^2(\mathbb{P})$. In order to compute $\mathbb{E}\, F(X)$, one writes for every simulation size $M \geq 1$,

$$\begin{aligned} \mathbb{E}\, F(X) &= \mathbb{E}\, F(\widehat{X}^x) + \mathbb{E}\,\big(F(X) - F(\widehat{X}^x)\big) \\ &= \underbrace{\mathbb{E}\, F(\widehat{X}^x)}_{(a)} + \underbrace{\frac{1}{M}\sum_{m=1}^{M} F\big(X^{(m)}\big) - F\big(\widehat{X^{(m)}}^x\big)}_{(b)} + R_{N,M}, \end{aligned} \tag{5.31}$$

where $X^{(m)}$, $m = 1, \dots, M$, are M independent copies of X, $\widehat{X^{(m)}}^x = \mathrm{Proj}_x(X^{(m)})$ and $R_{N,M}$ is a remainder term defined by (5.31). Term (a) can be computed by quantization and term (b) can be computed by a Monte Carlo simulation. Then, it is clear that

$$\|R_{N,M}\|_2 = \frac{\sigma\big(F(X) - F(\widehat{X}^x)\big)}{\sqrt{M}} \leq \frac{\|F(X) - F(\widehat{X}^x)\|_2}{\sqrt{M}} \leq [F]_{\mathrm{Lip}} \frac{\|X - \widehat{X}^x\|_2}{\sqrt{M}}$$

as $M \to +\infty$, where $\sigma(Y)$ denotes the standard deviation of a random variable Y. Furthermore,

$$\sqrt{M}\, R_{N,M} \xrightarrow{\mathcal{L}} \mathcal{N}\big(0; \mathrm{Var}(F(X) - F(\widehat{X}^x))\big) \quad \text{as} \quad M \to +\infty.$$

Consequently, if F is simply a Lipschitz continuous function and if $(\widehat{X}^{x^{(N)}})_{N\geq 1}$ is a sequence of *optimal quadratic quantizations of* X, then

$$\|R_{N,M}\|_2 \leq \frac{\|F(X) - F\big(\widehat{X}^{x^{(N)}}\big)\|_2}{\sqrt{M}} \leq C_{2,\delta}[F]_{\mathrm{Lip}} \frac{\sigma_{2+\delta}(X)}{\sqrt{M}N^{\frac{1}{d}}}, \tag{5.32}$$

where $C_{2,\delta}$ denotes the constant coming from the non-asymptotic version of Zador's Theorem (see Theorem 5.2(b)).

ℵ Practitioner's corner

As concerns practical implementation of this quantization-based variance reduction method, the main gap is the *nearest neighbor search* needed at each step to compute $\widehat{X^{(m)}}^x$ from $X^{(m)}$.

In one dimension, an (optimal) N-quantizer is usually directly obtained as a monotonic N-tuple with non-decreasing components and the complexity of a nearest number search on the real line based on a dichotomy procedure is approximately $\frac{\log N}{\log 2}$. Unfortunately, this one-dimensional setting is of little interest for applications.

In d dimensions, there exist nearest neighbor search procedures with a $O(\log N)$-complexity, once the N-quantizer has been given an appropriate tree structure (which costs $O(N \log N)$). The most popular tree-based procedure for nearest neighbor search is undoubtedly the K-d tree (see [99]). During the last ten years, several attempts to improve it have been carried out, among them, we may mention the Principal Axis Tree algorithm (see [207]). These procedures are efficient for quantizers with a large size N lying in a vector space with medium dimension (say up to 10).

An alternative to speed up the nearest neighbor search procedure is to restrict to product quantizers whose Voronoi cells are hyper-parallelepipeds. In such a case, the nearest neighbor search reduces to those on the d marginals with an approximate resulting complexity of $d\,\frac{\log N}{\log 2}$.

However, this nearest neighbor search procedure undoubtedly slows down the global procedure.

5.5.2 *Universal Stratified Sampling*

The main drawback of the preceding is the repeated use of nearest neighbor search procedures. Using a quantization-based stratification may be a way to take advantage of quantization to reduce the variance without having to implement such time consuming procedures. On the other hand, one important drawback of the regular stratification method as described in Sect. 3.5 is to depend on the function F, at least when concerned by the optimal choice for the allocation parameters q_i. Our aim is to show that quantization-based stratification has a uniform efficiency among the class of Lipschitz continuous functions. The first step is the universal stratified sampling for Lipschitz continuous functions detailed in the simple proposition below, where we use the notations introduced in Sect. 3.5. Also keep in mind that for a random vector $Y \in L^2_{\mathbb{R}^d}(\Omega, \mathcal{A}, \mathbb{P})$, $\|Y\|_2 = \left(\mathbb{E}\,|Y|^2\right)^{1/2}$ where $|\cdot|$ denotes the canonical Euclidean norm.

Proposition 5.4 (Universal stratification) *Let $X \in L^2_{\mathbb{R}^d}(\Omega, \mathcal{A}, \mathbb{P})$ and let $(A_i)_{i\in I}$ be a stratification of $\mathbb{R}^d$. For every $i \in I$, we define the local inertia of the random vector X on the stratum A_i by*

$$\sigma_i^2 = \mathbb{E}\left(|X - \mathbb{E}\,(X\,|X \in A_i)|^2\,|\,X \in A_i\right).$$

(a) Then, for every Lipschitz continuous function $F : (\mathbb{R}^d, |\cdot|) \to (\mathbb{R}^d, |\cdot|)$,

$$\forall\, i \in I, \qquad \sup_{[F]_{\mathrm{Lip}} \le 1} \sigma_{F,i} = \sigma_i, \tag{5.33}$$

where $\sigma_{F,i}$ is non-negative and defined by

$$\sigma_{F,i}^2 = \min_{a\in\mathbb{R}^d} \mathbb{E}\left(|F(X) - a|^2\,|\,X \in A_i\right) = \mathbb{E}\Big(|F(X) - \mathbb{E}\left(X\,|\,X \in A_i\right)|^2\,|\,X \in A_i\Big).$$

(b) Suboptimal choice: $q_i = p_i$.

$$\sup_{[F]_{\mathrm{Lip}} \le 1} \left(\sum_{i\in I} p_i \sigma_{F,i}^2\right) = \sum_{i\in I} p_i \sigma_i^2 = \left\|X - \mathbb{E}\left(X\,|\sigma(\{X \in A_i\},\ i \in I)\right)\right\|_2^2. \tag{5.34}$$

(c) Optimal choice of the q_i. *(see (3.10) for a closed form of the q_i)*

$$\begin{aligned} \sup_{[F]_{\mathrm{Lip}} \le 1} \left(\sum_{i\in I} p_i \sigma_{F,i}\right)^2 &= \left(\sum_{i\in I} p_i \sigma_i\right)^2 \\ &= \left\|X - \mathbb{E}\left(X\,|\sigma(\{X \in A_i\},\ i \in I)\right)\right\|_1^2. \end{aligned} \tag{5.35}$$

Remark. Any real-valued Lipschitz continuous function can be canonically seen as an $\mathbb{R}^d$-valued Lipchitz function, but then the above equalities (5.33)–(5.35) only hold as inequalities.

Proof. (a) Note that

$$\begin{aligned} \sigma_{F,i}^2 &= \mathrm{Var}\,(F(X)\,|\,X \in A_i) = \mathbb{E}\left((F(X) - \mathbb{E}\,(F(X)|X \in A_i))^2|\,X \in A_i\right) \\ &\le \mathbb{E}\left((F(X) - F(\mathbb{E}\,(X\,|X \in A_i)))^2|X \in A_i\right) \end{aligned}$$

owing to the very definition of conditional expectation as a minimizer with respect to the conditional distribution. Now using that F is Lipschitz continuous, it follows that

$$\sigma^2_{F,i} \le [F]^2_{\text{Lip}} \frac{1}{p_i} \mathbb{E}\left(\mathbf{1}_{\{X \in A_i\}} |X - \mathbb{E}(X\,|X \in A_i)|^2\right) = [F]^2_{\text{Lip}} \sigma^2_i.$$

The equalities in (b) and (c) straightforwardly follow from (a). Finally, it follows from Jensen's Inequality (or the monotonicity of conditional L^p-norms) that

$$\begin{aligned}\sum_{i=1}^N p_i \sigma_i &= \sum_{i=1}^N p_i \big[\mathbb{E}\left(|X - \mathbb{E}(X\,|X \in A_i)|^2 |X \in A_i\right)\big]^{\frac{1}{2}} \\ &\ge \big\|X - \mathbb{E}(X\,|\sigma(\{X \in A_i\}, i \in I))\big\|_1 .\end{aligned}$$

◇

5.5.3 A(n Optimal) Quantization-Based Universal Stratification: A Minimax Approach

Let $X \in L^2_{\mathbb{R}^d}(\Omega, \mathcal{A}, \mathbb{P})$ take at least N values with positive probability. The starting idea is to use the Voronoi diagram of an N-quantizer $x = (x_1, \dots, x_N)$ of X such that $\mathbb{P}(X \in C_i(x)) > 0$, $i = 1, \dots, N$, to design the strata in a stratification procedure. Firstly, this amounts to setting $I = \{1, \dots, N\}$ and

$$A_i = C_i(x), \;\; i \in I,$$

in the preceding. Then, for every $i \in \{1, \dots, N\}$, there exists a Borel function $\varphi(x_i, .) : [0, 1]^q \to \mathbb{R}^d$ such that

$$\varphi(x_i, U) \stackrel{d}{=} \mathcal{L}(X \mid \widehat{X}^x = x_i) = \frac{\mathbf{1}_{C_i(x)} \mathbb{P}_X(d\xi)}{\mathbb{P}\big(X \in C_i(x)\big)},$$

where $U \stackrel{d}{=} \mathcal{U}([0, 1]^r)$. Note that the dimension r is arbitrary: one may always assume that $r = 1$ by the fundamental theorem of simulation, but in order to obtain some closed forms for $\varphi(x_i, .)$, we are led to consider situations where $r \ge 2$ (or even infinite when considering a Von Neumann acceptance-rejection method).

Now let (ξ, U) be a pair of independent random vectors such that $\xi \stackrel{d}{=} \widehat{X}^x$ and $U \stackrel{d}{=} \mathcal{U}([0, 1]^r)$. Then, one checks that

$$\varphi(\xi, U) \stackrel{d}{=} X$$

so that one may assume without loss of generality that

$$X = \varphi(\widehat{X}^x, U), \quad U \stackrel{d}{=} \mathcal{U}\big([0, 1]^r\big), \quad \text{where } U \text{ and } \widehat{X}^x \text{are independent.}$$

In terms of implementation, as mentioned above we need a closed formula for the function φ which induces some stringent constraints on the choice of the N-quantizers. In particular, there is no reasonable hope to consider true optimal quadratic quantizers for that purpose. A reasonable compromise is to consider some optimal *product* quantization for which the function φ can easily be made explicit (see Sect. 3.5).

Let $\mathcal{A}_d(N)$ denote the family of all Borel partitions of $\mathbb{R}^d$ having at most N elements.

Proposition 5.5 (a) Suboptimal stratification *($p_i = q_i$). One has*

$$\inf_{(A_i)_{1\le i\le N}\in\mathcal{A}_d(N)}\ \sup_{[F]_{\rm Lip}\le 1}\left(\sum_{i\in I} p_i\sigma^2_{F,i}\right) = \min_{x\in(\mathbb{R}^d)^N}\left\|X-\widehat{X}^x\right\|_2^2.$$

(b) Optimal stratification. *One has*

$$\inf_{(A_i)_{1\le i\le N}\in\mathcal{A}_d(N)}\ \sup_{[F]_{\rm Lip}\le 1}\left(\sum_{i\in I} p_i\sigma_{F,i}\right) \ge \min_{x\in(\mathbb{R}^d)^N}\left\|X-\widehat{X}^x\right\|_1.$$

Proof. Let $A_i = C_i(x)$, $i = 1,\dots,N$. Then (5.34) and (5.35) rewritten in terms of quantization read

$$\sup_{[F]_{\rm Lip}\le 1}\left(\sum_{i\in I} p_i\sigma^2_{F,i}\right) \le \sum_{i\in I} p_i\sigma_i^2 = \left\|X-\mathbb{E}\,(X\,|\widehat{X}^x)\right\|_2^2 \tag{5.36}$$

and

$$\sup_{[F]_{\rm Lip}\le 1}\left(\sum_{i\in I} p_i\sigma_{F,i}\right)^2 \le \left(\sum_{i\in I} p_i\sigma_i\right)^2 \ge \left\|X-\mathbb{E}\,(X\,|\widehat{X}^x)\right\|_1^2 \tag{5.37}$$

where we used the obvious fact that $\sigma(\{X\in C_i(x)\}, i\in I) = \sigma(\widehat{X}^x)$.

(a) It follows from (5.34) that

$$\inf_{(A_i)_{1\le i\le N}}\left(\sum_{i\in I} p_i\sigma_i^2\right) = \inf_{(A_i)_{1\le i\le N}}\left\|X-\mathbb{E}\left(X\,|\sigma(\{X\in A_i\}, i\in I)\right)\right\|_2.$$

Now, it follows from Proposition 5.1(b) that, if $x^{(N)}$ denotes an optimal quantizer at level N for X (which has N pairwise distinct components),

$$\left\|X-\widehat{X}^{x^{(N)}}\right\|_2 = \min\left\{\left\|X-Y\right\|_2,\ Y:(\Omega,\mathcal{A},\mathbb{P})\to\mathbb{R}^d,\ |Y(\Omega)|\le N\right\}$$

and
$$\widehat{X}^{x^{(N)}} = \mathbb{E}\Big(X \,|\, \widehat{X}^{x^{(N)}}\Big).$$

Consequently, (5.36) completes the proof.

(b) It follows from (5.35) that

$$\sum_{i\in I} p_i\sigma_i \ge \big\|X - \mathbb{E}\,(X\,|\sigma(\{X\in A_i\}, i\in I)\big\|_1 \ge \big\|\mathrm{dist}\big(X, \Gamma_A\big)\big\|_1,$$

where $\Gamma_A := \big\{\mathbb{E}\,\big(X\,|\sigma(\{X\in A_i\}, i\in I)\big)(\omega),\ \omega\in\Omega\big\}$ has at most N elements. Now $\big\|\mathrm{dist}(X,\Gamma_A)\big\|_1 = \mathbb{E}\,\mathrm{dist}\big(X,\Gamma_A\big) = \big\|X - \widehat{X}^{x(A)}\big\|_1$, consequently,

$$\sum_{i\in I} p_i\sigma_i \ge \|X - \widehat{X}^{\Gamma_A}\|_1 = \min_{x\in(\mathbb{R}^d)^N} \big\|X - \widehat{X}^x\big\|_1. \qquad \diamond$$

As a conclusion, we see that the notion of universal stratification (with respect to Lipschitz continuous functions) and quantization are closely related since the variance reduction factor that can be obtained by such an approach is essentially ruled by the optimal quantization rate of the random vector X, that is $c_X N^{-\frac{1}{d}}$, according to Zador's Theorem (see Theorem 5.2).

One dimension

In this case the method applies straightforwardly provided both the distribution function $F_X(u) := \mathbb{P}(X \le u)$ of X (on $\mathbb{R}$) and its right continuous (canonical) inverse on $[0, 1]$, denoted by F_X^{-1} are computable.

We also need the additional assumption that the N-quantizer $x = (x_1, \ldots, x_N)$ satisfies the following continuity assumption

$$\mathbb{P}(X = x_i) = 0, \qquad i = 1, \ldots, N.$$

Note that this is always the case if X has a density. Then set

$$x_{i+\frac{1}{2}} = \frac{x_i + x_{i+1}}{2},\ i = 1, \ldots, N-1,\ x_{\frac{1}{2}} = -\infty,\ x_{N+\frac{1}{2}} = +\infty.$$

Then elementary computations show that, with $q = 1$,

$$\forall\, u\in[0,1],\quad \varphi_N(x_i, u) = F_X^{-1}\Big(F_X(x_{i-\frac{1}{2}}) + \big(F_X(x_{i+\frac{1}{2}}) - F_X(x_{i-\frac{1}{2}})\big)u\Big),\quad i = 1, \ldots, N. \tag{5.38}$$

Higher dimensions

We consider a random vector $X = (X^1, \ldots, X^d)$ whose marginals X^i are independent. This may appear as a rather stringent restriction in full generality although it is often possible to "extract" in a model an innovation with this correlation structure. At least in a Gaussian framework, such a reduction is always possible after an orthogonal diagonalization of its covariance matrix. One considers a *product quantizer* (see *e.g.* [225, 226]) defined as follows: for every $\ell \in \{1, \ldots, d\}$, let $x^{(N_\ell)} = (x_1^{(N_\ell)}, \ldots, x_{N_\ell}^{(N_\ell)})$ be an N_ℓ-quantizer of the marginal X^ℓ and set $N := N_1 \times \cdots \times N_d$. Then, define for every multi-index $\underline{i} := (i_1, \ldots, i_d) \in I := \prod_{\ell=1}^d \{1, \ldots, N_\ell\}$,

$$x_{\underline{i}} = \big(x_{i_1}^{(N_1)}, \ldots, x_{i_d}^{(N_d)}\big).$$

Then, one defines $\varphi_{N,X}(x_{\underline{i}}, u)$ by setting $q = d$ and

$$\varphi_{N,X}\big(x_{\underline{i}}, (u^1, \ldots, u^d)\big) = \Big(\varphi_{N,X_\ell}(x_{i_\ell}^{(N_\ell)}, u^\ell)\Big)_{1 \le \ell \le d}$$

where φ_{N,X_ℓ} is defined by (5.38).

For various numerical experiments, we refer to [70] in finite dimensions, or [192, 233] for an implementation on the Wiener space.

Chapter 6
Stochastic Approximation with Applications to Finance

6.1 Motivation

In Finance, one often faces some optimization problems or zero search problems. The former often reduce to the latter since, at least in a convex framework, minimizing a function amounts to finding a zero of its gradient. The most commonly encountered examples are the extraction of implicit parameters (implicit volatility of an option, implicit correlations for a single best-of-option or in the credit markets), the calibration, the optimization of an exogenous parameters for variance reduction (regression, importance sampling, etc). All these situations share a common feature: the involved functions all have a representation as an expectation, namely they read $h(y) = \mathbb{E}\, H(y, Z)$ where Z is a q-dimensional random vector. The aim of this chapter is to provide a toolbox – stochastic approximation – based on simulation to solve these optimization or zero search problems. It can be viewed as a non-linear extension of the Monte Carlo method.

Stochastic approximation can also be presented as a probabilistic extension of Newton–Raphson like zero search recursive procedures of the form

$$\forall\, n \geq 0, \qquad y_{n+1} = y_n - \gamma_{n+1} h(y_n) \qquad (0 < \gamma_n \leq \gamma_0), \tag{6.1}$$

where $h : \mathbb{R}^d \to \mathbb{R}^d$ is a continuous vector field satisfying a sub-linear growth assumption at infinity. Under some appropriate *mean-reverting* assumptions, one shows that such a procedure is bounded and eventually converges to a zero y_* of h. As an example, if one sets $\gamma_n = (J_h(y_{n-1}))^{-1}$ – where $J_h(y)$ denotes the Jacobian of h at y – the above recursion is just the regular Newton–Raphson procedure for zero search of the function h (one can also set $\gamma_n = \gamma\big(J_h(y_{n-1})\big)^{-1}$, $\gamma > 0$).

In one dimension, *mean-reversion* may be obtained by an increasing assumption made on the function h or, more simply, by assuming that $h(y)(y - y_*) > 0$ for every $y \neq y_*$: if this is the case, y_n is decreasing as long as $y_n > y_*$ and increasing whenever $y_n < y_*$. In higher dimensions, this assumption becomes $(h(y)\,|\,y - y_*) > 0$, $y \neq y_*$, and will be extensively called upon later.

G. Pagès, *Numerical Probability*, Universitext,
https://doi.org/10.1007/978-3-319-90276-0_6

More generally mean-reversion may follow from a monotonicity assumption on h in one (or higher) dimension and more generally follows from the existence of a so-called *Lyapunov function*. To introduce this notion, let us make a (light) connection with Ordinary Differential Equations (*ODEs*) and differential dynamical systems: thus, when $\gamma_n = \gamma > 0$, Eq. (6.1) is but the Euler scheme with step $\gamma > 0$ of the *ODE*

$$ODE_h \equiv \dot{y} = -h(y).$$

A Lyapunov function for ODE_h is a function $L : \mathbb{R}^d \to \mathbb{R}_+$ such that any solution $t \mapsto y(t)$ of the equation satisfies $t \mapsto L(y(t))$ is non-increasing as t increases. If L is differentiable this is equivalent to the condition $(\nabla L|h) \geq 0$ since

$$\frac{d}{dt}L(y(t)) = \big(\nabla L(y(t))|\dot{y}(t)\big) = -(\nabla L|h)(y(t)).$$

If such a Lyapunov function does exist (which is not always the case!), the system is said to be *dissipative*.

We essentially have two frameworks:

– the function L is identified *a priori*, it is the object of interest for optimization purposes, and one designs a function h *from* L, *e.g.* by setting $h = \nabla L$ (or possibly h positively proportional to ∇L, *i.e.* $h = \rho\nabla L$, where ρ is – at least – an everywhere positive function).

– The function naturally entering into the problem is h and one has to search for a Lyapunov function L (which may not exist). This usually requires a deep understanding of the problem from a dynamical point of view.

This duality also occurs in discrete time Stochastic Approximation Theory from its very beginning in the early 1950s (see [169, 253]).

As concerns the constraints on the Lyapunov function, due to the specificities of the discrete time setting, we will require some further regularity assumptions on ∇L, typically ∇L is Lipschitz continuous and $|\nabla L|^2 \leq C(1 + L)$ (essentially a quadratic growth property).

▶ **Exercises. 1.** Show that if a function $h : \mathbb{R}^d \to \mathbb{R}^d$ is non-decreasing in the following sense

$$\forall\, x,\ y \in \mathbb{R}^d, \qquad (h(y) - h(x)|y - x) \geq 0$$

and if $h(y_*) = 0$ then $L(y) = |y - y_*|^2$ is a Lyapunov function for ODE_h.

2. Assume furthermore that $\{h = 0\} = \{y_*\}$ and that h satisfies a sub-linear growth assumption: $|h(y)| \leq C(1 + |y|)$, $y \in \mathbb{R}^d$. Show that the sequence $(y_n)_{n \geq 0}$ defined by (6.1) converges toward y_*.

Now imagine that no straightforward access to numerical values of $h(y)$ is available but that h has an integral representation with respect to an $\mathbb{R}^q$-valued random vector Z, say

$$h(y) = \mathbb{E}\, H(y, Z), \quad H : \mathbb{R}^d \times \mathbb{R}^q \xrightarrow{Borel} \mathbb{R}^d, \quad Z \stackrel{d}{=} \mu, \tag{6.2}$$

satisfying $\mathbb{E}\,|H(y, Z)| < +\infty$ for every $y \in \mathbb{R}^d$. Assume that

– $H(y, z)$ is easy to compute for any pair (y, z),

– the distribution μ of Z can be simulated at a reasonable cost.

A first idea is to simply "randomize" the above zero search procedure (6.1) by using at each iterate a Monte Carlo simulation to approximate $h(y_n)$.

A more sophisticated idea is to try to do both simultaneously by using, on the one hand, $H(y_n, Z_{n+1})$ instead of $h(y_n)$ and, on the other hand, by letting the step γ_n go to 0 to asymptotically smoothen the chaotic effect induced by this "local" randomization. However, one should not allow γ_n to go to 0 too fast, so that an averaging effect occurs like in the Monte Carlo method. In fact, one should impose that $\sum_n \gamma_n = +\infty$, if only to ensure that the initial value Y_0 will be "forgotten" by the procedure.

Based on this heuristic analysis, we can reasonably hope that the $\mathbb{R}^d$-valued recursive procedure

$$\forall n \geq 0, \quad Y_{n+1} = Y_n - \gamma_{n+1} H(Y_n, Z_{n+1}), \tag{6.3}$$

where

$$(Z_n)_{n\geq 1} \text{ is an i.i.d. sequence with distribution } \mu \text{ defined on } (\Omega, \mathcal{A}, \mathbb{P})$$

and Y_0, defined on the same probability space, is independent of the sequence $(Z_n)_{n\geq 1}$, also converges to a zero y_* of h, under appropriate assumptions on both H and the gain sequence $\gamma = (\gamma_n)_{n\geq 1}$. We will call such a recursive procedure a *Markovian stochastic algorithm* or, more simply, a stochastic algorithm. There are more general classes of recursive stochastic algorithms but this framework is sufficient for our purpose.

The preceding can be seen as the "meta-theorem" of Stochastic Approximation since a large part of this theory is focused on making this algorithm converge toward its "target" (a zero of h). In this framework, the Lyapunov functions mentioned above are called upon to ensure the stability of the procedure.

$\triangleright$ *A toy-example: the Strong Law of Large Numbers.* As a first example note that the sequence of empirical means $(\overline{Z}_n)_{n\geq 1}$ of an i.i.d. sequence (Z_n) of integrable random variables satisfies

$$\bar{Z}_{n+1} = \bar{Z}_n - \frac{1}{n+1}(\bar{Z}_n - Z_{n+1}), \quad n \geq 0, \ \bar{Z}_0 = 0,$$

i.e. a stochastic approximation procedure of type (6.3) with $H(y, Z) = y - Z$ and $h(y) := y - z_*$ with $z_* = \mathbb{E}\, Z$ (so that $Y_n = \bar{Z}_n$). Then the procedure converges *a.s.* (and in L^1) to the unique zero z_* of h.

The (weak) rate of convergence of $(\bar{Z}_n)_{n\geq 1}$ is ruled by the *CLT* which may suggest that the generic rate of convergence of this kind of procedure is of the same type. Note that the deterministic counterpart of (6.3) with the same gain parameter, $y_{n+1} =$

$y_n - \frac{1}{n+1}(y_n - z_*)$, converges at a $\frac{1}{n}$-rate to z_* (and this is clearly not the optimal choice for γ_n in this deterministic framework).

However, if we do not know the value of the mean $z_* = \mathbb{E}\,Z$ but if we are able to simulate μ-distributed random vectors, the first recursive stochastic procedure can be easily implemented whereas the deterministic one cannot. The stochastic procedure we are speaking about is simply the regular Monte Carlo method!

▷ *A toy-model: extracting implicit volatility in a Black–Scholes model.* A second toy-example is the extraction of implicit volatility in a Black–Scholes model for a vanilla Call or Put option. In practice it is carried out by a deterministic Newton procedure (see *e.g.* [209]) since closed forms are available for both the premium and the *vega* of the option. But let us forget about that for a while to illustrate the basic principle of Stochastic Approximation. Let $x,\ K,\ T \in (0, +\infty)$, let $r \in \mathbb{R}$ and set for every $\sigma \in \mathbb{R}$,

$$X_t^{x,\sigma} = xe^{(r-\frac{\sigma^2}{2})t+\sigma W_t}, \quad t \geq 0.$$

We know that $\sigma \mapsto Put_{BS}(x, K, \sigma, r, T) = e^{-rt}\mathbb{E}\,(K - X_T^{x,\sigma})_+$ is an even function, increasing on $(0, +\infty)$, continuous with $\lim_{\sigma\to 0} Put_{BS}(x, K, \sigma, r, T) = (e^{-rt}K - x)_+$ and $\lim_{\sigma\to+\infty} Put_{BS}(x, K, \sigma, r, T) = e^{-rt}K$ (these bounds are model-free and can be directly derived by arbitrage arguments). Let $P_{market}(x, K, r, T) \in [(e^{-rt}K - x)_+, e^{-rt}K]$ be a consistent mark-to-market price for the Put option with maturity T and strike price K. Then the implied volatility $\sigma_{\text{impl}} := \sigma_{\text{impl}}(x, K, r, T)$ is defined as the unique positive solution to the equation

$$Put_{BS}(x, K, \sigma_{\text{impl}}, r, T) = P_{market}(x, K, r, T)$$

or, equivalently,

$$\mathbb{E}\left(e^{-rt}(K - X_T^{x,\sigma})_+ - P_{market}(x, K, r, T)\right) = 0.$$

This naturally suggests to devise the following stochastic algorithm to solve this equation numerically:

$$\sigma_{n+1} = \sigma_n - \gamma_{n+1}\left(\left(K - xe^{(r-\frac{\sigma_n^2}{2})T+\sigma_n\sqrt{T}Z_{n+1}}\right)_+ - P_{market}(x, K, r, T)\right)$$

where $(Z_n)_{n\geq 1}$ is an i.i.d. sequence of $\mathcal{N}(0; 1)$-distributed random variables and the step sequence $\gamma = (\gamma_n)_{n\geq 1}$ is for example given by $\gamma_n = \frac{c}{n}$ for some parameter $c > 0$. After a necessary "tuning" of the constant c (try $c = \frac{2}{x+K}$), one observes that

$$\sigma_n \longrightarrow \sigma_{\text{impl}} \quad a.s. \quad \text{as} \quad n \to +\infty.$$

▶ **Exercise.** Try it!

A priori, one might imagine that σ_n could converge toward $-\sigma_{\text{impl}}$ (which would not be a real problem) but it *a.s.* never happens because this negative solution is repulsive for the related ODE_h and "noisy". This is an important topic often referred to in the literature as "how stochastic algorithms never fall into noisy traps" (see [51, 191, 242], for example).

To conclude this introductory section, let us briefly return to the case where $h = \nabla L$ and $L(y) = \mathbb{E}\,\Lambda(y, Z)$ so that $\nabla L(y) = \mathbb{E}\,H(y, Z)$ with $H(y, z) := \frac{\partial \Lambda(y,z)}{\partial y}$. The function H is sometimes called a *local gradient* of L and the procedure (6.3) is known as a *stochastic gradient* procedure. When Y_n converges to some zero y_* of $h = \nabla L$ at which the algorithm is "noisy enough" – say *e.g.* $\mathbb{E}\,(H(y_*, Z)\,H(y_*, Z)^*) > 0$ is a definite symmetric matrix – then y_* is necessarily a *local minimum* of the potential L: y_* cannot be a *trap*. So, if L is strictly convex and $\lim_{|y|\to+\infty} L(y) = +\infty$, ∇L has a single zero y_* which is simply the global minimum of L: the *stochastic gradient* turns out to be a minimization procedure.

However, most recursive stochastic algorithms (6.3) are not stochastic gradients and the Lyapunov function, if it exists, is not naturally associated to the algorithm: finding a Lyapunov function to "stabilize" the algorithm (by bounding *a.s.* its paths, see the Robbins–Siegmund Lemma below) is often a difficult task which requires a deep understanding of the related ODE_h.

As concerns the rate of convergence, one must keep in mind that it is usually ruled by a *CLT* at a $1/\sqrt{\gamma_n}$-rate which can attain at most the $\sqrt{n}$-rate of the regular *CLT*. So, such a "toolbox" is clearly not competitive compared to a deterministic procedure, when available, but this rate should be compare to that of the Monte Carlo method (*i.e.* the *SLLN*) since their fields of application are similar: stochastic approximation is the natural extension of the Monte Carlo method to solve inverse or optimization problems related to functions having a representation as an expectation of simulable random functions.

Recently, several contributions (see [12, 196, 199]) have drawn the attention of the quants world to stochastic approximation as a tool for variance reduction, *implicitation* of parameters, model calibration, risk management…It is also used in other fields of finance like algorithmic trading as an *on-line* optimizing device for execution of orders (see *e.g.* [188, 189]). We will briefly discuss several (toy-)examples of application.

6.2 Typical *a.s.* Convergence Results

Stochastic Approximation theory provides various theorems which guarantee the *a.s.* and/or L^p-convergence of recursive stochastic approximation algorithms as defined by (6.3). We provide below a general (multi-dimensional) preliminary result known as the *Robbins–Siegmund Lemma* from which the main convergence results will be easily derived. However, though, strictly speaking, this result does not provide any direct *a.s.* convergence result.

In what follows the function H and the sequence $(Z_n)_{n\geq 1}$ are defined by (6.2) and h is the vector field from $\mathbb{R}^d$ to $\mathbb{R}^d$ defined by $h(y) = \mathbb{E}\, H(y, Z_1)$.

Theorem 6.1 (Robbins–Siegmund Lemma) *Let $h : \mathbb{R}^d \to \mathbb{R}^d$ and let $H : \mathbb{R}^d \times \mathbb{R}^q \to \mathbb{R}^d$ satisfy (6.2). Suppose that there exists a continuously differentiable function $L : \mathbb{R}^d \to \mathbb{R}_+$ satisfying*

$$\nabla L \text{ is Lipschitz continuous and } \quad |\nabla L|^2 \leq C(1+L) \tag{6.4}$$

for some real constant $C > 0$ such that h satisfies the mean-reverting assumption

$$(\nabla L|h) \geq 0. \tag{6.5}$$

Furthermore, suppose that H satisfies the following sub-linear growth assumption

$$\forall\, y \in \mathbb{R}^d, \quad \|H(y, Z)\|_2 \leq C\sqrt{1+L(y)} \tag{6.6}$$

(which implies $|h| \leq C\sqrt{1+L}$).

Let $\gamma = (\gamma_n)_{n\geq 1}$ be a sequence of positive real numbers satisfying the (so-called) decreasing step assumption

$$\sum_{n\geq 1} \gamma_n = +\infty \quad \text{and} \quad \sum_{n\geq 1} \gamma_n^2 < +\infty. \tag{6.7}$$

Finally, assume that Y_0 is independent of $(Z_n)_{n\geq 1}$ and $\mathbb{E}\, L(Y_0) < +\infty$.

Then, the stochastic algorithm defined by (6.3) satisfies the following five properties:

(i) $Y_n - Y_{n-1} \longrightarrow 0$ $\mathbb{P}$-a.s. and in $L^2(\mathbb{P})$ as $n \to +\infty$ (and $\sum_{n\geq 1} |\Delta Y_n|^2 < +\infty$ a.s.),

(ii) the sequence $(L(Y_n))_{n\geq 0}$ is $L^1(\mathbb{P})$-bounded,

(iii) $L(Y_n) \overset{a.s.}{\longrightarrow} L_\infty \in L^1(\mathbb{P})$ as $n \to +\infty$,

(iv) $\sum_{n\geq 1} \gamma_n (\nabla L|h)(Y_{n-1}) < +\infty$ a.s. as an integrable random variable,

(v) the sequence $M_n^\gamma = \sum_{k=1}^n \gamma_k \big(H(Y_{k-1}, Z_k) - h(Y_{k-1})\big)$ is an L^2-bounded square integrable martingale, and hence converges in L^2 and a.s.

Remarks and terminology. • The sequence $(\gamma_n)_{n\geq 1}$ is called a *step sequence* or a *gain parameter sequence*.

• If the function L satisfies Assumptions (6.4), (6.5), (6.6) and moreover $\lim_{|y|\to+\infty} L(y) = +\infty$, then L is called a *Lyapunov function* of the system like in Ordinary Differential Equation Theory.

• Note that Assumption (6.4) on L implies that $\nabla\sqrt{1+L}$ is bounded. Hence $\sqrt{L}$ has at most a linear growth so that L itself has at most a quadratic growth. This justifies the somewhat unexpected terminology "sub-linear growth" for Assumption (6.6).

• In spite of the standard terminology, the step sequence does not need to be decreasing in Assumption (6.7).

• A careful reading of the proof below shows that the assumption $\sum_{n\geq 1}\gamma_n = +\infty$ is not needed. However we leave it in the statement because it is dramatically useful for *any application* of this Lemma since it implies, combined with (iv), that

$$\underline{\lim}_n (\nabla L|h)(Y_{n-1}) = 0.$$

These assumptions are known as "Robbins–Siegmund assumptions".

• When $H(y,z) := h(y)$ (*i.e.* the procedure is noiseless), the above theorem provides a convergence result for the original deterministic procedure (6.1).

The key of the proof is the following quite classical convergence theorem for non-negative super-martingales (see [217]).

Theorem 6.2 *Let $(S_n)_{n\geq 0}$ be a non-negative super-martingale with respect to a filtration $(\mathcal{F}_n)_{n\geq 0}$, on a probability space $(\Omega, \mathcal{A}, \mathbb{P})$ (i.e. for every $n \geq 0$, $S_n \in L^1(\mathbb{P})$ and $\mathbb{E}\,(S_{n+1}|\mathcal{F}_n) \leq S_n$ a.s.), then, S_n converges $\mathbb{P}$-a.s. to an integrable (non-negative) random variable S_∞.*

For general convergence theorems for sub-, super- and true martingales we refer to any standard course on Probability Theory or, preferably, to [217].

Proof of Theorem 6.1. Set $\mathcal{F}_n := \sigma(Y_0, Z_1, \ldots, Z_n)$, $n \geq 1$, and for notational convenience $\Delta Y_n := Y_n - Y_{n-1}$, $n \geq 1$. It follows from the fundamental theorem of Calculus that there exists $\xi_{n+1} \in (Y_n, Y_{n+1})$ (geometric interval) such that

$$\begin{aligned}
L(Y_{n+1}) &= L(Y_n) + (\nabla L(\xi_{n+1})|\Delta Y_{n+1}) \\
&\leq L(Y_n) + (\nabla L(Y_n)|\Delta Y_{n+1}) + [\nabla L]_{\text{Lip}}|\Delta Y_{n+1}|^2 \\
&= L(Y_n) - \gamma_{n+1}(\nabla L(Y_n)|H(Y_n, Z_{n+1})) \qquad (6.8)\\
&\quad +[\nabla L]_{\text{Lip}}\gamma_{n+1}^2|H(Y_n, Z_{n+1})|^2 \\
&= L(Y_n) - \gamma_{n+1}(\nabla L(Y_n)|h(Y_n)) - \gamma_{n+1}(\nabla L(Y_n)|\Delta M_{n+1}) \qquad (6.9)\\
&\quad +[\nabla L]_{\text{Lip}}\gamma_{n+1}^2|H(Y_n, Z_{n+1})|^2,
\end{aligned}$$

where

$$\Delta M_{n+1} = H(Y_n, Z_{n+1}) - h(Y_n).$$

We aim at showing that $(\Delta M_n)_{n\geq 1}$ is a sequence of (square integrable) $(\mathcal{F}_n)$-martingale increments satisfying $\mathbb{E}\,(|\Delta M_{n+1}|^2 \mid \mathcal{F}_n) \leq \widetilde{C}(1 + L(Y_n))$ for an appropriate real constant $\widetilde{C} > 0$.

First, it is clear by induction that Y_n is $\mathcal{F}_n$-measurable for every $n \geq 0$ and so is ΔM_n owing to its very definition.

As a second step, note that $L(Y_n) \in L^1(\mathbb{P})$ and $H(Y_n, Z_{n+1}) \in L^2(\mathbb{P})$ for every index $n \geq 0$. This follows again by an induction based on (6.9): using that $|(a|b)| \leq \frac{1}{2}(|a|^2 + |b|^2)$, $a, b \in \mathbb{R}^d$, we first get

$$\mathbb{E}\,\big|\big(\nabla L(Y_n)|H(Y_n, Z_{n+1})\big)\big| \leq \frac{1}{2}\big(\mathbb{E}\,|\nabla L(Y_n)|^2 + \mathbb{E}\,|H(Y_n, Z_{n+1})|^2\big).$$

Now, Y_n being $\mathcal{F}_n$-measurable and Z_{n+1} being independent of $\mathcal{F}_n$,

$$\mathbb{E}\,\big(|H(Y_n, Z_{n+1})|^2 \,|\, \mathcal{F}_n\big) = \big[\mathbb{E}\,|H(y, Z_1)|^2\big]_{|y=Y_n}$$

so that

$$\begin{aligned}\mathbb{E}\,|H(Y_n, Z_{n+1})|^2 &= \mathbb{E}\big[\mathbb{E}\,\big(|H(Y_n, Z_{n+1})|^2 \,|\, \mathcal{F}_n\big)\big]\\ &= \mathbb{E}\Big[\big[\mathbb{E}\,|H(y, Z_1)|^2\big]_{|y=Y_n}\Big]\\ &\leq C^2\big(1 + \mathbb{E}\,L(Y_n)\big)\end{aligned}$$

owing to (6.6). Combined with (6.4) and plugged into the above inequality, this yields

$$\mathbb{E}\,\big|\big(\nabla L(Y_n)|H(Y_n, Z_{n+1})\big)\big| \leq C^2\big(1 + \mathbb{E}\,L(Y_n)\big).$$

By the same argument, we get

$$\mathbb{E}\,\big(H(Y_n, Z_{n+1}) \,|\, \mathcal{F}_n\big) = \big[\mathbb{E}\,\big(H(y, Z_1)\big)\big]_{|y=Y_n} = h(Y_n).$$

Plugging these bounds into (6.9), we derive that $L(Y_{n+1}) \in L^1(\mathbb{P})$. Consequently $\mathbb{E}\,(\Delta M_{n+1} \,|\, \mathcal{F}_n) = 0$. The announced inequality for $\mathbb{E}\,(|\Delta M_{n+1}|^2 \,|\, \mathcal{F}_n)$ holds with $\widetilde{C} = 2\,C^2$ owing to (6.6) and the inequality

$$|\Delta M_{n+1}|^2 \leq 2(|H(Y_n, Z_{n+1})|^2 + |h(Y_n)|^2).$$

At this stage, we derive from the fact that $\nabla L(Y_n)$ and $\Delta M_{n+1} \in L^2(\mathbb{P})$,

$$\mathbb{E}\,\big((\nabla L(Y_n)|\Delta M_{n+1}) \,|\, \mathcal{F}_n\big) = \big(\nabla L(Y_n) \,|\, \mathbb{E}\,(\Delta M_{n+1} \,|\, \mathcal{F}_n)\big) = 0.$$

Conditioning (6.9) with respect to $\mathcal{F}_n$ reads

$$\begin{aligned}\mathbb{E}\,\big(L(Y_{n+1}) \,|\, \mathcal{F}_n\big) + \gamma_{n+1}(\nabla L|h)(Y_n) &\leq L(Y_n) + C_{_L}\gamma_{n+1}^2\big(1 + L(Y_n)\big)\\ &\leq L(Y_n)(1 + C_{_L}\gamma_{n+1}^2) + C_{_L}\gamma_{n+1}^2,\end{aligned}$$

where $C_{_L} = C^2[\nabla L]_{\text{Lip}} > 0$. Then, adding the positive term

$$\sum_{k=1}^{n} \gamma_k (\nabla L|h)(Y_{k-1}) + C_{_L} \sum_{k\geq n+2} \gamma_k^2$$

on the left-hand side of the above inequality, adding $(1 + C_{_L}\gamma_{n+1}^2)$ times this term on the right-hand side and dividing the resulting inequality by $\prod_{k=1}^{n+1}(1 + C_{_L}\gamma_k^2)$ shows that the $\mathcal{F}_n$-adapted sequence

$$S_n = \frac{L(Y_n) + \sum_{k=0}^{n-1} \gamma_{k+1}(\nabla L|h)(Y_k) + C_{_L} \sum_{k\geq n+1} \gamma_k^2}{\prod_{k=1}^{n}(1 + C_{_L}\gamma_k^2)}, \quad n \geq 0,$$

is a (non-negative) super-martingale with $S_0 = L(Y_0) \in L^1(\mathbb{P})$. The fact that the added term is positive follows from the mean-reverting inequality $(\nabla L|h) \geq 0$. Hence $(S_n)_{n\geq 0}$ is $\mathbb{P}$-*a.s.* convergent toward a non-negative integrable random variable $S_{_\infty}$ by Theorem 6.2. Consequently, using that $\sum_{k\geq n+1} \gamma_k^2 \to 0$ as $n \to +\infty$, one gets

$$L(Y_n) + \sum_{k=0}^{n-1} \gamma_{k+1}(\nabla L|h)(Y_k) \xrightarrow{a.s.} \widetilde{S}_{_\infty} = S_{_\infty} \prod_{n\geq 1}(1 + C_{_L}\gamma_n^2) \in L^1(\mathbb{P}). \tag{6.10}$$

(*ii*) The super-martingale $(S_n)_{n\geq 0}$ being $L^1(\mathbb{P})$-bounded by $\mathbb{E}\, S_0 = \mathbb{E}\, L(Y_0) < +\infty$, one derives likewise that $(L(Y_n))_{n\geq 0}$ is L^1-bounded since

$$L(Y_n) \leq \left(\prod_{k=1}^{n}(1 + C_{_L}\gamma_k^2)\right) S_n, \quad n \geq 0,$$

and $\prod_{k\geq 1}(1 + C_{_L}\gamma_k^2) < +\infty$ owing to the decreasing step assumption (6.7) made on $(\gamma_n)_{n\geq 1}$.

(*iv*) Now, for the same reason, the series $\sum_{0\leq k\leq n-1} \gamma_{k+1}(\nabla L|h)(Y_k)$ (with non-negative terms) satisfies for every $n \geq 1$,

$$\mathbb{E}\left(\sum_{k=0}^{n-1} \gamma_{k+1}(\nabla L|h)(Y_k)\right) \leq \prod_{k=1}^{n}(1 + C_{_L}\gamma_k^2)\mathbb{E}\, S_0$$

so that, by the Beppo Levi monotone convergence Theorem for series with non-negative terms,

$$\mathbb{E}\left(\sum_{n\geq 0} \gamma_{n+1}(\nabla L|h)(Y_n)\right) < +\infty$$

so that, in particular,

$$\sum_{n\geq 0}\gamma_{n+1}(\nabla L|h)(Y_n) < +\infty \qquad \mathbb{P}\text{-}a.s.$$

and the series converges in L^1 to its $a.s.$ limit.

(iii) It follows from (6.10) that, $\mathbb{P}$-$a.s.$, $L(Y_n) \longrightarrow L_\infty$ as $n \to +\infty$, which is integrable since $\big(L(Y_n)\big)_{n\geq 0}$ is L^1-bounded.

(i) Note that, again by Beppo Levi's monotone convergence Theorem for series with non-negative terms,

$$\begin{aligned}\mathbb{E}\left(\sum_{n\geq 1}|\Delta Y_n|^2\right) = \sum_{n\geq 1}\mathbb{E}\,|\Delta Y_n|^2 &\leq \sum_{n\geq 1}\gamma_n^2\,\mathbb{E}\,|H(Y_{n-1},Z_n)|^2\\ &\leq C\sum_{n\geq 1}\gamma_n^2(1+\mathbb{E}\,L(Y_{n-1})) < +\infty\end{aligned}$$

so that $\mathbb{E}\,|\Delta Y_n|^2 \to 0$ and $\sum_{n\geq 1}|\Delta Y_n|^2 < +\infty$ $a.s.$ which in turns yields $\Delta Y_n = Y_n - Y_{n-1} \to 0$ $a.s.$

(v) We have $M_n^\gamma = \sum_{k=1}^n \gamma_k \Delta M_k$ so that M^γ is clearly an $(\mathcal{F}_n)$-martingale. Moreover,

$$\begin{aligned}\langle M^\gamma\rangle_n &= \sum_{k=1}^n \gamma_k^2\,\mathbb{E}\left(|\Delta M_k|^2\,|\mathcal{F}_{k-1}\right)\\ &\leq \sum_{k=1}^n \gamma_k^2\,\mathbb{E}\left(|H(Y_{k-1},Z_k)|^2\,|\mathcal{F}_{k-1}\right)\\ &= \sum_{k=1}^n \gamma_k^2\left[\mathbb{E}\,|H(y,Z_1)|^2\right]_{|y=Y_{k-1}},\end{aligned}$$

where we used in the last line that Z_k is independent of $\mathcal{F}_{k-1}$. Consequently, owing to (6.6) and (ii), one has

$$\mathbb{E}\,\langle M^\gamma\rangle_\infty < +\infty,$$

which in turn implies by Theorem 12.7 that M_n^γ converges $a.s.$ and in L^2. ◇

Corollary 6.1 (a) ROBBINS–MONRO ALGORITHM. *Assume that the mean function h of the algorithm is continuous and satisfies*

$$\forall\, y\in\mathbb{R}^d,\ y\neq y_*,\quad \big(y-y_*|h(y)\big) > 0. \tag{6.11}$$

Suppose furthermore that $Y_0 \in L^2(\mathbb{P})$ and that H satisfies

$$\forall\, y\in\mathbb{R}^d,\qquad \big\|H(y,Z)\big\|_2 \leq C(1+|y|).$$

Finally, assume that the step sequence $(\gamma_n)_{n\geq 1}$ satisfies (6.7). Then

$$\{h=0\}=\{y_*\}\quad and \quad Y_n \xrightarrow{a.s.} y_*.$$

The convergence also holds in every $L^p(\mathbb{P})$, $p \in (0,2)$ *(and* $\big(|Y_n - y_*|\big)_{n\geq 0}$ *is* L^2*-bounded).*

(b) STOCHASTIC GRADIENT $(h = \nabla L)$. *Let* $L : \mathbb{R}^d \to \mathbb{R}_+$ *be a differentiable function satisfying (6.4),* $\lim_{|y|\to+\infty} L(y) = +\infty$, *and* $\{\nabla L = 0\} = \{y_*\}$. *Assume the mean function of the algorithm is given by* $h = \nabla L$, *that the function* H *satisfies* $\mathbb{E}\,|H(y,Z)|^2 \leq C\big(1 + L(y)\big)$ *and that* $L(Y_0) \in L^1(\mathbb{P})$. *Assume that the step sequence* $(\gamma_n)_{n\geq 1}$ *satisfies (6.7). Then*

$$L(y_*) = \min_{\mathbb{R}^d} L \quad \textit{and} \quad Y_n \xrightarrow{a.s.} y_* \quad \textit{as} \quad n \to +\infty.$$

Moreover, $\nabla L(Y_n)$ *converges to* 0 *in every* $L^p(\mathbb{P})$, $p \in (0,2)$ *(and* $\big(L(Y_n)\big)_{n\geq 0}$ *is* L^1*-bounded so that* $\big(\nabla L(Y_n)\big)_{n\geq 0}$ *is* L^2*-bounded).*

Proof. (a) Assumption (6.11) implies that the mean-reverting assumption (6.5) is satisfied by the quadratic Lyapunov function $L(y) = |y - y_*|^2$ (which clearly satisfies (6.4)). The assumption on H is clearly the linear growth Assumption (6.6) for this function L. Consequently, it follows from the above Robbins–Siegmund Lemma that

$$|Y_n - y_*|^2 \longrightarrow L_\infty \in L^1(\mathbb{P}) \quad \text{and} \quad \sum_{n\geq 1} \gamma_n \big(h(Y_{n-1})|Y_{n-1} - y_*\big) < +\infty \quad \mathbb{P}\text{-}a.s.$$

and that $(|Y_n - y_*|^2)_{n\geq 0}$ is $L^1(\mathbb{P})$-bounded.

Let $\omega \in \Omega$ such that $|Y_n(\omega) - y_*|^2$ converges in $\mathbb{R}_+$ and the series $\sum_{n\geq 1} \gamma_n \big(Y_{n-1}(\omega) - y_* | h(Y_{n-1}(\omega))\big) < +\infty$. Since $\sum_{n\geq 1} \gamma_n \big(Y_{n-1}(\omega) - y_* | h(Y_{n-1}(\omega))\big) < +\infty$, it follows that

$$\underline{\lim}_n \big(Y_{n-1}(\omega) - y_* | h(Y_{n-1}(\omega))\big) = 0.$$

If $\underline{\lim}_n \big(Y_{n-1}(\omega) - y_* | h(Y_{n-1}(\omega))\big) > 0$, the convergence of the above series induces a contradiction with the fact that $\sum_{n\geq 1} \gamma_n = +\infty$. Let $\big(\phi(n,\omega)\big)_{n\geq 1}$ be a subsequence such that

$$\big(Y_{\phi(n,\omega)}(\omega) - y_* | h(Y_{\phi(n,\omega)}(\omega))\big) \longrightarrow 0 \quad \text{as} \quad n \to +\infty.$$

Now, $(Y_n(\omega))_{n\geq 0}$ being bounded, one may assume, up to one further extraction, still denoted by $\big(\phi(n,\omega)\big)_{n\geq 1}$, that $Y_{\phi(n,\omega)}(\omega) \to y_\infty = y_\infty(\omega)$. It follows by the continuity of h that $\big(y_\infty - y_* | h(y_\infty)\big) = 0$ which in turn implies that $y_\infty = y_*$. Now, since we know that $L(Y_n(\omega)) = |Y_n(\omega) - y_*|^2$ converges,

$$\lim_n \big|Y_n(\omega) - y_*\big|^2 = \lim_n \big|Y_{\phi(n,\omega)}(\omega) - y_*\big|^2 = 0.$$

Finally, for every $p\in(0,2)$, $(|Y_n-y_*|^p)_{n\geq 0}$ is $L^{\frac{2}{p}}(\mathbb{P})$-bounded, hence uniformly integrable. As a consequence the *a.s.* convergence holds in L^1, *i.e.* $Y_n\to y_*$ converges in $L^p(\mathbb{P})$.

It is clear from (6.11) that $\{h=0\}\subset\{y_*\}$. On the other hand, if $y=y_*=\varepsilon u$, $|u|=1,\varepsilon>0$, one has $\varepsilon\big(u|h(y_*+\varepsilon u)\big)>0$. Letting $\varepsilon\to 0$ implies that $\big(u|h(y_*)\big)\geq 0$ for every unitary vector u since h is continuous, which in turn implies (switching from u to $-u$) that $\big(u|h(y_*)\big)=0$. Hence $h(y_*)=0$ (otherwise $u=\frac{h(y_*)}{|h(y_*)|}$ yields a contradiction).

(*b*) One may apply the Robbins–Siegmund Lemma with L as a Lyapunov function since $(h|\nabla L)=|\nabla L|^2\geq 0$. The assumption on H is just the linear growth assumption (6.6). As a consequence

$$L(Y_n)\longrightarrow L_\infty\in L^1(\mathbb{P})\quad\text{and}\quad\sum_{n\geq 1}\gamma_n\big|\nabla L(Y_{n-1})\big|^2<+\infty\quad\mathbb{P}\text{-}a.s.$$

and $\big(L(Y_n)\big)_{n\geq 0}$ is $L^1(\mathbb{P})$-bounded. Let $\omega\in\Omega$ such that $L\big(Y_n(\omega)\big)$ converges in $\mathbb{R}_+$,

$$\sum_{n\geq 1}\gamma_n\big|\nabla L(Y_{n-1}(\omega))\big|^2<+\infty\quad\text{and}\quad Y_n(\omega)-Y_{n-1}(\omega)\to 0.$$

The same arguments as above show that

$$\underline{\lim}_n\big|\nabla L(Y_n(\omega))\big|^2=0.$$

From the convergence of $L(Y_n\big(\omega)\big)$ toward $L_\infty(\omega)$ and $\lim_{|y|\to+\infty}L(y)=+\infty$, one derives that $(Y_n(\omega))_{n\geq 0}$ is bounded. Then there exists a subsequence $\big(\phi(n,\omega)\big)_{n\geq 1}$ such that

$$Y_{\phi(n,\omega)}\to\widetilde{y},\quad\nabla L\big(Y_{\phi(n,\omega)}(\omega)\big)\to 0\quad\text{and}\quad L\big(Y_{\phi(n,\omega)}(\omega)\big)\to L_\infty(\omega)$$

as $n\to+\infty$.

Then $\nabla L(\widetilde{y})=0$ which implies $\widetilde{y}=y_*$ and $L_\infty(\omega)=L(y_*)$. Since L is non-negative, differentiable and goes to infinity at infinity, it attains its unique global minimum at y_*. In particular, $\{L=L(y_*)\}=\{\nabla L=0\}=\{y_*\}$. Consequently, the only possible limiting value for the bounded sequence $(Y_n(\omega))_{n\geq 1}$ is y_* since $L(Y_n)\to L(y_*)$, *i.e.* $Y_n(\omega)$ converges toward y_*.

The $L^p(\mathbb{P})$-convergence to 0 of $|\nabla L(Y_n)|$, $p\in(0,2)$, follows by the same uniform integrability argument as in (*a*). ◇

► **Exercises. 1.** Show that Claim (*a*) remains true if one only assumes that

$$y\longmapsto(h(y)\,|\,y-y_*)\ \text{ is lower semi-continuous.}$$

2. ***Non-homogeneous*** L^2***-strong law of large numbers by stochastic approximation***. Let $(Z_n)_{n\geq 1}$ be an i.i.d. sequence of square integrable random vectors. Let $(\gamma_n)_{n\geq 1}$ be a sequence of positive real numbers satisfying the decreasing step Assumption (6.7). Show that the recursive procedure defined by

$$Y_{n+1} = Y_n - \gamma_{n+1}(Y_n - Z_{n+1})$$

a.s. converges toward $y_* = \mathbb{E}\, Z_1$.

The above settings are in fact some special cases of a more general result, the so-called "pseudo-gradient setting", stated below. However its proof, in particular in a multi-dimensional setting, needs additional arguments, mainly the so-called *ODE* method (for Ordinary Differential Equation method) originally introduced by Ljung (see [200]). The underlying idea is to think of a stochastic algorithm as a perturbed Euler scheme with a decreasing step of the *ODE* $\dot{y} = -h(y)$. For an introduction to the *ODE* method, see Sect. 6.4.1; we also refer to classical textbooks on Stochastic Approximation like [39, 81, 180].

Theorem 6.3 (Pseudo-Stochastic Gradient) *Assume that L and h and the step sequence $(\gamma_n)_{n\geq 1}$ satisfy all the assumptions of the Robbins–Siegmund Lemma. Assume furthermore that*

$$\lim_{|y|\to+\infty} L(y) = +\infty \quad \textit{and} \quad (\nabla L|h) \textit{ is lower semi-continuous.}$$

Then, $\mathbb{P}(d\omega)$-a.s., there exists an $\ell = \ell(\omega) \geq 0$ and a connected component $\mathcal{C}_\infty(\omega)$ of $\big\{(\nabla L|h) = 0\big\} \cap \big\{L = \ell\big\}$ such that

$$\operatorname{dist}\big(Y_n(\omega), \mathcal{C}_\infty(\omega)\big) \longrightarrow 0 \quad \textit{as} \quad n \to +\infty.$$

In particular, if for every $\ell \geq 0$, $\{(\nabla L|h) = 0\} \cap \{L = \ell\}$ is locally finite ([1]), *then, $\mathbb{P}(d\omega)$-a.s., there exists an $\ell_\infty(\omega)$ such that Y_n converges toward a point of the set $\big\{(\nabla L|h) = 0\big\} \cap \big\{L = \ell_\infty(\omega)\big\}$.*

Proof (One-dimensional case). We consider an $\omega \in \Omega$ for which all the "*a.s.* conclusions" of the Robbins–Siegmund Lemma are true. Combining $Y_n(\omega) - Y_{n-1}(\omega) \to 0$ with the boundedness of the sequence $(Y_n(\omega))_{n\geq 0}$, one can show that the set $\mathcal{Y}_\infty(\omega)$ of the limiting values of $(Y_n(\omega))_{n\geq 0}$ is a connected compact set ([2]).

On the other hand, $\mathcal{Y}_\infty(\omega) \subset \big\{L = L_\infty(\omega)\big\}$ since $L(Y_n(\omega)) \to L_\infty(\omega)$. Furthermore, reasoning as in the proof of claim (*b*) of the above corollary shows that there exists a limiting value $y_* \in \mathcal{Y}_\infty(\omega)$ such that $\big(\nabla L(y_*)|h(y_*)\big) = 0$ so that $y_* \in \big\{(\nabla L|h) = 0\big\} \cap \big\{L = L_\infty(\omega)\big\}$.

---

[1]By *locally finite*, we mean "finite on every compact set".

[2]The method of proof is to first establish that $\mathcal{Y}_\infty(\omega)$ is "bien enchaîné" as a set. A subset $A \subset \mathbb{R}^d$ is "bien enchaîné" if for every $a,\ a' \in A$ and every $\varepsilon > 0$, there exists $p \in \mathbb{N}^*$, $b_0, b_1, \ldots, b_p \in A$ such that $b_0 = a$, $b_p = a'$, $|b_i - b_{i-1}| \leq \varepsilon$. Any connected set A is "bien enchaîné" and the converse is true if A is compact. What we need here is precisely this converse (see *e.g.* [13] for details).

At this stage, we assume that $d = 1$. Either $\mathcal{Y}_\infty(\omega) = \{y_*\}$ and the proof is complete, or $\mathcal{Y}_\infty(\omega)$ is a non-trivial compact interval as a compact connected subset of $\mathbb{R}$. The function L is constant on this interval, consequently its derivative L' is zero on $\mathcal{Y}_\infty(\omega)$ so that $\mathcal{Y}_\infty(\omega) \subset \big\{(\nabla L|h) = 0\big\} \cap \big\{L = L(y_*)\big\}$. Hence the conclusion. When $\big\{(\nabla L|h) = 0\big\} \cap \big\{L = \ell\big\}$ is locally finite, the conclusion is obvious since its connected components are reduced to single points. ◇

6.3 Applications to Finance

6.3.1 Application to Recursive Variance Reduction by Importance Sampling

This section was originally motivated by the seminal paper [12]. Finally, we followed the strategy developed in [199] which provides, in our mind, an easier to implement procedure. Assume we want to compute the expectation

$$\mathbb{E}\,\varphi(Z) = \int_{\mathbb{R}^d} \varphi(z) e^{-\frac{|z|^2}{2}} \frac{dz}{(2\pi)^{\frac{d}{2}}} \tag{6.12}$$

where $\varphi : \mathbb{R}^d \to \mathbb{R}$ is integrable with respect to the normalized Gaussian measure. In order to deal with a consistent problem, we assume throughout this section that

$$\mathbb{P}(\varphi(Z) \neq 0) > 0.$$

⊳ **Examples.** (a) A typical example is provided by an option pricing in a d-dimensional Black–Scholes model where, with the usual notations,

$$\varphi(z) = e^{-rt}\phi\left(\left(x_0^i\, e^{(r-\frac{\sigma_i^2}{2})T+\sigma_i\sqrt{T}(Az)_i}\right)_{1\le i\le d}\right),\quad x_0 = (x_0^1,\ldots,x_0^d) \in (0,+\infty)^d,$$

with A a lower triangular matrix such that the covariance matrix $R = AA^*$ has diagonal entries equal to 1 and ϕ a non-negative, continuous if necessary, payoff function. The dimension d corresponds to the number of underlying risky assets.

(b) Monte Carlo simulation of functionals of the Euler scheme of a diffusion (or Milstein scheme) appear as integrals with respect to a multivariate Gaussian vector. Then the dimension d can be huge since it corresponds to the product of the number of time steps by the number of independent Brownian motions driving the dynamics of the *SDE*.

Variance reduction by mean translation: first approach (see [12]**).**

A change of variable $z = \zeta + \theta$, for a fixed $\theta \in \mathbb{R}^d$, leads to

$$\mathbb{E}\,\varphi(Z) = e^{-\frac{|\theta|^2}{2}}\,\mathbb{E}\left(\varphi(Z+\theta)e^{-(\theta|Z)}\right). \tag{6.13}$$

Such a change of variable yields what is known as the Cameron–Martin formula, which can be seen here either as a somewhat elementary version of the Girsanov change of probability, or as the first step of an importance sampling procedure.

One natural way to optimize the computation by Monte Carlo simulation of $\mathbb{E}\,\varphi(Z)$ is to choose, among the above representations depending on the parameter $\theta \in \mathbb{R}^d$, the one with the lowest variance. This means solving, at least roughly, the following minimization problem

$$\min_{\theta\in\mathbb{R}^d}\Big[L(\theta) := e^{-|\theta|^2}\mathbb{E}\left(\varphi^2(Z+\theta)e^{-2(Z|\theta)}\right)\Big] \tag{6.14}$$

since $\mathrm{Var}\Big(e^{-\frac{|\theta|^2}{2}}\varphi(Z+\theta)e^{-(\theta|Z)}\Big) = L(\theta) - \big(\mathbb{E}\,\varphi(Z)\big)^2$.

A reverse change of variable shows that

$$L(\theta) = e^{\frac{|\theta|^2}{2}}\mathbb{E}\left(\varphi^2(Z)e^{-(Z|\theta)}\right). \tag{6.15}$$

Hence, if $\mathbb{E}\left(\varphi^2(Z)|Z|e^{a|Z|}\right) < +\infty$ for every $a \in (0, +\infty)$, one can always differentiate the function L on $\mathbb{R}^d$ owing to Theorem 2.2(b) with

$$\nabla L(\theta) = e^{\frac{|\theta|^2}{2}}\mathbb{E}\left(\varphi^2(Z)e^{-(\theta|Z)}(\theta - Z)\right),\ \theta \in \mathbb{R}^d. \tag{6.16}$$

Rewriting Eq. (6.15) as

$$L(\theta) = \mathbb{E}\left(\varphi^2(Z)e^{-\frac{|Z|^2}{2}}e^{\frac{1}{2}|Z-\theta|^2}\right) \tag{6.17}$$

clearly shows that L is *strictly convex* since $\theta \mapsto e^{\frac{1}{2}|\theta - z|^2}$ is strictly convex for every $z \in \mathbb{R}^d$ and $\mathbb{P}(\varphi(Z) > 0) > 0$. Furthermore, Fatou's Lemma implies $\lim_{|\theta|\to+\infty} L(\theta) = +\infty$.

Consequently, L has a unique global minimum θ_* which is also local, whence satisfies $\{\nabla L = 0\} = \{\theta_*\}$.

We now prove the classical lemma which shows that if L is strictly convex then $\theta \mapsto |\theta - \theta_*|^2$ is mean-reverting for ∇L (strictly, in the strengthened sense (6.11) of the Robbins–Monro framework).

Lemma 6.1 *(a) Let* $L : \mathbb{R}^d \to \mathbb{R}_+$ *be a differentiable convex function. Then*

$$\forall\,\theta,\ \theta' \in \mathbb{R}^d, \qquad (\nabla L(\theta) - \nabla L(\theta')|\theta - \theta') \geq 0.$$

If, furthermore, L is strictly convex, the above inequality is strict if $\theta \neq \theta'$.

(b) If L is twice differentiable and $D^2 L \geq \alpha I_d$ *for some real constant* $\alpha > 0$ *(in the sense that* $u^* D^2 L(\theta) u \geq \alpha |u|^2$, *for every* $\theta,\ u \in \mathbb{R}^d$*), then* $\lim_{|\theta|\to+\infty} L(\theta) = +\infty$ *and*

$$\forall\,\theta,\,\theta'\in\mathbb{R}^d,\qquad \begin{cases} (i)\;\;(\nabla L(\theta)-\nabla L(\theta')|\theta-\theta')\geq \alpha|\theta-\theta'|^2,\\[2ex] (ii)\;L(\theta')\geq L(\theta)+(\nabla L(\theta)\,|\,\theta'-\theta)+\tfrac{1}{2}\alpha|\theta'-\theta|^2.\end{cases}$$

Proof. (a) One introduces the differentiable function defined on the unit interval by

$$g(t)=L(\theta+t(\theta'-\theta))-L(\theta),\quad t\in[0,1].$$

The function g is convex and differentiable. Hence its derivative

$$g'(t)=(\nabla L(\theta+t(\theta'-\theta))|\theta'-\theta),$$

is non-decreasing so that $g'(1)\geq g'(0)$ which yields the announced inequality. If L is strictly convex, then $g'(1)>g'(0)$ (otherwise $g'(t)\equiv 0$ which would imply that L is affine on the geometric interval $[\theta,\theta']$).

(b) The function is twice differentiable under this assumption and

$$g''(t)=(\theta'-\theta)^*D^2L\big(\theta+t(\theta'-\theta)\big)(\theta'-\theta)\geq\alpha|\theta'-\theta|^2.$$

The conclusion follows by noting that $g'(1)-g'(0)\geq\inf_{s\in[0,1]}g''(s)$. Moreover, noting that $g(1)\geq g(0)+g'(0)+\frac{1}{2}\inf_{s\in[0,1]}g''(s)$ yields the inequality (ii). Finally, setting $\theta=0$ yields

$$L(\theta')\geq L(0)+\frac{1}{2}\alpha|\theta'|^2-|\nabla L(0)|\,|\theta'|\to+\infty\quad\text{as}\quad|\theta'|\to\infty. \qquad\diamond$$

This suggests (as noted in [12]) to consider the quadratic function V defined by $V(\theta):=|\theta-\theta_*|^2$ as a Lyapunov function instead of L defined in (6.15). Indeed, L is usually not essentially quadratic: as soon as $\varphi(z)\geq\varepsilon_0>0$, it is obvious that $L(\theta)\geq\varepsilon_0^2e^{|\theta|^2}$, but this exponential growth is also observed when φ is bounded away from zero outside a ball. Hence ∇L cannot be Lipschitz continuous either and, consequently, cannot be used as a Lyapunov function.

However, if one uses the representation of ∇L as an expectation derived from (6.17) by pathwise differentiation in order to design a stochastic gradient algorithm, namely considers the local gradient $H(\theta,z):=\varphi^2(z)e^{-\frac{|z|^2}{2}}\frac{\partial}{\partial\theta}\big(e^{\frac{1}{2}|z-\theta|^2}\big)$, a major difficulty remains: the convergence results in Corollary 6.1 do not apply, mainly because the linear growth assumption (6.6) in quadratic mean is not fulfilled by such a choice for H.

In fact, this "naive" procedure explodes at almost every implementation, as pointed out in [12]. This leads the author to introduce in [12] some variants of the algorithm involving repeated re-initializations – the so-called projections "à la Chen" – to force the stabilization of the algorithm and subsequently prevent explosion. The choice we make in the next section is different (and still other approaches are possible to circumvent this problem, see *e.g.* [161]).

An "unconstrained" approach based on a third change of variable (see [199])

The starting point is to find a new local gradient to represent ∇L in order to apply the above standard convergence results. We know, and already used above, that the Gaussian density is smooth by contrast with the payoff φ (at least in quantitative finance of derivative products): to differentiate L, we already switched the parameter θ from φ to the Gaussian density to take advantage of its smoothness by a change of variable. At this stage, we face the converse problem: we usually know what the behavior of φ at infinity is whereas we cannot efficiently control the behavior of $e^{-(\theta|Z)}$ inside the expectation as θ goes to infinity. So, it is natural to try to cancel this exponential term by plugging θ back in the payoff φ.

The first step is to make a new change of variable. Starting from (6.16), one gets

$$\begin{aligned}\nabla L(\theta) &= e^{\frac{|\theta|^2}{2}}\int_{\mathbb{R}^d}\varphi^2(z)(\theta-z)e^{-(\theta|z)-\frac{|z|^2}{2}}\frac{dz}{(2\pi)^{d/2}}\\ &= e^{|\theta|^2}\int_{\mathbb{R}^d}\varphi^2(\zeta-\theta)(2\theta-\zeta)e^{-\frac{|\zeta|^2}{2}}\frac{d\zeta}{(2\pi)^{d/2}}\quad (z:=\zeta-\theta),\\ &= e^{|\theta|^2}\mathbb{E}\left(\varphi^2(Z-\theta)(2\theta-Z)\right).\end{aligned}$$

Consequently, canceling the positive term $e^{|\theta|^2}$, we get

$$\nabla L(\theta)=0 \Longleftrightarrow \mathbb{E}\left(\varphi^2(Z-\theta)(2\theta-Z)\right)=0.$$

This suggests to work with the function inside the expectation (though not exactly a local gradient) up to an explicit appropriate multiplicative factor depending on θ to satisfy the linear growth assumption for the L^2-norm in θ.

From now on, we assume that there exist two positive real constants $a\geq 0$ and $C>0$ such that

$$0\leq\varphi(z)\leq Ce^{\frac{a}{2}|z|},\quad z\in\mathbb{R}^d. \tag{6.18}$$

► **Exercise.** (*a*) Show that under this assumption, $\mathbb{E}\left(\varphi^2(Z)|Z|e^{|\theta||Z|}\right)<+\infty$ for every $\theta\in\mathbb{R}^d$, which implies that (6.16) holds true.
(*b*) Show that in fact $\mathbb{E}\left(\varphi^2(Z)|Z|^m e^{|\theta||Z|}\right)<+\infty$ for every $\theta\in\mathbb{R}^d$ and every $m\geq 1$, which in turn implies that L is $\mathcal{C}^\infty$. In particular, show that for every $\theta\in\mathbb{R}^d$,

$$D^2L(\theta)=e^{\frac{|\theta|^2}{2}}\mathbb{E}\left(\varphi^2(Z)e^{-(\theta|Z)}\left(I_d+(\theta-Z)(\theta-Z)^t\right)\right)$$

(throughout this chapter, we adopt the notation ^{t} for transposition). Derive that

$$D^2L(\theta)\geq e^{\frac{|\theta|^2}{2}}\mathbb{E}\left(\varphi^2(Z)e^{-(\theta|Z)}\right)I_d>0$$

(in the sense of positive definite symmetric matrices) which proves again that L is strictly convex.

Taking Assumption (6.18) into account, we set

$$H_a(\theta, z) = e^{-a(|\theta|^2+1)^{\frac{1}{2}}} \varphi^2(z-\theta)(2\theta - z). \tag{6.19}$$

One checks that

$$\begin{aligned}\mathbb{E}\,|H_a(\theta, Z)|^2 &\le 2\,C^4 e^{-2a(|\theta|^2+1)^{\frac{1}{2}}}\,\mathbb{E}\left(e^{2a|Z|+2a|\theta|}(4|\theta|^2+|Z|^2)\right)\\ &\le 2\,C^4\left(4|\theta|^2\mathbb{E}\left(e^{2a|Z|}\right)+\mathbb{E}\left(e^{2a|Z|}|Z|^2\right)\right)\\ &\le C'(1+|\theta|^2)\end{aligned}$$

since $|Z|$ has a Laplace transform defined on the whole real line, which in turn implies $\mathbb{E}\left(e^{a|Z|}|Z|^r\right) < +\infty$ for every $a,\ r > 0$.

On the other hand, it follows that the resulting mean function h_a reads

$$h_a(\theta) = e^{-a(|\theta|^2+1)^{\frac{1}{2}}}\,\mathbb{E}\left(\varphi^2(Z-\theta)(2\,\theta - Z)\right)$$

or, equivalently,

$$h_a(\theta) = e^{-a(|\theta|^2+1)^{\frac{1}{2}} - |\theta|^2}\,\nabla L(\theta) \tag{6.20}$$

so that h_a is continuous, $(\theta - \theta_* | h_a(\theta)) > 0$ for every $\theta \ne \theta_*$ and $\{h_a = 0\} = \{\theta_*\}$.

Applying Corollary 6.1(a) (the Robbins–Monro Theorem), one derives that for any step sequence $\gamma = (\gamma_n)_{n\ge 1}$ satisfying (6.7), the sequence $(\theta_n)_{n\ge 0}$ recursively defined by

$$\theta_{n+1} = \theta_n - \gamma_{n+1} H_a(\theta_n, Z_{n+1}),\ \ n \ge 0, \tag{6.21}$$

where $(Z_n)_{n\ge 1}$ is an i.i.d. sequence with distribution $\mathcal{N}(0; I_d)$ defined on a probability space $(\Omega, \mathcal{A}, \mathbb{P})$ independent of the $\mathbb{R}^d$-valued random vector $\theta_0 \in L^2_{\mathbb{R}^d}(\Omega, \mathcal{A}, \mathbb{P})$, satisfies

$$\theta_n \xrightarrow{a.s.} \theta_* \qquad \text{as} \quad n \to +\infty.$$

Remarks. • The reason for introducing $(|\theta|^2 + 1)^{\frac{1}{2}}$ is that this function is explicit, behaves like $|\theta|$ at infinity and is also everywhere differentiable, which simplifies the discussion about the rate of convergence detailed further on.

• Note that *no regularity assumption is made on the payoff* φ.

• An alternative approach based on a large deviation principle but which needs some regularity assumption on the payoff φ is developed in [116]. See also [243].

• To prevent a possible "freezing" of the procedure, for example when the step sequence has been misspecified or when the payoff function is too anisotropic, one can replace the above procedure (6.21) by the following fully data-driven variant of the algorithm

$$\forall\, n \ge 0, \qquad \widetilde{\theta}_{n+1} = \widetilde{\theta}_n - \gamma_{n+1}\widetilde{H}_a(\widetilde{\theta}_n, Z_{n+1}), \qquad \widetilde{\theta}_0 = \theta_0, \tag{6.22}$$

where

$$\widetilde{H}_a(\theta, z) := \frac{\varphi^2(z-\theta)}{1+\varphi^2(-\theta)}(2\theta - z).$$

This procedure also converges *a.s.* under a sub-multiplicativity assumption on the payoff function φ (see [199]).

• A final – and often crucial – trick to boost the convergence when dealing with rare events, like for importance sampling purpose, is to "drive" a parameter from a "regular" value to the value that makes the event rare. Typically when trying to reduce the variance of a deep-out-of-the-money Call option like in the numerical illustrations below, a strategy can be to implement the above algorithm with *a slowly varying strike K_n from $K_0 = x_0$ to the "target" strike K* (see below) during the first half of iterations.

ℵ Practitioner's corner (On the weak rate of convergence)

This paragraph anticipates on Sects. 6.4.3 and 6.4.4. It can be skipped on a first reading. Assume that the step sequence has the following parametric form $\gamma_n = \dfrac{\alpha}{\beta + n}$, $n \geq 1$, and that $Dh_a(\theta_*)$ is positive in the following sense: all the eigenvalues of $Dh_a(\theta_*)$ have a positive real part. Then, the rate of convergence of θ_n toward θ_* is ruled by a *CLT* (at rate $\sqrt{n}$) if and only if (see Sect. 6.4.3)

$$\alpha > \frac{1}{2\Re e(\lambda_{a,\min})} > 0,$$

where $\lambda_{a,\min}$ is the eigenvalue of $Dh_a(\theta_*)$ with the lowest real part. Moreover, one shows that the theoretical best choice for α is $\alpha_{opt} := \frac{1}{\Re e(\lambda_{a,\min})}$. The asymptotic variance is made explicit, once again in Sect. 6.4.3. Let us focus now on $Dh_a(\theta_*)$.

Starting from the expression (6.20)

$$\begin{aligned} h_a(\theta) &= e^{-a(|\theta|^2+1)^{\frac{1}{2}} - |\theta|^2} \nabla L(\theta) \\ &= e^{-a(|\theta|^2+1)^{\frac{1}{2}} - \frac{|\theta|^2}{2}} \times \mathbb{E}\left(\varphi^2(Z)(\theta - Z)e^{-(\theta|Z)}\right) \qquad \text{by (6.16)} \\ &= g_a(\theta) \times \mathbb{E}\left(\varphi^2(Z)(\theta - Z)e^{-(\theta|Z)}\right). \end{aligned}$$

Then

$$Dh_a(\theta) = g_a(\theta)\mathbb{E}\left(\varphi^2(Z)e^{-(\theta|Z)}(I_d + ZZ^t - \theta Z^t)\right) + e^{-\frac{|\theta|^2}{2}} \nabla L(\theta) \otimes \nabla g_a(\theta)$$

(where $u \otimes v = [u_i v_j]_{1 \leq i,j \leq d}$). Using that $\nabla L(\theta_*) = h_a(\theta_*) = 0$, so that $h_a(\theta_*)\theta_*^t = 0$, we get

$$Dh_a(\theta_*) = g_a(\theta_*)\mathbb{E}\left(\varphi^2(Z)e^{-(\theta_*|Z)}(I_d + (Z - \theta_*)(Z - \theta_*)^t)\right).$$

Hence $Dh_a(\theta_*)$ is a definite positive symmetric matrix and its lowest eigenvalue $\lambda_{a,\min}$ satisfies

$$\lambda_{a,\min} \geq g_a(\theta_*)\mathbb{E}\left(\varphi^2(Z)e^{-(\theta_*|Z)}\right) > 0.$$

These computations show that if the behavior of the payoff φ at infinity is mis-evaluated, this leads to a bad calibration of the algorithm. Indeed, if one considers two real numbers a, a' satisfying (6.18) with $0 < a < a'$, then one checks with obvious notations that

$$\frac{1}{2\lambda_{a,\min}} = \frac{g_{a'}(\theta_*)}{g_a(\theta_*)}\frac{1}{2\lambda_{a',\min}} = e^{(a-a')(|\theta_*|^2+1)^{\frac{1}{2}}}\frac{1}{2\lambda_{a',\min}} < \frac{1}{2\lambda_{a',\min}}.$$

So the condition on α is more stringent with a' than with a. Of course, in practice, the user does not know these values (since she does not know the target θ_*), however she will be led to consider higher values of α than requested, which will lead to the deterioration of the asymptotic variance (see again Sect. 6.4.3).

These weak rate results seem to be of little help in practice since θ_* being unknown, also means that functions at θ_* are unknown. One way to circumvent this difficulty is to implement *Ruppert and Polyak's averaging principle*, described and analyzed in Sect. 6.4.4. First the procedure should be implemented with a slowly decreasing step of the form

$$\gamma_n = \frac{\alpha}{\beta + n^b}, \frac{1}{2} < b < 1, \alpha > 0, \beta \geq 0$$

and, as a second step, an averaging phase is added, namely set

$$\bar{\theta}_n = \frac{\theta_0 + \cdots + \theta_{n-1}}{n}, \quad n \geq 1.$$

Then $(\bar{\theta}_n)_{n\geq 1}$ satisfies a Central Limit Theorem at a rate $\sqrt{n}$ with an asymptotic variance corresponding to the optimal asymptotic variance obtained for the original algorithm $(\theta_n)_{n\geq 0}$ with theoretical optimal step sequences $\gamma_n = \frac{\alpha_{opt}}{\beta+n}$, $n \geq 1$.

The choice of the parameter β either for the original algorithm or in its averaged version does not depend on theoretical motivations. A heuristic rule is to choose it so that γ_n does not decrease too fast to avoid being "frozen" far from its target.

Adaptive implementation into the computation of $\mathbb{E}\,\varphi(Z)$

At this stage, like for the variance reduction by regression, we may follow two strategies – batch or adaptive – to reduce the variance.

▷ The *batch strategy*. This is the simplest and the most elementary strategy.

Phase 1: One first computes an hopefully good approximation of the optimal variance reducer, which we will denote by θ_{n_0} for a large enough n_0 that will remain fixed during the second phase devoted to the computation of $\mathbb{E}\,\varphi(Z)$. It is assumed that an n_0 sample of an i.i.d. sequence $(Z_m)_{1\leq m\leq n_0}$ of $\mathcal{N}(0; I_d)$-distributed random vectors.

Phase 2: As a second step, one implements a Monte Carlo simulation based on $\tilde{\varphi}(z)\varphi(z+\theta_{n_0})e^{(\theta_{n_0}|z)-\frac{|\theta_{n_0}|^2}{2}}$, *i.e.*

$$\mathbb{E}\,\varphi(Z) = \lim_M \frac{1}{M}\sum_{m=1}^{M}\varphi(Z_{m+n_0}+\theta_{n_0})e^{-(\theta_{n_0}|Z_{m+n_0})-\frac{|\theta_{n_0}|^2}{2}},$$

where $(Z_m)_{m\geq n_0+1}$ is an i.i.d. sequence of $\mathcal{N}(0;I_d)$-distributed random vectors independent of $(Z_m)_{1\leq m\leq n_0}$. This procedure satisfies a *CLT* with (conditional) variance $L(\theta_{n_0})-(\mathbb{E}\,\varphi(Z))^2$ (given θ_{n_0}).

▷ The *adaptive strategy*. This approach, introduced in [12], is similar to the adaptive variance reduction by regression presented in Sect. 3.2. The aim is to devise a procedure fully based on the simultaneous computation of the optimal variance reducer and $\mathbb{E}\,\varphi(Z)$ from the same sequence $(Z_n)_{n\geq 1}$ of i.i.d. $\mathcal{N}(0;I_d)$-distributed random vectors used in (6.21). To be precise, this leads us to devise the following adaptive estimator of $\mathbb{E}\,\varphi(Z)$:

$$\frac{1}{M}\sum_{m=1}^{M}\varphi(Z_m+\theta_{m-1})e^{-(\theta_{m-1}|Z_m)-\frac{|\theta_{m-1}|^2}{2}},\quad M\geq 1, \tag{6.23}$$

where the sequence $(\theta_m)_{m\geq 0}$ is obtained by iterating (6.21).

We will briefly prove that the above estimator is unbiased and convergent.

Let $\mathcal{F}_m := \sigma(\theta_0, Z_1,\dots,Z_m)$, $m\geq 0$, be the filtration of the (whole) simulation process. Using that θ_{m-1} is $\mathcal{F}_{m-1}$-adapted and Z_m is independent of $\mathcal{F}_{m-1}$ with the same distribution as Z, one derives classically that

$$\mathbb{E}\left(\varphi(Z_m+\theta_{m-1})e^{-(\theta_{m-1}|Z_m)-\frac{|\theta_{m-1}|^2}{2}}\,\Big|\,\mathcal{F}_{m-1}\right) = \left[\mathbb{E}\left(\varphi(Z+\theta)e^{-(\theta|Z)-\frac{|\theta|^2}{2}}\right)\right]_{|\theta=\theta_{m-1}}.$$

As a first consequence, the estimator defined by (6.23) is unbiased. Now let us define the $(\mathcal{F}_M)$-martingale

$$N_M := \sum_{m=1}^{M}\frac{\varphi(Z_m+\theta_{m-1})e^{-(\theta_{m-1}|Z_m)-\frac{|\theta_{m-1}|^2}{2}}-\mathbb{E}\,\varphi(Z)}{m}\mathbf{1}_{\{|\theta_{m-1}|\leq m\}},\quad M\geq 1.$$

It is clear that $(N_M)_{M\geq 1}$ has square integrable increments so that $N_M\in L^2(\mathbb{P})$ for every $M\in\mathbb{N}^*$ and

$$\mathbb{E}\left(\left(\varphi(Z_m+\theta_{m-1})e^{-(\theta_{m-1}\,|Z_m)-\frac{|\theta_{m-1}|^2}{2}}\right)^2\,\Big|\,\mathcal{F}_{m-1}\right)\mathbf{1}_{\{|\theta_{m-1}|\leq m\}} = L(\theta_{m-1})\mathbf{1}_{\{|\theta_{m-1}|\leq m\}}$$
$$\xrightarrow{a.s.} L(\theta_*)$$

as $m \to +\infty$, which in turn implies that

$$\langle N\rangle_\infty \le \Big(\sup_m L(\theta_m)\Big)\sum_{m\ge 1}\frac{1}{m^2} < +\infty \quad a.s.$$

Consequently (see Theorem 12.7 in the Miscellany Chapter), $N_{_M} \to N_\infty$ $\mathbb{P}$-$a.s.$ where N_∞ is an $a.s.$ finite random variable. Finally, Kronecker's Lemma (see Lemma 12.1) implies

$$\frac{1}{M}\sum_{m=1}^{M}\Big(\varphi(Z_m+\theta_{m-1})\,e^{-(\theta_{m-1}|Z_m)-\frac{|\theta_{m-1}|^2}{2}} - \mathbb{E}\,\varphi(Z)\Big)\mathbf{1}_{\{|\theta_{m-1}|\le m\}} \longrightarrow 0 \quad a.s.$$

as $M \to +\infty$. Since $\theta_m \to \theta_*$ $a.s.$ as $m \to +\infty$, $\mathbf{1}_{\{|\theta_{m-1}|\le m\}} = 1$ for large enough m so that

$$\frac{1}{M}\sum_{m=1}^{M}\varphi(Z_m+\theta_{m-1})\,e^{-(\theta_{m-1}|Z_m)-\frac{|\theta_{m-1}|^2}{2}} \longrightarrow \mathbb{E}\,\varphi(Z) \quad a.s. \quad \text{as} \quad M \to +\infty.$$

One can show, using the *CLT* theorem for triangular arrays of martingale increments (see [142] and Chap. 12, Theorem 12.8) that

$$\sqrt{M}\left(\frac{1}{M}\sum_{m=1}^{M}\varphi(Z_m+\theta_{m-1})\,e^{-(\theta_{m-1}|Z_m)-\frac{|\theta_{m-1}|^2}{2}} - \mathbb{E}\,\varphi(Z)\right) \xrightarrow{\mathcal{L}} \mathcal{N}\big(0,\sigma_*^2\big),$$

where $\sigma_*^2 = L(\theta_*) - \big(\mathbb{E}\,\varphi(Z)\big)^2$ is the minimal variance.

As set, this second approach seems more performing owing to its minimal asymptotic variance. For practical use, the verdict is more balanced and the batch approach turns out to be quite satisfactory.

NUMERICAL ILLUSTRATIONS. (a) *At-the-money Black–Scholes setting.* We consider a vanilla Call in a *B-S* model

$$X_T = x_0\,e^{(r-\frac{\sigma^2}{2})T+\sigma\sqrt{T}Z}, \quad Z \stackrel{d}{=} \mathcal{N}(0;1)$$

with the following parameters: $T=1$, $r=0.10$, $\sigma=0.5$, $x_0=100$, $K=100$. The Black–Scholes reference price of the Vanilla Call is 23.93.

The recursive optimization of θ was achieved by running *the data driven version* (6.22) with a sample $(Z_n)_n$ of size 10 000. A *first renormalization has been made prior to the computation*: we considered the equivalent problem (as far as variance reduction is concerned) where the starting value of the asset is 1 and the strike is the *moneyness* K/X_0. The procedure was initialized at $\theta_0 = 1$. (Using (3.4) would have led us to set $\theta_0 = -0.2$).

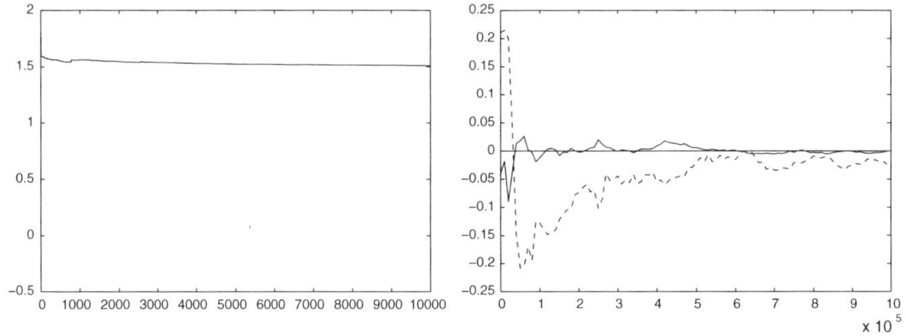

Fig. 6.1 *B-S* VANILLA CALL OPTION. $T = 1$, $r = 0.10$, $\sigma = 0.5$, $X_0 = 100$, $K = 100$. Left: convergence toward θ_* (up to $n = 10\,000$). Right: Monte Carlo simulation of size $M = 10^6$; dotted line: $\theta = 0$, solid line: $\theta = \theta_{10\,000} \simeq \theta_*$

We did not try to optimize the choice of the step γ_n following theoretical results on the weak rate of convergence, nor did we perform an averaging principle. We just applied the heuristic rule that if the function H (here H_a) takes its (usual) values within a few units, then choosing $\gamma_n = c\,\frac{20}{20+n}$ with $c \simeq 1$ (say $c \in [1/2, 2]$) leads to satisfactory performances of the algorithm.

The resulting value $\theta_{10\,000}$ was used in a standard Monte Carlo simulation of size $M = 10^6$ based on (6.13) and compared to a crude Monte Carlo simulation with $\theta = 0$. The numerical results are as follows:

– $\theta = 0$: 95 % Confidence interval = [23.92, 24.11] (pointwise estimate: 24.02).

– $\theta = \theta_{10\,000} \simeq 1.51$: 95 % Confidence interval = [23.919, 23.967] (pointwise estimate: 23.94).

The gain ratio in terms of standard deviations is $\frac{42.69}{11.01} = 3.88 \simeq 4$. This is observed on most simulations we made, however the convergence of θ_n may be more chaotic than displayed in Fig. 6.1 (left), where the convergence is almost instantaneous. The behavior of the two Monte Carlo simulations are depicted in Fig. 6.1 (right). The alternative original "parametrized" version of the algorithm ($H_a(\theta, z)$) with $a = 2\sigma\sqrt{T}$ yields quite similar results (when implemented with the same step and the same starting value).

FURTHER COMMENTS: As developed in [226], all of the preceding can be extended to non-Gaussian random vectors Z provided their distribution have a log-concave probability density p satisfying, for some positive ρ,

$$\log(p) + \rho|\cdot|^2 \text{ is convex.}$$

One can also replace the mean translation by other importance sampling procedures like those based on the Esscher transform. This has applications, *e.g.* when $Z = X_T$ is the value at time T of a process belonging to the family of subordinated Lévy processes (Lévy processes of the form $Z_t = W_{Y_t}$, where Y is a subordinator – an increasing Lévy process – independent of the standard Brownian motion W). For more insight on such processes, we refer to [40, 261].

6.3.2 Application to Implicit Correlation Search

We consider a 2-dimensional *B-S* toy model as defined by (2.2), *i.e.* $X_t^0 = e^{rt\cdot}$ (riskless asset) and

$$X_t^i = x_0^i e^{(r-\frac{\sigma_i^2}{2})t+\sigma_i W_t^i}, \quad x_0^i > 0, \quad i = 1, 2,$$

for the two risky assets, where $\langle W^1, W^2\rangle_t = \rho t$, $\rho \in [-1, 1]$ denotes the correlation between W^1 and W^2, that is, the correlation between the yields of the risky assets X^1 and X^2.

In this market, we consider a *best-of call* option defined by its payoff

$$\big(\max(X_T^1, X_T^2) - K\big)_+.$$

A market of such *best-of calls* is a market of the correlation ρ since the respective volatilities are obtained from the markets of vanilla options on each asset as implicit volatilities. In this 2-dimensional *B-S* setting, there is a closed formula for the premium involving the bi-variate standard normal distribution (see [159]), but what follows can be applied as soon as the asset dynamics – or their time discretization – can be simulated at a reasonable computational cost.

We will use a stochastic recursive procedure to solve the inverse problem in ρ

$$P_{BoC}(x_0^1, x_0^2, K, \sigma_1, \sigma_2, r, \rho, T) = P_{market} \text{ [Mark-to-market premium]}, \tag{6.24}$$

where

$$\begin{aligned} P_{BoC}(x_0^1, x_0^2, K, \sigma_1, \sigma_2, r, \rho, T) &:= e^{-rt}\mathbb{E}\Big(\big(\max(X_T^1, X_T^2) - K\big)_+\Big) \\ &= e^{-rt}\mathbb{E}\bigg(\Big(\max\Big(x_0^1 e^{\mu_1 T+\sigma_1\sqrt{T}Z_1}, x_0^2 e^{\mu_2 T+\sigma_2\sqrt{T}(\rho Z^1+\sqrt{1-\rho^2}Z^2)}\Big) - K\Big)_+\bigg), \end{aligned}$$

where $\mu_i = r - \frac{\sigma_i^2}{2}$, $i = 1, 2$, $Z = (Z^1, Z^2) \stackrel{d}{=} \mathcal{N}(0; I_2)$.

It is intuitive and easy to check (at least empirically by simulation) that the function $\rho \longmapsto P_{BoC}(x_0^1, x_0^2, K, \sigma_1, \sigma_2, r, \rho, T)$ is continuous and (strictly) decreasing on $[-1, 1]$. We assume that the market price is at least consistent, *i.e.* that $P_{market} \in [P_{BoC}(1), P_{BoC}(-1)]$ so that Eq. (6.24) in ρ has exactly one solution, say ρ_*. This example is not only a toy model because of its basic *B-S* dynamics, it is also due to the fact that, in such a model, more efficient deterministic procedures can be called upon, based on the closed form for the option premium. Our aim is to propose and illustrate below a general methodology for correlation search.

The most convenient way to prevent edge effects due to the fact that $\rho \in [-1, 1]$ is to use a trigonometric parametrization of the correlation by setting

$$\rho = \cos\theta, \;\; \theta \in \mathbb{R}.$$

At this stage, note that

$$\sqrt{1-\rho^2}\, Z^2 = |\sin\theta| Z^2 \stackrel{d}{=} (\sin\theta)\, Z^2$$

since $Z^2 \stackrel{d}{=} -Z^2$. Consequently, as soon as $\rho = \cos\theta$,

$$\rho Z^1 + \sqrt{1-\rho^2}\, Z^2 \stackrel{d}{=} (\cos\theta)\, Z^1 + (\sin\theta)\, Z^2$$

owing to the independence of Z^1 and Z^2.

In general, this introduces an over-parametrization, even inside $[0, 2\pi]$, since $\mathrm{Arccos}(\rho^*) \in [0, \pi]$ and $2\pi - \mathrm{Arccos}(\rho^*) \in [\pi, 2\pi]$ are both solutions to our zero search problem, but this is not at all a significant problem for practical implementation: a more careful examination would show that one of these two equilibrium points is "repulsive" and one is "attractive" for the procedure, see Sects. 6.4.1 and 6.4.5 for a brief discussion: this terminology refers to the status of an equilibrium for the *ODE* associated to a stochastic algorithm and the presence (or not) of noise. A noisy repulsive equilibrium cannot, *a.s.*, be the limit of a stochastic algorithm.

From now on, for convenience, we will just mention the dependence of the premium function in the variable θ, namely

$$\theta \longmapsto P_{BoC}(\theta) := e^{-rt}\mathbb{E}\left[\left(\max\left(x_0^1 e^{\mu_1 T + \sigma_1\sqrt{T}Z_1}, x_0^2 e^{\mu_2 T + \sigma_2\sqrt{T}\left((\cos\theta)\, Z^1 + (\sin\theta)\, Z^2\right)}\right) - K\right)_+\right].$$

The function P_{BoC} is a 2π-periodic continuous function. Extracting the implicit correlation from the market amounts to solving (with obvious notations) the equation

$$P_{BoC}(\theta) = P_{market} \qquad (\rho = \cos\theta),$$

where P_{market} is the quoted premium of the option (mark-to-market price). We need to slightly strengthen the consistency assumption on the market price, which is in fact necessary with almost any zero search procedure: we assume that P_{market} lies in the open interval

$$P_{market} \in \left(P_{BoC}(1), \max_{\theta} P_{BoC}(-1)\right)$$

i.e. that P_{market} is not an extremal value of P_{BoC}. So we are looking for a zero of the function h defined on $\mathbb{R}$ by

$$h(\theta) = P_{BoC}(\theta) - P_{market}.$$

This function admits a representation as an expectation given by

$$h(\theta) = \mathbb{E}\, H(\theta, Z),$$

where $H : \mathbb{R} \times \mathbb{R}^2 \to \mathbb{R}$ is defined for every $\theta \in \mathbb{R}$ and every $z = (z^1, z^2) \in \mathbb{R}^2$ by

$$H(\theta, z) = e^{-rt}\left(\max\left(x_0^1 e^{\mu_1 T + \sigma_1 \sqrt{T} z^1}, x_0^2 e^{\mu_2 T + \sigma_2 \sqrt{T}(z^1 \cos\theta + z^2 \sin\theta)}\right) - K\right)_+ - P_{market}$$

and $Z = (Z^1, Z^2) \stackrel{d}{=} \mathcal{N}(0; I_2)$.

Proposition 6.1 *Assume the above assumptions made on the P_{market} and the function P_{BoC}. If, moreover, the equation $P_{BoC}(\theta) = P_{market}$ has finitely many solutions on $[0, 2\pi]$, then the stochastic zero search recursive procedure defined by*

$$\theta_{n+1} = \theta_n - \gamma_{n+1} H(\theta_n, Z_{n+1}), \;\; \theta_0 \in \mathbb{R},$$

where $(Z_n)_{n\geq 1}$ is an i.i.d, $\mathcal{N}(0; I_2)$ distributed sequence and $(\gamma_n)_{n\geq 1}$ is a step sequence satisfying the decreasing step assumption (6.7), a.s. converges toward solution θ_* to $P_{BoC}(\theta) = P_{market}$.

Proof. For every $z \in \mathbb{R}^2$, $\theta \mapsto H(\theta, z)$ is continuous, 2π-periodic and dominated by a function $g(z)$ such that $g(Z) \in L^2(\mathbb{P})$ (g is obtained by replacing $z^1 \cos\theta + z^2 \sin\theta$ by $|z^1| + |z^2|$ in the above formula for H). One deduces that both the mean function h and $\theta \mapsto \mathbb{E}\, H^2(\theta, Z)$ are continuous and 2π-periodic, hence bounded.

The main difficulty in applying the Robbins–Siegmund Lemma is to find the appropriate Lyapunov function.

As the quoted value P_{market} is not an extremum of the function P, $\int_0^{2\pi} h_\pm(\theta)d\theta > 0$ where $h_\pm := \max(\pm h, 0)$. The two functions $h_\pm$ are 2π-periodic so that $\int_t^{t+2\pi} h_\pm(\theta)d\theta = \int_0^{2\pi} h_\pm(\theta)d\theta > 0$ for every $t > 0$. We consider any (fixed) solution θ_0 to the equation $h(\theta) = 0$ and two real numbers $\beta^\pm$ such that

$$0 < \beta^+ < \frac{\int_0^{2\pi} h_+(\theta)d\theta}{\int_0^{2\pi} h_-(\theta)d\theta} < \beta^-$$

and we set, for every $\theta \in \mathbb{R}$,

$$\ell(\theta) := h_+(\theta) - \beta^+ h_-(\theta)\mathbf{1}_{\{\theta \geq \theta_0\}} - \beta^- h_-(\theta)\mathbf{1}_{\{\theta \leq \theta_0\}}.$$

The function ℓ is clearly continuous, 2π-periodic "on the right" on $[\theta_0, +\infty)$ and "on the left" on $(-\infty, \theta_0]$. In particular, it is a bounded function. Furthermore, owing to the definition of $\beta^\pm$,

$$\int_{\theta_0-2\pi}^{\theta_0} \ell(\theta)d\theta < 0 \quad \text{and} \quad \int_{\theta_0}^{\theta_0+2\pi} \ell(\theta)d\theta > 0$$

so that

$$\lim_{\theta\to\pm\infty} \int_{\theta_0}^{\theta} \ell(u)du = +\infty.$$

As a consequence, there exists a real constant $C > 0$ such that the function

$$L(\theta) = \int_0^{\theta} \ell(u)du + C$$

is non-negative. Its derivative is given by $L' = \ell$ so that

$$L'h = \ell(h_+ - h_-) \geq (h_+)^2 + \beta^+(h_-)^2 \geq 0 \quad \text{and} \quad \left\{L'h = 0\right\} = \{h = 0\}.$$

It remains to prove that $L' = \ell$ is Lipschitz continuous. Calling upon the usual arguments to interchange expectation and differentiation (see Theorem 2.2(b)), one shows that the function P_{BoC} is differentiable at every $\theta \in \mathbb{R} \setminus 2\pi\mathbb{Z}$ with

$$P'_{BoC}(\theta) = \sigma_2\sqrt{T}\,\mathbb{E}\left(\mathbf{1}_{\{X_T^2(\theta)>\max(X_T^1,K)\}}X_T^2(\cos(\theta)Z^2 - \sin(\theta)Z^1)\right)$$

(with an obvious definition for $X_T^2(\theta)$). Furthermore,

$$\sup_{\theta\in\mathbb{R}\setminus 2\pi\mathbb{Z}} |P'_{BoC}(\theta)| \leq \sigma_2\sqrt{T}\,\mathbb{E}\left(x_0^2 e^{\mu_2 T+\sigma_2\sqrt{T}(|Z^1|+|Z^2|)}(|Z^2| + |Z^1|)\right) < +\infty$$

so that P_{Boc} is clearly Lipschitz continuous on the interval $[0, 2\pi]$, hence continuous on the whole real line by periodicity. Consequently h and $h_\pm$ are Lipschitz continuous, which implies in turn that ℓ is Lipschitz continuous as well.

Moreover, we know that the equation $P_{BoC}(\theta) = P_{market}$ has exactly two solutions on every interval of length 2π. Hence the set $\{h = 0\}$ is countable and locally finite, *i.e.* has a finite trace on any bounded interval.

One may apply Theorem 6.3 (for which we provide a self-contained proof in one dimension) to deduce that θ_n will converge toward a solution θ_* of the equation $P_{BoC}(\theta) = P_{market}$. ◇

Exercise. Show that P_{BoC} is continuously differentiable on the whole real line. [Hint: extend the derivative on $\mathbb{R}$ by continuity.]

Extend the preceding to any payoff $\varphi(X_T^1, X_T^2)$ where $\varphi : \mathbb{R}_+^2 \to \mathbb{R}_+$ is a Lipschitz continuous function. In particular, show without the help of differentiation that the corresponding function $\theta \mapsto P(\theta)$ is Lipschitz continuous.

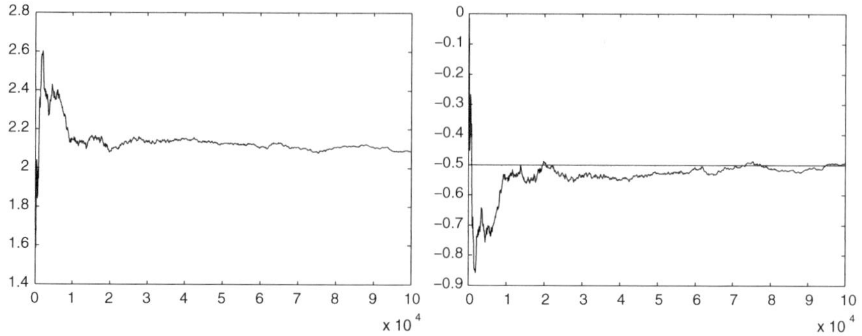

Fig. 6.2 *B-S* BEST-OF-CALL OPTION. $T = 1, r = 0.10, \sigma_1 = \sigma_2 = 0.30, X_0^1 = X_0^2 = 100, K = 100$. Left: convergence of θ_n toward a θ_* (up to $n = 100\,000$). Right: convergence of $\rho_n := \cos(\theta_n)$ toward -0.5

Table 6.1 *B-S* BEST-OF-CALL OPTION. $T = 1$, $r = 0.10$, $\sigma_1 = \sigma_2 = 0.30$, $X_0^1 = X_0^2 = 100$, $K = 100$. Convergence of $\rho_n := \cos(\theta_n)$ toward -0.5

n	$\rho_n := \cos(\theta_n)$
1000	−0.5606
10000	−0.5429
25000	−0.5197
50000	−0.5305
75000	−0.4929
100000	−0.4952

NUMERICAL EXPERIMENT. We set the model parameters to the following values

$$x_0^1 = x_0^2 = 100, \quad r = 0.10, \quad \sigma_1 = \sigma_2 = 0.30, \quad \rho = -0.50$$

and the payoff parameters

$$T = 1, \; K = 100.$$

The reference "Black–Scholes" price 30.75 is used as a mark-to-market price so that the target of the stochastic algorithm is $\theta_* = \text{Arccos}(-0.5) \mod \pi$. The stochastic approximation procedure parameters are

$$\theta_0 = 0, \quad n = 10^5.$$

The choice of θ_0 is "blind" on purpose. Finally, we set $\gamma_n = \frac{0.5}{n}$. No re-scaling of the procedure has been made in the example below (see Table 6.1 and Fig. 6.2).

► **Exercise** (*Yet another toy-example: extracting B-S-implied volatility by stochastic approximation*). Devise a similar procedure to compute the implied volatility in a standard Black–Scholes model (starting at $x > 0$ at $t = 0$ with interest rate r and maturity T).

(a) Show that the B-S premium $C_{BS}(\sigma)$ is even, increasing on $[0, +\infty)$ and continuous as a function of the volatility. Show that $\lim_{\sigma\to 0} C_{BS}(\sigma) = (x - e^{-rt}K)_+$ and $\lim_{\sigma\to+\infty} C_{BS}(\sigma) = x$.
(b) Deduce from (a) that for any mark-to-market price $P_{market} \in [(x - e^{-rt}K)_+, x]$, there is a unique (positive) B-S implicit volatility for this price.
(c) Consider, for every $\sigma \in \mathbb{R}$,

$$H(\sigma, z) = \chi(\sigma)\left(xe^{-\frac{\sigma^2}{2}T+\sigma\sqrt{T}z} - Ke^{-rt}\right)_+,$$

where $\chi(\sigma) = (1 + |\sigma|)e^{-\frac{\sigma^2}{2}T}$. Carefully justify this choice of H and implement the algorithm with $x = K = 100$, $r = 0.1$ and a market price equal to 16.73. Choose the step parameter of the form $\gamma_n = \frac{c}{x}\frac{1}{n}$, $n \geq 1$, with $c \in [0.5, 2]$ (this is simply a suggestion).

Warning. The above exercise is definitely a toy exercise! More efficient methods for extracting standard implied volatility are available (see *e.g.* [209], which is based on a Newton–Raphson zero search algorithm; a dichotomy approach is also very efficient).

► **Exercise** (*Extension to more general asset dynamics*). We now consider a pair of risky assets following two correlated local volatility models,

$$dX_t^i = X_t^i\big(rdt + \sigma_i(X_t)dW_t^i\big),\ X_0^i = x^i > 0,\ i = 1, 2,$$

where the functions $\sigma_i : \mathbb{R}_+^2 \to \mathbb{R}_+$ are bounded Lipschitz continuous functions and the Brownian motions W^1 and W^2 are correlated with correlation $\rho \in [-1, 1]$ so that $\langle W^1, W^2\rangle_t = \rho t$. (This ensures the existence and uniqueness of strong solutions for this SDE, see Chap. 7.)

Assume that we know how to simulate (X_T^1, X_T^2), either exactly, or at least as an approximation by an Euler scheme from a d-dimensional normal vector $Z = (Z^1, \ldots, Z^d) \stackrel{d}{=} \mathcal{N}(0; I_d)$.

Show that the above approach can be extended *mutatis mutandis*.

6.3.3 *The Paradigm of Model Calibration by Simulation*

Let $\Theta \subset \mathbb{R}^d$ be an open convex set of $\mathbb{R}^d$. Let

$$\begin{aligned} Y : (\Theta \times \Omega, \mathcal{B}or(\Theta) \otimes \mathcal{A}) &\longrightarrow (\mathbb{R}^p, \mathcal{B}or(\mathbb{R}^p)) \\ (\theta, \omega) &\longmapsto Y_\theta(\omega) = (Y_\theta^1(\omega), \ldots, Y_\theta^p(\omega)) \end{aligned}$$

be a random vector representative of p payoffs, "re-centered" by their mark-to-market price (see examples below). In particular, for every $i \in \{1, \dots, p\}$, $\mathbb{E}\, Y^i_\theta$ is representative of the error between the "theoretical price" obtained with parameter θ and the quoted price. To make the problem consistent we assume throughout this section that

$$\forall\, \theta \in \Theta, \quad Y_\theta \in L^1_{\mathbb{R}^p}(\Omega, \mathcal{A}, \mathbb{P}).$$

Let $S \in \mathcal{S}_+(p, \mathbb{R}) \cap GL(p, \mathbb{R})$ be a (positive definite) matrix. The resulting inner product is defined by

$$\forall\, u,\ v \in \mathbb{R}^p, \quad \langle u|v \rangle_S := u^* S v$$

and the associated Euclidean norm $|\cdot|_S$ by $|u|_S := \sqrt{\langle u|u \rangle_S}$.

A natural choice for the matrix S is a simple diagonal matrix $S = \mathrm{Diag}(w_1, \dots, w_p)$ with "weights" $w_i > 0$, $i = 1, \dots, p$.

The *paradigm* of model calibration is to find the parameter θ_* that minimizes the "aggregated error" with respect to the $|\cdot|_S$-norm. This leads to the following minimization problem

$$(C) \quad \equiv \quad \mathrm{argmin}_\theta \big|\mathbb{E}\, Y_\theta\big|_S = \mathrm{argmin}_\theta \frac{1}{2}\big|\mathbb{E}\, Y_\theta\big|^2_S.$$

Here are two simple examples to illustrate this somewhat abstract definition.

⊳ **Examples. 1.** *Black–Scholes model.* Let, for any $x, \sigma > 0$, $r \in \mathbb{R}$,

$$X^{x,\sigma}_t = x e^{(r - \frac{\sigma^2}{2})t + \sigma W_t}, \ t \ge 0,$$

where W is a standard Brownian motion. Then let $(K_i, T_i)_{i=1,\dots,p}$ be p "maturity-strike price" pairs. Set

$$\theta := \sigma, \qquad \Theta := (0, +\infty)$$

and

$$Y_\theta := \Big(e^{-rT_i}(X^{x,\sigma}_{T_i} - K_i)_+ - P_{market}(T_i, K_i)\Big)_{i=1,\dots,p}$$

where $P_{market}(T_i, K_i)$ is the mark-to-market price of the option with maturity $T_i > 0$ and strike price K_i.

2. *Merton model (mini-krach).* Now, for every $x, \sigma, \lambda > 0$, $a \in (0, 1)$, set

$$X^{x,\sigma,\lambda,a}_t = x\, e^{(r - \frac{\sigma^2}{2} + \lambda a)t + \sigma W_t}(1 - a)^{N_t}, \ t \ge 0,$$

where W is as above and $N = (N_t)_{t \ge 0}$ is a standard Poisson process with jump intensity λ. Set

$$\theta = (\sigma, \lambda, a), \qquad \Theta = (0, +\infty)^2 \times (0, 1)$$

and

$$Y_\theta = \left(e^{-rT_i}(X_{T_i}^{x,\sigma,\lambda,a} - K_i)_+ - P_{market}(T_i, K_i)\right)_{i=1\ldots,p}.$$

We will also have to make simulability assumptions on Y_θ and, if necessary, on its derivatives with respect to θ (see below). Otherwise our simulation-based approach would be meaningless.

At this stage, there are essentially two approaches that can be considered in order to solve this problem by simulation:

- A Robbins–Siegmund zero search approach of ∇L, which needs to have access to a *representation of the gradient* – assumed to exist – as an expectation of the function L.
- A more direct treatment based on the so-called *Kiefer–Wolfowitz procedure*, which is a variant of the Robbins–Siegmund approach based on a finite difference method (with decreasing step) which does not require the existence of a representation of ∇L as an expectation.

The Robbins–Siegmund approach

We make the following assumptions: for every $\theta_0 \in \Theta$,

$$(\mathrm{Cal}_{RZ}) \equiv \begin{cases} (i) & \mathbb{P}(d\omega)\!-\!a.s.,\ \theta \longmapsto Y_\theta(\omega) \text{ is differentiable} \\ & \text{at } \theta_0 \text{ with Jacobian } \partial_{\theta_0} Y_\theta(\omega), \\ (ii) & \exists\, U_{\theta_0},\ \text{neighborhood of } \theta_0 \text{ in } \Theta,\ \text{such that} \\ & \left(\frac{Y_\theta - Y_{\theta_0}}{|\theta-\theta_0|}\right)_{\theta\in U_{\theta_0}\setminus\{\theta_0\}} \text{is uniformly integrable.} \end{cases}$$

One checks – using the exercise "Extension to uniform integrability" which follows Theorem 2.2 – that $\theta \longmapsto \mathbb{E}\, Y_\theta$ is differentiable and that its Jacobian is given by

$$\partial_\theta \mathbb{E}\, Y_\theta = \mathbb{E}\, \partial_\theta Y_\theta.$$

Then, the function L is differentiable everywhere on Θ and its gradient (with respect to the canonical Euclidean norm) is given by

$$\forall\, \theta \in \Theta, \quad \nabla L(\theta) = \mathbb{E}\,(\partial_\theta Y_\theta)^t\, S\, \mathbb{E}\, Y_\theta = \mathbb{E}\left((\partial_\theta Y_\theta)^t\right) S\, \mathbb{E}\, Y_\theta.$$

At this stage we need a representation of $\nabla L(\theta)$ as an expectation. To this end, we construct, for every $\theta \in \Theta$, an independent copy $\widetilde{Y}_\theta$ of Y_θ defined as follows: we consider the product probability space $(\Omega^2, \mathcal{A}^{\otimes 2}, \mathbb{P}^{\otimes 2})$ and set, for every $(\omega, \widetilde{\omega}) \in \Omega^2$, $Y_\theta(\omega, \widetilde{\omega}) = Y_\theta(\omega)$ (the extension of Y_θ on Ω^2 still denoted by Y_θ) and $\widetilde{Y}_\theta(\omega, \widetilde{\omega}) = Y_\theta(\widetilde{\omega})$. It is straightforward by the product measure theorem that the two families $(Y_\theta)_{\theta\in\Theta}$ and $(\widetilde{Y}_\theta)_{\theta\in\Theta}$ are independent with the same distribution. From now on we will make the usual abuse of notation consisting in assuming that these two independent copies live on the probability space $(\Omega, \mathcal{A}, \mathbb{P})$.

Now, one can write

$$\forall\,\theta \in \Theta,\quad \nabla L(\theta) = \mathbb{E}\left((\partial_\theta Y_\theta)^t\right) S\, \mathbb{E}\,\widetilde{Y}_\theta$$
$$= \mathbb{E}\Big((\partial_\theta Y_\theta)^t\, S\, \widetilde{Y}_\theta\Big).$$

The standard situation, as announced above, is that Y_θ is a vector of payoffs written on d traded risky assets, re-centered by their respective quoted prices. The model dynamics of the d risky assets depends on the parameter θ, say

$$Y_\theta = \big(F_i(\theta, X_{T_i}(\theta))\big)_{i=1,\dots,p},$$

where the price dynamics $(X_t(\theta))_{t\geq 0}$ of the d traded assets is driven by a parametrized diffusion process

$$dX_t(\theta) = b(\theta, t, X_t(\theta))\, dt + \sigma(\theta, t, X_t(\theta))dW_t,\quad X_0(\theta) = x_0(\theta) \in \mathbb{R}^d,$$

where W is an $\mathbb{R}^q$-valued standard Brownian motion defined on a probability space $(\Omega, \mathcal{A}, \mathbb{P})$, b is an $\mathbb{R}^d$-valued parametrized function defined on $\Theta \times [0, T] \times \mathbb{R}^d$ and σ is an $\mathcal{M}(d, q)$-valued parametrized function defined on the same product space, both satisfying appropriate regularity assumptions.

The pathwise differentiability of Y_θ in θ needs that of $X_t(\theta)$ with respect to θ. This question is closely related to the θ-tangent process $\left(\frac{\partial X_t(\theta)}{\partial\theta}\right)_{t\geq 0}$ of $X(\theta)$. A precise statement is provided in Sect. 10.2.2, which ensures that if b and σ are smooth enough with respect to the variable θ, then such a θ-tangent process does exist and is a solution to a linear *SDE* (involving $X(\theta)$ in its coefficients).

Some differentiability properties are also required on the functions F_i in order to fulfill the above differentiability Assumption $(\mathrm{Cal}_{RZ})(i)$. As for model calibrations on vanilla derivative products performed in Finance, F_i is never everywhere differentiable – typically $F_i(y) := e^{-rT_i}(y - K_i)_+ - P_{market}(T_i, K_i)$ – but, if $X_t(\theta)$ has an absolutely continuous distribution (*i.e.* a probability density) for every time $t > 0$ and every $\theta \in \Theta$, then F_i only needs to be differentiable outside a Lebesgue negligible subset of $\mathbb{R}_+$. Finally, we can write formally

$$H\big(\theta, W(\omega)\big) := \big(\partial_\theta Y_\theta(\omega)\big)^t\, S\, \widetilde{Y}_\theta(\omega),$$

where W stands for an abstract random *innovation* taking values in an appropriate space. We denote by the capital letter W the innovation because when the underlying dynamics is a Brownian diffusion or its Euler–Maruyama scheme, it refers to a finite-dimensional functional of (two independent copies of) the $\mathbb{R}^q$-valued standard Brownian motion on an interval $[0, T]$: either (two independent copies of) $(W_{T_1}, \dots, W_{T_p})$ or (two independent copies of) the sequence $\big(\Delta W_{\frac{kT}{n}}\big)_{1\leq k\leq n}$ of Brownian increments with step $\frac{T}{n}$ over the interval $[0, T]$. Thus, these increments naturally appear in the simulation of the Euler scheme $\big(\bar{X}^n_{\frac{kT}{n}}(\theta)\big)_{0\leq k\leq n}$ of the process $\big(X_t(\theta)\big)_{t\in[0,T]}$ when the latter cannot be simulated directly (see Chap. 7 entirely devoted to the Euler scheme of Brownian diffusions). Of course, other situations may occur, especially when dealing with jump diffusions where W usually becomes the increment process of the driving Lévy process.

Nevertheless, we make the following reasonable meta-assumptions that

– the process W is simulable,

– the functional $H(\theta, w)$ can be easily computed for any input (θ, w).

Then, one may define recursively the following zero search algorithm for $\nabla L(\theta) = \mathbb{E}\, H(\theta, W)$, by setting

$$\theta_{n+1} = \theta_n - \gamma_{n+1} H(\theta_n, W^{n+1}),$$

where $(W^n)_{n\geq 1}$ is an i.i.d. sequence of copies of W and $(\gamma_n)_{n\geq 1}$ is a sequence of steps satisfying the usual decreasing step assumptions

$$\sum_n \gamma_n = +\infty \quad \text{and} \quad \sum_n \gamma_n^2 < +\infty.$$

In such a general framework, of course, one cannot ensure that the functions L and H will satisfy the basic assumptions needed to make stochastic gradient algorithms converge, typically

$$\nabla L \text{ is Lipschitz continuous} \quad \text{and} \quad \forall\, \theta \in \Theta, \quad \|H(\theta, \,.\,)\|_2 \leq C\big(1 + \sqrt{L(\theta)}\big)$$

or one of their numerous variants (see *e.g.* [39] for a large overview of possible assumptions). However, in many situations, one can make the problem fit into a converging setting by an appropriate change of variable on θ or by modifying the function L and introducing an appropriate explicit (strictly) positive "weight function" $\chi(\theta)$ that makes the product $\chi(\theta)H(\theta, W(\omega))$ fit with these requirements.

Despite this, the topological structure of the set $\{\nabla L = 0\}$ can be nontrivial, in particular disconnected. Nonetheless, as seen in Theorem 6.3, one can show, under natural assumptions, that

$$\theta_n \text{ converges to a connected component of } \{\chi|\nabla L|^2 = 0\} = \{\nabla L = 0\}.$$

The next step is that if ∇L has several zeros, they cannot all be local minima of L, especially when there are more than two of them (this is a consequence of the well-known Mountain-Pass Lemma, see [164]). Some are local maxima or saddle points of various kinds. These equilibrium points which are not local minima are called *traps*. An important fact is that, under some non-degeneracy assumptions on H at such a parasitic equilibrium point θ_∞ (typically $\mathbb{E}\, H(\theta_\infty, W)H(\theta_\infty, W)^t$ is positive definite at least in the direction of an unstable manifold of h at θ_∞), the algorithm will *a.s.* never converge toward such a "trap". This question has been extensively investigated in the literature in various settings for many years (see [33, 51, 95, 191, 242]).

A final problem may arise due to the incompatibility between the geometry of the parameter set Θ and the above recursive algorithm: to be really defined by the above recursion, we need Θ to be left stable by (almost) all the mappings $\theta \mapsto \theta - \gamma H(\theta, w)$, at least for γ small enough. If not the case, we need to introduce some constraints on the algorithm by projecting it onto Θ whenever θ_n skips outside Θ. This question was originally investigated in [54] when Θ is a convex set.

Once all these technical questions have been circumvented, we may state the following meta-theorem, which says that θ_n $a.s.$ converges toward a local minimum of L.

At this stage it is clear that calibration looks like quite a generic problem for stochastic optimization and that almost all difficulties arising in the field of Stochastic Approximation can be encountered when implementing such a (pseudo-)stochastic gradient to solve it.

The Kieffer–Wolfowitz approach

Practical implementations of the Robbins–Siegmund approach point out a specific technical difficulty: the random functions $\theta \mapsto Y_\theta(\omega)$ are not always pathwise differentiable (nor in the $L^r(\mathbb{P})$-sense, which could be enough). More important in some way, even if one shows that $\theta \mapsto \mathbb{E}\, Y(\theta)$ is differentiable, possibly by calling upon other techniques (log-likelihood method, Malliavin weights, etc) the resulting representation for $\partial_\theta Y(\theta)$ may turn out to be difficult to simulate, requiring much programming care, whereas the random vectors Y_θ can be simulated in a standard way. In such a setting, an alternative is provided by the Kiefer–Wolfowitz algorithm (K–W) which combines the recursive stochastic approximation principle with a finite difference approach to differentiation. The idea is simply to approximate the gradient ∇L by

$$\frac{\partial L}{\partial \theta_i}(\theta) \simeq \frac{L(\theta + \eta^i e_i) - L(\theta - \eta^i e_i)}{2\eta^i}, \quad 1 \le i \le p,$$

where $(e_i)_{1\le i\le p}$ denotes the canonical basis of $\mathbb{R}^p$ and $\eta = (\eta^i)_{1\le i\le p}$. This finite difference term has an integral representation given by

$$\frac{L(\theta + \eta^i e_i) - L(\theta - \eta^i e_i)}{2\eta_i} = \mathbb{E}\left(\frac{\Lambda(\theta + \eta^i, W) - \Lambda(\theta - \eta^i, W)}{2\eta^i}\right)$$

where, with obvious temporary notations,

$$\Lambda(\theta, W) := \langle Y(\theta, W), Y(\theta, W)\rangle_S = Y(\theta, W)^* \, S \, Y(\theta, W)$$

($Y(\theta, W)$ is related to the innovation W). Starting from this representation, we may derive a recursive updating formula for θ_n as follows

$$\theta^i_{n+1} = \theta^i_n - \gamma_{n+1} \frac{\Lambda(\theta_n + \eta^i_{n+1}, W^{n+1}) - \Lambda(\theta_n - \eta^i_{n+1}, W^{n+1})}{2\eta^i_{n+1}}, \quad 1 \le i \le p.$$

We reproduce below a typical convergence result for K–W procedures (see [39]) which is the natural counterpart of the stochastic gradient framework.

Theorem 6.4. *Assume that the function $\theta \mapsto L(\theta)$ is twice differentiable with a Lipschitz continuous Hessian. We assume that*

$$\theta \mapsto \|\Lambda(\theta, W)\|_2 \textit{ has (at most) linear growth}$$

and that the two step sequences respectively satisfy

$$\sum_{n\geq 1} \gamma_n = \sum_{n\geq 1} \eta_n^i = +\infty, \quad \sum_{n\geq 1} \gamma_n^2 < +\infty, \quad \eta_n \to 0, \quad \sum_{n\geq 1} \left(\frac{\gamma_n}{\eta_n^i}\right)^2 < +\infty.$$

Then, θ_n a.s. converges to a connected component of a $\{L = \ell\} \cap \{\nabla L = 0\}$ for some level $\ell \geq 0$.

A special case of this procedure in a linear framework is proposed in Sect. 10.1.2: the decreasing step finite difference method for greeks computation. The *traps* problem for the K–W algorithm (convergence toward a *local minimum* of L) has been more specifically investigated in [191].

Users must keep in mind that this procedure needs some care in the tuning of the step parameters γ_n and η_n. This may need some preliminary numerical experiments. Of course, all the recommendations made for the Robbins–Siegmund procedures remain valid. For more details on the K–W procedure we refer to [39].

6.3.4 Recursive Computation of the VaR *and the* CVaR *(I)*

For a more detailed introduction to Value-at-risk and Conditional Value-at-Risk (CVaR), see *e.g.* [254]. For a comprehensive overview on risk measures, see [92].

Theoretical background on Value-at-risk and Conditional Value-at-Risk

Let $X : (\Omega, \mathcal{A}, \mathbb{P}) \to \mathbb{R}$ be a random variable *representative of a loss*: $X \geq 0$ means a loss equal to X.

Definition 6.1. *The* Value at Risk *at level $\alpha \in (0, 1)$ is the (lowest) α-quantile of the distribution of X, i.e.*

$$\mathrm{VaR}_\alpha(X) := \inf\big\{\xi : \mathbb{P}(X \leq \xi) \geq \alpha\big\}. \tag{6.25}$$

The Value-at-Risk exists since $\lim_{\xi\to+\infty} \mathbb{P}(X \leq \xi) = 1$ and satisfies

$$\mathbb{P}\big(X < \mathrm{VaR}_\alpha(X)\big) \leq \alpha \leq \mathbb{P}\big(X \leq \mathrm{VaR}_\alpha(X)\big).$$

As soon as the distribution function F_X of X is continuous – or, equivalently, the distribution of X has no atom – the value at risk satisfies

$$\mathbb{P}\big(X \le \mathrm{VaR}_\alpha(X)\big) = \alpha. \tag{6.26}$$

Moreover, if F_X is also (strictly) increasing, then it is the unique solution of the above equation (otherwise it is the lowest one). In such a case, we will say that the *Value-at-Risk* is unique.

Roughly speaking α represents here *an alarm or warning level*, typically 0.95 or 0.99 like for confidence intervals. Beyond this level, the loss becomes unacceptable.

Unfortunately this measure of risk is not consistent, for several reasons which are discussed, *e.g.* by Föllmer and Schied in [92]. The main point is that it does not favor the diversification of portfolios to guard against the risk.

When $X \in L^1(\mathbb{P})$ with a continuous distribution, a consistent measure of risk is provided by the *Conditional Value-at-Risk* (at level α).

Definition 6.2. *Let $X \in L^1(\mathbb{P})$ with an atomless distribution. The* Conditional Value-at-Risk *at level $\alpha \in (0, 1)$ is defined by*

$$\mathrm{CVaR}_\alpha(X) := \mathbb{E}\big(X \,|\, X \ge \mathrm{VaR}_\alpha(X)\big). \tag{6.27}$$

Remark. Note that in the case of non-uniqueness of the α-quantile, the Conditional Value-at-risk is still well-defined since the above conditional expectation does not depend upon the choice of this α-quantile solution to (6.26).

► **Exercises. 1.** Assume that the distribution of X has no atom. Show that

$$\mathrm{CVaR}_\alpha(X) = \mathrm{VaR}_\alpha(X) + \frac{1}{1-\alpha}\int_{\mathrm{VaR}_\alpha(X)}^{+\infty} \mathbb{P}(X > u)\,du.$$

[Hint: use that for a non-negative r.v. Y, $\mathbb{E}\,Y = \displaystyle\int_0^{+\infty} \mathbb{P}(Y \ge y)\,dy$.]

2. Show that the conditional value-at-risk $\mathrm{CVaR}_\alpha(X)$ is a consistent measure of risk, *i.e.* that it satisfies the following three properties

- $\forall\,\lambda > 0, \quad \mathrm{CVaR}_\alpha(\lambda X) = \lambda \mathrm{CVaR}_\alpha(X)$.
- $\forall\, a \in \mathbb{R}, \quad \mathrm{CVaR}_\alpha(X + a) = \mathrm{CVaR}_\alpha(X) + a$.
- Let $X,\ Y \in L^1(\mathbb{P})$, $\mathrm{CVaR}_\alpha(X + Y) \le \mathrm{CVaR}_\alpha(X) + \mathrm{CVaR}_\alpha(Y)$.

The following formulation of the VaR_α and CVaR_α as solutions to an optimization problem is due to Rockafellar and Uryasev in [254].

Proposition 6.2. (The Rockafellar–Uryasev representation formula) *Let $X \in L^1(\mathbb{P})$ with a continuous distribution. The function $L : \mathbb{R} \to \mathbb{R}$ defined by*

$$L(\xi) = \xi + \frac{1}{1-\alpha}\,\mathbb{E}\,(X - \xi)_+$$

is convex and $\displaystyle\lim_{|\xi|\to+\infty} L(\xi) = +\infty$. Furthermore, L attains a minimum

$$\text{CVaR}_\alpha(X) = \min_{\xi \in \mathbb{R}} L(\xi) \geq \mathbb{E}\, X$$

at

$$\text{VaR}_\alpha(X) = \inf \text{argmin}_{\xi \in \mathbb{R}} L(\xi).$$

Proof. The function L is clearly convex and 1-Lipschitz continuous since both functions $\xi \mapsto \xi$ and $\xi \mapsto (x - \xi)_+$ are convex and 1-Lipschitz continuous for every $x \in \mathbb{R}$. As X has no atom, the function L is also differentiable on the whole real line with a derivative given for every $\xi \in \mathbb{R}$ by

$$L'(\xi) = 1 - \frac{1}{1-\alpha}\mathbb{P}(X > \xi) = \frac{1}{1-\alpha}\big(\mathbb{P}(X \leq \xi) - \alpha\big).$$

This follows from the interchange of differentiation and expectation allowed by Theorem 2.2(a) since $\xi \mapsto \xi + \frac{1}{1-\alpha}(X - \xi)_+$ is differentiable at a given ξ_0 on the event $\{X = \xi_0\}$, *i.e* $\mathbb{P}$-*a.s.* since X is atomless and, on the other hand, is Lipschitz continuous with Lipschitz continuous ratio $\frac{X}{1-\alpha} \in L^1(\mathbb{P})$. The second equality is obvious. Then L attains an absolute minimum at any solution ξ_α of the equation $\mathbb{P}(X > \xi_\alpha) = 1 - \alpha$, *i.e.* $\mathbb{P}(X \leq \xi_\alpha) = \alpha$. In particular, L attains a minimum at the value-at-risk which is the lowest such solution. Furthermore,

$$\begin{aligned}
L(\xi_\alpha) &= \xi_\alpha + \frac{\mathbb{E}\,((X - \xi_\alpha)_+)}{\mathbb{P}(X > \xi_\alpha)} \\
&= \frac{\xi_\alpha \mathbb{E}\mathbf{1}_{\{X > \xi_\alpha\}} + \mathbb{E}\,((X - \xi_\alpha)\mathbf{1}_{\{X > \xi_\alpha\}})}{\mathbb{P}(X > \xi_\alpha)} \\
&= \frac{\mathbb{E}\,(X\mathbf{1}_{\{X > \xi_\alpha\}})}{\mathbb{P}(X > \xi_\alpha)} \\
&= \mathbb{E}\,(X \mid \{X > \xi_\alpha\}).
\end{aligned}$$

The function L satisfies

$$\lim_{\xi \to +\infty} \frac{L(\xi)}{\xi} = \lim_{\xi \to +\infty} \left(1 + \frac{1}{1-\alpha}\mathbb{E}\,(X/\xi - 1)_+\right) = 1$$

and

$$\lim_{\xi \to +\infty} \frac{L(-\xi)}{\xi} = \lim_{\xi \to +\infty} \left(-1 + \frac{1}{1-\alpha}\mathbb{E}\,(X/\xi + 1)_+\right) = -1 + \frac{1}{1-\alpha} = \frac{\alpha}{1-\alpha}.$$

Hence, $\lim_{\xi \to \pm\infty} L(\xi) = +\infty$.

Finally, by Jensen's Inequality

$$L(\xi) \geq \xi + \frac{1}{1-\alpha}\big(\mathbb{E}\,X - \xi\big)_+.$$

One checks that the function on the right-hand side of the above inequality attains its minimum at its only break of monotonicity, *i.e.* when $\xi = \mathbb{E}\,X$. This completes the proof. ◇

Computing $\mathrm{VaR}_\alpha(X)$ **and** $\mathrm{CVaR}_\alpha(X)$ **by stochastic approximation**

▷ *First step: a stochastic gradient to compute the Value-at-risk.* The Rockafellar–Uryasev representation suggests to implement a stochastic gradient descent since the function L has a representation as an expectation

$$L(\xi) = \mathbb{E}\Big(\xi + \frac{1}{1-\alpha}(X-\xi)_+\Big).$$

Furthermore, if the distribution $\mathbb{P}_X$ has no atom, we know that, the function L being convex and differentiable with derivative $L' = \frac{F-\alpha}{1-\alpha}$ where $F = F_X$ denotes the c.d.f. of X. It satisfies

$$\forall\,\xi,\ \xi'\in\mathbb{R},\quad (L'(\xi)-L'(\xi'))(\xi-\xi') = \frac{1}{1-\alpha}\big(F(\xi)-F(\xi')\big)(\xi-\xi')\geq 0$$

and if the Value-at-Risk $\mathrm{CVaR}_\alpha(X)$ is the unique solution to $F(\xi)=\alpha$,

$$\forall\,\xi\in\mathbb{R},\quad \xi\neq \mathrm{VaR}_\alpha(X),$$
$$L'(\xi)\big(\xi-\mathrm{VaR}_\alpha(X)\big) = \frac{1}{1-\alpha}\big(F(\xi)-\alpha\big)\big(\xi-\mathrm{VaR}_\alpha(X)\big)>0.$$

Proposition 6.3 *Assume that* $X\in L^1(\mathbb{P})$ *with a unique Value-at-Risk* $\mathrm{VaR}_\alpha(X)$. *Let* $(X_n)_{n\geq 1}$ *be an i.i.d. sequence of random variables with the same distribution as* X, *let* $\xi_0\in L^1(\mathbb{P})$, *independent of* $(X_n)_{n\geq 1}$, *and let* $(\gamma_n)_{n\geq 1}$ *be a positive sequence of real numbers satisfying the decreasing step assumption (6.7). Then, the stochastic algorithm* $(\xi_n)_{n\geq 0}$ *defined by*

$$\xi_{n+1} = \xi_n - \gamma_{n+1}H(\xi_n, X_{n+1}),\quad n\geq 0, \tag{6.28}$$

where

$$H(\xi,x) := 1-\frac{1}{1-\alpha}\mathbf{1}_{\{x\geq\xi\}} = \frac{1}{1-\alpha}\big(\mathbf{1}_{\{x<\xi\}}-\alpha\big), \tag{6.29}$$

a.s. converges toward the Value-at-Risk, i.e.

$$\xi_n \xrightarrow{a.s.} \mathrm{VaR}_\alpha(X).$$

Furthermore, the sequence $(L(\xi_n))_{n\geq 0}$ *is* L^1*-bounded so that* $L(\xi_n)\to \mathrm{CVaR}_\alpha(X)$ *a.s. and in every* $L^p(\mathbb{P})$, $p\in(0,1]$.

Proof. First assume that $\xi_0\in L^2(\mathbb{P})$. The sequence $(\xi_n)_{n\geq 0}$ defined by (6.28) is the stochastic gradient related to the Lyapunov function $\widetilde{L}(\xi) = L(\xi)-\mathbb{E}\,X$, but owing to the convexity of L and the uniqueness of $\mathrm{VaR}_\alpha(X)$, it is more convenient to consider the quadratic Lyapunov function $\xi\mapsto\big(\xi-\mathrm{VaR}_\alpha(X)\big)^2$ and rely on Corollary 6.1(a) (Robbins–Monro theorem), once we have observed that the function

$(\xi, x) \mapsto H(\xi, x)$ is bounded by $\frac{\alpha}{1-\alpha}$ so that $\xi \mapsto \|H(\xi, X)\|_2$ is bounded as well. The conclusion follows directly from the Robbins–Monro theorem.

In the general case – $\xi_0 \in L^1(\mathbb{P})$ – one introduces the Lyapunov function $\widetilde{L}(\xi) = \frac{(\xi-\xi_\alpha)^2}{\sqrt{1+(\xi-\xi_\alpha)^2}}$ where we set $\xi_\alpha = \text{VaR}_\alpha(X)$ for convenience. The derivative of $\widetilde{L}$ is given by $\widetilde{L}'(\xi) = \frac{(\xi-\xi_\alpha)(2+(\xi-\xi_\alpha)^2)}{(1+(\xi-\xi_\alpha)^2)^{\frac{3}{2}}}$. One checks on the one hand that $\widetilde{L}'$ is Lipschitz continuous over the real line (*e.g.* because $\widetilde{L}''$ is bounded) and, on the other hand, $\{\widetilde{L}' = 0\} \cap \{L' = 0\} = \{L' = 0\} = \{\text{VaR}_\alpha(X)\}$. Then Theorem 6.3 (pseudo-Stochastic Gradient) applies and yields the announced conclusion (note that we need its one-dimensional version that we established in detail). ◇

► **Exercises. 1.** Show that if X has a bounded density f_X, then a direct application of the stochastic gradient convergence result (Corollary 6.1(*b*)) yields the announced result under the assumption $\xi_0 \in L^1(\mathbb{P})$. [Hint: show that L' is Lipschitz continuous.]

2. Under the additional assumption of Exercise 1, show that the mean function h satisfies $h'(x) = L''(x) = \frac{f_X(x)}{1-\alpha}$. Deduce a way to optimize the step sequence of the algorithm based on the *CLT* for stochastic algorithms stated further on in Sect. 6.4.3.

The second exercise is inspired by a simpler "α-quantile" approach. It leads to a more general *a.s.* convergence result for our algorithm, stated in the proposition below.

Proposition 6.4 *If $X \in L^p(\mathbb{P})$ for a $p > 0$ and is continuous with a unique value-at-risk and if $\xi_0 \in L^p(\mathbb{P})$, then the algorithm (6.28) a.s. converges toward* $\text{VaR}_\alpha(X)$.

Remark. The uniqueness of the value-at-risk can also be relaxed. The conclusion becomes that (ξ_n) *a.s.* converges to a random variable taking values in the "$\text{VaR}_\alpha(X)$ set": $\{\xi \in \mathbb{R} : \mathbb{P}(X \le \xi) = \alpha\}$ (see [29] for a statement in that direction).

▷ *Second step: adaptive computation of* $\text{CVaR}_\alpha(X)$. The main aim of this section is to compute on-line the $\text{CVaR}_\alpha(X)$. The fact that $L(\xi_n) \to \text{CVaR}_\alpha(X)$ is no practical help since the function L is not explicit. How can we proceed? The idea is to devise a *companion procedure* of the above stochastic gradient. Still set temporarily $\xi_\alpha = \text{VaR}_\alpha(X)$ for convenience. It follows from Proposition 6.3 and Césaro's averaging principle that $\dfrac{L(\xi_0) + \cdots + L(\xi_{n-1})}{n} \to \text{CVaR}_\alpha(X)$ *a.s.* and in L^1 since one has $L(\xi_{n-1}) \to \text{CVaR}_\alpha(X)$ *a.s.* and in L^1. In particular

$$\mathbb{E}\left(\frac{L(\xi_0) + \cdots + L(\xi_{n-1})}{n}\right) \longrightarrow \text{CVaR}_\alpha(X) \quad \text{as} \quad n \to +\infty.$$

On the other hand, we know that, for every $\xi \in \mathbb{R}$,

$$L(\xi) = \mathbb{E}\,\Lambda(\xi, X) \quad \text{where} \quad \Lambda(\xi, x) = \xi + \frac{(x-\xi)_+}{1-\alpha}.$$

Using that X_k and $(\xi_0, \xi_1, \dots, \xi_{k-1})$ are independent for every $k \geq 1$, one has

$$L(\xi_{k-1}) = \Big[\mathbb{E}\,\Lambda(\xi, X)\Big]_{|\xi=\xi_{k-1}} = \mathbb{E}\big[\Lambda(\xi_{k-1}, X_k)\,\big|\,\xi_{k-1}\big],\ k \geq 1,$$

so that

$$\mathbb{E}\left(\frac{\Lambda(\xi_0, X_1) + \cdots + \Lambda(\xi_{n-1}, X_n)}{n}\right) = \mathbb{E}\left(\frac{L(\xi_0) + \cdots + L(\xi_{n-1})}{n}\right)$$
$$\longrightarrow \mathrm{CVaR}_\alpha(X) \ \text{ as } \ n \to +\infty.$$

This suggests to consider the sequence $(C_n)_{n\geq 0}$ defined by

$$C_n = \frac{1}{n}\sum_{k=0}^{n-1} \Lambda(\xi_k, X_{k+1}), \quad n \geq 1, \quad C_0 = 0,$$

as a candidate to be an estimator of $\mathrm{CVaR}_\alpha(X)$. This sequence can clearly be recursively defined since, for every $n \geq 0$,

$$C_{n+1} = C_n - \frac{1}{n+1}\big(C_n - \Lambda(\xi_n, X_{n+1})\big). \tag{6.30}$$

Proposition 6.5 *Assume that $X \in L^{1+\rho}(\mathbb{P})$ for $\rho \in (0,1]$ and that $\xi_n \longrightarrow \mathrm{VaR}_\alpha(X)$ a.s. Then*

$$C_n \overset{a.s.}{\longrightarrow} \mathrm{CVaR}_\alpha(X) \quad as \quad n \to +\infty.$$

Proof. We will prove this claim in detail in the quadratic case $\rho = 1$. The proof in the general case relies on the Chow Theorem (see [81] or the second exercise right after the proof). First, one decomposes

$$C_n - L(\xi_\alpha) = \frac{1}{n}\sum_{k=0}^{n-1} L(\xi_k) - L(\xi_\alpha) + \frac{1}{n}\sum_{k=1}^{n} Y_k$$

with $Y_k := \Lambda(\xi_{k-1}, X_k) - L(\xi_{k-1})$, $k \geq 1$. It is clear that $\frac{1}{n}\sum_{k=0}^{n-1} L(\xi_k) - L(\xi_\alpha) \to 0$ as $n \to +\infty$ by Césaro's principle. As for the second term, we first note that

$$\Lambda(\xi, x) - L(\xi) = \frac{1}{1-\alpha}\Big((x-\xi)_+ - \mathbb{E}\,(X-\xi)_+\Big)$$

so that, $x \mapsto x_+$ being 1-Lipschitz continuous,

$$|\Lambda(\xi, x) - L(\xi)| \leq \frac{1}{1-\alpha}\mathbb{E}\,|X - x| \leq \frac{1}{1-\alpha}\big(\mathbb{E}\,|X| + |x|\big).$$

Consequently, for every $k \geq 1$,

$$\mathbb{E}\, Y_k^2 \leq \frac{2}{(1-\alpha)^2}\Big((\mathbb{E}\, X)^2 + \mathbb{E}\, X^2\Big).$$

We consider the filtration $\mathcal{F}_n := \sigma(\xi_0, X_1, \ldots, X_n)$. One checks that $(\xi_n)_{n\geq 1}$ is $(\mathcal{F}_n)$-adapted and that, for every $k \geq 1$,

$$\mathbb{E}\,(Y_k \,|\, \mathcal{F}_{k-1}) = \mathbb{E}\Big(\Lambda(\xi_{k-1}, X_k) \,|\, \mathcal{F}_{k-1}) - L(\xi_{k-1}\Big) = L(\xi_{k-1}) - L(\xi_{k-1}) = 0.$$

Hence, the sequence defined by $N_0 = 0$ and

$$N_n := \sum_{k=1}^{n} \frac{Y_k}{k}, \quad n \geq 1,$$

is a square integrable martingale with a predictable bracket process given by

$$\langle N \rangle_n = \sum_{k=1}^{n} \frac{\mathbb{E}\,(Y_k^2 \,|\, \mathcal{F}_{k-1})}{k^2}$$

so that

$$\mathbb{E}\,\langle N \rangle_\infty \leq \sup_n \mathbb{E}\, Y_n^2 \times \sum_{k\geq 1} \frac{1}{k^2} < +\infty.$$

Consequently $N_n \to N_\infty$ *a.s.* and in L^2 as $n \to +\infty$ (see Theorem 12.7(*b*)). Then the Kronecker Lemma (see Lemma 12.1) implies that

$$\frac{1}{n}\sum_{k=1}^{n} Y_k \longrightarrow 0 \quad a.s. \quad \text{as} \quad n \to +\infty$$

which finally implies that

$$C_n \longrightarrow \mathrm{CVaR}_\alpha(X) \quad a.s. \quad \text{as} \quad n \to +\infty.$$ ◇

Remark. For practical implementation, one may prefer to first estimate the $\mathrm{VaR}_\alpha(X)$ and, once it is done, use a regular Monte Carlo procedure to evaluate the $\mathrm{CVaR}_\alpha(X)$.

► **Exercises. 1.** Show that an alternative method to compute $\mathrm{CVaR}_\alpha(X)$ is to design the following recursive procedure

$$C_{n+1} = C_n - \gamma_{n+1}\big(C_n - \Lambda(\xi_n, X_{n+1})\big), \quad n \geq 0, \quad C_0 = 0, \tag{6.31}$$

where $(\gamma_n)_{n\geq 1}$ is the step sequence implemented in the algorithm (6.28) to compute $\mathrm{VaR}_\alpha(X)$.

2 (Proof of Proposition 6.5**).** Show that the conclusion of Proposition 6.5 remains valid if $X \in L^{1+\rho}(\mathbb{P})$. [Hint: rely on the Chow Theorem ([3]).]

ℵ Practitioner's corner

▷ WARNING!...TOWARD AN OPERATING PROCEDURE! As it is presented, the preceding is essentially a toy exercise for the following reason: in practice $\alpha \simeq 1$, the convergence of the above algorithm turns out to be slow and chaotic since $\mathbb{P}(X > \text{VaR}_\alpha(X)) = 1 - \alpha$ is close to 0. For a practical implementation on real life portfolios the above algorithm must be combined with an importance sampling transformation to "recenter" the simulation where things do happen. A realistic and efficient procedure is developed and analyzed in [29].

▷ A second practical improvement to the procedure is to make the level α vary slowly from, say, $\alpha_0 = \frac{1}{2}$ to the target level α during the grist part of the simulation.

▷ An alternative approach to this recursive algorithm is to invert the empirical measure of the innovations $(X_n)_{n\geq 1}$ (see [83]). This method is close to the one described above once rewritten in a recursive way (it corresponds to the step sequence $\gamma_n = \frac{1}{n}$).

6.3.5 *Stochastic Optimization Methods for Optimal Quantization*

Let $X : (\Omega, \mathcal{A}, \mathbb{P}) \to \mathbb{R}^d$ be a random vector taking at least $N \in \mathbb{N}^*$ values. Hence any optimal N-quantizer has N pairwise distinct components. We want to produce an optimal quadratic quantization of X at *level* N, *i.e.* to produce an N-quantizer which minimizes the quadratic quantization error as introduced in Chap. 5.

The starting point of numerical methods is that any optimal (or even locally optimal) quantizer $x = (x_1, \ldots, x_N)$ satisfies the stationarity equation as briefly recalled below.

Competitive Learning Vector Quantization

The *Competitive Learning Vector Quantization* algorithm – or *CLVQ* – is simply the stochastic gradient descent derived from the quadratic distortion function (see below). Unfortunately, this function, viewed as a potential to be minimized, does not fulfill any of the required assumptions to be a Lyapunov function in the sense of Theorem 6.1 (the Robbins–Siegmund Lemma) and Corollary 6.1(*b*) (Stochastic algorithm). So we cannot rely on these general results to ensure the *a.s.* convergence

[3] Let $(M_n)_{n\geq 0}$ be an $(\mathcal{F}_n, \mathbb{P})$-martingale null at 0 and let $\rho \in (0, 1]$; then

$$M_n \stackrel{a.s.}{\longrightarrow} M_\infty \quad \text{on} \quad \left\{ \sum_{n\geq 1} \mathbb{E}\left(|\Delta M_n|^{1+\rho} \,|\, \mathcal{F}_{n-1}\right) < +\infty \right\}.$$

of such a procedure. First, the potential does not go to infinity when the norm of the N-quantizer goes to infinity; secondly, the procedure is not well-defined when some components of the quantizers merge. Very partial results are known about the asymptotic behavior of this procedure (see *e.g.* [39, 224]) except in one dimension in the unimodal setting (log-concave density for the distribution to be quantized) where much faster deterministic Newton–Raphson like procedures can be implemented if the cumulative distribution and the first moment functions both have closed forms. But its use remains limited due to its purely one-dimensional feature.

However, these theoretical gaps (possible asymptotic "merge" of components or escape of components at infinity) are not observed in practical simulations. This fully justifies presenting it in detail.
Let us recall that the quadratic distortion function (see Definition 5.1.2) is defined as the squared quadratic mean-quantization error, *i.e.*

$$\forall x = (x_1, \dots, x_N) \in (\mathbb{R}^d)^N, \quad \mathcal{Q}_{2,N}(x) = \|X - \widehat{X}^{\Gamma_x}\|_2^2 = \mathbb{E}\, q_{2,N}(x, X),$$

where $\Gamma_x = \{x_1, \dots, x_N\}$ and the *local distortion* function $q_{2,N}(x, \xi)$ is defined by

$$\forall x \in (\mathbb{R}^d)^N, \; \forall \xi \in \mathbb{R}^d, \quad q_{2,N}(x, \xi) = \min_{1 \le i \le N} |\xi - x_i|^2 = \operatorname{dist}\big(\xi, \{x_1, \dots, x_N\}\big)^2.$$

Proposition 6.6 *The distortion function $\mathcal{Q}_{2,N}$ is continuously differentiable at N-tuples $x \in (\mathbb{R}^d)^N$ satisfying*

$$x \text{ has pairwise distinct components and } \mathbb{P}\Big(\bigcup_{1 \le i \le N} \partial C_i(x)\Big) = 0$$

with a gradient $\nabla \mathcal{Q}_{2,N} = \left(\frac{\partial \mathcal{Q}_{2,N}}{\partial x_i}\right)_{1 \le i \le N}$ given by

$$\frac{\partial \mathcal{Q}_{2,N}}{\partial x_i}(x) := \mathbb{E}\left(\frac{\partial q_{2,N}}{\partial x_i}(x, X)\right) = \int_{\mathbb{R}^d} \frac{\partial q_{2,N}}{\partial x_i}(x, \xi)\mathbb{P}_X(d\xi),$$

the local gradient *being given by*

$$\frac{\partial q_{2,N}}{\partial x_i}(x, \xi) := 2(x_i - \xi)\mathbf{1}_{\{\mathrm{Proj}_x(\xi) = x_i\}}, \quad 1 \le i \le N,$$

where Proj_x *denotes a (Borel) projection following the nearest neighbor rule on the* grid $\{x_1, \dots, x_N\}$.

Proof. First note that, as the N-tuple x has pairwise distinct components, all the interiors $\overset{\circ}{C}_i(x)$, $i = 1, \dots, N$, of the Voronoi cells induced by x are non-empty. Let $\xi \in \bigcup_{i=1}^{N} \overset{\circ}{C}_i(x) = \mathbb{R}^d \setminus \bigcup_{1 \le i \le N} \partial C_i(x)$. One has $\min_{i \ne j} \big||\xi - x_i| - |\xi - x_j|\big| > 0$ and

$$q_{2,N}(x,\xi) = \sum_{j=1}^{N} |x_j - \xi|^2 \mathbf{1}_{\{\xi \in \overset{\circ}{C}_j(x)\}}.$$

Now, if $x' \in (\mathbb{R}^d)^N$ satisfies $\max_{1\le i\le N} |x_i - x'_i| < \min_{i\neq j} \big||\xi - x_i| - |\xi - x_j|\big| \in (0,+\infty)$, then $\mathbf{1}_{\{\xi \in \overset{\circ}{C}_j(x)\}} = \mathbf{1}_{\{\xi \in \overset{\circ}{C}_j(x')\}}$ for every $i = 1,\dots,N$. Consequently,

$$\forall\, i \in \{1,\dots,N\},\quad \frac{\partial}{\partial x_i}\big(q_{2,N}(x,\xi)\big) = \mathbf{1}_{\{\xi \in \overset{\circ}{C}_i(x)\}} \frac{\partial |x_i - \xi|^2}{\partial x_i} = 2\,\mathbf{1}_{\{\xi \in \overset{\circ}{C}_i(x)\}}(x_i - \xi).$$

Hence, it follows from the assumption $\mathbb{P}\Big(\bigcup_{1\le i\le N} \partial C_i(x)\Big) = 0$ that $q_{2,N}(\,.\,,\xi)$ is $\mathbb{P}_X(d\xi)$-*a.s.* differentiable with a gradient $\nabla_x q_{2,N}(x,\xi)$ given by the above formula. On the other hand, for every $x, x' \in (\mathbb{R}^d)^N$, the function $\mathcal{Q}_{2,N}$ is locally Lipschitz continuous since

$$\begin{aligned}
\big|\mathcal{Q}_{2,N}(x') \;-\; &\mathcal{Q}_{2,N}(x)\big| \\
&\le \int_{\mathbb{R}^d} \big| \min_{1\le i\le N} |x_i - \xi| - \min_{1\le i\le N} |x'_i - \xi|\big| \Big(\min_{1\le i\le N} |x_i - \xi| \\
&\qquad\qquad + \min_{1\le i\le N} |x'_i - \xi|\Big) \mathbb{P}_X(d\xi) \\
&\le \max_{1\le i\le N} |x_i - x'_i| \int_{\mathbb{R}^d} \Big(\max_{1\le i\le N} \big(|x_i| + |\xi|\big) + \big(\min_{1\le i\le N} |x'_i| + |\xi|\big)\Big) \mathbb{P}_X(d\xi) \\
&\le C_X \max_{1\le i\le N} |x_i - x'_i|_\infty \Big(1 + \max_{1\le i\le N} |x_i| + \max_{1\le i\le N} |x'_i|\Big).
\end{aligned}$$

As a consequence of the local interchange Lebesgue differentiation Theorem 2.2(*a*), $\mathcal{Q}_{2,N}$ is differentiable at x with the announced gradient. In turn, the continuity of the gradient $\nabla \mathcal{Q}_{2,N}$ follows likewise from the Lebesgue continuity theorem (see *e.g.* [52]) ◇

Remarks. In fact, when $p > 1$, the L^p-distortion function $\mathcal{Q}_{p,N}$ with respect to a Euclidean norm is also differentiable at N-tuples having pairwise distinct components with gradient

$$\begin{aligned}
\nabla \mathcal{Q}_{p,N}(x) &= p\left(\int_{C_i(x)} \frac{x_i - \xi}{|x_i - \xi|} |x_i - \xi|^{p-1} \mu(d\xi)\right)_{1\le i\le N} \\
&= p\left(\mathbb{E}\left(\mathbf{1}_{\{X \in C_i(x)\}} \frac{x_i - X}{|x_i - X|} |x_i - X|^{p-1}\right)\right)_{1\le i\le N}.
\end{aligned}$$

An extension to the case $p \in (0,1]$ exists under appropriate continuity and integrability assumptions on the distribution μ so that $\mu(\{a\}) = 0$ for every a and the function $a \mapsto \int_{\mathbb{R}^d} |\xi - a|^{p-1}\mu(d\xi)$ remains bounded on compact sets of $\mathbb{R}^d$. A more general differentiation result exists for *strictly convex smooth* norms (see Lemma 2.5, p. 28 in [129]).

Remark. Note that if x is an optimal N-quantizer (and X is supported by at least N values), then the condition $\mathbb{P}\Big(\bigcup_{1\le i\le N}\partial C_i(x)\Big) = 0$ is always satisfied even if the distribution of X is not absolutely continuous and possibly assigns mass to hyperplanes. See Theorem 4.2 in [129] for a proof.

As emphasized in the introduction of this chapter, the gradient $\nabla\mathcal{Q}_{2,N}$ having an integral representation, it is formally possible to minimize $\mathcal{Q}_{2,N}$ using a stochastic gradient descent.

Unfortunately, it is easy to check that $\underline{\lim}_{|x|\to+\infty}\mathcal{Q}_{2,N}(x) < +\infty$, though $\underline{\lim}_{\min_i|x_i|\to+\infty}\mathcal{Q}_{2,N}(x) = +\infty$. Consequently, it is hopeless to apply the standard $a.s.$ convergence result for the stochastic gradient procedure from Corollary 6.1(b). But of course, we can still write it down formally and implement it.

▷ INGREDIENTS:

– A sequence $\xi^1,\ldots,\xi^n,\ldots$ of (simulated) independent copies of X,

– A $(0, 1)$-valued step sequence $(\gamma_n)_{n\ge1}$. One usually chooses the step in the parametric families $\gamma_n = \dfrac{c}{b+n} \downarrow 0$ (decreasing step satisfying (6.7)). Other choices are possible: slowly decreasing $\frac{c}{b+n^\vartheta}$, $\frac{1}{2} < \vartheta < 1$, in view of Ruppert and Polyak's averaging procedure, a small constant step $\gamma_n = \gamma_\infty \simeq 0$ ($\gamma_n \downarrow \gamma_\infty > 0$) to better explore the state space.

▷ THE STOCHASTIC GRADIENT DESCENT PROCEDURE.

Let $x^{[n]} = \big(x_1^{[n]},\ldots,x_N^{[n]}\big)$ denote the running N-quantizer at iteration n (keep in mind that the level remains N, fixed throughout the procedure). The procedure formally reads:

$$x^{[n+1]} = x^{[n]} - \gamma_{n+1}\nabla_x q_{2,N}\big(x^{[n]},\xi^{n+1}\big),\quad x^{[0]} \in (\mathbb{R}^d)^N,$$

where $x[0] = \big(x_1[0],\ldots,x_N[0]\big)$ is a starting value with pairwise distinct components in $\mathbb{R}^d$.

The updating of the current quantizer at time $n+1$ is performed as follows: let $x^{[n]} = (x_1^{[n]},\ldots,x_N^{[n]})$,

- *Competition phase* $(winner\ selection)$: $i^{[n+1]} \in \text{argmin}_i\big|x_i^{[n]} - \xi^{n+1}\big|$ (nearest neighbor search),
- *Learning phase* : $\begin{cases} x_{i^{[n+1]}}^{[n+1]} := \text{Dilatation}\big(\xi^{n+1}, 1-\gamma_{n+1}\big)(x_{i^{[n+1]}}^{[n]}\big),\\ x_i^{[n+1]} := x_i^{[n]},\ i \ne i^{[n+1]},\end{cases}$

where Dilatation(a,ρ) is the dilatation with center $a \in \mathbb{R}^d$ and coefficient $\rho \in (0, 1)$ defined by

$$\text{Dilatation}(a,\rho)(u) = a + \rho(a-u).$$

In case of conflict on the winner index one applies a prescribed rule (like selection at random of the true winner, uniformly among the winner candidates).

Warning! Note that choosing a $(0,1)$-valued step sequence $(\gamma_n)_{n\geq 1}$ is crucial to ensure that the learning phase is a dilatation with coefficient $\rho = 1-\gamma_{n+1}$ at iteration $n+1$.

One can easily check by induction that if $x^{[n]}$ has pairwise distinct components $x_i^{[n]}$, this is preserved by the learning phase. So, that the above procedure is well-defined. The name of the procedure – Competitive Learning Vector Quantization algorithm – is of course inspired by these two phases.

▷ HEURISTICS: $x^{[n]} \longrightarrow x^{(N)} \in \operatorname{argmin}_{x\in(\mathbb{R}^d)^N} \mathcal{Q}_{2,N}(x)$ as $n\to+\infty$, or, at least toward a local minima of $\mathcal{Q}_{2,N}$. (This implies that $x^{(N)}$ has pairwise distinct components.)

▷ ON- LINE COMPUTATION OF THE "COMPANION PARAMETERS": this phase is very important in view of numerical applications.

- *Weights* $p_i^{(N)} = \mathbb{P}\big(\widehat{X}^{x^{(N)}} = x_i^{(N)}\big),\ i=1,\ldots,N.$

  – Initialize: $p_i^{[0]} := 0,\ i=1,\ldots,N,$

  – Update: $$p_i^{[n+1]} := (1-\gamma_{n+1})p_i^{[n]} + \gamma_{n+1}\mathbf{1}_{\{i=i^{[n+1]}\}},\ n\geq 0. \tag{6.32}$$

  One has:
$$p_i^{[n]} \longrightarrow p_i^{(N)}\ a.s.\ \text{ on the event }\ \big\{x^{[n]}\to x^{(N)}\big\}\ \text{ as }\ n\to+\infty.$$

- *Distortion* $\mathcal{Q}_{2,N}(x^{(N)}) = \|X-\widehat{X}^{x^{(N)}}\|_2$: set

  – Initialize: $\mathcal{Q}_{2,N}^{[0]} := 0$

  – Update: $$\mathcal{Q}_{2,N}^{[n+1]} := (1-\gamma_{n+1})\mathcal{Q}_{2,N}^{[n]} + \gamma_{n+1}\big|x_{i^{[n+1]}}^{[n+1]} - \xi^{n+1}\big|^2,\ n\geq 0. \tag{6.33}$$

  One has
$$\mathcal{Q}_{2,N}^{[n]} \longrightarrow \mathcal{Q}_{2,N}\big(x^{(N)}\big)\ a.s.\ \text{ on the event }\ \big\{x^{[n]}\to x^{(N)}\big\}\ \text{ as }\ n\to+\infty.$$

Note that, since the ingredients involved in the above computations are those used in the competition phase (nearest neighbor search), there is (almost) no extra CPU time cost induced by this companion procedure. By contrast the nearest neighbor search is costly.

In some way the CLVQ algorithm can be seen as a Non-linear Monte Carlo Simulation devised to design an optimal skeleton of the distribution of X.

For partial theoretical results on the convergence of the CLVQ algorithm, we refer to [224] when X has a compactly supported distribution. To the best of our knowledge, no satisfactory theoretical convergence results are available when the distribution of X has an unbounded support. As for the convergence of the online adaptive companion procedures, the convergence proof relies on classical martingale arguments, we refer again to [224], but also to [20] (see also the exercise below).

► **Exercise.** (a) Replace the step sequence (γ_n) in (6.32) and (6.33) by $\tilde{\gamma}_n = \frac{1}{n}$, without modifying anything in the *CLVQ* procedure itself. Show that, if $x^{[n]} \to x^{(N)}$ *a.s.*, the resulting new procedures both converge toward their target. [Hint: Follow the lines of the convergence of the adaptive estimator of the CVaR in Sect. 6.3.4.]

(b) Extend this result to prove that the *a.s.* convergence holds on the event $\{x^{[n]} \to x^{(N)}\}$.

The *CLVQ* algorithm is recommended to obtain accurate results for small or medium levels N (less that 20) and medium dimensions d (less than 10).

The Randomized Lloyd I procedure

The randomized Lloyd I procedure described below is recommended when N is large and d is medium. We start again from the fact that if a function is differentiable at one of its local or global minima, then its gradient is zero at this point. Any global minimizer $x^{(N)}$ of the quadratic distortion function $\mathcal{Q}_{2,N}$ has pairwise distinct components and $\mathbb{P}$-negligible Voronoi cell boundaries as mentioned above. Consequently, owing to Proposition 6.6, the gradient of the quadratic distortion at such $x = x^{(N)}$ must be zero, *i.e.*

$$\frac{\partial \mathcal{Q}_{2,N}}{\partial x_i}(x,\xi) = 2\,\mathbb{E}\left((x_i - X)\mathbf{1}_{\{\mathrm{Proj}_x(X)=x_i\}}\right) = 0,\ \ 1 \le i \le N.$$

Moreover we know from Theorem 5.1.1(b) that $\mathbb{P}\big(X \in C_i(x)\big) > 0$ for every $i \in \{1,\ldots,N\}$ so that the above equation reads equivalently

$$x_i = \mathbb{E}\left(X \mid \{\widehat{X}^x = x_i\}\right),\ \ i = 1,\ldots,N. \tag{6.34}$$

This identity can also be rewritten more synthetically as a fixed point equality of the mapping $\widehat{X} \mapsto \mathbb{E}\left(X \mid \widehat{X}^x\right)$ since

$$\widehat{X}^x = \mathbb{E}\left(X \mid \widehat{X}^x\right) \tag{6.35}$$

since $\mathbb{E}\left(X \mid \widehat{X}^x\right) = \sum_{i=1}^{N} \mathbb{E}\left(X \mid \{\widehat{X}^x = x_i\}\right)\mathbf{1}_{\{\widehat{X}^x = x_i\}}$. Or, equivalently, but in a more tractable form, of the mapping

$$\Gamma \longmapsto \mathbb{E}\left(X \mid \widehat{X}^{\Gamma_x}\right)(\Omega) \tag{6.36}$$

defined on the set of subsets of $\mathbb{R}^d$ (grids) with (at most) N elements.

▷ Regular Lloyd's I procedure (definition). The Lloyd I procedure is simply the recursive procedure associated to the fixed point identity (6.34) (or (6.36)). In its generic form it reads, keeping the notation $x^{[n]}$ for the running N-quantizer at iteration $n \geq 0$:

$$x_i^{[n+1]} = \begin{cases} \mathbb{E}\left(X \mid \{\widehat{X}^{x^{[n]}} = x_i^{[n]}\}\right) \text{ if } \mathbb{P}\big(\widehat{X}^{x^{[n]}} = x_i^{[n]}\big) > 0, \\ x_i^{[n+1]} = x_i^{[n]} \qquad \text{ if } \mathbb{P}\big(\widehat{X}^{x^{[n]}} = x_i^{[n]}\big) = 0, \end{cases} \quad i = 1, \dots, N, \tag{6.37}$$

starting from an N-tuple $x^{[0]} \in (\mathbb{R}^d)^N$ with pairwise distinct components in $\mathbb{R}^d$.

We leave it as an exercise to show that this procedure is entirely determined by the *distribution* of the random vector X.

► **Exercise.** Prove that this recursive procedure only involves the distribution $\mu = \mathbb{P}_X$ of the random vector X.

The Lloyd I algorithm can be viewed as a two step procedure acting on random vectors as follows

$$\begin{cases} (i)\ \textit{Grid updating}: x^{[n+1]} = \widetilde{X}^{[n+1]}(\Omega) \ \text{ with } \ \widetilde{X}^{[n+1]} = \mathbb{E}\left(X \mid \widehat{X}^{x^{[n]}}\right), \\ (ii)\ \textit{Voronoi cells (weights) updating}: \widehat{X}^{x^{[n+1]}} \leftarrow \widetilde{X}^{[n+1]}. \end{cases}$$

The first step updates the grid, the second step re-assigns to each element of the grid its Voronoi cell, which can be also interpreted as a *weight* updating.

Proposition 6.6 *The Lloyd I algorithm makes the mean quadratic quantization error decrease, i.e.*

$$n \longmapsto \big\|X - \widehat{X}^{x^{[n]}}\big\|_2 \quad \textit{is non-increasing.}$$

Proof. It follows from the above decomposition of the procedure and the very definitions of nearest neighbor projection and conditional expectation as an orthogonal projector in $L^2(\mathbb{P})$ that, for every $n \in \mathbb{N}$,

$$\begin{aligned} \big\|X - \widehat{X}^{x^{[n+1]}}\big\|_2 &= \big\|\text{dist}(X, x^{[n+1]})\big\|_2 \\ &\leq \big\|X - \widetilde{X}^{x^{[n+1]}}\big\|_2 = \big\|X - \mathbb{E}\left(X \mid \widehat{X}^{x^{[n]}}\right)\big\|_2 \\ &\leq \big\|X - \widehat{X}^{x^{[n]}}\big\|_2. \end{aligned}$$

◇

Though attractive, this proposition is far from a convergence result for the Lloyd procedure since components of the running quantizer $x^{[n]}$ can *a priori* escape to infinity. In spite of a huge literature on the Lloyd I algorithm, also known as k-means in Statistics and Data Science, this question of its *a.s.* convergence toward a stationary, hopefully optimal, N-quantizer has so far only received partial answers. Let us cite [168], where the procedure is introduced and investigated, probably for the first time, in a one dimensional setting for unimodal distributions. More recently, as far as strongly continuous distribution are concerned ([4]), let us cite [79, 80] or [85], where *a.s.* convergence is established if X has a compactly supported density and [238],where the convergence is proved for unbounded strongly continuous distributions under an appropriate initialization of the procedure at level N depending on the lowest quantization error at level $N-1$ (see the splitting method in the Practitioner's corner below).

▷ THE RANDOMIZED LLOYD I PROCEDURE. It relies on the computation of $\mathbb{E}\,(X \mid \{\widehat{X}^x = x_i\}),\ 1 \le i \le N$, by a Monte Carlo simulation: if $\xi_1,\ldots,\xi_M,\ldots$ are independent copies of X,

$$\mathbb{E}\left(X \mid \widehat{X}^{x^{[n]}} = x_i^{[n]}\right) \simeq \frac{\sum_{1\le m\le M} \xi_m \mathbf{1}_{\{\widehat{\xi}_m^{x^{[n]}} = x_i^{[n]}\}}}{|\{1 \le m \le M,\ \widehat{\xi}_m^{x^{[n]}} = x_i^{[n]}\}|},$$

keeping in mind that the convergence holds when $M \to +\infty$. To be more precise, this amounts to setting at every iteration $n \ge 0$ and for every $i = 1,\ldots,N$,

$$x_i^{[n+1]} = \begin{cases} \dfrac{\sum_{1\le m\le M} \xi_m \mathbf{1}_{\{\widehat{\xi}_m^{x^{[n]}} = x_i^{[n]}\}}}{|\{1\le m\le M,\ \widehat{\xi}_m^{x^{[n]}} = x_i^{[n]}\}|} & \text{if } \{1 \le m \le M,\ \widehat{\xi}_m^{x^{[n]}} = x_i^{[n]}\} \ne \varnothing, \\ x_i^{[n]} & \text{otherwise}, \end{cases}$$

starting from $x(0) \in (\mathbb{R}^d)^N$ chosen to have pairwise distinct components in $\mathbb{R}^d$.

This randomized Lloyd procedure simply amounts to replacing the distribution $\mathbb{P}_X$ of X by the (random) empirical measure

$$\mu(\omega, d\xi)_M = \frac{1}{M}\sum_{m=1}^{M} \delta_{\xi_m(\omega)}(d\xi).$$

In particular, if we use the same sample $(\xi_m(\omega))_{1\le m\le M}$ at each iteration of the procedure, we still have the property that the procedure decreases a quantization error modulus (at level N) related to the distribution μ.

This suggests that the random i.i.d. sample $(\xi_m)_{m\ge 1}$ can also be replaced by deterministic copies obtained through a *QMC* procedure based on a representation of X of the form $X = \psi(U)$, $U \stackrel{d}{=} \mathcal{U}([0,1]^r)$.

ℵ Practitioner's corner

▷ *Splitting Initialization method II.* When computing quantizers of larger and larger sizes for the same distribution, a significant improvement of the method is to initialize the randomized Lloyd's procedure or the *CLVQ* at level $N+1$ by adding one component to the N-quantizer resulting from the previous execution of the procedure, at

[4] A Borel distribution μ on $\mathbb{R}^d$ is strongly continuous if $\mu(H) = 0$ for any hyperplane $H \subset \mathbb{R}^d$.

level N. To be more precise, one should initialize the procedure with the $N+1$-tuple $(x^{(N)}, \xi) \in (\mathbb{R}^d)^{N+1}$ where $x^{(N)}$ denotes the limiting value of the procedure at level N (assumed to exist, which is the case in practice). Such a protocol is known as the *splitting method.*

$\triangleright$ *Fast nearest neighbor search procedure(s) in* $\mathbb{R}^d$.

This is the key step in all stochastic procedures which intend to compute optimal (or at least "good") quantizers in higher dimensions. To speed it up, especially when d increases, is one of the major challenges of computer science.

– The Partial Distance Search paradigm (see [62]).

The nearest neighbor search in a Euclidean vector space can be reduced to the simpler problem to checking whether a vector $u = (u^1, \dots, u^d) \in \mathbb{R}^d$ is closer to 0 with respect to the canonical Euclidean distance than a given former "minimal record distance" $\delta_{\text{rec}} > 0$. The elementary "trick" is the following

$$
\begin{aligned}
(u^1)^2 \geq \delta^2_{\text{rec}} &\Longrightarrow |u| \geq \delta_{\text{rec}} \\
&\;\;\vdots \\
(u^1)^2 + \cdots + (u^\ell)^2 \geq \delta^2_{\text{rec}} &\Longrightarrow |u| \geq \delta_{\text{rec}} \\
&\;\;\vdots
\end{aligned}
$$

This is the simplest and easiest idea to implement but it seems that it is also the only one that still works as d increases.

– The *K-d tree* (Friedmann, Bentley, Finkel, 1977, see [99]): the principle is to store the N points of $\mathbb{R}^d$ in a tree of depth $O(\log(N))$ based on their coordinates on the canonical basis of $\mathbb{R}^d$.

– Further improvements are due to McNames (see [207]): the idea is to perform a pre-processing of the dataset of N points using a Principal Component Axis (*PCA*) analysis and then implement the K-d-tree method in the new orthogonal basis induced by the *PCA*.

Numerical optimization of quantizers for the normal distributions $\mathcal{N}(0; I_d)$ on $\mathbb{R}^d$, $d \geq 1$

The procedures that minimize the quantization error are usually stochastic (except in one dimension). The most famous ones are undoubtedly the so-called *Competitive Leaning Vector Quantization* algorithm (see [231] or [229]) and the Lloyd I procedure (see [106, 226, 231]) which have just been described and briefly analyzed above. More algorithmic details are also available on the website

`www.quantize.maths-fi.com`

For normal distributions a large scale optimization has been carried out based on a mixed *CLVQ*-Lloyd procedure. To be precise, grids have been computed for $d = 1$

up to 10 and $N = 1$ up to 5 000. Furthermore, several companion parameters have also been computed (still by simulation): weight, L^1-quantization error, (squared) L^2-quantization error (also known as distortion), local L^1 and L^2-pseudo-inertia of each Voronoi cell. All these grids can be downloaded on the above website.

Thus Fig. 6.3 depicts an optimal quadratic N-quantization of the bi-variate normal distribution $\mathcal{N}(0; I_2)$ with $N = 500$.

6.4 Further Results on Stochastic Approximation

This section is devoted to more advanced results on Stochastic Approximation and can be skipped on a first reading. Its first part deals with the connection between the asymptotic behavior of a stochastic algorithm with mean function h and that of the ordinary differential equation $ODE_h \equiv \dot{y} = -h(y)$ already introduced in the introduction. The second part is devoted to the main results about the rate of convergence of stochastic algorithms in both their original and averaged forms.

6.4.1 The Ordinary Differential Equation (ODE) Method

Toward the *ODE*

The starting idea – which goes back to Ljung in [200] – of the so-called *ODE* method is to consider a stochastic algorithm with mean function h as the perturbed discrete time Euler scheme with decreasing step of ODE_h. In this section, we will again

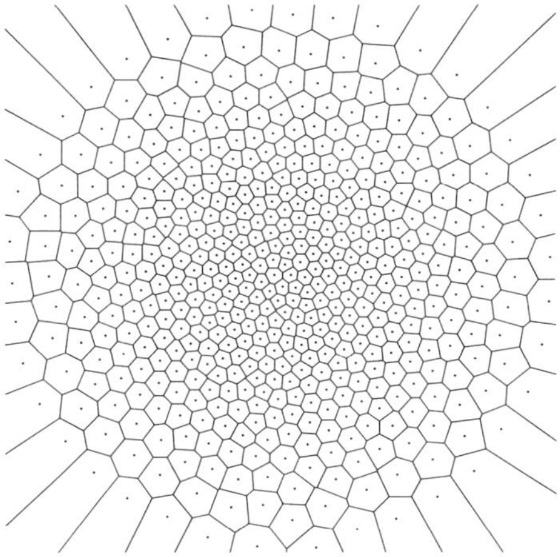

Fig. 6.3 An optimal quantization of the bi-variate normal distribution with size $N = 500$ (with J. Printems)

mainly deal with Markovian stochastic algorithms associated to an i.i.d. sequence $(Z_n)_{n\geq 1}$ of $\mathbb{R}^q$-valued innovations of the form (6.3), namely

$$Y_{n+1} = Y_n - \gamma_{n+1} H(Y_n, Z_{n+1})$$

where Y_0 is independent of the innovation sequence (all defined on the same probability space $(\Omega, \mathcal{A}, \mathbb{P})$), $H : \mathbb{R}^d \times \mathbb{R}^q \to \mathbb{R}^d$ is a Borel function such that $H(y, Z_1) \in L^2(\mathbb{P})$ for every $y \in \mathbb{R}^d$ and $(\gamma_n)_{n\geq 1}$ is sequence of step parameters.

We saw that this algorithm can be represented in a canonical way as follows:

$$Y_{n+1} = Y_n - \gamma_{n+1} h(Y_n) - \gamma_{n+1} \Delta M_{n+1}, \tag{6.38}$$

where $h(y) = \mathbb{E}\, H(y, Z_1)$ is the mean function of the algorithm. We define the filtration $\mathcal{F}_n = \sigma(Y_0, Z_1, \ldots, Z_n)$, $n \geq 0$. The sequence $(Y_n)_{n\geq 0}$ is $(\mathcal{F}_n)$-adapted and $\Delta M_n = \mathbb{E}\,(H(Y_{n-1}, Z_n) \mid \mathcal{F}_{n-1}) - h(Y_{n-1})$, $n \geq 1$, is a sequence of $(\mathcal{F}_n)$-martingale increments.

In the preceding we established that, under the assumptions of the Robbins–Siegmund Lemma (based on the existence of a "Lyapunov function", see Theorem 6.1 (ii) and (v)), the sequence $(Y_n)_{n\geq 0}$ is $a.s.$ bounded and that the martingale

$$M_n^\gamma = \sum_{k\geq 1} \gamma_k \Delta M_k \quad \text{is } a.s. \text{ convergent in } \mathbb{R}^d.$$

At this stage, to derive the $a.s.$ convergence of the algorithm itself in various settings (Robbins–Monro, stochastic gradient, one-dimensional stochastic gradient), we used direct *pathwise* arguments based on elementary topology. The main improvement provided by the *ODE* method is to provide more powerful tools from functional analysis derived from further investigations on the asymptotics of the "tail"-sequences $(Y_k(\omega))_{k\geq n}$, $n \geq 0$, assuming *a priori* that the whole sequence $(Y_n(\omega))_{n\geq 0}$ is bounded. The idea is to represent these tail sequences as stepwise constant càdlàg functions of the cumulative function of the steps. We will also need an additional assumption on the paths of the martingale $(M_n^\gamma)_{n\geq 1}$, however significantly less stringent than the above $a.s.$ convergence property. To keep on working in this direction, it is more convenient to temporarily abandon our stochastic framework and focus on a discrete time deterministic dynamics.

***ODE* method I**

Let us be more specific: first we consider a recursively defined sequence

$$y_{n+1} = y_n - \gamma_{n+1}\big(h(y_n) + \pi_{n+1}\big), \quad y_0 \in \mathbb{R}^d, \tag{6.39}$$

where $(\pi_n)_{n\geq 1}$ is a sequence of $\mathbb{R}^d$-valued vectors and $h : \mathbb{R}^d \to \mathbb{R}^d$ is a continuous Borel function.

We set $\Gamma_0 = 0$ and, for every integer $n \geq 1$,

$$\Gamma_n = \sum_{k=1}^{n} \gamma_k.$$

Then we define the stepwise constant càdlàg function $(y_t^{(0)})_{t \in \mathbb{R}_+}$ by

$$y_t^{(0)} = y_n \quad \text{if} \quad t \in [\Gamma_n, \Gamma_{n+1})$$

and the sequence of time shifted functions defined by

$$y_t^{(n)} = y_{\Gamma_n + t}^{(0)}, \ t \in \mathbb{R}_+.$$

Finally, we set for every $u \in \mathbb{R}_+$,

$$N(t) = \max \{k \ : \ \Gamma_k \le t\} = \min \{k \ : \ \Gamma_{k+1} > t\}$$

so that $N(t) = n$ if and only if $t \in [\Gamma_n, \Gamma_{n+1})$ (in particular, $N(\Gamma_n) = n$).

Expanding the recursive Equation (6.39), we get

$$y_n = y_0 - \sum_{k=1}^{n} \gamma_k h(y_{k-1}) - \sum_{k=1}^{n} \gamma_k \pi_k$$

which can be rewritten, for every $t \in [\Gamma_n, \Gamma_{n+1})$, as

$$\begin{aligned} y_t^{(0)} &= y_0^{(0)} - \sum_{k=1}^{n} \int_{\Gamma_{k-1}}^{\Gamma_k} h(\underbrace{y_s^{(0)}}_{=y_k}) ds - \sum_{k=1}^{n} \gamma_k \pi_k \\ &= y_0^{(0)} - \int_0^{\Gamma_n} h(y_s^{(0)}) ds - \sum_{k=1}^{n} \gamma_k \pi_k. \end{aligned}$$

As a consequence, for every $t \in \mathbb{R}_+$,

$$\begin{aligned} y_t^{(0)} &= y_0^{(0)} - \int_0^{\Gamma_{N(t)}} h(y_s^{(0)}) ds - \sum_{k=1}^{N(t)} \gamma_k \pi_k \\ &= y_0^{(0)} - \int_0^{t} h(y_s^{(0)}) ds + h(y_{N(t)}) \big(t - \Gamma_{N(t)}\big) - \sum_{k=1}^{N(t)} \gamma_k \pi_k. \end{aligned} \tag{6.40}$$

Then, using the very definition of the shifted function $y^{(n)}$ and taking advantage of the fact that $\Gamma_{N(\Gamma_n)} = \Gamma_n$, we derive, by subtracting (6.40) at times $\Gamma_n + t$ and Γ_n successively, that for every $t \in \mathbb{R}_+$,

$$\begin{aligned} y_t^{(n)} &= y_{\Gamma_n}^{(0)} + y_{\Gamma_n+t}^{(0)} - y_{\Gamma_n}^{(0)} \\ &= y_{\Gamma_n}^{(0)} - \int_{\Gamma_n}^{\Gamma_n+t} h(y_s^{(0)})ds + h(y_{N(t)})\big(t - \Gamma_{N(t)}\big) - \sum_{k=n+1}^{N(\Gamma_n+t)} \gamma_k \pi_k \\ &= y_0^{(n)} - \int_0^t h(y_s^{(n)})ds + R_t^{(n)} \end{aligned}$$

with $$R_t^{(n)} = h(y_{N(t+\Gamma_n)})\big(t + \Gamma_n - \Gamma_{N(t+\Gamma_n)}\big) - \sum_{k=n+1}^{N(\Gamma_n+t)} \gamma_k \pi_k,$$

keeping in mind that $y_0^{(n)} = y_{\Gamma_n}^{(0)} (= y_n)$. The term $R_n(t)$ is intended to behave as a remainder term as n goes to infinity. The next proposition establishes a first connection between the asymptotic behavior of the sequence of vectors $(y_n)_{n\geq 0}$ and that of the sequence of functions $(y^{(n)})_{n\geq 0}$.

Proposition 6.7 (*ODE* **method I**) *Assume that*

- $\mathbf{H}_1 \equiv$ *Both sequences* $(y_n)_{n\geq 0}$ *and* $(h(y_n))_{n\geq 0}$ *are bounded,*
- $\mathbf{H}_2 \equiv \forall n \geq 0,\ \gamma_n > 0,\ \lim_n \gamma_n = 0$ *and* $\sum_{n\geq 1} \gamma_n = +\infty$,
- $\mathbf{H}_3 \equiv \forall T \in (0, +\infty),\ \lim_n \sup_{t\in[0,T]} \left| \sum_{k=n+1}^{N(\Gamma_n+t)} \gamma_k \pi_k \right| = 0.$

Then:

(*a*) *The set* $\mathcal{Y}^\infty := \big\{$*limiting points of* $(y_n)_{n\geq 0}\big\}$ *is a compact connected set.*
(*b*) *The sequence* $(y^{(n)})_{n\geq 0}$ *is sequentially relatively compact* ([5]) *with respect to the topology of uniform convergence on compacts sets on the space* $\mathcal{B}(\mathbb{R}_+, \mathbb{R}^d)$ *of bounded functions from* $\mathbb{R}_+$ *to* $\mathbb{R}^d$ ([6]) *and all its limiting points lie in* $\mathcal{C}_b(\mathbb{R}_+, \mathcal{Y}^\infty)$.

Proof Let $M = \sup_{n\in\mathbb{N}} |h(y_n)|$.
(*a*) Let $T_0 = \sup_{n\in\mathbb{N}} \gamma_n < +\infty$. Then it follows from $\mathbf{H}_2$ that

$$|\gamma_{n+1}\pi_{n+1}| \leq \sup_{t\in[0,T_0]} \left| \sum_{k=n+1}^{N(\Gamma_n+t)} \gamma_k \pi_k \right| = 0.$$

It follows from (6.39) and $\mathbf{H}_3$ that

$$|y_{n+1} - y_n| \leq \gamma_{n+1} M + |\gamma_{n+1}\pi_{n+1}| \to 0 \quad \text{as} \quad n \to +\infty.$$

[5]In a metric space (E, d), a sequence $(x_n)_{n\geq 0}$ is sequentially relatively compact if from any subsequence one can extract a convergent subsequence with respect to the distance d.
[6]This topology is defined by the metric $d(f, g) = \sum_{k\geq 1} \frac{\min(\sup_{t\in[0,k]} |f(t)-g(t)|, 1)}{2^k}$.

As a consequence, the set $\mathcal{Y}^\infty$ is compact and *bien enchaîné* ([7]), hence connected.
(b) The sequence $\left(\int_0^{\cdot} h(y_s^{(n)})ds\right)_{n\geq 0}$ is uniformly Lipschitz continuous with Lipschitz continuous coefficient M since, for every $s,\ t \in \mathbb{R}_+,\ s \leq t$,

$$\left|\int_0^t h(y_u^{(n)})du - \int_0^s h(y_u^{(n)})du\right| \leq \int_s^t |h(y_u^{(n)})|du \leq M\,(t-s)$$

and $(y_0^{(n)})_{n\geq 0} = (y_n)_{n\geq 0}$ is bounded, hence it follows from the Arzela–Ascoli Theorem that the sequence of continuous $\left(y_0^{(n)} - \int_0^{\cdot} h(y_s^{(n)})ds\right)_{n\geq 0}$ is relatively compact in $\mathcal{C}(\mathbb{R}_+, \mathbb{R}^d)$ endowed with the topology, denoted by U_K, of the uniform convergence on compact intervals. On the other hand, for every $T \in (0, +\infty)$,

$$\begin{aligned}\sup_{t\in[0,T]}\left|y_t^{(n)} - y_0^{(n)} + \int_0^t h(y_s^{(n)})ds\right| &= \sup_{t\in[0,T]} |R_t^{(n)}| \\ &\leq M \sup_{k\geq n+1} \gamma_k + \sup_{t\in[0,T]}\left|\sum_{k=n+1}^{N(\Gamma_n+t)} \gamma_k \pi_k\right| \to 0 \text{ as } n \to +\infty\end{aligned}$$

owing to $\mathbf{H}_2$ and $\mathbf{H}_3$. Equivalently, this reads

$$y^{(n)} - \left(y_0^{(n)} - \int_0^{\cdot} h(y_s^{(n)})ds\right) \xrightarrow{U_K} 0 \quad \text{as} \quad n \to +\infty, \tag{6.41}$$

which implies that the sequence $\left(y^{(n)}\right)_{n\geq 0}$ is U_K-relatively compact. Moreover, both sequences $\left(y_0^{(n)} - \int_0^{\cdot} h(y_s^{(n)})ds\right)_{n\geq 0}$ and $\left(y^{(n)}\right)_{n\geq 0}$ have the same U_K-limiting values. In particular, these limiting functions are continuous with values in $\mathcal{Y}^\infty$. ◇

ODE **method II: flow invariance**

To state the first significant theorem on the so-called *ODE* method theorem, we introduce the *reverse* differential equation

$$ODE_h^* \equiv \dot{y} = h(y).$$

Theorem 6.5 **(ODE II)** *Assume* $\mathbf{H}_i$, $i = 1, 2, 3$, *hold and that the mean function* h *is continuous. Let* $y_0 \in \mathbb{R}^d$ *and* $\mathcal{Y}^\infty$ *be the set of limiting values of the sequence* $(y_n)_{n\geq 0}$ *recursively defined by* (6.39).
(a) *Any limiting function of the sequence* $(y^{(n)})_{n\geq 0}$ *is a* $\mathcal{Y}^\infty$*-valued solution of* $ODE_h \equiv \dot{y} = -h(y)$.

[7] A set A in a metric space (E, d) is "bien enchaîné" if for every $a, a' \in A$, and every $\varepsilon > 0$, there exists an integer $n \geq 1$ and $a_0, \ldots, a_n$ such that $a_0 = a$, $a_n = a'$ and $d(a_i, a_{i+1}) \leq \varepsilon$ for every $i = 0, \ldots, n-1$. Any connected set C in E is *bien enchaîné*. The converse is true if C is compact, see *e.g.* [13] for more details.

(b) Assume that ODE_h has a flow $\Phi_t(\xi)$ ([8]). Assume the existence of a flow for ODE_h^*. Then, the set $\mathcal{Y}^\infty$ is a compact, connected set, flow-invariant for both ODE_h and ODE_h^*.

Proof: (a) Given the above Proposition 6.7, the conclusion follows if we prove that any limiting function $y^{(\infty)} = U_K\text{-}\lim_n y^{(\varphi(n))}$ $(\varphi(n) \to +\infty)$ is a solution to ODE_h. For every $t \in \mathbb{R}_+$, $y_t^{(\varphi(n))} \to y_t^{(\infty)}$, hence $h\big(y_t^{(\varphi(n))}\big) \to h\big(y_t^{(\infty)}\big)$ since the function h is continuous. Then, by the Lebesgue dominated convergence theorem, one derives that for every $t \in \mathbb{R}_+$,

$$\int_0^t h\big(y_s^{(\varphi(n))}\big)ds \longrightarrow \int_0^t h\big(y_s^{(\infty)}\big)ds.$$

One also has $y_0^{(\varphi(n))} \to y_0^{(\infty)}$ so that finally, letting $\varphi(n) \to +\infty$ in (6.41), we obtain

$$y_t^{(\infty)} = y_0^{(\infty)} - \int_0^t h\big(y_s^{(\infty)}\big)ds.$$

One concludes by time differentiation since $h \circ y^{(\infty)}$ is continuous.

(b) Any $y_\infty \in \mathcal{Y}^\infty$ is the limit of a subsequence $y_{\varphi(n)}$ with $\varphi(n) \to +\infty$. Up to a new extraction, still denoted by $\varphi(n)$ for convenience, we may assume that $y^{(\varphi(n))} \to y^{(\infty)} \in \mathcal{C}(\mathbb{R}_+, \mathbb{R}^d)$ as $n \to +\infty$, uniformly on compact sets of $\mathbb{R}_+$. The function $y^{(\infty)}$ is a $\mathcal{Y}^\infty$-valued solution to ODE_h by (a) and $y^{(\infty)} = \Phi(y_0^{(\infty)})$ owing to the uniqueness assumption. This implies the invariance of $\mathcal{Y}^\infty$ under the flow of ODE_h.

For every $p \in \mathbb{N}$, we consider for large enough n, say $n \geq n_p$, the sequence $(y_{N(\Gamma_{\varphi(n)}-p)})_{n \geq n_p}$. It is clear by mimicking the proof of Proposition 6.7 that sequences of functions $(y^{(N(\Gamma_{\varphi(n)}-p))})_{n \geq n_p}$ are U_K-relatively compact. By a diagonal extraction procedure (still denoted by $\varphi(n)$), we may assume that, for every $p \in \mathbb{N}$,

$$y^{(N(\Gamma_{\varphi(n)}-p))} \xrightarrow{U_K} y^{(\infty),p} \quad \text{as} \quad n \to +\infty.$$

Since $y_{t+1}^{(N(\Gamma_{\varphi(n)}-p-1))} = y_t^{(N(\Gamma_{\varphi(n)}-p))}$ for every $t \in \mathbb{R}_+$ and every $n \geq n_{p+1}$, one has

$$\forall p \in \mathbb{N}, \quad \forall t \in \mathbb{R}_+, \quad y_{t+1}^{(\infty),p+1} = y_t^{(\infty),p}.$$

Furthermore, it follows from (a) that the functions $y^{(\infty),p}$ are $\mathcal{Y}^\infty$-valued solutions to ODE_h. One defines the function $\widetilde{y}^{(\infty)}$ by

$$\widetilde{y}_t^{(\infty)} = y_{p-t}^{(\infty),p}, \ t \in [p-1, p],$$

[8]For every $\xi \in \mathcal{Y}^\infty$, ODE_h admits a unique solution $(\Phi_t(\xi))_{t \in \mathbb{R}_+}$ starting at $\Phi_0(\xi) = \xi$.

which satisfies *a posteriori*, for every $p \in \mathbb{N}$, $\widetilde{y}_t^{(\infty)} = y_{p-t}^{(\infty),p}$, $t \in [0, p]$. This implies that $\widetilde{y}^{(\infty)}$ is a $\mathcal{Y}^\infty$-valued solution to ODE_h^* starting from y_∞ on $\cup_{p\geq 0}[0, p] = \mathbb{R}_+$. Uniqueness implies that $\widetilde{y}^{(\infty)} = \Phi_t^*(y_\infty)$, which completes the proof. $\diamond$

Remark. If uniqueness fails either for ODE_h or for ODE_h^*, one still has that $\mathcal{Y}^\infty$ is left invariant by ODE_h and ODE_h^* in the weaker sense that, for every $y_\infty \in \mathcal{Y}^\infty$, there exist $\mathcal{Y}^\infty$-valued solutions of ODE_h and ODE_h^*.

This property is the first step toward the deep connection between the asymptotic behavior of a recursive stochastic algorithm and its associated mean field ODE_h. Item (*b*) can be seen as a first criterion to direct possible candidates to a set of limiting values of the algorithm. Thus, any zero y_* of h, or equivalently equilibrium points of ODE_h, satisfies the requested invariance condition since $\Phi_t(y_*) = y_*$ for every $t \in \mathbb{R}_+$. No other single point can satisfy this invariance property. More generally, we have the following result.

Corollary 6.2 *If there are finitely many compact connected sets $\mathcal{X}_i$, $i \in I$ (I finite), two-sided invariant for ODE_h and ODE_h^*, then the sequence $(y_n)_{n\geq 0}$ converges toward one of these sets, i.e. there is an integer $i_0 \in I$ such that* $\text{dist}(y_n, \mathcal{X}_{i_0}) \to 0$ *as $n \to +\infty$.*

As an elementary example let us consider the *ODE*

$$\dot{y} = (1 - |y|)y + \varsigma y^\perp, \quad y_0 \in \mathbb{R}^2,$$

where $y = (y_1, y_2)$, $y^\perp = (-y_2, y_1)$ and ς is a positive real constant. Then the unit circle $C(0; 1)$ is clearly a connected compact set invariant under *ODE* and *ODE**. The singleton $\{0\}$ also satisfies this invariant property. In fact, $C(0; 1)$ is an attractor of *ODE* and $\{0\}$ is a repeller in the sense that the flow Φ of this *ODE* uniformly converges toward $C(0; 1)$ on every compact set $K \subset \mathbb{R}^2 \setminus \{0\}$.

We know that any recursive procedure with mean function h of the form $h(y_1, y_2) = (1 - |y|)y + \varsigma y^\perp$ satisfying $\mathbf{H}_i$, $i = 1, 2, 3$, will converge either toward $C(0, 1)$ or 0. But at this stage, we cannot eliminate the repulsive equilibrium $\{0\}$ (what happens if $y_0 = 0$ and the perturbation term sequence $\pi_n \equiv 0$).

Sharper characterizations of the possible set of limiting points of the sequence $(y_n)_{n\geq 0}$ have been established in close connection with the theory of perturbed dynamical systems. To be slightly more precise it has been shown that the set $\mathcal{Y}^\infty$ of limiting points of the sequence $(y_n)_{n\geq 0}$ is *internally chain recurrent* or, equivalently, contains no strict attractor for the dynamics of the *ODE*, *i.e.* as a subset $A \subset \mathcal{Y}^\infty$, $A \neq \mathcal{Y}^\infty$, such that $\phi_t(\xi)$ converges to A uniformly in $\xi \in \mathcal{X}^\infty$. Such results are beyond the scope of this monograph and we refer to [33] (see also [94] when uniqueness fails and the *ODE* has no flow) for an introduction to internal chain recurrence.

Unfortunately, such refined results are still not able to discriminate between these two candidates as a limiting set, though, as soon as π_n behaves like a non-degenerate noise when y_n is close to 0, it seems more likely that the algorithm converges toward the unit circle, like the flow of the ODE_h does (except when starting from 0). At this point probability comes back into the game: this intuition can be confirmed under the additional non-degeneracy assumption of the noise at 0 for the algorithm (the notion of "noisy trap"). Thus, if the procedure (6.39) is a generic path of a Markovian algorithm of the form (6.3) satisfying at 0

$$\mathbb{E}\, H(0, Z_1)H(0, Z_1)^t \text{ is symmetric positive definite,}$$

this generic path cannot converge toward 0 and will subsequently converge to $C(0; 1)$.

Practical aspects of assumption $\mathbf{H}_3$.

To make the connection with the original form of stochastic algorithms, we come back to hypothesis $\mathbf{H}_3$ in the following proposition. In particular, it emphasizes that this condition is less stringent than a standard convergence assumption on the series.

Proposition 6.8 *Assumption* $\mathbf{H}_3$ *is satisfied in two situations (or their combination):*
(a) Remainder term*: If* $\pi_n = r_n$ *with* $r_n \to 0$ *and* $\gamma_n \to 0$ *as* $n \to +\infty$*, then*

$$\sup_{t\in[0,T]}\Big|\sum_{n+1}^{N(\Gamma_n+t)}\gamma_k\pi_k\Big| \le \sup_{k\ge n+1}|\pi_k|(\Gamma_{N(\Gamma_n+T)} - \Gamma_n) \to 0 \quad as \quad n \to +\infty.$$

(b) Martingale perturbation term*: If* $\pi_n = \Delta M_n$, $n \ge 1$*, is a sequence of martingale increments and if the martingale* $M^\gamma = \sum_{k=1}^n \gamma_k \Delta M_k$ *a.s. converges in* $\mathbb{R}^d$*. Then, it satisfies a Cauchy condition so that*

$$\sup_{t\in[0,T]}\Big|\sum_{k=n+1}^{N(\Gamma_n+t)}\gamma_k\pi_k\Big| \le \sup_{\ell\ge n+1}\Big|\sum_{k=n+1}^{\ell}\gamma_k\pi_k\Big| \to 0 \quad a.s. \quad as \quad n \to +\infty.$$

(c) Mixed perturbation*: In practice, one often meets the combination of these rrsituations:*

$$\pi_n = \Delta M_n + r_n,$$

where r_n *is a remainder term which goes to* 0 *a.s. and* M_n^γ *is an a.s. convergent martingale.*

The *a.s.* convergence of the martingale $(M_n^\gamma)_{n\ge0}$ follows from the fact that $\sup_{n\ge1}\mathbb{E}\big((\Delta M_n)^2\,|\,\mathcal{G}_{n-1}\big) < +\infty$ *a.s.*, where $\mathcal{G}_n = \sigma(\Delta M_k,\ k = 1,\dots,n)$, $n \ge 1$ and $\mathcal{G}_0 = \{\varnothing, \Omega\}$. The *a.s.* convergence of $(M_n^\gamma)_{n\ge0}$ can even be relaxed for the martingale term. Thus we have the following classical results where $\mathbf{H}_3$ is satisfied while the martingale M_n^γ may diverge.

Proposition 6.9 (a) Métivier–Priouret criterion (see [213]). *Let* $(\Delta M_n)_{n\ge1}$ *be a sequence of martingale increments and let* $(\gamma_n)_{n\ge1}$ *be a sequence of non-negative steps satisfying* $\sum_n \gamma_n = +\infty$*. Then,* $\mathbf{H}_3$ *is a.s. satisfied with* $\pi_n = \Delta M_n$ *as soon as* *there exists a pair of Hölder conjugate exponents* $(p,q)\in(1,+\infty)^2$ *(i.e.* $\frac{1}{p}+\frac{1}{q}=1$*) such that*

$$\sup_n \mathbb{E}\,|\Delta M_n|^p<+\infty \quad \textit{and} \quad \sum_n \gamma_n^{1+\frac{q}{2}}<+\infty.$$

This allows for the use of steps of the form $\gamma_n\sim c_1 n^{-a}$, $a>\frac{2}{2+q}=\frac{2(p-1)}{3(p-1)+1}$.

(b) Exponential criterion (see *e.g.* [33]). *Assume that there exists a real number* $c>0$ *such that,*

$$\forall\,\lambda\in\mathbb{R},\quad \mathbb{E}\,e^{\lambda\Delta M_n}\le e^{c\frac{\lambda^2}{2}}.$$

Then, for every sequence $(\gamma_n)_{n\ge1}$ *such that* $\sum_{n\ge1} e^{-\frac{c}{\gamma_n}}<+\infty$, *Assumption* $\mathbf{H}_3$ *is satisfied with* $\pi_n=\Delta M_n$. *This allows for the use of steps of the form* $\gamma_n\sim c_1 n^{-a}$, $a>0$, *and* $\gamma_n=c_1(\log n)^{-1+a}$, $a>0$.

$\triangleright$ **Examples.** Typical examples where the sub-Gaussian assumption is satisfied are the following:

- $|\Delta M_n|\le K\in\mathbb{R}_+$ since, owing to Hoeffding's Inequality, $\mathbb{E}\,e^{\lambda\Delta M_n}\le e^{\frac{\lambda^2}{8}K^2}$ (see *e.g.* [193], see also the exercise that follows the proof of Theorem 7.4).
- $\Delta M_n\overset{d}{=}\mathcal{N}(0;\sigma_n^2)$ with $\sigma_n\le K$, so that $\mathbb{E}\,e^{\lambda\Delta M_n}\le e^{\frac{\lambda^2}{2}K^2}$.

The first case is very important since in many situations the perturbation term is a martingale term and is structurally bounded.

Application to an extended Lyapunov approach and pseudo-gradient

By relying on claim (a) in Proposition 6.7, one can also derive directly *a.s.* convergence results for an algorithm.

Proposition 6.10 (G-Lemma, see [94]) *Assume* $\mathbf{H}_i$, $i=1,2,3$. *Let* $G:\mathbb{R}^d\to\mathbb{R}_+$ *be a function satisfying*

$$(\mathcal{G})\equiv\Big(\lim_n y_n=y_\infty \ \textit{and}\ \lim_n G(y_n)=0\Big)\Longrightarrow G(y_\infty)=0.$$

Assume that the sequence $(y_n)_{n\ge0}$ *satisfies*

$$\sum_{n\ge0}\gamma_{n+1}G(y_n)<+\infty. \tag{6.42}$$

Then, there exists a connected component $\mathcal{X}^*$ *of the set* $\{G=0\}$ *such that* dist $(y_n,\mathcal{X}^*)=0$.

Remark. Any non-negative lower semi-continuous (l.s.c.) function $G:\mathbb{R}^d\to\mathbb{R}_+$ satisfies $(\mathcal{G})$ ([9]).

---

[9] A function $f:\mathbb{R}^d\to\mathbb{R}$ is lower semi-continuous if, for every $x\in\mathbb{R}^d$ and every sequence $x_n\to x$, $f(x)\le\underline{\lim}_n f(x_n)$.

Proof. First, it follows from Proposition (6.7) that the sequence $(y^{(n)}_n)_{n\geq 0}$ is U_K-relatively compact with limiting functions lying in $\mathcal{C}(\mathbb{R}_+, \mathcal{Y}^\infty)$ where $\mathcal{Y}^\infty$ still denotes the compact connected set of limiting values of $(y_n)_{n\geq 0}$.

Set, for every $y \in \mathbb{R}^d$,

$$\widetilde{G}(y) = \underline{\lim}_{x\to y} G(x) = \inf\left\{\underline{\lim}_{n} G(x_n),\ x_n \to y\right\}$$

so that $0 \leq \widetilde{G} \leq G$. The function $\widetilde{G}$ is the l.s.c. envelope of the function G, *i.e.* the highest l.s.c. function not greater than G. In particular, under Assumption $(\mathcal{G})$

$$\{G = 0\} = \{\widetilde{G} = 0\} \text{ is closed.}$$

First note that Assumption (6.42) reads

$$\int_0^{+\infty} G\big(y^{(0)}_s\big)ds < +\infty.$$

Let $y_\infty \in \mathcal{Y}^\infty$. Up to at most two extractions of subsequences, one may assume that $y^{(\varphi(n))} \to y^{(\infty)}$ for the U_K topology, where $y^{(\infty)}_0 = y_\infty$. It follows from Fatou's Lemma that

$$\begin{aligned}
0 \leq \int_0^{+\infty} \widetilde{G}\big(y^{(\infty)}_s\big)ds &= \int_0^{+\infty} \widetilde{G}\big(\lim_n y^{(\varphi(n))}_s\big)ds \\
&\leq \int_0^{+\infty} \underline{\lim}_n \widetilde{G}\big(y^{(\varphi(n))}_s\big)ds \quad \text{since } \widetilde{G} \text{ is l.s.c.} \\
&\leq \underline{\lim}_n \int_0^{+\infty} \widetilde{G}\big(y^{(\varphi(n))}_s\big)ds \quad \text{owing to Fatou's Lemma} \\
&\leq \underline{\lim}_n \int_0^{+\infty} G\big(y^{(\varphi(n))}_s\big)ds \\
&= \underline{\lim}_n \int_{\Gamma_{\varphi(n)}}^{+\infty} G\big(y^{(0)}_s\big)ds = 0.
\end{aligned}$$

Consequently, $\displaystyle\int_0^{+\infty} \widetilde{G}\big(y^{(\infty)}_s\big)ds = 0$, which implies that $\widetilde{G}\big(y^{(\infty)}_s\big) = 0$ ds-*a.s.* Now as the function $s \mapsto y^{(\infty)}_s$ is continuous it follows that $\widetilde{G}\big(y^{(\infty)}_0\big) = 0$ since $\widetilde{G}$ is l.s.c. This in turn implies $G\big(y^{\infty}_0\big) = 0$, *i.e.* $G\big(y_\infty\big) = 0$. As a consequence $\mathcal{Y}^\infty \subset \{G = 0\}$, which yields the result since on the other hand it is a connected set.
◇

Now we are in position to prove the convergence of stochastic pseudo-gradient procedures in the multi-dimensional case.

Proof of Theorem 6.3 (Pseudo-gradient). Each path of the stochastic algorithm fits into the *ODE* formalism (6.39) by setting $y_n = Y_n(\omega)$ and the perturbation $\pi_n = H\big(Y_{n-1}(\omega), Z_n(\omega)\big)$. Under the assumptions of the Robbins–Siegmund Lemma, we know that $\big(Y_n(\omega)\big)_{n\geq 0}$ is $\mathbb{P}(d\omega)$-*a.s.* bounded and $L\big(Y_n(\omega)\big) \to L_\infty(\omega)$. Combined with (claim (v)) of Theorem 6.1, this implies that $\mathbb{P}(d\omega)$-*a.s.*, Assumption $\mathbf{H}_i$, $i = 1, 2, 3$ are satisfied by $(y_n)_{n\geq 0}$. Consequently, Proposition 6.7 implies that the set $\mathcal{Y}^\infty(\omega)$ of limiting values of $(y_n)_{n\geq 0}$ is connected and compact. The former Proposition 6.10 applied to the l.s.c. function $G = (\nabla L\,|h)$ implies that $\mathcal{Y}^\infty(\omega) \subset \big\{(\nabla L\,|h) = 0\big\}$ as it is also contained in $\{L = \ell(\omega)\}$ with $\ell(\omega) = L_\infty(\omega)$. The conclusion follows. ◇

6.4.2 L^2-Rate of Convergence and Application to Convex Optimization

Proposition 6.11 *Let $(Y_n)_{n\geq 1}$ be a stochastic algorithm defined by (6.3) where the function H satisfies the quadratic linear growth assumption* (6.6), *namely*

$$\forall\, y \in \mathbb{R}^d, \qquad \|H(y, Z)\|_2 \leq C(1 + |y|).$$

Assume $Y_0 \in L^2(\mathbb{P})$ is independent of the i.i.d. sequence $(Z_n)_{n\geq 1}$. Assume there exists a $y_* \in \mathbb{R}^d$ and an $\alpha > 0$ such that the strong mean-reverting assumption

$$\forall\, y \in \mathbb{R}^d, \qquad (y - y_*\,|h(y)) > \alpha|y - y_*|^2 \tag{6.43}$$

holds. Finally, assume that the step sequence $(\gamma_n)_{n\geq 1}$ satisfies the usual decreasing step assumption (6.7) *and the additional assumption*

$$(G_\alpha) \equiv \overline{\lim_n}\left[a_n = \frac{1}{\gamma_{n+1}}\left(\frac{\gamma_n}{\gamma_{n+1}}\Big(1 - 2\,\alpha\,\gamma_{n+1}\Big) - 1\right)\right] = -\kappa^* < 0. \tag{6.44}$$

Then

$$Y_n \xrightarrow{a.s.} y_* \quad \textit{and} \quad \|Y_n - y_*\|_2 = O(\sqrt{\gamma_n}).$$

ℵ **Practitioner's corner.** • If $\gamma_n = \dfrac{\gamma_1}{n^\vartheta}$, $\frac{1}{2} < \vartheta < 1$, then (G_α) is satisfied for any $\alpha > 0$.
• If $\gamma_n = \dfrac{\gamma_1}{n}$, Condition (G_α) reads $\dfrac{1 - 2\,\alpha\,\gamma_1}{\gamma_1} = -\kappa^* < 0$ or equivalently $\gamma_1 > \dfrac{1}{2\,\alpha}$.

Proof of Proposition 6.11 The fact that $Y_n \to y_*$ *a.s.* is a straightforward consequence of Corollary 6.1 (Robbins–Monro framework). We also know that $\big(|Y_n - y_*|\big)_{n\geq 0}$ is L^2-bounded, so that, in particular $\big(Y_n\big)_{n\geq 0}$ is L^2-bounded. As concerns the quadratic rate of convergence, we re-start from the classical proof of Robbins–Siegmund's Lemma. Let $\mathcal{F}_n = \sigma(Y_0, Z_1, \ldots, Z_n)$, $n \geq 0$. Then

$$|Y_{n+1} - y_*|^2 = |Y_n - y_*|^2 - 2\gamma_{n+1}\big(H(Y_n, Z_{n+1})|Y_n - y_*\big) + \gamma_{n+1}^2|H(Y_n, Z_{n+1})|^2.$$

Since Y_n– is $\mathcal{F}_n$-measurable, hence independent of Z_{n+1}, we know that Likewise we get

$$\mathbb{E}\,|H(Y_n, Z_{n+1})|^2 = \mathbb{E}\big[\mathbb{E}\left(|H(Y_n, Z_{n+1})|^2|\,\mathcal{F}_n\right)\big] \le 2\,C^2(1 + \mathbb{E}\,|Y_n|^2) < +\infty.$$

Now, as $Y_n - y_*$ is $\mathcal{F}_n$-measurable, one has

$$\mathbb{E}\Big[(H(Y_n, Z_{n+1})|Y_n - y_*)\Big] = \mathbb{E}\Big[\big(\mathbb{E}\,(H(Y_n, Z_{n+1})|\mathcal{F}_n)|Y_n - y_*\big)\Big] = \mathbb{E}\,\big(h(Y_n)|Y_n - y_*\big).$$

This implies

$$\begin{aligned}\mathbb{E}\,|Y_{n+1} - y_*|^2 &= \mathbb{E}\,|Y_n - y_*|^2 - 2\gamma_{n+1}\mathbb{E}\,\big(h(Y_n)|Y_n - y_*\big) + \gamma_{n+1}^2\mathbb{E}\,|H(Y_n, Z_{n+1})|^2\\ &\le \mathbb{E}\,|Y_n - y_*|^2 - 2\gamma_{n+1}\mathbb{E}\,\big(h(Y_n)|Y_n - y_*\big) \; + 2\,\gamma_{n+1}^2 C^2\big(1 + \mathbb{E}\,|Y_n|^2\big)\\ &\le \mathbb{E}\,|Y_n - y_*|^2 - 2\,\alpha\,\gamma_{n+1}\mathbb{E}\,|Y_n - y_*|^2 + \gamma_{n+1}^2 C'\big(1 + \mathbb{E}\,|Y_n - y_*|^2\big)\end{aligned}$$

owing successively to the linear quadratic growth and the strong mean-reverting assumptions. Finally,

$$\mathbb{E}\,|Y_{n+1} - y_*|^2 = \mathbb{E}\,|Y_n - y_*|^2\Big(1 - 2\,\alpha\,\gamma_{n+1} + C'\gamma_{n+1}^2\Big) + C'\gamma_{n+1}^2.$$

If we set for every $n \ge 1$,

$$u_n = \frac{\mathbb{E}\,|Y_n - y_*|^2}{\gamma_n},$$

the above inequality can be rewritten using the expression for a_n,

$$\begin{aligned}u_{n+1} &\le u_n\frac{\gamma_n}{\gamma_{n+1}}\Big(1 - 2\,\alpha\,\gamma_{n+1} + C'\gamma_{n+1}^2\Big) + C'\gamma_{n+1}\\ &= u_n\Big(1 + \gamma_{n+1}(a_n + C'\gamma_n)\Big) + C'\gamma_{n+1}.\end{aligned}$$

Let n_0 be an integer such that, for every $n \ge n_0$, $a_n \le -\frac{3}{4}\kappa^*$ and $C'\gamma_n \le \frac{\kappa^*}{4}$. For these integers n, $1 - \frac{\kappa^*}{2}\gamma_{n+1} > 0$ and

$$u_{n+1} \le u_n\Big(1 - \frac{\kappa^*}{2}\gamma_{n+1}\Big) + C'\gamma_{n+1}.$$

Then, one derives by induction that

$$\forall\, n \ge n_0, \qquad 0 \le u_n \le \max\left(u_{n_0}, \frac{2C'}{\kappa^*}\right),$$

which completes the proof. ◇

► **Exercise.** Prove a similar result (under appropriate assumptions) for an algorithm of the form

$$Y_{n+1} = Y_n - \gamma_{n+1}\big(h(Y_n) + \Delta M_{n+1}\big),$$

where $h : \mathbb{R}^d \to \mathbb{R}^d$ is Borel continuous function and $(\Delta M_n)_{n\geq 1}$ a sequence of $\mathcal{F}_n$-martingale increments satisfying

$$|h(y)| \leq C(1 + |y - y_*|) \text{ and } \mathbb{E}\left(|\Delta M_{n+1}|^2 \,|\, \mathcal{F}_n\right) < C(1 + |Y_n|^2).$$

Application to α-convex optimization

Let $L : \mathbb{R}^d \to \mathbb{R}_+$ be a twice differentiable convex function with $D^2 L \geq \alpha I_d$, where $\alpha > 0$ (in the sense that $D^2 L(x) - \alpha I_d$ is a positive symmetric matrix for every $x \in \mathbb{R}^d$). Such a function is sometimes called *α-strictly convex*. Then, it follows from Lemma 6.1(*b*) that $\lim_{|y|\to+\infty} L(y) = +\infty$ and that, for every $x, \; y \in \mathbb{R}^d$,

$$\big(\nabla L(x) - \nabla L(y) | x - y\big) \geq \alpha |x - y|^2.$$

Hence L, being non-negative, attains its minimum at a point $y_* \in \mathbb{R}^d$. In fact, the above inequality straightforwardly implies that y_* is unique since $\{\nabla L = 0\}$ is clearly reduced to $\{y_*\}$.

Moreover, if we assume that ∇L is Lipschitz continuous, then for every $y, \; u \in \mathbb{R}^d$,

$$0 \leq u^* D^2 L(y) u = \lim_{t\to 0} \frac{\big(\nabla L(y + tu) - \nabla L(y) | u\big)}{t} \leq [\nabla L]_{\text{Lip}} |u|^2 < +\infty.$$

In that framework, the following proposition shows that the "value function" $L(Y_n)$ of stochastic gradient converges in L^1 at a $O\big(1/n\big)$-rate. This is an easy consequence of the above Proposition 6.11 and a well-known method in Statistics called *Δ-method.*

Note that, if L is α-strictly convex with a Lipschitz continuous gradient, then $L(y) \asymp |y|^2$

Proposition 6.12 *Let $L : \mathbb{R}^d \to \mathbb{R}_+$ be a twice differentiable strictly α-convex function with a Lipschitz continuous gradient hence satisfying $\alpha I_d \leq D^2 L \leq [\nabla L]_{\text{Lip}} I_d$ ($\alpha > 0$) in the sense of symmetric matrices.*

Let $(Y_n)_{n\geq 0}$ be a stochastic gradient descent associated to L, i.e. a stochastic algorithm defined by (6.3) *with a mean function $h = \nabla L$. Assume that the L^2-linear growth assumption* (6.6) *on the state function H, the independence assumptions on Y_0 and the innovation sequence $(Z_n)_{n\geq 1}$ from Proposition 6.11 all hold true. Finally, assume that the decreasing step assumption* (6.7) *and* (G_α) *are both satisfied by the sequence $(\gamma_n)_{n\geq 1}$. Then*

$$\mathbb{E}\, L(Y_n) - L(y_*) = \big\|L(Y_n) - L(y_*)\big\|_1 = O\big(\gamma_n\big).$$

Proof. It is clear that such a stochastic algorithm satisfies the assumptions of the above Proposition 6.11, especially the strong mean-reverting assumption (6.43), owing to the preliminaries on L that precede the proposition. By the fundamental theorem of Calculus, for every $n \geq 1$, there exists a $\Xi_n \in (Y_{n-1}, Y_n)$ (geometric interval in $\mathbb{R}^d$) such that

$$\begin{aligned} L(Y_n) - L(y_*) &= \big(\nabla L(y_*)|Y_n - y_*\big) + \frac{1}{2}(Y_n - y_*)^* D^2 L(\Xi_n)(Y_n - y_*) \\ &\leq \frac{1}{2}[\nabla L]_1 \left|Y_n - y_*\right|^2 \end{aligned}$$

where we used in the second line that $\nabla L(y_*) = 0$ and the above inequality. One concludes by taking the expectation in the above inequality and applying Proposition 6.11. ◇

► **Exercise.** Prove a similar result for a pseudo-stochastic gradient under appropriate assumptions on the mean function h.

6.4.3 *Weak Rate of Convergence: CLT*

We showed in Proposition 6.11 that under natural assumptions, including a strong mean-reverting assumption, a stochastic algorithm converges in L^2 to its target at a $\sqrt{\gamma_n}$-rate. Then, we saw that it suggests to use some steps of the form $\gamma_n = \frac{c}{n+b}$ provided c is not small (see Practitioner's corner that follows the proof of the proposition). Such a result corresponds to the Law of Large Numbers in quadratic mean and one may reasonably guess that a Central Limit Theorem can be established. In fact, the mean-reverting (to coercivity) assumption can be localized at the target y_*, leading a *CLT* at a $\sqrt{\gamma_n}$-rate, namely that $\frac{Y_n - y_*}{\sqrt{\gamma_n}}$ converges in distribution to some normal distribution involving the dispersion matrix $\Sigma_H(y_*) = \mathbb{E}\left[H(y_*, Z)H(y_*, Z)^t\right]$.

The *CLT* for Stochastic Approximation algorithms has given rise to an extensive literature starting from the pioneering work by Kushner [179] in the late 1970's (see also [49] for a result with Markovian innovations). We give here a result established by Pelletier in [241] whose main originality is its "locality" in the following sense: a *CLT* is shown to hold for the stable weak convergence, locally on the convergence set(s) of the algorithm to its equilibrium points. In particular, it solves the case of multi-target algorithms, which is often the case, *e.g.* for stochastic gradient descents associated to non-convex potentials. It could also be of significant help to elucidate the rate of convergence of algorithms with constraints or repeated projections like those introduced by Chen. Such is the case for Arouna's original adaptive variance reduction procedure for which a *CLT* has been established in [197] by a direct approach (see also [196] for a rigorous proof of the convergence).

Theorem 6.6 (**Pelletier [241]**) *We consider the stochastic procedure* $(Y_n)_{n\geq 0}$ *defined by (6.3). Let* $y_* \in \{h=0\}$ *be an equilibrium point. We make the following assumptions.*
(i) y_* is an attractor for ODE_h: *y_* is a "locally uniform attractor" for the* $ODE_h \equiv \dot{y} = -h(y)$ *in the following sense:*

$$h \text{ is differentiable at } y_* \text{ and all the eigenvalues of } Dh(y_*) \text{ have positive real parts.}$$

(ii) Regularity and growth control of H: *the function* H *satisfies the following regularity and growth control properties*

$$y \mapsto \mathbb{E}\big[H(y,Z)H(y,Z)^t\big] \text{ is continuous at } y_*$$

and

$$y \mapsto \mathbb{E}\,|H(y,Z)|^{2+\beta}, \text{ is locally bounded on } \mathbb{R}^d$$

for some $\beta > 0$.
(iii) Non-degenerate asymptotic variance: *Assume that the covariance matrix of* $H(y_*,Z)$

$$\Sigma_H(y_*) := \mathbb{E}\big[H(y_*,Z)H(y_*,Z)^t\big] \text{ is positive definite in } \mathcal{S}(d,\mathbb{R}). \tag{6.45}$$

(iv) Specification of the step sequence: *Assume that the step* γ_n *is of the form*

$$\gamma_n = \frac{c}{n^{\vartheta}+b}, \quad \frac{1}{2} < \vartheta \leq 1, \ b \geq 0, \ c > 0, \tag{6.46}$$

with

$$c > \frac{1}{2\,\Re e(\lambda_{min})} \quad \text{if} \quad \vartheta = 1, \tag{6.47}$$

where λ_{min} *denotes the eigenvalue of* $Dh(y_*)$ *with the lowest real part.*

Then, the a.s. convergence is ruled on the event $A_* = \{Y_n \to y_*\}$ *by the following stable Central Limit Theorem*

$$\sqrt{n^{\vartheta}}\,\big(Y_n - y_*\big) \stackrel{\mathcal{L}_{stably}}{\longrightarrow} \mathcal{N}(0; c\,\Sigma) \ \text{ on } A_*, \tag{6.48}$$

where $\Sigma := \displaystyle\int_0^{+\infty} e^{-s\left(Dh(y_*)^t - \frac{I_d}{2c_*}\right)}\Sigma_H(y_*)e^{-s\left(Dh(y_*) - \frac{I_d}{2c_*}\right)}ds$ *and* $c_* = c$ *if* $\vartheta = 1$ *and* $c_* = +\infty$ *otherwise.*

Note that the optimal rate is obtained when $\vartheta = 1$*, provided* c *satisfies* (6.47).

The *stable convergence in distribution* – denoted by "$\mathcal{L}_{stably}$" – mentioned in (6.48) means that there exists an extension $(\Omega', \mathcal{A}', \mathbb{P}')$ of $(\Omega, \mathcal{A}, \mathbb{P})$ and $Z : (\Omega', \mathcal{A}', \mathbb{P}') \to \mathbb{R}^d$ with $\mathcal{N}(0; I_d)$ distribution such that, for every bounded continuous function f

and every $A \in \mathcal{A}$,

$$\mathbb{E}\Big(\mathbf{1}_{A_*\cap A}\, f\big(\sqrt{n}(Y_n - y_*)\big)\Big) \longrightarrow \mathbb{E}\Big(\mathbf{1}_{A_*\cap A}\, f\big(\sqrt{c}\sqrt{\Sigma}\, Z\big)\Big) \quad \text{as} \quad n \to +\infty.$$

In fact, when the algorithm *a.s.* converges toward a unique target y^*, it has been shown in [203] (see Sect. 11.4) that one may assume that Z and $\mathcal{A}$ are independent so that, in particular, for every $A \in \mathcal{A}$,

$$\mathbb{E}\Big(f\big(\sqrt{n}(Y_n - y_*)\big) \,|\, A\Big) \longrightarrow \mathbb{E}\Big(f\big(\sqrt{c}\sqrt{\Sigma}\, Z\big)\Big) \quad \text{as} \quad n \to +\infty.$$

The proof is detailed for scalar algorithms but its extension to higher-dimensional procedures is standard, though slightly more technical. In the possibly multi-target setting, like in the above statement, it is most likely that such an improvement still holds true (though no proof is known to us). As $A_* \in \mathcal{A}$ is independent of Z, this would read, with the notation of the original theorem: if $\mathbb{P}(A_*) > 0$, for every A in $\mathcal{A}$ such that $\mathbb{P}(A \cap A_*) > 0$,

$$\mathbb{E}\Big(f\big(\sqrt{n}(Y_n - y_*)\big) \,|\, A \cap A_*\Big) \longrightarrow \mathbb{E}\Big(f\big(\sqrt{c}\sqrt{\Sigma}\, Z\big)\Big) \quad \text{as} \quad n \to +\infty.$$

Remarks. • It is clear that the best rate of convergence is obtained when $\vartheta = 1$, namely $\sqrt{n}$. With the restriction for practical implementation that the choice of the coefficient c is subject to a constraint involving an unknown lower bound (see Practitioner's corner below and Sect. 6.4.4 devoted to Ruppert and Polyak's averaging principle).

• When $\vartheta = 1$ and $0 < c \le \frac{1}{2\,\Re e(\lambda_{min})}$ other rates are obtained. Thus, if $c = \frac{1}{2\,\Re e(\lambda_{min})}$ and the maximal order of the Jordan blocks of $\lambda_{\min}$ is 1, the above weak convergence rate switch to $\sqrt{\frac{n}{\log n}}$ with a variance which can again be made explicit.
When $0 < c < \frac{1}{2\,\Re e(\lambda_{min})}$ and $\lambda_{\min}$ is real, still with Jordan blocks with order 1, under an additional assumption on the "differentiability rate" of h at y_*, one can show that there exists a non-zero $\mathbb{R}^d$-valued random vector Ξ such that

$$Y_n = y_* + n^{-c\lambda_{\min}}\big(\Xi + o(1)\big) \quad a.s. \ \text{as} \ n \to +\infty.$$

Similar results hold when $\lambda_{\min}$ is complex, possibly with Jordan order blocks greater than 1 or when $\gamma_n = o\big(\frac{1}{n}\big)$ but still satisfies the usual decreasing step Assumption (6.7) (like $\gamma_n = \frac{c}{\log(n+1)(n+b)}$).

ℵ Practitioner's corner

▷ *Optimal choice of the step sequence.* As mentioned in the theorem itself, the optimal weak rate is obtained for $\vartheta = 1$, provided the step sequence is of the form $\gamma_n = \frac{c}{n+b}$, with $c > \frac{1}{2\,\Re e(\lambda_{min})}$. The explicit expression for Σ – which depends on c as well – suggests that there exists an optimal choice of this parameter c minimizing the asymptotic variance of the algorithm.

We will deal with the one-dimensional case to get rid of some technicalities. If $d = 1$, then $\lambda_{\min} = h'(y_*)$, $\Sigma_H(y_*) = \mathrm{Var}\big(H(y_*, Z)\big)$ and a straightforward computation shows that

$$c\,\Sigma = \mathrm{Var}\big(H(y_*, Z)\big)\frac{c^2}{2c\,h'(y_*) - 1}.$$

This simple function of c attains its minimum on $(1/(2h'(y_*)), +\infty)$ at

$$c_{opt} = \frac{1}{h'(y_*)}$$

with a resulting asymptotic variance

$$\frac{\mathrm{Var}(H(y_*, Z))}{h'(y_*)^2}.$$

One shows that this is the lowest possible variance in such a procedure. Consequently, the best choice for the step sequence $(\gamma_n)_{n\geq 1}$ is

$$\gamma_n := \frac{1}{h'(y_*)n} \quad \text{or, more generally,} \quad \gamma_n := \frac{1}{h'(y_*)(n+b)},$$

where b can be tuned to "control" the step at the beginning of the simulation when n is small.

At this stage, one encounters the same difficulties as with deterministic procedures since y_* being unknown, $h'(y_*)$ is even "more" unknown. One can imagine to estimating this quantity as a companion procedure of the algorithm, but this turns out to be not very efficient. A more efficient approach, although not completely satisfactory in practice, is to implement the algorithm in its averaged version (see Sect. 6.4.4 below).

▷ *Exploration vs convergence rate.* However, one must keep in mind that this tuning of the step sequence is intended to optimize the rate of convergence of the algorithm during its final convergence phase. In real applications, this class of recursive procedures spends most of its time "exploring" the state space before getting trapped in some attracting basin (which can be the basin of a local minimum in the case of multiple critical points). The *CLT* rate occurs once the algorithm is trapped.

▷ *Simulated annealing.* An alternative to these procedures is to implement a simulated annealing procedure which "super-excites" the algorithm using an exogenous simulated noise in order to improve the efficiency of the exploring phase (see [103, 104, 241]). Thus, when the mean function h is a gradient ($h = \nabla L$), it finally converges – but only *in probability* – to the true minimum of the potential/Lyapunov function L. However, the final convergence rate is worse owing to the additional exciting noise which slows down the procedure in its convergence phase. Practitioners often use the above Robbins–Monro or stochastic gradient procedure with a sequence of steps $(\gamma_n)_{n\geq 1}$ which decreases to a positive limit $\underline{\gamma}$.

Proving a *CLT*

We now prove this *CLT* in the $1D$-framework when the algorithm *a.s.* converges toward a unique "target" y_*. Our method of proof is the so-called *SDE* method, which heavily relies on functional weak convergence arguments. We will have to admit few important results about weak convergence of processes, for which we will provide precise references. An alternative proof is possible based on the *CLT* for triangular arrays of martingale increments (see [142], see also Theorem 12.8 for a statement in the Miscellany Chapter). Such an alternative proof – in a one-dimensional setting – can be found in [203].

We propose below a proof of Theorem 6.6, dealing only with the regular weak convergence in the case of an *a.s.* convergence toward a unique equilibrium point y_*. The extension to a multi-target algorithm is not much more involved and we refer to the original paper [241]. Before getting onto the proof, we need to recall the discrete time Burkhölder–Davis–Gundy (*B.D.G.*) inequality (and the Marcinkiewicz–Zygmund inequality) for discrete time martingales. We refer to [263] (p. 499) for a proof and various developments.

Burkhölder–Davis–Gundy Inequality (discrete time) and Marcinkiewicz–Zygmund inequality. Let $p \in [1, +\infty)$. There exists universal real constants c_p and $C_p > 0$ such that, for every sequence $(\Delta M_n)_{n\geq 1}$ of $(\mathcal{F}_n)_{n\geq 1}$-martingale increments, for every $n \geq 1$,

$$c_p \left\| \sqrt{\sum_{k=1}^{n} (\Delta M_k)^2} \right\|_p \leq \left\| \max_{k=1,\dots,n} \left| \sum_{\ell=1}^{k} \Delta M_\ell \right| \right\|_p \leq C_p \left\| \sqrt{\sum_{k=1}^{n} (\Delta M_k)^2} \right\|_p . \tag{6.49}$$

If $p > 1$, one also has, for every $n \geq 1$

$$c_p \left\| \sqrt{\sum_{k=1}^{n} (\Delta M_k)^2} \right\|_p \leq \left\| \sum_{k=1}^{n} \Delta M_k \right\|_p \leq C_p \left\| \sqrt{\sum_{k=1}^{n} (\Delta M_k)^2} \right\|_p . \tag{6.50}$$

The left inequality in (6.50) remains true for $p = 1$ if the random variables $(\Delta M_n)_{n\geq 1}$ are independent. Then, the inequality takes the name of Marcinkiewicz–Zygmund inequality.

Proof of Theorem 6.4. We first introduce the augmented filtration of the innovations defined by $\mathcal{F}_n = \sigma(Y_0, Z_1, \dots, Z_n)$, $n \geq 0$. It is clear by induction that $(Y_n)_{n\geq 0}$ is $(\mathcal{F}_n)_{n\geq 0}$-adapted (in what follows we will sometimes use a random variable $Z \stackrel{d}{=} Z_1$). We first rewrite the recursion satisfied by the algorithm in its canonical form

$$Y_n = Y_{n-1} - \gamma_n\Big(h(Y_{n-1}) + \Delta M_n\Big),\ \ n \ge 1,$$

where

$$\Delta M_n = H(Y_{n-1}, Z_n) - h(Y_{n-1}),\ \ n \ge 1,$$

is a sequence of $\mathcal{F}_n$-martingale increments. The so-called *SDE* method is based on the same principle as the *ODE* method but with the quantity of interest

$$\Upsilon_n := \frac{Y_n - y_*}{\sqrt{\gamma_n}}, \quad n \ge 1.$$

This normalization is strongly suggested by the above L^2-convergence rate theorem. The underlying idea is to write a recursion on Υ_n which appears as an Euler scheme with decreasing step γ_n of an *SDE* having $\mathcal{N}\big(0;\sigma\big)$ as a stationary/steady regime.
STEP 1 (*Toward the SDE*). As announced, we assume that

$$Y_n \xrightarrow{a.s.} y_* \in \{h = 0\}.$$

We may assume (up to a change of variable resulting from by the translation $y \leftarrow y - y_*$) that

$$y_* = 0.$$

The differentiability of h at $y_* = 0$ reads

$$h(Y_n) = Y_n h'(0) + Y_n \eta(Y_n) \quad \text{with} \quad \lim_{y \to 0} \eta(y) = \eta(y_*) = 0.$$

Moreover the function η is locally bounded on the real line owing to the growth assumption made on $\|H(y,Z)\|_2$ which implies that h is locally bounded too owing to Jensen's inequality. For every $n \ge 1$, we have

$$\begin{aligned}
\Upsilon_{n+1} &= \sqrt{\frac{\gamma_n}{\gamma_{n+1}}}\,\Upsilon_n - \sqrt{\gamma_{n+1}}\Big(h(Y_n) + \Delta M_{n+1}\Big)\\
&= \sqrt{\frac{\gamma_n}{\gamma_{n+1}}}\,\Upsilon_n - \sqrt{\gamma_{n+1}}\,Y_n\Big(h'(0) + \eta(Y_n)\Big) - \sqrt{\gamma_{n+1}}\,\Delta M_{n+1}\\
&= \Upsilon_n - \Upsilon_n + \sqrt{\frac{\gamma_n}{\gamma_{n+1}}}\,\Upsilon_n - \sqrt{\gamma_{n+1}}\,\sqrt{\gamma_n}\,\Upsilon_n\Big(h'(0) + \eta(Y_n)\Big)\\
&\qquad - \sqrt{\gamma_{n+1}}\,\Delta M_{n+1}\\
&= \Upsilon_n - \gamma_{n+1}\Upsilon_n\left(\sqrt{\frac{\gamma_n}{\gamma_{n+1}}}\Big(h'(0) + \eta(Y_n)\Big) - \frac{1}{\gamma_{n+1}}\left(\sqrt{\frac{\gamma_n}{\gamma_{n+1}}} - 1\right)\right)\\
&\qquad - \sqrt{\gamma_{n+1}}\,\Delta M_{n+1}.
\end{aligned}$$

Assume that the sequence $(\gamma_n)_{n\ge1}$ is such that there exists a $c \in (0, +\infty]$ satisfying

$$\lim_n \left[a'_n = \frac{1}{\gamma_{n+1}}\left(\sqrt{\frac{\gamma_n}{\gamma_{n+1}}} - 1\right)\right] = \frac{1}{2c}.$$

Note that this implies $\lim_n \frac{\gamma_n}{\gamma_{n+1}} = 1$. One easily checks that this condition is satisfied by our two families of step sequences of interest since

– if $\gamma_n = \frac{c}{b+n}$, $c > 0$, $b \geq 0$, then $\lim_n a'_n = \dfrac{1}{2c} > 0$,

– if $\gamma_n = \frac{c}{n^{\vartheta}}$, $c > 0$, $\frac{1}{2} < \vartheta < 1$, then $\lim_n a'_n = 0$, *i.e.* $c = +\infty$.

Consequently, for every $n \geq 1$,

$$\Upsilon_{n+1} = \Upsilon_n - \gamma_{n+1}\Upsilon_n\Big(h'(0) - \frac{1}{2c} + \alpha_n^1 + \alpha_n^2\eta(Y_n)\Big) - \sqrt{\gamma_{n+1}}\Delta M_{n+1},$$

where $(\alpha_n^i)_{n\geq 1}$ $i = 1, 2$ are two deterministic sequences such that $\alpha_n^1 \to 0$ and $\alpha_n^2 \to 1$ as $n \to +\infty$.

Step 2 (*Localization(s)*). Since $Y_n \to 0$ *a.s.*, one can write the scenarii space Ω as follows

$$\forall\, \varepsilon > 0, \qquad \Omega = \bigcup_{N\geq 1} \Omega_{\varepsilon,N} \quad a.s. \quad \text{where} \quad \Omega_{\varepsilon,N} := \Big\{\sup_{n\geq N} |Y_n| \leq \varepsilon\Big\}.$$

Let $\varepsilon > 0$ and $N \geq 1$ be temporarily free parameters. We define the function $\widetilde{h} = \widetilde{h}_\varepsilon$ by

$$\forall\, y \in \mathbb{R}, \qquad \widetilde{h}(y) = h(y)\mathbf{1}_{\{|y|\leq\varepsilon\}} + Ky\mathbf{1}_{\{|y|>\varepsilon\}}$$

($K = K(\varepsilon)$ is also a parameter to be specified further on) and

$$\begin{cases} \widetilde{Y}_N^{\varepsilon,N} = Y_N \mathbf{1}_{\{|Y_N|\leq\varepsilon\}}, \\[2ex] \widetilde{Y}_{n+1}^{\varepsilon,N} = \widetilde{Y}_n^{\varepsilon,N} - \gamma_{n+1}\Big(\widetilde{h}_\varepsilon(\widetilde{Y}_n^{\varepsilon,N}) + \mathbf{1}_{\{|Y_n|\leq\varepsilon\}}\Delta M_{n+1}\Big), \ n \geq N. \end{cases}$$

It is straightforward to show by induction that, for every $\omega \in \Omega_{\varepsilon,N}$,

$$\forall\, n \geq N, \quad Y_n^{\varepsilon,N}(\omega) = Y_n(\omega).$$

To alleviate notation, we will drop the exponent ${}^{\varepsilon,N}$ in what follows and write $\widetilde{Y}_n$ instead of $\widetilde{Y}_n^{\varepsilon,N}$.

The mean function and $\mathcal{F}_n$-martingale increments associated to this new algorithm are

$$\widetilde{h} \quad \text{and} \quad \Delta\widetilde{M}_{n+1} = \mathbf{1}_{\left\{|Y_n|\leq\varepsilon\right\}}\Delta M_{n+1}, \ n \geq N,$$

which satisfy

$$\begin{aligned}\sup_{n\geq N}\mathbb{E}\left(|\Delta\widetilde{M}_{n+1}|^{2+\beta}\,|\mathcal{F}_n\right) &\leq 2^{1+\beta}\sup_{|\theta|\leq\varepsilon}\mathbb{E}\,|H(\theta,X)|^{2+\beta}\\ &\leq A(\varepsilon) < +\infty \quad a.s.\end{aligned} \tag{6.51}$$

In what follows, we will study the normalized error defined by

$$\widetilde{\Upsilon}_n := \frac{\widetilde{Y}_n}{\sqrt{\gamma_n}},\ \ n\geq N.$$

STEP 3 (*Specification of* ε *and* $K = K(\varepsilon)$). We start again from the differentiability of h at 0 (and $h'(0) > 0$),

$$h(y) = y\big(h'(0)+\eta(y)\big) \quad \text{with} \quad \lim_{y\to 0}\eta(y) = \eta(0) = 0.$$

– If $\gamma_n = \frac{c}{n+b}$ with $c > \frac{1}{2h'(0)}$ (and $b \geq 0$), we may choose $\rho = \rho(h) > 0$ small enough so that

$$c > \frac{1}{2h'(0)(1-\rho)}.$$

– If $\gamma_n = \frac{c}{(n+b)^{\vartheta}}$, $\frac{1}{2} < \vartheta < 1$ and, more generally, as soon as $\lim_n a'_n = 0$, any choice of $\rho\in(0,1)$ is possible. Now let $\varepsilon(\rho) > 0$ be such that $|y| \leq \varepsilon(\rho)$ implies $|\eta(y)| \leq \rho h'(0)$. It follows that

$$\theta h(y) = y^2\big(h'(0)+\eta(y)\big) \geq y^2(1-\rho)h'(0) \quad \text{if} \quad |y| \leq \eta(\rho).$$

Now we specify for the rest of the proof

$$\varepsilon = \varepsilon(\rho) \qquad \text{and} \qquad K = (1-\rho)h'(0) > 0.$$

As a consequence, the function $\widetilde{h}$ satisfies

$$\forall\, y\in\mathbb{R},\quad y\widetilde{h}(y) \geq Ky^2.$$

Consequently, since $(\gamma_n)_{n\geq 1}$ satisfies (G_α) with $\alpha = K$ and $\kappa^* = 2K - \frac{1}{c} > 0$, one derives following the lines of the proof of Proposition 6.11 (established for Markovian algorithms) that,

$$\|\widetilde{Y}_n\|_2 = O\big(\sqrt{\gamma_n}\big).$$

STEP 4 (The *SDE* method). First we apply Step 1 to our framework and we write

$$\widetilde{\Upsilon}_{n+1} = \widetilde{\Upsilon}_n - \gamma_{n+1}\widetilde{\Upsilon}_n\Big(h'(0) - \frac{1}{2c} + \widetilde{\alpha}^1_n + \widetilde{\alpha}^2_n\widetilde{\eta}(\widetilde{Y}_n)\Big) - \sqrt{\gamma_{n+1}}\,\Delta\widetilde{M}_{n+1}, \tag{6.52}$$

where $\widetilde{\alpha}_n^1 \to 0$ and $\widetilde{\alpha}_n^2 \to 1$ are two deterministic sequences and $\widetilde{\eta}$ is a bounded function.

At this stage, we want to re-write the above recursive equation in continuous time exactly like we did for the *ODE* method. To this end, we first set

$$\Gamma_n = \gamma_1 + \cdots + \gamma_n, \quad n \geq N,$$

and

$$\forall\, t \in \mathbb{R}_+, \quad N(t) = \min\big\{k \,:\, \Gamma_{k+1} \geq t\big\}, \quad \underline{t} = \Gamma_{N(t)}.$$

To alleviate the notation we also set $a = h'(0) - \frac{1}{2c} > 0$. We first define the càdlàg function $\widetilde{\Upsilon}^{(0)}$ on $[\Gamma_{_N}, +\infty)$ by setting

$$\widetilde{\Upsilon}^{(0)}_{(t)} = \widetilde{\Upsilon}_n, \ t \in [\Gamma_n, \Gamma_{n+1}), \ n \geq N,$$

so that, in particular, $\widetilde{\Upsilon}^{(0)}_{(t)} = \widetilde{\Upsilon}^{(0)}_{(\underline{t})}$. Following the strategy adopted for the *ODE* method, one expands the recursion (6.52) in order to obtain

$$\widetilde{\Upsilon}^{(0)}_{(\Gamma_n)} = \widetilde{\Upsilon}_{_N} - \int_{\Gamma_{_N}}^{\Gamma_n} \widetilde{\Upsilon}^{(0)}_{(t)}\big(a + \widetilde{\alpha}^1_{N(t)} + \widetilde{\alpha}^2_{N(t)}\widetilde{\eta}(\widetilde{Y}_{N(t)})\big)dt) - \sum_{k=N}^{n} \sqrt{\gamma_k}\,\Delta\widetilde{M}_k.$$

As a consequence, one has for every time $t \geq \Gamma_{_N}$,

$$\widetilde{\Upsilon}^{(0)}_{(t)} = \widetilde{\Upsilon}_{_N} - \int_{\Gamma_{_N}}^{\underline{t}} \widetilde{\Upsilon}^{(0)}_{s}\big(a + \widetilde{\alpha}^1_{N(s)} + \widetilde{\alpha}^2_{N(s)}\widetilde{\eta}(\widetilde{Y}_{N(s)})\big)ds - \sum_{k=N}^{N(t)} \sqrt{\gamma_k}\,\Delta\widetilde{M}_k. \tag{6.53}$$

Still like for the *ODE*, we are interested in the functional asymptotics at infinity of $\widetilde{\Upsilon}^{(0)}_{(t)}$, this time in a weak sense. To this end, we introduce for every $n \geq N$ and every $t \geq 0$, the sequence of time shifted functions, defined this time on $\mathbb{R}_+$:

$$\widetilde{\Upsilon}^{(n)}_{(t)} = \Upsilon^{(0)}_{(\Gamma_n+t)}, \quad t \geq 0, \ n \geq N.$$

It follows from (6.53) that these processes satisfy for every $t \geq 0$,

$$\widetilde{\Upsilon}^{(n)}_{(t)} = \widetilde{\Upsilon}_n - \underbrace{\int_{\Gamma_n}^{\underline{\Gamma_n+t}} \widetilde{\Upsilon}^{(0)}_{(s)}\big(a + \widetilde{\alpha}^1_{N(s)} + \widetilde{\alpha}^2_{N(s)}\widetilde{\eta}(\widetilde{Y}_{N(s)})\big)ds}_{=:\,\widetilde{A}^{(n)}_{(t)}} + \underbrace{\sum_{k=n+1}^{N(\Gamma_n+t)} \sqrt{\gamma_k}\,\Delta\widetilde{M}_k}_{=:\,\widetilde{M}^{(n)}_{(t)}}. \tag{6.54}$$

Step 5 (*Functional tightness*). At this stage, we need two fundamental results about functional weak convergence. The first is a criterion which implies the functional tightness of the distributions of a sequence of càdlàg processes $X^{(n)}$ (viewed as probability measures on the space $\mathbb{D}(\mathbb{R}_+, \mathbb{R})$ of càdlàg functions from $\mathbb{R}_+$ to $\mathbb{R}$). The second is an extension of Donsker's Theorem for sequences of martingales.

We recall the definition of the uniform continuity modulus defined for every function $f : \mathbb{R}_+ \to \mathbb{R}$ and $\delta, T > 0$ by

$$w(f, \delta, T) = \sup_{s,t\in[0,T],\ |s-t|\le\delta} |f(t) - f(s)|.$$

The terminology comes from the seminal property of this modulus: f is (uniformly) continuous over $[0, T]$ if and only if $\lim_{\delta\to 0} w(f, \delta, T) = 0$.

Theorem 6.7 (**C-tightness criterion, see [45], Theorem 15.5, p. 127**) *Let $(X^n_t)_{t\ge 0}$, $n \ge 1$, be a sequence of càdlàg processes null at $t = 0$.*

(a) If, for every $T > 0$ and every $\varepsilon > 0$,

$$\lim_{\delta\to 0} \overline{\lim_{n}}\, \mathbb{P}\big(w(X^{(n)}, \delta, T) \ge \varepsilon\big) = 0, \tag{6.55}$$

then the sequence $(X^n)_{n\ge 1}$ is C-tight in the following sense: from any subsequence $(X^{n'})_{n\ge 1}$ one may extract a subsequence $(X^{n''})_{n\ge 1}$ such that $X^{n''}$ converges in distribution toward a process X^∞ with respect to the weak topology on the space $\mathbb{D}(\mathbb{R}_+, \mathbb{R})$ induced by the topology of uniform convergence on compact sets ([10]) such that $\mathbb{P}(X^\infty \in C(\mathbb{R}_+, \mathbb{R})) = 1$.

(b) A criterion *(see [45], proof of Theorem 8.3 p. 56): If, for every $T > 0$ and every $\varepsilon > 0$,*

$$\lim_{\delta\to 0} \overline{\lim_{n}} \sup_{s\in[0,T]} \frac{1}{\delta} \mathbb{P}\Big(\sup_{s\le t\le s+\delta} |X^{(n)}_t - X^{(n)}_s| \ge \varepsilon\Big) = 0, \tag{6.56}$$

then the above condition (6.55) *in (a) is satisfied.*

The second theorem below provides a tightness criterion for a sequence of martingales based on the sequence of their bracket processes.

Theorem 6.8 (**Weak functional limit of a sequence of martingales, see [154]**) *Let $(M^n_{(t)})_{t\ge 0}$, $n \ge 1$, be a C-tight sequence of càdlàg (local) martingales, null at* 0*, with (existing) predictable bracket process $\langle M^n\rangle$. If*

$$\forall\, t \ge 0, \qquad \langle M^n\rangle_{(t)} \xrightarrow{a.s.} \sigma^2 t \quad \textit{as } n \to +\infty, \quad \sigma > 0,$$

then

$$M^n \xrightarrow{\mathcal{L}_{\mathbb{D}(\mathbb{R}_+,\mathbb{R})}} \sigma W$$

[10] Although this topology is not standard on this space, it is simply defined sequentially by $X^n \overset{(U)}{\Longrightarrow} X^\infty$ if for every bounded functional $F : \mathbb{D}(\mathbb{R}_+, \mathbb{R}) \to \mathbb{R}$, *continuous for the* $\|\cdot\|_{\sup}$*-norm*, $\mathbb{E}\,F(X^n) \to \mathbb{E}\,F(X^\infty)$.

where W denotes a standard Brownian motion ([11]).

Now we can apply these results to the processes $\widetilde{A}^{(n)}_{(t)}$ and $\widetilde{M}^{(n)}_{(t)}$.

First we aim at showing that the sequence of continuous processes $(\widetilde{A}^{(n)})_{n\ge 1}$ is C-tight. Since $\sup_{n\ge N, t\ge 0} |\widetilde{\eta}(\widetilde{Y}_{N(t)})| \le \|\widetilde{\eta}\|_{\sup} < +\infty$, there exists a real constant $C = C_{\|\widetilde{\eta}\|_{\sup}, \|\widetilde{\alpha}^1\|_{\sup}, \|\widetilde{\alpha}^2\|_{\sup}} > 0$ such that the sequence $(\widetilde{A}^{(n)})_{n\ge N}$ of time integrals satisfies for every $s \ge 0$ and every $\delta > 0$,

$$\sup_{s\le t\le s+\delta} \big|\widetilde{A}^{(n)}_{(t)} - \widetilde{A}^{(n)}_{(s)}\big| \le C \int_{\underline{s+\Gamma_n}}^{s+\delta+\Gamma_n} \big|\widetilde{\Upsilon}^{(0)}_{(u)}\big| du.$$

Hence, owing to the Schwarz Inequality,

$$\sup_{s\le t\le s+\delta} \big|\widetilde{A}^{(n)}_{(t)} - \widetilde{A}^{(n)}_{(s)}\big|^2 \le C^2\big(\underline{s+\delta+\Gamma_n} - \underline{s+\Gamma_n}\big) \int_{\underline{s+\Gamma_n}}^{s+\delta+\Gamma_n} \big|\widetilde{\Upsilon}^{(0)}_{(u)}\big|^2 du$$

so that

$$\begin{aligned}
\mathbb{E}\left[\sup_{s\le t\le s+\delta} |\widetilde{A}^{(0)}_{(t)} - \widetilde{A}^{(0)}_{(s)}|^2\right] &\le C^2 \sup_{n\ge N} \mathbb{E}\,|\widetilde{\Upsilon}_n|^2 \times \big(\underline{s+\delta+\Gamma_n} - \underline{s+\Gamma_n}\big)^2 \\
&\le C^2 \sup_{n\ge N} \|\widetilde{\Upsilon}_n\|_2^2 \times \big(\delta + \gamma_{N(s+\delta+\Gamma_n)+1}\big)^2.
\end{aligned}$$

Hence, for every $n \ge N$ and every $s \in [0, T]$,

$$\frac{1}{\delta}\,\mathbb{P}\Big(\sup_{s\le t\le s+\delta} |\widetilde{A}^{(n)}_{(t)} - \widetilde{A}^{(n)}_{(s)}| \ge \varepsilon\Big) \le \frac{C^2 \sup_n \|\widetilde{\Upsilon}_n\|_2^2 (\delta + \gamma_{N(T+\delta+\Gamma_n)+1})^2}{\delta\varepsilon}.$$

Noting that $\lim_{n\to+\infty} \gamma_{N(T+\delta+\Gamma_n)+1} = 0$, one derives that Criterion (6.56) is satisfied. Hence, the sequence $(\widetilde{A}^{(n)})_{n\ge N}$ is C-tight by applying the above Theorem 6.7.

Now, we deal with the martingales $\widetilde{M}^{(n)}$, $n \ge N$. Let us consider the filtration $\mathcal{F}^{(0)}_t = \mathcal{F}_n$, $t \in [\Gamma_n, \Gamma_{n+1})$. We define $\widetilde{M}^{(0)}$ by

$$\widetilde{M}^{(0)}_{(t)} = 0 \text{ if } t \in [0, \Gamma_N] \quad \text{and} \quad \widetilde{M}^{(0)}_{(t)} = \sum_{k=N}^{N(t)} \sqrt{\gamma_k}\,\Delta\widetilde{M}_{k+1} \ \text{ if } \ t \in [\Gamma_N, +\infty).$$

It is clear that $\big(\widetilde{M}^{(0)}_{(t)}\big)_{t\ge 0}$ is an $(\mathcal{F}^{(0)}_t)_{t\ge 0}$-martingale. Moreover, we know from (6.51) that $\sup_n \mathbb{E}\,|\Delta\widetilde{M}_n|^{2+\beta} \le A(\varepsilon) < +\infty$.

---

[11] This means that for every bounded functional $F : \mathbb{D}(\mathbb{R}_+, \mathbb{R}) \to \mathbb{R}$, measurable with respect to the σ-field spanned by finite projection $\alpha \mapsto \alpha(t)$, $t \in \mathbb{R}_+$, and continuous at every $\alpha \in \mathcal{C}(\mathbb{R}_+, \mathbb{R})$, one has $\mathbb{E}\,F(M^n) \to \mathbb{E}\,F(\sigma W)$ as $n \to +\infty$. In fact, this remains true for measurable functionals F which are $\mathbb{P}_{\sigma W}(d\alpha)$-*a.s.* continuous on $\mathcal{C}(\mathbb{R}_+, \mathbb{R})$, such that $\big(F(M^n)\big)_{n\ge 1}$ is uniformly integrable.

It follows from the Burkhölder–Davis–Gundy Inequality (6.49) that, for every $s \in [\Gamma_{_N}, +\infty)$,

$$\mathbb{E}\left[\sup_{s\le t\le s+\delta}\left|\widetilde{M}^{(0)}_{(t)}-\widetilde{M}^{(0)}_{(s)}\right|^{2+\beta}\right] \le C_\beta\,\mathbb{E}\left(\sum_{k=N(s)+1}^{N(s+\delta)}\gamma_k(\Delta\widetilde{M}_k)^2\right)^{1+\frac{\beta}{2}}$$

$$\le C_\beta\left(\sum_{k=N(s)+1}^{N(s+\delta)}\gamma_k\right)^{1+\frac{\beta}{2}}\mathbb{E}\left(\frac{\sum_{k=N(s)+1}^{N(s+\delta)}\gamma_k(\Delta\widetilde{M}_k)^2}{\sum_{k=N(s)+1}^{N(s+\delta)}\gamma_{+}+++k}\right)^{1+\frac{\beta}{2}}$$

$$\le C_\beta\left(\sum_{k=N(s)+1}^{N(s+\delta)}\gamma_k\right)^{1+\frac{\beta}{2}}\mathbb{E}\left(\frac{\sum_{k=N(s)+1}^{N(s+\delta)}\gamma_k|\Delta\widetilde{M}_k|^{2+\beta}}{\sum_{k=N(s)+1}^{N(s+\delta)}\gamma_k}\right)$$

$$\le C_\beta\left(\sum_{k=N(s)+1}^{N(s+\delta)}\gamma_k\right)^{\frac{\beta}{2}}\sum_{k=N(s)+1}^{N(s+\delta)}\gamma_k\,\mathbb{E}\,|\Delta\widetilde{M}_k|^{2+\beta},$$

where C_β is a positive real constant. One finally derives that, for every $s \in [\Gamma_{_N}, +\infty)$,

$$\mathbb{E}\left[\sup_{s\le t\le s+\delta}\left|M^{(0)}_{(t)}-M^{(0)}_{(s)}\right|^{2+\delta}\right] \le C_\delta A(\varepsilon)\left(\sum_{k=N(s)+1}^{N(s+\delta)}\gamma_k\right)^{1+\frac{\beta}{2}}$$

$$\le C_\delta A(\varepsilon)\Big(\delta+\sup_{k\ge N(s)+1}\gamma_k\Big)^{1+\frac{\beta}{2}}.$$

Noting that $\widetilde{M}^{(n)}_{(t)} = M^{(0)}_{\Gamma_n+t} - M^{(0)}_{\Gamma_n}$, $t \ge 0$, $n \ge N$, we derive

$$\forall\, n \ge N,\ \forall\, s \ge 0,\quad \mathbb{E}\sup_{s\le t\le s+\delta}\left|\widetilde{M}^{(n)}_{(t)}-\widetilde{M}^{(n)}_{(s)}\right|^{2+\beta} \le C'_\delta\Big(\delta+\sup_{k\ge N(\Gamma_n)+1}\gamma_k\Big)^{1+\frac{\beta}{2}}.$$

Then, by Markov's inequality, we have for every $\varepsilon > 0$ and $T > 0$,

$$\overline{\lim_n}\,\frac{1}{\delta}\sup_{s\in[0,T]}\mathbb{P}\Big(\sup_{s\le t\le s+\delta}\left|\widetilde{M}^{(n)}_{(t)}-\widetilde{M}^{(n)}_{(s)}\right| \ge \varepsilon\Big) \le C'_\delta\frac{\delta^{\frac{\beta}{2}}}{\varepsilon^{2+\beta}}.$$

The C-tightness of the sequence $(\widetilde{M}^{(n)})_{n\ge N}$ follows again from Theorem 6.7(b). Furthermore, for every $n \ge N$,

$$
\begin{aligned}
\langle \widetilde{M}^{(n)} \rangle_t &= \sum_{k=n+1}^{N(\Gamma_n+t)} \gamma_k\, \mathbb{E}\left((\Delta \widetilde{M}_k)^2 \mid \mathcal{F}_{k-1}\right) \\
&= \sum_{k=n+1}^{N(\Gamma_n+t)} \gamma_k \left(\left[\mathbb{E}\, H(y,Z)^2\right]_{|y=Y_{k-1}} - h(Y_{k-1})^2\right) \\
&\sim \left(\mathbb{E}\,[H(0,Z)^2] - h(0)^2\right) \sum_{k=n+1}^{N(\Gamma_n+t)} \gamma_k \quad \text{as} \quad n \to +\infty
\end{aligned}
$$

since $y \mapsto \mathbb{E}\left[H(y,Z)^2\right]$ and h are both continuous at $y_* = 0$ and $Y_k \to y_*$ as $k \to +\infty$. Using that $h(0) = 0$, it follows that

$$
\langle \widetilde{M}^{(n)} \rangle_t \longrightarrow \mathbb{E}\,[H(0,Z)^2] \times t \qquad \text{as} \quad n \to +\infty.
$$

Setting $\sigma^2 = \mathbb{E}\,[H(0,Z)^2]$, Theorem 6.8 then implies

$$
\widetilde{M}^{(n)} \overset{\mathcal{L}_{C(\mathbb{R}_+,\mathbb{R})}}{\longrightarrow} \sigma\, W^{(\infty)}
$$

where $W^{(\infty)}$ is a standard Brownian motion.

Step 6 (*Synthesis and conclusion*). The sequence of processes $\left(\widetilde{\Upsilon}^{(n)}_{(t)}\right)_{t\geq 0}$, $n \geq N$, satisfies, for every $n \geq N$,

$$
\forall\, t \geq 0, \quad \widetilde{\Upsilon}^{(n)}_{(t)} = \widetilde{\Upsilon}_n - \widetilde{A}^{(n)}_{(t)} - \widetilde{M}^{(n)}_{(t)}.
$$

Consequently, the sequence $\left(\widetilde{\Upsilon}^{(n)}\right)_{n\geq N}$ is C-tight since C-tightness is stable under addition, $(\widetilde{\Upsilon}_n)_{n\geq N}$ is tight (by L^2-boundedness) and $\left(\widetilde{A}^{(n)}\right)_{n\geq N}$, $\left(\widetilde{M}^{(n)}\right)_{n\geq N}$ are both C-tight.

The sequence of random variables $(\widetilde{\Upsilon}_n)_{n\geq N}$ is tight since it is L^2-bounded. Consequently, the sequence of processes $\left(\widetilde{\Upsilon}^{(n)}, \widetilde{M}^{(n)}\right)_{n\geq N}$ is C-tight as well.

Now let us elucidate the limit of $\left(\widetilde{A}^{(n)}_{(t)}\right)_{n\geq N}$

$$
\sup_{t\in[0,T]} \left| \widetilde{A}^{(n)}_t - a \int_{\Gamma_n}^{\Gamma_n+t} \widetilde{\Upsilon}^{(0)}_{(s)}\, ds \right| \leq a_n \int_0^T \left|\widetilde{\Upsilon}^{(n)}_{(s)}\right| ds
$$

where $a_n = \left[\sup_{k\geq n} |\widetilde{\alpha}^1_k| + \|\widetilde{\alpha}^2_{\cdot}\|_{\sup} \sup_{k\geq n} \left|\widetilde{\eta}(\widetilde{Y}_k)\right|\right]$. As $\widetilde{\eta}(\widetilde{Y}_n) \to 0$ *a.s.*, we derive that $a_n \to 0$ *a.s.*, whereas $\int_0^T \left|\widetilde{\Upsilon}^{(n)}_{(s)}\right| ds$ is L^1-bounded since

$$
\mathbb{E} \int_0^T \left|\widetilde{\Upsilon}^{(n)}_{(s)}\right| ds = \int_0^T \mathbb{E} \left|\widetilde{\Upsilon}^{(n)}_s\right| ds \leq \sup_{n\geq N} \|\widetilde{\Upsilon}_n\|_2 T < +\infty
$$

(we used that $\|\cdot|_1 \le \|\cdot\|_2$). Hence we obtain by Slutsky's Theorem $\sup_{t\in[0,T]}\left|\widetilde{A}^{(n)}_t - a\int_{\Gamma_n}^{\underline{\Gamma_n+t}} \widetilde{\Upsilon}^{(0)}_{(s)}ds\right| \to 0$ in probability. On the other hand,

$$\sup_{t\in[0,T]}\left|\int_{\underline{\Gamma_n+t}}^{\Gamma_n+t} \widetilde{\Upsilon}^{(0)}_{(s)}ds\right|^2 \le \sup_{t\in[0,T]}\left(\Gamma_n + t - \underline{\Gamma_n+t}\right)\int_{\Gamma_n}^{\Gamma_n+T} |\widetilde{\Upsilon}^{(0)}_{(s)}|ds$$
$$\le \sup_{k\ge n}\gamma_k \int_0^T |\widetilde{\Upsilon}^{(n)}_s|ds,$$

which implies likewise that $\sup_{t\in[0,T]}\left|\int_{\underline{\Gamma_n+t}}^{\Gamma_n+t} \widetilde{\Upsilon}^{(0)}_{(s)}ds\right|^2 \to 0$ in probability as $n\to+\infty$. Consequently, for every $T>0$,

$$\sup_{t\in[0,T]}\left|\widetilde{A}^{(n)}_t - a\int_0^t \widetilde{\Upsilon}^{(n)}_{(s)}ds\right| \to 0 \text{ in probability as } n\to+\infty. \tag{6.57}$$

Let $\left(\widetilde{\Upsilon}^{(\infty)}_{(t)}, \widetilde{M}^{(\infty)}_{(t)}\right)_{t\ge0}$ be a weak functional limiting value of $\left(\widetilde{\Upsilon}^{(n)}, \widetilde{M}^{(n)}\right)_{n\ge N}$ *i.e.* a weak limit along a subsequence $(\varphi(n))$. It follows from (6.54) that

$$\widetilde{\Upsilon}^{(n)} - \widetilde{\Upsilon}^{(n)}_{(0)} + a\int_0^{\cdot} \widetilde{\Upsilon}^{(n)}_{(s)}ds = \widetilde{M}^{(n)} - \left(\widetilde{A}^{(n)} - a\int_0^{\cdot}\widetilde{\Upsilon}^{(n)}_{(s)}ds\right).$$

Using that $(x,y)\mapsto\left(x\mapsto x - x(0) + a\int_0^{\cdot} x(s)ds, y\right)$ is clearly continuous for the $\|.\|_{\sup}$-norm and (6.57), it follows that

$$\Upsilon^{(\infty)}_{(t)} - \Upsilon^{(\infty)}_{(0)} + a\int_0^t \Upsilon^{(\infty)}_{(s)}ds = \sigma W^{(\infty)}_t, \quad t\ge 0.$$

This means that $\Upsilon^{(\infty)}$ is solution to the Ornstein–Uhlenbeck *SDE*

$$d\widetilde{\Upsilon}^{(\infty)}_{(t)} = -a\widetilde{\Upsilon}^{(\infty)}_{(t)}dt + \sigma\, dW^{(\infty)}_t \tag{6.58}$$

starting from a random variable $\widetilde{\Upsilon}^{(\infty)}_{(0)} \in L^2$ such that

$$\left\|\widetilde{\Upsilon}^{(\infty)}_{(0)}\right\|_2 \le \sup_{n\ge N}\|\widetilde{\Upsilon}_n\|_2.$$

This follows from the weak Fatou's Lemma for convergence in distribution (see Theorem 12.6(v)) since $\widetilde{\Upsilon}_{\varphi(n)} \stackrel{\mathcal{L}}{\to} \widetilde{\Upsilon}^{(\infty)}_{(0)}$. Let ν_0 be a weak limiting value of $(\widetilde{\Upsilon}_n)_{n\ge N}$ *i.e.* such that $\widetilde{\Upsilon}_{\psi(n)} \stackrel{(\mathbb{R})}{\Rightarrow} \nu_0$.

For every $t > 0$, one considers the sequence of integers $\psi_t(n)$ uniquely defined by

$$\Gamma_{\psi_t(n)} := \Gamma_{\psi(n)} - t.$$

Up to a new extraction, we may assume that we simultaneously have the convergence of

$$\widetilde{\Upsilon}^{(\psi(n))} \xrightarrow{\mathcal{L}} \widetilde{\Upsilon}^{(\infty,0)} \text{ starting from } \widetilde{\Upsilon}^{(\infty,0)}_{(0)} \stackrel{d}{=} \nu_0$$

and

$$\widetilde{\Upsilon}^{(\psi_t(n))} \xrightarrow{\mathcal{L}} \widetilde{\Upsilon}^{(\infty,-t)} \text{ starting from } \widetilde{\Upsilon}^{(\infty,-t)}_{(0)} \stackrel{d}{=} \nu_{-t}.$$

One checks by strong uniqueness of solutions of the above Ornstein–Uhlenbeck *SDE* (6.58) that

$$\widetilde{\Upsilon}^{(\infty,-t)}_{(t)} = \widetilde{\Upsilon}^{(\infty,0)}_{(0)}.$$

Now, let $(P_t)_{t\geq 0}$ denote the semi-group of the Ornstein–Uhlenbeck process defined on bounded Borel functions $f : \mathbb{R} \to \mathbb{R}$ by $P_t f(x) = \mathbb{E} f(X_t^x)$. From the preceding, for every $t \geq 0$,

$$\nu_0 = \nu_{-t} P_t.$$

Moreover, $(\nu_{-t})_{t\geq 0}$ is tight since it is L^2-bounded. Let $\nu_{-\infty}$ be a weak limiting value of ν_{-t} as $t \to +\infty$.

Let $\Upsilon^{\mu}_{(t)}$ denote a solution to (6.58) starting from a μ-distributed random variable independent of W. It is straightforward that its paths satisfy the confluence property

$$|\Upsilon^{\mu}_t - \Upsilon^{\mu'}_t| \leq |\Upsilon^{\mu}_0 - \Upsilon^{\mu'}_0| e^{-at}.$$

For every Lipschitz continuous function f with compact support,

$$\begin{aligned}
\left|\nu_{-\infty} P_t(f) - \nu_{-t} P^t(f)\right| &= \left|\mathbb{E} f(\Upsilon^{\nu_{-\infty}}_{(t)}) - \mathbb{E} f(\Upsilon^{\nu_{-t}}_{(t)})\right| \\
&\leq [f]_{\text{Lip}} \mathbb{E} \left|\Upsilon^{\nu_{-\infty}}_{(t)} - \Upsilon^{\nu_{-t}}_{(t)}\right| \\
&\leq [f]_{\text{Lip}} e^{-at} \mathbb{E} \left|\Upsilon^{\nu_{-\infty}}_{(0)} - \Upsilon^{\nu_{-t}}_{(0)}\right| \\
&\leq [f]_{\text{Lip}} e^{-at} \|\Upsilon^{\nu_{-\infty}}_{(0)} - \Upsilon^{\nu_{-t}}_{(0)}\|_2 \\
&\leq 2\,[f]_{\text{Lip}} e^{-at} \sup_{n\geq N} \|\widetilde{\Upsilon}_n\|_2 \\
&\longrightarrow 0 \quad \text{as} \quad t \to +\infty,
\end{aligned}$$

where we used in the penultimate line that $\sup_{n\geq N} \|\widetilde{\Upsilon}_n\|_2 < +\infty$

Consequently

$$\nu_0 = \lim_{t\to+\infty} \nu_{-\infty} P_t = \mathcal{N}\left(0; \frac{\sigma^2}{2a}\right).$$

We have proved that the distribution $\mathcal{N}\left(0;\frac{\sigma^2}{2a}\right)$ is the only possible limiting value, hence

$$\widetilde{\Upsilon}_n \xrightarrow{\mathcal{L}} \mathcal{N}\left(0;\frac{\sigma^2}{2a}\right) \quad \text{as} \quad n\to+\infty.$$

Now we return to Υ_n (prior to the localization). We just proved that for $\varepsilon=\varepsilon(\rho)$ and for every $N\geq 1$,

$$\widetilde{\Upsilon}_n^{\varepsilon,N} \xrightarrow{\mathcal{L}} \mathcal{N}\left(0;\frac{\sigma^2}{2a}\right) \quad \text{as} \quad n\to+\infty. \tag{6.59}$$

On the other hand, we already saw that $Y_n\to 0$ *a.s.* implies that $\Omega=\bigcup_{N\geq 1}\Omega_{\varepsilon,N}$ *a.s.* where $\Omega_{\varepsilon,N}=\left\{Y^{\varepsilon,N}=Y_n,\ n\geq N\right\}=\left\{\widetilde{\Upsilon}^{\varepsilon,N}=\Upsilon_n,\ n\geq N\right\}$. Moreover, the events $\Omega_{\varepsilon,N}$ are non-decreasing as N increases so that

$$\lim_{N\to+\infty}\mathbb{P}(\Omega_{\varepsilon,N})=1.$$

Owing to the localization principle, for every Borel bounded function f,

$$\forall\, n\geq N,\qquad \mathbb{E}\left|f(\Upsilon_n)-f(\Upsilon_n^{\varepsilon,N})\right|\leq 2\|f\|_\infty\mathbb{P}\big(\Omega^c_{\varepsilon,N}\big).$$

Combined with (6.59), if f is continuous and bounded, we get, for every $N\geq 1$,

$$\overline{\lim_n}\left|\mathbb{E}f(\Upsilon_n)-\mathbb{E}f\Big(\frac{\sigma}{\sqrt{2a}}\zeta\Big)\right|\leq 2\|f\|_\infty\mathbb{P}\big(\Omega^c_{\varepsilon,N}\big),$$

where $\zeta\stackrel{d}{=}\mathcal{N}(0;1)$. The result follows by letting N go to infinity and observing that for every bounded continuous function f

$$\lim_n \mathbb{E}f(\Upsilon_n)=\mathbb{E}f\Big(\frac{\sigma}{\sqrt{2a}}\zeta\Big)$$

i.e. $\Upsilon_n \xrightarrow{\mathcal{L}} \mathcal{N}\left(0;\frac{\sigma^2}{2a}\right)$ as $n\to+\infty$. ◇

6.4.4 The Averaging Principle for Stochastic Approximation

Practical implementations of recursive stochastic algorithms show that the convergence, although ruled by a *CLT*, is chaotic, even in the final convergence phase, except if the step is optimized to produce the lowest asymptotic variance. Of course, this optimal choice is not realistic in practice since it requires an *a priori* knowledge of what we are trying to compute.

The original motivation to introduce the averaging principle was to "smoothen" the behavior of a converging stochastic algorithm by considering the arithmetic mean of the past values *up to the n-th iteration* rather than the computed value at the n-th iteration. In fact, we will see that, if this averaging procedure is combined with the use of a "slowly decreasing" step parameter γ_n, one attains for free the best possible rate of convergence!

To be precise: let $(\gamma_n)_{n\geq 1}$ be a step sequence satisfying

$$\gamma_n = \frac{c}{n^{\vartheta} + b}, \quad \vartheta \in (1/2, 1), \ c > 0, \ b \geq 0.$$

Then, we implement the standard recursive stochastic algorithm (6.3) and set

$$\bar{Y}_n := \frac{Y_0 + \cdots + Y_{n-1}}{n}, \ n \geq 1.$$

Note that, of course, this empirical mean itself satisfies a recursive formula:

$$\forall\, n \geq 0, \quad \bar{Y}_{n+1} = \bar{Y}_n - \frac{1}{n+1}\big(\bar{Y}_n - Y_n\big), \quad \bar{Y}_0 = 0.$$

By Césaro's averaging principle it is clear that under the assumptions which ensure that $Y_n \to y_*$, one has

$$\bar{Y}_n \stackrel{a.s.}{\longrightarrow} y_* \quad \text{as} \quad n \to +\infty$$

as well. This is even true on the event $\{Y_n \to y_*\}$ (*e.g.* in the case of multiple targets). What is more unexpected is that, under natural assumptions (see Theorem 6.9 hereafter), the (weak) rate of this convergence is ruled by a *CLT*

$$\sqrt{n}\,(\bar{Y}_n - y_*) \stackrel{\mathcal{L}_{\text{stably}}}{\longrightarrow} \mathcal{N}(0; \Sigma_*) \quad \text{on the event} \quad \{Y_n \to y_*\},$$

where Σ_* is the *minimal possible asymptotic variance-covariance matrix*. Thus, if $d = 1$, $\Sigma_* = \frac{\text{Var}(H(y_*,Z))}{h'(y_*)^2}$ corresponding to the optimal choice of the constant c in the step sequence $\gamma_n = \frac{c}{n+b}$ in the *CLT* satisfied by the algorithm itself.

As we did for the *CLT* of the algorithm itself, we again state – and prove – this *CLT* for the averaged procedure for Markovian algorithms, though it can be done for more general recursive procedures of the form

$$Y_{n+1} = Y_n - \gamma_{n+1}\big(h(Y_n) + \Delta M_{n+1}\big),$$

where $(\Delta M_n)_{n\geq 1}$ is an $L^{2+\eta}$-bounded sequence of martingale increments. However, the adaptation to such a more general setting of what follows is an easy exercise.

Theorem 6.9 **(Ruppert and Polyak, see [245, 259], see also [81, 241, 246])** *Let $H : \mathbb{R}^d \times \mathbb{R}^q \to \mathbb{R}^d$ be a Borel function. Let $(Z_n)_{n\geq 1}$ be a sequence of i.i.d. $\mathbb{R}^q$-*

valued random vectors defined on a probability space $(\Omega, \mathcal{A}, \mathbb{P})$, *independent of* $Y_0 \in L^2_{\mathbb{R}^d}(\Omega, \mathcal{A}, \mathbb{P})$. *Then, we define the recursive procedure by*

$$Y_{n+1} = Y_n - \gamma_{n+1} H(Y_n, Z_{n+1}), \ \ n \geq 0.$$

Assume that, for every $y \in \mathbb{R}^d$, $H(y, Z) \in L^1(\mathbb{P})$ *so that the mean vector field* $h(y) = \mathbb{E}\, H(y, Z)$ *is well-defined (this is implied by (iii) below).*

We make the following assumptions:

(*i*) *The function h is zero at* y_* *and is "fast" differentiable at* y_*, *in the sense that*

$$\forall\, y \in \mathbb{R}^d, \quad h(y) = J_h(y_*)(y - y_*) + O(|y - y_*|^2),$$

where all eigenvalues of the Jacobian matrix $J_h(y_*)$ *of h at* y_* *have a (strictly) positive real part. (Hence* $J_h(y_*)$ *is invertible).*

(*ii*) *The algorithm* Y_n *converges toward* y_* *with positive probability.*

(*iii*) *There exists an* $\eta > 2$ *such that*

$$\forall\, K > 0, \ \sup_{|y| \leq K} \mathbb{E}\, |H(y, Z)|^{2+\eta} < +\infty. \tag{6.60}$$

(*iv*) *The mapping* $y \mapsto \mathbb{E}\big(H(y, Z)H(y, Z)^t\big)$ *is continuous at* y_*.
Set $\Sigma_* = \mathbb{E}\big(H(y_*, Z)H(y_*, Z)^t\big)$.

Then, if the step sequence is slowly decreasing of the form $\gamma_n = \frac{c}{n^{\vartheta}+b}$, $n \geq 1$, *with* $1/2 < \vartheta < 1$ *and* $c > 0$, $b \geq 0$, *the empirical mean sequence defined by*

$$\bar{Y}_n = \frac{Y_0 + \cdots + Y_{n-1}}{n}$$

satisfies the CLT with the optimal asymptotic variance, on the event $\{Y_n \to y_*\}$, *namely*

$$\sqrt{n}\,(\bar{Y}_n - y_*) \stackrel{\mathcal{L}_{stably}}{\longrightarrow} \mathcal{N}\big(0; J_h(y_*)^{-1}\Sigma_* J_h(y_*)^{-1}\big) \quad on \quad \{Y_n \to y_*\}.$$

Proof (partial). We will prove this theorem in the case of a scalar algorithm, that is we assume $d = 1$. Beyond this dimensional limitation, only adopted for notational convenience, we will consider a more restrictive setting than the one proposed in the above statement of the theorem. We refer for example to [81] for the general case. In addition (or instead) of the above assumption we assume that:

– the function H satisfies the linear growth assumption:

$$\|H(y, Z)\|_2 \leq C(1 + |\theta|),$$

– the mean function h satisfies the coercivity assumption (6.43) from Proposition 6.11 for some $\alpha > 0$ and has a Lipschitz continuous derivative.

At this point, note that the step sequences $(\gamma_n)_{n\geq 1}$ under consideration are non-increasing and all satisfy the Condition (G_α) of Proposition 6.11 (*i.e.* (6.44)) with α from (6.43). In particular, this implies that $Y_n \to y_*$ *a.s.* and $\|Y_n - y_*\|_2 = O(\sqrt{\gamma_n})$.

Without loss of generality we may assume that $y_* = 0$, by replacing Y_n by $Y_n - y_*$, $H(y, z)$ by $H(y_* + y, z)$, etc. We start from the canonical decomposition

$$\forall n \geq 0,\ Y_{n+1} = Y_n - \gamma_{n+1} h(Y_n) - \gamma_{n+1}\Delta M_{n+1},$$

where $\Delta M_{n+1} = H(Y_n, Z_{n+1}) - h(Y_n)$, $n \geq 0$, is a sequence of $\mathcal{F}_n$-martingale increments with $\mathcal{F}_n = \sigma(Y_0, Z_1, \ldots, Z_n)$, $n \geq 0$. As $h(0) = 0$ and h' is Lipschitz, one has for every $y \in \mathbb{R}$, $h(y) - h'(0)y = y^2\kappa(y)$ with $|\kappa(y)| \leq [h']_{\text{Lip}}$. Consequently, for every $k \geq 0$,

$$h'(0)Y_k = \frac{Y_k - Y_{k+1}}{\gamma_{k+1}} - \Delta M_{k+1} - Y_k^2\kappa(Y_k)$$

which in turn implies, by summing from $k = 0$ up to $n - 1$,

$$h'(0)\sqrt{n}\,\bar{Y}_n = -\frac{1}{\sqrt{n}}\sum_{k=1}^{n}\frac{Y_k - Y_{k-1}}{\gamma_k} - \frac{1}{\sqrt{n}}M_n - \frac{1}{\sqrt{n}}\sum_{k=0}^{n-1} Y_k^2\kappa(Y_k).$$

We will successively inspect the three sums on the right-hand side of the equation. First, by an Abel transform, we get

$$\sum_{k=1}^{n}\frac{Y_k - Y_{k-1}}{\gamma_k} = \frac{Y_n}{\gamma_n} - \frac{Y_0}{\gamma_1} + \sum_{k=2}^{n} Y_{k-1}\left(\frac{1}{\gamma_k} - \frac{1}{\gamma_{k-1}}\right).$$

Hence, using that the sequence $\left(\frac{1}{\gamma_n}\right)_{n\geq 1}$ is non-decreasing, we derive

$$\left|\sum_{k=1}^{n}\frac{Y_k - Y_{k-1}}{\gamma_k}\right| \leq \frac{|Y_n|}{\gamma_n} + \frac{|Y_0|}{\gamma_1} + \sum_{k=2}^{n} |Y_{k-1}|\left(\frac{1}{\gamma_k} - \frac{1}{\gamma_{k-1}}\right).$$

Taking expectation and using that $\mathbb{E}\,|Y_k| \leq \left\|Y_k\right\|_2 \leq C\sqrt{\gamma_k}$, $k \geq 1$, for some real constant $C > 0$, we get,

$$\begin{aligned}\mathbb{E}\left|\sum_{k=1}^{n}\frac{Y_k - Y_{k-1}}{\gamma_k}\right| &\leq \frac{\mathbb{E}\,|Y_n|}{\gamma_n} + \frac{\mathbb{E}\,|Y_0|}{\gamma_1} + \sum_{k=2}^{n}\mathbb{E}\,|Y_{k-1}|\left(\frac{1}{\gamma_k} - \frac{1}{\gamma_{k-1}}\right)\\ &\leq \frac{C}{\sqrt{\gamma_n}} + \frac{\mathbb{E}\,|Y_0|}{\gamma_1} + C\sum_{k=2}^{n}\sqrt{\gamma_{k-1}}\left(\frac{1}{\gamma_k} - \frac{1}{\gamma_{k-1}}\right)\end{aligned}$$

$$= \frac{C}{\sqrt{\gamma_n}} + \frac{\mathbb{E}\,|Y_0|}{\gamma_1} + C\sum_{k=2}^{n} \frac{1}{\sqrt{\gamma_{k-1}}}\left(\frac{\gamma_{k-1}}{\gamma_k} - 1\right).$$

Now, for every $k \geq 2$,

$$\frac{1}{\sqrt{\gamma_{k-1}}}\left(\frac{\gamma_{k-1}}{\gamma_k} - 1\right) = c^{-\frac{1}{2}}(k^{\vartheta} + b)^{\frac{1}{2}}\left(\frac{k^{\vartheta} + b}{(k-1)^{\vartheta} + b} - 1\right)$$
$$\sim b^{-\frac{1}{2}}\vartheta k^{\frac{\vartheta}{2}-1} \quad \text{as } k \to +\infty$$

so that

$$\sum_{k=2}^{n} \frac{1}{\sqrt{\gamma_{k-1}}}\left(\frac{\gamma_{k-1}}{\gamma_k} - 1\right) = O\left(n^{\frac{\vartheta}{2}}\right).$$

As a consequence, it follows from the obvious facts that $\lim_n n\gamma_n = +\infty$ and $\lim_n n^{\frac{\vartheta-1}{2}} = 0$, that

$$\lim_n \frac{1}{\sqrt{n}}\mathbb{E}\left|\sum_{k=1}^{n} \frac{Y_k - Y_{k-1}}{\gamma_k}\right| = 0,$$

i.e. $\frac{1}{\sqrt{n}}\sum_{k=1}^{n} \frac{Y_k - Y_{k-1}}{\gamma_k} \xrightarrow{L^1} 0$ as $n \to +\infty$.

The second term is the martingale M_n obtained by summing the increments ΔM_k. This is this martingale which will rule the global weak convergence rate. To analyze it, we rely on Lindeberg's Theorem 12.8 from the Miscellany Chapter, applied with $a_n = n$. First, if we set $\Sigma_H(y) = \mathbb{E}\left(H(y, Z) - h(y)\right)^2$, it is straightforward that condition (i) of Theorem 12.8 involving the martingale increments is satisfied since

$$\frac{\langle M\rangle_n}{n} = \frac{1}{n}\sum_{k=1}^{n} \Sigma_H(Y_{k-1}) \xrightarrow{a.s.} \Sigma_H(y_*),$$

owing to Césaro's Lemma and the fact that $Y_n \to y_*$ *a.s.* It remains to check the condition (ii) of Theorem 12.8, known as Lindeberg's condition. Let $\varepsilon > 0$. Then,

$$\mathbb{E}\left((\Delta M_k)^2 \mathbf{1}_{\{|\Delta M_k| \geq \varepsilon\sqrt{n}\}}\right) \leq \varsigma(Y_{k-1}, \varepsilon\sqrt{n}),$$

where $\varsigma(y, a) = \mathbb{E}\Big(\left(H(y, Z) - h(y)\right)^2 \mathbf{1}_{\{|H(y,Z)-h(y)| \geq a\}}\Big)$, $a > 0$. One shows, owing to Assumption (6.60), that h is bounded on compact sets which, in turn, implies that $\left((H(y, Z) - h(y))^2\right)_{|y| \leq K}$ makes up a uniformly integrable family. Hence $\varsigma(y, a) \to 0$ as $a \to 0$ for every $y \in \mathbb{R}^d$, uniformly on compact sets of $\mathbb{R}^d$. Since $Y_n \to y_*$ *a.s.*, $(Y_n)_{n\geq 0}$ is *a.s.* bounded, hence

$$\frac{1}{n}\sum_{k=1}^{n}\varsigma\big(Y_{k-1},\varepsilon\sqrt{n}\big)\le \max_{0\le k\le n-1}\varsigma\big(Y_k,\varepsilon\sqrt{n}\big)\to 0 \quad a.s. \quad \text{as} \quad n\to+\infty.$$

Hence Lindeberg's condition is satisfied. As a consequence,

$$\frac{1}{h'(0)}\frac{M_n}{\sqrt{n}}\xrightarrow{\mathcal{L}}\mathcal{N}\left(0;\frac{\Sigma_H(y_*)}{h'(0)^2}\right).$$

The third term can be handled as follows, at least in our strengthened framework. UnderAssumption (6.43) of Proposition 6.11, we know that $\mathbb{E}\,Y_n^2\le C\gamma_n$ for some real constant $C>0$ since the class of step sequences we consider satisfies the condition (G_α) (see Practitioner's corner after Proposition 6.11). Consequently,

$$\begin{aligned}\frac{1}{\sqrt{n}}\sum_{k=0}^{n-1}\mathbb{E}\,(Y_k^2|\kappa(Y_k)|) &\le [h']_{\text{Lip}}\sum_{k=0}^{n-1}\mathbb{E}\,Y_k^2=\frac{C\,[h']_{\text{Lip}}}{\sqrt{n}}\sum_{k=1}^{+\infty}\gamma_k\\ &\le\frac{Cc\,[h']_{\text{Lip}}}{\sqrt{n}}\left(\mathbb{E}\,Y_0^2+\sum_{k=1}^{n-1}k^{-\vartheta}\right)\\ &\le\frac{Cc\,[h']_{\text{Lip}}}{\sqrt{n}}\left(\mathbb{E}\,Y_0^2+\frac{n^{1-\vartheta}}{1-\vartheta}\right)\sim\frac{Cc\,[h']_{\text{Lip}}}{1-\vartheta}n^{\frac{1}{2}-\vartheta}\to 0\end{aligned}$$

as $n\to+\infty$. This implies that $\displaystyle\frac{1}{\sqrt{n}}\sum_{k=1}^{+\infty}Y_k^2|\kappa(Y_k)|\xrightarrow{L^1}0$. Slutsky's Lemma completes the proof. ◇

Remark. As far as the step sequence is concerned, we only used in the above proof that the step sequence $(\gamma_n)_{n\ge1}$ is non-increasing and satisfies the following three conditions

$$(G_\alpha),\qquad \sum_n\frac{\gamma_k}{\sqrt{k}}<+\infty\quad\text{and}\quad\lim_n\sqrt{n}\,\gamma_n=+\infty.$$

Indeed, we have seen in the former section that, in one dimension, the asymptotic variance $\frac{\Sigma_H^*}{h'(0)^2}$ obtained in the Ruppert-Polyak theorem is the *lowest possible asymptotic variance* in the *CLT* when specifying the step parameter in an optimal way ($\gamma_n=\frac{c_{opt}}{n+b}$). In fact, this discussion and its conclusions can be easily extended to higher dimensions (if one considers some matrix-valued step sequences) as emphasized, for example, in [81].

So, the Ruppert and Polyak averaging principle *performs as fast as the "fastest" regular stochastic algorithm* with no need to optimize the step sequence: the optimal asymptotic variance is realized for free!

ℵ **Practitioner's corner.** ▷ *How to choose* ϑ? If we carefully inspect the two reminder non-martingale terms in the above proof, we see that they converge in

L^1 to zero at explicit rates:

$$\frac{1}{\sqrt{n}}\sum_{k=1}^{n}\frac{Y_k - Y_{k-1}}{\gamma_k} = O_{L^1}\big(n^{\frac{\vartheta-1}{2}}\big) \quad\text{and}\quad \frac{1}{\sqrt{n}}\sum_{k=0}^{n-1} Y_k^2|\kappa(Y_k)| = O_{L^1}\big(n^{\frac{1}{2}-\vartheta}\big).$$

Hence the balance between these two terms is obtained by equalizing the two exponents $\frac{1}{2}-\vartheta$ and $\frac{\vartheta-1}{2}$ *i.e.*

$$\frac{1}{2}-\vartheta = \frac{\vartheta-1}{2} \iff \vartheta_{opt} = \frac{2}{3}.$$

See also [101].

▷ *When to start averaging?* In practice, one should not start the averaging at the true beginning of the procedure but rather wait for its stabilization, ideally once the "exploration/search" phase is finished. On the other hand, the compromise consisting in using a moving window (typically of length n after $2n$ iterations) does not yield the optimal asymptotic variance, as pointed out in [196].

► **Exercises. 1.** Test the above averaging principle on the former exercises and "numerical illustrations" by considering $\gamma_n = \gamma_1 n^{-\frac{2}{3}}$, $n \ge 1$, as suggested in the first item of the above Practitioner's corner. Compare with a direct approach with a step of the form $\widetilde{\gamma}_n = \frac{c'}{n+b'}$, with $c' > 0$ "large enough but not too large...", and $b' \ge 0$.

2. Show that, under the (stronger) assumptions that we considered in the proof of the former theorem, Proposition 6.12 holds true with the averaged algorithm $(\bar{Y}_n)_{n\ge 1}$, namely that

$$\mathbb{E}\, L(\bar{Y}_n) - L(y_*) = O\Big(\frac{1}{n}\Big).$$

6.4.5 *Traps (A Few Words About)*

In the presence of multiple equilibrium points, *i.e.* of multiple zeros of the *mean* function h, some of them turn out to be parasitic. This is the case for saddle points or local maxima of the potential function L in the framework of stochastic gradient descent. More generally any zero of h whose Jacobian $J_h(y_*)$ has at least one eigenvalue with non-positive real part is parasitic.

There is a wide literature on this problem which says, roughly speaking, that a noisy enough parasitic equilibrium point is *a.s.* not a possible limit point for a stochastic approximation procedure. Although natural and expected, such a conclusion is far from being straightforward to establish, as testified by the various works on the topic (see [191], see also [33, 81, 95, 242], etc). If the equilibrium is not noisy many situations may occur as illustrated by the two-armed bandit algorithm, whose zeros are all noiseless (see [184]).

To some extent local minima are parasitic too but this is another story and standard stochastic approximation does not provide satisfactory answers to this "second order problem" for which specific procedures like simulated annealing should be implemented, with the drawback of degrading the (nature and) rate of convergence.

Going deeper in this direction is beyond the scope of this monograph so we refer to the literature mentioned above and the references therein for more insight on this aspect of Stochastic Approximation.

6.4.6 *(Back to)* $\mathbf{VaR}_\alpha$ *and* $\mathbf{CVaR}_\alpha$ *Computation (II): Weak Rate*

We can apply both above *CLT*s to the VaR_α and $\mathrm{CVaR}_\alpha(X)$ algorithms (6.28) and (6.30). Since

$$h(\xi) = \frac{1}{1-\alpha}\Big(F(\xi) - \alpha\Big) \quad \text{and} \quad \mathbb{E}\,H(\xi, X)^2 = \frac{1}{(1-\alpha)^2}F(\xi)(1 - F(\xi)),$$

where F is the c.d.f. of X, one easily derives from Theorems 6.6 and 6.9 the following results.

Theorem 6.10 *Assume that* $\mathbb{P}_X = f(x)dx$, *where* f *is a probability density function continuous at* $\xi^*_\alpha = \mathrm{VaR}_\alpha(X)$.
(a) If $\gamma_n = \frac{c}{n^\vartheta + b}$, $\frac{1}{2} < \vartheta < 1$, $c > 0$, $b \ge 0$, *then*

$$n^{\frac{a}{2}}\left(\xi_n - \xi^*_\alpha\right) \xrightarrow{\mathcal{L}} \mathcal{N}\Big(0; \frac{c\alpha(1-\alpha)}{2f(\xi^*_\alpha)}\Big).$$

(b) If $\gamma_n = \frac{c}{n+b}$, $b \ge 0$, *and* $c > \frac{1-\alpha}{2f(\xi^*_\alpha)}$ *then*

$$\sqrt{n}\big(\xi_n - \xi^*_\alpha\big) \xrightarrow{\mathcal{L}} \mathcal{N}\Big(0; \frac{c^2\alpha}{2cf(\xi^*_\alpha) - (1-\alpha)}\Big),$$

so that the minimal asymptotic variance is attained with $c^*_\alpha = \frac{1-\alpha}{f(\xi^*_\alpha)}$ *with an asymptotic variance equal to* $\frac{\alpha(1-\alpha)}{f(\xi^*_\alpha)^2}$.
(c) Ruppert and Polyak's averaging principle: If $\gamma_n = \frac{c}{n^\vartheta + b}$, $\frac{1}{2} < \vartheta < 1$, $c > 0$, $b \ge 0$, *then*

$$\sqrt{n}\big(\xi_n - \xi^*_\alpha\big) \xrightarrow{\mathcal{L}} \mathcal{N}\Big(0; \frac{\alpha(1-\alpha)}{f(\xi^*_\alpha)^2}\Big).$$

The algorithm for the $\mathrm{CVaR}_\alpha(X)$ satisfies the same kind of *CLT*.

This result is not satisfactory because the asymptotic variance remains huge since $f(\xi^*_\alpha)$ is usually very close to 0 when α is close to 1. Thus if X has a normal distribution $\mathcal{N}(0;1)$, then it is clear that $\xi^*_\alpha \to +\infty$ as $\alpha \to 1$. Consequently,

$$1-\alpha = \mathbb{P}(X \geq \xi^*_\alpha) \sim \frac{f(\xi^*_\alpha)}{\xi^*_\alpha} \quad \text{as} \quad \alpha \to 1$$

so that

$$\frac{\alpha(1-\alpha)}{f(\xi^*_\alpha)^2} \sim \frac{1}{\xi^*_\alpha f(\xi^*_\alpha)} \to +\infty \quad \text{as} \quad \alpha \to 1.$$

This simply illustrates the "rare event" effect which implies that, when α is close to 1, the event $\{X_{n+1} > \xi_n\}$ is rare especially when ξ_n gets close to its limit $\xi^*_\alpha = \mathrm{VaR}_\alpha(X)$.

The way out is to add an importance sampling procedure to somewhat "re-center" the distribution around its $\mathrm{VaR}_\alpha(X)$. To proceed, we will take advantage of our recursive variance reduction by importance sampling described and analyzed in Sect. 6.3.1. This is the object of the next section.

6.4.7 $\mathbf{VaR}_\alpha$ *and* $\mathbf{CVaR}_\alpha$ *Computation (III)*

As emphasized in the previous section, the asymptotic variance of our "naive" algorithms for VaR_α and CVaR_α computation are not satisfactory, in particular when α is close to 1. To improve them, the idea is to mix the recursive data-driven variance reduction procedure introduced in Sect. 6.3.1 with the above algorithms.

First we make the (not so) restrictive assumption that the r.v. X, representative of a loss, can be represented as a function of a Gaussian normal vector $Z \stackrel{d}{=} \mathcal{N}(0; I_d)$, namely

$$X = \varphi(Z), \quad \varphi : \mathbb{R}^d \to \mathbb{R}, \quad \text{Borel function.}$$

Hence, for a level $\alpha \in (0,1]$, in a (temporarily) *static* framework (*i.e.* fixed $\xi \in \mathbb{R}$), the function of interest for variance reduction is defined by

$$\varphi_{\alpha,\xi}(z) = \frac{1}{1-\alpha}\Big(\mathbf{1}_{\{\varphi(z)\leq \xi\}} - \alpha\Big), \quad z \in \mathbb{R}^d.$$

So, still following Sect. 6.3.1 and taking advantage of the fact that $\varphi_{\alpha,\xi}$ is bounded, we design the following data driven procedure for the adaptive variance reducer (using the notations from this section),

$$\theta_{n+1} = \theta_n - \gamma_{n+1}\varphi_{\alpha,\xi}(Z_{n+1} - \theta_n)^2(2\theta_n - Z_{n+1}), \ n \geq 0,$$

so that $\mathbb{E}\,\varphi_{\alpha,\xi}(Z)$ can be computed adaptively by

$$\mathbb{E}\,\varphi_{\alpha,\xi}(Z) = e^{-\frac{|\theta|^2}{2}}\mathbb{E}\Big(\varphi_{\alpha,\xi}(Z)e^{-(\theta|Z)}\Big) = \lim_n \frac{1}{n}\sum_{k=1}^{n} e^{-\frac{|\theta_{k-1}|^2}{2}}\varphi_{\alpha,\xi}(Z_k+\theta_{k-1})e^{-(\theta_{k-1}|Z_k)}.$$

Considering now a dynamical version of these procedures in order to adapt ξ recursively leads us to design the following procedure:

$$\widetilde{\xi}_{n+1} = \widetilde{\xi}_n - \frac{\gamma_{n+1}}{1-\alpha}e^{-\frac{|\theta_n|^2}{2}}e^{-(\theta_n|Z_k)}\big(\mathbf{1}_{\{\varphi(Z_{n+1}+\theta_n)\le\widetilde{\xi}_n\}} - \alpha\big) \tag{6.61}$$

$$\widetilde{\theta}_{n+1} = \widetilde{\theta}_n - \gamma_{n+1}\big(\mathbf{1}_{\{\varphi(Z_{n+1})\le\widetilde{\xi}_n\}} - \alpha\big)^2(2\widetilde{\theta}_n - Z_{n+1})\big),\quad n\ge 1, \tag{6.62}$$

with an appropriate initialization (see the remark below). This procedure is *a.s.* convergent toward its target, denoted by $(\theta_\alpha, \xi_\alpha)$, and the averaged component $\big(\bar{\widetilde{\xi}}_n\big)_{n\ge0}$ of $(\widetilde{\xi}_n)_{n\ge0}$ satisfies a *CLT* (see [29]).

Theorem 6.11 **(Adaptive VaR computation with importance sampling)** (*a*) *If the step sequence satisfies the decreasing step assumption* (6.7)*, then*

$$(\widetilde{\xi}_n, \widetilde{\theta}_n) \stackrel{n\to+\infty}{\longrightarrow} (\xi_\alpha, \theta_\alpha) \quad a.s. \quad with \quad \xi_\alpha = \mathrm{VaR}_\alpha(X)$$

and

$$\theta_\alpha = \operatorname{argmin}_{\theta\in\mathbb{R}}\Big[V_{\alpha,\xi}(\theta) = e^{-|\theta|^2}\mathbb{E}\Big(\big(\mathbf{1}_{\{\varphi(Z+\theta)\le\xi\}} - \alpha\big)^2 e^{-2(\theta|Z)}\Big)\Big].$$

Note that $V_{\alpha,\xi}(0) = F(\xi)\big(1-F(\xi)\big)$.

(*b*) *If the step sequence satisfies* $\gamma_n = \frac{c}{n^\vartheta+b}$, $\frac{1}{2} < \vartheta < 1$, $b\ge 0$, $c>0$, *then*

$$\sqrt{n}\,\big(\bar{\widetilde{\xi}}_n - \xi_\alpha\big) \stackrel{\mathcal{L}}{\longrightarrow} \mathcal{N}\left(0; \frac{V_{\alpha,\xi_\alpha}(\theta_\alpha)}{f(\xi_\alpha)^2}\right) \quad as \quad n\to+\infty.$$

Remark. In practice it may be useful, as noted in [29], to make the level α slowly vary with n in (6.61) and (6.62), *e.g.* from 0.5 up to the requested level, usually close to 1. Otherwise, the procedure may freeze. The initialization of the procedure should be set accordingly.

6.5 From Quasi-Monte Carlo to Quasi-Stochastic Approximation

Plugging quasi-random numbers into a recursive stochastic approximation procedure instead of pseudo-random numbers is a rather natural idea given the performances of *QMC* methods for numerical integration. To the best of our knowledge, it goes back to the early 1990s in [186]. As expected, various numerical tests showed that it may significantly accelerate the convergence of the procedure like in Monte Carlo simulations.

In [186], this question is mostly investigated from a theoretical viewpoint. The main results are based on an extension of uniformly distributed sequences on unit hypercubes called *averaging systems*. The two main results are based, on the one hand, on a contraction assumption and on the other hand on a monotonicity assumption, which both require some stringent conditions on the function H. In the first setting, some *a priori* error bounds emphasize that *quasi-stochastic approximation* does accelerate the convergence rate of the procedure. Both results are one-dimensional, though the contracting setting could be easily extended to multi-dimensional procedures. Unfortunately, it turns out to be of little interest for practical applications.

In this section, we want to propose a more natural multi-dimensional framework in view of applications. First, we give the counterpart of the Robbins–Siegmund Lemma established in Sect. 6.2. It relies on a *pathwise Lyapunov function*, which remains a rather restrictive assumption. It emphasizes what kind of assumption is needed to establish theoretical results when using deterministic uniformly distributed sequences. A more detailed version, including several examples of applications, is available in [190].

Theorem 6.12 **(Robbins–Siegmund Lemma,** *QMC* **framework)**
(*a*) *Let* $h : \mathbb{R}^d \to \mathbb{R}^d$ *and let* $H : \mathbb{R}^d \times [0,1]^q \xrightarrow{Borel} \mathbb{R}^d$ *be such that*

$$\forall\, y \in \mathbb{R}^d, \quad h(y) = \mathbb{E}\big(H(y, U)\big), \quad where \quad U \stackrel{d}{=} \mathcal{U}([0,1]^q).$$

Suppose that

$$\{h = 0\} = \{y_*\}$$

and that there exists a differentiable function $L : \mathbb{R}^d \to \mathbb{R}_+$ *with a Lipschitz continuous gradient* ∇L*, satisfying*

$$|\nabla L| \le C_L \sqrt{1 + L},$$

such that H *fulfills the following* pathwise *mean-reverting assumption: the function* Φ^H *defined for every* $y \in \mathbb{R}^d$ *by*

$$\Phi^H(y) := \inf_{u \in [0,1]^q} \big(\nabla L(y) | H(y, u) - H(y_*, u)\big) \quad is\ l.s.c.\ and\ positive\ on\ \mathbb{R}^d \setminus \{y_*\}. \tag{6.63}$$

Furthermore, assume that

$$\forall\, y \in \mathbb{R}^d,\ \forall\, u \in [0,1]^q, \qquad |H(y, u)| \le C_H \sqrt{1 + L(y)} \tag{6.64}$$

(which implies that h is bounded) and that the function

$$u \mapsto H(y_*, u) \text{ has finite variation (in the measure sense).}$$

Let $\xi := (\xi_n)_{n\geq 1}$ be a uniformly distributed sequence over $[0,1]^q$ with low discrepancy, hence satisfying

$$\ell_n := \max_{1\leq k\leq n}\big(kD_k^*(\xi)\big) = O\big((\log n)^q\big).$$

Let $\gamma = (\gamma_n)_{n\geq 1}$ be a non-increasing sequence of gain parameters satisfying

$$\sum_{n\geq 1}\gamma_n = +\infty,\quad \gamma_n(\log n)^q \to 0 \quad \textit{and} \quad \sum_{n\geq 1}\max\big(\gamma_n - \gamma_{n+1}, \gamma_n^2\big)(\log n)^q < +\infty. \tag{6.65}$$

Then, the recursive procedure defined by

$$\forall\, n \geq 0,\quad y_{n+1} = y_n - \gamma_{n+1}H(y_n, \xi_{n+1}),\quad y_0 \in \mathbb{R}^d,$$

satisfies:

$$y_n \longrightarrow y_* \quad \textit{as} \quad n \to +\infty.$$

(*b*) *If $(y,u) \mapsto H(y,u)$ is continuous, then Assumption (6.63) reads*

$$\forall\, y \in \mathbb{R}^d \setminus \{y_*\},\ \forall\, u \in [0,1]^q,\quad (\nabla L(y)|H(y,u) - H(y_*,u)) > 0. \tag{6.66}$$

Proof. (*a*) STEP 1 (*Regular step*). The beginning of the proof is rather similar to the "regular" stochastic case except that we will use as a Lyapunov function

$$\Lambda = \sqrt{1+L}.$$

First note that $\nabla\Lambda = \frac{\nabla L}{2\sqrt{1+L}}$ is bounded (by the constant C_L) so that Λ is C_L-Lipschitz continuous. Furthermore, for every $x,\ y \in \mathbb{R}^d$,

$$\begin{aligned}
\big|\nabla\Lambda(y) - \nabla\Lambda(y')\big| &\leq \frac{|\nabla L(y) - \nabla L(y')|}{\sqrt{1+L(y)}} \\
&\quad + |\nabla L(y')|\left|\frac{1}{\sqrt{1+L(y)}} - \frac{1}{\sqrt{1+L(y')}}\right| \qquad (6.67)\\
&\leq [\nabla L]_{\text{Lip}}\frac{|y-y'|}{\sqrt{1+L(y)}} \\
&\quad + \frac{C_L}{\sqrt{1+L(y)}}\big(\sqrt{1+L(y)} - \sqrt{1+L(y')}\big) \\
&\leq [\nabla L]_{\text{Lip}}\frac{|y-y'|}{\sqrt{1+L(y)}} + \frac{C_L^2}{\sqrt{1+L(y)}}|y-y'| \\
&\leq C_\Lambda\frac{|y-y'|}{\sqrt{1+L(y)}} \qquad (6.68)
\end{aligned}$$

with $C_{\Lambda} = [\nabla L]_{\text{Lip}} + C_L^2$.

It follows, by using successively the fundamental theorem of Calculus applied to Λ between y_n and y_{n+1} and Hölder's Inequality, that there exists $\zeta_{n+1} \in (y_n, y_{n+1})$ (geometric interval) such that

$$\begin{aligned}\Lambda(y_{n+1}) &= \Lambda(y_n) - \gamma_{n+1}\big(\nabla\Lambda(y_n)\,|\,H(y_n, \xi_{n+1})\big)\\ &\quad + \gamma_{n+1}\big(\nabla\Lambda(y_n) - \nabla\Lambda(\zeta_{n+1})|H(y_n, \xi_{n+1})\big)\\ &\le \Lambda(y_n) - \gamma_{n+1}\big(\nabla\Lambda(y_n)\,|\,H(y_n, \xi_{n+1})\big)\\ &\quad + \gamma_{n+1}\big|\nabla\Lambda(y_n) - \nabla\Lambda(\zeta_{n+1})\big|\big|H(y_n, \xi_{n+1})\big|.\end{aligned}$$

Now, the above inequality (6.68) applied with $y = y_n$ and $y' = \zeta_{n+1}$ yields, knowing that $|\zeta_{n+1} - y_n| \le |y_{n+1} - y_n|$,

$$\begin{aligned}\Lambda(y_{n+1}) &\le \Lambda(y_n) - \gamma_{n+1}\big(\nabla\Lambda(y_n)\,|\,H(y_n, \xi_{n+1})\big) + \gamma_{n+1}^2\frac{C_{\Lambda}}{\sqrt{1+L(y_n)}}|H(y_n, \xi_{n+1})|^2\\ &\le \Lambda(y_n) - \gamma_{n+1}\big(\nabla\Lambda(y_n)\,|\,H(y_n, \xi_{n+1}) - H(y_*, \xi_{n+1})\big)\\ &\quad - \gamma_{n+1}(\nabla\Lambda(y_n)\,|\,H(y_*, \xi_{n+1})) + \gamma_{n+1}^2 C_{\Lambda}\Lambda(y_n).\end{aligned}$$

Then, using (6.64), we get

$$\begin{aligned}\Lambda(y_{n+1}) &\le \Lambda(y_n)\big(1 + C_{\Lambda}\gamma_{n+1}^2\big)\\ &\quad - \gamma_{n+1}\Phi^H(y_n) - \gamma_{n+1}\big(\nabla\Lambda(y_n)\,|\,H(y_*, \xi_{n+1})\big).\end{aligned} \tag{6.69}$$

Set, for every $n \ge 0$,

$$s_n := \frac{\Lambda(y_n) + \sum_{k=1}^n \gamma_k \Phi^H(y_{k-1})}{\prod_{k=1}^n (1 + C_{\Lambda}\gamma_k^2)}$$

with the usual convention $\sum_{\varnothing} = 0$. It follows from (6.63) that the sequence $(s_n)_{n\ge 0}$ is non-negative since all the terms involved in its numerator are non-negative.

Now (6.69) reads

$$\forall\, n \ge 0, \qquad 0 \le s_{n+1} \le s_n - \widetilde{\gamma}_{n+1}\big(\nabla\Lambda(y_n)\,|\,H\,(y_*, \xi_{n+1})\big) \tag{6.70}$$

where $\widetilde{\gamma}_n = \dfrac{\gamma_n}{\prod_{k=1}^n (1 + C_{\Lambda}\gamma_k^2)}$, $n \ge 1$.

STEP 2 (*QMC step*). Set for every $n \ge 1$,

$$m_n := \sum_{k=1}^n \gamma_k\big(\nabla\Lambda(y_{k-1})\,|\,H(y_*, \xi_k)\big) \qquad \text{and} \qquad S_n^* = \sum_{k=1}^n H(y_*, \xi_k).$$

First note that (6.65) combined with the Koksma–Hlawka Inequality (see Proposition (4.3)) imply

$$|S_n^*| \le C_\xi V\big(H(y_*,\,.)\big)(\log n)^q, \tag{6.71}$$

where $V\big(H(y_*,\,.)\big)$ denotes the variation in the measure sense of $H(y_*,\,.)$. An Abel transform yields (with the convention $S_0^* = 0$)

$$\begin{aligned} m_n &= \widetilde{\gamma}_n\big(\nabla\Lambda(y_{n-1})\,|\,S_n^*\big) - \sum_{k=1}^{n-1}\big(\widetilde{\gamma}_{k+1}\nabla\Lambda(y_k) - \widetilde{\gamma}_k\nabla\Lambda(y_{k-1})\,|\,S_k^*\big)\\ &= \underbrace{\widetilde{\gamma}_n\big(\nabla\Lambda(y_{n-1})\,|\,S_n^*\big)}_{(a)} - \underbrace{\sum_{k=1}^{n-1}\widetilde{\gamma}_k\big(\nabla\Lambda(y_k) - \nabla\Lambda(y_{k-1})\,|\,S_k^*\big)}_{(b)}\\ &\quad \underbrace{-\sum_{k=1}^{n-1}\Delta\widetilde{\gamma}_{k+1}\big(\nabla\Lambda(y_k)\,|\,S_k^*\big)}_{(c)}. \end{aligned}$$

We aim at showing that m_n converges in $\mathbb{R}$ toward a finite limit by inspecting the above three terms.

One gets, using that $\gamma_n \le \widetilde{\gamma}_n$,

$$|(a)| \le \gamma_n\|\nabla\Lambda\|_{\sup} O\big((\log n)^q\big) = O(\gamma_n(\log n)^q) \to 0 \quad \text{as} \quad n \to +\infty.$$

Owing to (6.68), the partial sum (b) satisfies

$$\begin{aligned} \sum_{k=1}^{n-1}\widetilde{\gamma}_k\big|\big(\nabla\Lambda(y_k) - \nabla\Lambda(y_{k-1})\,|\,S_k^*\big)\big| &\le C_{\Lambda}\sum_{k=1}^{n-1}\widetilde{\gamma}_k\gamma_k\frac{|H(y_{k-1},\xi_k)|}{\sqrt{1+L(y_{k-1})}}|S_k^*|\\ &\le C_{\Lambda}C_H V\big(H(y_*,\,.)\big)\sum_{k=1}^{n-1}\gamma_k^2(\log k)^q, \end{aligned}$$

where we used Inequality (6.71) in the second inequality.

Consequently the series $\sum_{k\ge1}\widetilde{\gamma}_k\big(\nabla L(y_k) - \nabla L(y_{k-1})\,|\,S_k^*\big)$ is (absolutely) convergent owing to Assumption (6.65).

Finally, one deals with term (c). First notice that

$$|\widetilde{\gamma}_{n+1} - \widetilde{\gamma}_n| \le \gamma_{n+1} - \gamma_n + C_{\Lambda}\gamma_{n+1}^2\gamma_n \le C'_{\Lambda}\max(\gamma_n^2, \gamma_{n+1} - \gamma_n)$$

for some real constant C'_{Λ}. One checks that the series (c) is also (absolutely) convergent owing to the boundedness of ∇L, Assumption (6.65) and the upper-bound (6.71) for S_n^*.

Then m_n converges toward a finite limit m_∞. This induces that the sequence $(s_n + m_n)_n$ is bounded below since $(s_n)_n$ is non-negative. Now, we know from (6.70) that $(s_n + m_n)$ is also non-increasing, hence convergent in $\mathbb{R}$, which in turn implies that the sequence $(s_n)_{n\geq 0}$ itself is convergent toward a finite limit. The same arguments as in the regular stochastic case yield

$$L(y_n) \longrightarrow L_\infty \text{ as } n \to +\infty \quad \text{and} \quad \sum_{n\geq 1} \gamma_n \Phi^H(y_{n-1}) < +\infty.$$

One concludes, still like in the stochastic case, that (y_n) is bounded and eventually converges toward the unique zero of Φ^H, *i.e.* y_*.

(b) is obvious. ◇

ℵ **Practitioner's corner** • The step assumption (6.65) includes all the step sequences of the form $\gamma_n = \frac{c}{n^\alpha}$, $\alpha \in (0, 1]$. Note that as soon as $q \geq 2$, the condition $\gamma_n(\log n)^q \to 0$ is redundant (it follows from the convergence of the series on the right owing to an Abel transform).

• One can replace the (slightly unrealistic) assumption on $H(y_*, \,.\,)$ by a more natural Lipschitz continuous assumption, provided one strengthens the step assumption (6.65) into

$$\sum_{n\geq 1} \gamma_n = +\infty, \quad \gamma_n(\log n)n^{1-\frac{1}{q}} \to 0$$

and

$$\sum_{n\geq 1} \max\big(\gamma_n - \gamma_{n+1}, \gamma_n^2\big)(\log n)n^{1-\frac{1}{q}} < +\infty.$$

This is a straightforward consequence of Proïnov's Theorem (Theorem 4.3), which implies that

$$|S_n^*| \leq C(\log n)\, n^{1-\frac{1}{q}}.$$

Note that the above new assumptions are satisfied by the step sequences $\gamma_n = \frac{c}{n^\rho}$, $1 - \frac{1}{q} < \rho \leq 1$.

• It is clear that the mean-reverting assumption on H is much more stringent in the *QMC* setting.

• It remains that theoretical spectrum of application of the above theorem is dramatically more narrow than the original one. However, from a practical viewpoint, one observes on simulations a very satisfactory behavior of such quasi-stochastic procedures, including the improvement of the rate of convergence with respect to the regular *MC* implementation.

► **Exercise.** We assume now that the recursive procedure satisfied by the sequence $(y_n)_{n\geq 0}$ is given by

$$\forall n \geq 0, \quad y_{n+1} = y_n - \gamma_{n+1}\big(H(y_n, \xi_{n+1}) + r_{n+1}\big), \quad y_0 \in \mathbb{R}^d,$$

where the sequence $(r_n)_{n\geq 1}$ is a perturbation term. Show that, if $\sum_{n\geq 1} \gamma_n r_n$ is a convergent series, then the conclusion of the above theorem remains true.

NUMERICAL EXPERIMENT: We reproduced here (without even trying to check any kind of assumption) the implicit correlation search recursive procedure tested in Sect. 6.3.2, implemented this time with a sequence of some quasi-random normal numbers, namely

$$(\zeta_n^1, \zeta_n^2) = \left(\sqrt{-2\log(\xi_n^1)}\sin(2\pi\xi_n^2), \sqrt{-2\log(\xi_n^1)}\cos(2\pi\xi_n^2)\right), \quad n \geq 1,$$

where $\xi_n = (\xi_n^1, \xi_n^2)$, $n \geq 1$, is simply a regular 2-dimensional Halton sequence (see Table 6.2 and Fig. 6.4).

Table 6.2 *B-S* BEST-OF-CALL OPTION. $T = 1$, $r = 0.10$, $\sigma_1 = \sigma_2 = 0.30$, $X_0^1 = X_0^2 = 100$, $K = 100$. Convergence of $\rho_n = \cos(\theta_n)$ toward a $\rho^* = toward - 0.5$ (up to $n = 100\,000$)

n	$\rho_n := \cos(\theta_n)$
1000	−0.4964
10000	−0.4995
25000	−0.4995
50000	−0.4994
75000	−0.4996
100000	−0.4998

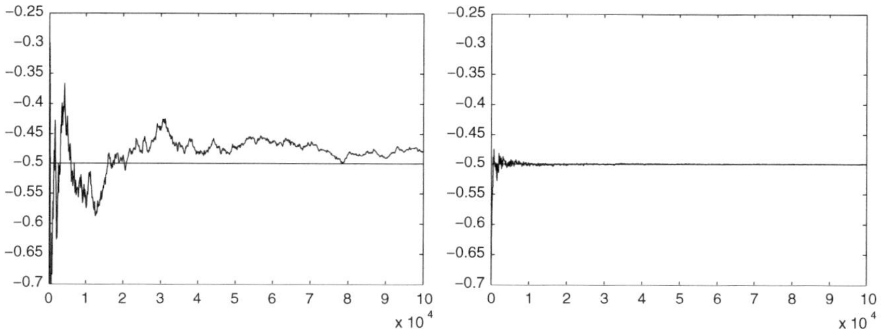

Fig. 6.4 *B-S* BEST-OF-CALL OPTION. $T = 1$, $r = 0.10$, $\sigma_1 = \sigma_2 = 0.30$, $X_0^1 = X_0^2 = 100$, $K = 100$. Convergence of $\rho_n = \cos(\theta_n)$ toward a $\rho^* = toward - 0.5$ (up to $n = 100\,000$). Left: *MC* implementation. Right: *QMC* implementation

6.6 Concluding Remarks

From a probabilistic viewpoint, many other moment or regularity assumptions on the Lyapunov function entail some *a.s.* convergence results. From the dynamical point of view, stochastic approximation is rather closely connected to the dynamics of the autonomous Ordinary Differential Equation (*ODE*) of its mean function, namely $\dot{y} = -h(y)$. However, the analysis of a stochastic algorithm cannot be completely "reduced" to that of its mean *ODE* as emphasized by several authors (see *e.g.* [33, 94]).

There is a huge literature on stochastic approximation, motivated by several fields: optimization, robotics, statistics, artificial neural networks and machine learning, self-organization an unsupervised learning, etc. For further insight on stochastic approximation, the main textbooks are probably [39, 81, 180] for prominently probabilistic aspects. One may read [33] for a dynamical system oriented point of view. For an occupation measure approach, one may also see [95].

The case of non-i.i.d. Markovian innovation with or without feedback is not treated in this chapter. This is an important topic for applications for which we mainly refer to [39], see also, among other more recent references on this framework, [93] which deals with a discontinuous mean dynamics.

Chapter 7
Discretization Scheme(s) of a Brownian Diffusion

The aim of this chapter is to investigate several discretization schemes of the (adapted) solution $(X_t)_{t\in[0,T]}$ to a d-dimensional Brownian Stochastic Differential Equation (*SDE*) formally reading

$$(SDE) \quad \equiv \quad dX_t = b(t, X_t)dt + \sigma(t, X_t)dW_t, \tag{7.1}$$

where $b : [0, T] \times \mathbb{R}^d \to \mathbb{R}^d$, $\sigma : [0, T] \times \mathbb{R}^d \to \mathcal{M}(d, q, \mathbb{R})$ are continuous functions (see the remark below), $W = (W_t)_{t\in[0,T]}$ denotes a q-dimensional standard Brownian motion defined on a probability space $(\Omega, \mathcal{A}, \mathbb{P})$ and $X_0 : (\Omega, \mathcal{A}, \mathbb{P}) \to \mathbb{R}^d$ is a random vector, independent of W. We assume that b and σ are Lipschitz continuous in x uniformly with respect to $t \in [0, T]$, *i.e.*

$$\forall\, t \in [0, T],\ \forall\, x,\ y \in \mathbb{R}^d, \quad \big|b(t, x) - b(t, y)\big| + \big\|\sigma(t, x) - \sigma(t, y)\big\| \le K|x - y|. \tag{7.2}$$

Note that $|\cdot|$ and $\|\,.\,\|$ may denote any norm on $\mathbb{R}^d$ and on $\mathcal{M}(d, q, \mathbb{R})$, respectively, in the above condition. However, without explicit mention, we will consider the canonical Euclidean and Fröbenius norms in what follows.

We consider the so-called *augmented* filtration of the *SDE* generated by X_0 and $\sigma(W_s,\ 0 \le s \le t)$, *i.e.*

$$\mathcal{F}_t := \sigma\big(X_0,\ \mathcal{N}_{\mathbb{P}},\ W_s,\ 0 \le s \le t\big),\ \ t \in [0, T], \tag{7.3}$$

where $\mathcal{N}_{\mathbb{P}}$ denotes the class of $\mathbb{P}$-negligible sets of $\mathcal{A}$ (*i.e.* all negligible sets if the σ-algebra $\mathcal{A}$ is $\mathbb{P}$-complete). When $X_0 = x_0 \in \mathbb{R}^d$ is deterministic, $\mathcal{F}_t = \mathcal{F}_t^W = \sigma\big(\mathcal{N}_{\mathbb{P}},\ W_s,\ 0 \le s \le t\big)$ is simply the augmented filtration of the Brownian motion W. One shows using Kolmogorov's 0-1 law (see [251] or [162]) that $(\mathcal{F}_t^W)$ is right continuous, *i.e.* $\mathcal{F}_t^W = \cap_{s>t}\mathcal{F}_s^W$ for every $t \in [0, T)$. The same holds for $(\mathcal{F}_t)$. Such

G. Pagès, *Numerical Probability*, Universitext,
https://doi.org/10.1007/978-3-319-90276-0_7

a combination of completeness and right continuity of a filtration is described as "usual conditions".

The following theorem shows both the existence and uniqueness of such an $(\mathcal{F}_t)$-adapted solution starting from X_0. Such stochastic processes are usually called Brownian diffusion processes or, more simply Brownian diffusions. We refer to [162], Theorem 2.9, p. 289, for a proof (among many other references).

Theorem 7.1 (Strong solution of (SDE)**)** *(see e.g. [162], Theorem 2.9, p. 289) Under the above assumptions on* b, σ, X_0 *and* W, *the SDE* (7.1) *has a unique* $(\mathcal{F}_t)$*-adapted solution* $X = (X_t)_{t\in[0,T]}$ *starting from* X_0 *at time* 0, *defined on the probability space* $(\Omega, \mathcal{A}, \mathbb{P})$, *in the following (integral) sense:*

$$\mathbb{P}\text{-}a.s. \quad \forall\, t \in [0, T], \quad X_t = X_0 + \int_0^t b(s, X_s)ds + \int_0^t \sigma(s, X_s)dW_s.$$

This solution has $\mathbb{P}$*-a.s. continuous paths.*

Notation. When $X_0 = x \in \mathbb{R}^d$, one denotes the solution of (SDE) on $[0, T]$ by X^x or $(X_t^x)_{t\in[0,T]}$.

Remarks. • A solution as described in the above theorem is known as a *strong* solution in the sense that it is defined on the probability space on which W lives.

• The global continuity assumption on b and σ can be relaxed to Borel measurability if we add the linear growth assumption

$$\forall\, t \in [0, T],\ \forall\, x \in \mathbb{R}^d, \quad |b(t, x)| + \|\sigma(t, x)\| \le K'(1 + |x|).$$

In fact, if b and σ are continuous this condition follows from (7.2) applied with (t, x) and $(t, 0)$, given the fact that $t \mapsto b(t, 0)$ is bounded on $[0, T]$ by continuity.

Moreover, still under this linear growth assumption, the Lipschitz assumption (7.2) can be relaxed to a *local Lipschitz* condition on b and σ, namely, for every $N \in \mathbb{N}^*$,

$$\exists K_N \in (0, +\infty), \forall\, t \in [0, T],\ \forall\, x,\ y \in B(0, N),$$
$$\big|b(t, x) - b(t, y)\big| + \big\|\sigma(t, x) - \sigma(t, y)\big\| \le K_N |x - y|.$$

• By adding the 0-th component t to X, *i.e.* by setting $Y_t := (t, X_t)$, one may sometimes assume that the (SDE) is autonomous, *i.e.* that the coefficients b and σ only depend on the space variable. This is often enough for applications, though it may induce some too stringent assumptions on the time variable in many theoretical results. Furthermore, when some *ellipticity* assumptions are required, this way of considering the equation no longer works since the equation $dt = 1dt + 0\,dW_t$ is completely degenerate (in terms of noise).

The above theorem admits an easy extension which allows us to define, like for *ODE*s, the *flow* of an *SDE*. Under the assumptions of Theorem 7.1, for every $t \in [0, T]$ and every $x \in \mathbb{R}^d$, there exists a unique $(\mathcal{F}^W_{t+s})_{s\in[t,T]}$-adapted solution $(X_s^{t,x})_{s\in[t,T]}$ to the above *SDE* (7.1) in the sense that

$$\mathbb{P}\text{-}a.s.\quad \forall\, s \in [t,T],\ X_t^{t,x} = x + \int_t^s b(u, X_u^{t,x})du + \int_t^s \sigma(u, X_u^{t,x})dW_u. \tag{7.4}$$

In fact, $(X_s^{t,x})_{s\in[t,T]}$ is adapted to the augmented filtration of the Brownian motion $W^{(t)} = W_{t+s} - W_t$, $s \in [t,T]$.

7.1 Euler–Maruyama Schemes

Except for some very specific equations, it is impossible to devise an exact simulation of the process X, even at a fixed time T. By exact simulation, we mean writing $X_T = \chi(U)$, $U \stackrel{d}{=} \mathcal{U}([0,1]^r)$, where $r \in \mathbb{N}^* \cup \{+\infty\}$, χ is an explicit function, defined on $[0,1]^r$ if $r < \infty$ and on $[0,1]^{(\mathbb{N}^*)}$ when $r = +\infty$. In fact, such exact simulation has been shown to be possible when $d = 1$ and $\sigma \equiv 1$, see [43], by an appropriate acceptance-rejection method. For a brief discussion on recent developments in this direction, we refer to the introduction of Chap. 9. Consequently, to approximate $\mathbb{E}\, f(X_T)$, or more generally $\mathbb{E}\, F\big((X_t)_{t\in[0,T]}\big)$ by a Monte Carlo method, one needs to approximate X by a process that can be simulated (at least at a fixed number of instants).

To this end, we will introduce three types of Euler schemes with step $\frac{T}{n}$ ($n \in \mathbb{N}^*$) associated to the *SDE* (7.1): the *discrete time* Euler scheme $\bar{X} = \big(\bar{X}_{\frac{kT}{n}}\big)_{0\le k\le n}$ with step $\frac{T}{n}$, its càdlàg stepwise constant extension known as the *stepwise constant* (Brownian) Euler scheme and the *continuous* or *genuine* (Brownian) Euler scheme.

7.1.1 The Discrete Time and Stepwise Constant Euler Schemes

The idea, like for *ODE*s in the deterministic framework, is to *freeze* the solution of the *SDE* between the regularly spaced discretization instants $\frac{kT}{n}$.

Discrete time Euler scheme

The *discrete time Euler scheme* with step $\frac{T}{n}$ is defined by

$$\bar{X}_{t_{k+1}^n} = \bar{X}_{t_k^n} + \frac{T}{n} b(t_k^n, \bar{X}_{t_k^n}) + \sigma(t_k^n, \bar{X}_{t_k^n})\sqrt{\frac{T}{n}}\, Z_{k+1}^n, \bar{X}_0 = X_0, \qquad k = 0,\dots,n-1, \tag{7.5}$$

where $t_k^n = \frac{kT}{n}$, $k = 0,\dots,n$ and $(Z_k^n)_{1\le k\le n}$ denotes a sequence of i.i.d. $\mathcal{N}(0; I_q)$-distributed random vectors given by

$$Z_k^n := \sqrt{\frac{n}{T}}\big(W_{t_k^n} - W_{t_{k-1}^n}\big), \quad k = 1,\dots,n.$$

We will often drop the superscript ^{n} in Z_k^n and write Z_k.

$\triangleright$ **Example (Black–Scholes).** The discrete time Euler scheme of a standard Black–Scholes dynamics $dX_t = X_t\big(rdt + \sigma dW_t\big)$, $X_0 = x_0 > 0$, reads with these notations

$$\bar{X}_{t^n_{k+1}} = \bar{X}_{t^n_k}\Big(1 + r\frac{T}{n} + \sigma\sqrt{\frac{T}{n}}\,Z^n_{k+1}\Big),\quad k = 0, \ldots, n-1,\quad \bar{X}_0 = X_0. \tag{7.6}$$

The stepwise constant Euler scheme

To alleviate he notation, from now on we write

$$\underline{t} := t^n_k \quad \text{if } t \in [t^n_k, t^n_{k+1}).$$

The *stepwise constant Euler scheme*, denoted by $(\widetilde{X}_t)_{t\in[0,T]}$ for convenience, is defined by

$$\widetilde{X}_t = \bar{X}_{\underline{t}},\quad t \in [0, T]. \tag{7.7}$$

7.1.2 *The Genuine (Continuous) Euler Scheme*

At this stage it is natural to extend the definition (7.5) of the Euler scheme at every instant $t \in [0, T]$ by interpolating the drift and the diffusion term in their own scale, that is, with respect to time for the drift and with respect to the Brownian motion for the diffusion coefficient, namely

$$\begin{aligned}\forall\, k \in \{0, \ldots, n-1\},\ \forall\, t \in [t^n_k, t^n_{k+1}),&\\ \bar{X}_t &= \bar{X}_{\underline{t}} + (t - \underline{t})b(\underline{t}, \bar{X}_{\underline{t}}) + \sigma(\underline{t}, \bar{X}_{\underline{t}})(W_t - W_{\underline{t}}),\\ \bar{X}_0 &= X_0.\end{aligned} \tag{7.8}$$

It is clear that $\lim_{t\to t^n_{k+1}, t<t^n_{k+1}} \bar{X}_t = \bar{X}_{t^n_{k+1}}$ since W has continuous paths. Consequently, thus defined, $(\bar{X}_t)_{t\in[0,T]}$ is an $\mathcal{F}_t$-adapted process with continuous paths.

The following proposition is the key property of the genuine (or continuous) Euler scheme.

Proposition 7.1 *Assume that b and σ are continuous functions on* $[0, T] \times \mathbb{R}^d$. *The genuine Euler scheme is a (continuous) Itô process* ([1]) *satisfying the pseudo-SDE with* frozen coefficients *satisfying*

$$\bar{X}_t = X_0 + \int_0^t b(\underline{s}, \bar{X}_{\underline{s}})ds + \int_0^t \sigma(\underline{s}, \bar{X}_{\underline{s}})dW_s,\ t \in [0, T]. \tag{7.9}$$

Proof. It is clear from (7.8), the recursive definition (7.5) at the discretization dates t^n_k and the continuity of b and σ that $\bar{X}_t \to \bar{X}_{t^n_{k+1}}$ as $t \to t^n_{k+1}$. Consequently, for every

[1] in the sense of Sect. 12.8 of the Miscellany Chapter.

$t \in [t_k^n, t_{k+1}^n]$,

$$\bar{X}_t = \bar{X}_{t_k^n} + \int_{t_k^n}^t b(\underline{s}, \bar{X}_{\underline{s}})ds + \int_{t_k^n}^t \sigma(\underline{s}, \bar{X}_{\underline{s}})dW_s$$

so that the conclusion follows by just concatenating the above identities between $t_0^n = 0, t_1^n, \ldots, t_k^n = \underline{t}$ and t. ◇

⊳ **Example (Black–Scholes).** The continuous time Euler scheme of a standard Black–Scholes dynamics reads, in these notation,

$$\forall\, t \in [0, T], \quad \bar{X}_t = \bar{X}_{\underline{t}}\Big(1 + r(t - \underline{t}) + \sigma(W_t - W_{\underline{t}})\Big), \quad \bar{X}_0 = X_0. \tag{7.10}$$

NOTATION: In the main statements, we will write $\bar{X}^n$ instead of $\bar{X}$ to recall the dependence of the Euler scheme in its step $\frac{T}{n}$. Idem for $\widetilde{X}$, etc.

Then, the main (classical) result is that under the assumptions on the coefficients b and σ mentioned above, $\sup_{t\in[0,T]} |X_t - \bar{X}_t|$ goes to zero in every $L^p(\mathbb{P})$, $0 < p < +\infty$, as $n \to +\infty$. Let us be more specific on this topic by providing error rates under slightly more stringent assumptions.

How to use this genuine scheme for practical simulation does not seem not obvious, at least not as obvious as the stepwise constant Euler scheme. However, it turns out to be an important method for improving the convergence rate of MC simulations, *e.g.* for option pricing. Using this scheme in simulations is possible for specific functionals F. It relies on the so-called *diffusion bridge method* and will be detailed further on in Chap. 8.

7.2 The Strong Error Rate and Polynomial Moments (I)

7.2.1 Main Results and Comments

We consider the *SDE* (7.1) and its Euler–Maruyama scheme(s) as defined by (7.5), (7.8). The first version of Theorem 7.2 below (including the second remark that follows) is mainly due to O. Faure in his PhD thesis (see [90]). An important preliminary step is to establish the existence of finite L^pmoments of the sup-norm of solutions when b and σ have linear growth, whenever X_0 itself lies in L^p.

Polynomial moments control

It is often useful to have at hand the following uniform bounds for the solution(s) of (SDE) and its Euler schemes, which first appears as a step of the proof of the rate but has many other applications. Thus it is an important step to prove the existence of global strong solutions to (SDE) (7.1) when b and σ are only locally Lipschitz continuous but both satisfy a linear growth assumption in the space variable x uniformly in $t \in [0, T]$.

Proposition 7.2 (Polynomial moments) *Assume that the coefficients b and σ of the SDE (7.1) are Borel functions simply satisfying the following linear growth assumption:*

$$\forall\, t \in [0, T],\ \forall\, x \in \mathbb{R}^d,\quad |b(t, x)| + \|\sigma(t, x)\| \le C(1 + |x|) \tag{7.11}$$

for some real constant $C = C_T > 0$ and a "horizon" $T > 0$. Then, for every $p \in (0, +\infty)$, there exists a universal positive real constant $\kappa_{p,d} > 0$ such that every strong solution $(X_t)_{t\in[0,T]}$ of (7.1) (if any) satisfies

$$\Big\| \sup_{t\in[0,T]} |X_t| \Big\|_p \le \kappa_{p,d} e^{\kappa_{p,d} C(C+1)T}(1 + \|X_0\|_p) \tag{7.12}$$

and, for every $n \ge 1$, the Euler scheme with step $\frac{T}{n}$ satisfies

$$\Big\| \sup_{t\in[0,T]} |\bar{X}^n_t| \Big\|_p \le \kappa_{p,d} e^{\kappa_p C(C+1)T}(1 + \|X_0\|_p). \tag{7.13}$$

Remarks. • The universal constant $\kappa_{p,d}$ is a numerical function of the *BDG* real constant $C^{BDG}_{p\vee 2,d}$.

• Less synthetic but more precise bounds can be obtained. For example, if $p \ge 2$, Inequality (7.55) established in the proof of the Proposition 7.6 reads, at time $t = T$,

$$\Big\| \sup_{t\in[0,T]} |X_t| \Big\|_p \le 2\, e^{(2+C(C^{BDG}_{d,p})^2)CT}\Big(\|X_0\|_p + C(T + C^{BDG}_{d,p}\sqrt{T})\Big)$$

which holds as an equality if $C = 0$ and $X_0 = 0$, unlike the above more synthetic general upper-bound (7.12).

One interesting consequence of this proposition is that, if b and σ are defined on $\mathbb{R}_+ \times \mathbb{R}^d$ and satisfy (7.11) with the same real constant C for every $T > 0$ and if there exists a global solution $(X_t)_{t\ge 0}$ of (SDE), then the above exponential control in T of its sup norm over $[0, T]$ established (7.12) holds true for every $T > 0$.

7.2.2 *Uniform Convergence Rate in $L^p(\mathbb{P})$*

First we introduce the following condition (H^β_T) which strengthens Assumption (7.2) by adding a time regularity assumption of the Hölder type, namely there exists a $\beta \in (0, 1]$ such that

$$(H_T^\beta) \equiv \begin{cases} \exists\, C_{b,\sigma,T} > 0 \;\text{ such that } \forall s, t \in [0,T],\; \forall\, x, y \in \mathbb{R}^d, \\ |b(t,x) - b(s,x)| + \|\sigma(t,x) - \sigma(s,x)\| \le C_{b,\sigma,T}\,(1+|x|)|t-s|^\beta \\ |b(t,x) - b(t,y)| + \|\sigma(t,x) - \sigma(t,y)\| \le C_{b,\sigma,T}\,|y-x|. \end{cases} \tag{7.14}$$

Theorem 7.2 (Strong Rate for the Euler scheme) (a) GENUINE EULER SCHEME. *Suppose the coefficients b and σ of the SDE (7.1) satisfy the above regularity condition (H_T^β) for a real constant $C_{b,\sigma,T} > 0$ and an exponent $\beta \in (0,1]$. Then the genuine Euler scheme $(\bar{X}_t^n)_{t\in[0,T]}$ converges toward $(X_t)_{t\in[0,T]}$ in every $L^p(\mathbb{P})$, $p > 0$, such that $X_0 \in L^p$, at a $O\big(n^{-(\frac{1}{2}\wedge\beta)}\big)$-rate. To be precise, there exists a universal constant $\kappa_p > 0$ only depending on p such that, for every $n \ge T$,*

$$\Big\| \sup_{t\in[0,T]} \big|X_t - \bar{X}_t^n\big| \Big\|_p \le K(p,b,\sigma,T)\big(1 + \|X_0\|_p\big)\left(\frac{T}{n}\right)^{\beta\wedge\frac{1}{2}} \tag{7.15}$$

where

$$K(p,b,\sigma,T) = \kappa_p C'_{b,\sigma,T} e^{\kappa_p(1+C'_{b,\sigma,T})^2 T}$$

and

$$C'_{b,\sigma,T} = C_{b,\sigma,T} + \sup_{t\in[0,T]} |b(t,0)| + \sup_{t\in[0,T]} \|\sigma(t,0)\| < +\infty. \tag{7.16}$$

(a') DISCRETE TIME EULER SCHEME. *In particular, (7.15) is satisfied when the supremum is restricted to discretization instants t_k^n, namely*

$$\Big\| \sup_{0\le k\le n} \big|X_{t_k} - \bar{X}_{t_k}^n\big| \Big\|_p \le K(p,b,\sigma,T)\big(1 + \|X_0\|_p\big)\left(\frac{T}{n}\right)^{\beta\wedge\frac{1}{2}}. \tag{7.17}$$

(a'') *If b and σ are defined on the whole $\mathbb{R}_+ \times \mathbb{R}^d$ and satisfy (H_T^β) with the same real constant $C_{b,\sigma}$* not depending on *T and if $b(\,.\,,0)$ and $\sigma(\,.\,,0)$ are bounded on $\mathbb{R}_+$, then $C'_{b,\sigma,T}$* does not depend on *T.*

This will be the case in the autonomous case, i.e. if $b(t,x) = b(x)$ and $\sigma(t,x) = \sigma(x)$, $t \in \mathbb{R}_+$, $x \in \mathbb{R}^d$, with b and σ Lipschitz continuous on $\mathbb{R}^d$.

(b) STEPWISE CONSTANT EULER SCHEME. ▷ *As soon as b and σ satisfy the linear growth assumption (7.11) with a real constant $L_{b,\sigma,T} > 0$, then, for every $p \in (0,+\infty)$ and every $n \ge T$,*

$$\begin{aligned}\Big\| \sup_{t\in[0,T]} \big|\bar{X}_t^n - \bar{X}_{\underline{t}}^n\big| \Big\|_p &\le \tilde{\kappa}_p e^{\tilde{\kappa}_p L_{b,\sigma,T} T}\big(1 + \|X_0\|_p\big)\sqrt{\frac{T(1+\log n)}{n}} \\ &= O\left(\sqrt{\frac{1+\log n}{n}}\right)\end{aligned}$$

where $\tilde{\kappa}_p > 0$ *is a positive real constant only depending on* p *(and increasing in* p*).*

$\triangleright$ *In particular, if* b *and* σ *satisfy the assumption* (H_T^β) *like in item* (a)*, then the stepwise constant Euler scheme* $(\widetilde{X}_t^n)_{t\in[0,T]}$ *converges toward* $(X_t)_{t\in[0,T]}$ *in every* $L^p(\mathbb{P})$*,* $p>0$*, such that* $X_0\in L^p$ *and for every* $n\geq T$*,*

$$\Big\|\sup_{t\in[0,T]}\big|X_t-\widetilde{X}_t^n\big|\Big\|_p \leq \widetilde{K}(p,b,\sigma,T)\big(1+\|X_0\|_p\big)\left(\sqrt{\frac{T(1+\log n)}{n}}+\left(\frac{T}{n}\right)^{\beta\wedge\frac{1}{2}}\right)$$
$$= O\left(\left(\frac{1}{n}\right)^\beta+\sqrt{\frac{1+\log n}{n}}\right)$$

where

$$\widetilde{K}(p,b,\sigma,T)=\tilde{\kappa}'_p(1+C'_{b,\sigma,T})e^{\tilde{\kappa}'_p(1+C'_{b,\sigma,T})^2T},$$

$\tilde{\kappa}'_p>0$ *is a positive real constant only depending on* p *(increasing in* p*) and* $C'_{b,\sigma,T}$ *is given by (7.16).*

WARNING! The complete and detailed proof of this theorem in its full generality, *i.e.* including the tracking of the constants, is postponed to Sect. 7.8. It makes use of stochastic calculus. A first version of the proof in the one-dimensional quadratic case is proposed in Sect. 7.2.3. However, owing to its importance for applications, the optimality of the upper-bound for the stepwise constant Euler scheme will be discussed right after the remarks below.

Remarks. • When $n\leq T$, the above explicit bounds still hold true with the same constants provided one replaces

$$2\left(\frac{T}{n}\right)^{\beta\wedge\frac{1}{2}}\ \text{by}\ \frac{1}{2}\left(\left(\frac{T}{n}\right)^\beta+\left(\frac{T}{n}\right)^{\frac{1}{2}}\right)$$

and

$$\sqrt{\frac{T}{n}(1+\log n)}\ \text{by}\ \frac{1}{2}\left(\sqrt{\frac{T}{n}(1+\log n)}+\frac{T}{n}\right).$$

• As a consequence, note that the time regularity exponent β rules the convergence rate of the scheme as soon as $\beta<1/2$. In fact, the method of proof itself will emphasize this fact: the idea is to use a Gronwall Lemma to upper-bound the error $X-\bar{X}$ in $L^p(\mathbb{P})$ by the $L^p(\mathbb{P})$-norm of the increments $X_s-X_{\underline{s}}$ or, equivalently, $\bar{X}_s-\bar{X}_{\underline{s}}$ as we will see later.

• If $b(t,x)$ and $\sigma(t,x)$ are globally Lipschitz continuous on $\mathbb{R}_+\times\mathbb{R}^d$ with Lipschitz continuous coefficient $C_{b,\sigma}$, one may consider the time t as a $(d+1)$-th spatial component of X and apply item (a'') of the above theorem directly.

The following corollary is a straightforward consequence of claims (*a*) of the above theorem (to be precise of (7.17)): it yields a (first) convergence rate for the pricing of "vanilla" European options (*i.e.* payoffs of the form $\varphi(X_T)$).

Corollary 7.1 *Let $\varphi : \mathbb{R}^d \to \mathbb{R}$ be an α-Hölder function for an exponent $\alpha \in (0, 1]$, i.e. a function such that $[\varphi]_\alpha := \sup_{x\neq y} \frac{|\varphi(x)-\varphi(y)|}{|x-y|^\alpha} < +\infty$. Then, there exists a real constant $C_{b,\sigma,T} \in (0, \infty)$ such that, for every $n \geq 1$,*

$$\big|\mathbb{E}\,\varphi(X_T) - \mathbb{E}\,\varphi(\bar{X}^n_T)\big| \leq [\varphi]_\alpha \mathbb{E}\,|X_T - \bar{X}^n_T|^\alpha \leq C_{b,\sigma,T}[\varphi]_\alpha \left(\frac{T}{n}\right)^{\frac{\alpha}{2}}.$$

We will see further on that this *weak error* ([2]) rate can be dramatically improved when b, σ and φ share higher regularity properties.

On the universality of the rate for the stepwise constant Euler scheme ($p \geq 2$)

Note that the rate in claim (*b*) of theorem is universal since, as we will see now, it holds as a sharp rate for the Brownian motion itself (here we deal with the case $d = 1$). Indeed, since W is its own genuine Euler scheme, $\bar{W}^n_t - \widetilde{W}^n_t = W_t - W_{\underline{t}}$ for every $t \in [0, T]$. Now,

$$\begin{aligned}\Big\|\sup_{t\in[0,T]} |W_t - W_{\underline{t}}|\Big\|_p &= \Big\|\max_{k=1,\dots,n} \sup_{t\in[t^n_{k-1},t^n_k)} |W_t - W_{t^n_{k-1}}|\Big\|_p \\ &= \sqrt{\frac{T}{n}}\Big\|\max_{k=1,\dots,n} \sup_{t\in[k-1,k)} |B_t - B_{k-1}|\Big\|_p\end{aligned}$$

where $B_t := \sqrt{\frac{n}{T}} W_{\frac{T}{n}t}$ is a standard Brownian motion owing to the scaling property. Hence

$$\Big\|\sup_{t\in[0,T]} |W_t - W_{\underline{t}}|\Big\|_p = \sqrt{\frac{T}{n}}\Big\|\max_{k=1,\dots,n} \zeta_k\Big\|_p$$

where the random variables $\zeta_k := \sup_{t\in[k-1,k)} |W_t - W_{k-1}|$ are i.i.d.
▷ *Lower bound.* Note that, for every $k \geq 1$,

$$\zeta_k \geq Z_k := |W_k - W_{k-1}|$$

since the Brownian motion (W_t) has continuous paths. The sequence $(Z_k)_{k\geq1}$ is i.i.d. as well, with the same distribution as $|W_1|$. Hence, the random variables Z_k^2 are still i.i.d. with a $\chi^2(1)$-distribution so that (see Exercises **1**. and **2**. below)

[2]The word "weak" refers here to the fact that the error $\big|\mathbb{E}\,\varphi(X_T) - \mathbb{E}\,\varphi(\bar{X}^n_T)\big|$ is related to the convergence of the distributions $\mathbb{P}_{\bar{X}^n_T}$ toward $\mathbb{P}_{X_T}$ as $n \to \infty$—*e.g.* weakly or in variation—whereas strong $L^p(\mathbb{P})$-convergence involves the joint distributions $\mathbb{P}_{(\bar{X}^n_T, X_T)}$.

$$\forall\, p\geq 2,\qquad \Big\|\max_{k=1,\dots,n}|Z_k|\Big\|_p=\sqrt{\|\max_{k=1,\dots,n} Z_k^2\|_{\frac{p}{2}}}\geq c_p\sqrt{\log n}.$$

Finally, one has

$$\forall\, p\geq 2,\qquad \Big\|\sup_{t\in[0,T]}|W_t-W_{\underline{t}}|\Big\|_p\geq c_p\sqrt{\frac{T}{n}}\sqrt{\log n}.$$

▷ *Upper bound.* To establish the upper-bound, we proceed as follows. First, note that

$$\zeta_1=\max\Big(\sup_{t\in[0,1)}W_t,\ \sup_{t\in[0,1)}(-W_t)\Big).$$

We also know that

$$\sup_{t\in[0,1)}W_t\stackrel{d}{=}|W_1|$$

(see *e.g.* [251], Reflection principle, p. 105). Hence using that, for every $a, b\geq 0$, $e^{(a\vee b)^2}\leq e^{a^2}+e^{b^2}$, that $\sup_{t\in[0,1)}W_t\geq 0$ and that $-W$ is also a standard Brownian motion, we derive that

$$\begin{aligned}\mathbb{E}\,e^{\theta\zeta_1^2}&\leq \mathbb{E}\,e^{\theta(\sup_{t\in[0,1)}W_t)^2}+\mathbb{E}\,e^{\theta(\sup_{t\in[0,1)}(-W_t))^2}\\&=2\,\mathbb{E}\,e^{\theta(\sup_{t\in[0,1)}W_t)^2}\\&=2\,\mathbb{E}\,e^{\theta W_1^2}=2\int_{\mathbb{R}}\exp\Bigg(-\frac{u^2}{2(\frac{1}{\sqrt{1-2\theta}})^2}\Bigg)\frac{du}{\sqrt{2\pi}}=\frac{2}{\sqrt{1-2\theta}}<+\infty\end{aligned}$$

as long as $\theta\in(0,\frac{1}{2})$. Consequently, it follows from Lemma 7.1 below applied with the sequence $(\zeta_n^2)_{n\geq 1}$, that

$$\forall\, p\in(0,+\infty),\qquad \Big\|\max_{k=1,\dots,n}\zeta_k\Big\|_p=\sqrt{\|\max_{k=1,\dots,n}\zeta_k^2\|_{\frac{p}{2}}}\leq C_{W,p}\sqrt{1+\log n},$$

i.e. for every $p\in(0,+\infty)$,

$$\Big\|\sup_{t\in[0,T]}|W_t-W_{\underline{t}}|\Big\|_p\leq C_{W,p}\sqrt{\frac{T}{n}\big(1+\log n\big)}. \tag{7.18}$$

Lemma 7.1 *Let $Y_1,\dots Y_n$ be non-negative random variables with the same distribution satisfying $\mathbb{E}\,(e^{\lambda Y_1})<+\infty$ for some $\lambda>0$. Then,*

$$\forall\, p\in(0,+\infty),\qquad \big\|\max(Y_1,\dots Y_n)\big\|_p\leq\frac{1}{\lambda}\big(\log n+C_{p,Y_1,\lambda}\big).$$

Proof: We may assume without loss of generality that $p \geq 1$ since the $\|\cdot\|_p$-norm is non-decreasing. First, assume $\lambda = 1$. Let $p \geq 1$. One sets

$$\varphi_p(x) = \big(\log(e^{p-1} + x)\big)^p - (p-1)^p, \quad x > 0.$$

The function φ_p is continuous, increasing, concave and one-to-one from $\mathbb{R}_+$ onto $\mathbb{R}_+$ (the term e^{p-1} is introduced to ensure the concavity). It follows that $\varphi_p^{-1}(y) = e^{((p-1)^p+y)^{1/p}} - e^{p-1} \leq e^{p-1+y^{1/p}}$ for every $y \geq 0$, owing to the elementary inequality $(u+v)^{\frac{1}{p}} \leq u^{\frac{1}{p}} + v^{\frac{1}{p}}$, $u,\ v \geq 0$. Hence,

$$\begin{aligned}\mathbb{E}\max_{k=1,\dots,n} Y_k^p &= \mathbb{E}\Big(\max_{k=1,\dots,n}\big(\varphi_p \circ \varphi_p^{-1}(Y_k^p)\big)\Big)\\ &= \mathbb{E}\Big(\varphi_p\Big(\max\big(\varphi_p^{-1}(Y_1^p),\dots,\varphi_p^{-1}(Y_n^p)\big)\Big)\Big)\end{aligned}$$

since φ_p is non-decreasing. Then Jensen's Inequality implies

$$\begin{aligned}\mathbb{E}\max_{k=1,\dots,n} Y_k^p &\leq \varphi_p\Big(\mathbb{E}\max_{k=1,\dots,n}\varphi_p^{-1}(Y_k^p)\Big)\\ &\leq \varphi_p\Big(\sum_{k=0}^{n}\mathbb{E}\,\varphi_p^{-1}(Y_k^p)\Big) = \varphi_p\big(n\,\mathbb{E}\,\varphi_p^{-1}(Y_1^p)\big)\\ &\leq \varphi_p\big(n\,\mathbb{E}\,e^{p-1+Y_1}\big) \leq \big(p-1+\log\big(1+n\,\mathbb{E}\,e^{Y_1}\big)\big)^p.\end{aligned}$$

Hence

$$\begin{aligned}\Big\|\max_{k=1,\dots,n} Y_k\Big\|_p &\leq \log\Big(1+n\,\mathbb{E}\,e^{Y_1}\Big) + p - 1 = \log n + \log\Big(\mathbb{E}\,e^{Y_1} + \frac{1}{n}\Big) + p - 1\\ &\leq \log n + C_{p,1,Y_1},\end{aligned}$$

where $C_{p,\lambda,Y_1} = \log\Big(\mathbb{E}\,e^{\lambda Y_1} + e^{p-1}\Big)$.

Let us return to the general case, *i.e.* $\mathbb{E}\,e^{\lambda Y_1} < +\infty$ for a $\lambda > 0$. Then

$$\begin{aligned}\Big\|\max(Y_1,\dots,Y_n)\Big\|_p &= \frac{1}{\lambda}\Big\|\max(\lambda Y_1,\dots,\lambda Y_n)\Big\|_p\\ &\leq \frac{1}{\lambda}\big(\log n + C_{p,\lambda,Z_1}\big).\end{aligned}$$

◇

► **Exercises. 1.** (*Completion of the proof of the above lower bound I*). Let Z be a non-negative random variable with distribution function $F(z) = \mathbb{P}(Z \leq z)$ and a continuous probability density function f. Assume that the survival function $\bar{F}(z) := \mathbb{P}(Z > z)$ satisfies: there exists $a\, c \in (0, +\infty)$ such that

$$\forall z \geq a, \quad \bar{F}(z) \geq c\, f(z). \tag{7.19}$$

Show that, if $(Z_n)_{n\geq 1}$ is i.i.d. with distribution $\mathbb{P}_Z(dz)$,

$$\forall p \geq 1, \quad \left\| \max(Z_1, \ldots, Z_n) \right\|_p \geq c \sum_{k=1}^{n} \frac{1 - F^k(a)}{k}$$
$$\geq c\Big(\log(n+1) + \log\big(1 - F(a)\big)\Big).$$

[Hint: one may assume $p = 1$. Establish the classical representation formula

$$\mathbb{E}\, U = \int_0^{+\infty} \mathbb{P}(U \geq u)du$$

for any non-negative random variable U and use some basic facts about Stieljès integral like $dF(z) = f(z)dz$, etc.]

2. (*Completion of the proof of the above lower bound II*). Show that the $\chi^2(1)$ distribution defined by its density on the real line $f(u) := \frac{e^{-\frac{u}{2}}}{\sqrt{2\pi u}} \mathbf{1}_{\{u>0\}}$ satisfies the above inequality (7.19). [Hint: use an integration by parts and usual comparison theorems for integrals to show that

$$\bar{F}(z) = \frac{2e^{-\frac{z}{2}}}{\sqrt{2\pi z}} - \int_z^{+\infty} \frac{e^{-\frac{u}{2}}}{u\sqrt{2\pi u}} du \sim \frac{2e^{-\frac{z}{2}}}{\sqrt{2\pi z}} \quad \text{as} \quad z \to +\infty\ .]$$

3. (*Euler scheme of the martingale Black–Scholes model*). Let $(Z_n)_{n\geq 1}$ be an i.i.d. sequence of $\mathcal{N}(0; 1)$-distributed random variables defined on a probability space $(\Omega, \mathcal{A}, \mathbb{P})$.

(a) Compute for every real number $a > 0$, the quantity $\mathbb{E}(1 + aZ_1)e^{aZ_1}$ as a function of a.

(b) Compute for every integer $n \geq 1$,

$$\left\| e^{a(Z_1+\cdots+Z_n) - n\frac{a^2}{2}} - \prod_{k=1}^{n}(1 + aZ_k) \right\|_2.$$

(c) Let $\sigma > 0$. Show the existence of two positive real constants c_i, $i = 1, 2$, such that

$$\left\| e^{(Z_1+\cdots+Z_n)/\sqrt{n} - \frac{1}{2}} - \prod_{k=1}^{n}\left(1 + \frac{Z_k}{\sqrt{n}}\right) \right\|_2 = \frac{\sigma^2 e^{\frac{\sigma^2}{2}}}{\sqrt{2n}} \Big(1 - \frac{\sigma^2(c_1\sigma^2 + c_2)}{n} + O(1/n^2)\Big).$$

(d) Deduce the exact convergence rate of the Euler scheme with step $\frac{T}{n}$ of the martingale Black–Scholes model

$$dX_t = \sigma X_t dW_t, \quad X_0 = x_0 > 0$$

as $n \to +\infty$. Conclude to the optimality of the rate established in Theorem 7.16(a') as a universal convergence rate of the Euler scheme (in a Lipschitz framework).

$A.s.$ **convergence rate(s)**

The last important result of this section is devoted to the $a.s.$ convergence of the Euler schemes toward the diffusion process with a first (elementary) approach to its rate of convergence.

Theorem 7.3 *If b and σ satisfy (H_T^β) for a $\beta \in (0, 1]$ and if X_0 is $a.s.$ finite, the continuous Euler scheme $\bar{X}^n = (\bar{X}_t^n)_{t\in[0,T]}$ $a.s.$ converges toward the diffusion X for the* sup-*norm over* $[0, T]$. *Furthermore, for every $\alpha \in \big[0, \beta \wedge \frac{1}{2}\big)$,*

$$n^\alpha \sup_{t\in[0,T]} |X_t - \bar{X}_t^n| \xrightarrow{a.s.} 0.$$

The proof follows from the L^p-convergence theorem by an approach “à la Borel–Cantelli”. The details are deferred to Sect. 7.8.6.

7.2.3 *Proofs in the Quadratic Lipschitz Case for Autonomous Diffusions*

We provide below a proof of both Proposition 7.2 and Theorem 7.2 in a simplified one dimensional, autonomous and quadratic ($p = 2$) setting. This means that $b(t, x) = b(x)$ *and* $\sigma(t, x) = \sigma(x)$ *are defined as Lipschitz continuous functions on the real line*. Then (SDE) admits a unique strong solution starting from X_0 on every interval $[0, T]$, which means that there exists a unique strong solution $(X_t)_{t\geq 0}$ starting from X_0.

Furthermore, we will not care about the structure of the real constants that come out, in particular no control in T is provided in this concise version of the proof. The complete and detailed proofs are postponed to Sect. 7.8.

Lemma 7.2 *(Gronwall's Lemma) Let $f : \mathbb{R}_+ \to \mathbb{R}_+$ be a Borel non-negative locally bounded function and let $\psi : \mathbb{R}_+ \to \mathbb{R}_+$ be a non-decreasing function satisfying*

$$(G) \quad \equiv \quad \forall\, t \geq 0, \quad f(t) \leq \alpha \int_0^t f(s)\, ds + \psi(t)$$

for a real constant $\alpha > 0$. Then

$$\forall\, t \geq 0, \qquad \sup_{0\leq s\leq t} f(s) \leq e^{\alpha t} \psi(t).$$

Proof. It is clear that the non-decreasing (finite) function $\varphi(t) := \sup_{0\le s\le t} f(s)$ satisfies (G) instead of f. Now the function $e^{-\alpha t}\int_0^t \varphi(s)\,ds$ has a right derivative at every $t \ge 0$ and

$$\left(e^{-\alpha t}\int_0^t \varphi(s)\,ds\right)'_r = e^{-\alpha t}\left(\varphi(t+) - \alpha\int_0^t \varphi(s)\,ds\right)$$
$$\le e^{-\alpha t}\psi(t_+)$$

where $\varphi(t_+)$ and $\psi(t_+)$ denote the right limits of φ and ψ at t. Then, it follows from the fundamental theorem of Calculus that the function

$$t \longmapsto e^{-\alpha t}\int_0^t \varphi(s)ds - \int_0^t e^{-\alpha s}\psi(s_+)\,ds \quad \text{is non-increasing and null at 0,}$$

hence, non-positive so that

$$\forall\, t \ge 0, \quad \int_0^t \varphi(s) \le e^{\alpha t}\int_0^t e^{-\alpha s}\psi(s+)\,ds.$$

Plugging this into the above inequality implies

$$\varphi(t) \le \alpha e^{\alpha t}\int_0^t e^{-\alpha s}\psi(s+)ds + \psi(t) = \alpha e^{\alpha t}\int_0^t e^{-\alpha s}\psi(s)\,ds + \psi(t)$$
$$\le \left(\alpha e^{\alpha t}\frac{1-e^{-\alpha t}}{\alpha} + 1\right)\psi(t) = e^{\alpha t}\psi(t),$$

where we used successively that a monotonic function is ds-$a.s.$ continuous and that ψ is non-decreasing. ◇

Now we recall the quadratic Doob Inequality which is needed in the proof (instead of the more sophisticated Burkhölder–Davis–Gundy required in the non-quadratic case).

Doob's Inequality (L^2 **case**) (see *e.g.* [183]).
(a) Let $M = (M_t)_{t\ge 0}$ be a continuous martingale with $M_0 = 0$. Then, for every $T > 0$,

$$\mathbb{E}\left(\sup_{t\in[0,T]} M_t^2\right) \le 4\,\mathbb{E}\,M_T^2 = 4\,\mathbb{E}\,\langle M\rangle_T.$$

(b) If M is simply a continuous *local* martingale with $M_0 = 0$, then, for every $T > 0$,

$$\mathbb{E}\left(\sup_{t\in[0,T]} M_t^2\right) \le 4\,\mathbb{E}\,\langle M\rangle_T.$$

Proof of Proposition 7.2 **(A first partial)**. We may assume without loss of generality that $\mathbb{E}\, X_0^2 < +\infty$ (otherwise the inequality is trivially fulfilled). Let $\tau_L := \min\big\{t : |X_t - X_0| \geq L\big\}$, $L \in \mathbb{N} \setminus \{0\}$ (with the usual convention $\min \varnothing = +\infty$). It is an $\mathcal{F}$-stopping time as the hitting time of a closed set by a process with continuous paths (see Sect. 7.8.2). Furthermore, for every $t \in [0, T]$,

$$|X_t^{\tau_L}| \leq L + |X_0|.$$

In particular, this implies that

$$\mathbb{E} \sup_{t \in [0,T]} |X_t^{\tau_L}|^2 \leq 2(L^2 + \mathbb{E}\, X_0^2) < +\infty.$$

Then,

$$\begin{aligned} X_t^{\tau_L} &= X_0 + \int_0^{t \wedge \tau_L} b(X_s) ds + \int_0^{t \wedge \tau_L} \sigma(X_s) dW_s \\ &= X_0 + \int_0^{t \wedge \tau_L} b(X_s^{\tau_L}) ds + \int_0^{t \wedge \tau_L} \sigma(X_s^{\tau_L}) dW_s \end{aligned}$$

owing to the local feature of both regular and stochastic integrals. The stochastic integral

$$M_t^{(L)} := \int_0^{t \wedge \tau_L} \sigma(X_s^{\tau_L}) dW_s$$

is a continuous local martingale null at zero with bracket process defined by

$$\langle M^{(L)} \rangle_t = \int_0^{t \wedge \tau_L} \sigma^2(X_s^{\tau_L}) ds.$$

Now, using that $t \wedge \tau_L \leq t$, we derive that

$$|X_t^{\tau_L}| \leq |X_0| + \int_0^t |b(X_s^{\tau_L})| ds + \sup_{s \in [0,t]} \big|M_s^{(L)}\big|,$$

which in turn immediately implies that

$$\sup_{s \in [0,t]} |X_s^{\tau_L}| \leq |X_0| + \int_0^t |b(X_s^{\tau_L})| ds + \sup_{s \in [0,t]} |M_s^{(L)}|.$$

The elementary inequality $(a + b + c)^2 \leq 3(a^2 + b^2 + c^2)$, $a, b, c \geq 0$, combined with the Schwarz Inequality successively yields

$$\begin{aligned}\sup_{s\in[0,t]} \left(X_s^{\tau_L}\right)^2 &\le 3\left(X_0^2 + \left(\int_0^t |b(X_s^{\tau_L})|ds\right)^2 + \sup_{s\in[0,t]} |M_s^{(L)}|^2\right)\\ &\le 3\left(X_0^2 + t\int_0^t |b(X_s^{\tau_L})|^2 ds + \sup_{s\in[0,t]} |M_s^{(L)}|^2\right).\end{aligned}$$

We know that the functions b and σ satisfy a linear growth assumption

$$|b(x)| + |\sigma(x)| \le C_{b,\sigma}(1+|x|), \quad x \in \mathbb{R},$$

as Lipschitz continuous functions. Then, taking expectation and using Doob's Inequality for the local martingale $M^{(L)}$ yields for an appropriate real constant $C_{b,\sigma,T} > 0$ (that may vary from line to line)

$$\begin{aligned}\mathbb{E}\left[\sup_{s\in[0,t]} (X_s^{\tau_L})^2\right] &\le 3\left(\mathbb{E}\,X_0^2 + T C_{b,\sigma}\int_0^t \left(1+\mathbb{E}\,|X_s^{\tau_L}|\right)^2 ds + \mathbb{E}\int_0^{t\wedge\tau_L} \sigma^2(X_s^{\tau_L})ds\right)\\ &\le C_{b,\sigma,T}\left(\mathbb{E}\,X_0^2 + \int_0^t (1+\mathbb{E}\,|X_s^{\tau_L}|)ds + \mathbb{E}\int_0^t (1+|X_s^{\tau_L}|)^2 ds\right)\\ &= C_{b,\sigma,T}\left(\mathbb{E}\,X_0^2 + \int_0^t (1+\mathbb{E}\,|X_s^{\tau_L}|^2)ds\right)\end{aligned}$$

where we used again (in the first inequality) that $\tau_L \wedge t \le t$. Finally, this can be rewritten as

$$\mathbb{E}\left[\sup_{s\in[0,t]} (X_s^{\tau_L})^2\right] \le C_{b,\sigma,T}\left(1+\mathbb{E}\,X_0^2 + \int_0^t \mathbb{E}\,(|X_s^{\tau_L}|^2)ds\right)$$

for a new real constant $C_{b,\sigma,T}$. Then, the Gronwall Lemma 7.2 applied to the bounded function $f_L(t) := \mathbb{E}\left(\sup_{s\in[0,t]}(X_s^{\tau_L})^2\right)$ at time $t = T$ implies

$$\mathbb{E}\left[\sup_{s\in[0,T]} (X_s^{\tau_L})^2\right] \le C_{b,\sigma,T}\left(1+\mathbb{E}\,X_0^2\right)e^{C_{b,\sigma,T}T}.$$

This holds or every $L \ge 1$. Now $\tau_L \uparrow +\infty$ $a.s.$ as $L \uparrow +\infty$ since $\sup_{0\le s\le t} |X_s| < +\infty$ for every $t \ge 0$ $a.s.$ Consequently,

$$\lim_{L\to+\infty} \sup_{s\in[0,T]} |X_s^{\tau_L}| = \sup_{s\in[0,T]} |X_s|.$$

Then Fatou's Lemma implies

$$\mathbb{E}\left[\sup_{s\in[0,T]} X_s^2\right] \le C_{b,\sigma,T}\left(1+\mathbb{E}\,X_0^2\right)e^{C_{b,\sigma,T}T} = C'_{b,\sigma,T}\left(1+\mathbb{E}\,X_0^2\right).$$

As for the Euler scheme, the proof follows closely the above lines, once we replace the stopping time τ_L by

$$\bar{\tau}_L = \min\big\{t \ :\ |\bar{X}_t - X_0| \geq L\big\}.$$

It suffices to note that, for every $s \in [0, T]$, $\sup_{u\in[0,s]} |\bar{X}_{\underline{u}}| \leq \sup_{u\in[0,s]} |\bar{X}_u|$. Then one shows that

$$\sup_{n\geq 1} \mathbb{E}\left[\sup_{s\in[0,T]} (\bar{X}^n_s)^2\right] \leq C_{b,\sigma,T}\big(1+\mathbb{E}\, X_0^2\big)e^{C_{b,\sigma,T}T}. \qquad \diamond$$

Proof of Theorem 7.2 **(simplified setting).** *(a) (Convergence rate of the continuous Euler scheme).* Combining the equations satisfied by X and its (continuous) Euler scheme yields

$$X_t - \bar{X}_t = \int_0^t \big(b(X_s) - b(\bar{X}_{\underline{s}})\big)ds + \int_0^t \big(\sigma(X_s) - \sigma(\bar{X}_{\underline{s}})\big)dW_s.$$

Consequently, using that b and σ are Lipschitz continuous, the Schwartz and Doob Inequalities lead to

$$\begin{aligned}
\mathbb{E}\left[\sup_{s\in[0,t]} |X_s - \bar{X}_s|^2\right] &\leq 2\,\mathbb{E}\left[\int_0^t [b]_{\text{Lip}}|X_s - \bar{X}_{\underline{s}}|ds\right]^2 \\
&\quad +2\,\mathbb{E}\left[\sup_{s\in[0,t]}\left(\int_0^s \big(\sigma(X_u) - \sigma(\bar{X}_{\underline{u}})\big)dW_u\right)^2\right] \\
&\leq 2\,[b]_{\text{Lip}}^2\mathbb{E}\left[\int_0^t |X_s - \bar{X}_{\underline{s}}|ds\right]^2 + 8\,\mathbb{E}\int_0^t \big(\sigma(X_u) - \sigma(\bar{X}_{\underline{u}})\big)^2 du \\
&\leq 2t[b]_{\text{Lip}}^2 \int_0^t \mathbb{E}\,|X_s - \bar{X}_{\underline{s}}|^2 ds + 8\,[\sigma]_{\text{Lip}}^2 \int_0^t \mathbb{E}\,|X_u - \bar{X}_{\underline{u}}|^2 du \\
&= C_{b,\sigma,T} \int_0^t \mathbb{E}\,|\,X_s - \bar{X}_{\underline{s}}|^2 ds \\
&\leq C_{b,\sigma,T} \int_0^t \mathbb{E}\,|X_s - \bar{X}_s|^2 ds + C_{b,\sigma,T} \int_0^t \mathbb{E}\,|\bar{X}_s - \bar{X}_{\underline{s}}|^2 ds \\
&\leq C_{b,\sigma,T} \int_0^t \mathbb{E}\left[\sup_{u\in[0,s]} |X_u - \bar{X}_u|^2\right] ds + C_{b,\sigma,T} \int_0^t \mathbb{E}\,|\bar{X}_s - \bar{X}_{\underline{s}}|^2 ds.
\end{aligned}$$

where $C_{b,\sigma,T}$ varies from line to line.

The function $f(t) := \mathbb{E}\big[\sup_{s\in[0,t]} |X_s - \bar{X}_s|^2\big]$ is locally bounded owing to Step 1. Consequently, it follows from Gronwall's Lemma 7.2 applied at $t = T$ that

$$\mathbb{E}\left[\sup_{s\in[0,T]} |X_s - \bar{X}_s|^2\right] \leq e^{C_{b,\sigma,T}T} C_{b,\sigma,T} \int_0^T \mathbb{E}\,|\bar{X}_s - \bar{X}_{\underline{s}}|^2 ds.$$

Now

$$\bar{X}_s - \bar{X}_{\underline{s}} = b(\bar{X}_{\underline{s}})(s - \underline{s}) + \sigma(\bar{X}_{\underline{s}})(W_s - W_{\underline{s}}) \tag{7.20}$$

so that, using Step 1 (for the Euler scheme) and the fact that $W_s - W_{\underline{s}}$ and $\bar{X}_{\underline{s}}$ are independent

$$\begin{aligned}
\mathbb{E}\,|\bar{X}_s - \bar{X}_{\underline{s}}|^2 &\le C_{b,\sigma}\left(\left(\frac{T}{n}\right)^2 \mathbb{E}\, b^2(\bar{X}_{\underline{s}}) + \mathbb{E}\,\sigma^2(\bar{X}_{\underline{s}})\,\mathbb{E}\,(W_s - W_{\underline{s}})^2\right)\\
&\le C_{b,\sigma}\Big(1 + \mathbb{E}\sup_{t\in[0,T]}|\bar{X}_t|^2\Big)\left(\left(\frac{T}{n}\right)^2 + \frac{T}{n}\right)\\
&\le C_{b,\sigma}\big(1 + \mathbb{E}\,X_0^2\big)\frac{T}{n}.
\end{aligned}$$

(*b*) *Stepwise constant Euler scheme*. We assume here—for pure convenience—that $X_0 \in L^4$. One derives from (7.20) and the linear growth assumption satisfied by b and σ (since they are Lipschitz continuous) that

$$\sup_{t\in[0,T]}|\bar{X}_t - \bar{X}_{\underline{t}}| \le C_{b,\sigma}\big(1 + \sup_{t\in[0,T]}|\bar{X}_t|\big)\Big(\frac{T}{n} + \sup_{t\in[0,T]}|W_t - W_{\underline{t}}|\Big)$$

so that

$$\Big\|\sup_{t\in[0,T]}|\bar{X}_t - \bar{X}_{\underline{t}}|\Big\|_2 \le C_{b,\sigma}\left\|\Big(1 + \sup_{t\in[0,T]}|\bar{X}_t|\Big)\left(\frac{T}{n} + \sup_{t\in[0,T]}|W_t - W_{\underline{t}}|\right)\right\|_2.$$

Now, note that if U and V are two real-valued random variables, the Schwarz Inequality implies

$$\|U\,V\|_2 = \sqrt{\|U^2V^2\|_1} \le \sqrt{\|U^2\|_2\|V^2\|_2} = \|U\|_4\|V\|_4.$$

Combining this with the Minkowski inequality in both resulting terms yields

$$\Big\|\sup_{t\in[0,T]}|\bar{X}_t - \bar{X}_{\underline{t}}|\Big\|_2 \le C_{b,\sigma}\Big(1 + \|\sup_{t\in[0,T]}|\bar{X}_t|\|_4\Big)\Big(T/n + \|\sup_{t\in[0,T]}|W_t - W_{\underline{t}}|\|_4\Big).$$

Now, as already mentioned in the first remark that follows Theorem 7.2,

$$\Big\|\sup_{t\in[0,T]}|W_t - W_{\underline{t}}|\Big\|_4 \le C_W\sqrt{\frac{T}{n}}\sqrt{1 + \log n}$$

which completes the proof... if we admit that $\Big\|\sup_{t\in[0,T]}|\bar{X}_t|\Big\|_4 \le C_{b,\sigma,T}\big(1 + \|X_0\|_4\big)$. ◇

Remarks. • The proof in the general L^p framework follows exactly the same lines, except that one replaces Doob's Inequality for a continuous (local) martingale $(M_t)_{t\geq 0}$ by the so-called Burkhölder–Davis–Gundy Inequality (see *e.g.* [251]) which holds for every exponent $p>0$ (only in the continuous setting)

$$\forall\, t\geq 0,\quad c_p\big\|\sqrt{\langle M\rangle_t}\big\|_p\leq \Big\|\sup_{s\in[0,t]}|M_s|\Big\|_p\leq C_p\,\big\|\sqrt{\langle M\rangle_t}\big\|_p=C_p\,\|\langle M\rangle_t\|_{\frac{p}{2}}^{\frac{1}{2}},$$

where c_p, C_p are positive real constants only depending on p. This general setting is developed in full detail in Sect. 7.8 (in the one-dimensional case to alleviate notation).

• In some so-called mean-reverting situations one may even get boundedness over $t\in(0,+\infty)$.

7.3 Non-asymptotic Deviation Inequalities for the Euler Scheme

The aim of this section is to establish non-asymptotic deviation inequalities for the Euler scheme in order to provide confidence intervals for the Monte Carlo method. We recall for convenience several important notations. Let $\|A\|=\big(\mathrm{Tr}(AA^*)\big)^{\frac{1}{2}}$ denote the Fröbenius norm of $A\in\mathcal{M}(d,q,\mathbb{R})$ (*i.e.* the canonical Euclidean norm on $\mathbb{R}^{dq}$) and let $|||A|||=\sup_{|x|=1}|Ax|$ be the operator norm of A (with respect to the canonical Euclidean norms on $\mathbb{R}^d$ and $\mathbb{R}^q$). Note that $|||A|||\leq\|A\|$.

We still consider the Brownian diffusion process solution to (7.1) on a probability space $(\Omega,\mathcal{A},\mathbb{P})$ with the same Lipschitz regularity assumptions made on the drift b and the diffusion coefficient σ. The q-dimensional driving Brownian motion is still denoted by W and the (augmented) natural filtration $(\mathcal{F}_t)_{t\in[0,T]}$ is still defined by (7.3). Furthermore, as the functions $b:[0,T]\times\mathbb{R}^d\to\mathbb{R}^d$ and $\sigma:[0,T]\times\mathbb{R}^d\to\mathcal{M}(d,q,\mathbb{R})$ satisfy a Lipschitz continuous assumption in x uniformly in $t\in[0,T]$, we define

$$[b]_{\mathrm{Lip}}=\sup_{t\in[0,T],\,x\neq y}\frac{|b(t,x)-b(t,y)|}{|x-y|}<+\infty \tag{7.21a}$$

and

$$[\sigma]_{\mathrm{Lip}}=\sup_{t\in[0,T],\,x\neq y}\frac{\|\sigma(t,x)-\sigma(t,y)\|}{|x-y|}<+\infty. \tag{7.21b}$$

The definition of the Brownian Euler scheme with step $\frac{T}{n}$, starting at X_0, is unchanged but to alleviate notation in this section, we will temporarily write $\bar{X}^n_k$ instead of $\bar{X}^n_{t^n_k}$. So we have: $\bar{X}^n_0=X_0$ and

$$\bar{X}^n_{k+1} = \bar{X}^n_k + \frac{T}{n}\, b(t^n_k, \bar{X}^n_k) + \sigma(t^n_k, \bar{X}^n_k)\sqrt{\frac{T}{n}} Z_{k+1},\ k = 0, \dots, n-1,$$

where $t^n_k = \frac{kT}{n}$, $k = 0, \dots, n$ and $Z_k = \sqrt{\frac{n}{T}}\big(W_{t^n_k} - W_{t^n_{k-1}}\big)$, $k = 1, \dots, n$ is an i.i.d. sequence of $\mathcal{N}(0; I_q)$ random vectors. When $X_0 = x \in \mathbb{R}^d$, we may denote occasionally by $(\bar{X}^{n,x}_k)_{0 \le k \le n}$ the Euler scheme starting at x.

The Euler scheme defines an $\mathbb{R}^d$-valued Markov chain with transitions $P_k(x, dy) = \mathbb{P}\big(\bar{X}^n_{k+1} \in dy \mid \bar{X}^n_k = x\big)$, $k = 0, \dots, n-1$, reading on bounded or non-negative Borel functions $f : \mathbb{R}^d \to \mathbb{R}$,

$$P_k(f)(x) = \mathbb{E}\Big(f(\bar{X}_{k+1}) \mid X_k = x\Big) = \mathbb{E}\, f\Big(\mathcal{E}_k(x, Z)\Big),\ k = 0, \dots, n-1,$$

where

$$\mathcal{E}_k(x, z) = x + \frac{T}{n}\, b(t^n_k, x) + \sigma(t^n_k, x)\sqrt{\frac{T}{n}}\, z,\ x \in \mathbb{R}^d,\ z \in \mathbb{R}^q,\ k = 0, \dots, n-1,$$

denotes the Euler scheme operator and $Z \stackrel{d}{=} \mathcal{N}(0; I_q)$. Then, set for every $k, \ell \in \{0, \dots, n-1\}$, $k < \ell$,

$$P_{k,\ell} = P_k \circ \dots \circ P_{\ell-1} \quad \text{and} \quad P_{\ell,\ell}(f) = f$$

so that we have, still for bounded or non-negative Borel functions f,

$$P_{k,\ell}(f)(x) = \mathbb{E}\Big(f(\bar{X}_\ell) \mid \bar{X}_k = x\Big).$$

We will need the following property satisfied by the Euler transitions.

Proposition 7.3 *Under the above Lipschitz assumption* (7.21a) *and* (7.21b) *on b and σ, the transitions P_k, $k = 0, \dots, n-1$, satisfy*

$$[P_k f]_{\mathrm{Lip}} \le [P_k]_{\mathrm{Lip}} [f]_{\mathrm{Lip}}, \quad k = 0, \dots, n-1,$$

with

$$\begin{aligned}[P_k]_{\mathrm{Lip}} &\le \left(\Big[I_d + \frac{T}{n} b(t^n_k, .)\Big]^2_{\mathrm{Lip}} + \frac{T}{n}[\sigma(t^n_k, .)]^2_{\mathrm{Lip}}\right)^{\frac{1}{2}} \\ &\le \left(1 + \frac{T}{n}\Big(C_{b,\sigma} + \frac{T}{n}\kappa_b\Big)\right)^{\frac{1}{2}}\end{aligned}$$

where

$$C_{b,\sigma} = 2[b]_{\mathrm{Lip}} + [\sigma]^2_{\mathrm{Lip}} \quad \textit{and} \quad \kappa_b = [b]^2_{\mathrm{Lip}}. \tag{7.22}$$

Proof. It is a straightforward consequence of the fact that

$$|P_k f(x) - P_k f(y)| \le [f]_{\text{Lip}} \Big\| x - y + \frac{T}{n}(b(t_k^n, x) - b(t_k^n, y)) + \sqrt{\frac{T}{n}}(\sigma(t_k^n, x) - \sigma(t_k^n, y))Z\Big\|_1$$
$$\le [f]_{\text{Lip}} \Big\| x - y + \frac{T}{n}(b(t_k^n, x) - b(t_k^n, y)) + \sqrt{\frac{T}{n}}(\sigma(t_k^n, x) - \sigma(t_k^n, y))Z\Big\|_2.$$

Now, one completes the proof by noting that

$$\Big\| x - y + \frac{T}{n}(b(t_k^n, x) - b(t_k^n, y)) + \sqrt{\frac{T}{n}}(\sigma(t_k^n, x) - \sigma(t_k^n, y))Z\Big\|_2^2$$
$$= \Big| x - y + \frac{T}{n}\big(b(t_k^n, x) - b(t_k^n, y)\big)\Big|^2 + \frac{T}{n}\big\|\sigma(t_k^n, x) - \sigma(t_k^n, y)\big\|^2. \quad \diamond$$

The key property is the following classical exponential inequality for the Gaussian measure (the proof below is due to Ledoux in [195]).

Proposition 7.4 *Let $g : \mathbb{R}^q \to \mathbb{R}$ be a Lipschitz continuous function (with respect to the canonical Euclidean measure on $\mathbb{R}^q$) and let Z be an $\mathcal{N}(0; I_q)$ distributed random vector. Then*

$$\forall\, \lambda \in \mathbb{R}, \quad \mathbb{E}\Big(e^{\lambda\left(g(Z) - \mathbb{E}\, g(Z)\right)}\Big) \le e^{\frac{\lambda^2}{2}[g]_{\text{Lip}}^2}. \tag{7.23}$$

Proof. STEP 1 (*Preliminaries*). We consider a standard Ornstein–Uhlenbeck process starting at $x \in \mathbb{R}^q$, being a solution to the stochastic differential equation

$$d\,\Xi_t^x = -\frac{1}{2}\,\Xi_t^x dt + dW_t, \;\; \Xi_0^x = x,$$

where W is a standard q-dimensional Brownian motion. One easily checks that this equation has a (unique) explicit solution on the whole real line given by

$$\Xi_t^x = x\, e^{-\frac{t}{2}} + e^{-\frac{t}{2}} \int_0^t e^{\frac{s}{2}}\, dW_s, \quad t \in \mathbb{R}_+.$$

This shows that $(\Xi_t^x)_{t\ge 0}$ is a Gaussian process such that

$$\mathbb{E}\,\Xi_t^x = x\, e^{-\frac{t}{2}}.$$

Using the Wiener isometry, we derive that the covariance matrix $\Sigma_{\Xi_t^x}$ of Ξ_t^x is given by

$$
\begin{aligned}
\Sigma_{\Xi_t^x} &= e^{-t}\left[\mathbb{E}\left(\int_0^t e^{\frac{s}{2}}\,dW_s^k\int_0^t e^{\frac{s}{2}}\,dW_s^\ell\right)\right]_{1\le k,\ell\le q}\\
&= e^{-t}\int_0^t e^s\,ds\,I_q\\
&= (1-e^{-t})I_q.
\end{aligned}
$$

(The time covariance structure of the process can be computed likewise but is of no use in this proof). As a consequence, for every Borel function $g:\mathbb{R}^q\to\mathbb{R}$ with polynomial growth,

$$
Q_t g(x) := \mathbb{E}\,g\big(\Xi_t^x\big) = \mathbb{E}\,g\Big(x\,e^{-t/2}+\sqrt{1-e^{-t}}Z\Big) \quad\text{where}\quad Z \stackrel{d}{=} \mathcal{N}(0;\,I_q). \tag{7.24}
$$

Hence, owing to the Lebesgue dominated convergence theorem,

$$
\lim_{t\to+\infty} Q_t g(x) = \mathbb{E}\,g(Z).
$$

Moreover, if g is differentiable with a gradient ∇g having polynomial growth,

$$
\nabla_x Q_t g(x) = e^{-\frac{t}{2}}\mathbb{E}\,\nabla g(\Xi_t^x). \tag{7.25}
$$

If g is twice continuously differentiable with a Hessian $D^2 g$ having polynomial growth, it follows from Itô's formula (see Sect. 12.8), and Fubini's Theorem that, for every $t\ge 0$,

$$
Q_t g(x) = g(x)+\int_0^t \mathbb{E}\big[(Lg)(\Xi_s^x)\big]ds = g(x)+\int_0^t Q_s(Lg)(x)ds, \tag{7.26}
$$

where L is the Ornstein–Uhlenbeck operator (infinitesimal generator of the above Ornstein–Uhlenbeck stochastic differential equation) which maps g to the function

$$
Lg:\xi\longmapsto Lg(\xi)=\frac{1}{2}\big(\Delta g(\xi)-(\xi|\nabla g(\xi))\big),
$$

where $\Delta g(\xi)=\sum_{1\le i\le q} g''_{x_i^2}(\xi)$ denotes the Laplacian of g and $\nabla g(\xi) = \begin{bmatrix} g'_{x_i}(\xi)\\ \vdots \\ g'_{x_q}(\xi)\end{bmatrix}$.

Now, if h and g are both twice differentiable with existing partial derivatives having polynomial growth, one has

$$
\mathbb{E}\big((\nabla g(Z)|\nabla h(Z)\big) = \sum_{k=1}^q \int_{\mathbb{R}^q} g'_{x_k}(z)h'_{x_k}(z)e^{-\frac{|z|^2}{2}}\frac{dz}{(2\pi)^{\frac{q}{2}}}.
$$

After noting that

$$\forall\, z = (z_1, \dots, z_q) \in \mathbb{R}^d, \quad \frac{\partial}{\partial z_k}\Big(h'_{x_k}(z)e^{-\frac{|z|^2}{2}}\Big) = e^{-\frac{|z|^2}{2}}\big(h''_{x_k^2}(z) - z_k h'_{x_k}(z)\big),$$
$$k = 1, \dots, q,$$

integrations by parts in each of the above q integrals yield the following identity

$$\mathbb{E}\,\big((\nabla g(Z)|\nabla h(Z)\big) = -2\,\mathbb{E}\,\big(g(Z)Lh(Z)\big). \tag{7.27}$$

One also derives from (7.26) and the continuity of $s \mapsto \mathbb{E}\,(Lg)(\Xi_s^x)$ that

$$\frac{\partial}{\partial t} Q_t g(x) = Q_t Lg(x).$$

In particular, $\big[\frac{\partial}{\partial t} Q_t g\big]_{|t=0}(x) = Q_0 Lg(x) = Lg(x)$. On the other hand, as $Q_t g(x)$ has partial derivatives with polynomial growth

$$\lim_{s\to 0} \frac{Q_s(Q_t g)(x) - Q_t g(x)}{s} = \Big[\frac{\partial}{\partial s} Q_s\Big]_{|s=0}(Q_t g)(x) = L Q_t g(x)$$

so that, finally, we get the classical identity

$$L Q_t g(x) = Q_t Lg(x). \tag{7.28}$$

STEP 2 (*The smooth case*). Let $g : \mathbb{R}^q \to \mathbb{R}$ be a continuously twice differentiable function with bounded existing partial derivatives such that $\mathbb{E}\, g(Z) = 0$. Let $\lambda \in \mathbb{R}$ be fixed. We define the function $H_\lambda : \mathbb{R}_+ \to \mathbb{R}_+$ by

$$H_\lambda(t) = \mathbb{E}\, e^{\lambda Q_t g(Z)}.$$

Let us first check that H_λ is well-defined by the above equality. Note that the function g is Lipschitz continuous since its gradient is bounded and $[g]_{\rm Lip} \le \|\nabla g\|_{\sup} = \sup_{\xi\in\mathbb{R}^q} |\nabla g(\xi)|$. It follows from (7.24) that, for every $z \in \mathbb{R}^q$,

$$\begin{aligned}
|Q_t g(z)| &\le [g]_{\rm Lip}\, e^{-t/2}|z| + |Q_t g(0)| \\
&\le [g]_{\rm Lip}|z| + \mathbb{E}\,|g(\sqrt{1-e^{-t}}Z)| \\
&\le [g]_{\rm Lip}|z| + |g(0)| + [g]_{\rm Lip}\mathbb{E}\,|Z| \\
&\le [g]_{\rm Lip}|z| + |g(0)| + [g]_{\rm Lip}\sqrt{q}.
\end{aligned}$$

This ensures the existence of H_λ since $\mathbb{E}\, e^{a|Z|} < +\infty$ for every $a \ge 0$. One shows likewise that, for every $z \in \mathbb{R}^q$ and every $t \in \mathbb{R}_+$,

$$
\begin{aligned}
|Q_t g(z)| &= |Q_t g(z) - \mathbb{E}\, g(Z)| \\
&\le [g]_{\mathrm{Lip}}\big(e^{-t/2}|z| + (1-\sqrt{1-e^{-t}})\mathbb{E}\,|Z|\big) \to 0 \quad \text{as} \quad t \to +\infty.
\end{aligned}
$$

One concludes by the Lebesgue dominated convergence theorem that

$$
\lim_{t\to+\infty} H_\lambda(t) = e^0 = 1.
$$

Furthermore, one shows, still using the same arguments, that H_λ is differentiable over $\mathbb{R}_+$ with a derivative given for every $t \in \mathbb{R}_+$ by

$$
\begin{aligned}
H'_\lambda(t) &= \lambda\, \mathbb{E}\left(e^{\lambda Q_t g(Z)} Q_t L g(Z)\right) \\
&= \lambda\, \mathbb{E}\left(e^{\lambda Q_t g(Z)} L Q_t g(Z)\right) \qquad \text{by (7.28)} \\
&= -\frac{\lambda}{2}\,\mathbb{E}\Big(\big(\nabla_z(e^{\lambda Q_t g(z)})_{|z=Z} | \nabla Q_t g(Z)\big)\Big) \qquad \text{by (7.27)} \\
&= -\frac{\lambda^2}{2} e^{-\frac{t}{2}} e^{-\frac{t}{2}}\,\mathbb{E}\Big(e^{\lambda Q_t g(Z)} |Q_t \nabla g(Z)|^2\Big) \qquad \text{by (7.25).}
\end{aligned}
$$

Consequently, as $\lim_{t\to+\infty} H_\lambda(t) = 1$,

$$
\begin{aligned}
H_\lambda(t) &= 1 - \int_t^{+\infty} H'_\lambda(s)ds \\
&\le 1 + \frac{\lambda^2}{2}\|\nabla g\|^2_{\sup} \int_t^{+\infty} e^{-s}\,\mathbb{E}\Big(e^{\lambda Q_s g(Z)}\Big)ds \\
&= 1 + K \int_t^{+\infty} e^{-s} H_\lambda(s)ds \quad \text{with} \quad K = \frac{\lambda^2}{2}\|\nabla g\|^2_{\sup}.
\end{aligned}
$$

One derives by induction, using that H_λ is non-increasing since its derivative is negative, that, for every integer $m \in \mathbb{N}^*$,

$$
H_\lambda(t) \le \sum_{k=0}^{m} K^k \frac{e^{-kt}}{k!} + H_\lambda(0) K^{m+1} \frac{e^{-(m+1)t}}{(m+1)!}.
$$

Letting $m \to +\infty$ finally yields

$$
H_\lambda(t) \le e^{K e^{-t}} \le e^K = e^{\frac{\lambda^2}{2}\|\nabla g\|^2_{\sup}}.
$$

One completes this step of the proof by applying the above inequality to the function $g - \mathbb{E}\, g(Z)$.

STEP 3 *(The Lipschitz continuous case)*. This step relies on an approximation technique which is closely related to sensitivity computation for options attached to non-regular payoffs (but in a situation where the Brownian motion plays the role of a pseudo-asset). Let $g : (\mathbb{R}^q, |\cdot|) \to (\mathbb{R}, |\cdot|)$ be a Lipschitz continuous function with Lipschitz coefficient $[g]_{\text{Lip}}$ and $\zeta \stackrel{d}{=} \mathcal{N}(0; I_q)$. One considers for every $\varepsilon > 0$,

$$g_\varepsilon(z) = \mathbb{E}\, g(z + \sqrt{\varepsilon}\,\zeta) = \int_{\mathbb{R}^q} f(u) e^{-\frac{|u-z|^2}{2\varepsilon}} \frac{du}{(2\pi\varepsilon)^{\frac{q}{2}}}.$$

It is clear that g_ε uniformly converges toward g on $\mathbb{R}^q$ since $|g_\varepsilon(z) - g(z)| \le \sqrt{\varepsilon}\,\mathbb{E}\,|\zeta|$ for every $z \in \mathbb{R}^q$. One shows likewise that g_ε is Lipschitz continuous and $[g_\varepsilon]_{\text{Lip}} \le [g]_{\text{Lip}}$. Moreover, the function g_ε is differentiable with gradient given by

$$\nabla g_\varepsilon(z) = -\frac{1}{\varepsilon}\int_{\mathbb{R}^q} f(u) e^{-\frac{|u-z|^2}{2\varepsilon}} (u - z) \frac{du}{(2\pi\varepsilon)^{\frac{q}{2}}} = \frac{1}{\sqrt{\varepsilon}}\mathbb{E}\left(g(z + \sqrt{\varepsilon}\zeta)\zeta\right).$$

It is always true that if a function $h : \mathbb{R}^q \to \mathbb{R}$ is differentiable and Lipschitz continuous then $\|\nabla h\|_{\sup} = [h]_{\text{Lip}}$ ([3]). Consequently,

$$\|\nabla f_\varepsilon\|_{\sup} = [f_\varepsilon]_{\text{Lip}} \le [f]_{\text{Lip}}.$$

The Hessian of f_ε is also bounded (by a constant depending on ε, but this has no consequence on the inequality of interest). One concludes by Fatou's Lemma

$$\mathbb{E}\, e^{\lambda f(Z)} \le \underline{\lim}_{\varepsilon \to 0} \mathbb{E}\, e^{\lambda f_\varepsilon(Z)} \le \underline{\lim}_{\varepsilon \to 0} e^{\frac{\lambda^2}{2}\|\nabla f_\varepsilon\|^2_{\sup}} \le e^{\frac{\lambda^2}{2}[f]^2_{\text{Lip}}}. \qquad \diamond$$

We are now in a position to state the main result of this section and its application to the design of confidence intervals (see also [100], where this result was proved by a slightly different method).

Theorem 7.4 *Assume* $|||\sigma|||_{\sup} := \sup_{(t,x)\in[0,T]\times\mathbb{R}^d} |||\sigma(t,x)||| < +\infty$. *Then, there exists a positive decreasing sequence* $K(b, \sigma, T, n) \in (0, +\infty)$, $n \ge 1$, *of real numbers such that, for every* $n \ge 1$ *and every Lipschitz continuous function* $f : \mathbb{R}^d \to \mathbb{R}$,

$$\forall\, \lambda \in \mathbb{R}, \quad \mathbb{E}\left(e^{\lambda\left(f(\bar{X}^n_T) - \mathbb{E}\, f(\bar{X}^n_T)\right)}\right) \le e^{\frac{\lambda^2}{2}|||\sigma|||^2_{\sup}[f]^2_{\text{Lip}} K(b,\sigma,T,n)}. \tag{7.29}$$

The choice $K(b, \sigma, T, n) = \dfrac{e^{(C_{b,\sigma} + \kappa_b \frac{T}{n})T}}{C_{b,\sigma}}$ *is admissible so that* $\lim_n \downarrow K(b, \sigma, T, n) = \dfrac{e^{C_{b,\sigma}T}}{C_{b,\sigma}}$, *where the real constant* $C_{b,\sigma}$ *is defined in (7.22) of Proposition 7.3.*

---

[3] In fact, for every $z, u \in \mathbb{R}^q$, $|u| = 1$, one has $|(\nabla h(z)|u)| = \lim_{\varepsilon\to 0} \varepsilon^{-1}|h(z + \varepsilon u) - h(z)| \le [h]_{\text{Lip}}$, whereas $|\nabla h(z)| = \sup_{|u|=1} |(\nabla h(z)|u)|$. Hence, $\|\nabla h\|_{\sup} = \sup_{z\in\mathbb{R}^q} |\nabla h(z)| \le [h]_{\text{Lip}}$. The reverse inequality is obvious since for every $z, z' \in \mathbb{R}^q$, $|h(z) - h(z')| = |(\nabla h(\zeta)|z' - z)| \le \|\nabla h\|_{\sup}|z' - z|$.

Application to the design of non-asymptotic confidence intervals

Let us briefly recall that such exponential inequalities yield deviation inequalities in the strong law of large numbers. Let $(\bar{X}^{n,\ell})_{\ell\geq 1}$ be independent copies of the discrete time Euler scheme $\bar{X}^n = (\bar{X}^n_{t^n_k})_{k=0,\dots,n}$ with step $\frac{T}{n}$ starting at $\bar{X}^n_0 = X_0$. Then, for every $\varepsilon > 0$ and every $\lambda > 0$, the Markov inequality and independence imply, for every integer $n \geq 1$,

$$\begin{aligned}\mathbb{P}\left(\frac{1}{M}\sum_{\ell=1}^{M} f(\bar{X}^{n,\ell}_T) - \mathbb{E}\, f(\bar{X}^n_T) > \varepsilon\right) &= \mathbb{P}\left(e^{\lambda\sum_{\ell=1}^{M}\left(f(\bar{X}^{n,\ell}_T)-\mathbb{E}\, f(\bar{X}^n_T)\right)} > e^{\lambda\varepsilon M}\right)\\ &\leq e^{-\lambda\varepsilon M}\,\mathbb{E}\, e^{\lambda\sum_{\ell=1}^{M}\left(f(\bar{X}^{n,\ell}_T)-\mathbb{E}\, f(\bar{X}^n_T)\right)}\\ &= e^{-\lambda\varepsilon M}\left(\mathbb{E}\, e^{\lambda\left(f(\bar{X}^n_T)-\mathbb{E}\, f(\bar{X}^n_T)\right)}\right)^M\\ &\leq e^{-\lambda\varepsilon M+\frac{\lambda^2}{2}M\|\!|\sigma\|\!|^2_{\sup}[f]^2_{\rm Lip}K(b,\sigma,T,n)}.\end{aligned}$$

The function $\lambda \mapsto -\lambda\varepsilon + \frac{\lambda^2}{2}\|\!|\sigma\|\!|^2_{\sup}[f]^2_{\rm Lip}K(b,\sigma,T,n)$ attains its minimum at

$$\lambda_{\min} = \frac{\varepsilon}{\|\!|\sigma\|\!|^2_{\sup}[f]^2_{\rm Lip}K(b,\sigma,T,n)}$$

so that, finally,

$$\mathbb{P}\left(\frac{1}{M}\sum_{\ell=1}^{M} f(\bar{X}^{n,\ell}_T) - \mathbb{E}\, f(\bar{X}^n_T) > \varepsilon\right) \leq e^{-\frac{\varepsilon^2 M}{2\|\!|\sigma\|\!|^2_{\sup}[f]^2_{\rm Lip}K(b,\sigma,T,n)}}.$$

Applying the above inequality to $-f$ yields by an obvious symmetry argument the two-sided deviation inequality

$$\forall\, n \geq 1,\quad \mathbb{P}\left(\left|\frac{1}{M}\sum_{\ell=1}^{M} f(\bar{X}^{n,\ell}_T) - \mathbb{E}\, f(\bar{X}^n_T)\right| > \varepsilon\right) \leq 2e^{-\frac{\varepsilon^2 M}{2\|\!|\sigma\|\!|^2_{\sup}[f]^2_{\rm Lip}K(b,\sigma,T,n)}}. \tag{7.30}$$

One easily derives from (7.30) confidence intervals provided (upper-bounds of) $\|\!|\sigma\|\!|$, $[f]^2_{\rm Lip}$ and $K(b, \sigma, T, n)$ are known.

The main feature of the above inequality (7.30), beyond the fact that it holds for possibly unbounded Lipschitz continuous functions f, is that the right-hand upper-bound does not depend on the time step $\frac{T}{n}$ of the Euler scheme. Consequently we can design confidence intervals for Monte Carlo simulations based on the Euler schemes *uniformly in the time discretization step* $\frac{T}{n}$.

Doing so, we can design non-asymptotic confidence intervals when computing $\mathbb{E}\, f(X_T)$ by a Monte Carlo simulation. We know that the bias is due to the discretization scheme and only depends on the step $\frac{T}{n}$: under appropriate assumptions on b, σ and f (see Sect. 7.6 for the first order expansion of the weak error), one has,

$$\mathbb{E}\, f(\bar{X}^n_{_T}) = \mathbb{E}\, f(X_{_T}) + \frac{c_1}{n^\alpha} + o(1/n).$$

Remark. Under the assumptions we make on b and σ, the Euler scheme converges $a.s.$ and in every L^p space (provided X_0 lies in L^p) for the sup norm over $[0, T]$ toward the diffusion process X, so that we deduce a similar result for independent copies X^ℓ of the diffusion itself, namely

$$\mathbb{P}\left(\left|\frac{1}{M}\sum_{\ell=1}^{M} f\big(X^\ell_{_T}\big) - \mathbb{E}\, f(X_T)\right| > \varepsilon\right) \le 2e^{-\frac{\varepsilon^2 M}{2\|\sigma\|^2_{\sup}[f]^2_{\rm Lip} K(b,\sigma,T)}}$$

with $K(b,\sigma,T) = \frac{e^{C_{b,\sigma}T}}{C_{b,\sigma}}$ and $C_{b,\sigma} = 2[b]_{\rm Lip} + [\sigma]^2_{\rm Lip}$.

Proof of Theorem 7.4. It follows from Proposition 7.4 that, for every Lipschitz continuous function $f : \mathbb{R}^d \to \mathbb{R}$, the Euler scheme operator satisfies

$$\forall\, \lambda \in \mathbb{R}, \quad \mathbb{E}\Big(e^{\lambda\left(f(\mathcal{E}_k(x,Z)) - P_k f(x)\right)}\Big) \le e^{\frac{\lambda^2}{2}\left(\sqrt{\frac{T}{n}}\|\sigma(t^n_k,x)\|\right)^2 [f]^2_{\rm Lip}}$$

since the function $z \mapsto f\big(\mathcal{E}(x,z)\big)$ is Lipschitz continuous from $\mathbb{R}^q$ to $\mathbb{R}$ with respect to the canonical Euclidean norm with a Lipschitz coefficient upper-bounded by $\sqrt{\frac{T}{n}}[f]_{\rm Lip}\|\sigma(t^n_k,x)\|$, where $\|\sigma(t^n_k,x)\|$ denotes the operator norm of $\sigma(t^n_k,x)$. Consequently, as $\|\sigma\|_{\sup} := \sup_{(t,x)\in[0,T]\times\mathbb{R}^d}\|\sigma(t,x)\|$,

$$\forall\, \lambda \in \mathbb{R}, \quad \mathbb{E}\left(e^{\lambda f(\bar{X}^n_{t^n_{k+1}})} \,\big|\, \mathcal{F}_{t^n_k}\right) \le e^{\lambda P_k f(\bar{X}^n_{t^n_k}) + \frac{\lambda^2}{2}\frac{T}{n}\|\sigma\|^2_{\sup}[f]^2_{\rm Lip}}.$$

Applying this inequality to $P_{k+1,n}f$ and taking expectation on both sides then yields

$$\forall\, \lambda \in \mathbb{R}, \quad \mathbb{E}\left(e^{\lambda P_{k+1,n} f(\bar{X}^n_{t^n_{k+1}})}\right) \le \mathbb{E}\left(e^{\lambda P_{k,n} f(\bar{X}^n_{t^n_k})}\right) e^{\frac{\lambda^2}{2}\frac{T}{n}\|\sigma\|^2_{\sup}[P_{k+1,n}f]^2_{\rm Lip}}$$

since $P_{k,n} = P_k \circ P_{k+1,n}$. By a straightforward backward induction from $k = n-1$ down to $k = 0$, combined with the fact that $P_{0,n}f(X_0) = \mathbb{E}\big(f(\bar{X}^n_T)|X_0\big)$, we obtain

$$\forall\, \lambda \in \mathbb{R}, \quad \mathbb{E}\left(e^{\lambda f(\bar{X}^n_T)}\right) \le \mathbb{E}\left(e^{\lambda \mathbb{E}\left(f(\bar{X}^n_T)|X_0\right)}\right) e^{\frac{\lambda^2}{2}\|\sigma\|^2_{\sup}\frac{T}{n}\sum_{k=0}^{n-1}[P_{k+1,n}f]^2_{\rm Lip}}.$$

First, note that by Jensen's Inequality applied to the convex function $e^{\lambda\,\cdot}$,

$$\mathbb{E}\left(e^{\lambda\,\mathbb{E}\left(f(\bar{X}^n_T)\,|X_0\right)}\right) \le \mathbb{E}\Big(\mathbb{E}\left(e^{\lambda f(\bar{X}^n_T)}\,|X_0\right)\Big) = \mathbb{E}\, e^{\lambda f(\bar{X}^n_T)}.$$

On the other hand, it is clear from their very definition that

$$[P_{k,n}]_{\rm Lip} \le \prod_{\ell=k}^{n-1} [P_\ell]_{\rm Lip}$$

(with the consistent convention that an empty product is equal to 1). Hence, owing to Proposition 7.3,

$$[P_{k,n}]_{\rm Lip} \le \Big(1 + C^{(n)}_{b,\sigma}\frac{T}{n}\Big)^{\frac{n-k}{2}} \quad \text{with} \quad C^{(n)}_{b,\sigma,T} = C_{b,\sigma} + \frac{T}{n}\kappa_b$$

so that

$$\begin{aligned}
\frac{T}{n}\sum_{k=1}^{n}[P_{k,n}f]^2_{\rm Lip} &= \frac{T}{n}\sum_{k=0}^{n-1}[P_{n-k,n}f]^2_{\rm Lip}\\
&= \frac{(1 + C^{(n)}_{b,\sigma}\frac{T}{n})^n - 1}{C^{(n)}_{b,\sigma}}[f]^2_{\rm Lip}\\
&\le \frac{1}{C_{b,\sigma}}e^{(C_{b,\sigma}+\kappa_b\frac{T}{n})T}[f]^2_{\rm Lip} = K(b,\sigma,T,n)[f]^2_{\rm Lip}.
\end{aligned}$$

◇

► **Exercise** (*Hoeffding Inequality and applications*). Let $Y : (\Omega, \mathcal{A}, \mathbb{P}) \to \mathbb{R}$ be a bounded centered random variable satisfying $|Y| \le A$ for some $A \in (0, +\infty)$.
(a) Show that, for every $\lambda > 0$,

$$e^{\lambda Y} \le \frac{1}{2}\Big(1 + \frac{Y}{A}\Big)e^{\lambda A} + \frac{1}{2}\Big(1 - \frac{Y}{A}\Big)e^{-\lambda A}.$$

(b) Deduce that, for every $\lambda > 0$,

$$\mathbb{E}\, e^{\lambda Y} \le \cosh(\lambda A) \le e^{\frac{\lambda^2 A^2}{2}}.$$

(c) Let $(X_k)_{k\ge 1}$ be an i.i.d. sequence of random vectors and $f : \mathbb{R}^d \to \mathbb{R}$ a bounded Borel function. Show that

$$\forall \varepsilon > 0,\ \forall M \ge 1,\quad \mathbb{P}\left(\left|\frac{1}{M}\sum_{k=1}^{M} f(X_k) - \mathbb{E}\, f(X)\right| > \varepsilon\right) \le 2\, e^{-\frac{\varepsilon^2 M}{2\|f\|^2_{\sup}}}.$$

The Monte Carlo method does depend on the dimension d

As a second step, it may seem natural to try to establish *deviation inequalities for the supremum of the Monte Carlo error over all Lipschitz continuous functions whose Lipschitz continuous coefficients are bounded by* 1, that is, for the L^1-Wasserstein distance, owing to the Monge–Kantorovich representation (see Sect. 7.2.2). For such results concerning the Euler scheme, we refer to [87].

At this point we want to emphasize that introducing this supremum *inside* the probability radically modifies the behavior of the Monte Carlo error and highlights a strong dependence on the structural dimension of the simulation. To establish this behavior, we will rely heavily on an argument from optimal vector quantization theory developed in Chap. 5.

Let $\mathrm{Lip}_1(\mathbb{R}^d, \mathbb{R})$ be the set of Lipschitz continuous functions from $\mathbb{R}^d$ to $\mathbb{R}$ with a Lipschitz continuous coefficient $[f]_{\mathrm{Lip}} \le 1$.

Let $X_\ell : (\Omega, \mathcal{A}, \mathbb{P}) \to \mathbb{R}^d$, $\ell \ge 1$, be independent copies of an integrable *d-dimensional random vector* X with distribution denoted by $\mathbb{P}_X$, independent of X for convenience. For every $\omega \in \Omega$ and every $M \ge 1$, let $\mu_M^X(\omega) = \frac{1}{M}\sum_{\ell=1}^M \delta_{X_\ell(\omega)}$ denote the empirical measure associated to $\big(X_\ell(\omega)\big)_{\ell\ge1}$. Then

$$W_1\big(\mu_M^X(\omega), \mathbb{P}_X\big) = \sup_{f\in \mathrm{Lip}_1(\mathbb{R}^d,\mathbb{R})} \left|\frac{1}{M}\sum_{\ell=1}^M f\big(X_\ell(\omega)\big) - \int_{\mathbb{R}^d} f(\xi)\mathbb{P}_X(d\xi)\right|$$

(the absolute values can be removed without damage since f and $-f$ simultaneously belong to $\mathrm{Lip}_1(\mathbb{R}^d, \mathbb{R})$). Now let us introduce the function defined for every $\xi \in \mathbb{R}^d$ by

$$f_{\omega,M}(\xi) = \min_{\ell=1,\dots,M} |X_\ell(\omega) - \xi| \ge 0.$$

It is clear from its very definition that $f_{\omega,M} \in \mathrm{Lip}_1(\mathbb{R}^d, \mathbb{R})$ owing to the elementary inequality

$$\left|\min_{1\le i\le M} a_i - \min_{1\le i\le M} b_i\right| \le \max_{1\le i\le M} |a_i - b_i|.$$

Then, for every $\omega \in \Omega$,

$$\begin{aligned}
W_1\big(\mu_M^X(\omega), \mathbb{P}_X\big) &\ge \left|\frac{1}{M}\sum_{\ell=1}^M \underbrace{f_{\omega,M}(X_\ell(\omega))}_{=0} - \int_{\mathbb{R}^d} f_{\omega,M}(\xi)\mathbb{P}_X(d\xi)\right| \\
&= \int_{\mathbb{R}^d} f_{\omega,M}(\xi)\mathbb{P}_X(d\xi) = \int_\Omega \min_{\ell=1,\dots,M} |X(\omega') - X_\ell(\omega)| d\mathbb{P}(\omega') \\
&\ge \inf_{(x_1,\dots,x_M)\in(\mathbb{R}^d)^M} \mathbb{E}\Big[\min_{1\le i\le M} |X - x_i|\Big] \\
&= \inf_{x\in(\mathbb{R}^d)^M} \big\|X - \widehat{X}^{\Gamma_x}\big\|_1.
\end{aligned}$$

The lower bound in the last line is just the optimal L^1-quantization error of the (distribution of the) random vector X at level M. It follows from Zador's Theorem (see the remark that follows Zador's Theorem 5.1.2 in Chap. 5 or [129]) that

$$\underline{\lim}_M\, M^{-\frac{1}{d}} \inf_{x\in(\mathbb{R}^d)^M} \big\|X - \widehat{X}^{\Gamma_x}\big\|_1 \ge J_{1,d}\|\varphi_X\|_{\frac{d}{d+1}},$$

where φ_X denotes the density of the nonsingular part of the distribution $\mathbb{P}_X$ of X with respect to the Lebesgue measure on $\mathbb{R}^d$, if it exists. The constant $J_{1,d} \in (0, +\infty)$ is a universal constant and the pseudo-norm $\|\varphi_X\|_{\frac{d}{d+1}}$ is finite as soon as $X \in L^{1+} = \cup_{\eta>0} L^{1+\eta}$. Furthermore, it is clear that as soon as the support of $\mathbb{P}_X$ is infinite, $e_{1,M}(X) > 0$. Combining these two inequalities we deduce that, for non-purely singular distributions (*i.e.* such that $\varphi_X \not\equiv 0$),

$$\underline{\lim}_M M^{-\frac{1}{d}} \sup_{f \in \mathrm{Lip}_1(\mathbb{R}^d, \mathbb{R})} \left| \frac{1}{M} \sum_{\ell=1}^{M} f(X_\ell) - \mathbb{E}\, f(X) \right| > 0.$$

This illustrates that the strong law of large numbers/Monte Carlo method is not as "dimension free" as is commonly admitted.

For recent results about the (non-asymptotic) behavior of $\mathbb{E}\, W_1(\mu_M^X, \mathbb{P}_X)$, we refer to [97]. It again emphasizes that the L^1-Wasserstein distance is not dimension free: in fact, the generic behavior of $\mathbb{E}\, W_1(\mu_M^X, \mathbb{P}_X)$ is $M^{-\frac{1}{d}}$ at least when $d \geq 3$.

7.4 Pricing Path-Dependent Options (I) (Lookback, Asian, etc)

Let us recall the notation $\mathbb{D}([0, T], \mathbb{R}^d) := \{\xi : [0, T] \to \mathbb{R}^d,\ \text{càdlàg}\}$. As a direct consequence of Theorem 7.2, if $F : \mathbb{D}([0, T], \mathbb{R}) \to \mathbb{R}$ is a Lipschitz continuous functional with respect to the sup norm, *i.e.* satisfies

$$|F(\xi) - F(\xi')| \leq [F]_{\mathrm{Lip}} \sup_{t \in [0,T]} \big|\xi(t) - \xi'(t)\big|,$$

then

$$\Big|\mathbb{E}\, F\big((X_t)_{t\in[0,T]}\big) - \mathbb{E}\, F\big((\bar{X}^n_{\underline{t}})_{t\in[0,T]}\big)\Big| \leq [F]_{\mathrm{Lip}} C_{b,\sigma,T} \sqrt{\frac{1 + \log n}{n}}$$

and

$$\Big|\mathbb{E}\, F\big((X_t)_{t\in[0,T]}\big) - \mathbb{E}\, F\big((\bar{X}^n_t)_{t\in[0,T]}\big)\Big| \leq C n^{-\frac{1}{2}}.$$

Typical example in option pricing. Assume that a one-dimensional diffusion process $X = (X_t)_{t\in[0,T]}$ models the dynamics of a single risky asset (we do not take into account here the consequences on the drift and diffusion coefficient term induced by the preservation of non-negativity nor the martingale property under a risk-neutral probability for the discounted asset…).

– The (partial) Lookback payoffs:

$$h_T := \left(X_T - \lambda \min_{t\in[0,T]} X_t\right)_+$$

where $\lambda = 1$ in the regular Lookback case and $\lambda > 1$ in the so-called "partial Lookback" case.

– Vanilla payoffs on supremum (like Calls and Puts) of the form

$$h_T = \varphi\Big(\sup_{t\in[0,T]} X_t\Big) \quad \text{or} \quad \varphi\Big(\inf_{t\in[0,T]} X_t\Big)$$

where φ is Lipschitz continuous on $\mathbb{R}_+$.

– Asian payoffs of the form

$$h_T = \varphi\left(\frac{1}{T-T_0}\int_{T_0}^{T} \psi(X_s)ds\right), \quad 0 \le T_0 < T,$$

where φ and ψ are Lipschitz continuous on $\mathbb{R}_+$. In fact, such Asian payoffs are even continuous with respect to the pathwise L^2-norm, *i.e.* $\|f\|_{L^2_T} := \sqrt{\int_0^T f^2(s)ds}$. In fact, the positivity of the payoffs plays no role here.

See Chap. 8 for improvements based on a weak error approach.

7.5 The Milstein Scheme (Looking for Better Strong Rates…)

Throughout this section we will consider an autonomous diffusion just for notational convenience ($b(t,\,.) = b$ and $\sigma(t,\,.) = \sigma$). The extension to general non-autonomous *SDE* s of the form (7.1) is straightforward (in particular it adds no further terms to the discretization scheme). The Milstein scheme has been originally designed (see. [214]) to produce a $O(1/n)$-error (in L^p), like…the Euler scheme for deterministic *ODE*s. However, in the framework of *SDE* s, such a scheme is of higher (second) order compared to the Euler scheme. In one dimension, its definition is simple and it can easily be implemented, provided the diffusion coefficient σ is differentiable (see below). In higher dimensions, several problems arise, of both a theoretical and practical (simulability) nature, making its use more questionable, especially when its performances are compared to the weak error rate satisfied by the Euler scheme (see Sect. 7.6).

7.5.1 The One Dimensional Setting

Let $X^x = (X^x_t)_{t\in[0,T]}$ denote the (autonomous) diffusion process starting at $x \in \mathbb{R}$ at time which is the 0 solution to the following *SDE* written in its integrated form

$$X^x_t = x + \int_0^t b(X^x_s)ds + \int_0^t \sigma(X^x_s)dW_s, \tag{7.31}$$

where we assume that the functions b and σ are twice differentiable with bounded existing derivatives (hence Lipschitz continuous).

The starting idea is to expand the solution X_t^x for small t in order to "select" the terms which go to zero at most as fast as t to 0 when $t \to 0$ (with respect to the $L^2(\mathbb{P})$-norm). Let us inspect the two integral terms successively. First,

$$\int_0^t b(X_s^x)ds = b(x)t + o(t) \quad \text{as} \quad t \to 0$$

since $X_t^x \to x$ as $t \to 0$ and b is continuous. Furthermore, as b is Lipschitz continuous and $\mathbb{E}\sup_{s\in[0,t]}|X_s^x - x|^2 \to 0$ as $t \to 0$ (see *e.g.* Proposition 7.7 further on), this expansion holds for the $L^2(\mathbb{P})$-norm as well (*i.e.* $o(t) = o_{L^2}(t)$).

As concerns the stochastic integral term, we keep in mind that $\mathbb{E}\,(W_{t+\Delta t} - W_t)^2 = \Delta t$ so that one may consider heuristically that, in a scheme, *a Brownian increment* $\Delta W_t := W_{t+\Delta t} - W_t$ *between* t *and* $t + \Delta t$ *behaves like* $\sqrt{\Delta t}$. Then, by Itô's Lemma (see Sect. 12.8), one has for every $s \in [0, t]$,

$$\sigma(X_s^x) = \sigma(x) + \int_0^s \Big(\sigma'(X_u^x)b(X_u^x) + \frac{1}{2}\sigma''(X_u^x)\sigma^2(X_u^x)\Big)du + \int_0^s \sigma'(X_u^x)\sigma(X_u^x)dW_u$$

so that

$$\int_0^t \sigma(X_s^x)dW_s = \sigma(x)W_t + \int_0^t\int_0^s \sigma(X_u^x)\sigma'(X_u^x)dW_u dW_s + O_{L^2}(t^{3/2}) \tag{7.32}$$

$$= \sigma(x)W_t + \sigma\sigma'(x)\int_0^t W_s dW_s + o_{L^2}(t) + O_{L^2}(t^{3/2}) \tag{7.33}$$

$$= \sigma(x)W_t + \frac{1}{2}\sigma\sigma'(x)(W_t^2 - t) + o_{L^2}(t) \tag{7.34}$$

since, by Itô's formula, $\int_0^t W_s dW_s = \frac{1}{2}(W_t^2 - t)$.

The $O_{L^2}(t^{3/2})$ in (7.32) comes from the fact that $u \mapsto \sigma'(X_u^x)b(X_u^x) + \frac{1}{2}\sigma''(X_u^x)\sigma^2(X_u^x)$ is $L^2(\mathbb{P})$-bounded in t in the neighborhood of 0 (note that b and σ have at most linear growth and use Proposition 7.2). Consequently, using Itô's fundamental isometry, the Fubini-Tonnelli Theorem and Proposition 7.2,

$$\begin{aligned}\mathbb{E}&\left[\int_0^t\int_0^s \Big(\sigma'(X_u^x)b(X_u^x)+\frac{1}{2}\sigma''(X_u^x)\sigma^2(X_u^x)\Big)du\, dW_s\right]^2\\ &= \mathbb{E}\int_0^t\left[\int_0^s \Big(\sigma'(X_u^x)b(X_u^x)+\frac{1}{2}\sigma''(X_u^x)\sigma^2(X_u^x)\Big)du\right]^2 ds\\ &\le C(1+x^4)\int_0^t\Big(\int_0^s du\Big)^2 ds = \frac{C}{3}(1+x^4)t^3.\end{aligned}$$

The $o_{L^2}(t)$ in Eq. (7.33) also follows from the combination of Itô's fundamental isometry (twice) and the Fubini-Tonnelli Theorem, which yields

$$\mathbb{E}\left(\int_0^t\int_0^s\left(\sigma\sigma'(X_u^x)-\sigma\sigma'(x)\right)dW_u dW_s\right)^2=\int_0^t\int_0^s\varepsilon(u)\,du\,ds,$$

where $\varepsilon(u)=\mathbb{E}\left(\sigma\sigma'(X_u^x)-\sigma\sigma'(x)\right)^2\to 0$ as $u\to 0$ by the Lebesgue dominated convergence Theorem. Finally note that, by scaling and using that $\mathbb{E}\,W_1^2=1$ and $\mathbb{E}\,W_1^4=3$,

$$\|W_t^2-t\|_2=t\|W_1^2-1\|_2=\sqrt{2}\,t$$

so that the second term in the right-hand side of (7.34) is exactly of order one. Then, X_t^x expands as follows

$$X_t^x=x+b(x)t+\sigma(x)W_t+\frac{1}{2}\sigma\sigma'(x)\big(W_t^2-t\big)+o_{L^2}(t).$$

Using the Markov property of the diffusion, one can reproduce the above reasoning on each time step $[t_k^n,t_{k+1}^n)$, *given the value of the scheme at time* t_k^n, this suggests to define the discrete time *Milstein* scheme $(\bar X_{t_k^n}^{mil,n})_{k=0,\ldots,n}$ with step $\frac{T}{n}$ as follows:

$$\begin{aligned}\bar X_0^{mil,n}&=X_0,\\ \bar X_{t_{k+1}^n}^{mil,n}&=\bar X_{t_k^n}^{mil,n}+\frac{T}{n}b\big(\bar X_{t_k^n}^{mil,n}\big)+\sigma\big(\bar X_{t_k^n}^{mil,n}\big)\sqrt{\frac{T}{n}}\,Z_{k+1}^n\\ &\quad+\frac{1}{2}\sigma\sigma'\big(\bar X_{t_k^n}^{mil,n}\big)\frac{T}{n}\Big((Z_{k+1}^n)^2-1\Big),\end{aligned}\tag{7.35}$$

where

$$Z_k^n=\sqrt{\frac{n}{T}}(W_{t_k^n}-W_{t_{k-1}^n}),\ k=1,\ldots,n.$$

Or, equivalently, by grouping the drift terms together,

$$\begin{aligned}\bar X_0^{mil,n}&=X_0,\\ \bar X_{t_{k+1}^n}^{mil,n}&=\bar X_{t_k^n}^{mil,n}+\Big(b\big(\bar X_{t_k^n}^{mil,n}\big)-\frac{1}{2}\sigma\sigma'\big(\bar X_{t_k^n}^{mil,n}\big)\Big)\frac{T}{n}+\sigma\big(\bar X_{t_k^n}^{mil,n}\big)\sqrt{\frac{T}{n}}\,Z_{k+1}^n\\ &\quad+\frac{1}{2}\sigma\sigma'\big(\bar X_{t_k^n}^{mil,n}\big)\frac{T}{n}(Z_{k+1}^n)^2.\end{aligned}\tag{7.36}$$

Like for the Euler scheme, the *stepwise constant Milstein scheme* is defined as

$$\widetilde X_t^{mil,n}=\bar X_{\underline{t}}^{mil,n},\quad t\in[0,T].\tag{7.37}$$

In what follows, when no ambiguity arises, we will often drop the superscript ^{n} in the notation of the Milstein scheme(s).

By interpolating the above scheme between the discretization times, *i.e.* freezing the coefficients of the scheme, we define the *continuous* or *genuine Milstein scheme* with step $\frac{T}{n}$ defined, with our standard notations $\underline{t}$, by

$$\begin{aligned}\bar{X}_0^{mil} &= X_0,\\ \bar{X}_t^{mil} &= \bar{X}_{\underline{t}}^{mil} + \big(b(\bar{X}_{\underline{t}}^{mil}) - \frac{1}{2}\sigma\sigma'(\bar{X}_{\underline{t}}^{mil})\big)(t-\underline{t}) + \sigma(\bar{X}_{\underline{t}}^{mil})(W_t - W_{\underline{t}})\\ &\quad + \frac{1}{2}\sigma\sigma'(\bar{X}_{\underline{t}}^{mil})(W_t - W_{\underline{t}})^2 \qquad (7.38)\end{aligned}$$

for every $t \in [0, T]$.

$\triangleright$ **Example (Black–Scholes model).** The discrete time Milstein scheme of a Black–Scholes model starting at $x_0 > 0$, with interest rate r and volatility $\sigma > 0$ over $[0, T]$ reads

$$\bar{X}_{t_{k+1}^n}^{mil} = \bar{X}_{t_k^n}^{mil}\left(1 + \Big(r - \frac{\sigma^2}{2}\Big)\frac{T}{n} + \sigma\sqrt{\frac{T}{n}}\,Z_{k+1}^n + \frac{\sigma^2 T}{2n}(Z_{k+1}^n)^2\right). \qquad (7.39)$$

The following theorem gives the rate of strong pathwise convergence of the Milstein scheme under slightly less stringent hypothesis than the usual $\mathcal{C}^1_{\text{Lip}}$-assumptions on b and σ.

Theorem 7.5 (Strong L^p-rate for the Milstein scheme) *(See e.g. [170]) Assume that b and σ are $\mathcal{C}^1$ on $\mathbb{R}$ with bounded, $\alpha_{b'}$ and $\alpha_{\sigma'}$-Hölder continuous first derivatives respectively, $\alpha_{b'}, \alpha_{\sigma'} \in (0, 1]$. Let $\alpha = \min(\alpha_{b'}, \alpha_{\sigma'})$.*

(a) Discrete time and genuine Milstein scheme. *For every $p \in (0, +\infty)$, there exists a real constant $C_{b,\sigma,T,p} > 0$ such that, for every $X_0 \in L^p(\mathbb{P})$, independent of the Brownian motion W, one has*

$$\Big\| \max_{0\le k\le n} \big|X_{t_k^n} - \bar{X}_{t_k^n}^{mil,n}\big| \Big\|_p \le \Big\| \sup_{t\in[0,T]} \big|X_t - \bar{X}_t^{mil,n}\big| \Big\|_p \le C_{b,\sigma,T,p}\big(1 + \|X_0\|_p\big)\Big(\frac{T}{n}\Big)^{\frac{1+\alpha}{2}}.$$

In particular, if b' and σ' are Lipschitz continuous

$$\Big\| \max_{0\le k\le n} \big|X_{t_k^n} - \bar{X}_{t_k^n}^{mil,n}\big| \Big\|_p \le \Big\| \sup_{t\in[0,T]} \big|X_t - \bar{X}_t^{mil,n}\big| \Big\|_p \le C_{p,b,\sigma,T}\big(1 + \|X_0\|_p\big)\frac{T}{n}.$$

(b) Stepwise constant Milstein scheme. *As concerns the stepwise constant Milstein scheme $\big(\widetilde{X}_t^{mil,n}\big)_{t\in[0,T]}$ defined by (7.37), one has (like for the Euler scheme!)*

$$\Big\| \sup_{t\in[0,T]} \big|X_t - \widetilde{X}_t^{mil,n}\big| \Big\|_p \le C_{b,\sigma,T}(1 + \|X_0\|_p)\sqrt{\frac{T}{n}(1 + \log n)}.$$

A detailed proof is provided in Sect. 7.8.8 in the case $X_0 = x \in \mathbb{R}$ and $p \in [2, +\infty)$.

Remarks. • As soon as the derivatives of b' and σ' are Hölder, the genuine Milstein scheme converges faster than the Euler scheme. Furthermore, if b and σ are $\mathcal{C}^1$ with Lipschitz continuous derivatives, the L^p-convergence rate of the genuine Milstein scheme is, as expected, $O\big(\frac{1}{n}\big)$.

• This $O(\frac{1}{n})$-rate obtained when b' and σ' are (bounded and) Lipschitz continuous should be compared to the *weak rate* investigated in Sect. 7.6, which is also $O(\frac{1}{n})$ for the approximation of $\mathbb{E}\, f(X_T)$ by $\mathbb{E}\, f(\bar{X}^n_T)$ (under sightly more stringent assumptions). Comparing the performances of both approaches should rely on numerical evidence and depends on the specified diffusion, function or step parameter.

• Claim (b) of the theorem shows that the stepwise constant Milstein scheme does not converge faster than the stepwise constant Euler scheme (except, of course, at the discretization times t^n_k). To convince yourself of this, just to think of the Brownian motion itself: in that case, $b \equiv 0$ and $\sigma' \equiv 0$, so that both the stepwise constant Milstein and Euler schemes coincide and consequently converge at the same rate! As a consequence, since it is the only simulable version of the Milstein scheme when dealing with *path-dependent functionals*, its use for the approximate computation (by Monte Carlo simulation) of the expectation of the form $\mathbb{E}\, F\big((X_t)_{t\in[0,T]}\big)$ should not provide better results than implementing the standard stepwise constant Euler scheme, as briefly described in Sect. 7.4.

By contrast, some functionals of the continuous Euler scheme can be simulated in an exact way: this is the purpose of Chap. 8 devoted to diffusion bridges. This is not the case for the Milstein scheme.

▶ **Exercises. 1.** *A.s. convergence of the Milstein scheme.* Derive from these L^p-rates an $a.s.$ rate of convergence for the Milstein scheme.

2. *Euler scheme of the Ornstein-Uhlenbeck process.* We consider on the one hand the sequence of random variables recursively defined by

$$Y_{k+1} = Y_k(1 + \mu\Delta) + \sigma\sqrt{\Delta}\, Z_{k+1},\ k \ge 0,\ Y_0 = 0,$$

where $\mu > 0$, $\Delta > 0$ are positive real numbers, and the Ornstein-Uhlenbeck process solution to the *SDE*

$$dX_t = \mu X_t dt + \sigma dW_t,\ X_0 = 0.$$

Set $t_k = k\Delta$, $k \ge 0$ and $Z_k = \frac{W_{t_k} - W_{t_{k-1}}}{\sqrt{\Delta}}$, $k \ge 1$.

(a) Show that, for every $k > 0$,

$$\mathbb{E}\,|Y_k|^2 = \frac{\sigma^2}{\mu}\,\frac{(1+\mu\Delta)^{2k} - 1}{2 + \mu\Delta}.$$

(b) Show that, for every $k \ge 0$, $X_{t_{k+1}} = e^{\mu\Delta} X_{t_k} + \sigma e^{\mu t_{k+1}} \displaystyle\int_{t_k}^{t_{k+1}} e^{-\mu s} dW_s$.

(c) Show that, for every $k \geq 0$, for every $\Delta_n > 0$

$$\begin{aligned}\mathbb{E}|Y_{k+1} - X_{t_{k+1}}|^2 \leq & (1 + \Delta_n)\, e^{2\mu\Delta}\mathbb{E}|Y_k - X_{t_k}|^2 \\ & + (1 + 1/\Delta_n)\, \mathbb{E}\,|Y_k|^2\big(e^{\mu\Delta} - 1 - \mu\Delta\big)^2 \\ & + \sigma^2 \int_0^{\Delta} (e^{\mu u} - 1)^2 du.\end{aligned}$$

In what follows we assume that $\Delta = \Delta_n = \frac{T}{n}$ where T is a positive real number and n is a non-zero integer which may vary. However, we keep on using the notation Y_k rather than $Y_k^{(n)}$.

(d) Show that, for every $k \in \{0, \ldots, n\}$, $\mathbb{E}\,|Y_k|^2 \leq \dfrac{\sigma^2 e^{2\mu T}}{2\mu}$.

(e) Deduce the existence of a real constant $C = C_{\mu,\sigma,T} > 0$ such that, for every $k \in \{0, \ldots, n-1\}$,

$$\mathbb{E}|Y_{k+1} - X_{t_{k+1}}|^2 \leq (1 + \Delta_n)e^{\mu\Delta_n}\mathbb{E}|Y_k - X_{t_k}|^2 + C\Delta_n^3.$$

Conclude that $\mathbb{E}|Y_{k+1} - X_{t_{k+1}}|^2 \leq C'\left(\frac{T}{n}\right)^2$ for some real constant $C' = C'_{\mu,\sigma,T} > 0$.

3. Revisit the above exercise when $\mu \in (-\frac{2}{\Delta}, 0)$? What can be said about the dependence of the real constants in T when $\mu \in (-\frac{2}{\Delta}, -\frac{1}{\Delta})$?

4. *Milstein scheme may preserve positivity.* We consider with the usual notations the scalar *SDE*

$$dX_t = b(t, X_t)dt + \sigma(t, X_t)dW_t, \;\; X_0 = x > 0, \;\; t \in [0, T]$$

with drift $b : [0, T] \times \mathbb{R} \to \mathbb{R}$ and diffusion coefficient $\sigma : [0, T] \times \mathbb{R} \to \mathbb{R}$, both assumed continuously differentiable in x on $[0, T] \times (0, +\infty)$. We do not prejudge the existence of a strong solution.

(a) Show that, if

$$\forall\, t \in [0, T], \;\; \forall\, \xi \in (0, +\infty), \;\; \begin{cases} (i) & \sigma(t, \xi) > 0, \;\; \sigma'_x(t, \xi) > 0, \\ (ii) & \dfrac{\sigma}{2\,\sigma'_x}(t, \xi) \leq \xi, \\ (iii) & \dfrac{\sigma\sigma'_x(t, \xi)}{2} \leq b(t, \xi), \end{cases} \tag{7.40}$$

then the discrete time Milstein scheme with step $\frac{T}{n}$ starting at $x > 0$ defined by (7.36) satisfies

$$\forall\, k \in \{0, \ldots, n\}, \;\; \bar{X}^{mil}_{t_k^n} > 0 \;\; a.s.$$

[Hint: decompose the right-hand side of Milstein scheme on (7.35) as the sum of three terms, one being a square.]
(b) Show that, under Assumption (7.40), the genuine Milstein scheme starting at $X > 0$ also satisfies $\bar{X}_t^{mil} > 0$ for every $t \in [0, T]$.
(c) Show that the Milstein scheme of a Black–Scholes model is positive if and only if $\sigma^2 \leq 2r$.

5. *Milstein scheme of a positive CIR process.* We consider a *CIR* process solution to the *SDE*

$$dX_t = \kappa(a - X_t)dt + \vartheta\sqrt{X_t}\,dW_t, \quad X_0 = x > 0, \ t \geq 0,$$

where $\kappa, a, \vartheta > 0$ and $\frac{\vartheta^2}{2\kappa a} \leq 1$. Such an *SDE* has a unique strong solution $(X_t^x)_{t\geq 0}$ living in $(0, +\infty)$ (see [183], Proposition 6.2.4, p.130). We set $Y_t = e^{\kappa t} X_t^x$, $t \geq 0$.
(a) Show that the Milstein scheme of the process $(Y_t)_{t\geq 0}$ is $a.s.$ positive.
(b) Deduce a way to devise a positive simulable discretization scheme for the *CIR* process (the convergence properties of this process are not requested).

6. *Extension.* We return to the setting of the above exercise **2.** We want to relax the assumption on the drift b. We still assume that the conditions (i)–(ii) in (7.40) on the function σ are satisfied and we formally set $Y_t = e^{\rho t} X_t^x$, $t \in [0, T]$, for some $\rho > 0$. Show that $(X_t^x)_{t\in[0,T]}$ is a solution to the above *SDE* if and only if $(Y_t)_{t\in[0,T]}$ is a solution to a stochastic differential equation

$$dY_t = \widetilde{b}(t, Y_t) + \widetilde{\sigma}(t, Y_t)dW_t, \quad Y_0 = x > 0,$$

where $\widetilde{b}, \widetilde{\sigma} : [0, T] \times \mathbb{R} \to \mathbb{R}$ are functions depending on b, σ and ρ to be determined.
(a) Show that $\widetilde{\sigma}$ still satisfies (i)–(ii) in (7.40).
(b) Deduce that if $\rho_0 = \sup_{\xi>0} \frac{1}{\xi}\Big(\frac{\sigma\sigma'_x(t,\xi)}{2} - b(t, \xi)\Big) < +\infty$, then for every $\rho \geq \rho_0$, the Milstein scheme of Y is $a.s.$ positive

These exercises enhance a major asset of the one-dimensional Milstein scheme, keeping in mind financial applications to pricing and hedging of derivatives: it may be used to preserve positivity. Obvious applications to the *CIR* model in fixed income (interest rates) or the Heston model for Equity derivatives.

7.5.2 Higher-Dimensional Milstein Scheme

Let us examine what the above expansion in small time becomes if the *SDE* (7.31) is modified to be driven by a 2-dimensional Standard Brownian motion $W = (W^1, W^2)$ (still with $d = 1$). It reads

$$dX_t^x = b(X_t^x)dt + \sigma_1(X_t^x)dW_t^1 + \sigma_2(X_t^x)dW_t^2, \quad X_0^x = x.$$

The same reasoning as that carried out above shows that the first order term $\sigma(x)\sigma'(x)\int_0^t W_s dW_s$ in (7.33) becomes

$$\sum_{i,j=1,2} \sigma_i'(x)\sigma_j(x)\int_0^t W_s^i dW_s^j.$$

In particular, when $i \neq j$, this term involves the two Lévy areas $\int_0^t W_s^1 dW_s^2$ and $\int_0^t W_s^2 dW_s^1$, linearly combined with, *a priori*, different coefficients.

If we return to the general setting of a d-dimensional diffusion driven by a q-dimensional standard Brownian motion, with (differentiable) drift $b : \mathbb{R}^d \to \mathbb{R}^d$ and diffusion coefficient $\sigma = [\sigma_{ij}] : \mathbb{R}^d \to \mathcal{M}(d, q, \mathbb{R})$, elementary though tedious computations lead us to define the (discrete time) Milstein scheme with step $\frac{T}{n}$ as follows:

$$\begin{aligned}
\bar{X}_0^{mil} &= X_0,\\
\bar{X}_{t_{k+1}^n}^{mil} &= \bar{X}_{t_k^n}^{mil} + \frac{T}{n} b(\bar{X}_{t_k^n}^{mil})\\
&\quad + \sigma(\bar{X}_{t_k^n}^{mil})\Delta W_{t_{k+1}^n} + \sum_{1\le i,j\le q} \partial\sigma_{.i}\sigma_{.j}(\bar{X}_{t_k^n}^{mil})\int_{t_k^n}^{t_{k+1}^n} (W_s^i - W_{t_k^n}^i) dW_s^j, \qquad (7.41)\\
&\qquad k = 0, \ldots, n-1,
\end{aligned}$$

where $\Delta W_{t_{k+1}^n} := W_{t_{k+1}^n} - W_{t_k^n} = \sqrt{\frac{T}{n}} Z_{k+1}^n$, $\sigma_{.i}(x)$ denotes the i-th column of the matrix σ and, for every $i, j \in \{1, \ldots, q\}$,

$$\forall\, x = (x^1, \ldots, x^d) \in \mathbb{R}^d, \quad \partial\sigma_{.i}\,\sigma_{.j}(x) := \sum_{\ell=1}^{d} \frac{\partial\sigma_{.i}}{\partial x^\ell}(x)\sigma_{\ell j}(x) \in \mathbb{R}^d. \qquad (7.42)$$

Remark. A more synthetic way to memorize this quantity is to note that it is the Jacobian matrix of the vector $\sigma_{.i}(x)$ applied to the vector $\sigma_{.j}(x)$.

The ability of simulating such a scheme entirely relies on the exact simulations

$$\left(W_{t_k^n}^1 - W_{t_{k-1}^n}^1, \ldots, W_{t_k^n}^q - W_{t_{k-1}^n}^q, \int_{t_{k-1}^n}^{t_k^n} (W_s^i - W_{t_k^n}^i) dW_s^j,\ i, j = 1, \ldots, q,\ i \neq j\right),$$
$$k = 1, \ldots, n,$$

i.e. of identical copies of the q^2-dimensional random vector

$$\left(W_t^1, \ldots, W_t^q, \int_0^t W_s^i dW_s^j, 1 \le i, j \le q, i \ne j\right)$$

(at $t = \frac{T}{n}$). To the best of our knowledge no convincing method (*i.e.* with a reasonable computational cost) to achieve this has been proposed so far in the literature (see however [102]).

The discrete time Milstein scheme can be successfully simulated when the tensors terms $\partial\sigma_{.i}\,\sigma_{.j}$ commute since in that case the Lévy areas disappear, as shown in the following proposition.

Proposition 7.5 (Commuting case) (*a*) *If the tensor terms* $\partial\sigma_{.i}\,\sigma_{.j}$ *commute, i.e. if*

$$\forall i,\ j \in \{1, \ldots, q\}, \qquad \partial\sigma_{.i}\,\sigma_{.j} = \partial\sigma_{.j}\,\sigma_{.i},$$

then the discrete time Milstein scheme reduces to

$$\bar{X}^{mil}_{t^n_{k+1}} = \bar{X}^{mil}_{t^n_k} + \frac{T}{n}\left(b(\bar{X}^{mil}_{t^n_k}) - \frac{1}{2}\sum_{i=1}^{q}\partial\sigma_{.i}\sigma_{.i}(\bar{X}^{mil}_{t^n_k})\right) + \sigma(\bar{X}^{mil}_{t^n_k})\Delta W_{t^n_{k+1}}$$
$$+\frac{1}{2}\sum_{1\le i,j\le q}\partial\sigma_{.i}\,\sigma_{.j}(\bar{X}^{mil}_{t^n_k})\Delta W^i_{t^n_{k+1}}\Delta W^j_{t^n_{k+1}}, \quad \bar{X}^{mil}_0 = X_0. \tag{7.43}$$

(*b*) *When* $q = 1$ *the commutation property is trivially satisfied.*

Proof. Let $i \ne j$ in $\{1, \ldots, q\}$. As $\partial\sigma_{.i}\,\sigma_{.j} = \partial\sigma_{.j}\,\sigma_{.i}$, both Lévy's areas involving W^i and W^j only appear through their sum in the scheme. Now, an integration by parts shows that

$$\int_{t^n_k}^{t^n_{k+1}}(W^i_s - W^i_{t^n_k})dW^j_s + \int_{t^n_k}^{t^n_{k+1}}(W^j_s - W^j_{t^n_k})dW^i_s = \Delta W^i_{t^n_{k+1}}\Delta W^j_{t^n_{k+1}}$$

since $(W^i_t)_{t\in[0,T]}$ and $(W^j_t)_{t\in[0,T]}$ are independent if $i \ne j$. The result follows noting that, for every $i \in \{1, \ldots, q\}$,

$$\int_{t^n_k}^{t^n_{k+1}}(W^i_s - W^i_{t^n_k})dW^i_s = \frac{1}{2}\left((\Delta W^i_{t^n_{k+1}})^2 - \frac{T}{n}\right). \qquad \diamond$$

$\triangleright$ **Example** (*Multi-dimensional Black–Scholes model*). We consider the standard multi-dimensional Black–Scholes model

$$dX^i_t = X^i_t\Big(r\,dt + \sum_{j=1}^{q}\sigma_{ij}dW^j_t\Big),\ X^i_0 = x^i_0 > 0,\ i = 1, \ldots, d,$$

where $W = (W^1, \ldots, W^q)$ a q-dimensional Brownian motion. Then, elementary computations yield the following expression for the tensor terms $\partial\sigma_{.i}\,\sigma_{.j}$

$$\forall i,\ j \in \{1, \ldots, q\}, \quad \partial \sigma_{.i}\, \sigma_{.j}(x) = \big[\sigma_{\ell i}\sigma_{\ell j} x^{\ell}\big]_{\ell=1,\ldots,d}$$

which obviously commute.

The rate of convergence of the Milstein scheme is formally the same in higher dimension as it is in one dimension: Theorem 7.5 remains true with a d-dimensional diffusion driven by a q-dimensional Brownian motion provided $b : \mathbb{R}^d \to \mathbb{R}^d$ and $\sigma : \mathbb{R}^d \to \mathcal{M}(d, q, \mathbb{R})$ are $\mathcal{C}^2$ with bounded existing partial derivatives.

Theorem 7.6 (Multi-dimensional discrete time Milstein scheme) *(See e.g. [170]) Assume that b and σ are $\mathcal{C}^1$ on $\mathbb{R}^d$ with bounded $\alpha_{b'}$ and $\alpha_{\sigma'}$-Hölder continuous existing partial derivatives, respectively. Let $\alpha = \min(\alpha_{b'}, \alpha_{\sigma'})$. Then, for every $p \in (0, +\infty)$, there exists a real constant $C_{p,b,\sigma,T} > 0$ such that for every $X_0 \in L^p(\mathbb{P})$, independent of the q-dimensional Brownian motion W, the error bound established in Theorem 7.5(a) remains valid.*

However, one should keep in mind that this strong rate result does not prejudge the ability to simulate this scheme. In a way, the most important consequence of this theorem concerns the Euler scheme.

Corollary 7.2 (Discrete time Euler scheme with constant diffusion coefficient) *If the drift $b \in \mathcal{C}^2(\mathbb{R}^d, \mathbb{R}^d)$ with bounded existing partial derivatives and if $\sigma(x) = \Sigma$ is constant, then the discrete time Euler and Milstein schemes coincide. As a consequence, the strong rate of convergence of the discrete time Euler scheme is, in that very specific case, $O(\frac{1}{n})$. Namely, for every $p \in (0, +\infty)$, there exists a real constant $C_{p,b,\sigma,T} > 0$*

$$\Big\| \max_{0 \le k \le n} |X_{t_k^n} - \bar{X}^n_{t_k^n}| \Big\|_p \le C_{p,b,\sigma,T} \frac{T}{n}\big(1 + \|X_0\|_p\big).$$

7.6 Weak Error for the Discrete Time Euler Scheme (I)

In many situations, like the pricing of "vanilla" European options, a discretization scheme $\bar{X}^n = (\bar{X}^n_t)_{t\in[0,T]}$ of a d-dimensional diffusion process $X = (X_t)_{t\in[0,T]}$ is introduced in order to compute by a Monte Carlo simulation an approximation $\mathbb{E}\, f(\bar{X}_T)$ of $\mathbb{E}\, f(X_T)$, *i.e.* only at a fixed (terminal) time. If one relies on the former *strong* rates of convergence, we get, as soon as $f : \mathbb{R}^d \to \mathbb{R}$ is *Lipschitz continuous* and b, σ satisfy (H_T^{β}) for $\beta \ge \frac{1}{2}$,

$$\begin{aligned}\Big|\mathbb{E}\, f(X_T) - \mathbb{E}\, f(\bar{X}^n_T)\Big| &\le [f]_{\text{Lip}}\mathbb{E}\,\big|X_T - \bar{X}^n_T\big| \le [f]_{\text{Lip}}\mathbb{E}\sup_{t\in[0,T]}\big|X_t - \bar{X}^n_t\big| \\ &= O\Big(\frac{1}{\sqrt{n}}\Big).\end{aligned}$$

In fact, the first inequality in this chain turns out to be highly non-optimal since it switches from a *weak* error (the difference only depending on the respective (marginal) distributions of X_T and $\bar{X}^n_T$) to a pathwise approximation $X_T - \bar{X}^n_T$ (4). To improve asymptotically the other two inequalities is hopeless since it has been shown (see the remark and comments in Sect. 7.8.6 for a brief introduction) that, under appropriate assumptions, $X_T - \bar{X}^n_T$ satisfies a central limit theorem at rate $n^{-\frac{1}{2}}$ with non-zero asymptotic variance. In fact, one even has a functional form of this central limit theorem for the whole process $(X_t - \bar{X}^n_t)_{t\in[0,T]}$ (see [155, 178]) still at rate $n^{-\frac{1}{2}}$. As a consequence, a rate faster than $n^{-\frac{1}{2}}$ in an L^1 sense would not be consistent with this central limit result.

Furthermore, numerical experiments confirm that the *weak rate of convergence* between the above two expectations is usually much faster than $n^{-\frac{1}{2}}$. This fact has long been known and has been extensively investigated in the literature, starting with the two seminal papers [270] by Talay–Tubaro and [24] by Bally–Talay, leading to an expansion of the time discretization error at an arbitrary accuracy when b and σ are smooth enough as functions. Two main settings have been investigated: when the function f is itself smooth and when the diffusion is "regularizing", *i.e.* propagates the regularizing effect of the driving Brownian motion (5) thanks to a non-degeneracy assumption on the diffusion coefficient σ, typically uniform ellipticity for σ (see below) or weaker assumption such as parabolic Hörmander (hypo-ellipticity) assumption (see [139] (6)).

The same kind of question has been investigated for specific classes of path-dependent functionals F of the diffusion X with some applications to exotic option pricing (see Chap. 8). These results, though partial and specific, often emphasize that the resulting *weak error rate* is the same as the strong rate derived from the Milstein scheme for these types of functionals, especially when they are Lipschitz continuous with respect to the sup-norm.

As a second step, we will show how the Richardson–Romberg extrapolation method provides a first procedure to take advantage of such weak rates, before fully exploiting a higher-order weak error rates expansion in Chap. 9 with the multilevel paradigm.

---

[4] *A priori* $\bar{X}^n_T$ and X_T could be defined on different probability spaces: recall the approximation of the Black–Scholes model by binomial models (see [59]).

[5] The regularizing property of the Brownian motion should be understood in its simplest form as follows: if f is a Borel bounded function on $\mathbb{R}^d$, then $f_\sigma(x) := \mathbb{E}\left(f(x+\sigma W_1)\right)$ is a $\mathcal{C}^\infty$ function for every $\sigma > 0$ and converges towards f as $\sigma \to 0$ in every L^p space, $p > 0$. This result is just a classical convolution result with a Gaussian kernel rewritten in a probabilistic form.

[6] Given in detail this condition is beyond the scope of this monograph. However, let us mention that if the column vectors $(\sigma_{.j}(x))_{1\le j\le q}$ span $\mathbb{R}^d$ for every $x \in \mathbb{R}^d$, this condition is satisfied. If not, the same spanning property is requested, after adding enough iterated Lie brackets of the coefficients of the *SDE*—re-written in a Stratonovich sense—and including the drift this time.

7.6.1 Main Results for $\mathbb{E}\, f(X_T)$: the Talay–Tubaro and Bally–Talay Theorems

We adopt the notations of the former Sect. 7.1, except that we still consider, for convenience, an autonomous version of the *SDE*, with initial condition $x \in \mathbb{R}^d$,

$$dX_t^x = b(X_t^x)dt + \sigma(X_t^x)dW_t, \quad X_0^x = x.$$

The notations $(X_t^x)_{t\in[0,T]}$ and $(\bar{X}_t^{n,x})_{t\in[0,T]}$ respectively denote the diffusion and the Euler scheme of the diffusion with step $\frac{T}{n}$ of the diffusion starting at x at time 0 (the superscript ^{n} will often be dropped).

The first result is the simplest result on the weak error, obtained under less stringent assumptions on b and σ.

Theorem 7.7 (see *[270]*) *Assume b and σ are four times continuously differentiable on $\mathbb{R}^d$ with bounded existing partial derivatives (this implies that b and σ are Lipschitz continuous). Assume $f : \mathbb{R}^d \to \mathbb{R}$ is four times differentiable with polynomial growth as well as its existing partial derivatives. Then, for every $x \in \mathbb{R}^d$,*

$$\mathbb{E}\, f(X_T^x) - \mathbb{E}\, f(\bar{X}_T^{n,x}) = O\left(\frac{1}{n}\right) \quad as \quad n \to +\infty. \tag{7.44}$$

Proof (partial). Assume $d = 1$ for notational convenience. We also assume that $b \equiv 0$, σ is bounded and f has bounded first four derivatives, for simplicity. The diffusion $(X_t^x)_{t\geq 0, x\in\mathbb{R}}$ is a homogeneous Markov process with transition semi-group $(P_t)_{t\geq 0}$ (see *e.g.* [162, 251] among other references) reading on Borel test functions g (*i.e.* bounded or non-negative)

$$P_t\, g(x) := \mathbb{E}\, g(X_t^x), \ t \geq 0, \ x \in \mathbb{R}.$$

On the other hand, the Euler scheme with step $\frac{T}{n}$ starting at $x \in \mathbb{R}$, denoted by $(\bar{X}_{t_k^n}^x)_{0\leq k\leq n}$, is a discrete time homogeneous Markov chain with transition reading on Borel test functions g

$$\bar{P}\, g(x) = \mathbb{E}\, g\left(x + \sigma(x)\sqrt{\frac{T}{n}}\, Z\right), \qquad Z \stackrel{d}{=} \mathcal{N}(0; 1).$$

To be more precise, this means for the diffusion process that, for any Borel test function g,

$$\forall\, s, t \geq 0, \quad P_t g(x) = \mathbb{E}\left(g(X_{s+t}) \,|\, X_s = x\right) = \mathbb{E}\, g(X_t^x)$$

and for its Euler scheme (still with $t_k^n = \frac{kT}{n}$)

$$\bar{P}\, g(x) = \mathbb{E}\left(g(\bar{X}^x_{t^n_{k+1}}) \mid \bar{X}^x_{t^n_k} = x\right) = \mathbb{E}\, g(\bar{X}^x_{\frac{T}{n}}),\ k = 0, \ldots, n-1.$$

Now, let us consider the four times differentiable function f. One gets, by the semi-group property satisfied by both P_t and $\bar{P}$,

$$\mathbb{E}\, f(X^x_T) = P_T f(x) = P^n_{\frac{T}{n}}(f)(x) \quad \text{and} \quad \mathbb{E}\, f(\bar{X}^x_T) = \bar{P}^n(f)(x).$$

Then, by writing the difference $\mathbb{E}\, f(X^x_T) - \mathbb{E}\, f(\bar{X}^x_T)$ in a telescopic way, switching from $P^n_{\frac{T}{n}}(f)$ to $\bar{P}^n(f)$, we obtain

$$\begin{aligned}\mathbb{E}\, f(X^x_T) - \mathbb{E}\, f(\bar{X}^x_T) &= \sum_{k=1}^{n} P^k_{\frac{T}{n}}(\bar{P}^{n-k} f)(x) - P^{k-1}_{\frac{T}{n}}(\bar{P}^{n-(k-1)} f)(x)\\ &= \sum_{k=1}^{n} P^{k-1}_{\frac{T}{n}}\big((P_{\frac{T}{n}} - \bar{P})(\bar{P}^{n-k} f)\big)(x). \qquad (7.45)\end{aligned}$$

This "domino" sum suggests two tasks to be accomplished:

– the first is to estimate precisely the asymptotic behavior of

$$P_{\frac{T}{n}} f(x) - \bar{P} f(x)$$

with respect to the step $\frac{T}{n}$ and the first (four) derivatives of the function g.

– the second is to control the (first four) derivatives of the functions $\bar{P}^\ell f$ with respect to the sup norm, uniformly in $\ell \in \{1, \ldots, n\}$ and $n \geq 1$, in order to propagate the above local error bound.

Let us deal with the first task. First, Itô's formula (see Sect. 12.8) yields

$$P_t f(x) := \mathbb{E}\, f(X^x_t) = f(x) + \underbrace{\mathbb{E} \int_0^t (f'\sigma)(X^x_s) dW_s}_{=0} + \frac{1}{2}\mathbb{E} \int_0^t f''(X^x_s)\sigma^2(X^x_s)\, ds,$$

where we use that $f'\sigma$ is bounded to ensure that the stochastic integral is a true martingale.

A Taylor expansion of $f\Big(x + \sigma(x)\sqrt{\frac{T}{n}}\, Z\Big)$ at x yields for the transition of the Euler scheme (after taking expectation)

$$
\begin{aligned}
\bar{P} f(x) &= \mathbb{E}\, f(\bar{X}^x_{_T}) \\
&= f(x) + f'(x)\sigma(x)\mathbb{E}\left(\sqrt{\frac{T}{n}}\, Z\right) + \frac{1}{2}(f''\sigma^2)(x)\mathbb{E}\left(\sqrt{\frac{T}{n}}\, Z\right)^2 \\
&\quad + f^{(3)}(x)\frac{\sigma^3(x)}{3!}\mathbb{E}\left(\sqrt{\frac{T}{n}}\, Z\right)^3 + \frac{\sigma^4(x)}{4!}\mathbb{E}\left(f^{(4)}(\xi)\left(\sqrt{\frac{T}{n}}\, Z\right)^4\right), \\
&\qquad\qquad\qquad\qquad\qquad\qquad\qquad\qquad \text{for a } \xi \in \left(x, \bar{X}_{_T}\right) \\
&= f(x) + \frac{T}{2n}(f''\sigma^2)(x) + \frac{\sigma^4(x)}{4!}\mathbb{E}\left(f^{(4)}(\xi)\left(\sqrt{\frac{T}{n}}\, Z\right)^4\right) \\
&= f(x) + \frac{T}{2n}(f''\sigma^2)(x) + \frac{\sigma^4(x)T^2}{4!n^2} c_n(f),
\end{aligned}
$$

where $|c_n(f)| \le 3\|f^{(4)}\|_{\sup}$. This follows from the well-known facts that $\mathbb{E}\, Z = \mathbb{E}\, Z^3 = 0$, $\mathbb{E}\, Z^2 = 1$ and $\mathbb{E}\, Z^4 = 3$. Consequently

$$
\begin{aligned}
P_{\frac{T}{n}} f(x) - \bar{P} f(x) = \frac{1}{2}\int_0^{\frac{T}{n}} \mathbb{E}\left((f''\sigma^2)(X^x_s) - (f''\sigma^2)(x)\right)ds& \\
- \frac{\sigma^4(x)T^2}{4!n^2} c_n(f).&
\end{aligned}
\tag{7.46}
$$

Applying again Itô's formula to the $\mathcal{C}^2$ function $\gamma := f''\sigma^2$ yields

$$
\mathbb{E}\left((f''\sigma^2)(X^x_s) - (f''\sigma^2)(x)\right) = \frac{1}{2}\mathbb{E}\left(\int_0^s \gamma''(X^x_u)\sigma^2(X^x_u)du\right)
$$

so that

$$
\forall\, s \ge 0, \qquad \sup_{x\in\mathbb{R}}\left|\mathbb{E}\left((f''\sigma^2)(X^x_s) - (f''\sigma^2)(x)\right)\right| \le \frac{s}{2}\|\gamma''\sigma^2\|_{\sup}.
$$

Elementary computations show that

$$
\|\gamma''\|_{\sup} \le C_\sigma \max_{k=2,3,4} \|f^{(k)}\|_{\sup},
$$

where C_σ depends on $\|\sigma^{(k)}\|_{\sup}$, $k = 0, 1, 2$, but not on f (with the standard convention $\sigma^{(0)} = \sigma$).

Consequently, we derive from (7.46) that

$$
\left|P_{\frac{T}{n}}(f)(x) - \bar{P}(f)(x)\right| \le C'_{\sigma,T} \max_{k=2,3,4} \|f^{(k)}\|_{\sup}\left(\frac{T}{n}\right)^2.
$$

The fact that the first derivative f' is not involved in these bounds is an artificial consequence of our assumption that $b \equiv 0$.

Now we switch to the second task. In order to plug this estimate in (7.45), we need to control the first four derivatives of $\bar{P}^{\ell} f$, $\ell = 1, \ldots, n$, uniformly with respect to k and n. In fact, we do not directly need to control the first derivative since $b \equiv 0$ but we will do it as a first example, illustrating the method in a simpler case.

Let us consider again the generic function f and its four bounded derivatives.

$$(\bar{P} f)'(x) = \mathbb{E}\left[f'\left(x + \sigma(x)\sqrt{\frac{T}{n}}\, Z\right)\left(1 + \sigma'(x)\sqrt{\frac{T}{n}}\, Z\right)\right]$$

so that

$$\begin{aligned}
|(\bar{P} f)'(x)| &\le \|f'\|_{\sup} \left\|1 + \sigma'(x)\sqrt{\frac{T}{n}}\, Z\right\|_1 \le \|f'\|_{\sup} \left\|1 + \sigma'(x)\sqrt{\frac{T}{n}}\, Z\right\|_2 \\
&= \|f'\|_{\sup} \sqrt{\mathbb{E}\left(1 + 2\sigma'(x)\sqrt{\frac{T}{n}}\, Z + \sigma'(x)^2 \frac{T}{n} Z^2\right)} \\
&= \|f'\|_{\sup} \sqrt{1 + (\sigma')^2(x)\frac{T}{n}} \le \|f'\|_{\sup}\left(1 + \sigma'(x)^2 \frac{T}{2n}\right)
\end{aligned}$$

since $\sqrt{1+u} \le 1 + \frac{u}{2}$, $u \ge 0$. Hence, we derive by induction that, for every $n \ge 1$ and every $\ell \in \{1, \ldots, n\}$,

$$\forall\, x \in \mathbb{R}, \quad \left|(\bar{P}^{\ell} f)'(x)\right| \le \|f'\|_{\sup}\big(1 + \sigma'(x)^2 T/(2n)\big)^{\ell} \le \|f'\|_{\sup}\, e^{\frac{\sigma'(x)^2 T}{2}},$$

where we used that $(1+u)^{\ell} \le e^{\ell u}$, $u \ge 0$. This yields

$$\|(\bar{P}^{\ell} f)'\|_{\sup} \le \|f'\|_{\sup}\, e^{\frac{\|\sigma'\|_{\sup}^2 T}{2}}.$$

Let us deal now with the second derivative,

$$\begin{aligned}
(\bar{P} f)''(x) &= \tfrac{d}{dx}\mathbb{E}\left[f'\left(x + \sigma(x)\sqrt{\tfrac{T}{n}}\, Z\right)\left(1 + \sigma'(x)\sqrt{\tfrac{T}{n}}\, Z\right)\right] \\
&= \mathbb{E}\left[f''\left(x + \sigma(x)\sqrt{\tfrac{T}{n}}\, Z\right)\left(1 + \sigma'(x)\sqrt{\tfrac{T}{n}}\, Z\right)^2\right] \\
&\quad + \mathbb{E}\left[f'\left(x + \sigma(x)\sqrt{\tfrac{T}{n}}\, Z\right)\sigma''(x)\sqrt{\tfrac{T}{n}}\, Z\right].
\end{aligned}$$

Now

$$\left|\mathbb{E}\left[f''\left(x+\sigma(x)\sqrt{\frac{T}{n}}\,Z\right)\left(1+\sigma'(x)\sqrt{\frac{T}{n}}\,Z\right)^2\right]\right| \le \|f''\|_{\sup}\left(1+\sigma'(x)^2\frac{T}{n}\right)$$

and, using that $f'\left(x+\sigma(x)\sqrt{\frac{T}{n}}\,Z\right)=f'(x)+f''(\zeta)\sigma(x)\sqrt{\frac{T}{n}}\,Z$ for some ζ, owing to the fundamental theorem of Calculus, we get

$$\left|\mathbb{E}\left(f'\left(x+\sigma(x)\sqrt{\frac{T}{n}}\,Z\right)\sigma''(x)\sqrt{\frac{T}{n}}\,Z\right)\right| \le \|f''\|_{\sup}\|\sigma\sigma''\|_{\sup}\mathbb{E}\,(Z^2)\frac{T}{n}$$

since $\mathbb{E}\,Z=0$. Hence

$$\forall\, x\in\mathbb{R},\quad |(\bar{P}f)''(x)|\le \|f''\|_{\sup}\left(1+\left(\|\sigma\sigma''\|_{\sup}+\|(\sigma')^2\|_{\sup}\right)\frac{T}{n}\right),$$

which implies the boundedness of $|(\bar{P}^{\ell}f)''(x)|$, $\ell=0,\dots,n-1$, $n\ge 1$.

The same reasoning yields the boundedness of all derivatives $(\bar{P}^{\ell}f)^{(i)}$, $i=1,2,3,4$, $\ell=1,\dots,n$, $n\ge 1$.

Now we can combine our local error bound with the control of the derivatives. Plugging these estimates into each term of (7.45) finally yields

$$\left|\mathbb{E}\,f(X^x_T)-\mathbb{E}\,f(\bar{X}^x_T)\right|\le C'_{\sigma,T}\max_{1\le\ell\le n,i=1,\dots,4}\|(\bar{P}^{\ell}f)^{(i)}\|_{\sup}\sum_{k=1}^{n}\frac{T^2}{n^2}\le C_{\sigma,T,f}\,T\frac{T}{n},$$

which completes the proof. ◇

► **Exercises. 1.** Complete the above proof by inspecting the case of higher-order derivatives ($k=3,4$).

2. Extend the proof to a (bounded) non-zero drift b.

If one assumes more regularity on the coefficients or some uniform ellipticity on the diffusion coefficient σ it is possible to obtain an expansion of the error at any order.

Theorem 7.8 (Weak error expansions) (a) Smooth function f (Talay–Tubaro's Theorem, see *[270]). Assume b and σ are infinitely differentiable with bounded partial derivatives. Assume $f:\mathbb{R}^d\to\mathbb{R}$ is infinitely differentiable with partial derivative having polynomial growth. Then, for every order $R\in\mathbb{N}^*$, the expansion*

$$(\mathcal{E}_{R+1})\;\equiv\;\mathbb{E}\,f(\bar{X}^{n,x}_T)-\mathbb{E}\,f(X^x_T)=\sum_{k=1}^{R}\frac{c_k}{n^k}+O(n^{-(R+1)}) \qquad (7.47)$$

as $n \to +\infty$, where the real valued coefficients $c_k = c_k(f, T, b, \sigma)$ depend on f, T, b and σ.

(b) Uniformly elliptic diffusion (Bally–Talay's Theorem, see [24]). *If b and σ are bounded, infinitely differentiable with bounded partial derivatives and if σ is uniformly elliptic, i.e.*

$$\forall\, x \in \mathbb{R}^d, \quad \sigma\sigma^*(x) \geq \varepsilon_0 I_d \quad \textit{for an } \varepsilon_0 > 0,$$

then the conclusion of (a) holds true for any bounded Borel function.

Method of proof for (a)**.** The idea is to to rely on the *PDE* method, *i.e.* considering the solution of the parabolic partial differential equation

$$\left(\frac{\partial}{\partial t} + L\right)(u)(t, x) = 0, \quad u(T, .) = f$$

where L defined by

$$(Lg)(x) = g'(x)b(x) + \frac{1}{2}g''(x)\sigma^2(x),$$

denotes the infinitesimal generator of the diffusion. It follows from the Feynman–Kac formula that (under some appropriate regularity assumptions)

$$u(0, x) = \mathbb{E}\, f(X^x_T).$$

Formally (in one dimension), the Feynman–Kac formula can be established as follows (see Theorem 7.11 for a more rigorous proof). Assuming that u is regular enough, *i.e.* $\mathcal{C}^{1,2}([0, T] \times \mathbb{R})$ to apply Itô's formula (see Sect. 12.8), then

$$\begin{aligned} f(X^x_T) &= u(T, X^x_T) \\ &= u(0, x) + \int_0^T \left(\frac{\partial}{\partial t} + L\right)(u)(t, X^x_t)dt + \int_0^T \partial_x u(t, X^x_t)\sigma(X^x_t)dW_t \\ &= u(0, x) + \int_0^T \partial_x u(t, X^x_t)\sigma(X^x_t)dW_t \end{aligned}$$

since u satisfies the above parabolic *PDE*. Assuming that $\partial_x u$ has polynomial growth, so that the stochastic integral is a true martingale, we can take expectation. Then, we introduce domino differences based on the Euler scheme as follows

$$\begin{aligned} \mathbb{E}\, f(\bar{X}^{n,x}_T) - \mathbb{E}\, f(X^x_T) &= \mathbb{E}\left(u(T, \bar{X}^{n,x}_T) - u(0, x)\right) \\ &= \sum_{k=1}^{n} \mathbb{E}\left(u(t^n_k, \bar{X}^{n,x}_{t^n_k}) - u(t^n_{k-1}, \bar{X}^{n,x}_{t^n_{k-1}})\right). \end{aligned}$$

The core of the proof consists in applying Itô's formula (to u, b and σ) to show that

$$\mathbb{E}\left(u(t_k^n, \bar{X}_{t_k^n}^{n,x}) - u(t_{k-1}^n, \bar{X}_{t_{k-1}^n}^{n,x})\right) = \frac{\mathbb{E}\,\phi(t_k^n, X_{t_k^n}^x)}{n^2} + o\left(\frac{1}{n^2}\right)$$

for some continuous function ϕ. Then, one derives (after new computations) that

$$\mathbb{E}\,f(X_T^x) - \mathbb{E}\,f(\bar{X}_T^{n,x}) = \frac{\mathbb{E}\int_0^T \phi(t, X_t^x)dt}{n} + o\left(\frac{1}{n}\right).$$

This approach will be developed in full detail in Sect. 7.8.9 where the theorem is rigorously proved.

Remarks. • The weak error expansion, alone or combined with strong error rates (in quadratic means), are major tools to fight against the bias induced by discretization schemes. This aspect is briefly illustrated below where we first introduce standard Richardson–Romberg extrapolation for diffusions. Wide classes of *multilevel* estimators especially designed to efficiently "kill" the bias while controlling the variance are introduced and analyzed in Chap. 9.

• A parametrix approach is presented in [172] which naturally leads to the higher-order expansion stated in Claim (*b*) of Theorem 7.8. The expansion is derived, in a uniformly elliptic framework, from an approximation result of the density of the diffusion by that of the Euler scheme.

• For extensions to less regular f—namely tempered distributions—in the uniformly elliptic case, we refer to [138].

• The last important information about weak error from the practitioner's viewpoint is that the weak error induced by the Milstein scheme has exactly the same order as that of the Euler scheme, *i.e.* $O(1/n)$. So the Milstein scheme seems at a first glance to be of little interest as long as one wishes to compute $\mathbb{E}\,f(X_T)$. However, we will see in Chap. 9 that combined with its fast strong convergence, it leads to unbiased-like multilevel estimators.

7.7 Bias Reduction by Richardson–Romberg Extrapolation (First Approach)

This section is a first introduction to bias reduction. Chapter 9 is entirely devoted to this topic and introduces more advanced methods like multilevel methods.

7.7.1 *Richardson–Romberg Extrapolation with Consistent Brownian Increments*

Bias-variance decomposition of the quadratic error in a Monte Carlo simulation

Let V be a vector space of continuous functions with linear growth satisfying $(\mathcal{E}_2)$ (the case of non-continuous functions is investigated in [225]). Let $f \in V$. For notational convenience, in view of what follows, we set $W^{(1)} = W$ and $X^{(1)} = X$ (including $X_0^{(1)} = X_0 \in L^2(\Omega, \mathcal{A}, \mathbb{P})$ throughout this section). A regular Monte Carlo simulation based on M independent copies $(\bar{X}_T^{(1)})^m$, $m = 1, \ldots, M$, of the Euler scheme $\bar{X}_T^{(1)}$ with step $\frac{T}{n}$ induces the following global (squared) quadratic error

$$\begin{aligned}\left\| \mathbb{E}\, f(X_T) - \frac{1}{M}\sum_{m=1}^{M} f\big((\bar{X}_T^{(1)})^m\big)\right\|_2^2 &= \big(\mathbb{E}\, f(X_T) - \mathbb{E}\, f(\bar{X}_T^{(1)})\big)^2 \\ &\quad + \left\| \mathbb{E}\, f(\bar{X}_T^{(1)}) - \frac{1}{M}\sum_{m=1}^{M} f\big((\bar{X}_T^{(1)})^m\big)\right\|_2^2 \\ &= \left(\frac{c_1}{n}\right)^2 + \frac{\mathrm{Var}\big(f(\bar{X}_T^{(1)})\big)}{M} + O(n^{-3}). \quad (7.48)\end{aligned}$$

The above formula is the bias-variance decomposition of the approximation error of the Monte Carlo estimator. The resulting quadratic error bound (7.48) emphasizes that this estimator does not take full advantage of the above expansion $(\mathcal{E}_2)$.

Richardson–Romberg extrapolation

To take advantage of the expansion, we will perform a Richardson–Romberg extrapolation. In this framework (originally introduced in the seminal paper [270]), one considers the strong solution $X^{(2)}$ of a "copy" of Eq. (7.1), driven by a second Brownian motion $W^{(2)}$ and starting from $X_0^{(2)}$ (independent of $W^{(2)}$ with the same distribution as $X_0^{(1)}$) both defined on the same probability space $(\Omega, \mathcal{A}, \mathbb{P})$ on which $W^{(1)}$ and $X_0^{(1)}$ are defined. One may always consider such a Brownian motion by enlarging the probability space Ω if necessary.

Then we consider the Euler scheme with *a twice smaller* step $\frac{T}{2n}$, denoted by $\bar{X}^{(2)}$, associated to $X^{(2)}$, *i.e.* starting from $X_0^{(2)}$ with Brownian increments built from $W^{(2)}$.

We assume from now on that $(\mathcal{E}_3)$ (as defined in (7.47)) holds for f to get more precise estimates but the principle would work with a function simply satisfying $(\mathcal{E}_2)$. Then combining the two time discretization error expansions related to $\bar{X}^{(1)}$ and $\bar{X}^{(2)}$, respectively, we get

$$\mathbb{E}\, f(X_T) = \mathbb{E}\big(2 f(\bar{X}_T^{(2)}) - f(\bar{X}_T^{(1)})\big) + \frac{c_2}{2\,n^2} + O(n^{-3}).$$

Then, the new global (squared) quadratic error becomes

$$\left\| \mathbb{E}\, f(X_T) - \frac{1}{M}\sum_{m=1}^{M} 2f\big((\bar{X}_T^{(2)})^m\big) - f\big((\bar{X}_T^{(1)})^m\big)\right\|_2^2$$
$$= \Big(\frac{c_2}{2n^2}\Big)^2 + \frac{\mathrm{Var}\big(2f(\bar{X}_T^{(2)}) - f(\bar{X}_T^{(1)})\big)}{M} + O(n^{-5}). \tag{7.49}$$

The structure of this quadratic error suggests the following natural question:

Is it possible to reduce the (asymptotic) time discretization error *without increasing the Monte Carlo error (at least asymptotically in n...)?*

Or, put differently: *to what extent is it possible to control the variance term* $\mathrm{Var}\big(2f(\bar{X}_T^{(2)}) - f(\bar{X}_T^{(1)})\big)$?

– *Lazy simulation.* If one adopts a somewhat "lazy" approach by using the pseudo-random number generator purely sequentially to simulate the two Euler schemes, this corresponds from a theoretical point of view to considering independent Gaussian white noises $(Z_k^{(1)})_k$ and $(Z_k^{(2)})_k$ to simulate the Brownian increments in both schemes and independent starting values or, equivalently, to assuming that $W^{(1)}$ and $W^{(2)}$ are two *independent Brownian motions* and that $X_0^{(1)}$ and $X_0^{(2)}$ are i.i.d. (square integrable) random variables. Then

$$\mathrm{Var}\big(2f(\bar{X}_T^{(2)}) - f(\bar{X}_T^{(1)})\big) = 4\mathrm{Var}\big(f(\bar{X}_T^{(2)})\big) + \mathrm{Var}\big(f(\bar{X}_T^{(1)})\big)$$
$$= 5\,\mathrm{Var}\big(f(\bar{X}_T)\big) \overset{n\to+\infty}{\longrightarrow} 5\,\mathrm{Var}\big(f(\bar{X}_T)\big).$$

In this approach the gain of one order on the bias (switch from $c_1 n^{-1}$ to $c_2 n^{-2}$) induces an increase of the variance by 5 and of the complexity by (approximately) 3.

– *Consistent simulation (of the Brownian increments).* If $W^{(i)} = W$ and $X_0^{(i)} = X_0 \in L^2(\mathbb{P})$, $i = 1, 2$, then

$$\mathrm{Var}\big(2f(\bar{X}_T^{(2)}) - f(\bar{X}_T^{(1)})\big) \overset{n\to+\infty}{\longrightarrow} \mathrm{Var}\big(2f(X_T) - f(X_T)\big) = \mathrm{Var}\big(f(X_T)\big)$$

since the Euler schemes $\bar{X}^{(i)}$, $i = 1, 2$, both converge in $L^2(\mathbb{P})$ to X.

This time, the same gain in terms of bias has no impact on the variance, at least for a refined enough scheme (n large). Of course, the complexity remains 3 times higher, but the pseudo-random number generator is less solicited by a factor $2/3$.

In fact, it is shown in [225] that this choice $W^{(i)} = W$, $X_0^{(i)} = X_0$, $i = 1, 2$, leading to *consistent Brownian increments for the two schemes*, is asymptotically optimal among all possible choices of Brownian motions $W^{(1)}$ and $W^{(2)}$. This result can be extended to Borel functions f when the diffusion is uniformly elliptic (and b, σ bounded, infinitely differentiable with bounded partial derivatives, see [225]).

ℵ Practitioner's corner

From a practical viewpoint, one first simulates an Euler scheme with step $\frac{T}{2n}$ using a white Gaussian noise $(Z_k^{(2)})_{k\geq 1}$, then one simulates the Gaussian white noise $Z^{(1)}$ of the Euler scheme with step $\frac{T}{n}$ by setting

$$Z_k^{(1)} = \frac{Z_{2k}^{(2)} + Z_{2k-1}^{(2)}}{\sqrt{2}}, \quad k \geq 1.$$

Numerical illustration. We wish to illustrate the efficiency of the Richardson–Romberg (*RR*) extrapolation in a somewhat extreme situation where the time discretization induces an important bias. To this end, we consider the Euler scheme of the Black–Scholes *SDE*

$$dX_t = X_t\big(rdt + \sigma dW_t\big)$$

with the following values for the parameters

$$X_0 = 100, \ r = 0.15, \ \sigma = 1.0, \ T = 1.$$

Note that such a volatility $\sigma = 100\%$ per year is equivalent to a 4 year maturity with volatility 50% (or 16 years with volatility 25%). A high interest rate is chosen accordingly. We consider the Euler scheme of this *SDE* with step $h = \frac{T}{n}$, namely

$$\bar{X}_{t_{k+1}} = \bar{X}_{t_k}\big(1 + rh + \sigma\sqrt{h}\, Z_{k+1}\big), \qquad \bar{X}_0 = X_0,$$

where $t_k = kh$, $k = 0, \ldots, n$ and $(Z_k)_{1\leq k\leq n}$ is a Gaussian white noise. We purposefully choose a coarse discretization step $n = 10$ so that $h = \frac{1}{10}$. One should keep in mind that, in spite of its virtues in terms of closed forms, both coefficients of the Black–Scholes *SDE* have linear growth so that it is quite a demanding benchmark, especially when the discretization step is coarse. We want to price a vanilla Call option with strike $K = 100$, *i.e.* to compute

$$C_0 = e^{-rt}\mathbb{E}\,(X_T - K)_+$$

using a crude Monte Carlo simulation and an *RR* extrapolation with consistent Brownian increments as described in the above practitioner's corner. The Black–Scholes reference premium is $C_0^{BS} = 42.9571$ (see Sect. 12.2). To equalize the complexity of the crude simulation and its *RR* extrapolated counterpart, we use M sample paths, $M = 2^k$, $k = 14, \ldots, 2^{26}$ for the *RR*-extrapolated simulation ($2^{14} \simeq 32\,000$ and $2^{26} \simeq 67\,000\,000$) and $3M$ for the crude Monte Carlo simulation. Figure 7.1 depicts the obtained results. The simulation is large enough so that, at its end, the observed error is approximately representative of the residual bias. The blue line (crude *MC*) shows the magnitude of the theoretical bias (close to 1.5) for such a coarse step whereas the red line highlights the improvement brought by the Richardson–

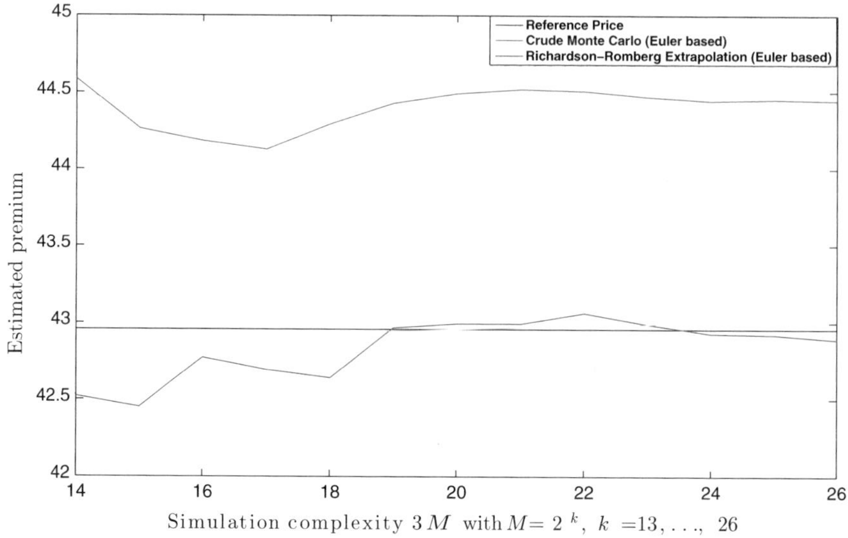

Fig. 7.1 CALL OPTION IN A B-S MODEL PRICED BY AN EULER SCHEME. $\sigma = 1.00$, $r = 0.15\%$, $T = 1$, $K = X_0 = 100$. Step $h = 1/10$ ($n = 10$). Black line: reference price; Red line: (Consistent) Richardson–Romberg extrapolation of the Euler scheme of size M; Blue line: Crude Monte Carlo simulation of size $3\,M$ of the Euler scheme (equivalent complexity)

Romberg extrapolation: the residual bias is approximately equal to 0.07, *i.e.* the bias is divided by more than 20.

▶ **Exercises. 1.** Let $X, Y \in L^2(\Omega, \mathcal{A}, \mathbb{P})$.
(a) Show that

$$\big|\mathrm{cov}(X, Y)\big| \le \sigma(X)\sigma(Y) \quad \text{and} \quad \sigma(X + Y) \le \sigma(X) + \sigma(Y)$$

where $\sigma(X) = \sqrt{\mathrm{Var}(X)} = \|X - \mathbb{E}\,X\|_2$ denotes the *standard-deviation* of X.
(b) Show that $\big|\sigma(X) - \sigma(Y)\big] \le \sigma(X - Y)$ and, for every $\lambda \in \mathbb{R}$, $\sigma(\lambda X) = |\lambda|\sigma(X)$.

2. Let X and $Y \in L^2(\Omega, \mathcal{A}, \mathbb{P})$ have the same distribution.

(a) Show that $\big|\sigma(X) - \sigma(Y)\big| \le \sigma(X - Y)$. Deduce that, for every $\alpha \in (-\infty, 0] \cup [1, +\infty)$,

$$\mathrm{Var}\big(\alpha X + (1 - \alpha)Y\big) \ge \mathrm{Var}(X).$$

(b) Deduce that consistent Brownian increments produce the Richardson-Romberg meta-scheme with the lowest asymptotically variance as n goes to infinity.

3. (ℵ *Practitioner's corner...*). (a) In the above numerical illustration, carry on testing the Richardson–Romberg extrapolation based on Euler schemes *versus* crude Monte Carlo simulation with steps $\frac{T}{n}$ and $\frac{T}{2n}$, $n = 5,\ 10,\ 20,\ 50$, respectively with

– independent Brownian increments,

– consistent Brownian increments.

(*b*) Compute an estimator of the variance of the estimators in both settings and compare the obtained results.

7.8 Further Proofs and Results

Throughout this section, we recall that $|\,.\,|$ always denotes for the canonical Euclidean norm on $\mathbb{R}^d$ and $\|A\| = \sqrt{\mathrm{Tr}(AA^*)} = \sqrt{\sum_{ij} a_{ij}^2}$ denotes for the Fröbenius norm of $A = [a_{ij}] \in \mathcal{M}(d, q, \mathbb{R})$. We will extensively use that, for every $u = (u^1, \dots, u^d) \in \mathbb{R}^d$, $|Au| \le \|A\|\,|u|$, which is an easy consequence of the Schwarz Inequality (in particular, $|||A||| \le \|A\|$). To alleviate notation, we will drop the exponent n in $t_k^n = \frac{kT}{n}$.

7.8.1 *Some Useful Inequalities*

On the non-quadratic case, Doob's Inequality is not sufficient to carry out the proof: we need the more general Burkhölder–Davis–Gundy Inequality. Furthermore, to get some real constants having the announced behavior as a function of T, we will also need to use the generalized Minkowski Inequality, (see [143]) established below in a probabilistic framework.

$\triangleright$ **The generalized Minkowski Inequality:** For any (bi-measurable) process $X = (X_t)_{t\ge 0}$ and for every $p \in [1, \infty)$,

$$\forall\, T \in [0, +\infty], \qquad \left\| \int_0^T X_t\, dt \right\|_p \le \int_0^T \|X_t\|_p\, dt. \tag{7.50}$$

Proof. First note that, owing to the triangle inequality $\left| \int_0^T X_s ds \right| \le \int_0^T |X_s| ds$, one may assume without loss of generality that X Is a non-negative process. If $p = 1$ the inequality is obvious. Assume now $p \in (1, +\infty)$. Let $T \in (0, +\infty)$ and let Y be a non-negative random variable defined on the same probability space as $(X_t)_{t\in[0,T]}$. Let $M > 0$. It follows from Fubini's Theorem and Hölder's Inequality that

$$\begin{aligned}
\mathbb{E}\left(\int_0^T (X_s \wedge M) ds\, Y \right) &= \int_0^T \mathbb{E}\big((X_s \wedge M)Y)\big)\, ds \\
&\le \int_0^T \|X_s \wedge M\|_p \|Y\|_q\, ds \qquad \text{with } q = \tfrac{p}{p-1} \\
&= \|Y\|_q \int_0^T \|X_s \wedge M\|_p ds.
\end{aligned}$$

The above inequality applied with $Y := \left(\int_0^T X_s \wedge M\, ds\right)^{p-1}$

$$\mathbb{E}\left(\int_0^T X_s \wedge M\, ds\right)^p \le \left[\mathbb{E}\left(\int_0^T X_s \wedge M ds\right)^p\right]^{1-\frac{1}{p}} \int_0^T \|X_s\|_p ds.$$

If $\mathbb{E}\left(\int_0^T X_s \wedge M_n\, ds\right)^p = 0$ for any sequence $M_n \uparrow +\infty$, the inequality is obvious since, by Beppo Levi's monotone convergence Theorem, $\int_0^T X_s ds = 0$ $\mathbb{P}$-$a.s.$ Otherwise, there is a sequence $M_n \uparrow +\infty$ such that all these integrals are non-zero (and finite since X is bounded by M and T is finite). Consequently, one can divide both sides of the former inequality to obtain

$$\forall n \ge 1, \qquad \left[\mathbb{E}\left(\int_0^T X_s \wedge M_n\, ds\right)^p\right]^{\frac{1}{p}} \le \int_0^T \|X_s\|_p ds.$$

Now letting $M_n \uparrow +\infty$ yields exactly the expected result owing to two successive applications of Beppo Levi's monotone convergence Theorem, the first with respect to the Lebesgue measure ds, the second with respect to $d\mathbb{P}$. When $T = +\infty$, the result follows by Fatou's Lemma by letting T go to infinity in the inequality obtained for finite T. ◇

▷ **The Burkhölder–Davis–Gundy Inequality (continuous time)** For every $p \in (0, +\infty)$, there exists two real constants $c_p^{BDG} > 0$ and $C_p^{BDG} > 0$ such that, for every *continuous* local martingale $(X_t)_{t\in[0,T]}$ null at 0,

$$c_p^{BDG} \left\| \sqrt{\langle X\rangle_T} \right\|_p \le \left\| \sup_{t\in[0,T]} |X_t| \right\|_p \le C_p^{BDG} \left\| \sqrt{\langle X\rangle_T} \right\|_p. \tag{7.51}$$

For a detailed proof based on a stochastic calculus approach, we refer to [251], p. 160. As, in this section, we are concerned with multi-dimensional continuous local martingale (stochastic integrals of matrices *versus* a q-dimensional Brownian motion), we need the following easy extension of the right inequality for d-dimensional local martingales $X_t = (X_t^1, \dots, X_t^d)$: set $\langle X\rangle_t = \sum_{i=1}^d \langle X^i\rangle_t$. Let $p \in [1, +\infty)$ and $C_{d,p}^{BDG} = d\, C_p^{BDG}$. Then

$$\left\| \sup_{t\in[0,T]} |X_t| \right\|_p \le C_{d,p}^{BDG} \left\| \sqrt{\langle X\rangle_T} \right\|_p. \tag{7.52}$$

In particular, if W is an $(\mathcal{F}_t)_{t\in[0,T]}$ standard Brownian motion on a filtered probability space $\left(\Omega, \mathcal{A}, (\mathcal{F}_t)_{t\in[0,T]}, \mathbb{P}\right)$ and $(H_t)_{t\in[0,T]} = ([H_t^{ij}])_{t\in[0,T]}$ is an $(\mathcal{F}_t)_{t\in[0,T]}$-

progressively measurable process having values in $\mathcal{M}(d, q, \mathbb{R})$ such that $\int_0^T \|H_t\|^2 dt < +\infty$ $\mathbb{P}$-$a.s.$, then the d-dimensional local martingale $\int_0^{\cdot} H_s dW_s$ satisfies

$$\left\| \sup_{[0,T]} \left| \int_0^t H_s \, dW_s \right| \right\|_p \leq C^{BDG}_{d,p} \left\| \sqrt{\int_0^T \|H_t\|^2 dt} \right\|_p . \tag{7.53}$$

7.8.2 Polynomial Moments (II)

Proposition 7.6 (a) *For every* $p \in (0, +\infty)$, *there exists a positive real constant* $\kappa'_p > 0$ *(increasing in* p*), such that, if* b, σ *satisfy:*

$$\forall t \in [0, T], \ \forall x \in \mathbb{R}^d, \quad |b(t, x)| + \|\sigma(t, x)\| \leq C(1 + |x|), \tag{7.54}$$

then, every strong solution of Equation (7.1) starting from the finite random vector X_0 *(if any), satisfies*

$$\forall\, p \in (0, +\infty), \qquad \left\| \sup_{s \in [0,T]} |X_s| \right\|_p \leq 2\, e^{\kappa'_p C T} \big(1 + \|X_0\|_p\big).$$

(b) *The same conclusion holds under the same assumptions for the continuous Euler scheme with step* $\frac{T}{n}$, $n \geq 1$, *as defined by (7.8),* with the same constant κ'_p *(which does not depend* n*), i.e.*

$$\forall\, p \in (0, +\infty), \ \forall n \geq 1, \qquad \left\| \sup_{s \in [0,T]} |\bar{X}^n_s| \right\|_p \leq 2\, e^{\kappa'_p C T} \big(1 + \|X_0\|_p\big).$$

Remarks. • Note that this proposition makes no assumption either on the existence of strong solutions to (7.1) or on some (strong) uniqueness assumption on a time interval or the whole real line. Furthermore, the inequality is meaningless when $X_0 \notin L^p(\mathbb{P})$.

• The case $p \in (0, 2)$ will be discussed at the end of the proof.

Proof. To alleviate the notation we assume from now on that $d = q = 1$.

(a) Step 1 (*The process: first reduction*). Assume $p \in [2, \infty)$. First we introduce for every integer $N \geq 1$ the stopping time $\tau_N := \inf\big\{t \in [0, T] : |X_t - X_0| > N\big\}$ (convention $\inf \varnothing = +\infty$). This is a stopping time since, for every $t \in \mathbb{R}_+$, $\{\tau_N < t\} = \bigcup_{r \in [0,t] \cap \mathbb{Q}} \{|X_r - X_0| > N\} \in \mathcal{F}_t$. Moreover, $\{\tau_N \leq t\} = \bigcap_{k \geq k_0} \left\{\tau_N < t + \frac{1}{k}\right\}$ for every

$k_0 \geq 1$, hence $\{\tau_N \leq t\} \in \bigcap_{k_0 \geq 1} \mathcal{F}_{t+\frac{1}{k_0}} = \mathcal{F}_{t+} = \mathcal{F}_t$ since the filtration is càd ([7]). Furthermore, $\sup_{t\in[0,T]} |X_t^{\tau_N}| \leq N + |X_0|$ so that the non-decreasing function f_N defined by $f_N(t) := \Big\| \sup_{s\in[0,t]} |X_{s\wedge\tau_N}| \Big\|_p$, $t \in [0,T]$, is bounded by $N + \|X_0\|_p$. On the other hand,

$$\sup_{s\in[0,t]} |X_{s\wedge\tau_N}| \leq |X_0| + \int_0^{t\wedge\tau_N} |b(s, X_s)|ds + \sup_{s\in[0,t]} \left| \int_0^{s\wedge\tau_N} \sigma(u, X_u)dW_u \right|.$$

It follows from successive applications of both the regular and the generalized Minkowski (7.50) Inequalities and of the *BDG* Inequality (7.53) that

$$\begin{aligned}
f_N(t) &\leq \|X_0\|_p + \int_0^t \|\mathbf{1}_{\{s\leq\tau_N\}} b(s, X_s)\|_p ds + C_{d,p}^{BDG} \left\| \sqrt{\int_0^{t\wedge\tau_N} \|\sigma(s, X_s)\|^2 ds} \right\|_p \\
&\leq \|X_0\|_p + \int_0^t \|b(s\wedge\tau_N, X_{s\wedge\tau_N})\|_p ds + C_{d,p}^{BDG} \left\| \sqrt{\int_0^t \|\sigma(s\wedge\tau_N, X_{s\wedge\tau_N})\|^2 ds} \right\|_p \\
&\leq \|X_0\|_p + C\int_0^t (1 + \|X_{s\wedge\tau_N}\|_p) ds + C_{d,p}^{BDG} C \left\| \sqrt{\int_0^t (1 + |X_{s\wedge\tau_N}|)^2 ds} \right\|_p \\
&\leq \|X_0\|_p + C\int_0^t (1 + \|X_{s\wedge\tau_N}\|_p) ds + C_{d,p}^{BDG} C \left\| \sqrt{t} + \sqrt{\int_0^t |X_{s\wedge\tau_N}|^2 ds} \right\|_p
\end{aligned}$$

where we used in the last line the Minkowski inequality on $L^2([0,T], dt)$ endowed with its usual Hilbert norm. Hence, the $L^p(\mathbb{P})$-Minkowski Inequality and the obvious identity $\|\sqrt{.}\|_p = \|\,.\,\|_{\frac{p}{2}}^{\frac{1}{2}}$ yield

$$f_N(t) \leq \|X_0\|_p + C\int_0^t (1 + \|X_{s\wedge\tau_N}\|_p) ds + C_{d,p}^{BDG} C \left(\sqrt{t} + \left\| \int_0^t |X_{s\wedge\tau_N}|^2 ds \right\|_{\frac{p}{2}}^{\frac{1}{2}} \right).$$

Now, as $\frac{p}{2} \geq 1$, the generalized $L^{\frac{p}{2}}(\mathbb{P})$-Minkowski Inequality (7.50) yields

[7] This holds true for any hitting time of an open set by an $\mathcal{F}_t$-adapted càd process.

$$
\begin{aligned}
f_N(t) &\le \|X_0\|_p + C\int_0^t (1+\|X_{s\wedge\tau_N}\|_p)ds \\
&\quad + C_{d,p}^{BDG} C\left(\sqrt{t} + \left[\int_0^t \big\||X_{s\wedge\tau_N}|^2\big\|_{\frac{p}{2}} ds\right]^{\frac12}\right)\\
&= \|X_0\|_p + C\int_0^t (1+\|X_{s\wedge\tau_N}\|_p)ds \\
&\quad + C_{d,p}^{BDG} C\left(\sqrt{t} + \left[\int_0^t \|X_{s\wedge\tau_N}\|_p^2 ds\right]^{\frac12}\right).
\end{aligned}
$$

Consequently, the function f_N satisfies

$$
f_N(t) \le C\left(\int_0^t f_N(s)ds + C_{d,p}^{BDG}\left(\int_0^t f_N^2(s)ds\right)^{\frac12}\right) + \psi(t),
$$

where

$$
\psi(t) = \|X_0\|_p + C\big(t + C_{d,p}^{BDG}\sqrt{t}\big).
$$

STEP 2. (*"À la Gronwall" Lemma*).

Lemma 7.3 ("À la Gronwall" Lemma) *Let* $f : [0,T] \to \mathbb{R}_+$ *and let* $\psi : [0,T] \to \mathbb{R}_+$ *be two non-negative non-decreasing functions satisfying*

$$
\forall\, t \in [0,T], \qquad f(t) \le A\int_0^t f(s)ds + B\left(\int_0^t f^2(s)ds\right)^{\frac12} + \psi(t),
$$

where A, B are two positive real constants. Then

$$
\forall\, t \in [0,T], \qquad f(t) \le 2\,e^{(2A+B^2)t}\,\psi(t).
$$

Proof. First, it follows from the elementary inequality $\sqrt{x\,y} \le \frac12\big(\frac{x}{B} + By\big)$, $x, y \ge 0$, $B > 0$, that

$$
\left(\int_0^t f^2(s)ds\right)^{\frac12} \le \left(f(t)\int_0^t f(s)ds\right)^{\frac12} \le \frac{f(t)}{2B} + \frac{B}{2}\int_0^t f(s)ds.
$$

Plugging this into the original inequality yields

$$
f(t) \le (2A + B^2)\int_0^t f(s)ds + 2\,\psi(t).
$$

Gronwall's Lemma 7.2 finally yields the announced result. ◇

STEP 3 (*Conclusion when* $p \in [2, +\infty)$). Applying the above generalized Gronwall Lemma to the functions f_N and ψ defined in Step 1 leads to

$$\forall\, t \in [0, T], \quad \Big\| \sup_{s\in[0,t]} |X_{s\wedge\tau_N}| \Big\|_p$$
$$= f_N(t) \le 2\, e^{(2+C(C^{BDG}_{d,p})^2)Ct}\Big(\|X_0\|_p + C(t + C^{BDG}_{d,p}\sqrt{t})\Big).$$

The sequence of stopping times τ_N is non-decreasing and converges toward τ_∞, taking values in $[0, T] \cup \{\infty\}$. On the event $\{\tau_\infty \le T\}$, $|X_{\tau_N} - X_0| \ge N$ so that $|X_{\tau_\infty} - X_0| = \lim_{N\to+\infty} |X_{\tau_N} - X_0| = +\infty$ since X_t has continuous paths. This is $a.s.$ impossible since $[0, T]$ is compact and $(X_t)_{t\ge 0}$ $a.s.$ has continuous paths. As a consequence $\tau_\infty = +\infty$, $a.s.$ which in turn implies that

$$\lim_N \sup_{s\in[0,t]} |X_{s\wedge\tau_N}| = \sup_{s\in[0,t]} |X_s| \qquad \mathbb{P}\text{-}a.s.$$

Then Fatou's Lemma implies, by letting N go to infinity, that, for every $t \in [0, T]$,

$$\Big\| \sup_{s\in[0,t]} |X_s| \Big\|_p \le \underline{\lim}_N \Big\| \sup_{s\in[0,t]} |X_{s\wedge\tau_N}| \Big\|_p$$
$$\le 2\, e^{(2+C(C^{BDG}_{d,p})^2)Ct}\Big(\|X_0\|_p + C(t + C^{BDG}_{d,p}\sqrt{t})\Big), \tag{7.55}$$

which yields, using that $\max(\sqrt{u}, u) \le e^u$, $u \ge 0$,

$$\forall\, t \in [0, T], \qquad \Big\| \sup_{s\in[0,t]} |X_s| \Big\|_p \le 2\, e^{(2+C(C^{BDG}_{d,p})^2)Ct}\Big(\|X_0\|_p + e^{Ct} + e^{C(C^{BDG}_{d,p})^2 t}\Big)$$
$$\le 2\, e^{(2+C(C^{BDG}_{d,p})^2)Ct}\big(e^{Ct} + e^{C(C^{BDG}_{d,p})^2 t}\big)(1 + \|X_0\|_p).$$

One derives the existence of a positive real constant $\kappa'_{p,d} > 0$, only depending on the *BDG* real constant $C^{BDG}_{p,d}$, such that

$$\forall\, t \in [0, T], \quad \Big\| \sup_{s\in[0,t]} |X_s| \Big\|_p \le \kappa'_{p,d}\, e^{\kappa'_p C(C+1)t}\big(1 + \|X_0\|_p\big).$$

STEP 4 (*Conclusion when* $p \in (0, 2)$). The extension can be carried out as follows: for every $x \in \mathbb{R}^d$, the diffusion process starting at x, denoted by $(X^x_t)_{t\in[0,T]}$, satisfies the following two obvious facts:

– the process X^x is $\mathcal{F}^W_t$-adapted, where $\mathcal{F}^W_t := \sigma(\mathcal{N}_{\mathbb{P}}, W_s,\ s \le t)$, $t \in [0, T]$.

– If X_0 is an $\mathbb{R}^d$-valued random vector defined on $(\Omega, \mathcal{A}, \mathbb{P})$, independent of W, then the process $X = (X_t)_{t\in[0,T]}$ starting from X_0 satisfies

$$X_t = X^{X_0}_t.$$

Consequently, using that $p \mapsto \|\cdot\|_p$ is non-decreasing, it follows that

$$\Big\| \sup_{s\in[0,t]} |X_s^x| \Big\|_p \le \Big\| \sup_{s\in[0,t]} |X_s^x| \Big\|_2 \le \kappa'_{2,d}\, e^{\kappa'_{2,d}C(C+1)t}\big(1+|x|\big).$$

Now

$$\mathbb{E}\left(\sup_{t\in[0,T]}|X_t|^p\right) = \int_{\mathbb{R}^d} \mathbb{P}_{X_0}(dx)\mathbb{E}\left(\sup_{t\in[0,T]}|X_t^x|^p\right)$$
$$\le 2^{(p-1)_+}(\kappa'_{2,d})^p e^{p\kappa'_{2,d}C(C+1)T}\big(1+\mathbb{E}\,|X_0|^p\big)$$

(where we used that $(u+v)^p \le 2^{(p-1)_+}(u^p+v^p)$, $u, v \ge 0$) so that

$$\Big\| \sup_{t\in[0,T]} |X_t| \Big\|_p \le 2^{(1-\frac{1}{p})_+}\kappa'_{2,d}\, e^{\kappa'_2 C(C+C)T}2^{(\frac{1}{p}-1)_+}\big(1+\|X_0\|_p\big)$$
$$= 2^{|1-\frac{1}{p}|}\kappa'_{2,d}\; e^{\kappa'_{2,d}CT}\big(1+\|X_0\|_p\big)$$
$$\le \kappa''_{2,d}\; e^{\kappa''_{2,d}CT}\big(1+\|X_0\|_p\big).$$

As concerns the SDE (7.1) itself, the same reasoning can be carried out only if (7.1) satisfies an existence and uniqueness assumption for any starting value X_0.

(*b*) (*Euler scheme*) The proof follows the same lines as above. One starts from the integral form (7.9) of the continuous Euler scheme and one introduces for every $n,\ N \ge 1$ the stopping times

$$\bar\tau_N = \bar\tau_N^n := \inf\big\{t\in[0,T] : |\bar X_t^n - X_0| > N\big\}.$$

To adapt the above proof to the continuous Euler scheme, we just need to note that

$$\forall\, s\in[0,t],\quad 0\le \underline{s}\le s \quad\text{and}\quad \big\|\bar X^n_{\underline{s}}\big\|_p \le \Big\| \sup_{s\in[0,t\wedge\bar\tau_N]} |\bar X_s| \Big\|_p. \qquad \diamond$$

7.8.3 L^p-Pathwise Regularity

Lemma 7.4 *Let $p\ge 1$ and let $(Y_t)_{t\in[0,T]}$ be an ($\mathbb{R}^d$-valued) Itô process defined on $(\Omega,\mathcal{A},\mathbb{P})$ by*

$$Y_t = Y_0 + \int_0^t G_s\,ds + \int_0^t H_s\,dW_s,\ \ t\in[0,T],$$

where G and H are $(\mathcal{F}_t)$-progressively measurable having values in $\mathbb{R}^d$ and $\mathcal{M}(d,q,\mathbb{R})$, respectively and satisfying $\int_0^T\big(|G_s|+\|H_s\|^2\big)ds < +\infty$ $\mathbb{P}$-a.s.

(a) For every $p\ge 2$, writing $\big\|H_t\big\|_p = \big\|\,\|H_t\|\,\big\|_p$ to alleviate notation,

$$\forall s, t \in [0, T], \quad \|Y_t - Y_s\|_p \le C_{d,p}^{BDG} \sup_{t\in[0,T]} \|H_t\|_p |t-s|^{\frac{1}{2}} + \sup_{t\in[0,T]} \|G_t\|_p |t-s|$$
$$\le \Big(C_{d,p}^{BDG} \sup_{t\in[0,T]} \|H_t\|_p + \sqrt{T} \sup_{t\in[0,T]} \|G_t\|_p\Big)|t-s|^{\frac{1}{2}}.$$

In particular, if $\sup_{t\in[0,T]} \|H_t\|_p + \sup_{t\in[0,T]} \|G_t\|_p < +\infty$, *the process* $t \mapsto Y_t$ *is Hölder with exponent* $\frac{1}{2}$ *from* $[0, T]$ *into* $L^p(\mathbb{P})$.
(*b*) *If* $p \in [1, 2)$, *then, for every* $s, t \in [0, T]$,

$$\|Y_t - Y_s\|_p \le C_{d,p}^{BDG} \Big\| \sup_{t\in[0,T]} \|H_t\| \Big\|_p |t-s|^{\frac{1}{2}} + \sup_{t\in[0,T]} \|G_t\|_p |t-s|$$
$$\le \Big(C_{d,p}^{BDG} \Big\| \sup_{t\in[0,T]} \|H_t\| \Big\|_p + \sqrt{T} \sup_{t\in[0,T]} \|G_t\|_p\Big)|t-s|^{\frac{1}{2}}.$$

Proof. (*a*) Let $0 \le s \le t \le T$. It follows from the standard and generalized Minkowski Inequalities and the *BDG* Inequality (7.53) applied to the stochastic integral $\big(\int_s^{s+u} H_r dW_r\big)_{u\ge 0}$ that

$$\begin{aligned}
\|Y_t - Y_s\|_p &\le \Big\| \int_s^t |G_u| du \Big\|_p + \Big\| \int_s^t H_u dW_u \Big\|_p \\
&\le \int_s^t \|G_u\|_p du + C_p^{BDG} \Bigg\| \sqrt{\int_s^t \|H_u\|^2 du} \Bigg\|_p \\
&\le \sup_{t\in[0,T]} \|G_t\|_p (t-s) + C_p^{BDG} \Big\| \int_s^t \|H_u\|^2 du \Big\|_{\frac{p}{2}}^{\frac{1}{2}} \\
&\le \sup_{t\in[0,T]} \|G_t\|_p (t-s) + C_p^{BDG} \sup_{u\in[0,T]} \big\| \|H_u\|^2 \big\|_{\frac{p}{2}}^{\frac{1}{2}} (t-s)^{\frac{1}{2}} \\
&= \sup_{t\in[0,T]} \|G_t\|_p (t-s) + C_p^{BDG} \sup_{u\in[0,T]} \big\| H_u \big\|_p (t-s)^{\frac{1}{2}}.
\end{aligned}$$

The second inequality simply follows from $|t-s| \le \sqrt{T}|t-s|^{\frac{1}{2}}$.
(*b*) If $p \in [1, 2]$, one simply uses that

$$\Big\| \int_s^t \|H_u\|^2 du \Big\|_{\frac{p}{2}}^{\frac{1}{2}} \le |t-s|^{\frac{1}{2}} \Big\| \sup_{u\in[0,T]} \|H_u\|^2 \Big\|_{\frac{p}{2}}^{\frac{1}{2}} = |t-s|^{\frac{1}{2}} \Big\| \sup_{u\in[0,T]} \|H_u\| \Big\|_p$$

and one concludes likewise. ◇

Remark. If H, G and Y are defined on the non-negative real line $\mathbb{R}_+$ and $\sup_{t\in\mathbb{R}_+} \big(\|G_t\|_p + \|H_t\|_p\big) < +\infty$, then $t \mapsto Y_t$ is locally $\frac{1}{2}$-Hölder on $\mathbb{R}_+$. If $H = 0$, the process is Lipschitz continuous on $[0, T]$.

Combining the above result for Itô processes (see Sect. 12.8) with those of Proposition 7.6 leads to the following result on pathwise regularity of the diffusion solution to (7.1) (when it exists) and the related Euler schemes.

Proposition 7.7 *If the coefficients b and σ satisfy the linear growth assumption (7.54) over $[0,T]\times\mathbb{R}^d$ with a real constant $C>0$, then the Euler scheme with step $\frac{T}{n}$ and any strong solution of (7.1) satisfy for every $p\ge 1$,*

$$\forall n\ge 1,\ \forall s,\,t\in[0,T],$$
$$\|X_t-X_s\|_p+\|\bar{X}^n_t-\bar{X}^n_s\|_p\le \kappa''_{p,d}Ce^{\kappa''_{p,d}C(C+1)T}(1+\sqrt{T})\big(1+\|X_0\|_p\big)|t-s|^{\frac{1}{2}}$$

where $\kappa''_{p,d}\in(0,+\infty)$ is a real constant only depending on $C^{BDG}_{p,d}$ (increasing in p).

Proof. As concerns the process X, this is a straightforward consequence of the above Lemma 7.4 by setting

$$G_t=b(t,X_t)\quad\text{and}\quad H_t=\sigma(t,X_t)$$

since

$$\max\Big(\big\|\sup_{t\in[0,T]}|G_t|\big\|_p,\big\|\sup_{t\in[0,T]}\|H_t\|\big\|_p\Big)\le C\big(1+\|\sup_{t\in[0,T]}|X_t|\|_p\big).$$

One specifies the real constant κ''_p using Proposition 7.6. ◇

7.8.4 L^p-Convergence Rate (II): Proof of Theorem 7.2

STEP 1 ($p\ge 2$). Set

$$\begin{aligned}\varepsilon_t&:=X_t-\bar{X}^n_t,\quad t\in[0,T]\\&=\int_0^t\Big(b(s,X_s)-b(\underline{s},\bar{X}_{\underline{s}})\Big)ds+\int_0^t\Big(\sigma(s,X_s)-\sigma(\underline{s},\bar{X}_{\underline{s}})\Big)dW_s\end{aligned}$$

so that

$$\sup_{s\in[0,t]}|\varepsilon_s|\le\int_0^t\big|(b(s,X_s)-b(\underline{s},\bar{X}_{\underline{s}})\big|ds+\sup_{s\in[0,t]}\left|\int_0^s(\sigma(u,X_u)-\sigma(\underline{u},\bar{X}_{\underline{u}}))dW_u\right|.$$

One sets for every $t\in[0,T]$,

$$f(t):=\big\|\sup_{s\in[0,t]}|\varepsilon_s|\big\|_p.$$

It follows from the (regular) Minkowski Inequality on $\big(L^p(\mathbb{P}),\|\cdot\|_p\big)$, *BDG* Inequality (7.53) and the generalized Minkowski inequality (7.50) that

$$\begin{aligned}
f(t) &\le \int_0^t \left\| b(s, X_s) - b(\underline{s}, \bar{X}_{\underline{s}}) \right\|_p ds + C_{d,p}^{BDG} \left\| \left[\int_0^t \left\| \sigma(s, X_s) - \sigma(\underline{s}, \bar{X}_{\underline{s}}) \right\|^2 ds \right]^{\frac{1}{2}} \right\|_p \\
&= \int_0^t \left\| b(s, X_s) - b(\underline{s}, \bar{X}_{\underline{s}}) \right\|_p ds + C_{d,p}^{BDG} \left\| \int_0^t \left\| \sigma(s, X_s) - \sigma(\underline{s}, \bar{X}_{\underline{s}}) \right\|^2 ds \right\|_{\frac{p}{2}}^{\frac{1}{2}} \\
&\le \int_0^t \left\| b(s, X_s) - b(\underline{s}, \bar{X}_{\underline{s}}) \right\|_p ds + C_{d,p}^{BDG} \left[\int_0^t \left\| \left\| \sigma(s, X_s) - \sigma(\underline{s}, \bar{X}_{\underline{s}}) \right\|^2 \right\|_{\frac{p}{2}} ds \right]^{\frac{1}{2}} \\
&= \int_0^t \left\| b(s, X_s) - b(\underline{s}, \bar{X}_{\underline{s}}) \right\|_p ds + C_{d,p}^{BDG} \left[\int_0^t \left\| \| \sigma(s, X_s) - \sigma(\underline{s}, \bar{X}_{\underline{s}}) \| \right\|_p^2 ds \right]^{\frac{1}{2}}.
\end{aligned}$$

Let us temporarily set $\tau_t^X = \big(1 + \| \sup_{t\in[0,t]} |X_s| \|\big)t$, $t \in [0, T]$. Using Assumption (H_T^β) (see (7.14)) and the Minkowski Inequality on $\big(L^2([0, T], dt), |\,.\,|_{L^2(dt)}\big)$ spaces, we get

$$\begin{aligned}
f(t) &\le C_{b,\sigma,T} \Bigg(\int_0^t \left((1 + \|X_s\|_p)(s - \underline{s})^\beta + \left\| X_s - \bar{X}_{\underline{s}} \right\|_p\right) ds \\
&\qquad + C_{d,p}^{BDG} \left[\int_0^t \left((1 + \|X_s\|_p)(s - \underline{s})^\beta + \left\| X_s - \bar{X}_{\underline{s}} \right\|_p\right)^2 ds \right]^{\frac{1}{2}} \Bigg) \\
&\le C_{b,\sigma,T} \Bigg(\int_0^t \left((1 + \|X_s\|_p)(s - \underline{s})^\beta + \left\| X_s - \bar{X}_{\underline{s}} \right\|_p\right) ds \\
&\quad + C_{d,p}^{BDG} \left[\Big(1 + \big\| \sup_{s\in[0,T]} |X_s| \big\|_p\Big) \left[\int_0^t s - \underline{s})^{2\beta} ds \right]^{\frac{1}{2}} + \left[\int_0^t \| X_s - \bar{X}_{\underline{s}} \|_p^2 ds \right]^{\frac{1}{2}} \right] \Bigg) \\
&\le C_{b,\sigma,T} \Bigg(\tau_t^X \int_0^t (s - \underline{s})^\beta ds + \int_0^t \left\| X_s - \bar{X}_{\underline{s}} \right\|_p ds \\
&\qquad + C_{d,p}^{BDG} \left[\tau_t^X \left[\int_0^t (s - \underline{s})^{2\beta} ds \right]^{\frac{1}{2}} + \left[\int_0^t \| X_s - \bar{X}_{\underline{s}} \|_p^2 ds \right]^{\frac{1}{2}} \right] \Bigg).
\end{aligned}$$

Now, using that $0 \le s - \underline{s} \le \frac{T}{n}$, we obtain

$$\begin{aligned}
f(t) \le C_{b,\sigma,T} \Bigg(&\left(1 + C_{d,p}^{BDG}\right) \left(\frac{T}{n}\right)^\beta \tau_t^X + \int_0^t \| X_s - \bar{X}_{\underline{s}} \|_p \, ds \\
&+ C_{d,p}^{BDG} \left[\int_0^t \| X_s - \bar{X}_{\underline{s}} \|_p^2 ds \right]^{\frac{1}{2}} \Bigg),
\end{aligned}$$

Now, noting that

$$\begin{aligned}\|X_s - \bar{X}_{\underline{s}}\|_p &\le \|X_s - X_{\underline{s}}\|_p + \|X_{\underline{s}} - \bar{X}_{\underline{s}}\|_p = \|X_s - X_{\underline{s}}\|_p + \|\varepsilon_{\underline{s}}\|_p\\ &\le \|X_s - X_{\underline{s}}\|_p + f(s),\end{aligned}$$

we derive

$$f(t) \le C_{b,\sigma,T}\left(\int_0^t f(s)ds + \sqrt{2}\,C_{d,p}^{BDG}\left(\int_0^t f(s)^2 ds\right)^{\frac{1}{2}} + \psi(t)\right) \tag{7.56}$$

where

$$\begin{aligned}\psi(t) := \left(1 + C_{d,p}^{BDG}\right)\left(\frac{T}{n}\right)^{\beta}\tau_t^X + \int_0^t \|X_s - X_{\underline{s}}\|_p\,ds&\\ + \sqrt{2}\,C_{d,p}^{BDG}\left(\int_0^t \|X_s - X_{\underline{s}}\|_p^2\,ds\right)^{\frac{1}{2}}&.\end{aligned} \tag{7.57}$$

STEP 2. It follows from Lemma 7.3 that

$$f(t) \le 2\,C_{b,\sigma,T}e^{2C_{b,\sigma,T}(1+C_{b,\sigma,T}C_{d,p}^{BDG})t}\psi(t). \tag{7.58}$$

Now, we will use the $L^p(\mathbb{P})$-path regularity of the diffusion process X established in Proposition 7.7 to provide an upper-bound for the function ψ. We first note that, as b and σ satisfy (H_T^β) with a positive real constant $C_{b,\sigma,T}$, they also satisfy the linear growth assumption (7.54) with

$$C'_{b,\sigma,T} := C_{b,\sigma,T} + \sup_{t\in[0,T]}\left(|b(t,0)| + \|\sigma(t,0)\|\right) < +\infty$$

since $b(\,.\,,0)$ and $\sigma(\,.\,,0)$ are β-Hölder hence bounded on $[0,T]$. Set for convenience $\widetilde{C}_{b,\sigma,T} = C'_{b,\sigma,T}(C'_{b,\sigma,T}+1)$. It follows from (7.57) and Proposition 7.7 that

$$\begin{aligned}\psi(t) &\le \left(1 + C_{d,p}^{BDG}\right)\left(\frac{T}{n}\right)^{\beta}\tau_t^X\\ &\quad + \kappa''_p C'_{b,\sigma}e^{\kappa''_p\widetilde{C}_{b,\sigma,T}t}\left(1+\|X_0\|_p\right)\left(1+\sqrt{t}\right)\left(\frac{T}{n}\right)^{\frac{1}{2}}\left(t + \sqrt{2}C_{d,p}^{BDG}\sqrt{t}\right)\\ &\le \left(1 + C_{d,p}^{BDG}\right)\tau_t^X\left(\frac{T}{n}\right)^{\beta}\\ &\quad + 2\left(1+\sqrt{2}C_{d,p}^{BDG}\right)(1+t)^2\kappa''_{p,d}C'_{b,\sigma}e^{\kappa''_{p,d}\widetilde{C}_{b,\sigma,T}t}\left(1+\|X_0\|_p\right)\left(\frac{T}{n}\right)^{\frac{1}{2}},\end{aligned}$$

where we used the inequality $(1+\sqrt{t})(t+\sqrt{2}C^{BDG}_{d,p}\sqrt{t}) \le 2(1+\sqrt{2}C^{BDG}_{d,p})(1+t)^2$ which can be established by inspecting the cases $0 \le t \le 1$ and $t \ge 1$.

Moreover, this time using Proposition 7.6(a), we derive that

$$\tau^X_t \le \big(1+\kappa'_{p,d}e^{\kappa'_{p,d}\widetilde{C}_{b,\sigma,T}T}\big)\big(1+\|X_0\|_p\big)t \le \tilde{\kappa}'_{p,d}e^{\tilde{\kappa}'_{p,d}\widetilde{C}_{b,\sigma,T}T}\big(1+\|X_0\|_p\big)t \quad (7.59)$$

where $\tilde{\kappa}'_{p,d} = 1+\kappa'_{p,d}$. Hence, plugging the right-hand side of (7.59) into the above inequality satisfied by ψ, we derive the existence of a real constant $\tilde{\kappa}_{p,d} > 0$, only depending on $C^{BDG}_{d,p}$, such that

$$\begin{aligned}\psi(t) &\le \tilde{\kappa}_{p,d}e^{\tilde{\kappa}_p\widetilde{C}_{b,\sigma,T}T}\big(1+\|X_0\|_p\big)t\left(\frac{T}{n}\right)^{\beta}\\ &\quad+\tilde{\kappa}_{p,d}C'_{b,\sigma}e^{\tilde{\kappa}_p\widetilde{C}_{b,\sigma}t}\big(1+\|X_0\|_p\big)(1+t)^2\left(\frac{T}{n}\right)^{\frac{1}{2}}\\ &\le \kappa_{p,d}C'_{b,\sigma}e^{\kappa_{p,d}(1+C'_{b,\sigma})^2t}\big(1+\|X_0\|_p\big)\left(\frac{T}{n}\right)^{\beta\wedge\frac{1}{2}},\end{aligned}$$

where we used $(1+u)^2 \le 2e^u$, $u \ge 0$, in the second line. The real constant $\kappa_{p,d}$ only depends on $\tilde{\kappa}_{p,d}$, hence on $C^{BDG}_{d,p}$. Finally, one plugs this bound into (7.58) at time T to get the announced upper-bound.

STEP 3 ($p \in (0,2)$). It remains to deal with the case $p \in [1,2)$. In fact, once we observe that Assumption (H^{β}_T) ensures the global existence and uniqueness of the solution X of (7.1) starting from a given random variable X_0 (independent of W), it can be solved following the approach developed in Step 4 of the proof of Proposition 7.6. We leave the details to the reader. ◇

Corollary 7.3 *(Lipschitz continuous framework) If b and σ satisfy Condition (H^1_T), i.e.*

$$\forall s,\, t \in [0,T],\ \forall x,\, y \in \mathbb{R}^d,\quad |b(s,x)-b(t,y)|+\|\sigma(s,x)-\sigma(t,y)\| \le C_{b,\sigma,T}\big(|t-s|+|x-y|\big),$$

then for every $p \in [1,\infty)$, there exists a real constant $\kappa_{p,d} > 0$ such that

$$\forall n \ge 1,\qquad \Big\|\sup_{t\in[0,T]}\big|X_t-\bar{X}^n_t\big|\Big\|_p \le \kappa_{p,d}C'_{b,\sigma}e^{\kappa_{p,d}(1+C'_{b,\sigma})^2t}\big(1+\|X_0\|_p\big)\left(\frac{T}{n}\right)^{\beta\wedge\frac{1}{2}}$$

where $C'_{b,\sigma,T} := C_{b,\sigma,T}+\sup_{t\in[0,T]}\big(|b(t,0)|+\|\sigma(t,0)\|\big) < +\infty$.

7.8.5 The Stepwise Constant Euler Scheme

The aim of this section is to prove in full generality Claim (b) of Theorem 7.2. We recall that the stepwise constant Euler scheme is defined by

$$\forall\, t \in [0, T], \quad \widetilde{X}_t := \bar{X}_{\underline{t}},$$

i.e. $\widetilde{X}_t = \bar{X}_{t_k}$, if $t \in [t_t, t_{k+1})$.

We saw in Sect. 7.2.1 that when $X = W$, a $\log n$ factor comes out in the *a priori* error bound. One must again keep in mind that this question is quite crucial, at least in higher dimensions, since the simulation of (functionals of) the genuine/continuous Euler scheme is not always possible (see Chap. 8) whereas the simulation of the stepwise constant Euler scheme is generally straightforward in any dimension, provided b and σ are known.

Proof of Theorem 7.2(b)**.** STEP 1 ($X_0 = x \in \mathbb{R}^d$). We may assume without loss of generality that $p \in [1, \infty)$ owing to the monotonicity of L^p-norms. Then

$$\bar{X}^n_t - \tilde{X}^n_t = \bar{X}^n_t - \bar{X}^n_{\underline{t}} = \int_{\underline{t}}^t b(\underline{s}, \bar{X}_{\underline{s}})ds + \int_{\underline{t}}^t \sigma(\underline{s}, \bar{X}_{\underline{s}})dW_s.$$

One derives that

$$\sup_{t\in[0,T]} \left|\bar{X}^n_t - \tilde{X}^n_t\right| \le \frac{T}{n} \sup_{t\in[0,T]} \left|b(\underline{t}, \bar{X}_{\underline{t}})\right| + \sup_{t\in[0,T]} \left|\sigma(\underline{t}, \bar{X}_{\underline{t}})(W_t - W_{\underline{t}})\right|. \tag{7.60}$$

Now, it follows from Proposition 7.6(b) that

$$\left\| \sup_{t\in[0,T]} \left|b(\underline{t}, \bar{X}^n_t)\right| \right\|_p \le 2e^{\kappa'_p C_{b,\sigma,T} T}\big(1 + |x|\big).$$

On the other hand, using the extended Hölder Inequality: for every $p \in (0, +\infty)$,

$$\forall\, r, s \ge 1, \ \frac{1}{r} + \frac{1}{s} = 1, \quad \|fg\|_p \le \|f\|_{rp}\ \|g\|_{sp},$$

with $r = s = 2$ (other choices are possible), leads to

$$\begin{aligned}\left\| \sup_{t\in[0,T]} |\sigma(\underline{t}, \bar{X}_{\underline{t}})(W_t - W_{\underline{t}})| \right\|_p &\le \left\| \sup_{t\in[0,T]} \left\|\sigma(\underline{t}, \bar{X}_{\underline{t}})\right\| \sup_{t\in[0,T]} |W_t - W_{\underline{t}}| \right\|_p \\ &\le \left\| \sup_{t\in[0,T]} \left\|\sigma(\underline{t}, \bar{X}_{\underline{t}})\right\| \right\|_{2p} \left\| \sup_{t\in[0,T]} |W_t - W_{\underline{t}}| \right\|_{2p}.\end{aligned}$$

Now, like for the drift b, one has

$$\left\| \sup_{t\in[0,T]} \left\|\sigma(\underline{t}, \bar{X}_{\underline{t}})\right\| \right\|_{2p} \leq 2e^{\kappa'_{2p} C_{b,\sigma,T} T}\big(1+|x|\big).$$

As concerns the Brownian term, one has

$$\left\| \sup_{t\in[0,T]} |W_t - W_{\underline{t}}| \right\|_{2p} \leq C_{W,2p}\sqrt{\frac{T}{n}(1+\log n)}$$

owing to (7.18) in Sect. 7.2.1. Finally, plugging these estimates into (7.60), yields

$$\begin{aligned}\left\| \sup_{t\in[0,T]} |\bar{X}^n_t - \widetilde{X}^n_t| \right\|_p &\leq 2\frac{T}{n}e^{\kappa'_p C_{b,\sigma,T} T}(1+|x|)\\ &\quad +2\,e^{\kappa'_{2p} C_{b,\sigma,T} T}(1+|x|)\times C_{W,2p}\sqrt{\frac{T}{n}(1+\log n)},\\ &\leq 2\big(C_{W,2p}+1\big)e^{\kappa'_{2p} C_{b,\sigma,T} T}\big(1+|x|\big)\left(\sqrt{\frac{T}{n}(1+\log n)}+\frac{T}{n}\right).\end{aligned}$$

The result follows by noting that $\sqrt{\frac{T}{n}\big(1+\log\, n\big)}+\frac{T}{n} \leq \big(1+\sqrt{T}\big)\sqrt{\frac{T}{n}\big(1+\log\, n\big)}$ for every integer $n \geq 1$ and by setting $\widetilde{\kappa}_p := 2\max\big((1+\sqrt{T})(C_{W,2p}+1),\, \kappa'_{2p}\big)$.

STEP 2 (*Random* X_0). When X_0 is no longer deterministic one uses that X_0 and W are independent so that, with obvious notations,

$$\mathbb{E}\sup_{t\in[0,T]}\big[|\bar{X}^{n,X_0}_t - \widetilde{X}^{n,X_0}_t|^p\big] = \int_{\mathbb{R}^d} \mathbb{P}_{X_0}(dx_0)\mathbb{E}\sup_{t\in[0,T]}\big[|\bar{X}^{n,x_0}_t - \widetilde{X}^{n,x_0}_t|^p\big],$$

which yields the announced result.

STEP 3 (*Combination of the upper-bounds*). This is a straightforward consequence of Claims (a) and (b). ◇

7.8.6 *Application to the a.s.-Convergence of the Euler Schemes and its Rate*

One can derive from the above L^p-rate of convergence an $a.s.$-convergence result. The main result is given in the following theorem (which extends Theorem 7.3 stated in the homogeneous Lipschitz continuous case).

Theorem 7.9 *If (H_T^β) holds and if X_0 is a.s. finite, the continuous Euler scheme $\bar{X}^n = (\bar{X}_t^n)_{t\in[0,T]}$ a.s. converges toward the diffusion X for the* sup-*norm over* $[0, T]$. *Furthermore, for every $\alpha \in [0, \beta \wedge \frac{1}{2})$,*

$$n^\alpha \sup_{t\in[0,T]} |X_t - \bar{X}_t^n| \xrightarrow{a.s.} 0.$$

The same convergence rate holds with the stepwise constant Euler scheme $(\widetilde{X}_t^n)_{t\in[0,T]}$.

Proof. We make no *a priori* integrability assumption on X_0. We rely on the localization principle at the origin. Let $N \in \mathbb{N}^*$; set $X_0^{(N)} := X_0 \mathbf{1}_{\{|X_0|\le N\}} + N \frac{X_0}{|X_0|} \mathbf{1}_{\{|X_0|>N\}}$ so that $|X_0^{(N)}| \le N$. Stochastic integration being a local operator, the solutions $(X_t^{(N)})_{t\in[0,T]}$ and $(X_t)_{t\in[0,T]}$ of the *SDE* (7.1) are equal on $\{X_0 = X_0^{(N)}\}$, namely on $\{|X_0| \le N\}$. The same property is obvious for the Euler schemes $\bar{X}^n$ and $\bar{X}^{n,(N)}$ starting from X_0 and $X_0^{(N)}$, respectively. For a fixed N, we know from Theorem 7.2 (a) that, for every $p \ge 1$,

$\exists\, C_{p,b,\sigma,\beta,T} > 0$ such that, $\forall\, n \ge 1$,

$$\mathbb{E}\Big(\sup_{t\in[0,T]} \big|\bar{X}_t^{(N)} - X_t^{(N)}\big|^p\Big) \le C_{p,b,\sigma,\beta,T} \left(\frac{T}{n}\right)^{p(\beta\wedge\frac{1}{2})} \big(1 + \|X_0\|_p\big)^p.$$

In particular,

$$\begin{aligned}
\mathbb{E}\Big(\mathbf{1}_{\{|X_0|\le N\}} \sup_{t\in[0,T]} \big|\bar{X}_t^{n,(N)} - X_t\big|^p\Big) &= \mathbb{E}\Big(\mathbf{1}_{\{|X_0|\le N\}} \sup_{t\in[0,T]} \big|\bar{X}_t^{n,(N)} - X_t^{(N)}\big|^p\Big) \\
&\le \mathbb{E}\Big(\sup_{t\in[0,T]} \big|\bar{X}_t^{n,(N)} - X_t^{(N)}\big|^p\Big) \\
&\le C_{p,b,\sigma,\beta,T}\big(1 + \|X_0^{(N)}\|_p\big)^p \left(\frac{T}{n}\right)^{p(\beta\wedge\frac{1}{2})}.
\end{aligned}$$

Let $\alpha \in (0, \beta \wedge \frac{1}{2})$ and let $p > \frac{1}{\beta\wedge\frac{1}{2}-\alpha}$. Then $\displaystyle\sum_{n\ge1} \frac{1}{n^{p(\beta\wedge\frac{1}{2}-\alpha)}} < +\infty$. Consequently, Beppo Levi's monotone convergence Theorem for series with non-negative terms implies

$$\begin{aligned}
&\mathbb{E}\Big(\mathbf{1}_{\{|X_0|\le N\}} n^{p\alpha} \sum_{n\ge1} \sup_{t\in[0,T]} \big|\bar{X}_t^n - X_t\big|^p\Big) \\
&\qquad \le C_{p,b,\sigma,\beta,T} N^p T^{p(\beta\wedge\frac{1}{2})} \sum_{n\ge1} n^{-p(\beta\wedge\frac{1}{2}-\alpha)} < +\infty.
\end{aligned}$$

Hence

$$\sum_{n\geq 1}\sup_{t\in[0,T]} n^{p\alpha}\big|\bar{X}^n_t - X_t\big|^p < +\infty, \quad \mathbb{P}\ a.s.$$

$$\text{on the event } \bigcup_{N\geq 1}\big\{|X_0|\leq N\big\} = \big\{X_0\in\mathbb{R}^d\big\} \overset{a.s.}{=} \Omega.$$

Finally, one gets:

$$\mathbb{P}\text{-}a.s. \qquad \sup_{t\in[0,T]}|\bar{X}^n_t - X_t| = o\left(\frac{1}{n^{\alpha}}\right) \quad \text{as} \quad n\to+\infty.$$

The proof for the stepwise constant Euler scheme follows exactly the same lines since an additional $\log n$ term plays no role in the convergence of the above series.

◇

Remarks and comments. • The above rate result strongly suggests that the critical index for the $a.s.$ rate of convergence is $\beta\wedge\frac{1}{2}$. The question is then: what happens when $\alpha=\beta\wedge\frac{1}{2}$? It is shown in [155, 178] that (when $\beta=1$), $\sqrt{n}(X_t-\bar{X}^n_t)\xrightarrow{\mathcal{L}}\Xi_t$, where $\Xi=(\Xi_t)_{t\in[0,T]}$ is a diffusion process driven by a Brownian motion $\widetilde{W}$ independent of W. This weak convergence holds in a functional sense, namely for the topology of the uniform convergence on $\mathcal{C}([0,T],\mathbb{R}^d)$. This process Ξ is not $\mathbb{P}$-$a.s.\equiv 0$ if $\sigma(x)\not\equiv 0$, even $a.s$ non-zero if σ never vanishes. The "weak functional" feature means first that we consider the processes as random variables taking values in their natural path space, namely the separable Banach space $\big(\mathcal{C}\big([0,T],\mathbb{R}^d\big),\|\,.\,\|_{\sup}\big)$. Then, one may consider the weak convergence of probability measures defined on the Borel σ-field of this space (see [45], Chap. 2 for an introduction). In particular, $\|.\|_{\sup}$ being trivially continuous,

$$\sqrt{n}\sup_{t\in[0,T]}|X_t-\bar{X}^n_t|\xrightarrow{\mathcal{L}}\sup_{t\in[0,T]}|\Xi_t|$$

which implies that, if Ξ is $a.s.$ non-zero, $\mathbb{P}$-$a.s.$, for every $\varepsilon>0$,

$$\overline{\lim_n}\, n^{\frac{1+\varepsilon}{2}}\sup_{t\in[0,T]}|X_t-\bar{X}^n_t| = +\infty.$$

(This easily follows either from the Skorokhod representation theorem; a direct approach is also possible.)

► **Exercise.** One considers the geometric Brownian motion $X_t=e^{-\frac{t}{2}+W_t}$ solution to

$$dX_t = X_t dW_t, \qquad X_0 = 1.$$

(a) Show that for every $n\geq 1$ and every $k\geq 0$,

$$\bar{X}_{t_k^n} = \prod_{\ell=1}^{k}\big(1+\Delta W_{t_\ell^n}\big) \quad \text{where } t_\ell^n = \frac{\ell T}{n},\ \Delta W_{t_\ell^n} = W_{t_\ell^n} - W_{t_{\ell-1}^n},\ \ell \geq 1.$$

(*b*) Show that

$$\forall \varepsilon > 0, \quad \overline{\lim_n}\, n^{\frac{1+\varepsilon}{2}} |X_T - \bar{X}_T^n| = +\infty \quad \mathbb{P}\text{-}a.s.$$

7.8.7 The Flow of an SDE, Lipschitz Continuous Regularity

If Assumption (7.2) holds, then for every $x \in \mathbb{R}^d$, there exists a unique solution, denoted by $(X_t^x)_{t\in[0,T]}$, to the *SDE* (7.1) defined on $[0, T]$ and starting from x. The mapping $(x, t) \mapsto X_t^x$ defined on $[0, T] \times \mathbb{R}^d$ is called the *flow* of the *SDE* (7.1). One defines likewise the flow of the Euler scheme(which always exists). We will now elucidate the regularity of these flows when Assumption (H_T^β) holds.

Theorem 7.10 *If the coefficients b and σ of (7.1) satisfy Assumption (H_T^β) for a real constant $C > 0$, then the unique strong solution $(X_t^x)_{t\in[0,T]}$ starting from $x \in \mathbb{R}^d$ on $[0, T]$ and the continuous Euler scheme $(\bar{X}^{n,x})_{t\in[0,T]}$ satisfy*

$\forall x, y \in \mathbb{R}^d, \forall n \geq 1$

$$\Big\| \sup_{t\in[0,T]} |X_t^x - X_t^y| \Big\|_p + \Big\| \sup_{t\in[0,T]} |\bar{X}_t^{n,x} - \bar{X}_t^{n,y}| \Big\|_p \leq 2\, e^{\kappa_3'(p,CT)} |x-y|,$$

where $\kappa_3'(p, u) = \big(2 + C(C_{d,p}^{BDG})^2\big)u,\ u \geq 0.$

Proof. We focus on the diffusion process $(X_t)_{t\in[0,T]}$. First note that if the above bound holds for some $p > 0$ then it holds true for any $p' \in (0, p)$ since the $\|\cdot\|_p$-norm is non-decreasing in p. Starting from

$$X_t^x - X_t^y = (x-y) + \int_0^t (b(s, X_s^x) - b(s, X_s^y))ds + \int_0^t \big(\sigma(s, X_s^x) - \sigma(s, X_s^y)\big)dW_s$$

one gets

$$\sup_{s\in[0,t]} |X_s^x - X_s^y| \leq |x-y| + \int_0^t |b(s, X_s^x) - b(s, X_s^y)|ds + \sup_{s\in[0,t]} \left| \int_0^s \big(\sigma(u, X_u^x) - \sigma(u, X_u^y)\big)dW_u \right|.$$

Then, setting for every $p \geq 2$, $f_p(t) := \Big\| \sup_{s\in[0,t]} |X_s^x - X_s^y| \Big\|_p$, it follows from the *BDG* Inequality (7.53) and the generalized Minkowski inequality (7.50) that

$$f_p(t) \le |x-y| + C\int_0^t \|X_s^x - X_s^y\|_p ds + C_{d,p}^{BDG}\left\|\sqrt{\int_0^t \|\sigma(s,X_s^x)-\sigma(s,X_s^y)\|^2 ds}\right\|_p$$

$$\le |x-y| + C\int_0^t \|X_s^x - X_s^y\|_p ds + C_{d,p}^{BDG}\, C\left\|\int_0^t |X_s^x - X_s^y|^2 ds\right\|_{\frac{p}{2}}^{\frac{1}{2}}$$

$$\le |x-y| + C\int_0^t \|X_s^x - X_s^y\|_p ds + C_{d,p}^{BDG}\, C\left(\int_0^t \left\|X_s^x - X_s^y\right\|_p^2 ds\right)^{\frac{1}{2}}.$$

Consequently, the function f_p satisfies

$$f_p(t) \le |x-y| + C\left(\int_0^t f_p(s)ds + C_{d,p}^{BDG}\left(\int_0^t f_p^2(s)ds\right)^{\frac{1}{2}}\right).$$

One concludes by the "à la Gronwall" Lemma 7.3 that

$$\forall\, t\in[0,T],\qquad f_p(t) \le e^{C(2+C(C_{d,p}^{BDG})^2)t}|x-y|.$$

The proof for the Euler scheme follows the same lines once we observe that $\underline{s} \in [0, s]$.

◇

7.8.8 The Strong Error Rate for the Milstein Scheme: Proof of Theorem 7.5

In this section, we return to the scalar case $d = q = 1$ and we prove Theorem 7.5. Throughout this section $C_{b,\sigma,p,T}$ and $K_{b,\sigma,p,T}$ are positive real constants that may vary from line to line.

First we note that the genuine (or continuous) Milstein scheme $(\bar{X}_t^{mil,n})_{t\in[0,T]}$, as defined by (7.38), can be written in an integral form as follows

$$\bar{X}_t^{mil,n} = X_0 + \int_0^t b(\bar{X}_{\underline{s}}^{mil,n})ds + \int_0^t \sigma(\bar{X}_{\underline{s}}^{mil,n})dW_s + \int_0^t\int_{\underline{s}}^s (\sigma\sigma')(\bar{X}_{\underline{u}}^{mil,n})dW_u dW_s \tag{7.61}$$

with our usual notation $\underline{t} = t_k^n = \frac{kT}{n}$ if $t\in[t_k^n, t_{k+1}^n)$ (so that $\underline{u} = \underline{s}$ if $u\in[\underline{s}, s]$). For notational convenience, we will also drop throughout this section the superscript ^{n} when no ambiguity arises.

(*a*) Step 1 (*Moment control*). Our first aim is to prove that the Milstein scheme has uniformly controlled moments at any order, namely that, for every $p\in(0,+\infty)$, there exists a real constant $C_{p,b,\sigma,T} > 0$ such that

$$\forall n \geq 1, \qquad \sup_{n\geq 1}\Big\| \sup_{t\in[0,T]} |\bar{X}_t^{mil,n}|\Big\|_p \leq C_{b,\sigma,T}\big(1+\|X_0\|_p\big). \tag{7.62}$$

We may assume without loss of generality that $X_0 \in L^p$ in throughout this step. Set

$$\bar{H}_s = \sigma(\bar{X}_{\underline{s}}^{mil}) + \int_{\underline{s}}^{s} (\sigma\sigma')(\bar{X}_{\underline{u}}^{mil})dW_u = \sigma(\bar{X}_{\underline{s}}^{mil}) + (\sigma\sigma')(\bar{X}_{\underline{s}}^{mil})(W_s - W_{\underline{s}})$$

so that

$$\bar{X}_t^{mil} = X_0 + \int_0^t b(\bar{X}_{\underline{s}}^{mil})ds + \int_0^t \bar{H}_s\, dW_s.$$

It follows from the boundedness of b' and σ' that b and σ satisfy a linear growth assumption.

We will follow the lines of the proof of Proposition 7.6, the specificity of the Milstein framework being that the diffusion coefficient is replaced by the process $\bar{H}_s$. So, our task is to control the term

$$\sup_{s\in[0,t]}\left|\int_0^{s\wedge\bar{\tau}_N} \bar{H}_u dW_u\right|$$

in L^p where $\bar{\tau}_N = \bar{\tau}_N^n := \inf\big\{t\in[0,T] : |\bar{X}_s^{mil,n} - X_0| > N\big\}, n, N \geq 1$.

First assume that $p \in [2,\infty)$. Since $\int_0^{t\wedge\bar{\tau}_N} \bar{H}_s dW_s$ is a continuous local martingale, it follows from the *BDG* Inequality (7.51) that

$$\left\|\sup_{s\in[0,t]}\left|\int_0^{s\wedge\bar{\tau}_N} \bar{H}_u dW_u\right|\right\|_p \leq C_p^{BDG}\left\|\int_0^{t\wedge\bar{\tau}_N} \bar{H}_s^2 ds\right\|_{\frac{p}{2}}^{\frac{1}{2}}.$$

Consequently, using the generalized Minkowski Inequality (7.50)

$$\begin{aligned}\left\|\sup_{s\in[0,t]}\left|\int_0^{s\wedge\bar{\tau}_N} \bar{H}_u dW_u\right|\right\|_p &\leq C_p^{BDG}\left[\int_0^t \|\mathbf{1}_{\{s\leq\bar{\tau}_N\}}\bar{H}_s\|_p^2 ds\right]^{\frac{1}{2}}\\ &= C_p^{BDG}\left[\int_0^t \|\mathbf{1}_{\{s\leq\bar{\tau}_N\}}\bar{H}_{s\wedge\bar{\tau}_N}\|_p^2 ds\right]^{\frac{1}{2}}.\end{aligned}$$

Now, for every $s \in [0,t]$,

$$\begin{aligned}\left\|\mathbf{1}_{\{s\le\bar\tau_N\}}\bar H_{s\wedge\bar\tau_N}\right\|_p &\le \left\|\sigma(\bar X^{mil}_{\underline{s}\wedge\bar\tau_N})\right\|_p + \left\|\mathbf{1}_{\{s\le\bar\tau_N\}}(\sigma\sigma')(\bar X^{mil}_{\underline{s}\wedge\bar\tau_N})(W_{s\wedge\bar\tau^N}-W_{\underline{s}\wedge\bar\tau_N})\right\|_p\\ &\le \left\|\sigma(\bar X^{mil}_{\underline{s}\wedge\bar\tau_N})\right\|_p + \left\|(\sigma\sigma')(\bar X^{mil}_{\underline{s}\wedge\bar\tau_N})(W_s-W_{\underline{s}})\right\|_p\\ &= \left\|\sigma(\bar X^{mil}_{\underline{s}\wedge\bar\tau_N})\right\|_p + \left\|(\sigma\sigma')(\bar X^{mil}_{\underline{s}\wedge\bar\tau_N})\right\|_p\left\|W_s-W_{\underline{s}}\right\|_p,\end{aligned}$$

where we used that $(\sigma\sigma')(\bar X^{mil}_{\underline{s}\wedge\bar\tau_N})$ is $\mathcal{F}_{\underline{s}}$-measurable, hence independent of $W_s - W_{\underline{s}}$. Using that σ and $\sigma\sigma'$ have at most linear growth since σ' is bounded, we derive

$$\left\|\mathbf{1}_{\{s\le\bar\tau_N\}}\bar H_{s\wedge\bar\tau_N}\right\|_p \le C_{b,\sigma,T}\big(1+\left\|\bar X^{mil}_{\underline{s}\wedge\bar\tau_N}\right\|_p\big).$$

Finally, following the lines of the first step of the proof of Proposition 7.6 leads to

$$\left\|\sup_{s\in[0,t]}\left|\int_0^{s\wedge\bar\tau_N}\bar H_u dW_u\right|\right\|_p \le C_{b,\sigma,T}C_p^{BDG}\left(\sqrt{t}+\left[\int_0^t\Big\|\sup_{u\in[0,s\wedge\bar\tau_N]}|\bar X^{mil}_u|\Big\|_p^2 ds\right]^{\frac12}\right).$$

Still following the lines of the proof Proposition 7.6, we include the step 4 to deal with the case $p\in(0,2)$.

Moreover, we get as a by-product that, for every $p>0$ and every $n\ge1$,

$$\left\|\sup_{t\in[0,T]}|\bar H_t|\right\|_p \le K_{b,\sigma,T,p}\big(1+\|X_0\|_p\big) < +\infty, \tag{7.63}$$

where $K_{b,\sigma,T,p}$ does not depend on the discretization step n. As a matter of fact, this follows from

$$\sup_{t\in[0,T]}|\bar H_t| \le C_{b,\sigma}\Big(1+\sup_{t\in[0,T]}|\bar X^{mil,n}_t|\Big)\Big(1+2\sup_{t\in[0,T]}|W_t|\Big)$$

so that, by the Schwarz Inequality when $p\ge 1/2$,

$$\left\|\sup_{t\in[0,T]}|\bar H_t|\right\|_p \le C_{b,\sigma}\Big(1+\sup_{n\ge1}\|\sup_{t\in[0,T]}|\bar X^{mil,n}_t|\|_{2p}\Big)\Big(1+2\|\sup_{t\in[0,T]}|W_t|\|_{2p}\Big),$$

where we used that $\|\cdot\|_{2p}$ is a norm in the right-hand side of the inequality. A similar bound holds when $p\in(0,1/2)$ since $\|1+V\|_{2p}\le 2^{\frac{1}{2p}}\big(1+\|V\|_{2p}\big)$ for any random variable V.

Now, by Lemma 7.4 devoted to the L^p-regularity of Itô processes, one derives the existence of a real constant $\kappa_{b,\sigma,p,T}\in(0,+\infty)$ (not depending on $n\ge1$) such that

$$\forall\, t\in[0,T],\ \forall\, n\ge1,\ \ \|\bar X^{mil,n}_t-\bar X^{mil,n}_{\underline{t}}\|_p \le \kappa_{b,\sigma,p,T}\big(1+\|X_0\|_p\big)\left(\frac{T}{n}\right)^{\frac12}. \tag{7.64}$$

STEP 2 (*Decomposition and analysis of the error when $p \in [2, +\infty)$, $X_0 = x \in \mathbb{R}^d$*). Set $\varepsilon_t := X_t - \bar{X}_t^{mil}$, $t \in [0, T]$, and

$$f_p(t) := \Big\| \sup_{s\in[0,t]} |\varepsilon_s| \Big\|_p, \quad t \in [0, T].$$

Using the diffusion equation and the continuous Milstein scheme one gets

$$\begin{aligned}
\varepsilon_t &= \int_0^t \big(b(X_s) - b(\bar{X}_{\underline{s}}^{mil})\big)ds + \int_0^t \big(\sigma(X_s) - \sigma(\bar{X}_{\underline{s}}^{mil})\big)dW_s \\
&\quad - \int_0^t \int_{\underline{s}}^s (\sigma\sigma')(\bar{X}_{\underline{u}}^{mil})dW_u dW_s \\
&= \int_0^t \big(b(X_s) - b(\bar{X}_s^{mil})\big)ds + \int_0^t \big(\sigma(X_s) - \sigma(\bar{X}_s^{mil})\big)dW_s \\
&\quad + \int_0^t \big(b(\bar{X}_s^{mil}) - b(\bar{X}_{\underline{s}}^{mil})\big)ds \\
&\quad + \int_0^t \Big(\sigma(\bar{X}_s^{mil}) - \sigma(\bar{X}_{\underline{s}}^{mil}) - (\sigma\sigma')(\bar{X}_{\underline{s}}^{mil})(W_s - W_{\underline{s}})\Big)dW_s.
\end{aligned}$$

First, one derives that

$$\begin{aligned}
\sup_{s\in[0,t]} |\varepsilon_s| &\le \|b'\|_{\sup} \int_0^t \sup_{u\in[0,s]} |\varepsilon_u| ds + \sup_{s\in[0,t]} \left| \int_0^s \big(\sigma(X_u) - \sigma(\bar{X}_u^{mil})\big)dW_u \right| \\
&\quad + \sup_{s\in[0,t]} \left| \int_0^s \big(b(\bar{X}_u^{mil}) - b(\bar{X}_{\underline{u}}^{mil})\big)du \right| \\
&\quad + \sup_{s\in[0,t]} \left| \int_0^s \Big(\sigma(\bar{X}_u^{mil}) - \sigma(\bar{X}_{\underline{u}}^{mil}) - (\sigma\sigma')(\bar{X}_{\underline{u}}^{mil})(W_u - W_{\underline{u}})\Big)dW_u \right|
\end{aligned}$$

so that, using twice the generalized Minkowski Inequality (7.50) and the *BDG* Inequality (7.51), one gets classically

$$\begin{aligned}
f_p(t) &\le \|b'\|_{\sup} \int_0^t f_p(s)ds + C_p^{BDG} \|\sigma'\|_{\sup} \sqrt{\int_0^t f_p(s)^2 ds} \\
&\quad + \underbrace{\left\| \sup_{s\in[0,t]} \left| \int_0^s \big(b(\bar{X}_u^{mil}) - b(\bar{X}_{\underline{u}}^{mil})\big)du \right| \right\|_p}_{(B)} \\
&\quad + \underbrace{\left\| \sup_{s\in[0,t]} \left| \int_0^s \Big(\sigma(\bar{X}_u^{mil}) - \sigma(\bar{X}_{\underline{u}}^{mil}) - (\sigma\sigma')(\bar{X}_{\underline{u}}^{mil})(W_u - W_{\underline{u}})\Big)dW_u \right| \right\|_p}_{(C)}.
\end{aligned}$$

Now using that b' is $\alpha_{b'}$-Hölder yields for every $u \in [0, T]$,

$$\begin{aligned} b(\bar{X}_u^{mil}) - b(\bar{X}_{\underline{u}}^{mil}) &= b'(\bar{X}_{\underline{u}}^{mil})(\bar{X}_u^{mil} - \bar{X}_{\underline{u}}^{mil}) + \rho_b(u)|\bar{X}_u^{mil} - \bar{X}_{\underline{u}}^{mil}|^{1+\alpha_{b'}} \\ &= b\, b'(\bar{X}_{\underline{u}}^{mil})(u - \underline{u}) + b'(\bar{X}_{\underline{u}}^{mil}) \int_{\underline{u}}^{u} \bar{H}_v dW_v \\ &\quad + \rho_b(u)\big|\bar{X}_u^{mil} - \bar{X}_{\underline{u}}^{mil}\big|^{1+\alpha_{b'}}, \end{aligned}$$

where $\rho_b(u)$ is defined by the above equation on the event $\{\bar{X}_u^{mil} \neq \bar{X}_{\underline{u}}^{mil}\}$ and is equal to 0 otherwise. This defines an $(\mathcal{F}_u)$-adapted process, bounded by the Hölder coefficient $[b']_{\alpha_{b'}}$ of b'. Using that for every $\xi \in \mathbb{R}$, $|bb'|(\xi) \leq \|b'\|_{\sup}\big(\|b'\|_{\sup} + |b(0)|\big)|\xi|$ and (7.64) yields

$$\begin{aligned} (B) \leq\ & \|b'\|_{\sup}\big(\|b'\|_{\sup} + |b(0)|\big)\Big\| \sup_{t\in[0,T]} |\bar{X}_t^{mil}|\Big\|_p \frac{T}{n} \\ & + [b']_{\alpha_{b'}} K_{b,\sigma,p,T}\big(1 + |x|\big) \left(\frac{T}{n}\right)^{\frac{1+\alpha_{b'}}{2}} \\ & + \Big\| \sup_{s\in[0,t]} \int_0^s \Big(b'(\bar{X}_{\underline{u}}^{mil}) \int_{\underline{u}}^{u} \bar{H}_v dW_v\Big) du \Big\|_p . \end{aligned}$$

The last term in the right-hand side of the above equation needs a specific treatment: a naive approach would yield a $\sqrt{\frac{T}{n}}$ term that would make the whole proof crash down. So we will transform the regular Lebesgue integral into a stochastic integral (hence a local martingale). This can be done either by a stochastic Fubini theorem, or in a more elementary way by an integration by parts.

Lemma 7.5 *Let $G : \Omega \times \mathbb{R} \to \mathbb{R}$ be an $(\mathcal{F}_t)_{t\in[0,T]}$-progressively measurable process such that $\int_0^T G_s^2 ds < +\infty$ a.s. Set $\bar{s} := \frac{kT}{n}$ if $s \in \Big[\frac{(k-1)T}{n}, \frac{kT}{n}\Big)$. Then for every $t \in [0, T]$,*

$$\int_0^t \left(\int_{\underline{s}}^s G_u dW_u\right) ds = \int_0^t (\bar{s} \wedge t - s) G_s dW_s .$$

Proof. For every $k = 1, \ldots, n$, an integration by parts yields

$$\int_{\frac{(k-1)T}{n}}^{\frac{kT}{n}} \left(\int_{\underline{s}}^s G_u dW_u\right) ds = \int_{\frac{(k-1)T}{n}}^{\frac{kT}{n}} \left(\int_{\frac{(k-1)T}{n}}^s G_u dW_u\right) ds = \int_{\frac{(k-1)T}{n}}^{\frac{kT}{n}} \Big(\frac{kT}{n} - s\Big) G_s dW_s .$$

Likewise, if $t \in \Big[\frac{(\ell-1)T}{n}, \frac{\ell T}{n}\Big)$, then $\displaystyle\int_{\frac{(\ell-1)T}{n}}^t \left(\int_{\underline{s}}^s G_u dW_u\right) ds = \int_{\frac{(\ell-1)T}{n}}^t (t - s) G_s \, dW_s$, which completes the proof by summing all the terms from $k = 1$ up to ℓ with the last one. ◇

We apply this lemma to the continuous adapted process $G_t = b'(\bar{X}^{mil}_{\underline{t}})\bar{H}_t$. We derive by standard arguments that

$$\begin{aligned}\left\|\sup_{s\in[0,t]}\int_0^s\Big(b'(\bar{X}^{mil}_{\underline{u}})\int_{\underline{u}}^u \bar{H}_v dW_v\Big)du\right\|_p &\le C_p^{BDG}\left(\int_0^T \|(\bar{t}\wedge T-t)b'(\bar{X}^{mil}_{\underline{t}})\bar{H}_t\|_p^2 dt\right)^{\frac{1}{2}}\\ &\le C_p^{BDG}\|b'\|_{\sup}\frac{T}{n}\left(\int_0^T\|\bar{H}_t\|_p^2 dt\right)^{\frac{1}{2}}\\ &\le C_{b,\sigma,p,T}\big(1+\|X_0\|_p\big)\frac{T}{n},\end{aligned}$$

where we used first that $0\le \bar{t}\wedge T - t\le \frac{T}{n}$ and then (7.63). Finally, one gets that

$$(B)\le C_{b,\sigma,p,T}\big(1+|x|\big)\left(\frac{T}{n}\right)^{1+\alpha_{b'}}.$$

We adopt a similar approach for the term (C). Elementary computations show that

$$\begin{aligned}&\sigma(\bar{X}^{mil}_u)-\sigma(\bar{X}^{mil}_{\underline{u}})-(\sigma\sigma')(\bar{X}^{mil}_{\underline{u}})(W_u-W_{\underline{u}})\\ &\quad=\sigma' b(\bar{X}^{mil}_{\underline{u}})\frac{T}{n}+\frac{1}{2}\sigma(\sigma')^2(\bar{X}^{mil}_{\underline{u}})\Big((W_u-W_{\underline{u}})^2-(u-\underline{u})\Big)\\ &\quad+\rho_\sigma(u)\big|\bar{X}^{mil}_u-\bar{X}^{mil}_{\underline{u}}\big|^{1+\alpha_{\sigma'}}\end{aligned}$$

where $\rho_\sigma(u)$ is an $(\mathcal{F}_u)$-adapted process bounded by the Hölder coefficient $[\sigma']_{\alpha_{\sigma'}}$ of σ'. Consequently, for every $p\ge 1$,

$$\begin{aligned}&\left\|\sigma(\bar{X}^{mil}_u)-\sigma(\bar{X}^{mil}_{\underline{u}})-(\sigma\sigma')(\bar{X}^{mil}_{\underline{u}})(W_u-W_{\underline{u}})\right\|_p\\ &\quad\le \left\|\sigma' b(\bar{X}^{mil}_{\underline{u}})\right\|_p(u-\underline{u})+\frac{1}{2}\left\|\sigma(\sigma')^2(\bar{X}^{mil}_{\underline{u}})\right\|_p\left\|(W_u-W_{\underline{u}})^2-(u-\underline{u})\right\|_p\\ &\qquad+[\sigma']_{\alpha_{\sigma'}}\left\|\big|\bar{X}^{mil}_u-\bar{X}^{mil}_{\underline{u}}\big|^{1+\alpha_{\sigma'}}\right\|^{1+\alpha_{\sigma'}}_{p(1+\alpha_{\sigma'})}\\ &\quad\le C_{b,\sigma,p,T}\big(1+|x|\big)\Big((u-\underline{u})+\|Z^2-1\|_p(u-\underline{u})+[\sigma']_{\alpha_{\sigma'}}(u-\underline{u})^{\frac{1+\alpha_{\sigma'}}{2}}\Big)\\ &\quad\le C_{b,\sigma,p,T}\big(1+|x|\big)\left(\frac{T}{n}\right)^{\frac{1+\alpha_{\sigma'}}{2}}.\end{aligned}$$

Now, owing to*BDG* Inequality, we derive, for every $p \geq 2$, that

$$(C) \leq C_p^{BDG}\left(\int_0^t \left\|\sigma(\bar{X}_u^{mil}) - \sigma(\bar{X}_{\underline{u}}^{mil}) - (\sigma\sigma')(\bar{X}_{\underline{u}}^{mil})(W_u - W_{\underline{u}})\right\|_p^2 du\right)^{\frac{1}{2}}$$
$$\leq C_p^{BDG} C_{b,\sigma,p,T}\big(1 + |x|\big)\left(\frac{T}{n}\right)^{\frac{1+\alpha_{\sigma'}}{2}}.$$

Finally, combining the upper-bounds for (B) and (C) leads to

$$f_p(t) \leq \|b'\|_{\sup} \int_0^t f_p(s)ds + C_p^{BDG}\|\sigma'\|_{\sup}\sqrt{\int_0^t f^2(s)ds} + C_{b,\sigma,p,T}(1 + |x|)\left(\frac{T}{n}\right)^{\frac{1+\alpha_{\sigma'}\wedge\alpha_{b'}}{2}}$$

so that, owing to the "à la Gronwall" Lemma 7.3, there exists a real constant

$$f_p(T) \leq C_{b,\sigma,p,T}(1 + |x|)\left(\frac{T}{n}\right)^{\frac{1+\alpha_{\sigma'}\wedge\alpha_{b'}}{2}}.$$

STEP 2 (*Extension to* $p \in (0, 2)$ *and random starting values* X_0). First one uses that $p \mapsto \|\cdot\|_p$ is non-decreasing to extend the above bound to $p \in (0, 2)$. Then, one uses that, if X_0 and W are independent, for any non-negative functional $\Phi : \mathcal{C}([0, T], \mathbb{R}^d)^2 \to \mathbb{R}_+$, one has with obvious notations

$$\mathbb{E}\Phi(X, \bar{X}^{mil}) = \int_{\mathbb{R}^d} \mathbb{P}_{X_0}(dx_0)\, \mathbb{E}\,\Phi(X^{x_0}, \bar{X}^{mil,x_0}).$$

Applying this identity with $\Phi(x, \bar{x}) = \sup_{t\in[0,T]} |x(t) - \bar{x}(t)|^p$, $x, \bar{x} \in \mathcal{C}([0, T], \mathbb{R})$, completes the proof.

(*b*) This second claim follows from the error bound established for the Brownian motion itself: as concerns the Brownian motion, both stepwise constant and continuous versions of the Milstein and the Euler scheme coincide. So a better convergence rate is hopeless. ◇

7.8.9 The Feynman–Kac Formula and Application to the Weak Error Expansion by the *PDE* Method

In this section we return to the purely scalar case $d = q = 1$, mainly for notational convenience, namely we consider the scalar version of (7.1), *i.e.*

$$dX_t = b(t, X_t)dt + \sigma(t, X_t)dW_t,$$

where W is a standard (scalar) Brownian motion defined on a probability space $(\Omega, \mathcal{A}, \mathbb{P})$ and X_0, defined on the same space, is independent of W. We make the following regularity assumption on b and σ:

$$\big(C_\infty\big) \equiv \; b, \sigma \; \in \mathcal{C}^\infty([0,T]\times\mathbb{R}) \quad \text{and} \quad \forall\, k_1,\, k_2 \in \mathbb{N},\; k_1+k_2 \geq 1,$$
$$\sup_{(t,x)\in[0,T]\times\mathbb{R}} \left|\frac{\partial^{k_1+k_2} b}{\partial t^{k_1}\partial x^{k_2}}(t,x)\right| + \left|\frac{\partial^{k_1+k_2} \sigma}{\partial t^{k_1}\partial x^{k_2}}(t,x)\right| < +\infty.$$

In particular, b and σ are Lipschitz continuous in $(t,x)\in[0,T]\times\mathbb{R}$. Thus, for every $t,\, t'\in[0,T]$ and every $x,\, x'\in\mathbb{R}$,

$$|b(t',x') - b(t,x)| \leq \sup_{(s,\xi)\in[0,T]\times\mathbb{R}} \left|\frac{\partial b}{\partial x}(s,\xi)\right| |x'-x| + \sup_{(s,\xi)\in[0,T]\times\mathbb{R}} \left|\frac{\partial b}{\partial t}(s,\xi)\right| |t'-t|.$$

Consequently, the *SDE* (7.1) always has a unique strong solution $(X_t)_{t\in[0,T]}$ starting from any $\mathbb{R}^d$-valued random vector X_0, independent of the Brownian motion W on $(\Omega, \mathcal{A}, \mathbb{P})$. Furthermore, as

$$|b(t,x)| + |\sigma(t,x)| \leq \sup_{t\in[0,T]} \big(|b(t,0)| + |\sigma(t,0)|\big) + C|x| \leq C'(1+|x|),$$

any such strong solution $(X_t)_{t\in[0,T]}$ satisfies (see Proposition 7.6):

$$\forall\, p\geq 1,\;\; X_0\in L^p(\mathbb{P}) \Longrightarrow \mathbb{E}\Big(\sup_{t\in[0,T]}|X_t|^p\Big) + \sup_n \mathbb{E}\Big(\sup_{t\in[0,T]}|\bar{X}^n_t|^p\Big) < +\infty. \quad (7.65)$$

We recall that the infinitesimal generator L of the diffusion reads on every function $g\in\mathcal{C}^{1,2}([0,T]\times\mathbb{R})$

$$Lg(t,x) = b(t,x)\frac{\partial g}{\partial x}(t,x) + \frac{1}{2}\sigma^2(t,x)\frac{\partial^2 g}{\partial x^2}(t,x).$$

As for the *source* function f appearing in the Feynman–Kac formula, which will also be the function of interest for the weak error expansion, we make the following regularity and growth assumption:

$$(F_\infty) \equiv \begin{cases} f \in \mathcal{C}^\infty(\mathbb{R},\mathbb{R}) \\ \text{and} \\ \forall\, k\in\mathbb{N},\; \exists\, r_k\in\mathbb{N},\; \exists\, C_k \in (0,+\infty),\;\; |f^{(k)}(x)| \leq C(1+|x|^{r_k}), \end{cases}$$

where $f^{(k)}$ denotes the k-th derivative of f.

The first result of the section is the fundamental link between Stochastic Differential Equations and Partial Differential Equations, namely the representation of the solution of the above parabolic *PDE* as the expectation of a marginal function of the *SDE* at its terminal time T whose infinitesimal generator is the second-order differential operator of the *PDE*. This representation is known as the Feynman–Kac formula.

Theorem 7.11 (Feynman–Kac formula) *Assume* (C_∞) *and* (F_∞) *hold.*
(a) *The parabolic PDE*

$$\frac{\partial u}{\partial t} + Lu = 0, \qquad u(T,\,.\,) = f \tag{7.66}$$

has a unique solution $u \in \mathcal{C}^\infty([0,T]\times\mathbb{R},\mathbb{R})$. *This solution satisfies*

$$\forall\, k \geq 0,\ \exists\, r(k,T) \in \mathbb{N} \ \text{ such that } \ \sup_{t\in[0,T]}\left|\frac{\partial^k u}{\partial x^k}(t,x)\right| \leq C_{k,T}\big(1+|x|^{r(k,T)}\big). \tag{7.67}$$

(b) Feynman–Kac formula*: If* $X_0 \in L^{r_0}(\mathbb{P})$, *the solution u admits the following representation*

$$\forall\, t \in [0,T],\quad u(t,x) = \mathbb{E}\big(f(X_T)\,|\,X_t = x\big) = \mathbb{E}\, f(X_T^{t,x}), \tag{7.68}$$

where $(X_s^{t,x})_{s\in[t,T]}$ *denotes the unique solution of the SDE (7.1) starting at x at time* t. *If, furthermore, the SDE is autonomous—namely* $b(t,x) = b(x)$ *and* $\sigma(t,x) = \sigma(x)$*—then*

$$\forall\, t \in [0,T],\quad u(t,x) = \mathbb{E}\, f\big(X_{T-t}^x\big).$$

NOTATION: To alleviate notation, we will use throughout this section the notations $\partial_x f, \partial_t f, \partial_{xt} f$, etc, for the partial derivatives instead of $\frac{\partial f}{\partial x}, \frac{\partial f}{\partial t}, \frac{\partial^2 u}{\partial x \partial t}, \ldots$

► **Exercise.** Combining the above bound for the spatial partial derivatives with $\partial_t u = -Lu$, show that

$$\forall\, k = (k_1,k_2) \in \mathbb{N}^2,\ \exists\, r(k,T) \in \mathbb{N} \ \text{ such that}$$

$$\sup_{t\in[0,T]}\left|\frac{\partial^{k_1+k_2} u}{\partial t^{k_1}\partial x^{k_2}}(t,x)\right| \leq C_{k,T}\big(1+|x|^{r(k,T)}\big).$$

Proof. (*a*) For this result, we refer to [2].
(*b*) Let u be the solution to the parabolic *PDE* (7.66). For every $t \in [0, T]$,

$$\begin{aligned} u(T, X_T) &= u(t, X_t) + \int_t^T \partial_t u(s, X_s)ds + \int_t^T \partial_x u(s, X_s)dX_s \\ &\quad + \frac{1}{2}\int_t^T \partial_{xx} u(s, X_s)d\langle X_s\rangle \\ &= u(t, X_t) + \int_t^T \big(\partial_t u + Lu\big)(s, X_s)ds + \int_t^T \partial_x u(s, X_s)\sigma(s, X_s)dW_s \\ &= u(t, X_t) + \int_t^T \partial_x u(s, X_s)\sigma(s, X_s)dW_s \end{aligned} \tag{7.69}$$

since u satisfies the *PDE* (7.66). Now the local martingale $M_t := \int_0^t \partial_x u(s, X_s)$ $\sigma(s, X_s)dW_s$ is a true martingale since

$$\langle M\rangle_t = \int_0^t \big(\partial_x u(s, X_s)\big)^2 \sigma^2(s, X_s)ds \le C\Big(1 + \sup_{s\in[0,T]} |X_s|^\theta\Big) \in L^1(\mathbb{P})$$

for an exponent $\theta \ge 0$. The above inequality follows from the assumptions on σ and the resulting growth properties of $\partial_x u$. The integrability is a consequence of Eq. (7.54). Consequently $(M_t)_{t\in[0,T]}$ is a true martingale and, using the assumption $u(T, .) = f$, we deduce that

$$\forall\, t \in [0, T], \qquad \mathbb{E}\big(f(X_T)\,|\,\mathcal{F}_t\big) = u(t, X_t).$$

The chain rule for conditional expectation implies,

$$\forall\, t \in [0, T], \qquad \mathbb{E}\big(f(X_T)\,|\,X_t\big) = u(t, X_t) \quad \mathbb{P}\text{-}a.s.$$

since X_t is $\mathcal{F}_t$-measurable. This shows that $u(t, .)$ is a regular version of the conditional expectation on the left-hand side.

If we rewrite (7.69) with the solution $(X_s^{t,x})_{s\in[0,T-t]}$, the same reasoning shows that

$$u(T, X_T^{t,x}) = u(t, x) + \int_t^T \partial_x u(s, X_s^{t,x})\sigma(s, X_s)dW_s$$

and taking expectation yields $u(t, x) = \mathbb{E}\, u(T, X_T^{t,x}) = \mathbb{E}\, f(X_T^{t,x})$.

When b and σ do not depend on t, one checks that $(X_{t+s}^{t,x})_{s\in[0,T-t]}$ is a solution to (7.1) starting at x at time 0, where W is replaced by the standard Brownian motion $W_s^{(t)} = W_{t+s} - W_t$, $s \in [0, T-t]$, whereas $(X_s^x)_{s\in[0,T-t]}$ is solution to (7.1) starting at x at time 0, driven by the original Brownian motion W. As b and σ are Lipschitz continuous, (7.1) satisfies a strong (*i.e.* pathwise) existence-uniqueness property. Following *e.g.* [251] (see Theorem (1.7), Chap. IX, p. 368), this implies *weak uniqueness*, *i.e.* that $(X^{t,x}_{t+s})_{s\in[0,T-t]}$ and $(X^x_s)_{s\in[0,T-t]}$ have the same distribution. Hence $\mathbb{E}\, f(X^{t,x}_T) = \mathbb{E}\, f(X^x_{T-t})$ which completes the proof. ◇

Remarks. • The proof of the Feynman–Kac formula itself (Claim (*b*)) only needs u to be $\mathcal{C}^{1,2}$ and b and σ to be continuous on $[0, T] \times \mathbb{R}$ and Lipschitz in x uniformly in $t \in [0, T]$.

• In the time homogeneous case $b(t, x) = b(x)$ and $\sigma(t, x) = \sigma(x)$, one can proceed by verification. Under smoothness assumption on b and σ, say $\mathcal{C}^2$ with bounded existing derivatives and Hölder second-order partial derivatives, one shows, using the tangent process of the diffusion, that the function $u(t, x)$ defined by $u(t, x) = \mathbb{E}\, f(X^x_{T-t})$ is $\mathcal{C}^{1,2}$ in (t, x). Then, the above claim (*b*) shows the existence of a solution to the parabolic PDE (7.66).

▶ **Exercise** (0-*order term*). If u is a $\mathcal{C}^{1,2}$-solution of the $PDE \equiv \partial_t u + Lu + ru = 0$, $u(T, .) = f$ where $r : [0, T] \times \mathbb{R} \to \mathbb{R}$ is a bounded continuous function, then show that, for every $t \in [0, T]$,

$$u(t, X_t) = \mathbb{E}\left(e^{\int_t^T r(s,X_s)ds} f(X_T)\right) \quad \mathbb{P}\text{-}a.s.$$

or, equivalently, that $u(t, x)$ is a regular version of the conditional expectation

$$u(t, X_t) = \mathbb{E}\left(e^{\int_t^T r(s,X_s)ds} f(X_T) \mid X_t = x\right) = \mathbb{E}\left(e^{\int_t^T r(s,X^{t,x}_s)ds} f(X^{t,x}_T)\right).$$

We may now pass to the second result of this section, the Talay–Tubaro weak error expansion theorem, stated here in the non-homogeneous case (but in one dimension).

Theorem 7.12 (Smooth case) *(Talay–Tubaro [270]): Assume that b and σ satisfy (C_∞), that f satisfies (F_∞) and that $X_0 \in L^{r_0}(\mathbb{P})$. Then, the weak error can be expanded at any order, namely*

$$\forall\, R \in \mathbb{N}^*,\quad \mathbb{E}\, f(\bar{X}^n_T) - \mathbb{E}\, f(X_T) = \sum_{k=1}^{R} \frac{c_k}{n^k} + O\Big(\frac{1}{n^{R+1}}\Big).$$

Remarks. • The result at a given order R also holds under weaker smoothness assumptions on b, σ and f (say $\mathcal{C}^{R+5}_b$, see [134]).

• Standard arguments show that the coefficients c_k in the above expansion are the first R terms of a sequence $(c_k)_{k\geq 1}$.

Proof ($R = 1$) Following the original approach developed in [270], we rely on the *PDE* method.

STEP 1 (*Representing and estimating* $\mathbb{E}\, f(X_T) - \mathbb{E}\, f(\bar{X}^n_T)$). It follows from the Feynman–Kac formula (7.68) and the terminal condition $u(T,.) = f$ that

$$\mathbb{E}\, f(X_T) = \int_{\mathbb{R}^d} \mathbb{E}\, f(X^x_T)\mathbb{P}_{X_0}(dx) = \int_{\mathbb{R}^d} u(0,x)\mathbb{P}_{X_0}(dx) = \mathbb{E}u(0, X_0)$$

and $\mathbb{E}\, f(\bar{X}^n_T) = \mathbb{E}\, u(T, \bar{X}^n_T)$. It follows that

$$\begin{aligned}\mathbb{E}\left(f(\bar{X}^n_T) - f(X_T)\right) &= \mathbb{E}\left(u(T, \bar{X}^n_T) - u(0, \bar{X}^n_0)\right)\\ &= \sum_{k=1}^{n} \mathbb{E}\left(u(t_k, \bar{X}^n_{t_k}) - u(t_{k-1}, \bar{X}^n_{t_{k-1}})\right).\end{aligned}$$

In order to evaluate the increment $u(t_k, \bar{X}^n_{t_k}) - u(t_{k-1}, \bar{X}^n_{t_{k-1}})$, we apply Itô's formula (see Sect. 12.8) between t_{k-1} and t_k to the function u and use that the Euler scheme satisfies the pseudo-*SDE* with "frozen" coefficients

$$d\bar{X}^n_t = b(\underline{t}, \bar{X}^n_{\underline{t}})dt + \sigma(\underline{t}, \bar{X}^n_{\underline{t}})dW_t.$$

Doing so, we obtain

$$\begin{aligned}u(t_k, \bar{X}^n_{t_k}) - u(t_{k-1}, \bar{X}^n_{t_{k-1}}) &= \int_{t_{k-1}}^{t_k} \partial_t u(s, \bar{X}^n_s)ds + \int_{t_{k-1}}^{t_k} \partial_x u(s, \bar{X}^n_s)d\bar{X}^n_s\\ &\quad + \frac{1}{2}\int_{t_{k-1}}^{t_k} \partial_{xx} u(s, \bar{X}^n_s)d\langle \bar{X}^n\rangle_s\\ &= \int_{t_{k-1}}^{t_k} (\partial_t + \bar{L})u(s, \underline{s}, \bar{X}^n_s, \bar{X}^n_{\underline{s}})ds\\ &\quad + \int_{t_{k-1}}^{t_k} \sigma(\underline{s}, \bar{X}^n_{\underline{s}})\partial_x u(s, \bar{X}^n_s)dW_s,\end{aligned}$$

where $\bar{L}$ is the "frozen" infinitesimal generator defined on functions $g \in \mathcal{C}^{1,2}([0,T]\times\mathbb{R})$ by

$$\bar{L}g(s, \underline{s}, x, \underline{x}) = b(\underline{s}, \underline{x})\partial_x g(s,x) + \frac{1}{2}\sigma^2(\underline{s}, \underline{x})\partial_{xx} g(s,x)$$

and $\partial_t g(s, \underline{s}, x, \underline{x}) = \partial_t g(s,x)$.

The bracket process of the local martingale $M_t = \int_0^t \partial_x u(s, \bar{X}^n_s)\sigma(\underline{s}, \bar{X}^n_{\underline{s}})dW_s$ is given for every $t \in [0,T]$ by

$$\langle M\rangle_T = \int_0^T \left(\partial_x\, u(s, \bar{X}_s)\right)^2 \sigma^2(\underline{s}, \bar{X}^n_{\underline{s}})ds.$$

Consequently, using that σ has (at most) linear growth in x, uniformly in $t \in [0, T]$, and the control (7.67) of $\partial_x u(s, x)$, we have

$$\langle M\rangle_T \le C\Big(1 + \sup_{t\in[0,T]} |\bar{X}^n_{\underline{t}}|^2 + \sup_{t\in[0,T]} |\bar{X}^n_t|^{r(1,T)+2}\Big) \in L^1(\mathbb{P})$$

since $\sup_{t\in[0,T]} |\bar{X}^n_t|$ lies in every $L^p(\mathbb{P})$. Hence, $(M_t)_{t\in[0,T]}$ is a true martingale, so that

$$\mathbb{E}\left(u(t_k, \bar{X}^n_{t_k}) - u(t_{k-1}, \bar{X}^n_{t_{k-1}})\right) = \mathbb{E}\left(\int_{t_{k-1}}^{t_k} (\partial_t + \bar{L})u(s, \underline{s}, \bar{X}^n_s, \bar{X}^n_{\underline{s}})ds\right)$$

(the integrability of the integral term follows from $\partial_t u = -Lu$, which ensures the polynomial growth of $(\partial_t + \bar{L})u(s, \underline{s}, x, \underline{x}))$ in x and $\underline{x}$ uniformly in $s, \underline{s}$. At this stage, the idea is to expand the above expectation into a term $\bar{\phi}(\underline{s}, \bar{X}^n_{\underline{s}})\frac{T}{n} + O\big((\frac{T}{n})^2\big)$. To this end we will again apply Itô's formula to $\partial_t u(s, \bar{X}^n_s)$, $\partial_x u(s, \bar{X}^n_s)$ and $\partial_{xx} u(s, \bar{X}^n_s)$, taking advantage of the regularity of u.

– *Term 1*. The function $\partial_t u$ being $\mathcal{C}^{1,2}([0, T] \times \mathbb{R})$, Itô's formula between $\underline{s} = t_{k-1}$ and s yields

$$\partial_t u(s, \bar{X}^n_s) = \partial_t u(\underline{s}, \bar{X}^n_{\underline{s}}) + \int_{\underline{s}}^{s} \Big(\partial_{tt} u(r, \bar{X}^n_r) + \bar{L}(\partial_t u)(r, \bar{X}^n_r)\Big)dr$$
$$+ \int_{\underline{s}}^{s} \sigma(\underline{r}, \bar{X}^n_{\underline{r}})\partial_{tx} u(r, \bar{X}^n_r)dW_r.$$

First, let us show that the local martingale term is the increment between $\underline{s}$ and s of a true martingale, denoted by $(M^{(1)}_t)_{t\in[0,T]}$ from now on. Note that $\partial_t u = -L u$ so that $\partial_{xt} u = -\partial_x L u$ is clearly a function with polynomial growth in x uniformly in $t \in [0, T]$ since

$$\left|\partial_x \left(b(t, x)\partial_x u + \frac{1}{2}\sigma^2(t, x)\partial_{xx} u\right)(t, x)\right| \le C\big(1 + |x|^{\theta_0}\big)$$

owing to (7.67). Multiplying this term by the function with linear growth $\sigma(t, x)$ preserves its polynomial growth. Consequently, $\big(M^{(1)}_t\big)_{t\in[0,T]}$ is a true martingale since $\mathbb{E}\,\langle M^{(1)}\rangle_T < +\infty$. On the other hand, using that $\partial_t u = -L u$ leads to

$$\partial_{tt} u(r, \bar{X}^n_r) + \bar{L}(\partial_t u)(r, \underline{r}, \bar{X}^n_r, \bar{X}^n_{\underline{r}}) = \partial_{tt} u(r, \bar{X}^n_r) - \bar{L} \circ Lu(r, \underline{r}, \bar{X}^n_r, \bar{X}^n_{\underline{r}})$$
$$=: \bar{\phi}^{(1)}(r, \underline{r}, \bar{X}^n_r, \bar{X}^n_{\underline{r}}),$$

where $\bar{\phi}^{(1)}$ satisfies for every $x, \ y \in \mathbb{R}$ and every $t, \ \underline{t} \in [0, T]$,

$$\big|\bar{\phi}^{(1)}(t, \underline{t}, x, \underline{x})\big| \le C_1\big(1 + |x|^{\theta_1} + |\underline{x}|^{\underline{\theta}_1}\big).$$

This follows from the fact that $\bar{\phi}^{(1)}$ is defined as a linear combination of products of $b, \partial_t b, \partial_x b, \partial_{xx} b, \sigma, \partial_t \sigma, \partial_x \sigma, \partial_{xx}\sigma, \partial_x u, \partial_{xx} u$ at (t, x) or $(\underline{t}, \underline{x})$ (with "$x = \bar{X}^n_r$" and "$\underline{x} = \bar{X}^n_{\underline{r}}$").

– *Term 2*. The function $\partial_x u$ being $\mathcal{C}^{1,2}$, Itô's formula yields

$$\partial_x u(s, \bar{X}^n_s) = \partial_x u(\underline{s}, \bar{X}^n_{\underline{s}}) + \int_{\underline{s}}^{s} \Big(\partial_{xt} u(r, \bar{X}^n_r) + \bar{L}(\partial_x u)(r, \underline{r}, \bar{X}^n_r, \bar{X}^n_{\underline{r}})\Big)\, dr + \int_{\underline{s}}^{s} \partial_{xx} u(r, \bar{X}^n_r)\sigma(\underline{r}, \bar{X}^n_{\underline{r}}) dW_r.$$

The stochastic integral is the increment of a true martingale (denoted by $(M^{(2)}_t)_{t\in[0,T]}$ in what follows) and using that $\partial_{xt} u = \partial_x(-L\, u)$, one shows likewise that

$$\begin{aligned}\partial_{tx} u(r, \bar{X}^n_r) + \bar{L}(\partial_x u)(r, \underline{r}, \bar{X}^n_r, \bar{X}^n_{\underline{r}}) &= \Big(\bar{L}(\partial_x u) - \partial_x(Lu)\Big)(r, \underline{r}, \bar{X}^n_r, \bar{X}^n_{\underline{r}}) \\ &= \bar{\phi}^{(2)}(r, \underline{r}, \bar{X}^n_r, \bar{X}^n_{\underline{r}}),\end{aligned}$$

where $(t, \underline{t}, x, \underline{x}) \mapsto \bar{\phi}^{(2)}(t, \underline{t}, x, \underline{x})$ has a polynomial growth in $(x, \underline{x})$ uniformly in $t, \underline{t} \in [0, T]$.

– *Term 3*. Following the same lines one shows that

$$\partial_{xx} u(s, \bar{X}^n_s) = \partial_{xx} u(\underline{s}, \bar{X}^n_{\underline{s}}) + \int_{\underline{s}}^{s} \bar{\phi}^{(3)}(r, \underline{r}, \bar{X}^n_r, \bar{X}^n_{\underline{r}})dr + M^{(3)}_s - M^{(3)}_{\underline{s}},$$

where $(M^{(3)}_t)_{t\in[0,T]}$ is a martingale and $\bar{\phi}^{(3)}$ has a polynomial growth in $(x, \underline{x})$ uniformly in $(t, \underline{t})$.

STEP 2 (*A first bound*). Collecting all the results obtained in Step 1 yields

$$\begin{aligned}u(t_k, \bar{X}^n_{t_k}) - u(t_{k-1}, \bar{X}^n_{t_{k-1}}) &= (\partial_t + L)(u)(t_{k-1}, \bar{X}^n_{t_{k-1}}) \\ &\quad + \int_{t_{k-1}}^{t_k}\int_{t_{k-1}}^{s} \bar{\phi}(r, \underline{r}, \bar{X}^n_r, \bar{X}^n_{\underline{r}})dr\, ds \\ &\quad + \int_{t_{k-1}}^{t_k}\Big[M^{(1)}_s - M^{(1)}_{t_{k-1}} + b(t_{k-1}, \bar{X}^n_{t_{k-1}})\big(M^{(2)}_s - M^{(2)}_{t_{k-1}}\big) \\ &\qquad + \frac{1}{2}\sigma^2(t_{k-1}, \bar{X}^n_{t_{k-1}})\big(M^{(3)}_s - M^{(3)}_{t_{k-1}}\big)\Big]ds \\ &\quad + M_{t_k} - M_{t_{k-1}}\end{aligned}$$

where

$$\bar{\phi}(r, \underline{r}, x,\ \underline{x}) = \bar{\phi}^{(1)}(r, x, \underline{x}) + b(\underline{r}, \underline{x})\bar{\phi}^{(2)}(r, x, \underline{x}) + \frac{1}{2}\sigma^2(\underline{r}, \underline{x})\bar{\phi}^{(3)}(r, x, \underline{x}).$$

Hence, the function $\bar{\phi}$ satisfies a polynomial growth assumption: there exists $\theta,\ \theta' \in \mathbb{N}$ such that

$$\forall\, t,\ \underline{t} \in [0,T],\ \forall\, x,\ \underline{x} \in \mathbb{R},\quad |\bar{\phi}(t,\underline{t},x,\underline{x})| \le C_{\bar{\phi}}(T)\big(1+|x|^{\theta}+|\underline{x}|^{\theta'}\big)$$

where $C_{\bar{\phi}}(T)$ can be chosen to be non-decreasing in T (if b and σ are defined on $[0,T'] \times \mathbb{R}$, $T' \ge T$ satisfying (C_∞) on it). The first term $(\partial_t + L)(u)(t_{k-1}, \bar{X}^n_{t_{k-1}})$ on the right-hand side of the above decomposition vanishes since $\partial_t u + Lu = 0$.

As concerns the third term, let us show that it has a zero expectation. One can use Fubini's Theorem since $\sup_{t\in[0,T]} |\bar{X}^n_t| \in L^p(\mathbb{P})$ for every $p > 0$ (this ensures the integrability of the integrand). Consequently

$$\begin{aligned}\mathbb{E}\Bigg[\int_{t_{k-1}}^{t_k}\Bigg[M^{(1)}_s - M^{(1)}_{t_{k-1}} &+ b(t_{k-1},\bar{X}^n_{t_{k-1}})(M^{(2)}_s - M^{(2)}_{t_{k-1}})\\ &+\frac{1}{2}\sigma^2(t_{k-1},\bar{X}^n_{t_{k-1}})(M^{(3)}_s - M^{(3)}_{t_{k-1}})\Bigg]ds\Bigg]\\ &= \int_{t_{k-1}}^{t_k} \mathbb{E}\,\big(M^{(1)}_s - M^{(1)}_{t_{k-1}}\big) + \mathbb{E}\Big(b(t_{k-1},\bar{X}^n_{t_{k-1}})\big(M^{(2)}_s - M^{(2)}_{t_{k-1}}\big)\Big)\\ &\quad + \frac{1}{2}\,\mathbb{E}\Big(\sigma^2(t_{k-1},\bar{X}^n_{t_{k-1}})\big(M^{(3)}_s - M^{(3)}_{t_{k-1}}\big)\Big)\,ds.\end{aligned}$$

Now, all the three expectations inside the integral are zero since the $M^{(k)}$, $k = 1, 2, 3$, are true martingales. Thus

$$\begin{aligned}\mathbb{E}\Big(b(t_{k-1},\bar{X}^n_{t_{k-1}})\big(M^{(2)}_s - M^{(2)}_{t_{k-1}}\big)\Big) &= \mathbb{E}\Big(\underbrace{b(t_{k-1},\bar{X}^n_{t_{k-1}})}_{\mathcal{F}_{t_{k-1}}\text{-measurable}}\underbrace{\mathbb{E}\,\big(M^{(2)}_s - M^{(2)}_{t_{k-1}}\mid \mathcal{F}_{t_{k-1}}\big)}_{=0}\Big)\\ &= 0,\end{aligned}$$

etc. Finally, the original expansion is reduced to

$$\mathbb{E}\,\big(u(t_k,\bar{X}^n_{t_k}) - u(t_{k-1},\bar{X}^n_{t_{k-1}})\big) = \int_{t_{k-1}}^{t_k}\int_{t_{k-1}}^{s} \mathbb{E}\,\bar{\phi}(r,\underline{r},\bar{X}^n_r,\bar{X}^n_{\underline{r}})\,dr\,ds \tag{7.70}$$

so that

$$\begin{aligned}\big|\mathbb{E}\,\big(u(t_k,\bar{X}^n_{t_k}) - u(t_{k-1},\bar{X}^n_{t_{k-1}})\big)\big| &\le \int_{t_{k-1}}^{t_k} ds \int_{t_{k-1}}^{s} dr\,\mathbb{E}\,\big(|\bar{\phi}(r,\underline{r},\bar{X}^n_r,\bar{X}^n_{\underline{r}})|\big)\\ &\le C_{\bar{\phi}}(T)\Big(1 + 2\,\mathbb{E}\Big[\sup_{t\in[0,T]}|\bar{X}^n_t|^{\theta\vee\theta'}\Big]\Big)\frac{(t_k - t_{k-1})^2}{2}\\ &\le C_{b,\sigma,f}(T)\left(\frac{T}{n}\right)^2,\end{aligned}$$

where, owing to Proposition 6.6, the function $C_{b,\sigma,f}(\cdot)$ only depends on T (in a non-decreasing manner). Summing over the terms for $k = 1, \ldots, n$ yields, keeping in mind that $\bar{X}_0^n = X_0$,

$$\begin{aligned}
\left|\mathbb{E}\, f(\bar{X}_{T}^n) - \mathbb{E}\, f(X_{T})\right| &= \left|\mathbb{E}\, u(T, \bar{X}_{T}^n) - \mathbb{E}\, u(0, \bar{X}_0^n)\right| \\
&\le \sum_{k=1}^{n} \left|\mathbb{E}\, u(t_k, \bar{X}_{t_k}^n) - \mathbb{E}\, u(t_{k-1}, \bar{X}_{t_{k-1}}^n)\right| \\
&\le n C_{b,\sigma,f}(T)\Big(\frac{T}{n}\Big)^2 \\
&= C'_{b,\sigma,f}(T)\frac{T}{n} \quad \text{with} \quad C'_{b,\sigma,f}(T) = T C_{b,\sigma,f}(T).
\end{aligned}$$

Let $k \in \{0, \ldots, n-1\}$. It follows from the preceding and the obvious equality $\frac{t_k}{k} = \frac{T}{n}$, $k = 1, \ldots, n$, that,

$$\forall\, k \in \{0, \ldots, n\}, \qquad \left|\mathbb{E}\, f(\bar{X}_{t_k}^n) - \mathbb{E}\, f(X_{t_k})\right| \le C'_{b,\sigma,f}(t_k)\frac{t_k}{k} \le C'_{b,\sigma,f}(T)\frac{T}{n}.$$

► **Exercise.** Compute an explicit (closed) form for the function $\bar{\phi}$.

STEP 3 (*First order expansion*). To obtain an expansion at order 1, one must return to the identity (7.70), namely

$$\mathbb{E}\left(u(t_k, \bar{X}_{t_k}^n) - u(t_{k-1}, \bar{X}_{t_{k-1}}^n)\right) = \int_{t_{k-1}}^{t_k}\int_{t_{k-1}}^{s} \mathbb{E}\, \bar{\phi}(r, \underline{r}, \bar{X}_r^n, \bar{X}_{\underline{r}}^n)\, dr\, ds.$$

This function $\bar{\phi}$ can be written explicitly as a polynomial of b, σ, u and (some of) their partial derivatives at (t, x) or $(\underline{t}, \underline{x})$. Consequently, if b, σ and f satisfy (C_∞) and (F_∞), respectively, one shows that the function $\bar{\phi}$ satisfies

(i) $\bar{\phi}$ is continuous in $(t, \underline{t}, x, \underline{x})$,

(ii) $\left|\partial^{m+m'}_{x^m \underline{x}^{m'}}\bar{\phi}(t, \underline{t}, x, \underline{x})\right| \le C_T\left(1 + |x|^{\chi(m,m',T)} + |\underline{x}|^{\chi'(m,m',T)}\right)$, $t,\ \underline{t} \in [0, T]$,

(iii) $\left|\partial^{m+m'}_{t^m \underline{t}^{m'}}\bar{\phi}(t, \underline{t}, x, \underline{x})\right| \le C_T\left(1 + |x|^{\theta(m,m',T)} + |\underline{x}|^{\theta'(m,m',T)}\right)$, $t,\ \underline{t} \in [0, T]$.

In fact, as above, a $\mathcal{C}^{1,2}$-regularity in (t, x) is in fact sufficient to get a second order expansion. We associate to $\bar{\phi}$ the function $\widetilde{\phi}$ defined by

$$\widetilde{\phi}(t, x) := \bar{\phi}(t, t, x, x)$$

(which is at least a $\mathcal{C}^{1,2}$-function and) whose time and space partial derivatives have polynomial growth in x uniformly in $t \in [0, T]$ owing to the above properties of $\bar{\phi}$.

The idea is once again to apply Itô's formula, this time to $\bar{\phi}(\,.\,, \underline{r}, \,.\,, \bar{X}^n_{\underline{r}})$. Let $r \in [t_{k-1}, t_k)$ (so that $\underline{r} = t_{k-1}$). Then,

$$\begin{aligned}
\bar{\phi}(r, \underline{r}, \bar{X}^n_r, \bar{X}^n_{\underline{r}}) &= \widetilde{\phi}(\underline{r}, \bar{X}^n_{\underline{r}}) + \int_{t_{k-1}}^{r} \partial_x \bar{\phi}(v, \underline{v}, \bar{X}^n_v, \bar{X}^n_{\underline{v}}) d\bar{X}^n_v \\
&\quad + \int_{t_{k-1}}^{r} \Big(\partial_t \bar{\phi}(v, \underline{r}, \bar{X}^n_v, \bar{X}^n_{\underline{r}}) + \frac{1}{2}\partial_{xx}\phi(v, r, \bar{X}^n_v, \bar{X}^n_{\underline{r}})\sigma^2(\underline{v}, \bar{X}^n_{\underline{v}})\Big) dv \\
&= \bar{\phi}(\underline{r}, \underline{v}, \bar{X}^n_{\underline{v}}, \bar{X}^n_{\underline{v}}) + \int_{t_{k-1}}^{r} \Big(\partial_t \bar{\phi}(v, \underline{v}, \bar{X}^n_v, \bar{X}^n_{\underline{v}}) + \bar{L}\,\bar{\phi}(\,.\,, \underline{v}, \,.\,, \bar{X}^n_{\underline{v}})(v, \underline{v}, \bar{X}^n_v, \bar{X}^n_{\underline{v}})\Big) dv \\
&\quad + \int_{t_{k-1}}^{r} \partial_x \bar{\phi}(v, \underline{v}, \bar{X}^n_v, \bar{X}^n_v)\sigma(\underline{v}, \bar{X}^n_{\underline{v}}) dW_v
\end{aligned}$$

where we used that the mute variable v satisfies $\underline{v} = \underline{r} = t_{k-1}$. The stochastic integral turns out to be the increment of a true square integrable martingale since

$$\sup_{v \in [0,T]} \Big|\partial_x \bar{\phi}(v, \underline{v}, \bar{X}^n_v, \bar{X}^n_{\underline{v}})\sigma(\underline{v}, \bar{X}^n_{\underline{v}})\Big| \le C\big(1 + \sup_{s \in [0,T]} |\bar{X}^n_s|^{\theta''}\big) \in L^2(\mathbb{P})$$

where $\theta'' \in \mathbb{N}$, owing to the above (ii) and the linear growth of σ. Then, Fubini's Theorem yields

$$\begin{aligned}
\mathbb{E}\Big(u(t_k, \bar{X}^n_{t_k}) - u(t_{k-1}, \bar{X}^n_{t_{k-1}})\Big) &= \frac{1}{2}\Big(\frac{T}{n}\Big)^2 \mathbb{E}\,\widetilde{\phi}(t_{k-1}, \bar{X}^n_{t_{k-1}}) \qquad (7.71) \\
&\quad + \mathbb{E}\int_{t_{k-1}}^{t_k}\int_{t_{k-1}}^{s}\int_{t_{k-1}}^{r}\Big(\partial_t \bar{\phi}(v, \underline{v}, \bar{X}^n_v, \bar{X}^n_{\underline{v}}) \\
&\qquad\qquad + \bar{L}\,\bar{\phi}(\,.\,, \underline{v}, \,.\,, \bar{X}^n_{\underline{v}})(v, \underline{v}, \bar{X}^n_v, \bar{X}^n_{\underline{v}})\Big) dv\, dr\, ds \\
&\quad + \underbrace{\int_{t_{k-1}}^{t_k}\int_{t_{k-1}}^{s} \mathbb{E}\,(N_r - N_{t_{k-1}}) dr\, ds}_{=0}.
\end{aligned}$$

Now,

$$\mathbb{E}\left(\sup_{v, r \in [0,T]} \big|\bar{L}\,\bar{\phi}(\,.\,, \underline{v}, \,.\,, \bar{X}^n_{\underline{v}})(v, \underline{v}, \bar{X}^n_v, \bar{X}^n_{\underline{v}})\big|\right) < +\infty$$

owing to (7.65) and the polynomial growth of b, σ, u and its partial derivatives. The same holds for $\partial_t \bar{\phi}(v, \underline{r}, \bar{X}^n_v, \bar{X}^n_{\underline{r}})$ so that

$$\left|\mathbb{E}\left[\int_{t_{k-1}}^{t_k}\int_{t_{k-1}}^{s}\int_{t_{k-1}}^{r}\Big(\partial_t \bar{\phi}(v, \underline{v}, \bar{X}^n_v, \bar{X}^n_{\underline{v}}) + (\,.\,, \underline{v}, \,.\,, \bar{X}^n_{\underline{v}})(v, \underline{v}, \bar{X}^n_v, \bar{X}^n_{\underline{v}})\Big) dv\, dr\, ds\right]\right| \le C_{b,\sigma,f,T}\frac{1}{3}\Big(\frac{T}{n}\Big)^3.$$

Summing from $k = 1$ to n yields

$$\mathbb{E}\left(f(\bar{X}^n_T) - f(X_T)\right) = \sum_{k=1}^{n} \mathbb{E}\left(u(t_k, \bar{X}^n_{t_k}) - u(t_{k-1}, \bar{X}^n_{t_{k-1}})\right)$$
$$= \frac{T}{2n}\mathbb{E}\Big(\int_0^T \widetilde{\phi}(\underline{s}, \bar{X}^n_{\underline{s}})ds\Big) + O\left(\left(\frac{T}{n}\right)^2\right).$$

In turn, for every $k \in \{0, \ldots, n\}$, the function $\widetilde{\phi}(t_k, \,.\,)$ satisfies Assumption (F_∞) with some bounds not depending on k (this in turn follows from the fact that the space partial derivatives of $\widetilde{\phi}$ have polynomial growth in x uniformly in $t \in [0, T]$). Consequently, by Step 2,

$$\max_{0\le k\le n} \left|\mathbb{E}\,\widetilde{\phi}(t_k, \bar{X}^n_{t_k}) - \mathbb{E}\,\widetilde{\phi}(t_k, X_{t_k})\right| \le \widetilde{C}'_{b,\sigma,f}(T)\frac{T}{n}$$

so that

$$\max_{0\le k\le n}\left|\mathbb{E}\left(\int_0^{t_k} \widetilde{\phi}(\underline{s}, \bar{X}^n_{\underline{s}})ds - \int_0^{t_k} \widetilde{\phi}(\underline{s}, X_{\underline{s}})ds\right)\right| = \max_{0\le k\le n}\left|\int_0^{t_k}\left(\mathbb{E}\,\widetilde{\phi}(\underline{s}, \bar{X}^n_{\underline{s}}) - \mathbb{E}\,\widetilde{\phi}(\underline{s}, X_{\underline{s}})\right)ds\right|$$
$$\le \widetilde{C}'_{b,\sigma,f}(T)\frac{T^2}{n}.$$

Applying Itô's formula to $\widetilde{\phi}(u, X_u)$ between $\underline{s}$ and s shows that

$$\widetilde{\phi}(s, X_s) = \widetilde{\phi}(\underline{s}, X_{\underline{s}}) + \int_{\underline{s}}^{s} (\partial_t + L)(\widetilde{\phi})(r, X_r)dr + \int_{\underline{s}}^{s} \partial_x\widetilde{\phi}(r, X_r)\sigma(X_r)dW_r,$$

which implies

$$\sup_{s\in[0,T]} \left|\mathbb{E}\,\widetilde{\phi}(s, X_s) - \mathbb{E}\,\widetilde{\phi}(\underline{s}, X_{\underline{s}})\right| \le \widetilde{C}''_{f,b,\sigma}(T)\frac{T}{n}.$$

Hence

$$\max_{0\le k\le n}\left|\mathbb{E}\Big(\int_0^{t_k} \widetilde{\phi}(\underline{s}, X_{\underline{s}})ds\Big) - \mathbb{E}\Big(\int_0^{t_k} \widetilde{\phi}(s, X_s)ds\Big)\right| \le \frac{\widetilde{C}'''_{b,\sigma,f}(T)}{n}. \tag{7.72}$$

Finally, combining all these bounds yields

$$\mathbb{E}\,u(t_k, \bar{X}^n_{t_k}) - \mathbb{E}\,u(t_{k-1}, \bar{X}^n_{t_{k-1}}) = \frac{T}{2n}\int_{t_{k-1}}^{t_k} \mathbb{E}\,\widetilde{\phi}(s, X_s)\,ds + O\Big(\frac{1}{n^2}\Big)$$

uniformly in $k \in \{0, \dots, n\}$. One concludes by summing from $k = 1$ to n that

$$\mathbb{E}\, f(\bar{X}^n_T) - \mathbb{E}\, f(X_T) = \frac{T}{2n} \int_0^T \mathbb{E}\, \widetilde{\phi}(s, X_s)\, ds + O\Big(\frac{1}{n^2}\Big), \tag{7.73}$$

which completes the proof of the first-order expansion. Note that the function $\widetilde{\phi}$ can be made explicit. ◇

One step beyond (*Toward* $R = 2$). To expand at a higher-order, say $R = 2$, one can proceed as follows: first we go back to (7.71) which can be re-written as

$$\mathbb{E}\Big(u(t_k, \bar{X}^n_{t_k}) - u(t_{k-1}, \bar{X}^n_{t_{k-1}})\Big) = \frac{1}{2}\Big(\frac{T}{n}\Big)^2 \mathbb{E}\, \widetilde{\phi}(t_{k-1}, \bar{X}^n_{t_{k-1}})$$
$$+\, \mathbb{E} \int_{t_{k-1}}^{t_k} \int_{t_{k-1}}^{s} \int_{t_{k-1}}^{r} \bar{\chi}(v, \underline{v}, \bar{X}^n_v, \bar{X}^n_{\underline{v}})\, dv\, dr\, ds,$$

where $\bar{\chi}$ and $\widetilde{\chi}(t, x) = \bar{\chi}(t, t, x, x)$ and their partial derivatives satisfy similar smoothness and growth properties as $\bar{\phi}$ and $\widetilde{\phi}$, respectively.

Then, one gets by mimicking the above computations

$$\mathbb{E}\Big(u(t_k, \bar{X}^n_{t_k}) - u(t_{k-1}, \bar{X}^n_{t_{k-1}})\Big) = \frac{1}{2}\Big(\frac{T}{n}\Big)^2 \mathbb{E}\, \widetilde{\phi}(t_{k-1}, \bar{X}^n_{t_{k-1}}) + \frac{1}{3}\Big(\frac{T}{n}\Big)^3 \mathbb{E}\, \widetilde{\chi}(t_{k-1}, \bar{X}^n_{t_{k-1}})$$
$$+ \mathbb{E} \int_{t_{k-1}}^{t_k} \int_{t_{k-1}}^{s} \int_{t_{k-1}}^{r} \int_{t_{k-1}}^{t} \bar{\psi}(v, \underline{v}, \bar{X}^n_v, \bar{X}^n_{\underline{v}}) dt\, dv\, dr\, ds,$$

where $\bar{\psi}$ still has the same structure, so that

$$\mathbb{E} \int_{t_{k-1}}^{t_k} \int_{t_{k-1}}^{s} \int_{t_{k-1}}^{r} \int_{t_{k-1}}^{t} \bar{\psi}(v, \underline{v}, \bar{X}^n_v, \bar{X}^n_{\underline{v}}) dt\, dv\, dr\, ds = O\bigg(\Big(\frac{T}{n}\Big)^4\bigg)$$

uniformly in $k \in \{0, \dots, n-1\}$. Consequently, still summing from $k = 1$ to n,

$$\mathbb{E}\, f(\bar{X}^n_T) - \mathbb{E}\, f(X_T) = \frac{c_{1,n}}{n} + \frac{c_{2,n}}{n^2} + O\bigg(\Big(\frac{T}{n}\Big)^3\bigg),$$

where $c_{1,n} = \dfrac{T}{2} \displaystyle\int_0^T \mathbb{E}\, \widetilde{\phi}(\underline{s}, \bar{X}^n_{\underline{s}}) ds$ and $c_{2,n} = \dfrac{T^2}{3} \displaystyle\int_0^T \mathbb{E}\, \widetilde{\chi}(\underline{s}, \bar{X}^n_{\underline{s}}) ds$. It is clear, since the funtion $\widetilde{\phi}$ and $\widetilde{\chi}$ are both continuous, that $c_{1,n} \to \dfrac{T}{2} \displaystyle\int_0^T \mathbb{E}\, \widetilde{\phi}(s, \bar{X}_s) ds$ and

$c_{2,n} \to \frac{T}{2}\int_0^T \mathbb{E}\,\widetilde{\chi}(s,\bar{X}_s)ds$. Moreover, extending to $\widetilde{\chi}$ the rate results established with $\widetilde{\phi}$, we even have that

$$c_{1,n} - \frac{T}{2}\int_0^T \mathbb{E}\,\widetilde{\phi}(s,X_s)ds = O\Big(\frac{1}{n}\Big) \quad \text{and} \quad c_{2,n} - \frac{T^2}{3}\int_0^T \mathbb{E}\,\widetilde{\chi}(s,X_s)ds = O\Big(\frac{1}{n}\Big).$$

To get a second-order expansion it remains to show that the term $O\big(\frac{1}{n}\big)$ in the convergence rate of $c_{1,n}$ is of the form $\frac{c_1'}{n} + O\big(\frac{1}{n^2}\big)$. This can be obtained by applying the results obtained for f in the proof of the first-order expansion to the functions $\widetilde{\phi}(t_k,\,.)$ with some uniformity in k (which is the technical point at this stage).

Further comments. In fact, to get higher-order expansions, this elementary approach, close to the original proof from [270], is not the most convenient. Other approaches have been developed since this seminal work on weak error expansions. Among them, let us cite [64], based on an elegant duality argument. A *parametrix approach* is presented in [172], which naturally leads to the higher-order expansion stated in the theorem. In fact, this paper is more connected with the Bally–Talay theorem since it relies, in a uniformly elliptic framework, on an approximation of the density of the diffusion by that of the Euler scheme.

ℵ **Practitioner's corner.** The original application of the first-order expansion, already mentioned in the seminal paper [270], is of course the Richardson–Romberg extrapolation introduced in Sect. 7.7. More recently, the first order expansion turned out to be one of the two key ingredients of the Multilevel Monte Carlo method (*MLMC*) introduced by M. Giles in [107] (see Sect. 9.5.2 in Chap. 9) and devised to efficiently kill the bias of a simulation while keeping the variance under control.

In turn, higher-order expansions are the key, first to Multistep Richardson–Romberg extrapolation (see [225]), then, recently to the weighted multilevel methods introduced in [198]. These two methods are also exposed in Chap. 9.

7.9 The Non-globally Lipschitz Case (A Few Words On)

The Lipschitz continuity assumptions made on b and σ to ensure the existence of a unique strong solution to (7.1) over the whole non-negative real line $\mathbb{R}_+$, hence on any interval $[0,T]$, can be classically replaced by a local Lipschitz assumption combined with a linear growth assumption. These assumptions read, respectively:

$(i)_{Liploc}$ *Local Lipschitz continuity assumption*. Assume that b and σ are Borel functions, locally Lipschitz continuous in (t,x) in the sense that

$$\begin{aligned}&\forall\, N \in \mathbb{N}^*,\ \exists\, L_N > 0 \text{ such that}\\ &\qquad \forall\, x,\ y \in B(0,N),\ \forall\, t \in [0,N],\\ &\qquad |b(t,x) - b(t,y)| + \|\sigma(t,x) - \sigma(t,y)\| \le L_N|x-y|\end{aligned}$$

(where $B(0,N)$ denotes the ball centered at 0 with radius N and $\|\cdot\|$ denotes, *e.g.* the Fröbenius norm of matrices).

$(ii)_{LinG}$ *Linear growth*

$$\exists\, C>0,\ \forall\, t\in\mathbb{R}_+,\ \forall\, x\in\mathbb{R}^d,\ |b(t,x)|\vee\|\sigma(t,x)\|\le C\big(1+|x|^2\big)^{\frac{1}{2}}.$$

However, although this less stringent condition makes it possible to consider typically oscillating coefficients, it does not allow to relax their linear growth behavior at infinity. In particular, it does not take into account the possible mean-reverting effect induced by the drifts, typically of the form $h(x)=-\text{sign}(x)|x|^p$, $p>1$.

In the autonomous case ($b(t,x)=b(x)$ and $\sigma(t,x)=\sigma(x)$), this phenomenon is investigated by Has'minskiĭ in [145] (Theorem 4.1, p. 84) where $(ii)_{LinG}$ is replaced by a *weak Lyapunov assumption.*

Theorem 7.13 *If* $b:\to\mathbb{R}^d$ *and* $\sigma:\mathbb{R}^d\to\mathcal{M}(d,q,\mathbb{R})$ *satisfy* $(i)_{Liploc}$ *(i.e. are locally Lipschitz continuous) and the following* weak *Lyapunov assumption:*
$(ii)_{WLyap}$ Weak Lyapunov assumption*: there exists a twice differentiable function* $V:\mathbb{R}^d\to\mathbb{R}_+$, $\mathcal{C}^2$ *with* $\lim_{|x|\to+\infty}V(x)=+\infty$ *and a real number* $\lambda\in\mathbb{R}$ *such that*

$$\begin{aligned}\forall\, x\in\mathbb{R}^d,\ LV(x) &:= \big(b(x)\,|\,\nabla V(x)\big)+\frac{1}{2}\text{Tr}\big(\sigma(x)D^2V(x)\sigma^*(x)\big)\\ &\le \lambda V(x),\end{aligned}\tag{7.74}$$

then, for every $x\in\mathbb{R}^d$, *the SDE* (7.1) *has a unique (strong) solution* $(X_t^x)_{t\ge 0}$ *on the whole non-negative real line starting from* x *at time* 0.

Remark. If $\lambda V(x)$ is replaced in (7.74) by $\lambda V(x)+\mu$ with $\lambda<0$ and $\mu\in\mathbb{R}$, the assumption becomes a standard *mean-reverting* or *Lyapunov* assumption and V is called a *Lyapunov function*. Then $(X_t)_{t\ge 0}$ admits a stationary regime such that $\nu=\mathbb{P}_{X_t}$, $t\in\mathbb{R}_+$, satisfies $\nu(V)<+\infty$.

A typical example of where $(i)_{Liploc}$ and $(ii)_{WLyap}$ are satisfied is the following. Let $p\in\mathbb{R}$ and set

$$b(x)=\kappa\big(1+|x|^2\big)^{\frac{p}{2}}x,\quad x\in\mathbb{R}^d.$$

The function b is clearly locally Lipschitz continuous but not Lipschitz continuous when $p>0$. Assume σ is locally Lipschitz continuous in x with linear growth in the sense of $(ii)_{LinG}$, then one easily checks with $V(x)=|x|^2+1$, that

$$\begin{aligned}LV(x) &= 2\big(b(x)\,|\,x\big)+\|\sigma(x)\|^2\\ &\le 2\kappa(1+|x|^2)^{\frac{p}{2}}|x|^2+C^2(1+|x|^2).\end{aligned}$$

- If $\kappa \geq 0$ and $p \leq 0$, then $(i)_{Liploc}$ and $(ii)_{LinG}$ are satisfied, as well as (7.74) (with $\lambda = 2\kappa + C^2$).
- If $\kappa < 0$, then (7.74) is satisfied by $\lambda = C^2$ for every p whereas $(ii)_{LinG}$ is not satisfied.

Remark. If $\kappa < 0$ and $p > 0$, then the mean-reverting assumption mentioned in the previous remark is satisfied with $\lambda \in (2\kappa, 0)$ with $\mu_\lambda = \sup_{x \in \mathbb{R}^d} \big[\big((C^2 - \lambda) + 2\kappa(1 + |x|^2)^{\frac{p}{2}}\big)|x|^2 + C^2\big]$. In fact, in that case the *SDE* admits a stationary regime.

This allows us to deal with *SDE*s like perturbed dissipative gradient equations, equations coming from Hamiltonian mechanics (see *e.g.* [267]), etc, where the drift shares mean-reverting properties but has a (non-linear) polynomial growth at infinity.

Unfortunately, this stability, or more precisely non-explosive property induced by the existence of a weak Lyapunov function as described above cannot be transferred to the regular Euler or Milstein schemes, which usually have an explosive behavior, depending on the step. Actually, this phenomenon can already be observed in a deterministic framework with *ODE*s but is more systematic with (true) *SDE*s since, unlike *ODE*s, no stability region usually exists depending on the starting value of the scheme.

A natural idea, inspired by the numerical analysis of *ODE*s leads to the introduction fully and partially drift-implicit Euler schemes, but also new classes of explicit approximation scarcely more complex than the regular Euler scheme. We will not go further in this direction but will refer, for example, to [152] for an in-depth study of the approximation of such *SDE*s, including moment bounds and convergence rates.

Chapter 8
The Diffusion Bridge Method: Application to Path-Dependent Options (II)

8.1 Theoretical Results About Time Discretization of Path-Dependent Functionals

In this section we deal with some "path-dependent" (European) options. Such contracts are characterized by the fact that their payoffs depend on the whole past of the underlying asset(s) between the origin $t = 0$ of the contract and its maturity T. This means that these payoffs are of the form $F\big((X_t)_{t\in[0,T]}\big)$, where F is a functional usually naturally defined from $\mathbb{D}([0,T],\mathbb{R}^d) \to \mathbb{R}_+$ (where $\mathbb{D}([0,T],\mathbb{R}^d)$ is the set of càdlàg functions $x : [0,T] \to \mathbb{R}^d$ ([1]) and $X = (X_t)_{t\in[0,T]}$ denotes the dynamics of the underlying asset). We still assume from now on that $X = (X_t)_{t\in[0,T]}$ is a solution to an $\mathbb{R}^d$-valued *SDE* of type (7.1):

$$dX_t = b(t, X_t)dt + \sigma(t, X_t)dW_t,$$

where X_0 is independent of the Brownian motion W.

The question of establishing weak error expansions, even at the first order, for families of functionals of a Brownian diffusion is even more challenging than for time marginal functions $f(\alpha(T))$, $\alpha \in \mathbb{D}([0,T],\mathbb{R}^d)$. In the recent years, several papers have provided such expansions for specific families of functionals F. These works were essentially motivated by the pricing of European path-dependent options, like Asian or Lookback options in one dimension, corresponding to functionals defined for every $\alpha \in \mathbb{D}([0,T],\mathbb{R}^d)$ by

$$F(\alpha) := f\left(\int_0^T \alpha(s)ds\right), \quad F(\alpha) := f\Big(\alpha(T), \sup_{t\in[0,T]} \alpha(t), \inf_{t\in[0,T]} \alpha(t)\Big)$$

[1]We need to define F on càdlàg functions in view of the stepwise constant Euler scheme, not to speak of jump diffusion driven by Lévy processes.

G. Pagès, *Numerical Probability*, Universitext,
https://doi.org/10.1007/978-3-319-90276-0_8

and, in higher dimensions, to barrier options with functionals of the form

$$F(\alpha) = f\big(\alpha(T)\big)\mathbf{1}_{\{\tau_D(\alpha)>T\}},$$

where D is an open domain of $\mathbb{R}^d$ and $\tau_D(\alpha) := \inf\{s \in [0, T] : \alpha(s) \text{ or } \alpha(s-) \notin D\}$ is the exit time from D by the generic càdlàg path α ([2]). In both frameworks, f is usually at least Lipschitz continuous. Let us quote from the literature two well-known examples of results (in a homogeneous framework, *i.e.* $b(t,x) = b(x)$ and $\sigma(t,x) = \sigma(x)$).

– The following theorem is established in [120].

Theorem 8.1 *(a) If the domain D is bounded and has a smooth enough boundary (in fact $\mathcal{C}^3$), if $b \in \mathcal{C}^3(\mathbb{R}^d, \mathbb{R}^d)$, $\sigma \in \mathcal{C}^3\big(\mathbb{R}^d, \mathcal{M}(d,q,\mathbb{R})\big)$, σ uniformly elliptic on D (i.e. $\sigma\sigma^*(x) \geq \varepsilon_0 I_d$, $\varepsilon_0 > 0$), then, for every bounded Borel function f vanishing in a neighborhood of ∂D,*

$$\mathbb{E}\Big(f(X_T)\mathbf{1}_{\{\tau_D(X)>T\}}\Big) - \mathbb{E}\Big(f(\widetilde{X}^n_T)\mathbf{1}_{\{\tau_D(\widetilde{X}^n)>T\}}\Big) = O\left(\frac{1}{\sqrt{n}}\right) \quad \text{as} \quad n \to +\infty \quad (8.1)$$

where $\widetilde{X}$ denotes the stepwise constant Euler scheme.
(b) If, furthermore, b and σ are $\mathcal{C}^5$ and D is a half-space, then the genuine Euler scheme $\bar{X}^n$ *satisfies*

$$\mathbb{E}\Big(f(X_T)\mathbf{1}_{\{\tau_D(X)>T\}}\Big) - \mathbb{E}\Big(f(\bar{X}^n_T)\mathbf{1}_{\{\tau_D(\bar{X}^n)>T\}}\Big) = O\left(\frac{1}{n}\right) \quad \text{as} \quad n \to +\infty. \quad (8.2)$$

Note, however, that these assumptions are unfortunately not satisfied by standard barrier options (see below).

– It is suggested in [264] (with a rigorous proof when $X = W$) that if $b, \sigma \in \mathcal{C}^4_b(\mathbb{R}, \mathbb{R})$, σ is uniformly elliptic and $f \in \mathcal{C}^{4,2}_{pol}(\mathbb{R}^2)$ (existing partial derivatives with polynomial growth), then

$$\mathbb{E}\Big(f(X_T, \max_{t\in[0,T]} X_t)\Big) - \mathbb{E}\Big(f(\widetilde{X}^n_T, \max_{0\leq k\leq n} \widetilde{X}^n_{t_k})\Big) = O\left(\frac{1}{\sqrt{n}}\right) \quad \text{as} \quad n \to +\infty. \quad (8.3)$$

A similar improvement – $O(\frac{1}{n})$ rate – as above can be expected (but still remains a conjecture) when replacing $\widetilde{X}$ by the continuous Euler scheme $\bar{X}^n$, namely

$$\mathbb{E}\Big(f(X_T, \max_{t\in[0,T]} X_t)\Big) - \mathbb{E}\Big(f(\bar{X}^n_T, \max_{t\in[0,T]} \bar{X}^n_t)\Big) = O\left(\frac{1}{n}\right) \quad \text{as} \quad n \to +\infty.$$

[2]When α is continuous or stepwise constant and càdlàg, $\tau_D(\alpha) := \inf\{s \in [0,T] : \alpha(s) \notin D\}$.

More recently, relying on new techniques based on transport inequalities and the Wasserstein metric, a partial result toward this conjecture has been established in [5] with an error of $O(n^{-(\frac{2}{3}-\eta)})$ for every $\eta > 0$.

If we forget about the regularity assumptions, the formal intersection between these two classes of path-dependent functionals (or non-negative payoffs) is not empty since the payoff of a barrier option with domain $D = (-\infty, L)$ can be written

$$f(X_T)\mathbf{1}_{\{\tau_D(X)>T\}} = F\Big(X_T, \sup_{t\in[0,T]} X_t\Big) \quad \text{with} \quad F(x, y) = f(x)\mathbf{1}_{\{y<L\}}.$$

Unfortunately, such a function g is never a smooth function so that, if the second result is true it does not solve the first one.

For other results concerning these weak error expansions for functionals, we refer to [17, 122, 123, 160] and the references therein. By contrast with the "vanilla case", these results are somewhat disappointing since they point out that the weak error obtained with the stepwise constant Euler scheme is not significantly better than the strong error since the only gain is a $\sqrt{1 + \log n}$ factor. The positive side is that we can reasonably hope that using the genuine Euler scheme will again yield the $O(1/n)$-rate in the first-order expansion of the time discretization error, provided we know how to simulate the functional of interest of this scheme.

8.2 From Brownian to Diffusion Bridge: How to Simulate Functionals of the Genuine Euler Scheme

To take advantage of the above rates and the reasonable guess we may have about higher-order expansions, we do not need to simulate the genuine Euler scheme itself (which is meaningless) but some specific functionals of this scheme like the maximum, the minimum, time integrals, etc, between two time discretization instants t_k^n and t_{k+1}^n, *given the (simulated) values of the discrete time Euler scheme*. This means bridging this discrete time Euler scheme into its genuine extension. First, we deal with the standard Brownian motion itself.

8.2.1 The Brownian Bridge Method

We still denote by $(\mathcal{F}_t^W)_{t\geq 0}$ the (completed) natural filtration of a standard Brownian motion W. We begin with a quick study of the standard Brownian bridge between 0 and T.

Proposition 8.1 *Let $W = (W_t)_{t\geq 0}$ be a standard Brownian motion.*
(a) Let $T > 0$. Then, the so-called standard Brownian bridge on $[0, T]$ *defined by*

$$Y_t^{W,T} := W_t - \frac{t}{T} W_T, \quad t \in [0, T], \tag{8.4}$$

is an $\mathcal{F}_T^W$-measurable centered Gaussian process, independent of $(W_{T+s})_{s \geq 0}$, whose distribution is characterized by its covariance structure

$$\mathbb{E}\left(Y_s^{W,T} Y_t^{W,T}\right) = s \wedge t - \frac{st}{T} = \frac{(s \wedge t)(T - s \vee t)}{T}, \qquad 0 \leq s, t \leq T.$$

(b) Let $T_0, T_1 \in (0, +\infty)$, $T_0 < T_1$. Then

$$\mathcal{L}\left((W_t)_{t \in [T_0, T_1]} \mid W_s, s \notin (T_0, T_1)\right) = \mathcal{L}\left((W_t)_{t \in [T_0, T_1]} \mid W_{T_0}, W_{T_1}\right)$$

so that $(W_t)_{t \in [T_0, T_1]}$ and $(W_s)_{s \notin (T_0, T_1)}$ are independent given (W_{T_0}, W_{T_1}). Moreover, this conditional distribution is given by

$$\mathcal{L}\left((W_t)_{t \in [T_0, T_1]} \mid W_{T_0} = x, W_{T_1} = y\right) = \mathcal{L}\left(x + \frac{t - T_0}{T_1 - T_0}(y - x) + (Y_{t-T_0}^{B, T_1 - T_0})_{t \in [T_0, T_1]}\right),$$

where B is a generic standard Brownian motion.

Proof (a) The process $Y^{W,T}$ is clearly centered since W is. Elementary computations based on the covariance structure of the standard Brownian Motion

$$\mathbb{E}\, W_s W_t = s \wedge t, \quad s, t \in [0, T],$$

show that, for every $s, t \in [0, T]$,

$$\begin{aligned}
\mathbb{E}\,(Y_t^{W,T} Y_s^{W,T}) &= \mathbb{E}\, W_t W_s - \frac{s}{T}\mathbb{E}\, W_t W_T - \frac{t}{T}\mathbb{E}\, W_s W_T + \frac{st}{T^2}\mathbb{E} W_T^2 \\
&= s \wedge t - \frac{st}{T} - \frac{ts}{T} + \frac{ts}{T} \\
&= s \wedge t - \frac{ts}{T} = \frac{(s \wedge t)(T - s \vee t)}{T}.
\end{aligned}$$

Let $\mathbb{G}_W = \overline{\operatorname{span}\left\{W_t, t \geq 0\right\}}^{L^2(\mathbb{P})}$ be the closed vector subspace of $L^2(\Omega, \mathcal{A}, \mathbb{P})$ spanned by the Brownian motion W. Since it is a centered Gaussian process, it is a well-known fact that independence and absence of correlation coincide in $\mathbb{G}_W$. The process $Y^{W,T}$ belongs to this space by construction. Likewise one shows that, for every $u \geq T$, $\mathbb{E}\,(Y_t^{W,T} W_u) = 0$ so that $Y_t^{W,T} \perp \operatorname{span}(W_u, u \geq T)$. Consequently, $Y^{W,T}$ is independent of $(W_{T+u})_{u \geq 0}$.
(b) First note that for every $t \in [T_0, T_1]$,

$$W_t = W_{T_0} + W_{t-T_0}^{(T_0)},$$

where $W_t^{(T_0)} = W_{T_0+t} - W_{T_0}$, $t \geq 0$, is a standard Brownian motion, independent of $\mathcal{F}_{T_0}^W$. Rewriting (8.4) for $W^{(T_0)}$ leads to

$$W_s^{(T_0)} = \frac{s}{T_1 - T_0} W_{T_1-T_0}^{(T_0)} + Y_s^{W^{(T_0)},T_1-T_0}.$$

Plugging this identity into the above equality at time $s = t - T_0$ leads to

$$W_t = W_{T_0} + \frac{t - T_0}{T_1 - T_0}(W_{T_1} - W_{T_0}) + Y_{t-T_0}^{W^{(T_0)},T_1-T_0}. \tag{8.5}$$

It follows from (a) that the process $\widetilde{Y} := (Y_{t-T_0}^{W^{(T_0)},T_1-T_0})_{t\in[T_0,T_1]}$ is a Gaussian process, measurable with respect to $\mathcal{F}_{T_1-T_0}^{W^{(T_0)}}$ by (a), hence it is independent of $\mathcal{F}_{T_0}^W$ since $W^{(T_0)}$ is. Consequently, $\widetilde{Y}$ is independent of $(W_t)_{t\in[0,T_0]}$. Furthermore, $\widetilde{Y}$ is independent of $(W_{T_1-T_0+u}^{(T_0)})_{u\geq 0}$ by (a). Hence it is $L^2(\mathbb{P})$-orthogonal to $W_{T_1+u} - W_{T_0}$, $u \geq 0$, in $\mathbb{G}_W$. As it is also orthogonal to W_{T_0} in $\mathbb{G}_W$ since W_{T_0} is $\mathcal{F}_{T_0}^W$-measurable, it is orthogonal to $W_{T_1+u} = W_{T_1-T_0+u}^{(T_0)} + W_{T_0}$, $u \geq 0$. Which in turn implies independence of $\widetilde{Y}$ and $(W_{T_1+u})_{u\geq 0}$ since all these random variables lie in $\mathbb{G}_W$.

Finally, the same argument – in $\mathbb{G}_W$, no correlation implies independence – implies that $\widetilde{Y}$ is independent of $(W_s)_{s\in\mathbb{R}_+\setminus(T_0,T_1)}$ *i.e.* of the σ-field $\sigma(W_s,\ s \notin (T_0, T_1))$. The end of the proof follows from the above identity (8.5) and the exercises below. ◇

► **Exercises. 1.** Let X, Y, Z be three random vectors defined on a probability space $(\Omega, \mathcal{A}, \mathbb{P})$ taking values in $\mathbb{R}^{k_X}$, $\mathbb{R}^{k_Y}$ and $\mathbb{R}^{k_Z}$, respectively. Assume that Y and (X, Z) are independent. Show that for every bounded Borel function $f : \mathbb{R}^{k_Y} \to \mathbb{R}$ and every Borel function $g : \mathbb{R}^{k_X} \to \mathbb{R}^{k_Y}$,

$$\mathbb{E}\big(f(g(X) + Y) \mid (X, Z)\big) = \mathbb{E}\big(f(g(X) + Y) \mid X\big).$$

Deduce that $\mathcal{L}\big(g(X) + Y \mid (X, Z)\big) = \mathcal{L}\big(g(X) + Y \mid X\big)$.

2. Deduce from the previous exercise that

$$\mathcal{L}\big((W_t)_{t\in[T_0,T_1]} \mid (W_t)_{t\notin(T_0,T_1)}\big) = \mathcal{L}\big((W_t)_{t\in[T_0,T_1]} \mid (W_{T_0}, W_{T_1})\big).$$

[Hint: consider the finite-dimensional marginal distributions $(W_{t_1}, \dots, W_{t_n})$ given $(X, W_{s_1}, \dots, W_{s_p})$, where $t_i \in [T_0, T_1]$, $i = 1, \dots, n$, $X = (W_{T_0}, W_{T_1})$ and $s_j \in [T_0, T_1]^c$, $j = 1, \dots, p$, then use the decomposition (8.5).]

3. The conditional distribution of $(W_t)_{t\in[T_0,T_1]}$ given W_{T_0}, W_{T_1} is that of a Gaussian process, hence it can also be characterized by its expectation and covariance structure. Show that they are given respectively by

$$\mathbb{E}\left(W_t \mid W_{T_0} = x, W_{T_1} = y\right) = \frac{T_1 - t}{T_1 - T_0}x + \frac{t - T_0}{T_1 - T_0}y, \quad t \in [T_0, T_1],$$

and

$$\text{Cov}\left(W_t, W_s \mid W_{T_0} = x, W_{T_1} = y\right) = \frac{(T_1 - t)(s - T_0)}{T_1 - T_0}, \quad s \le t, \; s, t \in [T_0, T_1].$$

8.2.2 The Diffusion Bridge (Bridge of the Genuine Euler Scheme)

Now we return to the (continuous) Euler scheme defined by (7.8).

Proposition 8.2 (Bridge of the Euler scheme) *Assume that* $\sigma(t, x) \neq 0$ *for every* $t \in [0, T]$, $x \in \mathbb{R}$.
(a) The processes $(\bar{X}^n_t)_{t \in [t^n_k, t^n_{k+1}]}$, $k = 0, \ldots, n-1$, *are conditionally independent given the* σ*-field* $\sigma(\bar{X}^n_{t^n_k}, \; k = 0, \ldots, n)$.
(b) Furthermore, for every $k \in \{0, \ldots, n\}$, *the conditional distribution*

$$\mathcal{L}\Big((\bar{X}^n_t)_{t \in [t^n_k, t^n_{k+1}]} \mid \bar{X}^n_{t^n_\ell} = x_\ell, \; \ell = 0, \ldots, n\Big) = \mathcal{L}\Big((\bar{X}^n_t)_{t \in [t^n_k, t^n_{k+1}]} \mid \bar{X}^n_{t^n_k} = x_k, \bar{X}_{t^n_{k+1}} = x_{k+1}\Big)$$
$$= \mathcal{L}\left(\left(x_k + \frac{n(t - t^n_k)}{T}(x_{k+1} - x_k) + \sigma(t^n_k, x_k) Y^{B, T/n}_{t - t^n_k}\right)_{t \in [t^n_k, t^n_{k+1}]}\right)$$

where $(Y^{B, T/n}_s)_{s \in [0, T/n]}$ *is a Brownian bridge (related to a generic Brownian motion* B*) as defined by (8.4). The distribution of this Gaussian process (sometimes called a diffusion bridge) is entirely characterized by:*

– *its expectation function* $\left(x_k + \dfrac{n(t - t^n_k)}{T}(x_{k+1} - x_k)\right)_{t \in [t^n_k, t^n_{k+1}]}$

and

– *its covariance operator* $\sigma^2(t^n_k, x_k) \dfrac{(s \wedge t - t^n_k)(t^n_{k+1} - s \vee t)}{t^n_{k+1} - t^n_k}$, $s, t \in [t^n_k, t^n_{k+1}]$.

Proof. Elementary computations show that for every $t \in [t^n_k, t^n_{k+1}]$,

$$\bar{X}^n_t = \bar{X}^n_{t^n_k} + \frac{t - t^n_k}{t^n_{k+1} - t^n_k}(\bar{X}^n_{t^n_{k+1}} - \bar{X}^n_{t^n_k}) + \sigma(t^n_k, \bar{X}^n_{t^n_k}) Y^{W^{(t^n_k)}, T/n}_{t - t^n_k}$$

(keeping in mind that $t^n_{k+1} - t^n_k = T/n$). Consequently, the conditional independence claim will follow if the processes $\left(Y^{W^{(t^n_k)}, T/n}_t\right)_{t \in [0, T/n]}$, $k = 0, \ldots, n-1$, are independent given $\sigma\big(\bar{X}^n_{t^n_\ell}, \; \ell = 0, \ldots, n\big)$. Now, it follows from the assumption on the diffusion coefficient σ that

$$\sigma\big(\bar{X}^n_{t^n_\ell}, \; \ell = 0, \ldots, n\big) = \sigma\big(X_0, \; W_{t^n_\ell}, \; \ell = 1, \ldots, n\big).$$

So we have to establish the conditional independence of the processes $\big(Y_t^{W^{(t_k^n)},T/n}\big)_{t\in[0,T/n]}$, $k=0,\dots,n-1$, given $\sigma(X_0,\, W_{t_k^n},\, k=1,\dots,n)$ or, equivalently, given $\sigma(W_{t_k^n},\, k=1,\dots,n)$, since X_0 and W are independent (note that all the above bridges are $\mathcal{F}_T^W$-measurable). First observe that all the bridges $(Y_t^{W^{(t_k^n)},T/n})_{t\in[0,T/n]}$, $k=0,\dots,n-1$ and W live in the same Gaussian space $\mathbb{G}_W$. We know from Proposition 8.1(a) that each bridge $(Y_t^{W^{(t_k^n)},T/n})_{t\in[0,T/n]}$ is independent of both $\mathcal{F}_{t_k^n}^W$ and $\sigma(W_{t_{k+1}^n+s}-W_{t_k^n},\, s\geq 0)$. Hence, it is independent in particular, of all $\sigma(\{W_{t_\ell^n},\, \ell=1,\dots,n\})$ again because $\mathbb{G}_W$ is a Gaussian space. On the other hand, all bridges are independent since they are built from independent Brownian motions $(W_t^{(t_k^n)})_{t\in[0,T/n]}$. Hence, the bridges $\big(Y_t^{W^{(t_k^n)},T/n}\big)_{t\in[0,T/n]}$, $k=0,\dots,n-1$, are i.i.d. and independent of $\sigma\big(W_{t_k^n},\, k=1,\dots,n\big)$ and consequently of $\sigma\big(X_0,\, W_{t_k^n},\, k=1,\dots,n\big)$.

Now $\bar{X}^n_{t_k^n}$ is $\sigma\big(\{W_{t_\ell^n},\, \ell=1,\dots,k\}\big)$-measurable, consequently are has

$$\sigma\big(\bar{X}^n_{t_k^n},\, W_{t_k^n},\, \bar{X}^n_{t_{k+1}^n}\big)\subset\sigma\big(\{W_{t_\ell^n},\, \ell{=}1,\dots,n\}\big)$$

so that $\big(Y_t^{W^{(t_k^n)},T/n}\big)_{t\in[0,T/n]}$ is independent of $\big(\bar{X}^n_{t_k^n},\, W_{t_k^n},\, \bar{X}^n_{t_{k+1}^n}\big)$. The conclusion follows. ◇

Now we know the distribution of the genuine Euler scheme between two successive discretization times t_k^n and t_{k+1}^n conditionally to the Euler scheme at its discretization times. Now, we are in position to simulate some functionals of the continuous Euler scheme, namely its supremum.

Proposition 8.3 *The distribution of the supremum of the Brownian bridge starting at* 0 *and arriving at* y *at time* T*, defined by* $Y_t^{W,T,y}=\frac{t}{T}y+W_t-\frac{t}{T}W_T$ *on* $[0,T]$*, is given by*

$$\mathbb{P}\Big(\sup_{t\in[0,T]}Y_t^{W,T,y}\leq z\Big)=\begin{cases}1-\exp\big(-\frac{2}{T}z(z-y)\big) & \textit{if } z\geq\max(y,0),\\ 0 & \textit{if } z\leq\max(y,0).\end{cases}$$

Proof. The key is to have in mind that the distribution of $Y^{W,T,y}$ is that of the conditional distribution of W given $W_T=y$. So, we can derive the result from an expression of the joint distribution of $\big(\sup_{t\in[0,T]}W_t,\, W_T\big)$, *e.g.* from

$$\mathbb{P}\Big(\sup_{t\in[0,T]}W_t\geq z,\; W_T\leq y\Big).$$

It is well-known from the symmetry principle that, for every $z\geq\max(y,0)$,

$$\mathbb{P}\Big(\sup_{t\in[0,T]}W_t\geq z,\; W_T\leq y\Big)=\mathbb{P}(W_T\geq 2z-y).$$

We briefly reproduce the proof for the reader's convenience. If $z = 0$, the result is obvious since $W_T \stackrel{d}{=} -W_T$. If $z > 0$, one introduces the hitting time $\tau_z := \inf\{s > 0 : W_s = z\}$ of $[z, +\infty)$ by W (convention $\inf \varnothing = +\infty$). This is a $(\mathcal{F}_t^W)$-stopping time since $[z, +\infty)$ is a closed set and W is a continuous process (this uses that $z > 0$). Furthermore, τ_z is $a.s.$ finite since $\overline{\lim}_{t\to+\infty} W_t = +\infty$ $a.s.$ Consequently, still by continuity of its paths, $W_{\tau_z} = z$ $a.s.$ and $W_{\tau_z+t} - W_{\tau_z}$ is independent of $\mathcal{F}^W_{\tau_z}$. As a consequence, for every $z \geq \max(y, 0)$, using that $W_{\tau_z} = z$ on the event $\{\tau_z \leq T\}$,

$$\begin{aligned}
\mathbb{P}\Big(\sup_{t\in[0,T]} W_t \geq z,\ W_t \leq y\Big) &= \mathbb{P}\big(\tau_z \leq T,\ W_T - W_{\tau_z} \leq y - z\big)\\
&= \mathbb{P}\big(\tau_z \leq T,\ -(W_T - W_{\tau_z}) \leq y - z\big)\\
&= \mathbb{P}\big(\tau_z \leq T,\ W_T \geq 2z - y\big)\\
&= \mathbb{P}\big(W_T \geq 2z - y\big)
\end{aligned}$$

since $2z - y \geq z$. Consequently, one may write for every $z \geq \max(y, 0)$,

$$\mathbb{P}\Big(\sup_{t\in[0,T]} W_t \geq z,\ W_T \leq y\Big) = \int_{2z-y}^{+\infty} h_T(\xi)\,d\xi \quad \text{with} \quad h_T(\xi) = \frac{e^{-\frac{\xi^2}{2T}}}{\sqrt{2\pi T}}.$$

Hence, since the involved functions are differentiable one has

$$\begin{aligned}
\mathbb{P}\Big(\sup_{t\in[0,T]} W_t \geq z \,|\, W_T = y\Big) &= \lim_{\eta\to0} \frac{\Big(\mathbb{P}(W_T \geq 2z - (y+\eta)) - \mathbb{P}(W_T \geq 2z - y)\Big)/\eta}{\Big(\mathbb{P}(W_T \leq y+\eta) - \mathbb{P}(W_T \leq y)\Big)/\eta}\\
&= \frac{h_T(2z-y)}{h_T(y)} = e^{-\frac{(2z-y)^2-y^2}{2T}} = e^{-\frac{2z(z-y)}{2T}}.
\end{aligned}$$

◇

Corollary 8.1 *Let $\lambda > 0$ and let $x, y \in \mathbb{R}$. If $Y^{W,T}$ denotes the standard Brownian bridge of W between 0 and T, then for every $z \in \mathbb{R}$,*

$$\mathbb{P}\left(\sup_{t\in[0,T]} \Big(x + (y-x)\frac{t}{T} + \lambda Y_t^{W,T}\Big) \leq z\right) = \begin{cases} 1 - \exp\Big(-\frac{2}{T\lambda^2}(z-x)(z-y)\Big) & if\ z \geq \max(x,y), \\ 0 & if\ z < \max(x,y). \end{cases} \tag{8.6}$$

Proof. First note that

$$x + (y-x)\frac{t}{T} + \lambda Y_t^{W,T} = \lambda x' + \lambda\left(y'\frac{t}{T} + Y_t^{W,T}\right)$$

with $x' = x/\lambda$, $y' = (y-x)/\lambda$.

Then, the result follows by the previous proposition, using that for any real-valued random variable ξ, every $\alpha \in \mathbb{R}$ and every $\beta \in (0, +\infty)$,

$$\mathbb{P}(\alpha + \beta\xi \le z) = \mathbb{P}\Big(\xi \le \frac{z-\alpha}{\beta}\Big) = 1 - \mathbb{P}\Big(\xi > \frac{z-\alpha}{\beta}\Big). \qquad \diamond$$

▶ **Exercise.** Show using $-W \stackrel{d}{=} W$ that

$$\mathbb{P}\left(\inf_{t\in[0,T]}\Big(x + (y-x)\frac{t}{T} + \lambda Y_t^{W,T}\Big) \le z\right) = \begin{cases} \exp\big(-\frac{2}{T\lambda^2}(z-x)(z-y)\big) & \text{if } z \le \min(x,y), \\ 1 & \text{if } z > \min(x,y). \end{cases}$$

8.2.3 *Application to Lookback Style Path-Dependent Options*

In this section, we focus on Lookback style options (including general barrier options), *i.e.* exotic options related to payoffs of the form $h_T := f\big(X_T, \sup_{t\in[0,T]} X_t\big)$.

We want to compute an approximation of $e^{-rt}\mathbb{E}\, h_T$ using a Monte Carlo simulation based on the continuous time Euler scheme, *i.e.* we want to compute $e^{-rt}\mathbb{E}\, f\big(\bar{X}_T^n, \sup_{t\in[0,T]} \bar{X}_t^n\big)$. We first note, owing to the chaining rule for conditional expectation, that

$$\mathbb{E}\, f\big(\bar{X}_T^n, \sup_{t\in[0,T]} \bar{X}_t^n\big) = \mathbb{E}\big[\mathbb{E}\,(f(\bar{X}_T^n, \sup_{t\in[0,T]} \bar{X}_t^n) \mid \bar{X}_{t_\ell}^n, \ell = 0, \dots, n)\big].$$

We derive from Proposition 8.2 that

$$\mathbb{E}\,\big(f(\bar{X}_T^n, \sup_{t\in[0,T]} \bar{X}_t^n) \mid \bar{X}_{t_\ell}^n = x_\ell, \ell = 0, \dots, n\big) = f\big(x_n, \max_{0\le k\le n-1} M_{x_k, x_{k+1}}^{n,k}\big)$$

where, owing to Proposition 8.3,

$$M_{x,y}^{n,k} := \sup_{t\in[0,T/n]}\Big(x + \frac{nt}{T}(y-x) + \sigma(t_k^n, x) Y_t^{W^{(t_k^n)}, T/n}\Big)$$

are independent. This can also be interpreted as the random variables $M^{n,k}_{\bar{X}^n_{t^n_k},\bar{X}^n_{t^n_{k+1}}}$, $k = 0,\ldots,n-1$, being conditionally independent given $\bar{X}^n_{t^n_k}$, $k = 0,\ldots,n$. Following Corollary 8.1, the distribution function $G^{n,k}_{x,y}$ of $M^{n,k}_{x,y}$ is given by

$$G^{n,k}_{x,y}(z) = \left[1 - \exp\left(-\frac{2n}{T\sigma^2(t^n_k,x)}(z-x)(z-y)\right)\right]\mathbf{1}_{\{z\geq \max(x,y)\}}, \quad z \in \mathbb{R}.$$

Then, the inverse distribution simulation rule (see Proposition 1.1) yields that

$$\sup_{t\in[0,T/n]}\left(x + \frac{t}{\frac{T}{n}}(y-x) + \sigma(t^n_k,x)Y_t^{W^{(t^n_k)},T/n}\right) \stackrel{d}{=} (G^{n,k}_{x,y})^{-1}(U), \quad U \stackrel{d}{=} \mathcal{U}([0,1])$$
$$\stackrel{d}{=} (G^{n,k}_{x,y})^{-1}(1-U), \tag{8.7}$$

where we used that $U \stackrel{d}{=} 1-U$. To determine $(G^{n,k}_{x,y})^{-1}$ (at $1-u$), it remains to solve the equation $G^{n,k}_{x,y}(z) := 1-u$ under the constraint $z \geq \max(x,y)$, *i.e.*

$$1 - \exp\left(-\frac{2n}{T\sigma^2(t^n_k,x)}(z-x)(z-y)\right) = 1-u, \quad z \geq \max(x,y),$$

or, equivalently,

$$z^2 - (x+y)z + xy + \frac{T}{2n}\sigma^2(t^n_k,x)\log(u) = 0, \quad z \geq \max(x,y).$$

The above equation has two solutions, the solution below satisfying the constraint. Consequently,

$$(G^n_{x,y})^{-1}(1-u) = \frac{1}{2}\left(x + y + \sqrt{(x-y)^2 - 2T\sigma^2(t^n_k,x)\log(u)/n}\right).$$

Finally,

$$\mathcal{L}\Big(\max_{t\in[0,T]}\bar{X}^n_t \mid \{\bar{X}^n_{t_k} = x_k,\ k = 0,\ldots,n\}\Big) = \mathcal{L}\Big(\max_{0\leq k\leq n-1}(G^n_{x_k,x_{k+1}})^{-1}(1-U_{k+1})\Big),$$

where $(U_k)_{1\leq k\leq n}$ are i.i.d. and uniformly distributed random variables over the unit interval.

Pseudo-code for Lookback style options

We assume for the sake of simplicity that the interest rate r is 0. By Lookback style options we mean the class of options whose payoff involve possibly both $\bar{X}^n_T$ and the maximum of (X_t) over $[0,T]$, *i.e.*

$$\mathbb{E}\, f\big(\bar{X}^n_T, \sup_{t\in[0,T]} \bar{X}^n_t\big).$$

Regular Call on maximum is obtained by setting $f(x, y) = (y - K)_+$, the (maximum) Lookback option by setting $f(x, y) = y - x$ and the (maximum) partial lookback $f_\lambda(x, y) = (y - \lambda x)_+$ $\lambda > 1$.

We want to compute a Monte Carlo approximation of $\mathbb{E}\, f\big(\bar{X}^n_T, \sup_{t\in[0,T]} \bar{X}^n_t\big)$ using the continuous Euler scheme. We reproduce below a pseudo-script to illustrate how to use the above result on the conditional distribution of the maximum of the Brownian bridge.

- Set $S^f = 0$.

`for` $m = 1$ `to` M

  - Simulate a path of the discrete time Euler scheme and set $x_k := \bar{X}^{n,(m)}_{t^n_k}$, $k = 0, \dots, n$.
  - Simulate $\Xi^{(m)} := \max_{0\le k\le n-1} (G^n_{x_k,x_{k+1}})^{-1}(1 - U^{(m)}_k)$, where $(U^{(m)}_k)_{1\le k\le n}$ are i.i.d. with $\mathcal{U}([0, 1])$-distribution.
  - Compute $f(\bar{X}^{n,(m)}_T, \Xi^{(m)})$.
  - Compute $S^f_m := f(\bar{X}^{n,(m)}_T, \Xi^{(m)}) + S^f_{m-1}$.

`end.`(m)

- Eventually,

$$\mathbb{E}\, f\big(\bar{X}^n_T, \sup_{t\in[0,T]} \bar{X}^n_t\big) \simeq \frac{S^f_M}{M}$$

for large enough M ([3]).

Once one can simulate $\sup_{t\in[0,T]} \bar{X}^n_t$ (and its minimum, see exercise below), it is easy to price by simulation the exotic options mentioned in the former section (Lookback, options on maximum) but also the barrier options, since one can decide whether or not the *continuous* Euler scheme strikes a barrier (up or down). The Brownian bridge is also involved in the methods designed for pricing Asian options.

► **Exercise.** (a) Show that the distribution of the infimum of the Brownian bridge $(Y^{W,T,y}_t)_{t\in[0,T]}$ starting at 0 and arriving at y at time T is given by

$$\mathbb{P}\Big(\inf_{t\in[0,T]} Y^{W,T,y}_t \le z\Big) = \begin{cases} \exp\big(-\frac{2}{T}z(z-y)\big) & \text{if } z \le \min(y, 0), \\ 1 & \text{if } z \ge \min(y, 0). \end{cases}$$

---

[3]…Of course one needs to compute the empirical variance (approximately) given by

$$\frac{1}{M}\sum_{m=1}^{M} f(\Xi^{(m)})^2 - \left(\frac{1}{M}\sum_{m=1}^{M} f(\Xi^{(m)})\right)^2$$

in order to design a confidence interval, without which the method is simply nonsense….

(b) Derive a formula similar to (8.7) for the conditional distribution of the minimum of the continuous Euler scheme using now the inverse distribution functions

$$(F^{n,k}_{x,y})^{-1}(u) = \frac{1}{2}\left(x + y - \sqrt{(x-y)^2 - 2\,T\sigma^2(t^n_k, x)\log(u)/n}\right),\quad u \in (0,1),$$

of the random variable $\inf_{t\in[0,T/n]}\left(x + \frac{t}{\frac{T}{n}}(y-x) + \sigma(x)Y^{W,T/n}_t\right)$.

Warning! The above method is not appropriate for simulating the joint distribution of the $(n+3)$-tuple $\left(\bar{X}^n_{t^n_k},\ k = 0,\ldots,n,\ \inf_{t\in[0,T]}\bar{X}^n_t,\ \sup_{t\in[0,T]}\bar{X}^n_t\right)$.

8.2.4 Application to Regular Barrier Options: Variance Reduction by Pre-conditioning

By *regular barrier options* we mean barrier options having a constant level as a barrier. An up-and-out Call is a typical example of such options with a payoff given by

$$h_T = (X_T - K)_+\mathbf{1}_{\{\sup_{t\in[0,T]} X_t \le L\}}$$

where K denotes the strike price of the option and L $(L > K)$ its barrier.

In practice, the “Call” part is activated at T only if the process (X_t) hits the barrier $L \le K$ between 0 and T. In fact, as far as simulation is concerned, this “Call part” can be replaced by any Borel function f such that both $f(X_T)$ and $f(\bar{X}^n_T)$ are integrable (this is always true if f has polynomial growth owing to Proposition 7.2). Note that these so-called barrier options are in fact a sub-class of generalized maximum Lookback options having the specificity that the maximum only shows up through an indicator function.

Then, one may derive a general weighted formula for $\mathbb{E}\left(f(\bar{X}^n_T)\mathbf{1}_{\{\sup_{t\in[0,T]}\bar{X}^n_t\le L\}}\right)$, which is an approximation of $\mathbb{E}\left(f(X_T)\mathbf{1}_{\{\sup_{t\in[0,T]} X_t\le L\}}\right)$.

Proposition 8.4 (Up-and-Out Call option)

$$\mathbb{E}\left(f(\bar{X}^n_T)\mathbf{1}_{\left\{\sup_{t\in[0,T]}\bar{X}^n_t\le L\right\}}\right) = \mathbb{E}\left[f(\bar{X}^n_T)\mathbf{1}_{\{\max_{0\le k\le n}\bar{X}^n_{t_k}\le L\}}\prod_{k=0}^{n-1}\left(1 - e^{-\frac{2n}{T}\frac{(\bar{X}^n_{t_k}-L)(\bar{X}^n_{t_{k+1}}-L)}{\sigma^2(t^n_k,\bar{X}^n_{t_k})}}\right)\right]. \tag{8.8}$$

Proof of Equation (8.8). This formula is a typical application of pre-conditioning described in Sect. 3.4. We start from the chaining rule for conditional expectations:

$$\mathbb{E}\left(f(\bar{X}^n_T)\mathbf{1}_{\{\sup_{t\in[0,T]}\bar{X}^n_t\le L\}}\right) = \mathbb{E}\left[\mathbb{E}\left(f(\bar{X}^n_T)\mathbf{1}_{\{\sup_{t\in[0,T]}\bar{X}^n_t\le L\}}\mid \bar{X}^n_{t_k},\ k = 0,\ldots,n\right)\right].$$

Then we use the conditional independence of the bridges of the genuine Euler scheme given the values $\bar{X}^n_{t_k}$, $k = 0, \ldots, n$, established in Proposition 8.2. It follows that

$$\begin{aligned}
\mathbb{E}\left(f(\bar{X}^n_T)\mathbf{1}_{\{\sup_{t\in[0,T]}\bar{X}^n_t \le L\}}\right) &= \mathbb{E}\left(f(\bar{X}^n_T)\mathbb{P}\Big(\sup_{t\in[0,T]}\bar{X}^n_t \le L \mid \bar{X}_{t_k},\ k=0,\ldots,n\Big)\right)\\
&= \mathbb{E}\left(f(\bar{X}^n_T)\prod_{k=1}^{n} G^n_{\bar{X}^n_{t_k},\bar{X}^n_{t_{k+1}}}(L)\right)\\
&= \mathbb{E}\left[f(\bar{X}^n_T)\mathbf{1}_{\left\{\max_{0\le k\le n}\bar{X}^n_{t_k}\le L\right\}}\prod_{k=0}^{n-1}\left(1-e^{-\frac{2n}{T}\frac{(\bar{X}^n_{t_k}-L)(\bar{X}^n_{t_{k+1}}-L)}{\sigma^2(t^n_k,\bar{X}^n_{t_k})}}\right)\right]. \qquad \diamond
\end{aligned}$$

Furthermore, we know that *the random variable in the right-hand side always has a lower variance* since it is a conditional expectation of the random variable in the left-hand side, namely

$$\begin{aligned}
\mathrm{Var}\left(f(\bar{X}^n_T)\mathbf{1}_{\left\{\max_{0\le k\le n}\bar{X}_{t_k}\le L\right\}}\prod_{k=0}^{n-1}\left(1-e^{-\frac{2n}{T}\frac{(\bar{X}^n_{t_k}-L)(\bar{X}^n_{t_{k+1}}-L)}{\sigma^2(t^n_k,\bar{X}^n_{t_k})}}\right)\right)&\\
\le \mathrm{Var}\left(f(\bar{X}^n_T)\mathbf{1}_{\left\{\sup_{t\in[0,T]}\bar{X}^n_t\le L\right\}}\right)&.
\end{aligned}$$

► **Exercises. 1.** *Down-and-Out option.* Show likewise that for every Borel function $f \in L^1(\mathbb{R}, \mathbb{P}_{\bar{X}^n_T})$,

$$\mathbb{E}\left(f(\bar{X}^n_T)\mathbf{1}_{\left\{\inf_{t\in[0,T]}\bar{X}^n_t\ge L\right\}}\right) = \mathbb{E}\left(f(\bar{X}^n_T)\mathbf{1}_{\left\{\min_{0\le k\le n}\bar{X}^n_{t_k}\ge L\right\}}e^{-\frac{2n}{T}\sum_{k=0}^{n-1}\frac{(\bar{X}^n_{t_k}-L)(\bar{X}^n_{t_{k+1}}-L)}{\sigma^2(t^n_k,\bar{X}^n_{t_k})}}\right) \tag{8.9}$$

and that the expression in the second expectation has a lower variance.

2. Extend the above results to barriers of the form

$$L(t) := e^{at+b},\ a,\ b \in \mathbb{R}.$$

Remark Formulas like (8.8) and (8.9) can be used to produce quantization-based cubature formulas (see [260]).

8.2.5 Asian Style Options

The family of Asian options is related to payoffs of the form

$$h_T := h\left(\int_0^T X_s ds\right)$$

where $h : \mathbb{R}_+ \to \mathbb{R}_+$ is a non-negative Borel function. This class of options may benefit from a specific treatment to improve the rate of convergence of its time discretization. This may be viewed as a consequence of the continuity of the functional $f \mapsto \int_0^T f(s)ds$ from $(L^1([0,T], dt), \|\cdot\|_1)$.

What follows mostly comes from [187], where this problem has been extensively investigated.

▷ *Approximation phase:* Let

$$X_t^x = x\exp(\mu t + \sigma W_t), \quad \mu = r - \frac{\sigma^2}{2}, \quad x > 0.$$

Then

$$\int_0^T X_s^x ds = \sum_{k=0}^{n-1} \int_{t_k^n}^{t_{k+1}^n} X_s^x ds = \sum_{k=0}^{n-1} X_{t_k^n}^x \int_0^{T/n} \exp(\mu s + \sigma W_s^{(t_k^n)}) ds.$$

So, we need to approximate the time integrals coming out in the r.h.s of the above equation. Let B be a standard Brownian motion. It is clear by a scaling argument already used in Sect. 7.2.2 that $\sup_{s\in[0,T/n]}|B_s|$ is proportional to $\sqrt{T/n}$ in L^p, $p \in [1,\infty)$. Although it is not true in the a.s. sense, owing to a missing log log term coming from the *LIL*, we may write "almost" rigorously

$$\forall s \in [0, T/n], \qquad \exp(\mu s + \sigma B_s) = 1 + \mu s + \sigma B_s + \frac{\sigma^2}{2} B_s^2 + \text{"}O(n^{-\frac{3}{2}})\text{"}.$$

Hence

$$\begin{aligned}\int_0^{T/n} \exp(\mu s + \sigma B_s) ds &= \frac{T}{n} + \frac{\mu}{2}\frac{T^2}{n^2} + \sigma \int_0^{T/n} B_s ds + \frac{\sigma^2}{2}\frac{T^2}{2n^2} \\ &\quad + \frac{\sigma^2}{2}\int_0^{T/n} (B_s^2 - s) ds + \text{"}O(n^{-\frac{5}{2}})\text{"}, \\ &= \frac{T}{n} + \frac{r}{2}\frac{T^2}{n^2} + \sigma \int_0^{T/n} B_s ds \\ &\quad + \frac{\sigma^2}{2}\left(\frac{T}{n}\right)^2 \int_0^1 (\widetilde{B}_u^2 - u) du + \text{"}O(n^{-\frac{5}{2}})\text{"},\end{aligned}$$

where $\widetilde{B}_u = \sqrt{\frac{T}{n}} B_{\frac{T}{n}u}$, $u \in [0,1]$, is a standard Brownian motion on the unit interval since, combining a scaling and a change of variable,

$$\int_0^{T/n} (B_s^2 - s)ds = \left(\frac{T}{n}\right)^2 \int_0^1 (\widetilde{B}_u^2 - u)du.$$

Owing to the fact that the random variable $\int_0^1 (\widetilde{B}_u^2 - u)du$ is centered and that, when B is replaced successively by $(W_t^{(t_k^n)})_{t\in[0,\frac{T}{n}]}$, $k = 0, \ldots, n-1$, the resulting random variables are independent hence i.i.d. one can in fact consider that the contribution of this term is $O(n^{-\frac{5}{2}})$. To be more precise, the random variable $\int_0^1 ((\widetilde{W}_u^{(t_k^n)})^2 - u)\,du$ is independent of $\mathcal{F}_{t_k^n}^W$, $k = 0, \ldots, n-1$, so that

$$\begin{aligned}\left\|\sum_{k=0}^{n-1} X_{t_k^n}^x \int_0^1 \big((\widetilde{W}_u^{(t_k^n)})^2 - u\big)\,du\right\|_2^2 &= \sum_{k=0}^{n-1} \left\|X_{t_k^n}^x \int_0^1 \big((\widetilde{W}_u^{(t_k^n)})^2 - u\big)\,du\right\|_2^2 \\ &= \sum_{k=0}^{n-1} \left\|X_{t_k^n}^x\right\|_2^2 \left\|\int_0^1 \big((\widetilde{W}_u^{(t_k^n)})^2 - u\big)\,du\right\|_2^2 \\ &\le n\left(\frac{T}{n}\right)^4 \left\|\int_0^1 \big(B_u^2 - u\big)\,du\right\|_2^2 \|X_T^x\|_2^2\end{aligned}$$

since $(X_t^x)^2$ is a sub-martingale. As a consequence

$$\left\|\sum_{k=0}^{n-1} X_{t_k^n}^x \int_0^1 \big((\widetilde{W}_u^{(t_k^n)})^2 - u\big)\,du\right\|_2 = O(n^{-\frac{3}{2}}),$$

which justifies to considering "conventionally" the contribution of each term to be $O(n^{-\frac{5}{2}})$, *i.e.* negligible. This leads us to use the following approximation

$$\int_0^{T/n} \exp\big(\mu s + \sigma W_s^{(t_k^n)}\big)ds \simeq I_k^n := \frac{T}{n} + \frac{r}{2}\frac{T^2}{n^2} + \sigma \int_0^{T/n} W_s^{(t_k^n)} ds.$$

▷ *Simulation phase:* Now, it follows from Proposition 8.2 applied to the Brownian motion (which is its own genuine Euler scheme), that the n-tuple of processes $\big(W_t^{(t_k^n)}\big)_{t\in[0,T/n]}$, $k = 0, \ldots, n-1$, are independent processes given $\sigma(W_{t_k^n}, k = 1, \ldots, n)$ with conditional distribution given by

$$\begin{aligned}\mathcal{L}\Big((W_t^{(t_\ell^n)})_{t\in[0,T/n]} \mid W_{t_k^n} = w_k,\, k = 1, \ldots, n\Big) &= \mathcal{L}\Big((W_t^{(t_\ell^n)})_{t\in[0,T/n]} \mid W_{t_\ell^n} = w_\ell,\, W_{t_{\ell+1}^n} = w_{\ell+1}\Big) \\ &= \mathcal{L}\left(\left(\frac{nt}{T}(w_{\ell+1} - w_\ell) + Y_t^{W,T/n}\right)_{t\in[0,T/n]}\right)\end{aligned}$$

for every $\ell \in \{0, \ldots, n-1\}$.

Consequently, the random variables $\int_0^{T/n} W_s^{(t_\ell^n)} ds$, $\ell = 0, \ldots, n-1$, are conditionally independent given $\sigma(W_{t_k^n}, k = 1, \ldots, n)$ with a Gaussian conditional distribution $\mathcal{N}\Big(m(W_{t^n_\ell}, W_{t^n_{\ell+1}}); \sigma^2\Big)$ having

$$m(w_\ell, w_{\ell+1}) = \int_0^{T/n} \frac{nt}{T}(w_{\ell+1} - w_\ell)dt = \frac{T}{2n}(w_{\ell+1} - w_\ell)$$

(with $w_0 = 0$) as a mean and a (deterministic) variance $\sigma^2 = \mathbb{E}\left(\int_0^{T/n} Y_s^{W,T/n} ds\right)^2$. We can use the exercise below for a quick computation of this quantity based on stochastic calculus. The computation that follows is more elementary and relies on the distribution of the Brownian bridge between 0 and $\frac{T}{n}$:

$$\begin{aligned}
\mathbb{E}\left(\int_0^{T/n} Y_s^{W,T/n} ds\right)^2 &= \int_{[0,T/n]^2} \mathbb{E}\,(Y_s^{W,T/n} Y_t^{W,T/n})\, ds\, dt \\
&= \frac{n}{T}\int_{[0,T/n]^2} (s \wedge t)\left(\frac{T}{n} - (s \vee t)\right) ds\, dt \\
&= \left(\frac{T}{n}\right)^3 \int_{[0,1]^2} (u \wedge v)(1 - (u \vee v))\, du\, dv \\
&= \frac{1}{12}\left(\frac{T}{n}\right)^3.
\end{aligned}$$

► **Exercise.** Use stochastic calculus to show directly that

$$\mathbb{E}\left(\int_0^T Y_s^{W,T} ds\right)^2 = \mathbb{E}\left(\int_0^T W_s ds - \frac{T}{2}W_T\right)^2 = \int_0^T \left(\frac{T}{2} - s\right)^2 ds = \frac{T^3}{12}.$$

Now we are in a position to write a pseudo code for a Monte Carlo simulation.

Pseudo-code for the pricing of an Asian option with payoff $h\left(\int_0^T X_s^x ds\right)$

`for` $m := 1$ `to` M

- Simulate the Brownian increments $\Delta W^{(m)}_{t^n_{k+1}} \stackrel{d}{:=} \sqrt{\frac{T}{n}} Z_k^{(m)}$, $k = 1, \ldots, n$, $Z_k^{(m)}$ i.i.d. with distribution $\mathcal{N}(0; 1)$; set
  – $w_k^{(m)} := \sqrt{\frac{T}{n}}\left(Z_1^{(m)} + \cdots + Z_n^{(m)}\right)$,
  – $x_k^{(m)} := x \exp\left(\mu t_k^n + \sigma w_k^{(m)}\right)$, $k = 0, \ldots, n$.
- Simulate *independently* $\zeta_k^{(m)}$, $k = 1, \ldots, n$, i.i.d. with distribution $\mathcal{N}(0; 1)$ and set

$$I_k^{n,(m)} \stackrel{d}{:=} \frac{T}{n} + \frac{r}{2}\left(\frac{T}{n}\right)^2 + \sigma\left(\frac{T}{2n}(w_{k+1} - w_k) + \frac{1}{\sqrt{12}}\left(\frac{T}{n}\right)^{\frac{3}{2}} \zeta_k^{(m)}\right), \qquad k = 0, \ldots, n-1.$$

- Compute

$$h_T^{(m)} =: h\left(\sum_{k=0}^{n-1} x_k^{(m)} I_k^{n,(m)}\right)$$

`end.`(m)

$$\text{Premium} \simeq e^{-rt}\frac{1}{M}\sum_{m=1}^{M} h_T^{(m)}.$$

`end.`

▷ *Time discretization error estimates:* Set

$$A_T = \int_0^T X_s^x ds \qquad \text{and} \qquad \bar{A}_T^n = \sum_{k=0}^{n-1} X_{t_k^n}^x I_k^n.$$

This scheme, which is clearly simulable but closely dependent on the Black–Scholes model in its present form, induces the following time discretization error, established in [187].

Proposition 8.5 (Time discretization error) *For every* $p \geq 2$,

$$\| A_T - \bar{A}_T^n \|_p = O(n^{-\frac{3}{2}})$$

so that, if $g : \mathbb{R}^2 \to \mathbb{R}$ *is Lipschitz continuous, then*

$$\| g(X_T^x, A_T) - g(X_T^x, \bar{A}_T^n) \|_p = O\big(n^{-\frac{3}{2}}\big).$$

Remark. The main reason for not considering higher-order expansions of $\exp(\mu t + \sigma B_t)$ is that we are not able to simulate at a reasonable cost the triplet $\big(B_t, \int_0^t B_s ds, \int_0^t B_s^2 ds\big)$, which is no longer a Gaussian vector and, consequently, $\big(\int_0^t B_s ds, \int_0^t B_s^2 ds\big)$ given B_t.

Chapter 9
Biased Monte Carlo Simulation, Multilevel Paradigm

WARNING: Even more than in other chapters, we recommend the reader to carefully read the "Practitioner's corner" sections to get more information on practical aspects in view of implementation. In particular, many specifications of the structure parameters of the multilevel estimators are specified *a priori* in various Tables, but these specifications admit variants which are discussed and detailed in the Practitioner's corner sections.

9.1 Introduction

In the first chapters of this monograph, we explored how to efficiently implement Monte Carlo simulation method to compute various quantities of interest which can be represented as the expectation of a random variable. So far, we focused focused on the numerous variance reduction techniques.

One major field of application of Monte Carlo simulation, in Quantitative Finance, but also in Physics, Biology and Engineering Sciences, is to take advantage of the representation of solutions of various classes of *PDE*s as expectations of functionals of solutions of Stochastic Differential Equations (*SDE*). The most famous (and simplest) example of such a situation is provided by the Feynman–Kac representation of the solution of the parabolic $PDE \equiv \frac{\partial u}{\partial t} + Lu = f$, where $Lu = \big(b|\nabla_x u\big)(x) + \frac{1}{2}\mathrm{Tr}\big(a(x)D^2u(x)\big)$, $u(T,\,.) = f$ with $b : \mathbb{R}^d \to \mathbb{R}^d$ and $a : \mathbb{R}^d \to \mathcal{M}(d,\mathbb{R})$ established in Theorem 7.11 of Chap. 7. Under appropriate assumptions on a, b and f, a solution u exists and we can represent u by $u(t,x) = \mathbb{E}\, f(X^x_{T-t})$, where $(X^x_t)_{t\in[0,T]}$ is the Brownian diffusion, *i.e.* the unique solution to $dX^x_t = b(X^x_t)dt + \sigma(X^x_t)dW_t$, $X^x_0 = x$, where $a(\xi) = \sigma\sigma^*(\xi)$. This connection can be extended to some specific path-dependent functionals of the diffusion, typically of the form $f\big(X^x_T, \sup_{t\in[0,T]} X^x_t\big)$, $f\big(X^x_T, \inf_{t\in[0,T]} X^x_t\big)$, $f\Big(X^x_T, \int_0^T \varphi(X^x_t)dt\Big)$. Thus,

G. Pagès, *Numerical Probability*, Universitext,
https://doi.org/10.1007/978-3-319-90276-0_9

in the latter case, one can introduce the 2-dimensional degenerate *SDE* satisfied by the pair (X_t^x, Y_t) where $Y_t = \int_0^t \varphi(X_s)ds$

$$dX_t = b(t, X_t)dt + \sigma(t, X_t)dW_t, \ X_0 = x, \quad dY_t = \varphi(X_t)dt, \ Y_0 = 0.$$

Then, one can solve the "companion" parabolic *PDE* associated to its infinitesimal generator. See *e.g.* [2], where other examples in connection with pricing and hedging of financial derivatives under a risk neutral assumption are connected with the corresponding *PDE* derived by the Feynman–Kac formula (see Theorem 7.11). Numerical methods based on numerical schemes adapted to these *PDE*s are proposed. However these *PDE* based approaches are efficient only in low dimensions, say $d \leq 3$. In higher dimensions, the Feynman–Kac representation is used in the reverse sense: one computes $u(0, x) = \mathbb{E} f(X_T^x)$ by a Monte Carlo simulation rather than trying to numerically solve the *PDE*. As far as derivative pricing and hedging is concerned this means that *PDE* methods are efficient, *i.e.* they outperform Monte Carlo simulation, until dimension $d = 3$, but beyond, the Monte Carlo (or quasi-Monte Carlo) method has no credible alternative, except for very specific models. When dealing with risk estimation where the structural dimension is always high, there is even less alternative.

As we saw in Chap. 7, the exact simulation of such a diffusion process, at fixed horizon T or "through" functionals of its paths, is not possible in general. A noticeable exception is the purely one-dimensional setting ($d = q = 1$ with our notations) where Beskos and Roberts proposed in [42] (see also [43]) such an exact procedure for vectors $(X_{t_1}, \dots, X_{t_n})$, $t_k = \frac{kT}{n}$, $k = 1, \dots, n$, based on an acceptance-rejection method involving a Poisson process. Unfortunately, this method admits no significant extension in higher dimensions.

More recently, various groups, first in [9, 18], then in [149] and in [240]) developed a kind of "weak exact" simulation method for random variables of the form $f(X_T)$ in the following sense: they exhibit a simulable random variable Ξ satisfying $\mathbb{E}\,\Xi = \mathbb{E} f(X_T)$ and $\text{Var}(\Xi) < +\infty$. Beyond the assumptions made on b and σ, the approach is essentially limited to this type of "vanilla" payoff (depending on the diffusion at one fixed time T) and seems, from this point of view, less flexible than the multilevel methods developed in what follows. These methods, as will be seen further on, rely, in the case of diffusions, on the simulation of discretization schemes which are intrinsically biased. However, we have at hand two kinds of information: the strong convergence rate of such schemes toward the diffusion itself in the L^p-spaces (see the results for the Euler and Milstein schemes in Chap. 7) on the one hand and some expansions of the weak error, *i.e.* of the bias on the other hand. We briefly saw in Sect. 7.7 how to use such a weak error expansion to devise a Richardson–Romberg extrapolation to (partially) kill this bias. We will first improve this approach by introducing the multistep Richardson–Romberg extrapolation, which takes advantage of higher-order expansions of the weak error. Then we will present two avatars of the multilevel paradigm. The multilevel paradigm introduced by M. Giles in [107] (see also [148]) combines both weak error expansions and a strong approximation rate to get rid of the bias as quickly as possible. We present two natural frameworks for multilevel methods, the original one due to Giles, when a first order expansion of the weak error is available, and one combined with multistep Richardson–Romberg extrapolation, which takes advantage of a higher-order expansion.

The first section develops a convenient abstract framework for biased Monte Carlo simulation which will allow us to present and analyze in a unified way various approaches and improvements, starting from crude Monte Carlo simulation, standard and multistep Richardson–Romberg extrapolation, the multilevel paradigm in both its weighted and unweighted forms and, finally, the randomized multilevel methods. On our way, we illustrate how this framework can be directly applied to time stepping problems – mainly for Brownian diffusions at this stage – but also to optimize the so-called *Nested Monte Carlo* method. The object of the Nested Monte Carlo method is to compute by simulation of functions of conditional expectations which rely on the nested combination of *inner* and *outer* simulations. This is an important field of investigation of modern simulation, especially in actuarial sciences to compute quantities like the Solvency Capital Requirement (*SCR*) (see *e.g.* [53, 76]). Numerical tests are proposed to highlight the global efficiency of these multilevel methods. Several exercises will hopefully help to familiarize the reader with the basic principles of multilevel simulation but also some recent more sophisticated improvements like antithetic meta-schemes. At the very end of the chapter, we return to the weak exact simulation and its natural connections with randomized multilevel simulation.

9.2 An Abstract Framework for Biased Monte Carlo Simulation

The quantity of interest is

$$I_0 = \mathbb{E}\, Y_0,$$

where $Y_0 : (\Omega, \mathcal{A}, \mathbb{P}) \to \mathbb{R}$ is a square integrable real random variable whose simulation cannot be performed at a reasonable computational cost (or complexity). We assume that, nevertheless, it can be approximated, in a sense made more precise later by a family of random variables $(Y_h)_{h\in\mathcal{H}}$ as $h \to 0$, defined on the same probability space, where $\mathcal{H} \subset (0, +\infty)$ is a set of *bias* parameters such that $\mathcal{H} \cup \{0\}$ is a compact set and

$$\forall\, n \in \mathbb{N}^*, \quad \frac{\mathcal{H}}{n} \subset \mathcal{H}.$$

Note that $\mathcal{H}$ being nonempty, it contains a sequence $\frac{h}{n} \to 0$ so that 0 is a limiting point of $\mathcal{H}$. As far as approximation of Y_0 by Y_h is concerned, the least we can ask is that the induced "weak error" converges to 0, *i.e.*

$$\mathbb{E}\, Y_h \to \mathbb{E}\, Y_0 \quad \text{as} \quad h \to 0. \tag{9.1}$$

However, stronger assumptions will be needed in the sequel, in terms of weak but also strong (L^2) approximation. Our ability to specify the simulation complexity of Y_h for every $h \in \mathcal{H}$ will be crucial when selecting such a family for practical implementation. Waiting for some more precise specifications, we will simply assume that

$$Y_h \text{ can be simulated at a computational cost of } \kappa(h),$$

which we suppose to be "acceptable" by contrast with that of Y_0, where $\kappa(h)$ is more or less representative of the number of floating point operations performed to simulate one (pseudo-)realization of Y_h from (finitely many pseudo-)random numbers. We will assume that $h \mapsto \kappa(h)$ is a decreasing function of h and that $\lim_{h\to 0} \kappa(h) = +\infty$ in order to model the fact that the closer Y_h is to Y_0, the higher the simulation complexity of Y_h is, keeping in mind that Y_0 cannot be simulated at a reasonable cost. It is often convenient to specify h based on this complexity by assuming that $\kappa(h)$ is inverse linear in h, *i.e.* $\kappa(h) = \frac{\bar{\kappa}}{h}$.

To make such an approximate simulation method acceptable, we need to control at least the error induced by the computation of $\mathbb{E}\, Y_h$ by a Monte Carlo simulation, *i.e.* by considering $\widehat{(Y_h)}_M = \frac{1}{M}\sum_{k=1}^{M} Y_h^k$, where $(Y_h^k)_{k\geq 1}$ is a sequence of independent copies of Y_h. For that purpose, all standard ways to measure this error (quadratic norm, *CLT* to design confidence intervals or *LIL*) rely on the existence of a variance for Y_h, that is, $Y_h \in L^2$. Furthermore, except in very specific situations, one further natural constraint on the $(Y_h)_{h\in\mathcal{H}}$ is that $(\mathrm{Var}(Y_h))_{h\in\mathcal{H}}$ remains bounded as $h \to 0$ in $\mathcal{H}$ but, in practice, this is not enough and we often require the more precise

$$\mathrm{Var}(Y_h) \to \mathrm{Var}(Y_0) \quad \text{as } h \to 0 \tag{9.2}$$

or, equivalently, under (9.1) (convergence of expectations)

$$\mathbb{E}\, Y_h^2 \to \mathbb{E}\, Y_0^2 \quad \text{as } h \to 0.$$

In some sense, the term "approximation" to simply denote the convergence of the first two moments is a language abuse, but we will see that Y_0 is usually a functional $F(X)$ of an underlying "structural stochastic process" X and Y_h is defined as the same functional of a discretization scheme $F(\widetilde{X}^n)$, so that the above two convergences are usually the consequence of the weak convergence of $\widetilde{X}^n$ toward X in distribution with respect to the topology on the functional space of the paths of the processes X and $\widetilde{X}^n$, typically the space $\mathbb{D}([0, T], \mathbb{R}^d)$ of *right continuous left limited functions on* $[0, T]$.

Let us give two typical examples where this framework will apply.

⊳ **Examples.**

1. *Discretization scheme of a diffusion (Euler).* Let $X = (X_t)_{t\in[0,T]}$ be a Brownian diffusion with Lipschitz continuous drift $b(t, x)$ and diffusion coefficient $\sigma(t, x)$, both defined on $[0, T] \times \mathbb{R}^d$ and having values in $\mathbb{R}^d$ and $\mathcal{M}(d, q, \mathbb{R})$, respectively, associated to a q-dimensional Brownian motion W defined on a probability space

$(\Omega, \mathcal{A}, \mathbb{P})$, independent of X_0. Let $(\widetilde{X}^n_t)_{t\in[0,T]}$ and $(\bar{X}^n_t)_{t\in[0,T]}$ denote the càdlàg stepwise constant and genuine (continuous) Euler schemes with step $\frac{T}{n}$, respectively, as defined in Chap. 7. Let $F : \mathbb{D}([0,T], \mathbb{R}^d) \to \mathbb{R}$ be a ($\mathbb{P}_X$-$a.s.$) continuous functional. Then, set

$$\mathcal{H} = \left\{\frac{T}{n},\ n \in \mathbb{N}^*\right\}, \quad h = \frac{T}{n}, \quad Y_h = F\big((\widetilde{X}^n_t)_{t\in[0,T]}\big) \text{ and } F\big((\bar{X}^n_t)_{t\in[0,T]}\big)$$

respectively. Note that the complexity of the simulation is proportional to the number n of time discretization steps of the scheme, *i.e.* $\kappa(h) = \frac{\bar{\kappa}}{h}$.

2. *Nested Monte Carlo.* The aim of the so-called *Nested Monte Carlo* simulation is to compute expectations involving a conditional expectation, *i.e.* of the form

$$I_0 = \mathbb{E}\, Y_0 \quad \text{with} \quad Y_0 = f\big(\mathbb{E}\,(F(X, Z) \mid X)\big),$$

where X, Z are two independent random vectors defined on a probability space $(\Omega, \mathcal{A}, \mathbb{P})$ having values in $\mathbb{R}^d$ and $\mathbb{R}^q$, respectively, and $F : \mathbb{R}^d \times \mathbb{R}^q \to \mathbb{R}$ is a Borel function such that $F(X, Z) \in L^1(\mathbb{P})$. We assume that both X and Z can be simulated at a reasonable computational cost. However, except in very specific situations, one cannot directly simulate exact independent copies of Y_0 due to the presence of a conditional expectation. Taking advantage of the independence of X and Z, one has

$$Y_0 = f\big([\mathbb{E}\, F(x, Z)]_{|x=X}\big).$$

This suggests to first simulate M independent copies X_m, $m = 1, \dots, M$, of X and, for each such copy, to estimate $\big[\mathbb{E}\, F(x, Z)\big]_{|x=X_m}$ by an *inner Monte Carlo simulation* of size $K \in \mathbb{N}^*$ by

$$\frac{1}{K}\sum_{k=1}^{K} F(X_m, Z^m_k), \quad \text{where} \quad (Z^m_k)_{k=1,\dots,K} \quad \text{are i.i.d. copies of } Z, \text{ independent of } X,$$

finally leading us to consider the estimator

$$\widehat{I}_{M,K} = \frac{1}{M}\sum_{m=1}^{M} f\left(\frac{1}{K}\sum_{k=1}^{K} F(X_m, Z^m_k)\right).$$

The usual terminology used by practitioners to discriminate between the two sources of simulations in the above naive nested estimator is the following: the simulations of the random variables Z^m_k indexed by k running from 1 to K for each fixed m are called *inner simulations* whereas the simulations of the random variables X_m indexed by m running from 1 to M are called the *outer simulations*.

This can be inserted in the above biased framework by setting, an integer $K_0 \geq 1$ being fixed,

$$\mathcal{H} = \left\{\frac{1}{K_0 k},\ k \in \mathbb{N}^*\right\},\quad Y_0 = f\big([\mathbb{E}\, F(x, Z)]_{|x=X}\big) \tag{9.3}$$

and

$$Y_h = f\left(\frac{1}{K}\sum_{k=1}^{K} F(X, Z_k)\right),\ h = \frac{1}{K} \in \mathcal{H},\ (Z_k)_{k=1,\ldots,K} \text{ i.i.d. copies of } Z \tag{9.4}$$

independent of X.

If f is continuous with linear growth, it is clear by the *SLLN* under the above assumptions that $Y_h \to Y_0$ in L^1, so that $\mathbb{E}\, Y_h \to \mathbb{E}\, Y_0$. If, furthermore, $F(X, Z) \in L^2(\mathbb{P})$, then $\text{Var}(Y_h) \to \text{Var}(Y_0)$ as $h \to 0$ in $\mathcal{H}$ (*i.e.* as $K \to +\infty$). Consequently, if K and M go to infinity, then $\widehat{I}_{M,K}$ will converge toward its target $f\big(\mathbb{E}\,(F(X, Z) \mid X)\big)$ as $M, K \to +\infty$ in a proper way, but, it is clearly a biased estimator at finite range.

Note that the complexity of the simulation of Y_h as defined above is proportional to the length K of this inner Monte Carlo simulation, *i.e.* again of the form $\kappa(h) = \frac{\bar{\kappa}}{h}$.

Two main settings are extensively investigated in nested Monte Carlo simulation: the case of smooth functions f (at least Hölder) on the one hand and that of indicator functions of intervals. This second setting is very important as it corresponds to the computation of quantiles of a conditional expectation, and, in a second stage, of quantities like Value-at-Risk which is the related inverse problem. Thus, it is a major issue in actuarial sciences to compute the *Solvency Capital Requirement* (*SCR*) which appears, technically speaking, exactly as a quantile of a conditional expectation. To be more precise, this conditional expectation is the value of the available capital at a maturity T ($T = 30$ years or even more) given a short term future (say 1 year). For a first example of computation of *SCR* based on a nested Monte Carlo simulation, we refer to [31]. See also [76].

Our aim in this chapter is to introduce methods which reduce the bias efficiently while keeping the variance under control so as to satisfy a prescribed global quadratic error with the smallest possible "budget", that is, with the lowest possible complexity. This leads us to model a Monte Carlo simulation as a constrained optimization problem. This amounts to minimizing the (computational) cost of the simulation for a prescribed quadratic error. We first develop this point of view for a crude Monte Carlo simulation to make it clear and, more importantly, to have a reference. To this end, we will also need to switch to more precise versions of (9.1) and (9.2), as will be emphasized below. Other optimization approaches can be adopted which turn out to be equivalent *a posteriori*.

9.3 Crude Monte Carlo Simulation

The starting point is the *bias-variance decomposition* of the error induced by using independent copies $(Y_h^k)_{k\geq 1}$ of Y_h to approximate the quantity of interest $I_0 = \mathbb{E}\, Y_0$ by the Monte Carlo estimator

$$\widehat{I}_M = \frac{1}{M}\sum_{k=1}^{M} Y_h^k,$$

namely

$$\big\|\mathbb{E}\, Y_0 - \widehat{I}_M\big\|_2^2 = \underbrace{\big(\mathbb{E}\, Y_0 - \mathbb{E}\, Y_h\big)^2}_{\text{squared bias}} + \underbrace{\frac{\mathrm{Var}(Y_h)}{M}}_{\text{Monte Carlo variance}}. \tag{9.5}$$

This decomposition is a straightforward consequence of the identity

$$\mathbb{E}\, Y_0 - \frac{1}{M}\sum_{k=1}^{M} Y_h^k = \big(\mathbb{E}\, Y_0 - \mathbb{E}\, Y_h\big) \overset{\perp}{+} \left(\mathbb{E}\, Y_h - \frac{1}{M}\sum_{k=1}^{M} Y_h^k\right).$$

The two terms are orthogonal in L^2 since $\mathbb{E}\left(\mathbb{E}\, Y_h - \frac{1}{M}\sum_{k=1}^{M} Y_h^k\right) = 0$.

Definition 9.1. *The squared quadratic error* $\|I_0 - \widehat{I}_M\|_2^2 = \mathbb{E}\,(I_0 - \widehat{I}_M)^2$ *is called the* Mean Squared Error *(MSE) of the estimator whereas the quadratic error itself* $\|I_0 - \widehat{I}_M\|_2$ *is called the* Rooted Mean Squared Error *(RMSE).*

Our aim in what follows is to minimize the cost of a Monte Carlo simulation for a prescribed (upper-bound of the) $RMSE = \varepsilon > 0$, assumed to be small, and in any case in $(0, 1]$ in non-asymptotic results. The idea is to make a balance between the bias term and the variance term by an appropriate choice of the parameter h and the size M of the simulation in order to achieve this prescribed error bound at a minimal cost. To this end, we will strengthen our bias assumption (9.1) by assuming a *weak rate* of convergence or first order weak expansion error

$$\big(WE\big)_1^{\alpha} \;\equiv\; \mathbb{E}\, Y_h = \mathbb{E}\, Y_0 + c_1 h^{\alpha} + o(h^{\alpha}) \tag{9.6}$$

where $\alpha > 0$. We assume that this first order expansion is consistent, *i.e.* that $c_1 \neq 0$.

Remark. Note that under the assumptions of Theorem 7.8, (a) and (b), the above assumption holds true in the framework of diffusion approximation by the Euler scheme with $\alpha = 1$ when $Y_h = f(\bar{X}_T^n)$ (with $h = \frac{T}{n}$). For path-dependent functionals, we saw in Theorem 8.1 that α drops down (at least) to $\alpha = \frac{1}{2}$ in situations involving the indicator function of an exit time, like for barrier options and the genuine Euler scheme.

As for the variance, we make no significant additional assumption at this stage, except that, if $\mathrm{Var}(Y_h) \to \mathrm{Var}(Y_0)$, we may reasonably assume that

$$\bar{\sigma}^2 = \sup_{h \in \mathcal{H}} \mathrm{Var}(Y_h) < +\infty,$$

keeping in mind that, furthermore, we can replace $\bar{\sigma}^2$ by $\mathrm{Var}(Y_0)$ in asymptotic results when $h \to 0$.

Consequently, we can upper-bound the MSE of $\widehat{I}_M$ by

$$MSE(\widehat{I}_M) \le c_1^2 h^{2\alpha} + \frac{\bar{\sigma}^2}{M} + o\big(h^{2\alpha}\big).$$

On the other hand, we make the natural assumption (see the above examples) that the simulation of (one copy of) Y_h has complexity

$$\kappa(h) = \frac{\bar{\kappa}}{h}. \tag{9.7}$$

We denote by $\mathrm{Cost}(\widehat{I}_M)$ the global complexity – or computational cost – of the simulation of the estimator $\widehat{I}_M$. It is clearly given by

$$\mathrm{Cost}(\widehat{I}_M) = M\kappa(h)$$

since a Monte Carlo estimator is linear. Consequently, our minimization problem reads

$$\inf_{RMSE(\widehat{I}_M) \le \varepsilon} \mathrm{Cost}(\widehat{I}_M) = \inf_{RMSE(\widehat{I}_M) \le \varepsilon} M\kappa(h). \tag{9.8}$$

If we replace $MSE(\widehat{I}_M)$ by its above upper-bound and neglect the term $o(h^{2\alpha})$, this problem can be re-written by the following approximate one:

$$\inf\left\{\frac{M\bar{\kappa}}{h},\ h \in \mathcal{H},\ M \in \mathbb{N}^*,\ c_1^2 h^{2\alpha} + \frac{\bar{\sigma}^2}{M} \le \varepsilon^2\right\}. \tag{9.9}$$

Now, note that

$$\frac{M\bar{\kappa}}{h} = \frac{M\bar{\kappa}\,\bar{\sigma}^2/M}{h\,\bar{\sigma}^2/M} = \frac{\bar{\kappa}\,\bar{\sigma}^2}{h(\bar{\sigma}^2/M)},$$

where the MSE constraint reads $\bar{\sigma}^2/M \le \varepsilon^2 - c_1^2 h^{2\alpha}$. It is clear that, in order to minimize the above ratio, this constraint should be saturated, so that

$$\inf\left\{\frac{M\bar\kappa}{h},\ h\in\mathcal{H},\ M\in\mathbb{N}^*,\ c_1^2h^{2\alpha}+\frac{\bar\sigma^2}{M}\le\varepsilon^2\right\}\le\inf_{0<h<(\varepsilon/|c_1|)^{\frac{1}{\alpha}}}\frac{\bar\kappa\bar\sigma^2}{h(\varepsilon^2-c_1^2h^{2\alpha})} \tag{9.10}$$

$$=\frac{\bar\kappa\bar\sigma^2}{\sup_{0<h<(\varepsilon/c_1)^{\frac{1}{\alpha}}}h(\varepsilon^2-c_1^2h^{2\alpha})}$$

$$=\frac{(1+2\alpha)^{1+\frac{1}{2\alpha}}}{2\alpha}\bar\kappa\,\bar\sigma^2\,\frac{|c_1|^{\frac{1}{\alpha}}}{\varepsilon^{2+\frac{1}{\alpha}}}, \tag{9.11}$$

where the infimum in the right-hand side of (9.10) is in fact attained at

$$h_{\min}(\varepsilon)=\left(\frac{\varepsilon}{\sqrt{1+2\alpha}\,|c_1|}\right)^{\frac{1}{\alpha}}.$$

Then one sets

$$h^*(\varepsilon)=\text{ nearest lower neighbor in }\mathcal{H}\text{ of }h_{\min}(\varepsilon),$$

which lies in $\mathcal{H}$ but is possibly slightly suboptimal. Thus, if $\mathcal{H}$ is of the form $\mathcal{H}=\left\{\frac{\mathbf{h}}{n},\ n\ge 1\right\}$,

$$h^*(\varepsilon)=\mathbf{h}\lceil\mathbf{h}/h_{\min}(\varepsilon)\rceil^{-1}. \tag{9.12}$$

We derive the simulation size by plugging $h_{\min}(\varepsilon)$ into the saturated constraint $c_1^2h^{2\alpha}+\frac{\bar\sigma^2}{M}=\varepsilon^2$, which yields after elementary computations

$$M=M^*(\varepsilon)=\left\lceil\left(1+\frac{1}{2\alpha}\right)\frac{\bar\sigma^2}{\varepsilon^2}\right\rceil.$$

Taking advantage of the fact that $\mathrm{Var}(Y_h)$ converges to $\mathrm{Var}(Y_0)$, one may improve the above minimal complexity as $\varepsilon\to 0$ (and make more rigorous the above optimization by taking into account the term $o(h^{2\alpha})$ that we neglected). In fact, one easily checks that

$$\inf_{RMSE(\widehat{I}_M)\le\varepsilon}\mathrm{Cost}(\widehat{I}_M)\lesssim\frac{(1+2\alpha)^{1+\frac{1}{2\alpha}}}{2\alpha}\bar\kappa\,\mathrm{Var}(Y_0)|c_1|^{\frac{1}{\alpha}}\frac{1}{\varepsilon^{2+\frac{1}{\alpha}}}\quad\text{as }\ \varepsilon\to 0. \tag{9.13}$$

This optimization procedure should be compared to a *virtual* unbiased simulation by averaging independent copies of Y_0 if the simulation cost κ_0 of Y_0 is finite. In such a case, $MSE(\widehat{I}_M)=\frac{\mathrm{Var}(Y_0)}{M}$, so that $M^*(\varepsilon)=\frac{\mathrm{Var}(Y_0)}{\varepsilon^2}$ and the resulting complexity reads

$$\inf_{RMSE(\widehat{I}_M)\le\varepsilon}\mathrm{Cost}=\kappa_0M^*(\varepsilon)=\frac{\kappa_0\mathrm{Var}(Y_0)}{\varepsilon^2}\quad\text{and}\quad M^*(\varepsilon)=\frac{\mathrm{Var}(Y_0)}{\varepsilon^2}. \tag{9.14}$$

The main conclusion of this introductory section is the following: the challenge of reducing (killing...) the bias is to switch from the exponent $2 + \frac{1}{\alpha}$ to an exponent as close to 2 as possible. This is a quite important question, as highlighted by the following simple illustration based on (9.11) or (9.13):

- To increase the accuracy of an unbiased simulation by a factor of 2, one needs a 4 times longer simulation.
- If the simulation is biased and $\big(WE\big)_1^\alpha$ holds with $\alpha = 1$, one needs an 8 times longer simulation to increase the accuracy by a factor of 2.
- If the simulation is biased and $\big(WE\big)_1^\alpha$ holds with $\alpha = 1/2$, one needs a 16 times longer simulation to increase the accuracy by a factor of 2.

9.4 Richardson–Romberg Extrapolation (II)

Richardson–Romberg extrapolation based on weak error expansion has already been briefly introduced in the framework of diffusions and their discretization schemes in Sect. 7.7.

9.4.1 General Framework

We assume that the weak error expansion of $\mathbb{E}\, Y_h$ in the $(h^{\alpha\ell})_{\ell\geq 1}$ scale holds at the second order, *i.e.*

$$\big(WE\big)_2^\alpha \;\equiv\; \mathbb{E}\, Y_h = \mathbb{E}\, Y_0 + c_1 h^\alpha + c_2 h^{2\alpha} + o\big(h^{2\alpha}\big), \tag{9.15}$$

and that (9.2) holds, *i.e.* $\mathrm{Var}(Y_h) \to \mathrm{Var}(Y_0)$ as well as $\bar{\sigma}^2 = \sup_{h\in\mathcal{H}} \mathrm{Var}(Y_h) < +\infty$.

Remark. The implementation of Richardson–Romberg extrapolation only requires $\big(WE\big)_1^\alpha$. We assumed $\big(WE\big)_2^\alpha$ to enhance its best possible performance. The existence of such an expansion (already tackled in the diffusion framework) will be discussed further on in Sect. 9.5.1, *e.g.* for nested Monte Carlo simulation.

One deduces by linearly combining (9.15) with h and $\frac{h}{2}$ that

$$2^\alpha\, \mathbb{E}\, Y_{\frac{h}{2}} - \mathbb{E}\, Y_h = (2^\alpha - 1)\mathbb{E}\, Y_0 + (2^{-\alpha} - 1)c_2\, h^{2\alpha} + o\big(h^{2\alpha}\big). \tag{9.16}$$

This suggests a natural way to modify the estimator of $I_0 = \mathbb{E}\, Y_0$ to reduce the bias: set

$$\widetilde{Y}_h = \frac{2^\alpha Y_{\frac{h}{2}} - Y_h}{2^\alpha - 1}, \qquad h \in \mathcal{H}.$$

It is clear from (9.16) that $(\widetilde{Y}_h)_{h\in\mathcal{H}}$ satisfies $\big(WE\big)_1^{2\alpha}$. Without additional assumptions $\widetilde{Y}_h$ does not satisfy Assumption (9.2) (convergence of the variance), however its variance remains bounded if that of Y_h does since

$$\begin{aligned}\mathrm{Var}(\widetilde{Y}_h) = \sigma(\widetilde{Y}_h)^2 &\le \frac{1}{(2^\alpha-1)^2}\big(2^\alpha\sigma(Y_{\frac{h}{2}})+\sigma(Y_h)\big)^2\\ &\le \Big(\frac{2^\alpha+1}{2^\alpha-1}\Big)^2\bar{\sigma}^2 \end{aligned} \tag{9.17}$$

with the notations of the former section. Also note that the simulation cost $\tilde{\kappa}(h)$ of $\widetilde{Y}_h$ satisfies $\tilde{\kappa}(h)\le\kappa(h)+2\kappa(h)=\frac{3\bar{\kappa}}{h}$; note that for some applications like nested Monte Carlo one may obtain a slower increase of the complexity since $\tilde{\kappa}(h)=\kappa(h)\vee(2\kappa(h))=\frac{2\bar{\kappa}}{h}$. This leads us to introduce, for every $h\in\mathcal{H}$, the *Richardson–Romberg estimator*

$$\widetilde{I}_M^{RR}=\widetilde{I}_M^{RR}(h,M)=\frac{1}{M}\sum_{k=1}^{M}\widetilde{Y}_h^k=\frac{1}{M(2^\alpha-1)}\sum_{k=1}^{M}2^\alpha Y_{\frac{h}{2}}^k-Y_h^k,\;\; M\ge 1,$$

where $(Y_h^k, Y_{\frac{h}{2}}^k)_{k\ge1}$ are independent copies of $(Y_h, Y_{\frac{h}{2}})$. Applying the results of the former section to the family $(\widetilde{Y}_h)_{h\in\mathcal{H}}$ with $\tilde{\alpha}=2\alpha$, $\tilde{\kappa}(h)$, $\tilde{c}_1=-(1-2^{-\alpha})c_2$ and the upper-bound (9.17) of $\sigma(\widetilde{Y}_h)$, we deduce from (9.11) that

$$\inf_{\|\widetilde{I}_M^{RR}-I_0\|_2\le\varepsilon}\mathrm{Cost}(\widetilde{I}_M^{RR})\le\frac{(1+4\alpha)^{1+\frac{1}{4\alpha}}}{4\alpha}\frac{3(2^\alpha+1)^2[(1-2^{-\alpha})|c_2|]^{\frac{1}{2\alpha}}}{(2^\alpha-1)^2}\frac{\bar{\kappa}\,\bar{\sigma}^2}{\varepsilon^{2+\frac{1}{2\alpha}}}.$$

In view of the exponent $2+\frac{1}{2\alpha}$ of ε (the prescribed *RMSE*) in the above minimal Cost function, it turns out that we are halfway toward an unbiased simulation.

However, the brute force upper-bound established for $\mathrm{Var}(\widetilde{Y}_h)$ is not satisfactory since it relies on the triangle inequality for the standard deviation, so we will try to improve it by investigating the internal structure of the family $(Y_h)_{h\in\mathcal{H}}$. (In what follows all $\lim_{h\to0}$, $\overline{\lim}_{h\to0}$, etc, should always be understood for $h\in\mathcal{H}$.)

If we apply the reverse triangle inequality to $\widetilde{Y}_h$, we know that

$$\mathrm{Var}(\widetilde{Y}_h)\ge\frac{1}{(2^\alpha-1)^2}(2^\alpha\sigma(Y_h)-\sigma(Y_{\frac{h}{2}}))^2$$

so that, under Assumption (9.2),

$$\underline{\lim}_{h\to0}\mathrm{Var}(\widetilde{Y}_h)\ge\frac{(2^\alpha-1)^2}{(2^\alpha-1)^2}\sigma(Y_0)^2=\mathrm{Var}(Y_0). \tag{9.18}$$

On the other hand, by Schwartz's Inequality, $\mathrm{Cov}(Y_h, Y_{\frac{h}{2}}) \le \sqrt{\mathrm{Var}(Y_h)\mathrm{Var}(Y_{\frac{h}{2}})}$, which implies

$$\overline{\lim_{h\to 0}}\, \mathrm{Cov}(Y_h, Y_{\frac{h}{2}}) \le \mathrm{Var}(Y_0). \tag{9.19}$$

Bi-linearity of the covariance operator yields

$$\mathrm{Var}(2^\alpha Y_{\frac{h}{2}} - Y_h) = 2^{2\alpha}\mathrm{Var}(Y_{\frac{h}{2}}) - 2^{1+\alpha}\mathrm{Cov}(Y_h, Y_{\frac{h}{2}}) + \mathrm{Var}(Y_h)$$

so that

$$\overline{\lim_{h\to 0}}\, \mathrm{Var}(\widetilde{Y}_h) = \frac{2^{2\alpha}+1}{(2^\alpha-1)^2}\mathrm{Var}(Y_0) - \frac{2^{1+\alpha}}{(2^\alpha-1)^2}\underline{\lim_{h\to 0}}\, \mathrm{Cov}(Y_h, Y_{\frac{h}{2}}).$$

Combining the lower bound (9.18) and the upper bound (9.19) with the former equality shows that:

$$\begin{aligned}\lim_{h\to 0} \mathrm{Var}(\widetilde{Y}_h) = \mathrm{Var}(\widetilde{Y}_0) \text{ iff } & \overline{\lim_{h\to 0}}\, \mathrm{Var}(\widetilde{Y}_h) \le \mathrm{Var}(\widetilde{Y}_0)\\ \text{iff } & \underline{\lim_{h\to 0}}\, \mathrm{Cov}(Y_h, Y_{\frac{h}{2}}) \ge \mathrm{Var}(Y_0)\\ \text{iff } & \lim_{h\to 0} \mathrm{Cov}(Y_h, Y_{\frac{h}{2}}) = \mathrm{Var}(Y_0).\end{aligned}$$

As a consequence, the best choice, if possible, is to select a family $(Y_h)_{h\in\mathcal{H}}$ of *approximators* satisfying

$$\lim_{h\to 0} \mathrm{Cov}(Y_h, Y_{\frac{h}{2}}) = \mathrm{Var}(Y_0).$$

But, then

$$\begin{aligned}\left\|Y_h - Y_{\frac{h}{2}}\right\|_2^2 &= \mathrm{Var}(Y_h - Y_{\frac{h}{2}}) + \left(\mathbb{E}\,(Y_h - Y_{\frac{h}{2}})\right)^2\\ &= \mathrm{Var}(Y_h) - 2\,\mathrm{Cov}(Y_h, Y_{\frac{h}{2}}) + \mathrm{Var}(Y_{\frac{h}{2}}) + \left(\mathbb{E}\,(Y_h - Y_{\frac{h}{2}})\right)^2\\ &\to \mathrm{Var}(Y_0)(1-2+1) + 0^2 = 0 \quad \text{as} \quad h\to 0 \quad \text{in} \quad \mathcal{H},\end{aligned}$$

i.e.

$$Y_h - Y_{\frac{h}{2}} \xrightarrow{L^2} 0 \quad \text{as} \quad h\to 0 \quad \text{in} \quad \mathcal{H}. \tag{9.20}$$

This strongly suggests, in order to better control the variance of the estimator, to consider a family $(Y_h)_{h\in\mathcal{H}}$ which *strongly* converges toward Y_0 in quadratic norm (which in turn will imply that $\left\|Y_h - Y_{\frac{h}{2}}\right\|_2 \to 0$) that is

$$Y_h \xrightarrow{L^2} Y_0 \quad \text{as} \quad h\to 0 \quad \text{in} \quad \mathcal{H}. \tag{9.21}$$

Warning (the temptation of laziness)! It should be emphasized that the *lazy approach* consisting in considering Y_h and $Y_{\frac{h}{2}}$ to be independent leads to

$$\mathrm{Var}(\widetilde{Y}_h) = \frac{2^{2\alpha}\mathrm{Var}(Y_h) + \mathrm{Var}(Y_{\frac{h}{2}})}{(2^\alpha - 1)^2} \xrightarrow{h\to 0} \frac{2^{2\alpha}+1}{(2^\alpha-1)^2}\mathrm{Var}(Y_0).$$

Thus the variance of $\widetilde{Y}_h$ is approximately multiplied by $\frac{2^{2\alpha}+1}{(2^\alpha-1)^2}$ compared to $\mathrm{Var}(Y_h)$ when h is small! This corresponds to a factor of 5 when $\alpha = 1$ and a $\frac{3}{3-2\sqrt{2}} \simeq 17.5$ factor when $\alpha = \frac{1}{2}$. This huge increase of the variance becomes a major obstacle to the implementation of the method, except possibly when ε is very small. In any case, this choice is never competitive with that of a coherent family of approximators $(Y_h)_{h\in\mathcal{H}}$.

⊳ **Examples. 1.** In the framework of Euler discretization schemes for Brownian diffusions, when $Y_h = F\big((\widetilde{X}^n_t)_{t\in[0,T]}\big)$ or $Y_h = F\big((\bar{X}^n_t)_{t\in[0,T]}\big)$ with F a $\|\cdot\|_{\sup}$-Lipschitz continuous, the strong convergence assumption (9.21) amounts to the convergence in L^2 of $\sup_{t\in[0,T]}|X_t - \widetilde{X}^n_t|$ or $\sup_{t\in[0,T]}|X_t - \bar{X}^n_t|$ to 0 established in Chap. 7 (Theorem 7.2).

2. In the nested Monte Carlo framework as described by (9.3) and (9.4) at the beginning of the chapter, (9.21) is an easy consequence of the *SLLN* and Fubini's Theorem when the function f is Lipschitz continuous and $F(X, Z) \in L^2$.

▶ **Exercise.** Prove the statement about nested the Monte Carlo method in the preceding example. Extend the result to the case where f is ρ-Hölder under appropriate assumptions on $F(X, Z)$.

As a conclusion to this section, let us mention that it may happen that (9.20) holds and (9.21) fails. This is possible if the rate of convergence in (9.20) is not fast enough. Such a situation is observed with the (weak) approximation of the geometrical Brownian motion by a binomial tree (see [32]).

9.4.2 ℵ Practitioner's Corner

Brownian diffusions

Let $X = (X_t)_{t\in[0,T]}$ be the Brownian diffusion solution to the *SDE* with Lipschitz continuous drift $b(t, x)$ and diffusion coefficient $\sigma(t, x)$, driven by a q-dimensional Brownian motion on a probability space $(\Omega, \mathcal{A}, \mathbb{P})$ (with X_0 independent of W). The Euler scheme with step $h = \frac{T}{n}$ – so that $\mathcal{H} = \big\{\frac{T}{n},\ n \geq 1\big\}$ – reads:

$$\bar{X}^n_{t^n_{k+1}} = \bar{X}^n_{t^n_k} + \frac{T}{n}b(t^n_k, \bar{X}^n_{t^n_k}) + \sqrt{\frac{T}{n}}\sigma(t^n_k, \bar{X}^n_{t^n_k})U^n_{k+1}, \quad \bar{X}^n_0 = X_0,$$

where $t_k^n = \frac{kT}{n}$ and $U_k^n = \sqrt{\frac{n}{T}}(W_{t_k^n} - W_{t_{k-1}^n})$, $k = 1, \dots, n$. We consider now the Euler scheme with step $\frac{T}{2n}$, designed with the same Brownian motion. We have

$$\bar{X}^{2n}_{t^{2n}_{k+1}} = \bar{X}^{2n}_{t^{2n}_k} + \frac{T}{2n} b(t_k^{2n}, \bar{X}^{2n}_{t^{2n}_k}) + \sqrt{\frac{T}{2n}}\,\sigma(t_k^{2n}, \bar{X}^{2n}_{t^{2n}_k}) U^{2n}_{k+1}, \quad \bar{X}^{2n}_0 = X_0,$$

where $t_k^{2n} = \frac{kT}{2n}$ and $U_k^{2n} = \sqrt{\frac{2n}{T}}(W_{t_k^{2n}} - W_{t_{k-1}^{2n}})$, $k = 1, \dots, 2n$. Hence, it is clear by their very definition that

$$U_k^n = \frac{U^{2n}_{2k-1} + U^{2n}_{2k}}{\sqrt{2}}, \quad k = 1, \dots, n. \tag{9.22}$$

The *joint simulation* of these two schemes can be simply performed as follows:

- First simulate $2n$ independent pseudo-random numbers U_k^{2n}, $k = 1, \dots, 2n$, with distribution $\mathcal{N}(0; I_q)$;
- then compute the U_k^n, $k = 1, \dots, n$ using (9.22).

An alternative method is to:

- first simulate the Brownian increments of the coarse scheme with step $\frac{T}{n}$,
- then use the recursive simulation of the Brownian motion to simulate the increments of the refined scheme with step $\frac{T}{2n}$. To be precise, a straightforward application of the Brownian bridge method (see Proposition 8.1, applied with $W_t^{(\frac{T}{n})} = W_{\frac{T}{n}+t} - W_{\frac{T}{n}}$ between 0 and $\frac{T}{n}$) yields that

  $$\mathcal{L}\big(W_{\frac{(2k+1)T}{2n}} - W_{\frac{kT}{n}} \mid W_{\frac{kT}{n}} = x,\ W_{\frac{(k+1)T}{n}} = y\big) = \mathcal{N}\Big(\frac{y-x}{2}; \frac{T}{4n}\Big)$$

  so that

  $$\mathcal{L}\big(W_{\frac{(2k+1)T}{2n}} \mid W_{\frac{kT}{n}} = x,\ W_{\frac{(k+1)T}{n}} = y\big) = \mathcal{N}\Big(\frac{x+y}{2}; \frac{T}{4n}\Big).$$

Note that the joint simulation of (functionals of) two genuine (continuous) Euler schemes with respective steps $\frac{T}{n}$ and $\frac{T}{2n}$ cannot be performed in an as elementarily way: the joint simulation of the diffusion bridge method for both schemes remains, to our knowledge, an open problem.

Nested Monte Carlo

– *Weak error expansion* (f *smooth*). In the nested Monte Carlo framework defined in (9.4) and (9.3), the question of the existence of a first- or second-order weak error expansion, with $\alpha = 1$, remains reasonably elementary when the function f is smooth enough. Thus, if f is five times differentiable with bounded existing partial derivatives and $F(X, Z) \in L^5$, then $(WE)_2^\alpha$ holds with $\alpha = 1$. This is a special case of an expansion at order R established in [198].

– *Weak error expansion* (f *indicator function*). When $f = \mathbf{1}_{[a,+\infty)}$ or, more generally, any indicator function of an interval, establishing the existence of a first-order weak error expansion is a much more involved task, first achieved in [128] to our knowledge. It relies on smoothness assumptions on the joint law of (Ξ_0, Ξ_h). In [113], a weak error expansion at order $R \geq 2$, that is, $\big(WE\big)^{\alpha}_{R}$, still with $\alpha = 1$, is established by a duality method (see also Sect. 9.6.2 for a more precise statement).

– *Complexity.* Temporarily assume $\mathbf{h} = 1$ – *i.e.* $K_0 = 1$ – for simplicity. Given the form of $Y_h = f\Big(\frac{1}{K}\sum_{k=1}^{K} F(X, Z_k)\Big)$, $(Z_k)_{k\geq 1}$ i.i.d. the expression of the complexity of $\widetilde{Y}_h$ differs from the generic case and can be slightly improved (if the computation of one value of the function f is neglected) since one just needs to simulate $\big(F(X, Z_k)\big)_{k=1,\ldots,2K}$ to simulate both Y_h and $Y_{\frac{h}{2}}$. Consequently, the complexity of the computation of $\widetilde{Y}_h$ in this framework is

$$\widetilde{\kappa}(h) = \bar{\kappa}\, 2K = \frac{2\bar{\kappa}}{h} \qquad \text{(rather than } \tfrac{3\bar{\kappa}}{h}\text{)}.$$

9.4.3 Going Further in Killing the Bias: The Multistep Approach

In this section, we extend the above Richardson–Romberg method by introducing a multistep approach based on a higher-order extension of the weak error $\mathbb{E}\, Y_h - \mathbb{E}\, Y_0$. The technique described in this section should be understood as a foundation for the multilevel methods rather than a direct approach since it will be outperformed by the weighted multilevel methods. In particular, the weights introduced in this section will be used in the next one devoted to the multilevel paradigm.

Throughout this section, we assume that the following holds for some integer $R \geq 2$

$$\big(WE\big)^{\alpha}_{R} \;\equiv\; \mathbb{E}\, Y_h = \mathbb{E}\, Y_0 + \sum_{r=1}^{R} c_r h^{\alpha r} + o\big(h^{\alpha R}\big). \tag{9.23}$$

As for all Taylor-like expansions of this type, the coefficients c_r are unique and do not depend on $R \geq r$.

We also make the strong convergence assumption (9.21) ($Y_h \to Y_0$ in L^2) as well, though it is not mandatory in this setting.

Definition 9.2. *An increasing* R*-tuple* $\underline{n} = (n_1, \ldots, n_R)$ *of positive integers satisfying* $1 = n_1 < n_2 < \cdots < n_R$ *is called a vector of* refiners. *For every* $h \in \mathcal{H}$*, the resulting sub-family of approximators is denoted by* $Y_{\underline{n},h} := \big(Y_{\frac{h}{n_i}}\big)_{i=1,\ldots,R}$.

The driving idea is, a vector $\underline{n}$ of refiners being fixed, to extend the Richardson–Romberg estimator by searching for a linear combination of the components of $\mathbb{E}\, Y_{\frac{h}{n_r}}$ which kills the resulting bias up to order $R-1$, relying on the expansion $\big(WE\big)^{\alpha}_{R}$ and then deriving the multistep estimator accordingly. To determine this linear combination, whose coefficients hopefully will not depend on $h \in \mathcal{H}$, we consider a generic R-tuple of *weights* $\mathbf{w}^R = (\mathbf{w}_1^R, \dots, \mathbf{w}_R^R) \in \mathbb{R}^R$.

To alleviate notation we will drop the superscript ^{R} and write $\mathbf{w}$ instead of $\mathbf{w}^R$ when no ambiguity arises. Idem for the components $\mathbf{w}$. Then, owing to $\big(WE\big)^{\alpha}_{R}$,

$$\sum_{j=1}^{R} \mathbf{w}_j \mathbb{E}\, Y_{\frac{h}{n_j}} = \Big(\sum_{j=1}^{R} \mathbf{w}_j\Big)\mathbb{E}\, Y_0 + \sum_{j=1}^{R} \mathbf{w}_j \sum_{r=1}^{R} c_r \frac{h^{\alpha r}}{n_j^{\alpha}} + o\Big(h^{\alpha R}\Big)$$

$$= \Big(\sum_{j=1}^{R} \mathbf{w}_j\Big)\mathbb{E}\, Y_0 + \sum_{r=1}^{R} c_r h^{\alpha r}\left[\sum_{j=1}^{R} \frac{\mathbf{w}_j}{n_j^{\alpha r}}\right] + o\Big(h^{\alpha R}\Big), \quad (9.24)$$

where we interchanged the two sums in the second line. If w is a solution to the system

$$\begin{cases} \displaystyle\sum_{j=1}^{R} \mathbf{w}_j = 1 \\ \displaystyle\sum_{j=1}^{R} \frac{\mathbf{w}_j}{n_j^{\alpha r}} = 0, \quad r = 1, \dots, R-1, \end{cases}$$

or, equivalently,

$$\sum_{j=1}^{R} \frac{\mathbf{w}_j}{n_j^{\alpha(r-1)}} = 0^{r-1}, \quad r = 1, \dots, R, \quad (9.25)$$

then (9.24) reads

$$\sum_{j=1}^{R} \mathbf{w}_j\, \mathbb{E}\, Y_{\frac{h}{n_j}} = \widetilde{\mathbf{W}}^{(R)}_{R+1} c_R h^{\alpha R} + o\Big(h^{\alpha R}\Big),$$

where $\widetilde{\mathbf{W}}^{(R)}_{R+1}$ is defined by

$$\widetilde{\mathbf{W}}^{(R)}_{R+1} = \sum_{j=1}^{R} \frac{\mathbf{w}_j}{n_j^{\alpha R}}. \quad (9.26)$$

Like for $\mathbf{w}$, we will often write $\widetilde{\mathbf{W}}_{R+1}$ instead of $\widetilde{\mathbf{W}}^{(R)}_{R+1}$ in what follows. The linear system (9.25) is obviously of Vandermonde type, namely

$$\mathrm{Vand}(x_1, \dots, x_R)\, \mathbf{w} = [c^{i-1}]_{i=1:R},$$

where the Vandermonde matrix attached to an R-tuple $(x_1, \dots, x_R) \in (0, +\infty)^R$ is defined by $\mathrm{Vand}(x_1, \dots, x_R) = [x_j^{i-1}]_{i,j=1:R}$ and $c \in \mathbb{R}$. In our case, $x_i = n_i^{-\alpha}$, $i = 1, \dots, R$ and $c = 0$. As a consequence, Cramer's formulas yield

$$\mathbf{w}_i = \frac{\det[\mathrm{Vand}(n_1^{-\alpha}, \dots, n_{i-1}^{-\alpha}, 0, n_{i+1}^{-\alpha}, \dots, n_R^{-\alpha})]}{\det[\mathrm{Vand}[(n_1^{-\alpha}, \dots, n_i^{-\alpha}, \dots, n_R^{-\alpha})]}, \quad i = 1, \dots, R.$$

As a consequence, the weight vector solution to (9.25) has a closed form since it is classical background that

$$\forall\, x_1, \dots, x_R > 0, \quad \det[\mathrm{Vand}(x_1, \dots, x_i, \dots, x_R)] = \prod_{1 \le i < j \le R} (x_j - x_i).$$

Synthetic formulas are given in the following proposition for $\mathbf{w}$ and $\widetilde{\mathbf{W}}_{R+1}$.

Proposition 9.1 (*a*) *For every fixed integer $R \ge 2$, the weight vector* $\mathbf{w}$ *admits a closed form given by*

$$\forall i \in \{1, \dots, R\}, \quad \mathbf{w}_i = \frac{n_i^{\alpha(R-1)}}{\prod_{j \ne i}(n_i^\alpha - n_j^\alpha)} = (-1)^{R-i} \frac{1}{\prod_{j \ne i} |1 - (n_j/n_i)^\alpha|}. \tag{9.27}$$

(*b*) *Furthermore,*

$$\widetilde{\mathbf{W}}_{R+1} = \frac{(-1)^{R-1}}{\underline{n}!^\alpha} \quad \textit{where} \quad \underline{n}! = \prod_{i=1}^{R} n_i. \tag{9.28}$$

Proof. (*a*) Temporarily set $x_i = n_i^{-\alpha}$, $i = 1, \dots, R$. The above Cramer formula reads, after canceling the terms not containing the index i which appear simultaneously in both products at the top and at the bottom of the ratio,

$$\begin{aligned}
\mathbf{w}_i &= \frac{\prod_{1 \le k < i}(x_k - 0) \times \prod_{i < \ell \le R}(0 - x_\ell)}{\prod_{1 \le k < i}(x_k - x_i) \times \prod_{i < \ell \le R}(x_i - x_\ell)} \\
&= \prod_{1 \le k \le R, k \ne i} \frac{x_k}{x_k - x_i} = \frac{x_i^{-(R-1)}}{\prod_{k \ne i}(x_i^{-1} - x_k^{-1})} \\
&= \frac{n_i^{\alpha(R-1)}}{\prod_{k \ne i}(n_i^\alpha - n_k^\alpha)} = (-1)^{R-i} \frac{n_i^{\alpha(R-1)}}{\prod_{k \ne i} |n_i^\alpha - n_k^\alpha|},
\end{aligned}$$

where we used in the last line that the refiners are increasing.

Furthermore, setting $X = 0$ in the elementary decomposition of the rational ratio

$$\frac{1}{\prod_{i=1}^R (X - \frac{1}{x_i})} = \sum_{i=1}^R \frac{1}{\prod_{k \ne i}(\frac{1}{x_i} - \frac{1}{x_k})} \frac{1}{(X - \frac{1}{x_i})}$$

yields the elementary identity

$$(-1)^{R-1}\prod_{i=1}^{R} x_i = \sum_{i=1}^{R} \frac{x_i}{\prod_{k\neq i}(\frac{1}{x_i}-\frac{1}{x_k})}$$
$$= \sum_{i=1}^{R} \frac{x_i^R \prod_{k\neq i} x_k}{\prod_{k\neq i}(x_k - x_i)} = \sum_{i=1}^{R} x_i^R\, \mathbf{w}_i.$$

Replacing x_i by its value $n_i^{-\alpha}$ for every index i leads to the announced result since, by the very definition (9.26) of $\widetilde{\mathbf{W}}_{R+1}$,

$$\widetilde{\mathbf{W}}_{R+1} = \sum_{i=1}^{R} \frac{\mathbf{w}_i}{n_i^{\alpha}} = (-1)^{R-1}\sum_{i=1}^{R} x_i^R\, \mathbf{w}_i = (-1)^{R-1}\prod_{i=1}^{R} x_i = \frac{(-1)^{R-1}}{\underline{n}!^{\alpha}}. \qquad \diamond$$

As a straightforward consequence of the former proposition, we get with the weights given by (9.27),

$$\sum_{i=1}^{R} \mathbf{w}_i\, \mathbb{E}\, Y_{\frac{h}{n_i}} = \mathbb{E}\, Y_0 + (-1)^{R-1}\frac{c_R}{\underline{n}!^{\alpha}} h^{\alpha R} + o\big(h^{\alpha R}\big). \tag{9.29}$$

Remark. The main point to be noticed is the universality of these weights, which do not depend on $h \in \mathcal{H}$, but only on the chosen vector of refiners $\underline{n}$ and the exponent α of the weak error expansion.

These computations naturally suggest the following definition for a family of multistep estimators of *depth* R.

Definition 9.3. (Multistep estimator) *The family of multistep estimators of depth R associated to the vector of refiners $\underline{n}$ is defined for every $h \in \mathcal{H}$ and every simulation size $M \geq 1$ by*

$$\widehat{I}_M^{RR} = \widehat{I}_M^{RR}(R, \underline{n}, h, M) = \frac{1}{M}\sum_{k=1}^{M}\sum_{i=1}^{R} \mathbf{w}_i\, Y^k_{\frac{h}{n_i}} \tag{9.30}$$
$$= \frac{1}{M}\sum_{i=1}^{R} \mathbf{w}_i \sum_{k=1}^{M} Y^k_{\frac{h}{n_i}}, \tag{9.31}$$

where $Y^k_{\underline{n},h}$ are independent copies of $Y_{\underline{n},h} = \big(Y_{\frac{h}{n_i}}\big)_{i=1,\dots,R}$ and the weight vector $\mathbf{w}$ is given by (9.27).

The parameter R is called the depth *of the estimator.*

Remark. If $R = 2$ and $\underline{n} = (1, 2)$ the multistep estimator is just the regular Richardson–Romberg estimator introduced in the previous section.

► **Exercises. 1.** *Weights of interest.* Keep in mind that the formulas below, especially those in (b), will be extensively used in what follows.

(a) Show that, if $n_i = i$ and $\alpha = 1$, then, the depth $R \geq 2$ being fixed,

$$\mathbf{w}_i = (-1)^{R-i}\frac{i^R}{i!(R-i)!}, \quad i = 1, \dots, R \quad \text{and} \quad \widetilde{\mathbf{W}}_{R+1} = \frac{(-1)^{R-1}}{R!}. \tag{9.32}$$

(*b*) Show that if $n_i = N^{i-1}$ (N integer, $N \geq 2$) and $\alpha \in (0, +\infty)$, then

$$\begin{aligned}\mathbf{w}_i &= (-1)^{R-i}\frac{1}{\prod_{1\leq j\leq R, j\neq i}|1 - N^{-\alpha(i-j)}|}\\ &= (-1)^{R-i}\frac{N^{-\frac{\alpha}{2}(R-j)(R-j+1)}}{\prod_{j=1}^{i-1}(1-N^{-\alpha j})\prod_{j=1}^{R-j}(1-N^{-\alpha j})}\end{aligned} \tag{9.33}$$

and

$$\widetilde{\mathbf{W}}_{R+1} = \frac{(-1)^{R-1}}{N^{\alpha\frac{R(R-1)}{2}}}.$$

2. *When* $c_1 = 0$. Assume that $\big(WE\big)_R^\alpha$ holds with $c_1 = 0$

$$\big(WE\big)_R^\alpha \equiv \mathbb{E}\,Y_h = \mathbb{E}\,Y_0 + \sum_{r=2}^{R} c_r h^{\alpha r} + o\big(h^{\alpha R}\big).$$

Prove the existence and determine an $(R-1)$-tuple of weights $\big(\mathbf{w}_1^\dagger, \dots, \mathbf{w}_{R-1}^\dagger\big)$ and a coefficient $\widetilde{\mathbf{W}}_R^\dagger$ such that

$$\sum_{i=1}^{R-1}\mathbf{w}_i^\dagger\,\mathbb{E}\,Y_{\frac{h}{n_i}} = \widetilde{\mathbf{W}}_R^\dagger c_R h^{\alpha R} + o(h^{\alpha R}).$$

[Hint: One can proceed either by mimicking the general case or by an appropriate re-scaling of the regular weights at order $R-1$.]

Let us briefly analyze the basic properties of this estimator.

– *Bias*. The preceding shows that, if $\big(WE\big)_R^\alpha$ holds, then the bias of $\widehat{I}_M^{RR}$ is independent of M, and reads

$$\mathbb{E}\,\widehat{I}_M^{RR} - \mathbb{E}\,Y_0 = (-1)^{R-1}c_R\left(\frac{h^R}{\underline{n}!}\right)^\alpha + o\big(h^{\alpha R}\big).$$

– *Complexity*. The (unitary) complexity (*i.e.* for $M = 1$) is given by

$$\kappa^{RR}(h) = \bar{\kappa}\Big(\frac{n_1}{h} + \dots + \frac{n_r}{h} + \dots + \frac{n_R}{h}\Big) = \bar{\kappa}\,\frac{|\underline{n}|}{h},$$

where $|\underline{n}| = n_1 + \dots + n_R$ is the ℓ^1-norm of the refiner vector $\underline{n}$.

Note that we neglect the multiplication by the weights in this evaluation. The first reason is that they are pre-computed off-line, the second is that the computation of $\widehat{I}^{RR}_{M}$ defined in (9.30) is performed in practice by (9.31), which only requires R multiplications.

– *Variance.* As expected

$$\operatorname{Var}\big(\widehat{I}^{RR}_{M}\big) = \frac{\operatorname{Var}\Big(\sum_{1\le i\le R} \mathbf{w}_i\, Y_{\frac{h}{n_i}}\Big)}{M}.$$

► **Exercises. 1.** *Analysis of the Richardson–Romberg estimator.* Let $I_0 = \mathbb{E}\, Y_0$. Show by mimicking the analysis of the crude Monte Carlo simulation (carried out under assumption $\big(WE\big)^{\alpha}_{2}$) that, if $\mathcal{H} = \big\{\frac{\mathbf{h}}{n},\, n \ge 1\big\}$,

– $\big(WE\big)^{\alpha}_{R}$ holds

and

– $\|Y_h - Y_0\|_2 \to 0$ as $h \to 0$ in $\mathcal{H}$,

then the Multistep Richardson–Romberg estimator of I_0 satisfies

$$\inf_{\substack{h\in\mathcal{H}\\ \|\widehat{I}^{RR}_{M}-I_0\|_2<\varepsilon}} \operatorname{Cost}\big(\widehat{I}^{RR}_{M}\big) \sim \left(\frac{(1+2\alpha R)^{1+\frac{1}{2\alpha R}}}{2\alpha R}\right) |c_{_R}|^{\frac{1}{\alpha R}} \operatorname{Var}\left(\sum_{i=1}^{R} \mathbf{w}_i\, Y_{\frac{h}{n_i}}\right) \frac{|\underline{n}|}{\underline{n}!^{\frac{1}{R}}}\, \frac{1}{\varepsilon^{2+\frac{1}{\alpha R}}}$$

$$\sim \left(\frac{(1+2\alpha R)^{1+\frac{1}{2\alpha R}}}{2\alpha R}\right) |c_{_R}|^{\frac{1}{\alpha R}} \operatorname{Var}(Y_0) \frac{|\underline{n}|}{\underline{n}!^{\frac{1}{R}}}\, \frac{1}{\varepsilon^{2+\frac{1}{\alpha R}}} \quad \text{as } \varepsilon \to 0$$

where the optimal parameters $h^*(\varepsilon)$ and $M^*(\varepsilon)$ (simulation size) are given by

$$h^*(\varepsilon) = \mathbf{h}\lceil \mathbf{h}/h_{\rm opt}(\varepsilon)\rceil^{-1} \in \mathcal{H} \;\text{ with }\; h_{\rm opt}(\varepsilon) = \frac{\varepsilon^{\frac{1}{\alpha R}}}{(1+2\alpha R)^{\frac{1}{2\alpha R}}\, |c_{_R}|^{\frac{1}{\alpha R}}}$$

and

$$M^*(\varepsilon) = \left\lceil \left(1+\frac{1}{2\alpha R}\right) \frac{\operatorname{Var}(Y_{h^*(\varepsilon)})}{\varepsilon^2} \right\rceil.$$

(As defined, $h^*(\varepsilon)$ is the nearest lower neighbor of $h_{opt}(\varepsilon)$ in $\mathcal{H}$.)

2. Show that

(a) $\frac{|\underline{n}|}{\underline{n}!^{\frac{1}{R}}} \ge R$.

(b) If $n_i = i,\ i = 1,\dots,R$, $|\underline{n}| = \frac{R(R+1)}{2}$ and $\underline{n}!^{\frac{1}{R}} = n!^{\frac{1}{R}} \sim \frac{R}{e}$ as $R \uparrow \infty$ so that $\frac{|\underline{n}|}{\underline{n}!^{\frac{1}{R}}} \sim \frac{e(R+1)}{2}$.

(c) If $n_i = N^{i-1}$ ($N \in \mathbb{N}$, $N \ge 2$), then $\frac{|\underline{n}|}{\underline{n}!^{\frac{1}{R}}} \sim N^{\frac{R-1}{2}}$.

(d) Show that the choice $n_i = i,\ i = 1,\dots,R$, for the refiners may be not optimal to minimize $\frac{|\underline{n}|}{\underline{n}!^{\frac{1}{R}}}$ [Hint: test $n_i = i+1,\ i = 2,\dots,R$].

The conclusions that can be drawn from the above exercises are somewhat contradictory concerning the asymptotic cost of the Multistep estimator $\big(\widehat{I}^{RR}_M\big)_{M\geq 1}$ to achieve an $RMSE$ of ε:

- At a first glance, it does fill the gap between the crude Monte Carlo rate – $\varepsilon^{-2-\frac{1}{\alpha}}$ – and the virtual unbiased Monte Carlo simulation – ε^{-2} – since $\frac{1}{\alpha R} \to 0$ as R grows to infinity.
- When looking more carefully at the formula, when implemented with explicit refiners (see exercise 1), it appears (see exercise 2) that this asymptotic cost contains the factor $\frac{|\underline{n}|}{\underline{n}!^{\frac{1}{R}}}$, which increases, at least linearly in R, when R grows to infinity. This suggests that the increase of the depth of the multistep simulation will strongly impact the complexity of the global simulation, making it of little interest when ε is not dramatically small. This is verified by numerical experiments.

The multilevel paradigm described below will allow us to overcome this problem.

ℵ Practitioner's corner

Much of the information provided in this section will be useful in Sect. 9.5.1 that follows, devoted to weighted multilevel methods.

▷ ***The refiners and the weights***. As suggested in the preceding Exercise 2, the choice $n_i = i$ is close to optimality to minimize $\frac{|\underline{n}|}{\underline{n}!^{\frac{1}{R}}}$. The other somewhat hidden condition to obtain the announced asymptotic behavior of the complexity as $\varepsilon \to 0$ is the strong convergence assumption $Y_h \to Y_0$ in L^2.

⋖ **Two examples of weights.**
(a) $\alpha = 1, n_i = i, i = 1, \ldots, R$:

$$R = 2 : \mathbf{w}_1^{(R)} = -1, \quad \mathbf{w}_2^{(R)} = 2.$$

$$R = 3 : \mathbf{w}_1^{(R)} = \frac{1}{2}, \quad \mathbf{w}_2^{(R)} = -4, \quad \mathbf{w}_3^{(R)} = \frac{9}{2}.$$

$$R = 4 : \mathbf{w}_1^{(R)} = -\frac{1}{6}, \quad \mathbf{w}_2^{(R)} = 4, \quad \mathbf{w}_3^{(R)} = -\frac{27}{2}, \quad \mathbf{w}_4^{(R)} = \frac{32}{3}.$$

(b) $\alpha = \frac{1}{2}, n_i = i, i = 1, \ldots, R$:

$$R = 2 : \mathbf{w}_1^{(R)} = -(1+\sqrt{2}), \qquad \mathbf{w}_2^{(R)} = \sqrt{2}(1+\sqrt{2}).$$

$$R = 3 : \mathbf{w}_1^{(R)} = \frac{\sqrt{3}-\sqrt{2}}{2\sqrt{2}-\sqrt{3}-1}, \; \mathbf{w}_2^{(R)} = -2\frac{\sqrt{3}-1}{2\sqrt{2}-\sqrt{3}-1},$$

$$\mathbf{w}_3^{(R)} = 3\frac{\sqrt{2}-1}{2\sqrt{2}-\sqrt{3}-1}.$$

$$R = 4 : \mathbf{w}_1^{(R)} = -\frac{(1+\sqrt{2})(1+\sqrt{3})}{2}, \qquad \mathbf{w}_2^{(R)} = 4\Big(\frac{3}{2}+\sqrt{2}\Big)(\sqrt{3}+\sqrt{2}),$$

$$\mathbf{w}_3^{(R)} = -\frac{3}{2}(\sqrt{3}+\sqrt{2})(2+\sqrt{3})(3+\sqrt{3}), \qquad \mathbf{w}_4^{(R)} = 4(2+\sqrt{2})(2+\sqrt{3}).$$

▷ ***Nested Monte Carlo***. Once again, given the form of the approximator Y_h, the complexity of the procedure is significantly smaller in this case than announced in the general framework. In fact,

$$\kappa^{RR}(h) = \frac{n_R}{h}$$

so that, revisiting Exercise **1.**,

$$\inf_{\substack{h\in\mathcal{H}\\ \|\widehat{I}_M^{RR}-I_0\|_2<\varepsilon}} \mathrm{Cost}(\widehat{I}_M^{RR}) \sim \left(\frac{(1+2\alpha R)^{1+\frac{1}{2\alpha R}}}{2\alpha R}\right)|c_R|^{\frac{1}{\alpha R}}\,\mathrm{Var}(Y_0)\frac{1}{\underline{n}!^{\frac{1}{R}}}\frac{1}{\varepsilon^{2+\frac{1}{\alpha R}}} \quad \text{as } \varepsilon \to 0.$$

What should be noticed is that, when $n_i = i$, $i = 1, \ldots, R$, the ratio $\frac{|\underline{n}|}{\underline{n}!^{\frac{1}{R}}}$ (which grows at least like R) is replaced in that framework by $\frac{1}{\underline{n}!^{\frac{1}{R}}} \sim \frac{e}{R}$ as $R \uparrow +\infty$.

▷ ***Brownian diffusions***. Among the practical aspects to be dealt with, the most important one is undoubtedly to simulate (independent copies of) the vector $Y_{\underline{n},h}$ when $Y_0 = F\big(X\big)$, $X = (X_t)_{t\in[0,T]}$ is a Brownian diffusion and $Y_h = F\big(\widetilde{X}^n\big)$, where F is a functional defined on $\mathbb{D}([0,T],\mathbb{R}^d)$ and $\widetilde{X}^n = (\widetilde{X}^n_t)_{t\in[0,T]}$ is the (càdlàg) stepwise constant Euler (or Milstein) scheme with step $\frac{T}{n}$ of the underlying *SDE*, especially when $\underline{n} = (1, \ldots, R)$.

Like for standard Richardson–Romberg extrapolation, one has the choice to simulate the Brownian increments in a consistent way as requested either by using an abacus starting from the most refined scheme or by a recursive simulation starting from the coarsest increment. The structure of the refiners ($n_i = i$, $i = 1, \ldots, R$) makes the task significantly more involved than for the standard Romberg extrapolation.

We refer to [225] for an algorithm adapted to these refined schemes which spares time and complexity in the simulation of these Brownian increments.

9.5 The Multilevel Paradigm

The multilevel paradigm was originally introduced by M. Giles in [107] (see also [148]) in a framework based on a first order expansion of the weak error. This original approach is developed in Sect. 9.5.2, but, to ensure continuity with the previous section devoted to multistep estimators, we made the choice to introduce the *weighted* version of this multilevel paradigm as developed in [198]. It relies on higher-order expansions $\big(WE\big)^{\alpha}_{R}$ of the weak error. In a second step, we will discuss Giles' regular version, which only requires $\big(WE\big)^{\alpha}_{1}$, and compare the respective performances of these two classes of estimators.

Throughout this section, unless explicitly mentioned (*e.g.* in exercises), the refiners have a geometric structure, namely

$$n_i = N^{i-1},\ i = 1, \ldots, R. \tag{9.34}$$

Nevertheless, we will still use the synthetic notation n_i.

We will also assume, though it is not mandatory, that the parameter set $\mathcal{H}$ is of the form

$$\mathcal{H} = \left\{ \frac{\mathbf{h}}{n},\ n \geq 1 \right\} \quad \text{where} \quad \mathbf{h} \in (0, +\infty).$$

9.5.1 Weighted Multilevel Setting

The basic principle of the multilevel paradigm is to split the Monte Carlo simulation into two parts: a first part made of a *coarse* level based on an estimator of $Y_h = Y_{\frac{h}{n_1}}$ with a not too small h, that will achieve the most part of the job to compute $I_0 = \mathbb{E}\, Y_0$ with a bias $\mathbb{E}\, Y_h$, and a second part made of several correcting *refined* levels relying on increments $Y_{\frac{h}{n_i}} - Y_{\frac{h}{n_{i-1}}}$ to correct the former bias. These increments being small – since both $Y_{\frac{h}{n_i}}$ and $Y_{\frac{h}{n_{i-1}}}$ are close to Y_0 – have small variance and need smaller simulated samples to perform the expected bias correction. By combining these increments in an appropriate way, one can almost "kill" the bias while keeping under control the variance and the complexity of the simulation under control.

STEP 1 (*Killing the bias*). Let us assume

$$\big(WE\big)_R^\alpha \quad \equiv \quad \mathbb{E}\, Y_h = \mathbb{E}\, Y_0 + \sum_{r=1}^{R} c_r h^{\alpha r} + o\big(h^{\alpha R}\big). \tag{9.35}$$

Then, starting again from the weighted combination (9.24) of the elements of $Y_{\underline{n},h}$ with the weight vector $\mathbf{w}$ given by (9.27), one gets by an Abel transform

$$\begin{aligned}
\sum_{i=1}^{R} \mathbf{w}_i\, \mathbb{E}\, Y_{\frac{h}{n_i}} &= \mathbf{W}_1 \mathbb{E}\, Y_h + \sum_{i=2}^{R} \mathbf{W}_i \Big(\mathbb{E}\, Y_{\frac{h}{n_i}} - \mathbb{E}\, Y_{\frac{h}{n_{i-1}}} \Big) \\
&= \mathbf{W}_1 \mathbb{E}\, Y_h + \sum_{i=2}^{R} \mathbf{W}_i \mathbb{E}\Big(Y_{\frac{h}{n_i}} - Y_{\frac{h}{n_{i-1}}} \Big) \\
&= \mathbb{E} \left[\underbrace{Y_h^{(1)}}_{\text{coarse level}} + \sum_{i=2}^{R} \mathbf{W}_i \underbrace{\Big(Y^{(i)}_{\frac{h}{n_i}} - Y^{(i)}_{\frac{h}{n_{i-1}}} \Big)}_{\text{refined level } i \geq 2} \right]
\end{aligned} \tag{9.36}$$

where

$$\mathbf{W}_i = \mathbf{W}_i^{(R)} = \mathbf{w}_i + \cdots + \mathbf{w}_R,\ i = 1, \ldots, R, \tag{9.37}$$

(note that $\mathbf{W}_1 = \mathbf{w}_1 + \cdots + \mathbf{w}_R = 1$) and, in the last line, the families $(Y^{(i)}_{\underline{n},h})_{i=1,\ldots,R}$ are independent copies of $Y_{\underline{n},h}$. This last point at this stage may seem meaningless but, doing so, we draw the attention of the reader, in view of the future variance computations ([1]).

As the weights satisfy the Vandermonde system (9.25), we derive from (9.36) that

$$\mathbb{E}\,Y_h^{(1)} + \sum_{i=2}^{R} \mathbf{W}_i\,\mathbb{E}\Big(Y^{(i)}_{\frac{h}{n_i}} - Y^{(i)}_{\frac{h}{n_{i-1}}}\Big) = \mathbb{E}\,Y_0 + \widetilde{\mathbf{W}}_{R+1} c_R h^{\alpha R} + o\big(h^{\alpha R}\big), \tag{9.38}$$

where $\widetilde{\mathbf{W}}_{R+1} = \dfrac{(-1)^{R-1}}{N^{\frac{R(R-1)}{2}\alpha}}$ since $n_i = N^{i-1}$.

STEP 2 (*Multilevel estimator*). As already noted, $\mathbb{E}\,Y_h^{(1)} \simeq I_0$ and $Y_h^{(1)}$ is close to (a copy of) Y_0 so that it has no reason to be small,whereas $Y^{(i)}_{\frac{h}{n_i}} - Y^{(i)}_{\frac{h}{n_{i-1}}} \simeq 0$ since both quantities are supposed to be close to (copies of) Y_0. The underlying idea to devise the (weighted) multilevel estimator is to calibrate the R different sizes M_i of sample paths assigned to each level $i = 1, \ldots, R$ so that $M_1 + \cdots + M_R = M$. This leads to a first definition of *Multilevel Richardson–Romberg* estimators or *weighted Multilevel* estimators attached to $(Y_h)_{h\in\mathcal{H}}$.

They are defined, for every $h \in \mathcal{H}$ and every $M \in \mathbb{N}$, $M \geq R$ where $M_1, \ldots, M_R \in \mathbb{N}^*$ and $M_1 + \cdots + M_R = M$, by

$$\widehat{I}^{ML2R}_{M_1,\ldots,M_R} = \frac{1}{M_1}\sum_{k=1}^{M_1} Y_h^{(1),k} + \sum_{i=2}^{R}\frac{\mathbf{W}_i}{M_i}\sum_{k=1}^{M_i}\Big(Y^{(i),k}_{\frac{h}{n_i}} - Y^{(i),k}_{\frac{h}{n_{i-1}}}\Big),$$

where $\big(Y^{(i),k}_{\underline{n},h}\big)$, $i = 1, \ldots, R$, $k \geq 1$, are i.i.d. copies of $Y_{\underline{n},h} := \big(Y_{\frac{h}{n_i}}\big)_{i=1,\ldots,R}$ and the weight vector $(\mathbf{W}_i)_{i=1,\ldots,R}$ is given by (9.37).

However, in view of the optimization of the M_i, it is more convenient to introduce the formal framework of stratification: we re-write the above multilevel estimator as a stratified estimator, *i.e.* we set $q_i = \frac{M_i}{M}$, $i = 1, \ldots, R$, so that

$$\widehat{I}^{ML2R}_{M} := \frac{1}{M}\left[\sum_{k=1}^{M_1}\frac{Y_h^{(1),k}}{q_1} + \sum_{i=2}^{R}\frac{\mathbf{W}_i}{q_i}\sum_{k=1}^{M_i}\Big(Y^{(i),k}_{\frac{h}{n_i}} - Y^{(i),k}_{\frac{h}{n_{i-1}}}\Big)\right].$$

[1] Other choices for the correlation structure of the families $Y^{(i)}_{\underline{n},h}$ could *a priori* be considered, *e.g.* some negative correlations between successive levels, but this would cause huge simulation problems since the control of the correlation between two families seems difficult to monitor *a priori*.

Conversely, if

$$q = (q_i)_{1 \le i \le R} \in \mathcal{S}_R := \Big\{ (u_i)_{1 \le i \le R} \in (0,1)^R : \sum_{1 \le i \le R} u_i = 1 \Big\}$$

(so $\mathcal{S}_R$ denotes here the "open" simplex of $[0,1]^R$), the above estimator is well defined as soon as $M \ge \frac{1}{\min_i q_i}$ $(\ge R)$ by setting $M_i = \lfloor q_i M \rfloor \ge 1$, $i = 1, \dots, R$. Then $M_1 + \cdots + M_R$ is not equal to M but lies in $\{M - R + 1, \dots, M\}$. This leads us to the following slightly different formal definition.

Definition 9.4. (Weighted Multilevel estimators) *The family of weighted – or Richardson–Romberg – multilevel (ML2R) estimators attached to $(Y_h)_{h \in \mathcal{H}}$ is defined as follows: for every $h \in \mathcal{H}$, every $q = (q_i)_{1 \le i \le R} \in \mathcal{S}_R$ and every integer $M \ge 1$,*

$$\widehat{I}_M^{ML2R} = \widehat{I}_M^{ML2R}(q, h, R, \underline{n}) := \frac{1}{M_1} \sum_{k=1}^{M_1} Y_h^{(1),k} + \sum_{i=2}^{R} \frac{\mathbf{W}_i}{M_i} \sum_{k=1}^{M_i} \Big(Y_{\frac{h}{n_i}}^{(i),k} - Y_{\frac{h}{n_{i-1}}}^{(i),k} \Big), \tag{9.39}$$

(convention the $\frac{1}{0}\sum_{k=1}^{0} = 0$) where $M_i = \lfloor q_i M \rfloor$, $i = 1, \dots, R$, $\big(Y_{\underline{n},h}^{(i),k}\big)$, $i = 1, \dots, R$, $k \ge 1$, are i.i.d. copies of $Y_{\underline{n},h} := \big(Y_{\frac{h}{n_i}}\big)_{i=1,\dots,R}$ and the weight vector $(\mathbf{W}_i)_{i=1,\dots,R}$ is given by (9.37).

Note that, as soon as $M \ge M(q) := (\min_i q_i)^{-1}$, all $M_i = \lfloor q_i M \rfloor \ge 1$, $i = 1, \dots, R$. This condition on M will be implicitly assumed in what follows so that no level is empty.

Remark. The short notation $\widehat{I}_M^{ML2R}$ is convenient but clearly abusive since these estimators depend on the whole set of parameters $(h, q, \underline{n}, R)$ (coarse bias parameter, vector of paths allocation across the levels, vector of refiners and depth). Note that, owing to our parametrization of $\underline{n}$, we could replace $\underline{n}$ by its *root* N in the list of parameters of the estimator.

We are now in a position to evaluate the basic characteristics of this estimator.

▷ *Bias*. For every $M \ge M(q) := \frac{1}{\min_i q_i}$, each $M_i \ge 1$ so that the estimator satisfies,

$$\mathrm{Bias}\big(\widehat{I}_M^{ML2R}\big) = \mathrm{Bias}\big(\widehat{I}_1^{RR}\big) = (-1)^{R-1} c_R \Big(\frac{h^R}{\underline{n}!}\Big)^{\alpha} + o\Big(\Big(\frac{h^R}{\underline{n}!}\Big)^{\alpha}\Big) \tag{9.40}$$

owing to (9.38). Of course, by linearity, this bias does not depend on M.

▷ *Variance*. Taking advantage of the mutual independence of $(Y_{\underline{n},h}^{(j)})_{i=1,\dots,R}$ across the levels, one straightforwardly computes the variance of $\widehat{I}_M^{ML2R}$: for every $M \ge M(q)$, we have

$$\begin{aligned}\mathrm{Var}(\widehat{I}_{M}^{ML2R}) &= \frac{\mathrm{Var}(Y_h^1)}{M_1} + \sum_{i=2}^{R} \mathbf{W}_i^2 \frac{\mathrm{Var}\big(Y^i_{\frac{h}{n_i}} - Y^i_{\frac{h}{n_{i-1}}}\big)}{M_i} \\ &= \frac{1}{M}\left(\frac{\mathrm{Var}(Y_h)}{q_1} + \sum_{i=2}^{R} \mathbf{W}_i^2 \frac{\mathrm{Var}\big(Y_{\frac{h}{n_i}} - Y_{\frac{h}{n_{i-1}}}\big)}{q_i}\right) \qquad (9.41)\\ &= \frac{1}{M}\sum_{i=1}^{R}\frac{\sigma^2_{\mathrm{w}}(i,h)}{q_i}, \qquad (9.42)\end{aligned}$$

where we set

$$\sigma^2_{\mathrm{w}}(1,h) = \mathrm{Var}(Y_h) \ \text{ and } \ \sigma^2_{\mathrm{w}}(i,h) = \mathbf{W}_i^2\mathrm{Var}\Big(Y_{\frac{h}{n_i}} - Y_{\frac{h}{n_{i-1}}}\Big),\ i = 2,\dots,R.$$

▷ *Complexity.* The complexity of such an estimator is *a priori* given — or at least dominated – by

$$\begin{aligned}\mathrm{Cost}(\widehat{I}_{M}^{ML2R}) &= \bar{\kappa}\left(\frac{M_1}{h} + \sum_{i=2}^{R} M_i\Big(\frac{n_i}{h} + \frac{n_{i-1}}{h}\Big)\right)\\ &\le \frac{\bar{\kappa}M}{h}\Big(q_1 + \sum_{i=2}^{R} q_i(n_i + n_{i-1})\Big)\\ &= \frac{\bar{\kappa}M}{h}\Big(q_1 + \big(1+N^{-1}\big)\sum_{i=2}^{R} q_i n_i\Big), \qquad (9.43)\end{aligned}$$

where we used that $M_i = \lfloor Mq_i \rfloor \le Mq_i$ and the specified form $n_i = N^{i-1}$, $i = 1,\dots,R$, (see (9.34)) of the refiners in the penultimate line (and below). If we set

$$\kappa_1 = 1,\ \kappa_i = \big(1+N^{-1}\big)n_i,\ i = 2,\dots,R,$$

then, if we neglect the error induced by replacing $M_i = \lfloor Mq_i \rfloor$ by Mq_i, we may set

$$\mathrm{Cost}(\widehat{I}_{M}^{ML2R}) = \frac{\bar{\kappa}M}{h}\sum_{i=1}^{R}\kappa_i q_i. \qquad (9.44)$$

► **Exercise.** Show that in the case of nested Monte Carlo where $Y_h = f\Big(\frac{1}{K}\sum_{k=1}^{K} F(X, Z_k)\Big)$, the complexity reads (with the same approximation $M_i = Mq_i$ as above)

$$\mathrm{Cost}(\widehat{I}_{M}^{ML2R}) = \bar{\kappa}\Big(\frac{M_1}{h} + \sum_{i-2}^{R} M_i\frac{n_i}{h}\Big) = \frac{\bar{\kappa}M}{h}\Big(q_1 + \sum_{i=2}^{R} q_i n_i\Big) \qquad (9.45)$$

so that, for such a nested simulation, one may set $\kappa_i = n_i$, $i = 1, \ldots, R$.

STEP 3 (*Keeping the variance and the complexity under control*). Like for the analysis of the crude and Monte Carlo multistep estimators, the aim, at this stage, is to minimize the simulation complexity (or cost in short) of the whole simulation for a prescribed $RMSE\ \varepsilon > 0$. This amounts to solving, at least approximately, the following optimization problem

$$\inf_{RMSE(\widehat{I}_M^{ML2R}) \le \varepsilon} \mathrm{Cost}(\widehat{I}_M^{ML2R}). \tag{9.46}$$

The same formal manipulations as in the multistep framework show that

$$\mathrm{Cost}(\widehat{I}_M^{ML2R}) = \frac{\mathrm{Cost}(\widehat{I}_M^{ML2R})\mathrm{Var}(\widehat{I}_M^{ML2R})}{\mathrm{Var}(\widehat{I}_M^{ML2R})} = \frac{\mathrm{Effort}(\widehat{I}_M^{ML2R})}{\mathrm{Var}(\widehat{I}_M^{ML2R})}. \tag{9.47}$$

Plugging the respective expressions (9.42) and (9.44) of the variance and the complexity into their product yields the following expression of the effort (after simplifying by M)

$$\mathrm{Effort}(\widehat{I}_M^{ML2R}) = \mathrm{Cost}(\widehat{I}_M^{ML2R})\mathrm{Var}(\widehat{I}_M^{ML2R}) = \frac{\bar{\kappa}}{h}\left(\sum_{i=1}^{R} \kappa_i q_i\right)\left(\sum_{i=1}^{R} \frac{\sigma_{\mathrm{w}}^2(i,h)}{q_i}\right) \tag{9.48}$$

so that the effort *does not depend upon* the size M of the simulation in the sense that

$$\mathrm{Effort}(\widehat{I}_M^{ML2R}) = \mathrm{Cost}(\widehat{I}_{M_q}^{ML2R})\mathrm{Var}(\widehat{I}_{M_q}^{ML2R}) = \mathrm{Effort}(\widehat{I}_{M_q}^{ML2R}),$$

where $M_q = \lceil 1/\min_i q_i \rceil$. Such a property is universal among Monte Carlo estimators.

The bias-variance decomposition of the weighted multilevel estimator $\widehat{I}_M^{ML2R}$ and the fact that the $RMSE$ constraint should be saturated to maximize the denominator of the ratio in (9.47) allow us to reformulate our minimization problem as

$$\inf_{\|\widehat{I}_M^{ML2R} - I_0\|_2 \le \varepsilon} \mathrm{Cost}(\widehat{I}_M^{ML2R}) = \inf_{|\mathrm{Bias}(\widehat{I}_M^{ML2R})| < \varepsilon} \left[\frac{\mathrm{Effort}(\widehat{I}_M^{ML2R})}{\varepsilon^2 - \mathrm{Bias}(\widehat{I}_M^{ML2R})^2}\right] \tag{9.49}$$

keeping in mind that the right-hand side *does not depend on M when $M \ge M_q$*.

This reformulation suggests to consider the above optimization problem as a function of two of the three free parameters, namely q and h, the depth R being fixed (the root N has a special status). Note that the right-hand side of (9.49) seemingly no longer depends on the simulation size M. In fact, this size is determined further on in (9.53) as a function of the *RMSE* ε and the optimized parameters $q^*(\varepsilon)$ and $h^*(\varepsilon)$. We propose a sub-optimal solution for (9.49) divided in two steps:

– first minimizing the effort for a fixed $h \in \mathcal{H}$, *i.e.* the numerator of the above ratio,

– then minimizing the cost in the bias parameter $h \in H$, that is, maximize the denominator of the ratio (and plug the resulting optimized h in numerator).

Doing so, it is clear that we will only get sub-optimal solutions to the original optimization problem since the numerator depend on h. Nevertheless, the computations become tractable and lead to closed forms for our optimized parameters q and h.

▷ *Minimizing the effort.*

This phase follows from the elementary lemma below, a straightforward application of the Schwartz Inequality and its equality case (see [52]).

Lemma 9.1. *Let $a_i > 0$, $b_i > 0$ and $q \in \mathcal{S}_R$. Then*

$$\left[\sum_{i=1}^{R} \frac{a_i}{q_i}\right]\left[\sum_{i=1}^{R} b_i q_i\right] \geq \left[\sum_{i=1}^{R} \sqrt{a_i b_i}\right]^2$$

and equality holds if and only if $q_i = \frac{\sqrt{a_i b_i^{-1}}}{q^\dagger}$, $i = 1, \ldots, R$, *with* $q^\dagger = q^\dagger(a, b) = \sum_{i=1}^{R} \sqrt{a_i b_i^{-1}}$.

Applying this to (9.48), we derive the solution to the effort minimization problem:

$$\text{argmin}_{q \in \mathcal{S}_R} \text{Effort}(\widehat{I}_1^{ML2R}) = q^* = \left(\frac{\sigma_{\text{w}}(i, h)}{q^{*,\dagger}(h)\sqrt{\kappa_i}}\right)_{i=1,\ldots,R} \tag{9.50}$$

with $q^{*,\dagger}(h) = \sum_{1 \leq i \leq R} \frac{\sigma_{\text{W}}(i,h)}{\sqrt{\kappa_i}}$. The resulting minimal effort reads

$$\min_{q \in \mathcal{S}_R} \text{Effort}(\widehat{I}_1^{ML2R}) = \frac{\bar{\kappa}}{h}\left(\sum_{i=1}^{R} \sqrt{\kappa_i}\, \sigma_{\text{w}}(i, h)\right)^2.$$

However, this formal approach cannot be implemented as such in practice since the true values of $\sigma_{\text{w}}(i, h)$ are usually unknown and should be replaced by an upper-bound (see (9.58) and (9.61) further on).

▷ *Minimizing the resulting cost.*

If we now compile our results on the above effort minimization, the cost formulation (9.44) and the bias expansion (9.40), the global optimization problem $\inf_{RMSE(\widehat{I}_M^{ML2R}) \leq \varepsilon} \text{Cost}(\widehat{I}_M^{ML2R})$ is “dominated”, if we neglect the second-order term in the bias, by

$$\inf_{h \in \mathcal{H}\,:\, c_R^2\left(\frac{h^R}{\underline{n}!}\right)^{2\alpha} < \varepsilon^2} \bar{\kappa}\, \frac{\left(\sum_{i=1}^{R} \sqrt{\kappa_i}\, \sigma_{\text{w}}(i, h)\right)^2}{h\left(\varepsilon^2 - c_R^2\left(\frac{h^R}{\underline{n}!}\right)^{2\alpha}\right)}.$$

We adopt the suboptimal strategy consisting in maximizing the denominator of this ratio, where we want to plug the resulting $h^*(\varepsilon)$ into the whole ratio. Then, we temporarily forget about the constraint $h \in \mathcal{H}$: it reads

$$\sup_{h\in\mathcal{H}:0<h<\left(\frac{\varepsilon\, \underline{n}!^{\alpha}}{|c_R|}\right)^{\frac{1}{R\alpha}}} \left[h\left(\varepsilon^2 - c_R^2\left(\frac{h^R}{\underline{n}!}\right)^{2\alpha}\right)\right] \le \sup_{0<h<\left(\frac{\varepsilon\, \underline{n}!^{\alpha}}{|c_R|}\right)^{\frac{1}{R\alpha}}} \left[h\left(\varepsilon^2 - c_R^2\left(\frac{h^R}{\underline{n}!}\right)^{2\alpha}\right)\right].$$

Elementary optimization in h shows that

$$\sup_{h:\,c_R^2\left(\frac{h^R}{\underline{n}!}\right)^{2\alpha}<\varepsilon^2} h\left(\varepsilon^2 - c_R^2\left(\frac{h^R}{\underline{n}!}\right)^{2\alpha}\right) = \varepsilon^2 \frac{2\alpha R}{(1+2\alpha R)^{1+\frac{1}{2\alpha R}}}\left(\frac{\varepsilon}{|c_R|}\right)^{\frac{1}{\alpha R}} \underline{n}!^{\frac{1}{R}}$$

is in fact a maximum attained at

$$\widetilde{h}(\varepsilon) = \left(\frac{\varepsilon}{|c_R|}\right)^{\frac{1}{\alpha R}} \frac{\underline{n}!^{\frac{1}{R}}}{(1+2\alpha R)^{\frac{1}{2\alpha R}}}. \tag{9.51}$$

This leads us to set

$$h^*(\varepsilon) = \text{lower nearest neighbor of } \widetilde{h}(\varepsilon) \text{ in } \mathcal{H} = \mathbf{h}\left\lceil \mathbf{h}/\widetilde{h}(\varepsilon)\right\rceil^{-1}.$$

When $\varepsilon \to 0$, $\widetilde{h}(\varepsilon) \to 0$ and $h^*(\varepsilon) \sim \widetilde{h}(\varepsilon)$ but, of course, it remains *a priori* suboptimal at finite range so that

$$\inf_{RMSE(\widehat{I}_M^{ML2R})\le\varepsilon} \text{Cost}^{ML2R}(h, M, q) \le \bar{\kappa}\frac{(1+2\alpha R)^{1+\frac{1}{2\alpha R}}}{2\alpha R}\frac{|c_R|^{\frac{1}{\alpha R}}}{\underline{n}!^{\frac{1}{R}}}\frac{\left(\sum_{1\le i\le R}\sqrt{\kappa_i}\,\sigma_{\text{W}}(i, h^*(\varepsilon))\right)^2}{\varepsilon^{2+\frac{1}{\alpha R}}}. \tag{9.52}$$

▷ *Computing the size $M^*(\varepsilon)$ of the simulation.*

The global size $M^*(\varepsilon)$ of the simulation is obtained by saturating the constraint $\|\widehat{I}_M^{ML2R} - I_0\|_2 \le \varepsilon$ with $h = \widetilde{h}(\varepsilon)$, using that $\|\widehat{I}_M^{ML2R} - I_0\|_2^2 = \frac{\text{Var}\left(\widehat{I}_1^{ML2R}\right)}{M^*(\varepsilon)} + \text{Bias}\left(\widehat{I}_M^{ML2R}\right)^2$. We get

$$M^*(\varepsilon) = \frac{\text{Var}\left(\widehat{I}_1^{ML2R}\right)}{\varepsilon^2 - \text{Bias}\left(\widehat{I}_M^{ML2R}\right)^2}.$$

Plugging the value of the variance (9.42) and the bias (9.40) into this equation finally yields

$$M^*(\varepsilon) = \left\lceil \left(1+\frac{1}{2\alpha R}\right)\frac{\text{Var}\left(\widehat{I}_1^{ML2R}\right)}{\varepsilon^2}\right\rceil = \left\lceil \left(1+\frac{1}{2\alpha R}\right)\frac{q^{*,\dagger}\sum_{1\le i\le R}\sqrt{\kappa_i}\,\sigma_{\text{W}}(i, h^*(\varepsilon))}{\varepsilon^2}\right\rceil. \tag{9.53}$$

One may re-write this formula in a more tractable or convenient way (see Table 9.1 in Practitioner's corner further on) by replacing κ_i, $\sigma_{\mathrm{w}}(i,h)$ and $h^*(\varepsilon)$ by their available expressions or bounds.

Remark. If we assume furthermore that $Y_h \to Y_0$ in L^2 as $\varepsilon \to 0$, then $h^*(\varepsilon) \to 0$ so that $\left(\sum_{1\le i\le R} \kappa_i \sigma_{\mathrm{w}}(i, h^*(\varepsilon))\right)^2 \to \mathrm{Var}(Y_0)$ since $\sigma_{\mathrm{w}}(i, h^*(\varepsilon)) \to 0$ for $i = 2, \ldots, R$, $\sigma_{\mathrm{w}}(1, h^*(\varepsilon)) \to \mathrm{Var}(Y_0)$ and $\mathbf{W}_1 = 1$. Finally, we get the following asymptotic result

$$\inf_{\|\widehat{I}_M^{ML2R} - I_0\|_2 \le \varepsilon} \mathrm{Cost}^{ML2R}(h, M, q) \precsim \frac{(1+2\alpha R)^{1+\frac{1}{2\alpha R}}}{2\alpha R} \frac{|c_{_R}|^{\frac{1}{\alpha R}}}{\underline{n}!^{\frac{1}{R}}} \frac{\mathrm{Var}(Y_0)}{\varepsilon^{2+\frac{1}{\alpha R}}} \quad \text{as} \quad \varepsilon \to 0. \tag{9.54}$$

Temporary conclusions

– The first good news that can be drawn from this asymptotic result is that we still have the term $\varepsilon^{2+\frac{1}{\alpha R}}$, which shows that, the larger the depth R is, the closer we get to an unbiased simulation.

– The second good news is that the ratio $\frac{|\underline{n}|}{\underline{n}!^{\frac{1}{R}}}$ ($\ge R$) that appeared in the multistep framework and caused problems at fixed *RMSE* $\varepsilon > 0$ is now replaced by $\frac{1}{\underline{n}!^{\frac{1}{R}}} = \frac{1}{N^{\frac{R-1}{2}}}$ which goes to 0 as $R \to +\infty$.

– However this last result (9.54) remains purely asymptotic as $\varepsilon \to 0$. In view of practical implementation, it is not yet satisfactory since we plan to design an estimator $\widehat{I}_M^{ML2R}$ for a fixed prescribed $RMSE$ ε. In particular, at this stage, it seems not to be impacted by the convergence rate of the variance of Y_h to that of Y_0, but this is only an artifact induced by the asymptotic approach.

Fortunately, we have not yet taken advantage of the last free parameter of the problem, namely the depth R of the estimator.

STEP 4 (*Final optimization* ($R = R(\varepsilon) \to +\infty$)). In this last phase of the optimization process, we start from the non-asymptotic optimized upper-bound (9.52) of $\mathrm{Cost}\big(\widehat{I}_M^{ML2R}\big)$. Our aim at this stage is to specify the depth R of the estimator as a function of the *RMSE* ε, once the above optimized specifications of both the effort and the step have been performed. To achieve this final task we need to have a more precise non-asymptotic control on the optimized effort.

$\triangleright$ *Non-asymptotic control of the effort (or* **when strong convergence comes back into the game***).*

We temporarily return to a generic parameter $h \in \mathcal{H}$ (to alleviate notation). To control the effort of the estimator w.r.t. the depth R, the (strong) L^2-convergence of $Y_h - Y_{\frac{h}{2}}$ to 0 (combined with $\mathbb{E}\, Y_h \to 0$) as $h \to 0$ is not precise enough: we need a rate of convergence of $\mathrm{Var}(Y_h - Y_{\frac{h}{2}})$ as $h \to 0$ in $\mathcal{H}$, in order to control the weighted standard deviation terms $\sigma_{\mathrm{w}}(i,h) = |\mathbf{W}_i|\,\sigma\Big(Y_{\frac{h}{n_i}} - Y_{\frac{h}{n_{i-1}}}\Big)$.

To this end, we will make the following "optimistic" (or "demanding") hypothesis: we assume from now on that there exists $\beta > 0$ and $V_1 > 0$ such that

$$(VE)_\beta \quad \equiv \quad \forall\, h,\, h' \in \mathcal{H} \cup \{0\}, \quad \mathrm{Var}(Y_h - Y_{h'}) \le V_1 |h - h'|^{\beta}. \tag{9.55}$$

This choice is justified by the following facts: in the diffusion framework ($Y_h = f\big(\bar{X}_T\big)$ or $F\big((\bar{X}^n_t)_{t\in[0,T]}\big)$, $\bar{X}^n$ discretization scheme and $h = \frac{T}{n}$), Assumption $(VE)_\beta$ is consistent with the asymptotic variance of the Central Limit Theorem established in [34], namely a weak convergence $h^{-\frac{\beta}{2}}\big(Y_h - Y_{\frac{h}{N}}\big) \to \left(\frac{N-1}{N}\right)^{\frac{\beta}{2}} \Sigma_h\, Z,\, Z \stackrel{d}{=} \mathcal{N}(0;1)$ (when $\beta = 1$). This is confirmed by various empirical evidences not reproduced here for other values of β. In the nested Monte Carlo framework, $(VE)_\beta$ is also satisfied (see Proposition 9.2 further on) through the elementary inequality $\mathrm{Var}\big(Y_h - Y_{h'}\big) \le \big\|Y_h - Y_{h'}\big\|_2^2 \le V_1|h-h'|^\beta$. However, it is also clear that this bound is not the most accessible one in many situations: thus in a diffusion framework, one has more naturally access to a control of $Y_h - Y_0$ in quadratic norm, relying on results like those established in Chap. 7 devoted to time discretization schemes of *SDE*s. In fact, what follows can be easily adapted to a less sharp framework where one has only access to a control of the form

$$\forall\, h \in \mathcal{H}, \quad \big\|Y_h - Y_0\big\|_2^2 \le V_1 h^\beta.$$

This is, for example, the framework adopted in [198] (see also the next exercise).

Let us briefly discuss this assumption from a technical viewpoint. First, note that $(VE)_\beta$ *is clearly implied* by

$$(SE)_\beta \quad \equiv \quad \forall\, h,\, h' \in \mathcal{H}\cup\{0\}, \quad \big\|Y_h - Y_{h'}\big\|_2^2 \le V_1|h-h'|^\beta \tag{9.56}$$

which is more directly connected to the quadratic strong rate of convergence of Y_h toward Y_0. It even holds possibly with the same V_1 since $\mathrm{Var}(Y_h - Y_{h'}) \le \big\|Y_h - Y_{h'}\big\|_2^2$.

Moreover, if both $\big(WE\big)_1^\alpha$ and $(VE)_\beta$ hold, then there exists a real constant $\widetilde{V}_1 \in (0,+\infty)$ such that

$$\|Y_h - Y_{h'}\|_2^2 = \mathrm{Var}(Y_h - Y_{h'}) + \big(\mathbb{E}\,Y_h - \mathbb{E}\,Y_{h'}\big)^2 \le \widetilde{V}_1 h^{(2\alpha)\wedge\beta},$$

i.e. $(SE)_{(2\alpha)\wedge\beta}$ holds with $\widetilde{V}_1$. Conversely, if $(SE)_\beta$ and $\big(WE\big)_1^\alpha$ hold, it follows from the Schwartz Inequality that $2\alpha \ge \beta$.

The control of the effort under this new hypothesis is straightforward: it follows from Assumption $(VE)_\beta$ that, for every $i \in \{2,\dots,R\}$,

$$\sigma_{\mathbf{w}}(i,h) = |\mathbf{W}_i|\sigma\big(Y_{\frac{h}{n_i}} - Y_{\frac{h}{n_{i-1}}}\big) \tag{9.57}$$

$$\le |\mathbf{W}_i|\sqrt{V_1}\left|\frac{h}{n_i} - \frac{h}{n_{i-1}}\right|^{\frac{\beta}{2}}$$

$$= |\mathbf{W}_i|\sqrt{V_1}h^{\frac{\beta}{2}}\left|\frac{1}{n_i} - \frac{1}{n_{i-1}}\right|^{\frac{\beta}{2}} \tag{9.58}$$

$$\le |\mathbf{W}_i|\sqrt{V_1}\,h^{\frac{\beta}{2}}N^{-(i-1)\frac{\beta}{2}}(N-1)^{\frac{\beta}{2}},$$

where we used in the last line the specific form of the refiners $n_i = N^{i-1}$. On the coarse level,

$$\sigma_{\mathrm{w}}(1,h) = \sigma(Y_h) = \sqrt{\mathrm{Var}(Y_h)}.$$

As

$$\kappa_i = n_i + n_{i-1} = n_i\Big(1+\frac{1}{N}\Big), \quad i \in \{2,\dots,R\}, \tag{9.59}$$

we finally obtain, after setting

$$\theta = \theta_h = \sqrt{\frac{V_1}{\mathrm{Var}(Y_h)}}, \tag{9.60}$$

$$q_1^* = \frac{1}{q^{\dagger,*}}, \quad q_i^* = \frac{\theta h^{\frac{\beta}{2}} |\mathbf{W}_i| \left|\frac{h}{n_i} - \frac{h}{n_{i-1}}\right|^{\frac{\beta}{2}}}{q^{\dagger,*}\sqrt{n_i+n_{i-1}}}, \quad i = 2,\dots,R \tag{9.61}$$

and

$$\begin{aligned}\left(\sum_{i=1}^{R}\sqrt{\kappa_i}\,\sigma_{\mathrm{w}}(i,h)\right)^2 &\le \mathrm{Var}(Y_h)\left(1+\theta h^{\frac{\beta}{2}}\sum_{i=2}^{R}|\mathbf{W}_i|\big(n_{i-1}^{-1}-n_i^{-1}\big)^{\frac{\beta}{2}}\sqrt{n_i+n_{i-1}}\right) \\ &= \mathrm{Var}(Y_h)\left(1+\theta h^{\frac{\beta}{2}}(1+N^{-1})^{\frac{1}{2}}(N-1)^{\frac{\beta}{2}}\sum_{i=2}^{R}|\mathbf{W}_i| N^{(i-1)\frac{1-\beta}{2}}\right)^2,\end{aligned} \tag{9.62}$$

keeping in mind that $(\mathbf{W}_i)_{1\le i\le R} = (\mathbf{W}_i^{(R)})_{1\le i\le R}$ is an R-tuple, not the first R terms of a sequence.

Remarks. • If the two processes $Y_{\frac{h}{n_{i-1}}}$ and $Y_{\frac{h}{n_i}}$ do not remain "close" in a pathwise sense, the multilevel estimator may "diverge": while it maintains its performance as a *bias killer*, it loses all its variance control abilities as observed, for example, in [153] with the regular multilevel estimator presented in the next section. The same effect can be reproduced with this weighted estimator. Typically with diffusions, this means that the two discretization schemes involved in each level i are based on the same driving Brownian motion W^i (these W^i being independent).

• However, in some specific situations, one may even get rid of the strong convergence itself, see *e.g.* [32] for such an approach for diffusion processes where an approximation of the underlying diffusion process is performed by a kind of binomial tree which preserves the performances of the multi-level paradigm, though not converging strongly. This idea is applied to deal with Lévy-driven diffusions processes whose discretization schemes involve a wienerization of the small jumps.

► **Exercise** (*Effort control under quadratic convergence, see [198].*) In practice, the simplest or the most commonly known information available about a family $(Y_h)_{h\in\mathcal{H}}$ of approximations of Y_0 concerns the quadratic rate of convergence of Y_h toward Y_0, namely that there exists $\beta > 0$, $\bar{V}_1 > 0$, such that the following property holds

$$(SE)^0_\beta \quad \equiv \quad \forall\, h \in \mathcal{H}, \quad \big\| Y_h - Y_0 \big\|_2^2 \le \bar{V}_1 h^\beta. \tag{9.63}$$

(a) Show that, under $(SE)^0_\beta$ (and for refiners of the form $n_i = N^{i-1}$)

$$\sigma_{\mathbf{W}}(i,h) \le |\mathbf{W}_i| \sqrt{\bar{V}_1} h^{\frac{\beta}{2}} N^{-(i-1)\frac{\beta}{2}} \big(1 + N^{\frac{\beta}{2}}\big), \; i = 2, \ldots, R.$$

[Hint: use the sub-additivity of standard deviation.]

(b) Show that, if $\kappa_i = n_i(1+\frac{1}{N})$ for $i \in \{2, \ldots, R\}$ (and $\kappa_1 = 1$), one finally obtains

$$\left(\sum_{i=1}^R \sqrt{\kappa_i}\, \sigma_{\mathbf{W}}(i,h)\right)^2 \le \operatorname{Var}(Y_h)\left(1 + \bar{\theta}_h h^{\frac{\beta}{2}} (1+N^{-1})^{\frac{1}{2}} (N^{\frac{\beta}{2}} + 1) \sum_{i=2}^R |\mathbf{W}_i| N^{(i-1)\frac{1-\beta}{2}}\right)^2$$

where $\bar{\theta}_h = \sqrt{\frac{\bar{V}_1}{\operatorname{Var}(Y_h)}}$.

(c) Show that on the coarse level $\sigma_{\mathbf{W}}(1,h) \le \sigma(Y_0) + \sqrt{\bar{V}_1} h^{\frac{\beta}{2}}$. Deduce the less sharp inequality

$$\left(\sum_{i=1}^R \sqrt{\kappa_i}\, \sigma_{\mathbf{W}}(i,h)\right)^2 \le \operatorname{Var}(Y_h)\left(1 + \bar{\theta}_0 h^{\frac{\beta}{2}} (1+N^{-1})^{\frac{1}{2}} (N^{\frac{\beta}{2}} + 1) \sum_{i=1}^R |\mathbf{W}_i| N^{(i-1)\frac{1-\beta}{2}}\right)^2$$

where $\bar{\theta}_0 = \sqrt{\frac{\bar{V}_1}{\operatorname{Var}(Y_0)}}$.

▷ *Optimization of the depth* $R = R(\varepsilon)$.

First, we need to make one last additional assumption, namely that the weak error expansion holds true at any order and that the coefficients c_r do not go too fast toward infinity as $r \to +\infty$, namely

$$(WE)^\alpha_\infty \quad \equiv \quad \begin{cases} (WE)^\alpha_R \text{ holds for every depth } R \ge 2 \\ \text{and} \\ \widetilde{c}_\infty = \overline{\lim}_{R\to+\infty} |c_R|^{\frac{1}{R}} \in (0, +\infty) \end{cases} \tag{9.64}$$

(the case $\widetilde{c}_\infty = 0$ can be handled in what follows by considering $\widetilde{c}_\infty = \eta > 0$ to be arbitrarily small, though it will provide suboptimal results).

The key point at this stage is to note that, for a fixed *RMSE* $\varepsilon > 0$, the complexity formally goes to 0 as $R \to +\infty$ owing to (9.52) since $\frac{(1+2\alpha R)^{1+\frac{1}{2\alpha R}}}{2\alpha R} \to 1$ and $|c_R|^{\frac{1}{\alpha R}}$ remains bounded whereas $\underline{n}!^{\frac{1}{R}} = N^{\frac{R-1}{2}} \uparrow +\infty$. Moreover, the complexity depends on ε as $\varepsilon^{-(2+\frac{1}{\alpha R})}$ so that, the larger R is, the more this complexity behaves like that of an unbiased estimator (in ε^{-2}). These facts strongly suggest to consider R as large as possible, under the constraint that $h^*(\varepsilon)$ lies in $\mathcal{H}$ *and remains close to* $\widetilde{h}(\varepsilon)$, which is equivalent to $\widetilde{h}(\varepsilon) \le \mathbf{h}$.

This argument is all the more true if ε is small but remains heuristic at this stage and cannot be considered as a rigorous optimization: such a procedure may turn out to be suboptimal, especially if the prescribed *RMSE* is not small enough. An alternative for practical implementation is to numerically minimize the upper-bound of the cost on the right-hand side of (9.52), once the parameter θ has been estimated (see Practitioner's corner further on). One reason for computing a closed formula for the depth R and the other parameters of the Richardson–Romberg estimator is to obtain asymptotic bounds on the complexity (see Theorem 9.1).

The inequality $\widetilde{h}(\varepsilon) \le \mathbf{h}$ reads, owing to (9.51) and after taking the log and temporarily considering R as a real number,

$$R(R-1) - \frac{2}{\alpha \log N}\log\Big(\frac{1}{\varepsilon}\Big) - \frac{2R}{\log N}\log\Big(\mathbf{h}\,|c_{_R}|^{\frac{1}{\alpha R}}\Big) - \frac{\log(1+2\alpha R)}{\alpha \log N} \le 0.$$

At this stage, we need to consider approximations if we want to get a closed form at any order R to be able to carry out an asymptotic study of the behavior of our estimator.

First, we assume that $|c_{_R}|^{\frac{1}{\alpha R}} \simeq \widetilde{c}_{\infty}^{\frac{1}{\alpha}}$ (which is true if R is large enough under our assumptions...). Then, the above inequality reads

$$R^2 - R\left(1 + \frac{2\log\big(\mathbf{h}\,\widetilde{c}_{\infty}^{\frac{1}{\alpha}}\big)}{\log N}\right) - \frac{2}{\alpha \log N}\log\left(\frac{\sqrt{1+2\alpha\, R}}{\varepsilon}\right) \le 0.$$

Solving this inequality as if $\frac{2}{\alpha \log N}\log\Big(\frac{\sqrt{1+2\alpha\,R}}{\varepsilon}\Big)$ is a constant term yields

$$R \le \varphi_\epsilon(R) := \frac{1}{2} + \frac{\log\big(\mathbf{h}\,\widetilde{c}_{\infty}^{\frac{1}{\alpha}}\big)}{\log N} + \sqrt{\left(\frac{1}{2} + \frac{\log\big(\mathbf{h}\,\widetilde{c}_{\infty}^{\frac{1}{\alpha}}\big)}{\log N}\right)^2 + \frac{2\log\big(\frac{\sqrt{1+2\alpha\,R}}{\varepsilon}\big)}{\alpha \log N}}.$$

Note that φ_ε is increasing in R. The highest admissible depth $R^*(\varepsilon)$ is the highest integer R which satisfies the above equality. If we set $R_0 = 1$ and $R_{k+1} = \varphi_\varepsilon(R_k)$, then $R_k \uparrow R_\infty$ and

$$R^*(\varepsilon) = \lfloor R_\infty \rfloor. \tag{9.65}$$

Although the above convergence is almost instantaneous (two or three iterations are enough) a simpler "closed form" is to set set $R^*(\varepsilon) := \lceil \varphi_\epsilon(2) \rceil$, *i.e*

$$R^*(\varepsilon) = \left\lceil \frac{1}{2} + \frac{\log(\mathbf{h}\,\widetilde{c}_{\infty}^{\frac{1}{\alpha}})}{\log(N)} + \sqrt{\left(\frac{1}{2} + \frac{\log(\mathbf{h}\,\widetilde{c}_{\infty}^{\frac{1}{\alpha}})}{\log(N)}\right)^2 + \frac{2\log(\frac{\sqrt{1+4\alpha}}{\varepsilon})}{\alpha\log(N)}}\,\right\rceil. \tag{9.66}$$

On the way to an asymptotic analysis of the optimized weighted multilevel estimators, the first quantity to be investigated is clearly the family of weight vectors

$\mathbf{W}^{(R)} = \left(\mathbf{W}_i^{(R)}\right)_{i=1,\ldots,R}$ as R grows. The following lemma shows that they remain uniformly bounded as R grows.

Lemma 9.2 *Let $\alpha > 0$ and the associated weights $\left(\mathbf{W}_j^{(R)}\right)_{i=1,\ldots,R}$ be defined by (9.37) with $n_i = N^{i-1}$.*

(a) Let $a_\ell = \prod_{1\le k\le \ell-1} (1 - N^{-k\alpha})^{-1}$, $\ell \ge 1$ and $b_\ell = (-1)^\ell N^{-\frac{\alpha}{2}\ell(\ell+1)}$.

$$\mathbf{W}_j^{(R)} = \sum_{\ell=j}^{R} a_\ell a_{R-\ell+1} b_{R-\ell} = \sum_{\ell=0}^{R-j} a_{R-\ell} a_{\ell+1} b_\ell.$$

(b) The sequence $\alpha_\ell \uparrow a_\infty = \prod_{1\le k\le \ell-1}(1 - N^{-k\alpha})^{-1}$ as $\ell \uparrow +\infty$ and $\widetilde{B}_\infty = \sum_{\ell=0}^{+\infty} |b_\ell| < +\infty$ so that the weights $\mathbf{W}_j^{(R)}$ are uniformly bounded and

$$\forall R \in \mathbb{N}^*, \forall j \in \{1, \ldots, R\}, \qquad |\mathbf{W}_j^{(R)}| \le a_\infty^2 \widetilde{B}_\infty < +\infty. \tag{9.67}$$

Proof. (a) By re-writing (9.33) we get $\mathbf{w}_j = \mathbf{w}_j^{(R)} = a_j b_{R-j} a_{R-j+1}$ so that

$$\mathbf{W}_j^{(R)} = \sum_{\ell=j}^{R} a_\ell b_{R-\ell} a_{R-\ell+1} = \sum_{\ell=0}^{R-j} a_{R-\ell} b_\ell a_{\ell+1}.$$

(b) It is clear that $a_\ell \uparrow a_\infty < +\infty$ and that the series with general term b_ℓ is absolutely convergent, since $\widetilde{B}_\infty = \sum_{\ell\ge1} N^{-\frac{\alpha}{2}\ell(\ell+1)} < +\infty$. Hence

$$\forall R \in \mathbb{N}^*, \forall j \in \{1, \ldots, R\}, \qquad \left|\mathbf{W}_j^{(R)}\right| \le a_\infty^2 \widetilde{B}_\infty.$$ ◇

Now, we have to inspect the behavior of $\sum_{i=1}^{R} \sqrt{\kappa_i}\, \sigma_{\mathrm{w}}\big(i, h^*(\varepsilon)\big)$ depending on the value of β. Given (9.48), there are three different cases:

- $\beta \in (1, +\infty)$: *fast strong approximation* (corresponding, for example, to the use of the Milstein scheme for vanilla payoffs in a local volatility model),
- $\beta = 1$: *regular strong approximation* (corresponding, for example, to the use of the Euler scheme for vanilla or lookback payoffs in a local volatility model),
- $\beta \in (0, 1)$: *slow strong approximation* (corresponding, for example, to the use of the Euler scheme for payoffs including barriers in a local volatility model or the computation of quantile for risk measure purposes).

Combining all of the preceding and elementary though slightly technical computations (see [198] for details), lead to the following theorem.

Theorem 9.1 (**Weighted** *ML2R* **estimator**) *Let* $n_i = N^{i-1}$, $i \geq 1$. *Assume* $(VE)_\beta$ *and* $\big(WE\big)^\alpha_\infty$ *(see (9.64)) and let* $h = \mathbf{h}$ *and let* $\theta = \theta_\mathbf{h} = \sqrt{\frac{V_1}{\mathrm{Var}(Y_\mathbf{h})}}$ *(see (9.60)).*

(a) The family of ML2R estimators satisfies, as $\varepsilon \to 0$,

$$\inf_{\substack{h\in\mathcal{H},\, q\in\mathcal{S}_R,\, R,\, M\geq 1,\\ \|\widehat{I}_M^{ML2R}-I_0\|_2\leq\varepsilon}} \mathrm{Cost}\big(\widehat{I}_M^{ML2R}\big) \precsim K^{ML2R}\frac{\mathrm{Var}(Y_\mathbf{h})}{\varepsilon^2}\times\begin{cases}1 & \textit{if } \beta>1,\\ \log(1/\varepsilon) & \textit{if } \beta=1,\\ e^{\frac{1-\beta}{\sqrt{\alpha}}\sqrt{2\log(1/\varepsilon)\log(N)}} & \textit{if } \beta<1.\end{cases}$$

where $K^{ML2R} = K^{ML2R}(\alpha, \beta, N, \mathbf{h}, \widetilde{c}_\infty, V_1)$.

(b) When $\beta < 1$ *the* optimal rate *is achieved with* $N = 2$.

(c) The rates in the right-hand side in (a) are achieved by setting in the definition (9.39) of $\widehat{I}_M^{ML2R}$, $h = \mathbf{h}$, $q = q^*(\varepsilon)$, $R = R^*(\varepsilon)$ *and* $M = M^*(\varepsilon)$. *Closed forms are available for* $q^*(\varepsilon)$, $R^*(\varepsilon)$, $M^*(\varepsilon)$ *(see Table 9.1 in Practitioner's corner hereafter and the comments that follow for variants).*

Comments. • If $\beta > 1$ (fast strong approximation), the weighted Multilevel estimator (*ML2R*) behaves in terms of rate like an unbiased estimator, namely ε^{-2}. Any further gain will rely on the reduction of the (optimal) constant $K^{ML2R}(\alpha, \beta, N, \theta, \mathbf{h}, \widetilde{c}_\infty, V_1)$.
• If $\beta \in (0, 1]$, the exact unbiased rate cannot be achieved. However, it is almost attained, at least in a polynomial scale, since

$$\inf_{\substack{h\in\mathcal{H},\, q\in\mathcal{S}_R,\, R,\, M\geq 1\\ \|\widehat{I}_M^{ML2R}-I_0\|_2\leq\varepsilon}} \mathrm{Cost}\big(\widehat{I}_M^{ML2R}\big) = o(\varepsilon^{-\eta}),\ \forall\,\eta>0.$$

We can sum up Theorem 9.1 by noting that

ML2R estimators make it always possible to carry out quasi-unbiased simulations, whatever the "strong error" rate parameter β *is.*

• Closed forms for the constants $K^{ML2R}(\alpha, \beta, N, \theta, \mathbf{h}, \widetilde{c}_\infty, V_1)$ are given in [198] in the slightly different setting where $(VE)_\beta$ is replaced by $(SE)_\beta$ (see (9.56)). Under $(VE)_\beta$ such formulas are left as an exercise (see Exercise **1.** after the proof of the theorem).

• Under a "sharp" version of $(VE)_\beta$, a refined version of Lemma 9.2 (see Exercise **2.**(a)-(b) after the proof of the theorem) and few additional conditions, a strong law of large numbers as well as a *CLT* can be established for the optimized weighted Multilevel Richardson–Romberg estimator (see [112]). Based on this refined version of Lemma 9.2, one may also obtain better constants $K^{ML2R}(\alpha, \beta, N, \theta, \mathbf{h}, \widetilde{c}_\infty, V_1)$ than those obtained in [198], see [113] and Exercise **2.**(c).
• The theorem could be stated using $\mathrm{Var}(Y_0)$ by simply checking that

$$\operatorname{Var}(Y_{\mathbf{h}}) \le \operatorname{Var}(Y_0)\big(1+\theta_0\mathbf{h}\big)^2 \quad \text{with} \quad \theta_0 = \sqrt{\frac{V_1}{\operatorname{Var}(Y_0)}} \le \frac{\theta}{1-\theta\,\mathbf{h}}.$$

Proof of Theorem 9.1. The starting point of the proof is the upper-bound (9.52) of the complexity of *ML2R* estimator. We will analyze its behaviour as $\varepsilon \to 0$ when the parameters h and R are optimized (see Table 9.1) *i.e.* set at $h = \mathbf{h}$ and $R = R^*(\varepsilon)$.

First note that, with our choice of geometric refiners $n_i = N^{i-1}$, $\underline{n}!^{\frac{1}{R}} = N^{\frac{R-1}{2}}$. Then, if we denote $\mathbf{W}^* = \sup_{R\ge 1}\max_{1\le r\le R}\mathbf{W}_i^{(R)}$, we know that $\mathbf{W}^* < +\infty$ owing to Lemma 9.2. Hence, if we denote the infimum of the complexity of the *ML2R* estimators as defined in the theorem statement by

$$\operatorname{Cost}_{opt} = \inf_{\substack{h\in\mathcal{H},\, q\in\mathcal{S}_R,\, R,\, M\ge 1,\\ \|\widehat{I}_M^{ML2R}-I_0\|_2\le\varepsilon}} \operatorname{Cost}\big(\widehat{I}_M^{ML2R}\big),$$

it follows from (9.52) that

$$\begin{aligned}\varepsilon^2\operatorname{Cost}_{opt} &\le \varepsilon^2\operatorname{Cost}^{ML2R}\big(\mathbf{h}, M^*(\varepsilon), q^*(\varepsilon), N\big)\\ &\le \bar{\kappa}\frac{(1+2\alpha R)^{1+\frac{1}{2\alpha R}}}{2\alpha R}\,\frac{|c_R|^{\frac{1}{\alpha R}}}{\underline{n}!^{\frac{1}{R}}}\,\frac{\operatorname{Effort}_{opt}^{ML2R}}{\varepsilon^{\frac{1}{\alpha R}}}.\end{aligned}$$

where $\operatorname{Effort}_{opt}^{ML2R}$ is given by setting $h = \mathbf{h}$ in (9.48), namely

$$\operatorname{Effort}_{opt}^{ML2R} = \operatorname{Var}(Y_{\mathbf{h}})K_1(\beta,\theta,\mathbf{h},V_1)\left(\sum_{i=1}^{R}|\mathbf{W}_i|(n_{i-1}^{-1}-n_i^{-1})^{\frac{\beta}{2}}\sqrt{n_i+n_{i-1}}\right)^2.$$

Now, specifying in this inequality the values of the refiners n_i, yields

$$\operatorname{Effort}_{opt}^{ML2R} \le \operatorname{Var}(Y_{\mathbf{h}})K_1'(\beta, N,\theta,\mathbf{h},V_1)\left(\sum_{i=1}^{R}|\mathbf{W}_i|N^{\frac{1-\beta}{2}}\right)^2.$$

Keeping in mind that $\lim_{\varepsilon\to 0}R^*(\varepsilon) = +\infty$ owing to (9.66), one deduces that $\lim_{R\to+\infty}\frac{(1+2\alpha R)^{1+\frac{1}{2\alpha R}}}{2\alpha R} = 1$ and, using $(WE)_\infty^\alpha$, that $\lim_{R\to+\infty}|c_R|^{\frac{1}{\alpha R}} = \widetilde{c}_\infty^{\frac{1}{\alpha}} < +\infty$. Moreover, note that

$$R^*(\varepsilon) = \left\lceil \sqrt{\frac{2\log(1/\varepsilon)}{\alpha\log N}} + r^* + O\left(\frac{1}{\sqrt{\log(1/\varepsilon)}}\right)\right\rceil \quad \text{as} \quad \varepsilon\to 0$$

where $r^* = r^*(\alpha, N, \mathbf{h})$. In particular, one deduces that

$$\begin{aligned}
N^{-\frac{R^*(\varepsilon)-1}{2}}\varepsilon^{-\frac{1}{\alpha R^*(\varepsilon)}} &= \sqrt{N}\exp\left(-\frac{\log N}{2}R^*(\varepsilon)+\frac{\log(1/\varepsilon)}{\alpha R^*(\varepsilon)}\right)\\
&\le N^{\frac{1}{2}}\exp\left(-\sqrt{\frac{\log N\log(1/\varepsilon)}{2\alpha}}+\sqrt{\frac{\log N\log(1/\varepsilon)}{2\alpha}}-\frac{r^*}{2}\log N+O\Big(\big(\log(1/\varepsilon)\big)^{-\frac{1}{2}}\Big)\right)\\
&= K_2(\alpha,\beta,N,\theta,\mathbf{h},\widetilde{c}_\infty)\,\ell(\varepsilon)\quad\text{with}\quad \ell(\varepsilon)\to 1\quad\text{as}\quad\varepsilon\to 0.
\end{aligned}$$

At this stage, it is clear that the announced rate results will follow from the asymptotic behaviour of

$$\left(\sum_{i=1}^{R^*(\varepsilon)}|\mathbf{W}_i|N^{\frac{1-\beta}{2}}\right)^2\quad\text{as}\quad\varepsilon\to 0,$$

depending on the value of β. Elementary computations yield

$$\sum_{i=1}^{R}|\mathbf{W}_i|N^{\frac{1-\beta}{2}}\le\mathbf{W}^*\begin{cases}\dfrac{N^{R\frac{1-\beta}{2}}}{N^{\frac{1-\beta}{2}}-1} & \text{if } \beta<1\\[2ex] R & \text{if } \beta=1\\[2ex] \dfrac{N^{\frac{1-\beta}{2}}}{1-N^{\frac{1-\beta}{2}}} & \text{if } \beta>1.\end{cases}$$

Now we are in a position to inspect the three possible settings.

– $\beta>1$: It follows from what precedes that $\overline{\lim}_{\varepsilon\to 0}\,\varepsilon^2\,\mathrm{Cost}_{opt}<+\infty$.

– $\beta=1$: As $R^*(\varepsilon)\sim\sqrt{\frac{2\log(1/\varepsilon)}{\alpha\log N}}$ when $\varepsilon\to 0$, it follows that $R^*(\varepsilon)^2\sim\frac{2\log(1/\varepsilon)}{\alpha\log N}$ so that

$$\overline{\lim_{\varepsilon\to 0}}\,\frac{\varepsilon^2}{\log(1/\varepsilon)}\mathrm{Cost}_{opt}<+\infty.$$

– $\beta<1$: The conclusion follows from the fact that

$$\begin{aligned}
N^{(R^*(\varepsilon)-1)(1-\beta)} &\le N^{\beta-1}\exp\left((1-\beta)\log N\left[\sqrt{\frac{2\log(1/\varepsilon)}{\alpha\log N}}+O\left(\big(\log(1/\varepsilon)\big)^{-\frac{1}{2}}\right)\right]\right)\\
&\le K_1''(\theta,\mathbf{h},\alpha,\beta,N)\exp\left(\frac{1-\beta}{\sqrt{\alpha}}\sqrt{2\log N\log(1/\varepsilon)}\right)\lambda(\varepsilon)
\end{aligned}$$

with $\lim_{\varepsilon\to 0}\lambda(\varepsilon)=1$. One concludes that

$$\overline{\lim_{\varepsilon\to 0}}\,\varepsilon^2\exp\left(-\frac{1-\beta}{\sqrt{\alpha}}\sqrt{2\log N\log(1/\varepsilon)}\right)\mathrm{Cost}_{opt}<+\infty.$$

◇

► **Exercises. 1.** *Explicit constants.* Determine explicit values for the constant $K^{ML2R}(\alpha, \beta, N, \theta, \mathbf{h}, \widetilde{c}_{\infty})$ depending on β. [Hint: Inspect carefully the above proof and use Lemma 9.2.]

2. *Quest for better constants.* With the notation of Lemma 9.2 let $\alpha > 0$ and the associated weights $\big(\mathbf{W}_j^{(R)}\big)_{j=1,\dots,R}$ be as given in (9.37).

(a) For every $\gamma > 0$ and for every integer $N \geq 2$,

$$\lim_{R \to +\infty} \sum_{j=2}^{R} \left|\mathbf{W}_j^R\right| N^{-\gamma(j-1)} = \frac{1}{N^{\gamma} - 1}.$$

(b) Let $(v_j)_{j \geq 1}$ be a bounded sequence of positive real numbers. Let $\gamma \in \mathbb{R}$ and assume that $\lim_j v_j = 1$ when $\gamma \geq 0$. Then the following limits hold: for every $N \geq 2$,

$$\sum_{j=2}^{R} \left|\mathbf{W}_j^R\right| N^{\gamma(j-1)} v_j \overset{R \to +\infty}{\sim} \begin{cases} \sum_{j \geq 2} N^{\gamma(j-1)} v_j < +\infty & \text{for } \gamma < 0, \\ R & \text{for } \gamma = 0, \\ N^{\gamma R} a_{\infty} \sum_{j \geq 1} \left|\sum_{\ell=0}^{j-1} b_{\ell}\right| N^{-\gamma j} & \text{for } \gamma > 0. \end{cases}$$

(c) Use these results to improve the values of the constants $K^{ML2R}(\alpha, \beta, N, \theta, \mathbf{h}, \widetilde{c}_{\infty}, V_1)$ in Theorem 9.1 compared to those derived by relying on Lemma 9.2 in the former exercise.

Remark. These sharper results on the weights are needed to establish the *CLT* satisfied by the *ML2R* estimators as $\varepsilon \to 0$ (see [112]).

ℵ Practitioner's corner I: Specification and calibration of the estimator parameters

We assume in this section that Assumptions $(WE)_{\infty}^{\alpha}$ and $(SE)_{\beta}$ are in force. The coarsest parameter $\mathbf{h}$ is also assumed to be specified by exogenous considerations or constraints. The specification and calibration phase of the parameters of the *ML2R* estimators for a prescribed *RMSE* $\varepsilon > 0$ (quadratic error) is two-fold:

– As a first step, for a given root $N \geq 2$, we can specify all the parameters or calibrate all the parameters from the theoretical knowledge we have of the pair of characteristics (α, β). The details are given further on. One must keep in mind that we first determine $R = R^*(\varepsilon)$, then the weights $\mathbf{W}^{(R)}$ and the optimal bias parameter $h = h^*(\varepsilon)$. These only depend on (α, β) (except for $\widetilde{c}_{\infty}$, see the discussion further on). Then, we calibrate V_1 and $\mathrm{Var}(Y_h)$ (hence $\theta = \theta_h$), which are more "payoff" dependent. Finally, we derive the optimal allocation policy $q(\varepsilon) = (q_i(\varepsilon))_{1 \leq i \leq R}$ and the global size or budget of the simulation $M^*(\varepsilon)$.

All these parameters are listed in Table 9.1 (we dropped all superscripts * inside the Table to alleviate notation).

Since our way to determine the optimal depth $R^*(\varepsilon)$ consists in saturating the constraint $h^*(\varepsilon) \leq \mathbf{h}$ into an equality, it follows that the analytic expression for $h^*(\varepsilon)$

as a projection of $h_{\max}(\varepsilon)$ on $\mathcal{H}$, namely

$$h^*(\varepsilon) = \mathbf{h}\Big/\Big\lceil \mathbf{h}(1+2\alpha R)^{\frac{1}{2\alpha R}} \widetilde{c}_{\infty}^{\frac{1}{\alpha}} \varepsilon^{-\frac{1}{\alpha R}} N^{-\frac{R-1}{2}} \Big\rceil,$$

eventually boils down to $h^*(\varepsilon) = \mathbf{h}$.

– As a second step, we will roughly perform a minimization of the complexity based on the upper-bound (9.54) of $\mathrm{Cost}(\widehat{I}_M^{ML2R})$.

Remark. Our specifications, especially those of the allocation policy and of the global size of the estimator (obtained from formula (9.52)) rely on the fact that the complexity of the simulation of $Y_{\frac{h}{n_i}} - Y_{\frac{h}{n_{i-1}}}$ is dominated by $\frac{\bar{\kappa}}{h}(n_i + n_{i-1})$, which typically corresponds to the case of diffusion. See the comments further on for variants, especially for nested Monte Carlo where the complexity is lower (see the end of the paragraph devoted to nested Monte Carlo in Practitioner's corner III).

Let us start with the *structure* parameters, namely α, the resulting weights $\mathbf{W}^{(R)}$, β and the coarsest bias parameter $\mathbf{h}$.

▷ ***Structure parameters of the*** *ML2R* ***estimator: root, weights and bias parameter*** **h**

– For a weak error characteristic $\alpha > 0$, the resulting weights $(\mathbf{W}_i^{(R)})_{i=1,\dots,R}$ are given by

$$\mathbf{W}_i^{(R)} = \mathbf{w}_i^{(R)} + \cdots + \mathbf{w}_R^{(R)} \quad \text{where} \quad \mathbf{w}_i^{(R)} = (-1)^{R-i} \frac{1}{\prod_{j \neq i} |1 - N^{-\alpha(i-j)}|}, \quad i = 1, \dots, R$$

(the formula for $\mathbf{w}_i^{(R)}$ is established in (9.33)). Keep in mind that $\mathbf{W}_1^{(R)} = 1$.

– For a given root $N \in \mathbb{N}$, $N \geq 2$, the refiners are fixed as follows: $n_i = N^{i-1}$, $i = 1, \dots, R$.

– When $\beta \in (0, 1)$, the optimal choice for the root is $N = 2$.

– The choice of the coarsest bias parameter **h** is usually naturally suggested by the model under consideration. As all the approaches so far have been developed for an absolute error, it seems natural to choose **h** to be lower or close to one in practice: in particular, in a diffusion framework with horizon $T \gg 1$, we will not choose $\mathbf{h} = T$ but rather $T/2$ or $T/3$, etc; likewise, for nested Monte Carlo if the variance of the inner simulation is too large, the parameter $K_0 = 1/\mathbf{h}$ will be chosen strictly greater than 1 (say equal to a few units). See further on in this section for other comments.

▷ ***How to calibrate*** $\widetilde{c}_{\infty}$ **(and its connections with h and** $R^*(\varepsilon)$**)**

First, it is not possible to include an estimation of $\widetilde{c}_{\infty}$ in a pre-processing phase – beyond the fact that $R^*(\varepsilon)$ depends on $\widetilde{c}_{\infty}$ – since the natural statistical estimators are far too unstable. On the other hand, to the best of our knowledge, under $\big(WE\big)_{\infty}^{\alpha}$, there are no significant results about the behavior of the coefficients c_r as $r \to +\infty$, whatever the assumptions on the model are.

– If $\mathbf{h} = 1$, then $h \in \mathcal{H}$ can be considered as small and we observe that if the coefficients c_r of the weak error expansion have polynomial growth, *i.e.* $|c_r| = O(r^a)$

Table 9.1 Parameters of the ML2R estimator (standard framework)

$\underline{n}$	$n_i = N^{i-1},\ i = 1,\ldots,R \qquad$ (convention: $n_0 = n_0^{-1} = 0$)
$R = R^*(\varepsilon)$	$\left\lceil \frac{1}{2} + \frac{\log(\widetilde{c}_\infty^{\frac{1}{\alpha}}\mathbf{h})}{\log(N)} + \sqrt{\left(\frac{1}{2} + \frac{\log(\widetilde{c}_\infty^{\frac{1}{\alpha}}\mathbf{h})}{\log(N)}\right)^2 + 2\frac{\log\left(\frac{\sqrt{1+4\alpha}}{\varepsilon}\right)}{\alpha\log(N)}}\right\rceil$ (see also (9.66) and (9.69))
$h = h^*(\varepsilon)$	$\mathbf{h}$ (does not depend upon ε)
$q = q^*(\varepsilon)$	$q_1(\varepsilon) = \frac{1}{q_\varepsilon^\dagger},\ q_j(\varepsilon) = \frac{\theta h^{\frac{\beta}{2}}\,\lvert\mathbf{W}_j^{(R)}(N)\rvert\left(n_{j-1}^{-1} - n_j^{-1}\right)^{\frac{\beta}{2}}}{q_\varepsilon^\dagger\sqrt{n_{j-1}+n_j}},\ j = 2,\ldots,R,$ with $q_\varepsilon^\dagger$ s.t. $\sum_{j=1}^R q_j(\varepsilon) = 1$ and $\theta = \theta_h = \sqrt{\frac{V_1}{\mathrm{Var}(Y_h)}}$
$M = M^*(\varepsilon)$	$\left\lceil \left(1 + \frac{1}{2\alpha R}\right)\frac{\mathrm{Var}(Y_h)\,q_\varepsilon^\dagger\left(1 + \theta h^{\frac{\beta}{2}}\sum_{j=2}^R \lvert\mathbf{W}_j^{(R)}(N)\rvert\left(n_{j-1}^{-1} - n_j^{-1}\right)^{\frac{\beta}{2}}\sqrt{n_{j-1}+n_j}\right)}{\varepsilon^2}\right\rceil$

as $r \to \infty$, then $\widetilde{c}_\infty = \lim_{r\to+\infty} |c_r|^{\frac{1}{r}} = 1 = \mathbf{h}$ so that, from a practical point of view, $|c_R|^{\frac{1}{R}} \simeq 1$ as $R = R(\varepsilon)$ grows.

– If $\mathbf{h} \neq 1$, it is natural to perform a change of unit to return to the former situation, *i.e.* express everything with respect to $\mathbf{h}$. It reads

$$\mathbb{E}\left[\frac{Y_h}{\mathbf{h}}\right] = \mathbb{E}\left[\frac{Y_0}{\mathbf{h}}\right] + \sum_{r=1}^R c_r \mathbf{h}^{\alpha r-1}\left(\frac{h}{\mathbf{h}}\right)^{\alpha r} + \mathbf{h}^{\alpha R-1} o\left(\left(\frac{h}{\mathbf{h}}\right)^{\alpha R}\right).$$

If we consider that this expansion should behave as the above normalized one, it is natural to "guess" that $|c_R h^{\alpha R-1}|^{\frac{1}{R}} \to 1$ as $R \to +\infty$, *i.e.*

$$\widetilde{c}_\infty = \mathbf{h}^{-\alpha}. \tag{9.68}$$

Note that this implies that $\widetilde{c}_\infty^{\frac{1}{\alpha}}\mathbf{h} = 1$, which dramatically simplifies the expression of $R(\varepsilon)$ in Table 9.1 since several terms involving $\log(\widetilde{c}_\infty^{\frac{1}{\alpha}}\mathbf{h})$ vanish. The resulting formula reads

$$R^*(\varepsilon) = \left\lceil \frac{1}{2} + \sqrt{\frac{1}{4} + 2\frac{\log\left(\frac{\sqrt{1+4\alpha}}{\varepsilon}\right)}{\alpha\log(N)}}\right\rceil \tag{9.69}$$

which still grows as $O\left(\sqrt{\log(1/\varepsilon)}\right)$ as $\varepsilon \to 0$.

Various numerical experiments (see [113, 114]) show that the *ML2R* estimator, implemented with this choice $|c_R|^{\frac{1}{R}} \simeq \mathbf{h}^{-\alpha}$, is quite robust under variations of the coefficients c_r.

At this stage, for a prescribed *RMSE*, $R(\varepsilon)$ can be computed. Note that, given the value of $R(\varepsilon)$ (and the way it has been specified), one has

$$h^*(\varepsilon) = \mathbf{h}.$$

$\triangleright$ ***Payoff-dependent simulation parameters: estimation of*** $\mathrm{Var}(Y_{\mathbf{h}})$, V_1 ***and*** θ

Such an estimation is crucial since $\mathrm{Var}(Y_{\mathbf{h}})$ and $\theta = \sqrt{\frac{V_1}{\mathrm{Var}(Y_{\mathbf{h}})}}$ both appear in the computation of the optimal allocation policy $\big(q_i(\varepsilon)\big)_{1\le i\le R(\varepsilon)}$ and in the global size of the *ML2R* estimator. A (rough) statistical pre-processing is subsequently mandatory.

– *Estimation of* $\mathrm{Var}(Y_{\mathbf{h}})$. Taking into account our optimization procedure, one always has $h^*(\varepsilon) = \mathbf{h}$ (see Table 9.1.) One can without damage perform a rough estimation of $\mathrm{Var}(Y_{\mathbf{h}})$. Namely

$$\mathrm{Var}(Y_{\mathbf{h}}) \simeq \frac{1}{m}\sum_{k=1}^{m}(Y^k_{\mathbf{h}} - \overline{Y}_{\mathbf{h},m})^2 \quad \text{where} \quad \overline{Y}_{\mathbf{h},m} = \frac{1}{m}\sum_{k=1}^{m} Y^k_{\mathbf{h}}. \tag{9.70}$$

– *Estimation of* V_1. Under $(VE)_\beta$, for every $h_0,\ h_0' \in \mathcal{H}$, $h_0' < h_0$,

$$\mathrm{Var}\big(Y_{h_0} - Y_{h_0'}\big) \le V_1\big|h_0 - h_0'\big|^\beta.$$

This suggests to choose a not too small $h_0,\ h_0'$ in $\mathcal{H}$ to estimate V_1 based on the formula

$$V_1 \simeq \widehat{V}_1(h_0, h') = |h_0 - h_0'|^{-\beta}\mathrm{Var}\big(Y_{h_0} - Y_{h_0'}\big).$$

It remains to estimate the variance by its usual estimator, which yields

$$V_1 \simeq \widehat{V}_1(h_0, h_0', m) = \frac{|h_0 - h_0'|^{-\beta}}{m}\sum_{k=1}^{m}(Y^k_{h_0} - Y^k_{h_0'} - \overline{Y}_{h_0,h_0',m})^2,$$

where

$$\overline{Y}_{h_0,h_0',m} = \frac{1}{m}\sum_{k=1}^{m} Y^k_{h_0} - Y^k_{h_0'}. \tag{9.71}$$

We make a rather *conservative choice* by giving the priority to fulfilling the prescribed error ε at the cost of a possible small increase in complexity. This led us to consider in all the numerical experiments that follow

$$h_0 = \frac{\mathbf{h}}{5} \quad \text{and} \quad h_0' = \frac{h_0}{2} = \frac{\mathbf{h}}{10}$$

so that $|h_0 - h_0'|^{-\beta} = 10^\beta \mathbf{h}^{-\beta}$.

These choices turn out to be quite robust and provide satisfactory results in all investigated situations. They have been adopted in all the numerical experiments reproduced in this chapter.

▶ **Exercise.** Determine a way to estimate $\bar{V}_1$ in (9.63) when $(SE)^0_\beta$ is the only available strong error rate.

▷ ***Calibration of the root*** N

This corresponds to the second phase described at the beginning of this section.

– First, let us recall that *if* $\beta \in (0, 1)$, *then the best root is always* $N = 2$.

– When $\beta \geq 1$ – keeping in mind that R will never go beyond 10 or 12 for common values of the prescribed error ε, and once the parameters $\mathrm{Var}(Y_0)$ and V_1 have been estimated – it is possible to compute the upper-bound (9.54) of $\mathrm{Cost}(\widehat{I}^{ML2R}_M)$ for various values of N and select the minimizing root.

ℵ Practitioner's corner II: Design of a confidence interval at level α

We temporarily adopt in this section the slight abuse of notation $\widehat{I}^{ML2R}_{M(\varepsilon)}$ to denote the *ML2R* estimator designed to satisfy the prescribed *RMSE* ε.

The asymptotic behavior of *ML2R* estimators as the *RMSE* $\varepsilon \to 0$, especially the *CLT*, has been investigated in several papers (see [34, 35, 75], among others) in various frameworks (Brownian diffusions, Lévy-driven diffusions,...). In [112], a general approach is proposed. It is established, under a sharp version of $(VE)_\beta$ and, when $\beta \leq 1$, an additional uniform integrability condition that both weighted and regular Multilevel estimators are ruled by *CLT* at rate ε as the *RMSE* $\varepsilon \to 0$. By a sharp version of $(VE)_\beta$, we mean

$$\lim_{h\to 0,\, h\in\mathcal{H}} h^{-\beta}\left\|Y_h - Y_{\frac{h}{N}}\right\|_2^2 = v_1\left(1 - \frac{1}{N}\right)^\beta$$

where $v_1 \in (0, V_1]$.

As a consequence, it is possible to design a confidence interval under the assumption that this asymptotic Gaussian behavior holds true. The reasoning is as follows: we start from the fact that

$$\left\|\widehat{I}^{ML2R}_{M(\varepsilon)} - I_0\right\|_2^2 = \mathrm{Bias}^2\big(\widehat{I}^{ML2R}_{M(\varepsilon)}\big) + \mathrm{Var}\big(\widehat{I}^{ML2R}_{M(\varepsilon)}\big) = \varepsilon,$$

keeping in mind that $I_0 = \mathbb{E}\, Y_0$. Then, both $\big|\mathrm{Bias}\big(\widehat{I}^{ML2R}_{M(\varepsilon)}\big)\big|$ and the standard deviation $\sigma\big(\widehat{I}^{ML2R}_{M(\varepsilon)}\big)$ are dominated by ε. Let q_c denote the quantile at confidence level $c \in (0, 1)$ for the normal distribution ([2]). If ε is small enough,

$$\mathbb{P}\Big(\widehat{I}^{ML2R}_{M(\varepsilon)} \in \big[\mathbb{E}\,\widehat{I}^{ML2R}_{M(\varepsilon)} - q_c\sigma\big(\widehat{I}^{ML2R}_{M(\varepsilon)}\big),\, \mathbb{E}\,\widehat{I}^{ML2R}_{M(\varepsilon)} + q_c\sigma\big(\widehat{I}^{ML2R}_{M(\varepsilon)}\big)\big]\Big) \simeq c.$$

On the other hand, $\mathbb{E}\,\widehat{I}^{ML2R}_{M(\varepsilon)} = I_0 + \mathrm{Bias}\big(\widehat{I}^{ML2R}_{M(\varepsilon)}\big)$ so that

[2] c is used here to avoid confusion with the exponent α of the weak error expansion.

$$\Big[\mathbb{E}\,\widehat{I}^{ML2R}_{M(\varepsilon)} - q_c\sigma\big(\widehat{I}^{ML2R}_{M(\varepsilon)}\big),\ \mathbb{E}\,\widehat{I}^{ML2R}_{M(\varepsilon)} + q_c\sigma\big(\widehat{I}^{ML2R}_{M(\varepsilon)}\big)\Big]$$
$$\subset \Big[I_0 - \Big(\big|\mathrm{Bias}\big(\widehat{I}^{ML2R}_{M(\varepsilon)}\big)\big| + q_c\sigma\big(\widehat{I}^{ML2R}_{M(\varepsilon)}\big)\Big),\ I_0 + \big|\mathrm{Bias}\big(\widehat{I}^{ML2R}_{M(\varepsilon)}\big)\big| + q_c\sigma\big(\widehat{I}^{ML2R}_{M(\varepsilon)}\big)\Big]. \tag{9.72}$$

Now it is clear that the maximum of $u + q_c v$ under the constraints $u,\ v \geq 0$, $u^2 + v^2 = \varepsilon^2$ is equal to $\max(1, q_c)\varepsilon$. Combining these facts eventually leads to the following (conservative) confidence interval:

$$\mathbb{P}\Big(\widehat{I}^{ML2R}_{M(\varepsilon)} \in \big[I_0 - \max(1, q_c)\varepsilon,\ I_0 - \max(1, q_c)\varepsilon\big]\Big) \simeq c. \tag{9.73}$$

Note that in general c is small so that $q_c > 1$. A sharper (*i.e.* narrower) confidence interval at a given confidence level c can be obtained by estimating on-line the empirical variance of the estimator based on Eq. (9.41), namely

$$\widehat{\sigma}\big(\widehat{I}^{ML2R}_{M(\varepsilon)}\big) \simeq \frac{1}{\sqrt{M}}\left[\frac{1}{M_1}\sum_{k=1}^{M_1}\Big(Y_h^{(1),k} - \Big[\overline{Y_h^{(1)}}\Big]_{M_1}\Big)^2 + \sum_{i=2}^{R}\frac{\mathbf{W}_i^2}{M_i}\sum_{k=1}^{M_i}\Big(Y^{(i),k}_{\frac{h}{n_i}} - Y^{(i),k}_{\frac{h}{n_{i-1}}} - \Big[\overline{Y^{(i)}_{\frac{h}{n_i}} - Y^{(i)}_{\frac{h}{n_{i-1}}}}\Big]_{M_i}\Big)^2\right]^{\frac{1}{2}}$$

where $\Big[\overline{Y}\Big]_M = \frac{1}{M}\sum_{k=1}^{M} Y^k$ denotes the empirical mean of i.i.d. copies Y^k of the inner variable Y, $M_i = q_i(\varepsilon)M(\varepsilon)$, etc. Then, one evaluates the squared bias by setting

$$\widehat{\mathrm{Bias}}\big(\widehat{I}^{ML2R}_{M(\varepsilon)}\big) \simeq \sqrt{\varepsilon^2 - \mathrm{Var}\big(\widehat{I}^{ML2R}_{M}\big)}.$$

Then plugging these estimates into (9.72), *i.e.* replacing the bias $\mathrm{Bias}\big(\widehat{I}^{ML2R}_{M(\varepsilon)}\big)$ and the standard-deviation $\sigma\big(\widehat{I}^{ML2R}_{M(\varepsilon)}\big)$ by their above estimates $\widehat{\mathrm{Bias}}\big(\widehat{I}^{ML2R}_{M(\varepsilon)}\big)$ and $\widehat{\sigma}\big(\widehat{I}^{ML2R}_{M(\varepsilon)}\big)$, respectively, in the magnitude $\big|\mathrm{Bias}\big(\widehat{I}^{ML2R}_{M(\varepsilon)}\big)\big| + q_c\sigma\big(\widehat{I}^{ML2R}_{M(\varepsilon)}\big)$ of the theoretical confidence interval yields the expected empirical confidence interval at level c

$$I_{\varepsilon,c} = \Big[I_0 - \Big(\big|\widehat{\mathrm{Bias}}\big(\widehat{I}^{ML2R}_{M(\varepsilon)}\big)\big| + q_c\widehat{\sigma}\big(\widehat{I}^{ML2R}_{M(\varepsilon)}\big)\Big),\ I_0 + \big|\widehat{\mathrm{Bias}}\big(\widehat{I}^{ML2R}_{M(\varepsilon)}\big)\big| + q_c\widehat{\sigma}\big(\widehat{I}^{ML2R}_{M(\varepsilon)}\big)\Big].$$

ℵ Practitioner's corner III: The assumptions $(VE)_\beta$ and $\big(WE\big)^\alpha_R$

In practice, one often checks $(VE)_\beta$ as a consequence of $(SE)_\beta$, as illustrated below.

▷ ***Brownian diffusions (discretization schemes)***

Our task here is essentially to simulate two discretization schemes with step h and h/N in a coherent way, *i.e.* based on increments of the same Brownian motion. A discretization scheme with step $h = \frac{T}{n}$ formally reads

$$X^{(n)}_{kh} = \Phi\big(X^{(n)}_{(k-1)h}, \sqrt{h}\,U^{(n)}_k\big),\ k = 1, \ldots, n,\quad \bar{X}^n_0 = X_0,$$

where $U_k^{(n)} = \frac{W_{kh}-W_{k(h-1)}}{\sqrt{h}} \stackrel{d}{=} \mathcal{N}(0; I_q)$ are i.i.d. The most elementary way to jointly simulate the coarse scheme $X^{(n)}$ and the fine scheme $X^{(Nn)}$ is to start by simulating a path $(X^{(Nn)}_{k\frac{h}{N}})_{k=0,\dots nN}$ of the fine scheme using $\big(U_k^{(nN)}\big)_{k=1,\dots,nN}$ and then simulate the path $(X^{(n)}_{kh})_{k=0,\dots,n}$ using the sequence $(U_k^{(n)})_{k=1,\dots,n}$ defined by

$$U_k^{(n)} = \frac{\sum_{\ell=1}^{N} U^{(nN)}_{(k-1)N+\ell}}{\sqrt{N}},\ k = 1, \dots, N.$$

– *Condition* $(VE)_\beta$ *or* $(SE)^0_\beta$. For discretization schemes of Brownian diffusions, the bias parameter is given by $h = \frac{T}{n} \in \mathcal{H} = \Big\{\frac{\mathbf{h}}{n},\ n \geq 1\Big\}$. If $Y_h = F\big((\bar{X}^n_t)_{t\in[0,T]}\big)$, $F\big((\tilde{X}^n_t)_{t\in[0,T]}\big)$ (Euler schemes) or $F\big((\tilde{X}^{n,mil}_t)_{t\in[0,T]}\big)$ (genuine Milstein scheme).

The easiest condition to check is $(SE)^0_\beta$, since it can usually be established by combining the (Hölder/Lipschitz) regularity of the functional F with respect to the $\|\cdot\|_{\sup}$-norm with the L^2-rate of convergence of the (simulable) discretization scheme under consideration ($\sqrt{h}$ for the discrete time Euler scheme at time T, $\sqrt{h\log(1/h)}$ for the sup-norm and h for the Milstein scheme).

The condition $(VE)_\beta$ is not as easy to establish. However, as far as the Euler scheme is concerned, when $h \in \mathcal{H}$ and $h' = \frac{h}{N}$, it can be derived from a functional Central Limit Theorem for the difference $\bar{X}^n - \bar{X}^{nN}$, which weakly converges at a rate $\sqrt{n\frac{N}{N-1}}$ toward a diffusion process free of N as $n \to +\infty$. Combined with appropriate regularity and growth assumptions on the function f or the functional F, this yields that

$$\big\|Y_h - Y_{\frac{h}{N}}\big\|_2^2 \leq V_i\Big|h - \frac{h}{N}\Big|^\beta \quad \text{with} \quad \beta = 1.$$

For more details, we refer to [34] and to [112] for the application to path-dependent functionals.

– *Condition* $\big(WE\big)^\alpha_R$. As for the Euler scheme(s), we refer to Theorem 7.8 (for 1-marginals) and Theorem 8.1 for functionals. For 1-marginals $F(x) = f\big(x(T)\big)$, the property $(\mathcal{E}_{R+1})$ implies $\big(WE\big)^1_R$. When F is a "true" functional, the available results are partial since no higher-order expansion of the weak error is established and $\big(WE\big)^\alpha_R$, usually with $\alpha = \frac{1}{2}$ (like for barrier options) or 1 (like for lookback payoffs), should still be considered as a conjecture, though massively confirmed by numerical experiments.

▷ ***Nested Monte Carlo***

We retain the notation introduced at the beginning of the chapter. For convenience, we write

$$\Xi_0 = \mathbb{E}\big(F(X,Z)\,|\,X\big) \quad \text{and} \quad \Xi_h = \frac{1}{K}\sum_{k=1}^{K} F(X, Z_k),\ h = \frac{1}{K} \in \mathcal{H} = \Big\{\frac{1}{K},\ K \in K_0\mathbb{N}^*\Big\},$$

so that $Y_0 = f(\Xi_0)$ and $Y_h = f\big(\Xi_h\big)$.

– *Condition* $(VE)_\beta$ *or* $(SE)_\beta$. The following proposition yields criteria for $(SE)_\beta$ to be fulfilled.

Proposition 9.2 (Strong error) (a) Lipschitz continuous function f, quadratic case. *Assume* $F(X,Z)\in L^2$. *If* f *is Lipschitz continuous, then,*

$$\forall\, h,\, h'\in \mathcal{H}\cup\{0\},\quad \big\|Y_h - Y_{h'}\big\|_2^2 \le V_1|h-h'|,$$

where $V_1 = [f]^2_{\rm Lip}\big\|F(X,Z)-\mathbb{E}\,(F(X,Z)\,|\,X)\big\|_2^2 \le [f]^2_{\rm Lip}{\rm Var}\big(F(X,Z)\big)$ *so that* $(SE)_\beta$ *(hence* $(VE)_\beta$*) is satisfied with* $\beta=1$ *by setting* $h'=0$.

(b) Lipschitz continuous function f, L^p-case. *If, furthermore,* $F(X,Z)\in L^p$, $p\ge 2$, *then*

$$\forall\, h,\, h'\in \mathcal{H}\cup\{0\},\quad \big\|Y_h - Y_{h'}\big\|_p^p \le V_{\frac p2}|h-h'|^{\frac p2}, \tag{9.74}$$

where $V_{\frac p2} = \big(2[f]_{\rm Lip}C_p^{MZ}\big)^p\big\|F(X,Z)-\mathbb{E}\,\big(F(X,Z)\,|\,X\big)\big\|_p^p$ *and* C_p^{MZ} *is the universal constant in the right-hand side of the Marcinkiewicz–Zygmund (or B.D.G.) Inequality (6.49).*

(c) Indicator function $f=\mathbf{1}_{(-\infty,a]}$, $a\in\mathbb{R}$. *Assume* $F(X,Z)\in L^p$ *for some* $p\ge 2$. *Assume that the distributions of* Ξ_0, Ξ_h, $h\in(0,h_0]\cap\mathcal{H}$ $(h_0\in\mathcal{H})$, *are absolutely continuous with uniformly bounded densities* g_{Ξ_h}. *Then, for every* $h\in(0,h_0]\cap\mathcal{H}$,

$$\Big\|\mathbf{1}_{\{\Xi_h\le a\}} - \mathbf{1}_{\{\Xi_{h'}\le a\}}\Big\|_2^2 \le C_{X,Y,p}|h-h'|^{\frac{p}{2(p+1)}},$$

where

$$C_{X,Y,p} = 2^{\frac{p}{p+1}}\Big(p^{-\frac{p}{p+1}} + p^{\frac{1}{p+1}}\Big)\Big(\sup_{h\in(0,h_0]\cap\mathcal{H}}\|g_{\Xi_h}\|_{\sup}+1\Big)^{\frac{p}{p+1}}\big(2\,C_p^{MZ}\big\|F(X,Z)-\mathbb{E}\,\big(F(X,Z)\,|\,Y\big)\big\|_p\big)^{\frac{p}{p+1}}.$$

Hence, Assumption $(SE)_\beta$ *holds with* $\beta = \frac{p}{2(p+1)}\in\big(0,\frac12\big)$.

Remark. The boundedness assumption made on the densities g_{Ξ_h} of Ξ_h in the above Claim (c) may look unrealistic. However, if the density g_{Ξ_0} of Ξ_0 is bounded and the assumptions of Theorem 9.2 (b) further on – to be precise, in the item devoted to weak error expansion $\big(WE\big)_R^\alpha$ – are satisfied for $R=0$, then

$$g_{\Xi_h}(\xi)\le g_{\Xi_0}(\xi)+o(h^{\frac12})\quad \text{uniformly in } \xi\in\mathbb{R}^d.$$

Consequently, there exists an $h_0=\frac{1}{K_0}\in\mathcal{H}$ such that, for every $h\in(0,h_0]$,

$$\forall\,\xi\in\mathbb{R}^d,\quad g_{\Xi_h}(\xi)\le g_{\Xi_0}(\xi)+1.$$

Proof. (a) Except for the constant, this claim is a special case of Claim (b) when $p = 2$. We leave the direct proof as an exercise.

(b) Set $\widetilde{F}(\xi, z) = F(\xi, z) - \mathbb{E}\, F(\xi, Z)$, $\xi \in \mathbb{R}^d$, $z \in \mathbb{R}^q$, and assume that $h = \frac{1}{K}$ and $h' = \frac{1}{K'} > 0$, $1 \le K \le K'$, $K, K' \in \mathbb{N}^*$. First note that, X and Z being independent, Fubini's theorem implies

$$\big\|Y_h - Y_{h'}\big\|_p^p \le [f]_{\rm Lip}^p \int_{\mathbb{R}^d} \mathbb{P}_X(d\xi)\, \mathbb{E}\left|\frac{1}{K}\sum_{k=1}^{K} \widetilde{F}(\xi, Z_k) - \frac{1}{K'}\sum_{k=1}^{K'} \widetilde{F}(\xi, Z_k)\right|^p.$$

Then, for every $\xi \in \mathbb{R}^d$, it follows from Minkowski's Inequality that

$$\left\|\frac{1}{K}\sum_{k=1}^{K} \widetilde{F}(\xi, Z_k) - \frac{1}{K'}\sum_{k=1}^{K'} \widetilde{F}(\xi, Z_k)\right\|_p \le |h - h'|\Big\|\sum_{k=1}^{K} \widetilde{F}(\xi, Z_k)\Big\|_p + h'\Big\|\sum_{k=K+1}^{K'} \widetilde{F}(\xi, Z_k)\Big\|_p.$$

Applying the Marcinkiewicz–Zygmund Inequality to both terms on the right-hand side of the above inequality yields

$$\begin{aligned}\left\|\frac{1}{K}\sum_{k=1}^{K} \widetilde{F}(\xi, Z_k) - \frac{1}{K'}\sum_{k=1}^{K'} \widetilde{F}(\xi, Z_k)\right\|_p &\le |h - h'|C_p^{MZ}\left\|\sum_{k=1}^{K} \widetilde{F}(\xi, Z_k)^2\right\|_{\frac{p}{2}}^{\frac{1}{2}} + h'C_p^{MZ}\left\|\sum_{k=K+1}^{K'} \widetilde{F}(\xi, Z_k)^2\right\|_{\frac{p}{2}}^{\frac{1}{2}}\\ &\le |h - h'|K^{\frac{1}{2}}\big\|\widetilde{F}(\xi, Z)\big\|_p + h'(K' - K)^{\frac{1}{2}}\big\|\widetilde{F}(\xi, Z)\big\|_p\end{aligned}$$

where we again used (twice) Minkowski's Inequality in the last line. Finally, for every $\xi \in \mathbb{R}^d$,

$$\begin{aligned}\left\|\frac{1}{K}\sum_{k=1}^{K} \widetilde{F}(\xi, Z_k) - \frac{1}{K'}\sum_{k=1}^{K'} \widetilde{F}(\xi, Z_k)\right\|_p &\le C_p^{MZ}\big\|\widetilde{F}(\xi, Z)\big\|_p\left((h - h')\frac{1}{\sqrt{h}} + h'\Big(\frac{1}{h'} - \frac{1}{h}\Big)^{\frac{1}{2}}\right)\\ &= C_p^{MZ}\big\|\widetilde{F}(\xi, Z)\big\|_p (h - h')^{\frac{1}{2}}\left(\Big(1 - \frac{h'}{h}\Big)^{\frac{1}{2}} + \Big(\frac{h'}{h}\Big)^{\frac{1}{2}}\right)\\ &\le 2\, C_p^{MZ}\big\|\widetilde{F}(\xi, Z)\big\|_p (h - h')^{\frac{1}{2}}.\end{aligned}$$

Plugging this bound into the above equality yields

$$\begin{aligned}\big\|Y_h - Y_{h'}\big\|_p^p &\le (2\, C_p^{MZ})^p \int_{\mathbb{R}^d} \mathbb{P}_Y(d\xi)\big\|\widetilde{F}(\xi, Z)\big\|_p^p (h - h')^{\frac{p}{2}}\\ &= (2\, C_p^{MZ})^p \big\|F(X, Z) - \mathbb{E}\,(F(X, Z)|\, X)\big\|_p^p (h - h')^{\frac{p}{2}}\\ &\le \big(4\, C_p^{MZ})^p \big\|F(X, Z)\big\|_p^p (h - h')^{\frac{p}{2}}.\end{aligned}$$

If $h' = 0$, we get

$$\begin{aligned}\mathbb{E}\,|Y_h - Y_0|^p &\le [f]_{\mathrm{Lip}}^p \mathbb{E}\left|\left(\frac{1}{K}\sum_{k=1}^{K} F(X, Z_k)\right) - \mathbb{E}\left(F(X, Z)\,|X\right)\right|^p \\ &= [f]_{\mathrm{Lip}}^p \int_{\mathbb{R}^d} \mathbb{E}\left|\frac{1}{K}\sum_{k=1}^{K} \widetilde{F}(\xi, Z_k)\right|^p \mathbb{P}_X(d\xi)\end{aligned}$$

since $\mathbb{E}\,(F(X,Z)\,|X) = \big[\mathbb{E}\,F(\xi, Z)\big]_{|\xi=X}$ *a.s.* and the result follows as above.
(c) The proof of this claim relies on Claim (b) and the following elementary lemma.

Lemma 9.3 *Let Ξ and Ξ' be two real-valued random variables lying in $L^p(\Omega, \mathcal{A}, \mathbb{P})$, $p \ge 1$, with densities g_Ξ and $g_{\Xi'}$, respectively. Then, for every $a \in \mathbb{R}$,*

$$\left\|\mathbf{1}_{\{\Xi'\le a\}} - \mathbf{1}_{\{\Xi\le a\}}\right\|_2^2 \le \left(p^{-\frac{p}{p+1}} + p^{\frac{1}{p+1}}\right)\left(\|g_\Xi\|_{\sup} + \|g_{\Xi'}\|_{\sup}\right)^{\frac{p}{p+1}} \left\|\Xi - \Xi'\right\|_p^{\frac{p}{p+1}}. \tag{9.75}$$

Proof. Let $\lambda > 0$ be a free parameter. Note that

$$\begin{aligned}\left\|\mathbf{1}_{\{\Xi\le a\}} - \mathbf{1}_{\{\Xi'\le a\}}\right\|_2^2 &= \mathbb{P}\big(\Xi \le a \le \Xi'\big) + \mathbb{P}\big(\Xi' \le a \le \Xi\big) \\ &\le \mathbb{P}\big(\Xi \le a,\ \Xi' \ge a+\lambda\big) + \mathbb{P}\big(\Xi \le a \le \Xi' \le a+\lambda\big) \\ &\quad + \mathbb{P}\big(\Xi' \le a,\ \Xi \ge a+\lambda\big) + \mathbb{P}\big(\Xi' \le a \le \Xi \le a+\lambda\big) \\ &\le \mathbb{P}(\Xi' - \Xi \ge \lambda) + \mathbb{P}(\Xi - \Xi' \ge \lambda) \\ &\quad + \mathbb{P}(\Xi' \in [a, a+\lambda]) + \mathbb{P}(\Xi \in [a, a+\lambda]) \\ &= \mathbb{P}\big(|\Xi' - \Xi| \ge \lambda\big) + \mathbb{P}(\Xi \in [a, a+\lambda]) + \mathbb{P}(\Xi' \in [a, a+\lambda]) \\ &\le \frac{\mathbb{E}\,|\Xi' - \Xi|^p}{\lambda^p} + \lambda\Big(\|g_\Xi\|_{\sup} + \|g_{\Xi'}\|_{\sup}\Big).\end{aligned}$$

A straightforward optimization in λ yields the announced result. ◇

At this stage, Claim (c) becomes straightforward: plugging the upper-bound of the densities and (9.74) into Inequality (9.75) of Lemma 9.3 applied with $\Xi = \Xi_h$ and $\Xi' = \Xi_{h'}$ completes the proof. ◇

► **Exercise.** (a) Prove Claim (a) of the above proposition with the announced constant.
(b) Show that if f is ρ-Hölder, $\rho \in (0, 1]$ and $p \ge \frac{2}{\rho}$, then

$$\forall\, h, h' \in \mathcal{H}, \quad \|Y_h - Y_{h'}\|_p \le [f]_{\rho, Hol} \|F(X,Z) - \mathbb{E}\,(F(X,Z)\,|\,X)\|_{p\rho}^{\rho}\, |h - h'|^{\frac{\rho}{2}}.$$

– *Condition* $\big(WE\big)_R^\alpha$. For this expansion we again need to distinguish between the smooth case and the case of indicator functions. In the non-regular case where f is an indicator function, we will need a smoothness assumption on the joint distribution of (Ξ_0, Ξ_h) that will be formulated as a smoothness assumption on the pair $(\Xi_0, \Xi_h -$

Ξ_0). The first expansion result in that direction was established in [128]. The result in Claim (*b*) below is an extension of this result in the sense that if $R = 1$, our assumptions coincide with theirs.

We denote by φ_0 the natural regular (Borel) version of the conditional mean function of $F(X, Z)$ given X, namely

$$\varphi_0(\xi) = \mathbb{E}\, F(\xi, Z) = \mathbb{E}\big(F(X, Z)\,|\, X = \xi\big), \quad \xi \in \mathbb{R}.$$

In particular, $Y_0 = \varphi_0(\Xi_0)$ $\mathbb{P}$-*a.s*.

Theorem 9.2 (Weak error expansion) (*a*) Smooth setting. *Let* $R \in \mathbb{N}^*$. *Assume* $F(X, Z) \in L^{2R+1}(\mathbb{P})$ *and let* $f : \mathbb{R} \to \mathbb{R}$ *be a* $2R + 1$ *times differentiable function with bounded derivatives* $f^{(k)}$, $k = R + 1, \ldots, 2R + 1$. *Then, there exists real coefficients* $c_1(f), \ldots, c_R(f)$ *such that*

$$\forall h \in \mathcal{H}, \quad \mathbb{E}\, Y_h = \mathbb{E}\, Y_0 + \sum_{r=1}^{R} c_r(f) h^r + O(h^{R+1/2}). \tag{9.76}$$

(*b*) Density function and smooth joint density. *Assume that* $F(X, Z) \in L^{2R+1}(\mathbb{P})$ *for some* $R \in \mathbb{N}$ *and that* $d = 1$. *Assume the distribution* $(\Xi_0, \Xi_h - \Xi_0)$ *has a smooth density with respect to the Lebesgue measure on* $\mathbb{R}^2$. *Let* g_{Ξ_0} *be the density of* Ξ_0 *and let* $g_{\Xi_0|\Xi_h-\Xi_0=\tilde{\xi}}$ *be (a regular version of) the conditional density of* Ξ_0 *given* $\Xi_h - \Xi_0 = \tilde{\xi}$. *Assume that the functions* φ_0, g_{Ξ_0}, $g_{\Xi_0|\Xi_h-\Xi_0=\tilde{\xi}}$, $\tilde{\xi} \in \mathbb{R}$, *are* $2R + 1$ *times differentiable. Assume that* φ_0 *is monotonic (hence one-to-one) with a derivative* m' *which is never* 0. *If* $\sup_{h\in\mathcal{H},\,\tilde{\xi},\,\xi\in\mathbb{R}} |g^{(2R+1)}_{\Xi_0|\Xi_h-\Xi_0=\tilde{\xi}}(\xi)| < +\infty$, *then*

$$g_{\Xi_h}(\xi) = g_{\Xi_0}(\xi) + g_{\Xi_0}(\xi) \sum_{r=1}^{R} \frac{h^r}{r!} P_r(\xi) + O(h^{R+\frac{1}{2}}) \quad \textit{uniformly in } \xi \in \mathbb{R}, \tag{9.77}$$

where the functions P_r *are* $2(R - r) + 1$ *times differentiable functions*, $r = 1, \ldots, R$.
(*c*) Indicator function and smooth joint density. *Let* G_{Ξ_h} *and* G_{Ξ_0} *denote the c.d.f. functions of* Ξ_h *and* Ξ_0, *respectively. Under the assumptions of* (*b*), *if, furthermore,* $\sup_{h\in\mathcal{H},\,\tilde{\xi},\,\xi\in\mathbb{R}} |g^{(2R)}_{\Xi_0|\Xi_h-\Xi_0=\tilde{\xi}}(\xi)| < +\infty$ *and* $\lim_{\xi\to-\infty} g^{(2R)}_{\Xi_0|\Xi_h-\Xi_0=\tilde{\xi}}(\xi) = 0$ *for every* $\tilde{\xi} \in \mathbb{R}$, *then, one also has for every* $\xi \in \mathbb{R}$,

$$G_{\Xi_h}(\xi) = G_{\Xi_0}(\xi) + \sum_{r=1}^{R} \frac{h^r}{r!} \mathbb{E}\left(P_r(\Xi_0)\mathbf{1}_{\{\Xi_0\le\xi\}}\right) + O(h^{R+\frac{1}{2}}). \tag{9.78}$$

If $f = f_a = \mathbf{1}_{(-\infty,a]}$, $a \in \mathbb{R}$, *then* $\mathbb{E}\, Y_h = G_{\Xi_h}(a)$, $h \in \mathcal{H} \cup \{0\}$, *so that*

$$\mathbb{E}\,Y_h = \mathbb{E}\,Y_0 + \sum_{r=1}^{R} \frac{h^r}{r!}\mathbb{E}\left(P_r(\Xi_0)\mathbf{1}_{\{\Xi_0 \le a\}}\right) + O(h^{R+\frac{1}{2}}).$$

We will admit these results whose proofs turn out to be too specific and technical for this textbook. A first proof of Claim (a) can be found in [198]. The whole theorem is established in [113, 114] with closed forms for the coefficients $c_r(f)$ and the functions P_r involving the derivatives of f (for the coefficients $c_r(f)$), the conditional mean φ_0, the densities g_{Ξ_h}, $g_{\Xi_0-\Xi_h}$, their derivatives and the so-called partial *Bell polynomials* (see [69], p. 307). Claim (c) derives from (b) by integrating with respect to ξ and it can be extended to any bounded Borel function $\Psi : \mathbb{R} \to \mathbb{R}$ instead of $\mathbf{1}_{(-\infty,a]}$.

– *Complexity of nested Monte Carlo.* When computing the *ML2R* estimator in nested Monte Carlo simulation, one has, at each fine level $i \ge 2$,

$$Y_{\frac{h}{n_i}} - Y_{\frac{h}{n_{i-1}}} = f\left(\frac{1}{n_i K}\sum_{k=1}^{n_i K} F(X, Z_k)\right) - f\left(\frac{1}{n_{i-1} K}\sum_{k=1}^{n_{i-1} K} F(X, Z_k)\right)$$

with the same terms Z_k inside each instance of the function f. Hence, if we neglect the computational cost of f itself, the complexity of the fine level i reads

$$\bar{\kappa}\, n_i\, K = \bar{\kappa}\,\frac{n_i}{h}.$$

As a consequence, it is easily checked that $\sqrt{n_{j-1} + n_j}$ should be replaced by $\sqrt{n_j}$ in the above Table 9.1.

9.5.2 Regular Multilevel Estimator (Under First Order Weak Error Expansion)

What is called here a "regular" multilevel estimator is the original multilevel estimator introduced by M. Giles in [107] (see also [148] in a numerical integration framework and [167] for the Statistical Romberg extrapolation). The bias reduction is obtained by an appropriate stratification based on a simple "domino" or cascade property described hereafter. This simpler weightless structure makes its analysis possible under a first-order weak error expansion (and the β-control of the strong convergence similar to that introduced for *ML2R* estimators). We will see later in Theorem 9.3 the impact on the performances of this family of estimators compared to the *ML2R* family detailed in Theorem 9.1 of the previous section.

The regular Multilevel estimator

We assume that there exists a first-order weak error expansion, namely $(WE)_1^{\alpha}$

$$\big(WE\big)_1^\alpha \equiv \mathbb{E}\,Y_h = \mathbb{E}\,Y_0 + c_1 h^\alpha + o\big(h^\alpha\big). \tag{9.79}$$

Like for the *ML2R* estimator, we assume the refiners are powers of a root N: $n_i = N^{i-1}$, $i = 1, \ldots, L$, where L will denote throughout this section the depth of the regular multilevel estimator, following [107].

Note that, except for the final step – the depth optimization of L – most of what follows is similar to what we just did in the weighted framework, provided one sets $R = L$ and $\mathbf{W}_i^{(L)} = 1$, $i = 1, \ldots, L$.

STEP 1 (*Killing the bias*). Let $L \geq 2$ be an integer which will be the *depth* of the regular multilevel estimator. Then

$$\mathbb{E}\,Y_{\frac{h}{n_L}} - \mathbb{E}\,Y_0 = c_1\left(\frac{h}{n_{_L}}\right)^\alpha + o\left(\left(\frac{h}{n_{_L}}\right)^\alpha\right).$$

Now introducing artificially a telescopic sum, we get

$$\begin{aligned}
\mathbb{E}\,Y_{\frac{h}{n_L}} &= \mathbb{E}\,Y_h + \Big(\mathbb{E}\,Y_{\frac{h}{2}} - \mathbb{E}\,Y_h\Big) + \cdots + \Big(\mathbb{E}\,Y_{\frac{h}{n_L}} - \mathbb{E}\,Y_{\frac{h}{n_{L-1}}}\Big)\\
&= \mathbb{E}\,Y_h + \sum_{i=2}^{L} \mathbb{E}\,Y_{\frac{h}{n_i}} - Y_{\frac{h}{n_{i-1}}}\\
&= \mathbb{E}\Bigg[\underbrace{Y_h}_{\text{coarse level}} + \sum_{i=2}^{L} \underbrace{Y_{\frac{h}{n_i}} - Y_{\frac{h}{n_{i-1}}}}_{\text{refined level } i}\Bigg].
\end{aligned}$$

We will assume again that the $Y_{\frac{h}{n_i}} - Y_{\frac{h}{n_{i-1}}}$ variables at each level are sampled *independently*, *i.e.*

$$\mathbb{E}\,Y_0 = \mathbb{E}\left[Y_h^{(1)} + \sum_{i=2}^{L} Y_{\frac{h}{n_i}}^{(i)} - Y_{\frac{h}{n_{i-1}}}^{(i)}\right] - c_1\left(\frac{h}{n_{_L}}\right)^\alpha + o\left(\left(\frac{h}{n_{_L}}\right)^\alpha\right),$$

where the families $Y_{\underline{n},h}^{(i)}$, $i = 1, \ldots, L$, are independent.

STEP 2 (*Regular Multilevel estimator*). As already noted in the previous section, $Y_h^{(1)}$ has no reason to be small since it is close to a copy of Y_0, whereas, by contrast, $Y_{\frac{h}{n_i}}^{(i)} - Y_{\frac{h}{n_{i-1}}}^{(i)}$ is approximately 0 as the difference of two variables approximating Y_0. The seminal idea of the multilevel paradigm, introduced by Giles in [107], is to dispatch the L different simulation sizes M_i across the levels $i = 1, \ldots, L$ so that $M_1 + \cdots + M_{_L} = M$. This leads to the following family of (regular) multilevel estimators attached to $(Y_h)_{h\in\mathcal{H}}$ and the refiners vector $\underline{n}$:

$$\widehat{I}^{MLMC}_{M,M_1,\ldots,M_L} := \frac{1}{M_1}\sum_{k=1}^{M_1} Y_h^{(1),k} + \sum_{i=2}^{L}\frac{1}{M_i}\sum_{k=1}^{M_i}\Big(Y^{(i),k}_{\frac{h}{n_i}} - Y^{(i),k}_{\frac{h}{n_{i-1}}}\Big),$$

where $M_1,\ldots,M_L \in \mathbb{N}^*$, $M_1+\cdots+M_L = M$, $\big(Y^{(i),k}_{\underline{n},h}\big)_{i=1,\ldots,L,k\geq 1}$ are i.i.d. copies of $Y_{\underline{n},h} := \big(Y_{\frac{h}{n_i}}\big)_{i=1,\ldots,L}$.

As for weighted *ML2R* estimators, it is more convenient to re-write the above multilevel estimator as a stratified estimator by setting $q_i = \frac{M_i}{M}$, $i = 1,\ldots,L$. This leads us to the following formal definition.

Definition 9.5 *The family of (regular) multilevel estimators attached to* $(Y_h)_{h\in\mathcal{H}}$ *and the refiners vector* $\underline{n}$ *is defined, as follows: for every* $h\in\mathcal{H}$, *every* $q = (q_i)_{1\leq i\leq R}\in\mathcal{S}_R$ *and every integer* $M\geq 1$,

$$\widehat{I}^{MLMC}_{M} = \widehat{I}^{MLMC}_{M}(q,h,L,\underline{n}) = \frac{1}{M}\left[\sum_{k=1}^{M_1}\frac{Y_h^{(1),k}}{q_1} + \sum_{i=2}^{L}\sum_{k=1}^{M_i}\frac{1}{q_i}\Big(Y^{(i),k}_{\frac{h}{n_i}} - Y^{(i),k}_{\frac{h}{n_{i-1}}}\Big)\right] \tag{9.80}$$

(convention the $\frac{1}{0}\sum_{k=1}^{0} = 0$*), where* $M_i = \lfloor q_i M\rfloor$, $i = 1,\ldots,R$, $\big(Y^{(i),k}_{\underline{n},h}\big)$, $i=1,\ldots,R$, $k\geq 1$, *are i.i.d. copies of* $Y_{\underline{n},h} := \big(Y_{\frac{h}{n_i}}\big)_{i=1,\ldots,R}$.

This definition is similar to that of *ML2R* estimators where all the weights $\mathbf{W}_i$ would have been set to 1. Note that, as soon as $M\geq M(q) := (\min_i q_i)^{-1}$, all $M_i = \lfloor q_i M\rfloor \geq 1$, $i=1,\ldots,R$. This condition on M will be implicitly assumed in what follows so that no level is empty.

▷ *Bias.* The telescopic/domino structure of the estimator implies, regardless of the sizes $M_i\geq 1$, that

$$\mathrm{Bias}\big(\widehat{I}^{MLMC}_{M}\big) = \mathrm{Bias}\big(\widehat{I}^{RR}_{1}\big) = \mathbb{E}\,Y_{\frac{h}{n_L}} - \mathbb{E}\,Y_0 = c_1\left(\frac{h}{n_L}\right)^{\alpha} + o\left(\left(\frac{h}{n_L}\right)^{\alpha}\right). \tag{9.81}$$

▷ *Variance.* Taking advantage of the independence of the levels, one straightforwardly computes the variance of $\widehat{I}^{MLMC}_{M}$

$$\begin{aligned}\mathrm{Var}\big(\widehat{I}^{MLMC}_{M}\big) &= \frac{\mathrm{Var}(Y_h^{(1)})}{M_1} + \sum_{i=2}^{L}\frac{\mathrm{Var}\Big(Y^{(i)}_{\frac{h}{n_i}} - Y^{(i)}_{\frac{h}{n_{i-1}}}\Big)}{M_i}\\ &= \frac{1}{M}\left(\frac{\mathrm{Var}(Y_h^1)}{q_1} + \sum_{i=2}^{L}\frac{\mathrm{Var}\Big(Y^{(i)}_{\frac{h}{n_i}} - Y^{(i)}_{\frac{h}{n_{i-1}}}\Big)}{q_i}\right)\\ &= \frac{1}{M}\sum_{i=1}^{L}\frac{\sigma^2(i,h)}{q_i},\end{aligned} \tag{9.82}$$

where $\sigma^2(1,h) = \mathrm{Var}\big(Y_h^{(1)}\big)$ and $\sigma^2(i,h) = \mathrm{Var}\Big(Y^{(i)}_{\frac{h}{n_i}} - Y^{(i)}_{\frac{h}{n_{i-1}}}\Big)$, $i = 2, \ldots, L$.

$\triangleright$ *Complexity.* The complexity, or simulation cost, of the *MLMC* estimator is clearly the same as that of its weighted counterpart investigated in the previous section, *i.e.* is *a priori* given — or at least dominated – by

$$\mathrm{Cost}(\widehat{I}_M^{MLMC}) = \frac{\bar{\kappa} M}{h} \sum_{i=1}^{L} \kappa_i q_i, \tag{9.83}$$

where $\kappa_1 = 1$, $\kappa_i = \big(1 + N^{-1}\big)n_i$, $i = 2, \ldots, L$.

As usual, at this stage, the aim is to minimize the cost or complexity of the whole simulation for a prescribed $RMSE$ $\varepsilon > 0$, *i.e.* solving

$$\inf_{RMSE(\widehat{I}_M^{MLMC}) \le \varepsilon} \mathrm{Cost}(\widehat{I}_M^{MLMC}). \tag{9.84}$$

Following the lines of what was done in the weighted multilevel *ML2R* estimator, we check that this problem is equivalent to solving

$$\inf_{|\mathrm{Bias}(\widehat{I}_M^{MLMC})| < \varepsilon} \frac{\mathrm{Effort}(\widehat{I}_M^{MLMC})}{\varepsilon^2 - \mathrm{Bias}\big(\widehat{I}_M^{MLMC}\big)^2},$$

where, as soon as $M \ge M(q)$,

$$\begin{aligned}\mathrm{Effort}(\widehat{I}_M^{MLMC}) &= \kappa(\widehat{I}_M^{MLMC})\mathrm{Var}(\widehat{I}_M^{MLMC}) \\ &= \frac{\bar{\kappa}}{h}\left(\sum_{i=1}^{L} \frac{\sigma^2(i,h)}{q_i}\right)\left(\sum_{i=1}^{L} \kappa_i q_i\right).\end{aligned} \tag{9.85}$$

Note that neither the effort, nor the bias depend on M provided $M \ge M(q)$.

We still proceed by first minimizing the effort for a fixed h (and R) as a function of the vector $q \in \mathcal{S}_L$ and then minimizing the denominator of the above ratio cost in the bias parameter $h \in \mathcal{H}$.

$\triangleright$ *Minimizing the effort.* Lemma 9.1 yields the solution to the minimization of the effort, namely

$$\mathrm{argmin}_{q \in \mathcal{S}_L} \mathrm{Effort}(\widehat{I}_M^{MLMC}) = q^* = \left(\frac{\sigma(i,h)}{q^{*,\dagger}(h)\sqrt{\kappa_i}}\right)_{i=1,\ldots,L}$$

with $q^{*,\dagger}(h) = \displaystyle\sum_{1 \le i \le L} \frac{\sigma(i,h)}{\sqrt{\kappa_i}}$ and a resulting minimal effort reading

$$\min_{q\in\mathcal{S}_L} \text{Effort}(\widehat{I}_M^{MLMC}) = \frac{\bar{\kappa}}{h}\left(\sum_{i=1}^{L}\sqrt{\kappa_i}\,\sigma(i,h)\right)^2.$$

▷ *Minimizing the resulting cost.* Compiling the above results, the cost minimization

$$\inf_{RMSE(\widehat{I}_M^{MLMC})\le\varepsilon} \text{Cost}(\widehat{I}_M^{MLMC})$$

reads, if we neglect the second-order term in the bias expansion (9.81),

$$\inf_{\left|\text{Bias}(\widehat{I}_M^{MLMC})\right|\le\varepsilon} \frac{\bar{\kappa}\left(\sum_{1\le i\le L}\sqrt{\kappa_i}\,\sigma(i,h)\right)^2}{h\left(\varepsilon^2 - c_1^2\left(\frac{h}{n_L}\right)^{2\alpha}\right)}.$$

As for the weighted framework, we adopt a (slightly) suboptimal strategy that is solves

$$\begin{aligned}\sup_{\left|\text{Bias}(\widehat{I}_M^{MLMC})\right|\le\varepsilon}\left[h\left(\varepsilon^2 - c_1^2\left(\frac{h}{n_L}\right)^{2\alpha}\right)\right] &= \sup_{h\in\mathcal{H}:\ h\le\frac{\varepsilon n_L}{|c_1|}}\left[h\left(\varepsilon^2 - c_1^2\left(\frac{h}{n_L}\right)^{2\alpha}\right)\right]\\ &\le \sup_{h:0<h\le\frac{\varepsilon n_L}{|c_1|}}\left[h\left(\varepsilon^2 - c_1^2\left(\frac{h}{n_L}\right)^{2\alpha}\right)\right].\end{aligned}$$

This equivalence can be made rigorous (see [198] for more details) as well as the fact that it suffices (at least asymptotically as $\varepsilon\to 0$) to maximize the denominator.

First note that

$$\sup_{0<h<\frac{\varepsilon n_L}{|c_1|}} h\left(\varepsilon^2 - c_1^2\left(\frac{h}{n_L}\right)^{2\alpha}\right) = \varepsilon^2\frac{2\alpha}{(1+2\alpha)^{1+\frac{1}{2\alpha}}}\left(\frac{\varepsilon}{|c_1|}\right)^{\frac{1}{\alpha}} n_L$$

attains a maximum at

$$\widetilde{h}(\varepsilon) = \left(\frac{\varepsilon}{|c_1|}\right)^{\frac{1}{\alpha}}\frac{n_L}{(1+2\alpha)^{\frac{1}{2\alpha}}} \tag{9.86}$$

so that we are led to set

$$h^*(\varepsilon) = \text{lower nearest neighbor of } \widetilde{h}(\varepsilon) \text{ in } \mathcal{H} = \mathbf{h}\lceil \mathbf{h}/\widetilde{h}(\varepsilon)\rceil^{-1}. \tag{9.87}$$

When $\varepsilon\to 0$, $\widetilde{h}(\varepsilon)\to 0$ and $h^*(\varepsilon)\sim\widetilde{h}(\varepsilon)$, but of course it remains *a priori* suboptimal at finite range so that

$$\inf_{RMSE(\widehat{I}_M^{MLMC})\le\varepsilon} \mathrm{Cost}(\widehat{I}_M^{MLMC}) \le \bar{\kappa}\frac{(1+2\alpha)^{1+\frac{1}{2\alpha}}}{2\alpha}\frac{|c_1|^{\frac{1}{\alpha}}}{n_{_L}}\frac{\left(\sum_{1\le i\le L}\sqrt{\kappa_i}\,\sigma(i,h^*(\varepsilon))\right)^2}{\varepsilon^{2+\frac{1}{\alpha}}}. \tag{9.88}$$

STEP 3 (*Final optimization* $L = L(\varepsilon) \to +\infty$). This step is devoted to the minimization of the upper-bound (9.88) with respect to the depth L under the additional assumption $(VE)_\beta$ (see (9.55)), which provides a non-asymptotic control of the variance of the Y_h.

$\triangleright$ *Non-asymptotic control of the effort.* Following the lines of the weighted framework, after replacing all the weights $\mathbf{W}_i$ by 1, we get for every $h \in \mathcal{H}$, keeping in mind that $\kappa_i = n_i(1+\frac{1}{N})$ for $i \in \{2,\dots,L\}$,

$$\begin{aligned}\left(\sum_{i=1}^{L}\sqrt{\kappa_i}\,\sigma(i,h)\right)^2 &\le \mathrm{Var}(Y_h)\left(1+\theta h^{\frac{\beta}{2}}(1+N^{-1})^{\frac{1}{2}}(N-1)^{\frac{\beta}{2}}\sum_{i=2}^{L}N^{(i-1)\frac{1-\beta}{2}}\right)^2\\ &= \mathrm{Var}(Y_h)\left(1+\theta h^{\frac{\beta}{2}}a_{_N}\frac{1-N^{\frac{1-\beta}{2}(L-1)}}{1-N^{\frac{1-\beta}{2}}}\right)^2\end{aligned} \tag{9.89}$$

with

$$a_{_N} = (1+N^{-1})^{\frac{1}{2}}(N-1)^{\frac{\beta}{2}}N^{\frac{1-\beta}{2}} \quad \text{and the convention} \quad \frac{1-1^{L-1}}{1-1} = L-1 \tag{9.90}$$

corresponding to the case $\beta = 1$ ([3])

$\triangleright$ *Optimization of the depth* $L = L(\varepsilon)$. One checks that the upper-bound (9.88) of the complexity formally goes to 0 as $L \uparrow +\infty$ for a fixed $\varepsilon > 0$ since $n_{_L} = N^{L-1} \uparrow +\infty$. As in the weighted setting, the depth L is limited by the constraint on the bias parameter $h^*(\varepsilon) \le \mathbf{h}$. This constraint implies, owing to (9.87), that

$$L \le L_{\max}(\varepsilon) = 1 + \frac{\log\left(\left(\frac{|c_1|}{\varepsilon}\sqrt{1+2\alpha}\right)^{\frac{1}{\alpha}}\mathbf{h}\right)}{\log(N)}.$$

However, *unlike the ML2R estimator*, the exponent $2+\frac{1}{\alpha}$ of the (prescribed) *RMSE* ε appearing in the upper-bound of the complexity does not depend on the depth L, so it seems natural to try to go deeper in the optimization. This can be achieved in different manners (see *e.g.* [141]) depending on what is assumed to be a variable or a fixed parameter. We shall carry on with our approach which lets the bias parameter h vary in $\mathcal{H}$.

Based on the formula (9.88) of the complexity, the expression of the effort (9.89) and using the formula (9.86) of $\widetilde{h}(\varepsilon)$ (before projection on $\mathcal{H}$), the minimization problem

[3]Note that in the nested Monte Carlo framework, $\kappa_i = n_i$, so that $a_{_N} = (N-1)^{\frac{\beta}{2}}N^{\frac{1-\beta}{2}}$.

$$\min_{1\le L\le L_{\max}(\varepsilon)} \frac{\left(\sum_{1\le i\le L}\sqrt{\kappa_i}\sigma(i,\widetilde{h}(\varepsilon))\right)^2}{n_L} \tag{9.91}$$

is dominated by

$$\sup_{h\in\mathcal{H}}\mathrm{Var}(Y_h)\min_{1\le L\le L_{\max}}\left(N^{-\frac{L-1}{2}}+\theta\left(\frac{\varepsilon}{|c_1|\sqrt{1+2\alpha}}\right)^{\frac{\beta}{2\alpha}}a_N N^{-\frac{1-\beta}{2}(L-1)}\frac{1-N^{\frac{1-\beta}{2}(L-1)}}{1-N^{\frac{1-\beta}{2}}}\right)^2, \tag{9.92}$$

still with our convention when $\beta = 1$, where a_N is given by (9.90).

In (9.92) the function to be minimized in L is not decreasing (where L is viewed as a real variable) so that saturating the constraint on L, *e.g.* by the condition $\widetilde{h}(\varepsilon)\le \mathbf{h}$, may not be optimal.

Lemma 9.4 *The function of L, viewed as a* $[1,+\infty)$*-valued real variable, to be minimized in* (9.92)*, attains its minimum at*

$$\Lambda^*(\varepsilon)=1+\frac{1}{\log N}\left(\frac{1}{\alpha}\log\left(\frac{|c_1|\sqrt{1+2\alpha}}{\varepsilon}\right)+\frac{2}{\beta}\log\left(\frac{N^{\frac{1-\beta}{2}}-1}{\frac{1-\beta}{2}}\,\frac{1}{2\theta(1+N^{-1})^{\frac12}(N-1)^{\frac{\beta}{2}}N^{\frac{1-\beta}{2}}}\right)\right)_+ \tag{9.93}$$

with the convention $\dfrac{N^{\frac{1-\beta}{2}}-1}{\frac{1-\beta}{2}}=\log N$ *if* $\beta=1$.

► **Exercise.** Prove the lemma. [Hint: consider directly the function between brackets and inspect the three cases $\beta\in(0,1)$, $\beta=1$ and $\beta>1$.]

At this stage we define the optimized depth of the *MLMC* simulation for a prescribed *RMSE* ε by

$$L^*(\varepsilon)=\lceil \Lambda^*(\varepsilon)\wedge L_{\max}(\varepsilon)\rceil. \tag{9.94}$$

▷ *Optimal size of the simulation.* The global size $M^*(\varepsilon)$ of the simulation is obtained by saturating the *RMSE* constraint, *i.e.* $\|\widehat{I}_M^{MLMC}-I_0\|_2\le\varepsilon$ with $h=\widetilde{h}(\varepsilon)$, using that $\|\widehat{I}_M^{MLMC}-I_0\|_2^2=\frac{\mathrm{Var}(\widehat{I}_1^{MLMC})}{M^*(\varepsilon)}+\mathrm{Bias}(\widehat{I}_M^{MLMC})^2$. We get

$$M^*(\varepsilon)=\frac{\mathrm{Var}(\widehat{I}_1^{MLMC})}{\varepsilon^2-\mathrm{Bias}(\widehat{I}_M^{MLMC})^2}.$$

Plugging the value of the variance (9.82) and the bias (9.81) into this equation finally yields

$$M^*(\varepsilon)=\left\lceil\left(1+\frac{1}{2\alpha}\right)\frac{\mathrm{Var}(\widehat{I}_1^{MLMC})}{\varepsilon^2}\right\rceil=\left\lceil\left(1+\frac{1}{2\alpha}\right)\frac{q^{*,\dagger}\sum_{1\le i\le L(\varepsilon)}\sqrt{\kappa_i}\,\sigma(i,h^*(\varepsilon))}{\varepsilon^2}\right\rceil \tag{9.95}$$

which can in turn be re-written in an implementable form (see Table 9.2 in Practitioner's corner) with the available formulas or bounds for κ_i and $\sigma(i,h)$ and $h^*(\varepsilon)$.

At this stage, it remains to inspect again the same three cases, depending on the parameter β ($\beta > 1$, $\beta = 1$ and $\beta \in (0, 1)$) to obtain the following theorem.

Theorem 9.3 (*MLMC* **estimator, see [107, 198]**) *Let* $n_i = N^{i-1}$, $i = 1, \dots, L$, $N \in \mathbb{N} \setminus \{0, 1\}$. *Assume* $(SE)_\beta$ *and* $\big(WE\big)_1^\alpha$. *Let* $\theta = \theta_{\mathbf{h}} = \sqrt{\frac{V_1}{\mathrm{Var}(Y)}}$.
(a) The MLMC estimator $(\widehat{I}_M^{MLMC})_{M\geq 1}$ *satisfies, as* $\varepsilon \to 0$,

$$\inf_{\substack{h\in\mathcal{H},\, q\in\mathcal{S}_L,\, L,\, M\geq 1\\ \|\widehat{I}_M^{MLMC} - I_0\|_2^2\leq\varepsilon^2}} \mathrm{Cost}\big(\widehat{I}_M^{MLMC}\big) \precsim K^{MLMC}\frac{\mathrm{Var}(Y_{\mathbf{h}})}{\varepsilon^2} \times \begin{cases} 1 & \text{if } \beta > 1,\\ \big(\log(1/\varepsilon)\big)^2 & \text{if } \beta = 1,\\ \varepsilon^{-\frac{1-\beta}{\alpha}} & \text{if } \beta < 1,\end{cases}$$

where $K^{MLMC} = K^{MLMC}(\alpha, \beta, \theta, \mathbf{h}, c_1, V_1)$.
(b) The rates in the right-hand side in (a) are achieved by setting in the definition (9.80) *of the estimator* $\widehat{I}_M^{MLMC}$, $h = h^*(\varepsilon, L^*(\varepsilon))$, $q = q^*(\varepsilon)$, $R = R^*(\varepsilon)$ *and* $M = M^*(\varepsilon)$.

Closed forms for the depth $L^*(\varepsilon)$, *the bias parameter* $h^* = h^*(\varepsilon, L^*(\varepsilon)) \in \mathcal{H}$, *the optimal allocation vector* $q^*(\varepsilon)$ *and the global size* $M^*(\varepsilon)$ *of the simulation are reported in Table 9.2 below (see Practitioner's corner hereafter for these formulas and their variants).*

Remarks. • The *MLMC* estimator achieves the rate ε^{-2} of an unbiased simulation as soon as $\beta > 1$ (fast strong rate) under lighter conditions than *ML2R* since it only requires a first-order weak error expansion.

• When $\beta = 1$, the cost of the *MLMC* estimator for a prescribed $RMSE$ is $o(\varepsilon^{-2+\eta})$ for every $\eta > 0$ but it is slower by a factor $\log\left(\frac{1}{\varepsilon}\right)$ than the weighted *ML2R* Multilevel estimator. Keep in mind than for $\varepsilon = 1\text{bp} = 10^{-4}$, $\log\left(\frac{1}{\varepsilon}\right) \simeq 6.9$.

• When $\beta \in (0, 1)$, the simulation cost is no longer close to unbiased simulation since $\varepsilon^{-\frac{1-\beta}{\alpha}} \to +\infty$ as $\varepsilon \to 0$ at a polynomial rate while the weighted *ML2R* estimator remains close to the unbiased framework, see Theorem 9.1 and the comments that follow. In that setting, which is great interest in derivative pricing since it corresponds, for example, to barrier options or γ-hedging, the weighted *ML2R* clearly outperforms the regular *MLMC* estimator.

Proof of Theorem 9.3. The announced rates can be obtained by saturating the value of h at $\mathbf{h}$ *i.e.* by replacing $\Lambda^*(\varepsilon)$ in (9.94) by $+\infty$. Then the proof follows the lines of that of Theorem 9.1. At some place one has to use that

$$\begin{aligned}\mathrm{Var}(Y_{h^*(\varepsilon)}) &\leq \big(\sigma(Y_{h^*(\varepsilon)} - Y_{\mathbf{h}}) + \sigma(Y_{\mathbf{h}})\big)^2\\ &\leq \mathrm{Var}(Y_{\mathbf{h}})\big(1 + \sqrt{V_1}(\mathbf{h} - h^*(\varepsilon))^{1/2}\big)^2 \leq \mathrm{Var}(Y_{\mathbf{h}})\big(1 + \sqrt{V_1}\mathbf{h}^{1/2}\big)^2.\end{aligned}$$

Details are left to the reader. ◇

Table 9.2 Parameters of the MLMC estimator (standard framework)

$\underline{n}$	$n_i = N^{i-1},\ i = 1, \ldots, L$ (convention $n_0 = n_0^{-1} = 0$)
$L = L^*(\varepsilon)$	$\min\left(1 + \left\lceil \dfrac{\log\left((1+2\alpha)^{\frac{1}{2\alpha}} \left(\frac{\lvert c_1\rvert}{\varepsilon}\right)^{\frac{1}{\alpha}} \mathbf{h}\right)}{\log(N)} \right\rceil, \lceil \Lambda(\varepsilon) \rceil\right)$
$\Lambda(\varepsilon)$	$\left(\dfrac{\log\left((1+2\alpha)^{\frac{1}{2\alpha}} \left(\frac{\lvert c_1\rvert}{\varepsilon}\right)^{\frac{1}{\alpha}}\right)}{\log(N)} + \dfrac{2}{\beta \log N} \log\left(\dfrac{N^{\frac{1-\beta}{2}}-1}{\frac{1-\beta}{2}} \dfrac{1}{2\theta N^{\frac{1-\beta}{2}} (1+N^{-1})^{\frac{1}{2}} (N-1)^{\frac{\beta}{2}}}\right)\right)_+$ with the convention $\frac{N^0-1}{0} = \log N$ when $\beta = 1$.
$h = h^*(\varepsilon)$	$\mathbf{h} \Big/ \left\lceil \mathbf{h}(1+2\alpha)^{\frac{1}{2\alpha}} \left(\dfrac{\lvert c_1\rvert}{\varepsilon}\right)^{\frac{1}{\alpha}} N^{-(L-1)} \right\rceil$
$q = q^*(\varepsilon)$	$q_1(\varepsilon) = \dfrac{1}{q_\varepsilon^\dagger},\quad q_j(\varepsilon) = \dfrac{\theta h^{\frac{\beta}{2}} \left(n_{j-1}^{-1} - n_j^{-1}\right)^{\frac{\beta}{2}}}{q_\varepsilon^\dagger \sqrt{n_{j-1} + n_j}},\quad j = 2, \ldots, L,$ with $q_\varepsilon^\dagger$ s.t. $\sum_{j=1}^{L} q_j(\varepsilon) = 1$ and $\theta = \sqrt{\dfrac{V_1}{\mathrm{Var}(Y_h)}}$.
$M = M^*(\varepsilon)$	$\left\lceil \left(1 + \dfrac{1}{2\alpha}\right) \dfrac{\mathrm{Var}(Y_h)\, q_\varepsilon^\dagger \left(1 + \theta h^{\frac{\beta}{2}} \sum_{j=2}^{L} \left(n_{j-1}^{-1} - n_j^{-1}\right)^{\frac{\beta}{2}} \sqrt{n_{j-1} + n_j}\right)}{\varepsilon^2} \right\rceil$

ℵ Practitioner's corner

Most recommendations are similar to that made for the *ML2R* estimator. We focus in what follows on specific features of the regular *MLMC* estimators. The parameters have been designed based on a complexity of fine level i given by $\kappa(n_i + n_{i-1})$. When this complexity is of the form κn_i (like in nested Monte Carlo), the terms $\sqrt{n_j + n_{j-1}}$ should be replaced by $\sqrt{n_j}$.

▷ ***Payoff dependent parameters for MLMC estimator***

We set $n_i = N^{i-1}, i = 1, \ldots, L$. The values of the (optimized) parameters are listed in Table 9.2 (* superscripts are often removed for simplicity).

Let us make few additional remarks:

- Note that, by contrast with the *ML2R* framework, there is no *a priori* reason to choose $N = 2$ as a root when $\beta \in (0, 1)$.
- The value of the depth $L^*(\varepsilon)$ proposed in Table 9.2 is sharper than the value $\widetilde{L}^*(\varepsilon) = 1 + \left\lceil \log\left((1+2\alpha)^{\frac{1}{2\alpha}} \left(\frac{\lvert c_1\rvert}{\varepsilon}\right)^{\frac{1}{\alpha}} \mathbf{h}\right) / \log(N) \right\rceil$ given in [198]. However, the impact of this sharper optimized depth on the numerical simulations remains limited and $\widetilde{L}^*(\varepsilon)$ is a satisfactory choice in practice (it corresponds to setting

$\Lambda^*(\varepsilon) = +\infty$): in fact, in most situations, one observes that $L^*(\varepsilon) = \widetilde{L}^*(\varepsilon)$. In any case the optimized depth grows as $O\Big(\log(1/\varepsilon)\Big)$ when $\varepsilon \to 0$.

▷ ***Numerical estimation of the parameters*** $\mathrm{Var}(Y_h)$ $(h = h^*(\varepsilon))$, V_1 ***and*** θ

– The same remark about possible variants of the complexity of the simulation of $Y_{\frac{h}{n_i}} - Y_{\frac{h}{n_{i-1}}}$ can be made for the regular *MLMC* estimator with the same impact on the computation of the allocation vector $q(\varepsilon)$.

– Like with the *ML2R* estimator, one needs to estimate $\mathrm{Var}(Y_h)$, this time with $h = h^*(\varepsilon)$, and the constant V_1 in $(VE)_\beta$ in order to have a proxy of $\mathrm{Var}(Y_{h^*(\varepsilon)})$ and $\theta = \sqrt{\frac{V_1}{\mathrm{Var}(Y_{h^*(\varepsilon)})}}$, both used to specify the parameters reported in Table 9.2. The estimation procedures are the same as that developed in Sect. 9.5.1 in Practitioner's corner I (see (9.70) and (9.71)).

▷ **Calibration of the parameter** c_1

This is an opportunity of improvement offered by the *MLMC* estimator: if the coefficient α is known, *e.g.* by theoretical means, it is possible to estimate the coefficient c_1 in the first-order weak error expansion $\big(WE\big)_1^\alpha$. First we approximate by considering the following proxies, defined for $h_0, h_0' \in \mathcal{H}$, $0 < h_0' < h_0$, by

$$\widehat{c}_1(h_0, h_0') = \frac{\mathbb{E}\,Y_{h_0} - \mathbb{E}\,Y_{h_0'}}{h_0^\alpha - (h_0')^\alpha} \simeq c_1$$

if h_0 and h_0' are small (but not too small). Nevertheless, like for the estimation of V_1 (see Practitioner's corner I in Sect. 9.5.1), we adopt a conservative strategy which naturally leads us to consider not too small h_0 and h_0', namely, *like for the ML2R estimator,*

$$h_0 = \frac{\mathbf{h}}{5} \quad \text{and} \quad h_0' = \frac{h_0}{2} = \frac{\mathbf{h}}{10}.$$

Then, it remains to replace $\mathbb{E}\left(Y_{h_0} - Y_{\frac{h_0'}{2}}\right)$ by its usual estimator to obtain the (obviously biased) estimator of the form

$$\widehat{c}_1(h_0, h_0', m) = \frac{(h_0)^{-\alpha}}{(1-2^{-\alpha})}\frac{1}{m}\sum_{k=1}^{m} Y_{h_0}^k - Y_{h_0'}^k,$$

where $\left(Y_{h^*(\varepsilon)}^k, Y_{\frac{h^*(\varepsilon)}{2}}^k\right)_{k\geq 1}$ are i.i.d. copies of $\left(Y_{h^*(\varepsilon)}, Y_{\frac{h^*(\varepsilon)}{2}}\right)$ and $m \ll M$ (m is *a priori* small with respect to the size M of the master simulation).

One must be aware that this opportunity to estimate c_1 (which has no counterpart for *MLL2R*) is also a factor of instability for the *MLMC* estimators, if c_1 is not estimated or not estimated with enough accuracy. From this point of view the *MLMC* estimators are less robust than the *ML2R* estimators.

▷ **Design of confidence intervals**

The conditions of application and the way to design confidence intervals for the *MLMC* estimators when the *RMSE* $\varepsilon \to 0$ is the same as that described in Practitioner's corner II for weighted multilevel estimators in Sect. 9.5.1 (see Eq. (9.73)).

9.5.3 Additional Comments and Provisional Remarks

▷ *Heuristic recommendations for Brownian diffusions.* We can give few additional recommendations on how to implement the above Multilevel methods for Brownian diffusions. It is clear in view of the results of Chaps. 7 and 8 that:

- for vanilla Lipschitz payoffs, one may apply the method with $\alpha = 1$ and $\beta = 1$;
- for path-dependent payoffs of lookback type (without barrier), the same holds true;
- for barrier-like payoffs, $\beta = \frac{1}{2}$ is appropriate in practice, if the diffusion has polynomial moments at any order and $\alpha = \frac{1}{2}$.

But these recommendations are partly heuristic (or correspond to critical cases), especially for path-dependent functionals.

▷ *Extension to Lévy-driven diffusions.* The weak expansion error at order 1 for Lévy-driven diffusion processes discretized by an Euler scheme was established in [74] in view of the implementation of a multilevel simulation. For other results about the discretization schemes of Lévy-driven diffusion processes, we refer to classical papers like [157, 250].

▷ *More about the strong convergence assumption* $(VE)_\beta$. A careful reading of the chapter emphasizes that the assumptions

$$(VE)_\beta \;\equiv\; \mathrm{Var}(Y_h - Y_{h'}) \le V_1 |h - h'|^\beta \quad \text{or} \quad (SE)_\beta \;\equiv\; \|Y_h - Y_{h'}\|_2^2 \le V_1 |h - h'|^\beta$$

are the key to controlling the effort of the multilevel estimator after the optimization of the allocation vector $(q_i)_i$, see *e.g.* (9.57) still in the weighted Multilevel section. Using the seemingly more natural $(SE)_\beta^0$: $\|Y_h - Y_0\|_2^2 \le \bar{V}_1 h^\beta$ (see [198]) yields a less efficient design of the multilevel estimators, not to speak of the fact that V_1 is easier to estimate in practice than $\bar{V}_1$. But the virtue of this choice goes beyond this calibrating aspect. In fact, as emphasized in [32], assumptions $(VE)_\beta$ and $(SE)_\beta$ may be satisfied in situations where Y_h *does not converge in* L^2 *toward* Y_0 because the underlying scheme itself does not converge toward the continuous time process. This is the case for the binomial tree (see *e.g.* [59]) toward the Black–Scholes model. More interestingly, the author takes advantage of this situation to show that the Euler discretization scheme of a Lévy driven diffusion can be *wienerized* and still satisfies $(SE)_\beta$ for a larger β: the *wienerization* of the small jumps of a Lévy process consists in replacing the increments of the compensated small enough jump process over a time interval $[t, t + \delta t]$ by increments $c(B_{t+\Delta t} - B_t)$, where B is a standard Brownian motion and c is chosen to equalize the variances (B is independent of the existing Brownian and the large jump components of the Lévy process, if any). The wienerization trick was introduced in [14, 66] and makes this modified Euler scheme simulable, which is usually not the case for the standard Euler scheme if the process jumps infinitely often on each nontrivial time interval with non-summable jumps.

$\triangleright$ *Nested Monte Carlo.* As for the strong convergence $(SE)_\beta$ and weak error expansion $\big(WE\big)^\alpha_R$, we refer to Practitioner's corner II, Sect. 9.5.1.

$\triangleright$ *Limiting behavior.* The asymptotics of multilevel estimators, especially the *CLT*, has been investigated in various frameworks (Brownian diffusions, Lévy-driven diffusions, etc) in several papers (see [34, 35, 75]). See also [112] for a general approach where both weighted and regular Multilevel estimators are analyzed: a *SLLN* and a *CLT* are established for multilevel estimators as the *RMSE* $\varepsilon \to 0$.

9.6 Antithetic Schemes (a Quest for $\beta > 1$)

9.6.1 *The Antithetic Scheme for Brownian Diffusions: Definition and Results*

We consider our standard Brownian diffusion *SDE* (7.1) with drift b and diffusion coefficient σ driven by a q-dimensional standard Brownian motion W defined on a probability space $(\Omega, \mathcal{A}, \mathbb{P})$ with $q \geq 2$. In such a framework, we saw that the Milstein scheme cannot be simulated efficiently due to the presence of Lévy areas induced by the rectangular terms coming out in the second-order term of the scheme. This seems to make impossible it to reach the unbiased setting "$\beta > 1$" since the Euler scheme corresponds, as we saw above, to the critical case $\beta = 1$ (scheme of order $\frac{1}{2}$).

First we introduce a *truncated Milstein scheme* with step $h = \frac{T}{n}$. We start from the definition of the discrete time Milstein scheme as defined by (7.41). The scheme is not considered as simulable when $q \geq 2$ because of the Lévy areas

$$\int_{t_k^n}^{t_{k+1}^n} (W_s^i - W_{t_k^n}^i)dW_s^j$$

for which no efficient method of simulation is known when $i \neq j$. On the other hand a simple integration by parts shows that, for every $i \neq j$ and every $k = 0, \dots, n-1$,

$$\int_{t_k^n}^{t_{k+1}^n} (W_s^i - W_{t_k^n}^i)dW_s^j + \int_{t_k^n}^{t_{k+1}^n} (W_s^j - W_{t_k^n}^j)dW_s^i = (W_{t_{k+1}^n}^i - W_{t_k^n}^i)(W_{t_{k+1}^n}^j - W_{t_k^n}^j),$$

whereas $\displaystyle\int_{t_k^n}^{t_{k+1}^n} (W_s^i - W_{t_k^n}^i)dW_s^i = \frac{1}{2}\big((\Delta W_{t_{k+1}^n}^i - W_{t_k^n}^i)^2 - T/n\big)$ for $i = 1, \dots, q$.

The idea is to symmetrize the "rectangular" Lévy areas, *i.e.* to replace them by half of their sum with their symmetric counterpart, namely $\frac{1}{2}(W^i_{t^n_{k+1}} - W^i_{t^n_k})(W^j_{t^n_{k+1}} - W^j_{t^n_k})$. This yields the *truncated Milstein scheme* defined as follows:

$$\begin{aligned}\breve{X}^n_0 &= X_0,\\ \breve{X}^n_{t^n_{k+1}} &= \breve{X}^n_{t^n_k} + h\, b\big(t^n_k, \breve{X}^n_{t^n_k}\big) + \sqrt{h}\,\sigma\big(t^n_k, \breve{X}^n_{t^n_k}\big)\\ &\quad + \sum_{1\le i\le j\le q} \breve{\sigma}_{ij}(\breve{X}^n_{t^n_k})\big(\Delta W^i_{t^n_{k+1}} \Delta W^j_{t^n_{k+1}} - \mathbf{1}_{\{i=j\}} h\big),\end{aligned} \tag{9.96}$$

where $t^n_k = \frac{kT}{n}$, $\Delta W_{t^n_{k+1}} = W_{t^n_{k+1}} - W_{t^n_k}$, $k = 0, \ldots, n-1$ and

$$\begin{aligned}\breve{\sigma}_{ij}(x) &= \frac{1}{2}\big(\partial\sigma_{.i}\, \sigma_{.j}(x) + \partial\sigma_{.j}\, \sigma_{.i}(x)\big),\ 1 \le i < j \le q,\\ \breve{\sigma}_{ii}(x) &= \frac{1}{2}\partial\sigma_{.i}\, \sigma_{.i}(x),\ 1 \le i \le q\end{aligned}$$

(where $\partial\sigma_{.i}\, \sigma_{.j}$ is defined by (7.42)).

This scheme can clearly be simulated, but the analysis of its strong convergence rate, *e.g.* in L^2 when $X_0 \in L^2$, shows a behavior quite similar to the Euler scheme, *i.e.*

$$\Big\| \max_{k=0,\ldots,n} \big|\breve{X}^n_{t^n_k} - X_{t^n_k}\big| \Big\|_2 \le C_{b,\sigma,T}\sqrt{\frac{T}{n}}\big(1 + \|X_0\|_2\big)$$

if b and σ are $\mathcal{C}^1_{\mathrm{Lip}}$. In terms of weak error it also behaves like the Euler scheme: under similar smoothness assumptions on b, σ and f, or ellipticity on σ, it satisfies a first-order expansion in $h = \frac{T}{n}$: $\mathbb{E}\, f(X_T) - \mathbb{E}\, f(\breve{X}^n_T) = c_1 h + O(h^2)$, *i.e.* $\big(WE\big)^1_1$ (see [110]). Under additional assumptions, it is expected that higher-order weak error expansion $\big(WE\big)^1_R$ holds true.

Rather than trying to search for a higher-order simulable scheme, the idea introduced in [110] is to combine several such truncated Milstein schemes at different scales in order to make the fine level i behave *as if* the scheme satisfies $(SE)_\beta$ for a $\beta > 1$. To achieve this, the fine scheme of the level is duplicated into two fine schemes with the same step but based on *swapped* Brownian increments. Let us be more specific: the above scheme with step $h = \frac{T}{n}$ can be described as a homogeneous Markov chain associated to the mapping $\widetilde{\mathcal{M}} = \widetilde{\mathcal{M}}_{b,\sigma} : \mathbb{R}^d \times \mathcal{H} \times \mathbb{R}^q \to \mathbb{R}^d$ as follows

$$\breve{X}^n_{t^n_{k+1}} = \widetilde{\mathcal{M}}\big(\breve{X}^n_{t^n_k}, h, \Delta W_{t^n_{k+1}}\big),\ k = 0, \ldots, n-1,\ \breve{X}^n_0 = X_0.$$

We consider a first scheme $\big(\breve{X}^{2n,[1]}_{t^{2n}_k}\big)_{k=0,\ldots,2n}$ with step $\frac{h}{2} = \frac{T}{2n}$ which reads on two time steps

$$\breve{X}^{2n,[1]}_{t^{2n}_{2k+1}} = \widetilde{\mathcal{M}}\Big(\breve{X}^{2n,[1]}_{t^{2n}_{2k}}, \frac{h}{2}, \Delta W_{t^{2n}_{2k+1}}\Big),$$
$$\breve{X}^{2n,[1]}_{t^{2n}_{2(k+1)}} = \widetilde{\mathcal{M}}\Big(\breve{X}^{2n,[1]}_{t^{2n}_{2k+1}}, \frac{h}{2}, \Delta W_{t^{2n}_{2(k+1)}}\Big),\ k = 0, \dots, n-1.$$

We consider a second scheme with step $\frac{T}{2n}$, denoted by $\big(\breve{X}^{2n,[2]}_{t^{2n}_k}\big)_{k=0,\dots,2n}$, identical to the first one, except that the Brownian increments are *pairwise swapped*, *i.e.* the increments $\Delta W_{t^{2n}_{2k+1}}$ and $\Delta W_{t^{2n}_{2(k+1)}}$ are *swapped*. It reads

$$\breve{X}^{2n,[2]}_{t^{2n}_{2k+1}} = \widetilde{\mathcal{M}}\Big(\breve{X}^{2n,[2]}_{t^{2n}_{2k}}, \frac{h}{2}, \Delta W_{t^{2n}_{2(k+1)}}\Big),$$
$$\breve{X}^{2n,[2]}_{t^{2n}_{2(k+1)}} = \widetilde{\mathcal{M}}\Big(\breve{X}^{2n,[2]}_{t^{2n}_{2k+1}}, \frac{h}{2}, \Delta W_{t^{2n}_{2k+1}}\Big),\ k = 0, \dots, n-1.$$

The following theorem, established in [110] in the autonomous case ($b(t,x) = b(x)$, $\sigma(t,x) = \sigma(x)$), makes precise the fact that it produces a meta-scheme satisfying $(SE)_\beta$ with $\beta = 2$ (and a root $N = 2$) in a multilevel scheme.

Theorem 9.4 *Assume* $b \in \mathcal{C}^2(\mathbb{R}^d, \mathbb{R}^d)$, $\sigma \in \mathcal{C}^2(\mathbb{R}^d, \mathcal{M}(d,q,\mathbb{R}))$ *with bounded existing partial derivatives of* b, σ, $\breve{\sigma}$ *(up to order 2).*

(*a*) Smooth payoff. *Let* $f \in \mathcal{C}^2(\mathbb{R}^d, \mathbb{R})$ *also with bounded existing partial derivatives (up to order 2). Then, there exists a real constant* $C_{b,\sigma,T} > 0$

$$\Big\| f(\breve{X}^n_T) - \frac{1}{2}\big(f(\breve{X}^{2n,[1]}_T) + f(\breve{X}^{2n,[2]}_T)\big)\Big\|_2 \le C_{b,\sigma,T}\frac{T}{n}\big(1 + \|X_0\|_2\big), \tag{9.97}$$

i.e. $(SE)_\beta$ *is satisfied with* $\beta = 2$.

(*b*) Almost smooth payoff. *Assume that* f *is Lipschitz continuous on* $\mathbb{R}^d$ *such that its first two order partial derivatives exist outside a Lebesgue negligible set* $\mathcal{N}_0$ *of* $\mathbb{R}^d$ *and are uniformly bounded. Assume moreover that the diffusion* $(X_t)_{t\in[0,T]}$ *satisfies*

$$\overline{\lim_{\varepsilon\to 0}}\, \varepsilon^{-1}\mathbb{P}\Big(\inf_{z\in\mathcal{N}_0} |X_T - z| \le \varepsilon\Big) < +\infty. \tag{9.98}$$

Then

$$\Big\| f(\breve{X}^n_T) - \frac{1}{2}\big(f(\breve{X}^{2n,[1]}_T) + f(\breve{X}^{2n,[2]}_T)\big)\Big\|_2 \le C_{b,\sigma,T,\eta}\Big(\frac{T}{n}\Big)^{\frac{3}{4}-\eta}\big(1 + \|X_0\|_2\big) \tag{9.99}$$

for every $\eta \in (0, \frac{3}{4})$, *i.e.* $(SE)_\beta$ *is satisfied for every* $\beta \in (0, \frac{3}{2})$.

For detailed proofs, we refer to Theorems 4.10 and 5.2 in [110]. At least in a one-dimensional framework, the above condition (9.98) is satisfied if the distribution of X_T has a bounded density.

$\aleph$ **Practitioner's corner: Application to multilevel estimators with root** $N = 2$

The principle is rather simple at this stage. Let us deal with the simplest case of a vanilla payoff (or 1-marginal) $Y_0 = f(X_T)$ and assume that f is such that $(SE)_\beta$ is satisfied for some $\beta \in (1, 2]$.

▷ *Coarse level.* On the coarse (first) level $i = 1$, we simply set $Y_h^{(1)} = f\big(\check{X}_T^{n,(1)}\big)$, where the truncated Milstein scheme is driven by a q-dimensional Brownian motion $W^{(1)}$. Keep in mind that this scheme is of order $\frac{1}{2}$ so that coarse and fine levels will not be ruled by the same β.

▷ *Fine levels.* At each fine level $i \geq 2$, the regular difference $Y_{\frac{h}{n_i}}^{(i)} - Y_{\frac{h}{n_{i-1}}}^{(i)}$ (with $n_i = 2^{i-1}$ and $h \in \mathcal{H}$), is replaced by

$$\frac{1}{2}\big(Y_{\frac{h}{n_i}}^{[1],(i)} + Y_{\frac{h}{n_i}}^{[2],(i)}\big) - Y_{\frac{h}{n_{i-1}}}^{(i)},$$

where $Y_{\frac{h}{n_{i-1}}}^{(i)} = f\big(\check{X}_T^{nN^{i-2},(i)}\big)$, $Y_{\frac{h}{n_i}}^{[r],(i)} = f\big(\check{X}_T^{nN^{i-1},[r],(i)}\big)$, $r = 1, 2$. At each level, the three truncated antithetic Milstein schemes are driven by the same q-dimensional Brownian motions $W^{(i)}$, $i = 1, \ldots, R$ (or L, depending on the implemented type of multilevel meta-scheme). Owing to Theorem 9.4, it is clear that, under appropriate assumptions,

$$\Big\|\frac{1}{2}\big(Y_{\frac{h}{n_i}}^{[1],(i)} + Y_{\frac{h}{n_i}}^{[2],(i)}\big) - Y_{\frac{h}{n_{i-1}}}^{(i)}\Big\|_2^2 \leq V_1\Big(\frac{h}{n_i}\Big)^\beta \quad \text{with} \quad \beta = \Big(\frac{3}{2}\Big)^- \text{ or } \beta = 2.$$

As a second step, the optimization of the simulation parameter can be carried out following the lines of the case $\alpha = 1$ and $\beta = \frac{3}{2}$ or $\beta = 2$. However, the first level is ruled at a strong rate $\beta = 1$ which slightly modifies the definition of the sample allocations q_i in Table 9.1 for the weighted Multilevel estimator. Moreover, the complexity of fine level i is now of the form $\bar{\kappa}\,(2n_i + n_{i-1})/h$ instead of $\bar{\kappa}(n_i + n_{i-1})/h$ so that $\sqrt{n_i + n_{i-1}}$ should be replaced *mutatis mutandis* by $\sqrt{2n_i + n_{i-1}}$ in the same Table. Table 9.2 for regular Multilevel should of course be modified accordingly.

▷ *Parameter modifications for the antithetic estimator*:

1. New value for V_0: now $V_0 = \mathbb{E}\, Y_h = \mathbb{E}\, f(\check{X}_T^n)$ (were $h = h(\varepsilon)$ has been optimized of course).
2. Updated parameters of the estimator: see Table 9.3.

9.6.2 *Antithetic Scheme for Nested Monte Carlo (Smooth Case)*

For the sake of simplicity, we assume in this section that the root $N = 2$, but what follows can be extended to any integer $N \geq 2$ (see *e.g.* [113]). In a regular multilevel Nested Monte Carlo simulation, the fine level $i \geq 2$ – with or without weight – relies

Table 9.3 Parameters of the *ML2R* estimator (standard antithetic diffusion framework)

N	2 (in our described setting)
$R = R^*(\varepsilon)$	$\left\lceil \frac{1}{2} + \frac{\log(\widetilde{c}_\infty^{\frac{1}{\alpha}}\mathbf{h})}{\log(N)} + \sqrt{\left(\frac{1}{2} + \frac{\log(\widetilde{c}_\infty^{\frac{1}{\alpha}}\mathbf{h})}{\log(N)}\right)^2 + 2\frac{\log\left(\frac{\sqrt{1+4\alpha}}{\varepsilon}\right)}{\alpha\log(N)}}\right\rceil$
$h = h^*(\varepsilon)$	$\mathbf{h}$
$\underline{n}$	$n_i = N^{i-1},\ i = 1, \ldots, R$ (convention: $n_0 = n_0^{-1} := 0$)
$q = q^*(\varepsilon)$	$q_1(\varepsilon) = \frac{1}{q_\varepsilon^\dagger},\ q_j(\varepsilon) = \frac{\theta h^{\frac{\beta}{2}} \lvert\mathbf{W}_j^{(R)}(N)\rvert\left(n_{j-1}^{-1} - n_j^{-1}\right)^{\frac{\beta}{2}}}{q_\varepsilon^\dagger\sqrt{n_{j-1} + 2n_j}},\ j = 2, \ldots, R,$ with $q_\varepsilon^\dagger$ s.t. $\sum_{j=1}^{R} q_j(\varepsilon) = 1,\ \theta = \sqrt{\frac{V_1}{\breve{V}_h}}$ and $\breve{V}_h = \mathrm{Var}(\breve{Y}_h)$
$M = M^*(\varepsilon)$	$\left\lceil \left(1 + \frac{1}{2\alpha R}\right) \frac{\breve{V}_h q_\varepsilon^\dagger \left(1 + \theta h^{\frac{\beta}{2}} \sum_{j=2}^{R} \lvert\mathbf{W}_j^{(R)}(N)\rvert\left(n_{j-1}^{-1} - n_j^{-1}\right)^{\frac{\beta}{2}}\sqrt{n_{j-1} + 2n_j}\right)}{\varepsilon^2} \right\rceil$

on quantities of the form

$$Y_{\frac{h}{2}} - Y_h = f\left(\frac{1}{2K}\sum_{k=1}^{2K} F(X, Z_k)\right) - f\left(\frac{1}{K}\sum_{k=1}^{K} F(X, Z_k)\right)$$

where $h = \frac{1}{K} = h_i = \frac{1}{2^{i-2}K_0} \in \mathcal{H}$. If f is Lipschitz continuous, we know (see Proposition 9.2(*a*)) that $\left\|Y_h - Y_{\frac{h}{2}}\right\|_2^2 \le [f]_{\mathrm{Lip}} \frac{\mathrm{Var}\left(F(X,Z_1)\right)}{2} h$ so that $(SE)_\beta$ and, subsequently, $(VE)_\beta$ is satisfied with $\beta = 1$.

The *antithetic version of the fine level* $Y_h - Y_{\frac{h}{2}}$ that we present below seems to have been introduced independently in [55, 61, 140]; see also [109, 114]. It consists in replacing the estimator Y_h in a smart enough way by a random variable $\widetilde{Y}_h$ with the same expectation as Y_h, designed from the same X and the same innovations Z_k but closer to $Y_{\frac{h}{2}}$ in $L^2(\mathbb{P})$. This leads us to set

$$\widetilde{Y}_h = \frac{1}{2}\left[f\left(\frac{1}{K}\sum_{k=1}^{K} F(X, Z_k)\right) + f\left(\frac{1}{K}\sum_{k=K+1}^{2K} F(X, Z_k)\right)\right]. \tag{9.100}$$

It is obvious that $\mathbb{E}\,\widetilde{Y}_h = \mathbb{E}\,Y_h$, $h \in \mathcal{H}$. Now we can substitute in the fine levels of the Multilevel estimators (with or without weights) the quantities $Y_{\frac{h}{2}} - \widetilde{Y}_h$ in place of the quantities of the form $Y_{\frac{h}{2}} - Y_h$ *without modifying their bias.*

Proposition 9.3 *Assume that f is differentiable with a ρ-Hölder derivative f', $\rho \in (0,1]$, and that $F(X, Z_1) \in L^{2(1+\rho)}(\mathbb{P})$. Then, there exists a real constant $V_1 = V_1(f, \rho, F, X, Z)$ such that*

$$\forall\, h \in \mathcal{H}, \quad \big\|\widetilde{Y}_h - Y_{\frac{h}{2}}\big\|_2^2 \le V_1 h^{1+\rho}. \tag{9.101}$$

As a consequence, weighted and unweighted multilevel estimators (where Y_h is replaced by $\widetilde{Y}_h$), can be implemented with $\beta = 1 + \rho > 1$.

Proof. One derives from the elementary inequality

$$\Big| f\Big(\frac{a+b}{2}\Big) - \frac{1}{2}\big(f(a) + f(b)\big)\Big| \le [f']_{\rho,Hol}\frac{|b-a|^{1+\rho}}{2^\rho}$$

that

$$\begin{aligned}
\big\|\widetilde{Y}_h - Y_{\frac{h}{2}}\big\|_2^2 &\le \frac{[f']^2_{\rho,Hol}}{4^\rho}\mathbb{E}\left|\frac{1}{K}\sum_{k=1}^{K} F(X, Z_k) - \frac{1}{K}\sum_{k=K+1}^{2K} F(X, Z_k)\right|^{2(1+2\rho)} \\
&= \frac{[f']^2_{\rho,Hol}}{4^\rho}\frac{1}{K^{2(1+\rho)}}\mathbb{E}\left|\sum_{k=1}^{K} F(X, Z_k) - F(X, Z_{K+k})\right|^{2(1+\rho)} \\
&\le \frac{[f']^2_{\rho,Hol}}{4^\rho}\frac{C_{2(1+\rho)}}{K^{2(1+\rho)}}\mathbb{E}\left[\sum_{k=1}^{K}\big(F(X, Z_k) - F(X, Z_{K+k})\big)^2\right]^{1+\rho} \\
&= \frac{[f']^2_{\rho,Hol}}{4^\rho}\frac{C_{2(1+\rho)}}{K^{2(1+\rho)}}\left\|\sum_{k=1}^{K}\big(F(X, Z_k) - F(X, Z_{K+k})\big)^2\right\|_{1+\rho}^{1+\rho}
\end{aligned}$$

owing to the discrete time *B.D.G.* Inequality (see (6.49)) applied to the sequence $\big(F(X, Z_k) - F(X, Z_{K+k})\big)_{k\ge 1}$ of $\sigma(X, Z_1, \ldots, Z_n)$-adapted martingale increments. Then,
Minkowski's Inequality yields

$$\begin{aligned}
\big\|\widetilde{Y}_h - Y_{\frac{h}{2}}\big\|_2^2 &\le \frac{[f']^2_{\rho,Hol}}{4^\rho}\frac{C_{2(1+\rho)}}{K^{2(1+\rho)}}K^{1+\rho}\big\|F(X, Z_1) - F(X, Z_{K+1})\big\|_{2(1+\rho)}^{2(1+\rho)} \\
&= \frac{V_1}{K^{1+\rho}}.
\end{aligned}$$

This completes the proof. ◇

► **Exercises. 1.** Extend the above antithetic multilevel nested Monte Carlo method to locally ρ-Hölder continuous functions $f : \mathbb{R}^d \to \mathbb{R}$ satisfying

$$\forall\, x, y \in \mathbb{R}^d, \qquad \big|f(x) - f(y)\big| \le [f]_{\rho,loc}|x-y|^\rho\Big(1 + |x|^r + |y|^r\Big).$$

2. Update Table 9.3 to take into account the specific complexity of the antithetic nested Monte Carlo method.

3. Extend the antithetic approach to any root $N \geq 2$ by replacing Y_h in the generic expression $Y_{\frac{h}{N}} - Y_h$ of a fine level with root N by

$$\widetilde{Y}_h = \frac{1}{N}\sum_{n=1}^{N} f\left(\frac{1}{K}\sum_{k=1}^{K} F(X, Z_{(n-1)K+k})\right).$$

Remark (on convexity). If furthermore f is convex, it follows from the preceding that

$$\mathbb{E}\, Y_h = \mathbb{E}\, \widetilde{Y}_h \geq \mathbb{E}\, Y_{\frac{h}{N}}.$$

We know that $Y_h \to Y_0 = f\big(\mathbb{E}\left(F(X, Z_1) \,|\, X\right)\big)$ in $L^2(\mathbb{P})$ as $h \to 0$ in $\mathcal{H}$. It follows that, for every fixed $h \in \mathcal{H}$,

$$\mathbb{E}\, Y_h \geq \mathbb{E}\, Y_{\frac{h}{N^k}} \to \mathbb{E}\, Y_0 \quad \text{as} \quad k \to +\infty.$$

Consequently, for the regular *MLMC* estimator (with $h = h^*(\varepsilon)$, optimized bias parameter), one has (see (9.81)):

$$\mathbb{E}\, \widehat{I}_M^{MLMC} = \mathbb{E}\, Y_{\frac{h}{N^{L-1}}} \geq \mathbb{E}\, Y_0 = I_0$$

i.e. the regular *MLMC* estimator is *upper-biased*. This allows us to produce asymmetrical narrower confidence intervals (see Practitioner's corner in Sect. 9.5.2) since the bias $\mathbb{E}\, \widehat{I}_M^{MLMC} - I_0$ is always non-negative.

9.7 Examples of Simulation

9.7.1 *The Clark–Cameron System*

We consider the 2-dimensional *SDE* (in its integral form)

$$X_t^1 = \mu t + W_t^1, \quad X_t^2 = \sigma \int_0^t X_s^1\, dW_s^2, \quad t \in [0, T], \tag{9.102}$$

where $W = (W^1, W^2)$ is a 2-dimensional standard Brownian motion. This *SDE*, known as the *Clark–Cameron oscillator*, is an example of diffusion for which the strong convergence rate of the Euler scheme with step $\frac{T}{n}$ is exactly $O\Big(\sqrt{\frac{T}{n}}\Big)$. It means that the choice $\beta = 1$ is optimal. For this reason, it is part of the numerical

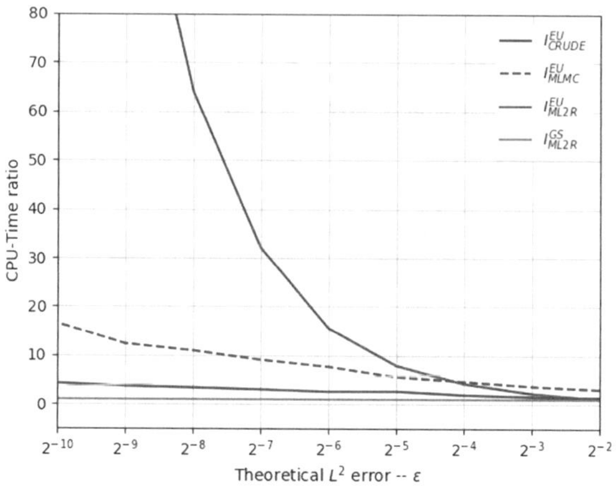

Fig. 9.1 CLARK–CAMERON OSCILLATOR: CPU-time ratio w.r.t. CPU-time of ML2R (in seconds) as a function of the prescribed RMSE ε (with D. Giorgi)

experiments carried out in [110] on weighted and unweighted multilevel estimators (see also [6] where other experiments are carried out with this model).

We want to compute

$$P(\mu, \sigma, T) = 10\,\mathbb{E}\cos(X_T^2)$$

for the following values of the parameters: $\mu = 1, \sigma = 1, T = 1$. Actually, a closed form exists for $\mathbb{E}\cos(X_T^2)$ for any admissible values of $\mu, \sigma >$ and $T > 0$, namely

$$\mathbb{E}\,\cos(X_T^2) = \frac{e^{-\frac{\mu^2 T}{2}\left(1-\frac{\tanh(\sigma T)}{\sigma T}\right)}}{\sqrt{\cosh(\sigma T)}}. \tag{9.103}$$

This yields the reference value

$$P(1, 1, 1) \simeq 7.14556.$$

We defer to Sect. 12.11 in the Miscellany chapter the proof of this identity (also used in [113] as a reference value).

We implemented the following schemes with their correspondences in terms of characteristic exponents (α, β):

- Crude Euler scheme: $\alpha = \beta = 1$ (I^{EU}_{CRUDE}),
- *MLMC* with Euler scheme: $\alpha = \beta = 1$ (I^{EU}_{MLMC}),
- *ML2R* with Euler scheme: $\alpha = \beta = 1$ (I^{EU}_{ML2R}),
- *ML2R* with the Giles–Szpruch antithetic Milstein scheme: $\alpha = \beta = 2$ (I^{GS}_{ML2R}).

We reproduce in Fig. 9.1 below the graph $\varepsilon \longmapsto$ CPU-time $\times\, \varepsilon^2$ where ε denotes the theoretical (prescribed) *RMSE* (log-scale).

This graphic highlights that Multilevel estimators make it possible to achieve accuracies which are simply out of reach with regular Monte Carlo estimators.

9.7.2 Option Pricing

We propose in this paragraph first a graphical comparison between crude Monte Carlo and multilevel methods, then an internal comparison between weighted and unweighted multilevel estimators, based on two types of options: a vanilla option (Call) and a barrier option both in a Black–Scholes model. We recall that the Black–Scholes dynamics for a traded risky asset reads:

$$dX_t = X_t\big(r\,dt + \sigma\, dW_t\big),\ \ X_0 = x_0. \tag{9.104}$$

This model can be simulated in an exact way at fixed time $t \in [0, T]$ since the above *SDE* has a closed solution $X_t = x_0 e^{(r-\frac{\sigma^2}{2})t+\sigma W_t}$.

All the simulations of this section have been performed by processing a C++ script on a Mac Book Pro (2.7 GHz intel Core i5).

Vanilla Call $(\alpha = \beta = 1)$

The (discounted) Vanilla Call payoff is given by $\varphi(X_T) = e^{-rT}\big(X_T - K\big)_+$. As for the model parameters we set:

$$x_0 = 50,\ r = 0.5,\ \sigma = 0.4,\ T = 1,\ K = 40.$$

Note that in order to produce a significant bias in the implemented discretization schemes, the interest purposefully been set at an unrealistic value of 50 %. The premium of this option (see Sect. 12.2 in the Miscellany chapter) has a closed form given by the Black–Scholes formula, namely the premium is given by

$$I_0 = \text{Call}_0 \simeq 25.9308.$$

– The selected time discretization schemes are the Euler scheme (see (7.6)) for which $\beta = 1$), and the Milstein scheme (see (7.39)) for which $\beta = 2$. Both are simulable since we are in one dimension and share the weak error exponent $\alpha = 1$.

– The constant $\widetilde{c}_\infty$ is set at 1 in the *ML2R* estimator and c_1 is roughly estimated in a preliminary phase to design the *MLMC* estimator (see Practitioner's corner of Sect. 9.5.2).

Though of no practical interest, one must have in mind that the Black–Scholes *SDE* is quite a demanding benchmark to test the efficiency of discretization schemes because both its drift and diffusion coefficients go to infinity as fast as possible under a linear growth control assumption. This is all the more true given the high interest rate.

All estimators are designed following the specifications detailed in the Practitioner's corners sections of this chapter (including the crude unbiased Monte Carlo simulation based on the exact simulation of X_T). As for (weighted and unweighted) multilevel estimators, we set $\mathbf{h} = 1$ and estimate the "payoff-dependent" parameters following the recommendations made in Practitioner's corner I (see (9.70) and (9.71) with $h_0 = \mathbf{h}/5$ and $h_0' = h_0/2$):

$$\text{Euler scheme: } \mathrm{Var}(Y_{\mathbf{h}}) \simeq 137.289, \; V_1 \simeq 40.2669 \quad \text{and} \quad \theta \simeq 0.5416.$$
$$\text{Milstein scheme: } \mathrm{Var}(Y_{\mathbf{h}}) \simeq 157.536, \; V_1 \simeq 1130.899 \quad \text{and} \quad \theta \simeq 2.6793.$$

The parameters of the multilevel estimators have been settled following the values in Tables 9.1 and 9.2. The crude Monte Carlo simulation has been implemented using the parameter values given in Sect. 9.3.

We present below tables of numerical results for crude Monte Carlo, *MLMC* and *ML2R* estimators for both Euler and Milstein schemes. Crude Monte Carlo has been run for target *RMSE* running from $\varepsilon = 2^{-k}$, $k = 1, \ldots, 6$ and multilevel estimators for *RMSE* running from $\varepsilon = 2^{-k}$, $k = 1, \ldots, 8$. We observe globally that our estimators fulfill the prescribed *RMSE* so that it is relevant to compare the CPU computation times.

Crude versus Multilevel estimators: See Tables 9.4, 9.5, 9.6, 9.7, 9.8 and 9.9. We observe at a 2^{-6} level for the Euler scheme that *MLMC* is 45 times faster than crude *MC* and *ML2R* is 130 times faster than crude *MC*. As for the Milstein scheme these ratios attain 166 and 255, respectively. This illustrates in a striking way that multilevel methods simply make possible simulations that would be unreasonable to undertake otherwise (in fact, that is why we limit ourselves to this lower range of *RMSE*).

ML2R versus MLMC estimators: See Tables 9.6, 9.7, 9.8 and 9.9. This time we make the comparison at an *RMSE* level of $\varepsilon = 2^{-8}$. As for the Euler scheme we observe that the *ML2R* is 4.62 times faster than the *MLMC* estimator whereas, with the Milstein scheme, this ratio is still 1.58. Note that, being in a setting $\beta = 2 > 1$, both estimators are assumed to behave like unbiased estimators which means that, in the case of *ML2R*, the constant C in the asymptotic rate $C\varepsilon^{-2}$ of the complexity seemingly remains lower with *ML2R* than with *MLMC*.

In the following figures, the performances (vertices) of the estimators are depicted against the *empirical RMSE* $\widetilde{\varepsilon}$ (abscissas). The label above each point of the graph indicates the *target RMSE* ε. The empirical *RMSE*, based on the bias-variance decomposition, has been computed *a posteriori* by performing 250 independent trials for each estimator. The performances of the estimators are measured and compared in various ways, mixing CPU time and $\widetilde{\varepsilon}$. Details are provided for each figure.

Table 9.4 MC Euler

l_s	Epsilon	CPU Time (s)	Emp. RMSE
1	0.50000	0.00339	0.50518
2	0.25000	0.02495	0.28324
3	0.12500	0.19411	0.14460
4	0.06250	1.50731	0.06995
5	0.03125	11.81181	0.03611
6	0.01562	93.40342	0.01733

Table 9.5 MC Milstein

l_s	Epsilon	CPU Time (s)	Emp. RMSE
1	0.50000	0.00454	0.46376
2	0.25000	0.03428	0.25770
3	0.12500	0.26948	0.12421
4	0.06250	2.12549	0.06176
5	0.03125	16.92783	0.03180
6	0.01562	135.14195	0.01561

Table 9.6 MLMC Euler

l_s	Epsilon	CPU Time (s)	Emp. RMSE
1	0.50000	0.00079	0.55250
2	0.25000	0.00361	0.27192
3	0.12500	0.01729	0.15152
4	0.06250	0.08959	0.06325
5	0.03125	0.41110	0.03448
6	0.01562	2.05944	0.01644
7	0.00781	13.08539	0.00745
8	0.00391	60.58989	0.00351

Table 9.7 MLMC Milstein

l_s	Epsilon	CPU Time (s)	Emp. RMSE
1	0.50000	0.00079	0.42039
2	0.25000	0.00296	0.19586
3	0.12500	0.01193	0.10496
4	0.06250	0.04888	0.05278
5	0.03125	0.19987	0.02871
6	0.01562	0.81320	0.01296
7	0.00781	3.29153	0.00721
8	0.00391	13.25764	0.00331

Table 9.8 ML2R Euler

l_s	Epsilon	CPU Time (s)	Emp. RMSE
1	0.50000	0.00048	0.61429
2	0.25000	0.00183	0.31070
3	0.12500	0.00798	0.12511
4	0.06250	0.03559	0.06534
5	0.03125	0.16060	0.03504
6	0.01562	0.71820	0.01680
7	0.00781	3.17699	0.00855
8	0.00391	13.09290	0.00363

Table 9.9 ML2R Milstein

l_s	Epsilon	CPU Time (s)	Emp. RMSE
1	0.50000	0.00070	0.36217
2	0.25000	0.00231	0.20908
3	0.12500	0.00819	0.10206
4	0.06250	0.03171	0.05154
5	0.03125	0.13266	0.02421
6	0.01562	0.52949	0.01269
7	0.00781	2.11587	0.00696
8	0.00391	8.45961	0.00332

The graphics are produced with the following conventions:

– Each method has its distinctive label: $\times$ for *MLMC*, $+$ for *ML2R*, $\circ$ for crude *MC* on discretization schemes and $\odot$-dashed line for exact simulation (when possible).

– Euler scheme-based estimators are depicted with solid lines whereas Milstein scheme-based estimators are depicted with dashed lines and the exact simulation with a dotted line.

It is important to keep in mind that *an estimator satisfies the prescribed error RSMSE* $\varepsilon = 2^{-\ell}$ *if its labels are aligned vertically above the corresponding abscissa* $2^{-\ell}$.

In Fig. 9.2 the performance measure is simply the average CPU time of one estimator (in seconds) for each of the five tested estimators, depicted in log-log scale.

These graphics confirm that both multilevel estimators satisfy the assigned *RMSE* constraint ε in the sense that the empirical *RMSE* $\widetilde{\varepsilon} \leq \varepsilon$.

It also stresses the tremendous gain provided by multilevel estimators (note that when the absolute accuracy is 2^{-8}, it corresponds to a relative accuracy of $0.015\,\%$.)

Figure 9.3 is based on the same simulations as Fig. 9.2. The performances are now measured through the empirical *MSE* $\times$ CPU time = $\widetilde{\varepsilon}^2 \times$ CPU time. Figure 9.3 highlights in the critical case $\beta = 1$ that the computational cost of *MLMC* estimators increases faster by a $\log\big(1/\varepsilon\big)$ factor than *ML2R* multilevel estimators.

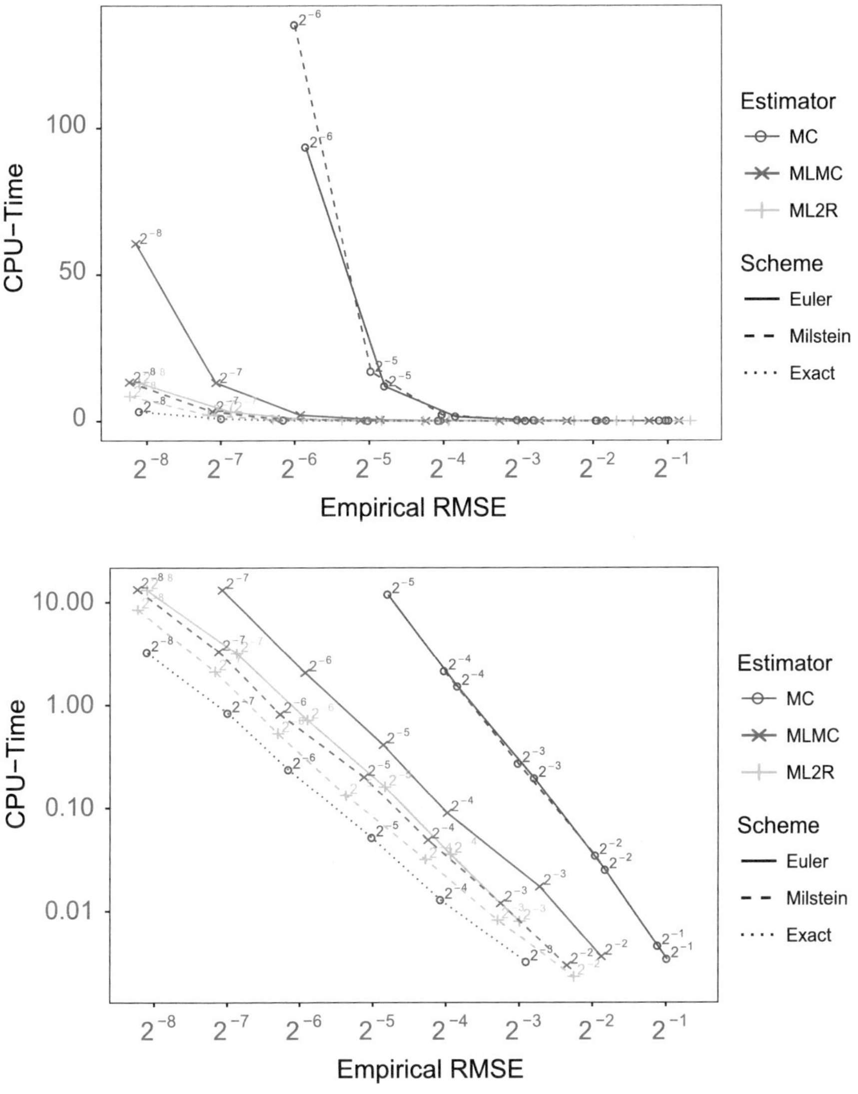

Fig. 9.2 CALL OPTION PRICING: CPU-time as a function of the empirical RMSE $\widetilde{\varepsilon}$. Top: log-regular scale; Bottom: log-log scale (with D. Giorgi and V. Lemaire)

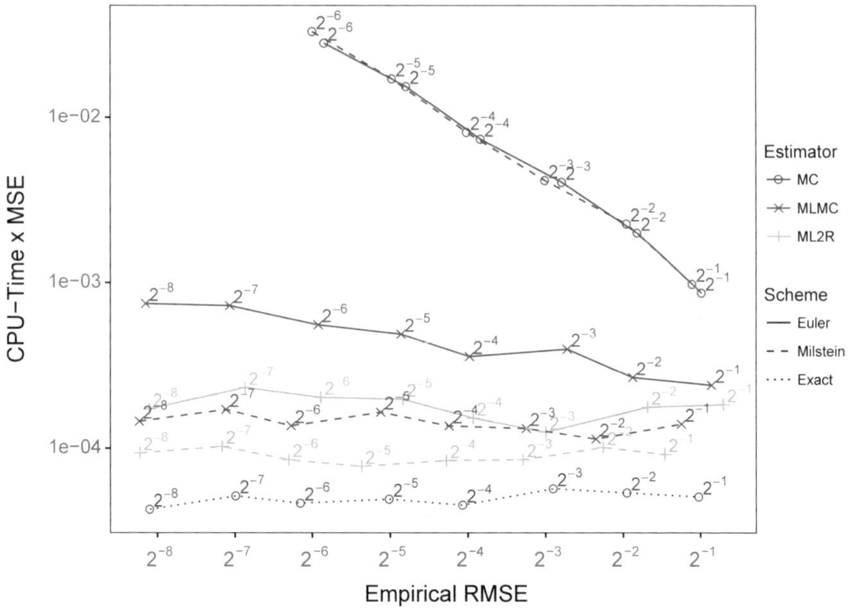

Fig. 9.3 Call option pricing by Euler and Milsteinschemes: MSE × CPU-time as a function of the empirical RMSE $\widetilde{\varepsilon}$, log-log- scale. (With D. Giorgi and V. Lemaire.)

Figure 9.3 illustrates for the Milstein scheme the "virtually" unbiased setting ($\beta = 2$) obtained here with the Multilevel Milstein scheme. We verify that multilevel curves remain approximately flat but lie significantly above the true unbiased simulation which, as expected, remains unrivaled. The weighted multilevel seems to be still faster in practice.

For more simulations involving path-dependent options, we refer to [107, 109] for regular multilevel and to [112, 198] for a comparison between weighted and regular Multilevel estimators on other options like Lookback options (note that the specifications are slightly more "conservative" in these last two papers than those given here, whereas in Giles' papers, the estimators designed are in an adaptive form). Now we will briefly explore a path-dependent option, namely a barrier option for which the characteristic exponent β is lower than 1, namely $\beta = 0.5$ (in $(SE)^0_\beta$) and α is most likely also equal to 0.5. We still consider a Black–Scholes dynamics to have access to a closed form for the option premium.

Barrier option ($\alpha = \beta = 0.5$)

We consider an *Up-and-Out Call option* to illustrate the case $\beta = 0.5 < 1$ and $\alpha = 0.5$. This path-dependent option with strike K and barrier $B > K$ is defined by its functional payoff

$$\varphi(x) = e^{-rT}(x(T) - K)_+ \mathbf{1}_{\{\max_{t\in[0,T]} x(t) \le B\}}, \quad x \in \mathcal{C}([0, T], \mathbb{R}).$$

The parameters of the Black–Scholes model are set as follows:

$$x_0 = 100, \quad r = 0, \quad \sigma = 0.15, \quad \text{and} \quad T = 1.$$

With $K = 100$ and $B = 120$, the price obtained by the standard closed-form solution is

$$I_0 = \text{Up-and-Out Call}_0 = 1.855225.$$

We consider here a simple (and highly biased) approximation of $\max_{t\in[0,T]} X_t$ by $\max_{k\in\{1,\dots,n\}} \bar{X}_{kh}$ $(h = \frac{T}{n})$ *without any "help" from a Brownian bridge approximation, as investigated in Sects.* 8.2.1 *and* 8.2.3.

By adapting the computations carried out in Lemma 9.3 with the convergence rate for the sup norm of the discrete time Euler scheme, we obtain that $\beta = 0.5^-$ for $(SE)^0_\beta$ (*i.e.* $(SE)^0_\beta$ holds for every $\beta \in (0, \frac{1}{2})$ but *a priori* not for $\beta = 0.5$, due to the $\log h$ term coming out in the quadratic strong convergence rate). The design of the multilevel estimators is adapted to $(SE)^0_\beta$ for both estimators.

As for the weak error rate, the first-order with $\alpha = 0.5$ has been established in [17]. But assuming that $R = +\infty$ with $\alpha = 0.5$, *i.e.* that the weak error expansion holds at any order in that scale, stands as a pure conjecture at higher-orders.

The rough estimation of payoff dependent parameters yields with $\mathbf{h} = 1$, $\text{Var}(Y_{\mathbf{h}}) \simeq 30.229$ and $V_1 \simeq 11.789$ (with $h_0 = \mathbf{h}/5$, $h'_0 = h_0/2$) so that $\theta \simeq 0.6245$.

Figure 9.4 highlights that since $\beta = 0.5$, we observe that the function (CPU-time)$\times\varepsilon^2$ increases much faster for *MLMC* than *ML2R* as ε goes to 0. This agrees with the respective theoretical asymptotic rates for both estimators since the computational cost of *ML2R* estimators remains a $o(\varepsilon^{-(2+\eta)})$ for every $\eta > 0$, even for $\beta < 1$, whereas the performances of the *MLMC* estimator behaves like $o\big(\varepsilon^{-(2+\frac{1-\beta}{\alpha})}\big)$.

In fact, in this highly biased example with slow strong convergence, the ratio *MLMC*/*ML2R* in terms of CPU time, as a function of the prescribed $\widetilde{\varepsilon} = 2^{-k}$, attains approximately 45 when $k = 6$.

Remark. The performances of the *ML2R* estimator on this option tends to support the conjecture that an expansion of the weak error at higher-orders exists.

9.7.3 Nested Monte Carlo

As an example we will consider two settings. First, a compound option ([4]) (here a Call-on-Put option) and, as a second example, a quantile of a Call option (both in a Black–Scholes model). In the first example we will implement both regular and antithetic multilevel nested estimators.

Compound option (Put-on-Call) $(\alpha = \beta = 1)$

[4]More generally, a compound option is an option on (the premium of) an option.

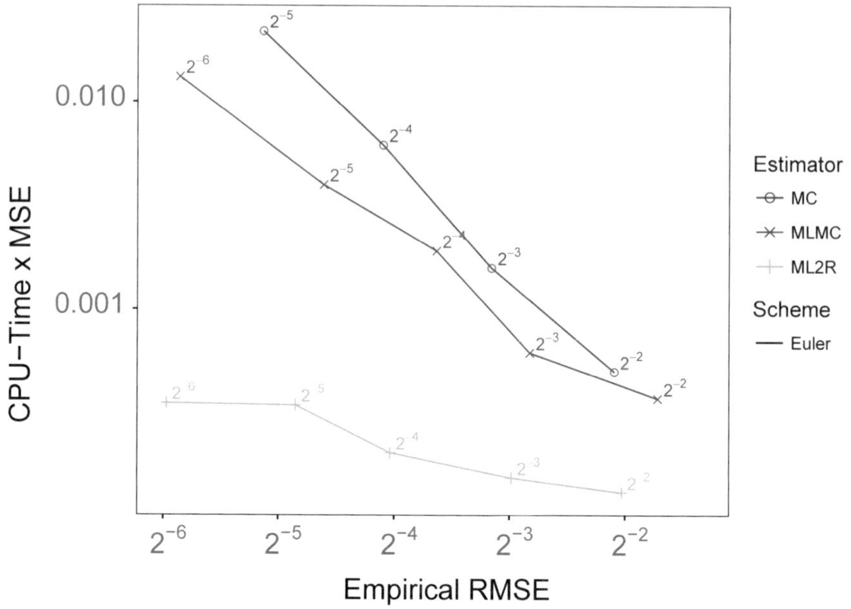

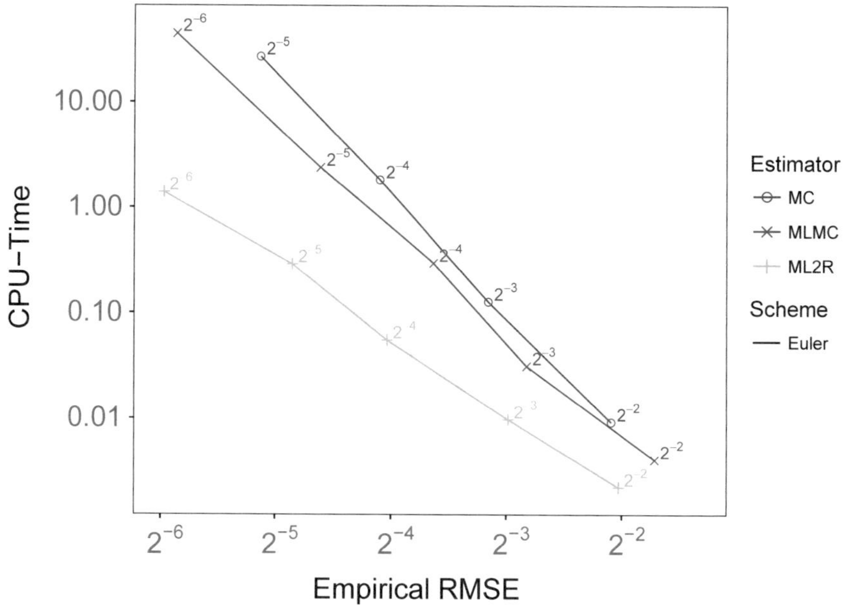

Fig. 9.4 Barrier option in a Black–Scholes model. Top: $\widetilde{\varepsilon}^2 \times$ CPU-time ($y-$ axis, log-scale) as a function of $\widetilde{\varepsilon}$ ($x-$ axis, $\log_2$ scale) for MLMC and ML2R estimators. Bottom: CPU-time (y–axis, log-scale) as a function of $\widetilde{\varepsilon}$ (x–axis, $\log_2$ scale). (With V. Lemaire.)

A compound option being simply an option on an option, its payoff involves the value of another option. We consider here the example of a European style *Put-on-Call*, still in a Black–Scholes model, with parameters (r, σ) (see (9.104) above). We consider a Call of maturity T_2 and strike K_2. At date T_1, $T_1 < T_2$, the holder of the option has the right to sell it at (strike) price K_1. The payoff of such a *Put-on-Call* option reads

$$e^{-rT_1}\Big(K_1 - e^{-r(T_2-T_1)}\mathbb{E}\left((X_{T_2} - K_2)_+ \,|\, X_{T_1}\right)\Big)_+.$$

To comply with the nested multilevel framework ([5]), we set here $\mathcal{H} = \{1/k,\ k \in k_0\mathbb{N}^*\}$ with $k_0 = 2$ and

$$Y_0 = f\Big(\mathbb{E}\left((X_{T_2} - K_2)_+ \,|\, X_{T_1}\right)\Big),\quad Y_{\frac{1}{k}} = f\left(\frac{1}{k}\sum_{\ell=1}^{k}\left(F(X_{T_1}, Z^\ell) - K_2\right)_+\right),\ \frac{1}{k} \in \mathcal{H},$$

where $f(x) = e^{-rT_1}\left(K_1 - e^{-r(T_2-T_1)}x\right)_+$, $(Z^k)_{k\geq 1}$ is an i.i.d. sequence of $\mathcal{N}(0; 1)$-distributed random variables and F is such that

$$X_{T_2} = F(X_{T_1}, Z) = X_{T_1} e^{(r-\frac{\sigma^2}{2})(T_2-T_1)+\sigma\sqrt{T_2-T_1}Z},\quad Z \stackrel{d}{=} \mathcal{N}(0; 1).$$

Note that, in these experiments, the underlying process $(X_t)_{t\in[0,T_2]}$ *is not discretized in time* since it can be simulated in an exact way. The bias error is exclusively due to the inner Monte Carlo estimator of the conditional expectation.

The parameters of the Black–Scholes dynamics $(X_t)_{t\in[0,T_2]}$ are

$$x_0 = 100,\quad r = 0.03,\quad \sigma = 0.2,$$

whereas those of the *Put-on-Call* payoff are $T_1 = 1/12$, $T_2 = 1/2$ and $K_1 = 6.5$, $K_2 = 100$.

To compute a reference price in spite of the absence of closed form we proceed as follows: we first note that, by the (homogeneous) Markov property, the Black–Scholes formula at time T_1 reads (see Sect. 12.2)

$$e^{-r(T_2-T_1)}\mathbb{E}\left((X_{T_2} - K_2)_+ \,|\, X_{T_1}\right) = \mathrm{Call}_{BS}(X_{T_1}, K_2, r, \sigma, T_2 - T_1)$$

(see Sect. 12.2) so that $Y_0 = g(X_{T_1})$ where

$$g(x) = f\Big(\left(K_1 - \mathrm{Call}_{BS}(x, K_2, r, \sigma, T_2 - T_1)\right)_+\Big).$$

This can be computed by a standard *unbiased* Monte Carlo simulation. To reduce the variance one may introduce the control variate

[5] Due to a notational conflict with the strike prices K_1, K_2, etc, we temporarily modify our standard notations for denoting inner simulations and the bias parameter set $\mathcal{H}$.

$$\Xi = \big(K_1 - e^{-r(T_2-T1)}(X_{T_2} - K_1)_+\big)_+ - \mathbb{E}\big(\big(K_1 - e^{-r(T_2-T1)}(X_{T_2} - K_1)_+\big)_+\big).$$

This leads, with the numerical values of the current example, to the reference value

$$\mathbb{E}\,Y_0 \simeq 1.39456 \quad \text{and} \quad \big|\mathbb{E}\,Y_0 - 1.39456\big| \leq 2^{-15} \text{ at a 95\%-confidence level.}$$

▶ **Exercise.** Check the above reference value using a large enough Monte Carlo simulation with the above control variate. How would you still speed up this simulation?

We tested the following four estimators:

- Regular *MLMC* and *ML2R*,
- Antithetic *MLMC* and *ML2R*.

Theorem 9.2 (a) suggests, though f is simply Lipschitz continuous, that the weak error expansion holds with $\alpha = 1$ and Proposition 9.2(a) shows that the strong error rate is satisfied with $\beta = 1$ for both regular multilevel estimators.

As for the implementation of the antithetic nested estimator on this example, it seems *a priori* totally inappropriate since the derivative f' of f is not even defined everywhere and has subsequently no chance to be (Hölder-)continuous. However, empirical tests suggest that we may assume that (9.101) (the variant of $(SE)_\beta$ involving the antithetic $\widetilde{Y}$ term defined in (9.100)) holds with $\beta = 1.5$, which led us to implement this specification for both multilevel estimators.

For the implementation we set $K_0 = 2$ (at least two inner simulations), *i.e.* $\mathbf{h} = \frac{1}{2}$.
A rough computation of the payoff-dependent parameters yields:

– Regular nested: $\mathrm{Var}(Y_{\mathbf{h}}) \simeq 7.938$, $V_1 \simeq 23.943$ (still with $h_0 = \mathbf{h}/5$ and $h_0' = h_0/2$) so that $\theta \simeq 1.74$.

– Antithetic: $\mathrm{Var}(Y_{\mathbf{h}}) \simeq 7.867$, $V_1 \simeq 14.689$ so that $\theta \simeq 1.3664$.

Note in Fig. 9.5 that, for a prescribed *RMSE* of 2^{-8}, *ML2R* is faster than *MLMC* as a function of the empirical *RMSE* $\widetilde{\varepsilon}$ by a factor of approximately 4 for regular nested simulations (8.24 s *versus* 33.98 s) and 2 for antithetic simulations (2.26 s *versus* 4.37 s). The gain with respect to a crude Monte Carlo simulation at the same accuracy (68.39 s) is beyond a factor of 30.

Digital option on a call option $(\beta = 0.5^-, \alpha = 1)$

Now we are interested in a quantile of the premium at time T_1 of a *call* option with maturity T_2 in the same model, namely to compute

$$\mathbb{P}\big(e^{-r(T_2-T_1)}\mathbb{E}\,(X_{T_2} - K_2)_+ \,|\, X_{T_1}) \geq K_1\big) = \mathbb{E}\left(\mathbf{1}_{\left\{e^{-r(T_2-T_1)}\mathbb{E}\,(X_{T_2}-K_2)_+ \,|\, X_{T_1})\geq K_1\right\}}\right).$$

Such a value can still be seen as a compound option, namely the value of a digital option on the premium at time T_1 of a call of maturity T_2. We kept the above parameters for the Black–Scholes model and the underlying call option but also for the digital component, namely $K_1 = 6.5$, $T_1 = 1/2$; $K_2 = 100$ and $T_2 = 1/2$.

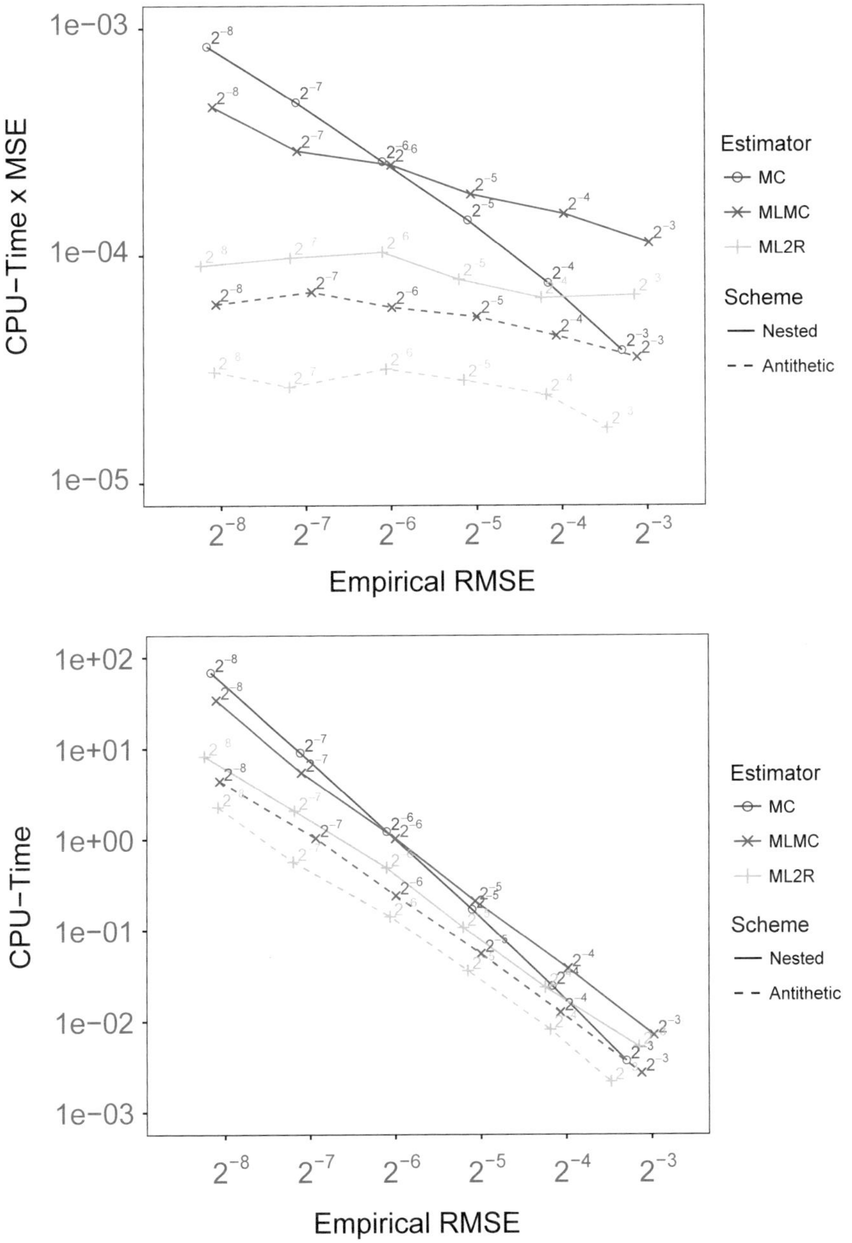

Fig. 9.5 Compound option in a Black- - Scholes model: Top: $\widetilde{\varepsilon}^2 \times$ CPU-time (y–axis, regular scale) as a function of the empirical RMSE $\widetilde{\varepsilon}$ (x-axis, log scale). Bottom: CPU-time in log-log scale. (With D. Giorgi and V. Lemaire)

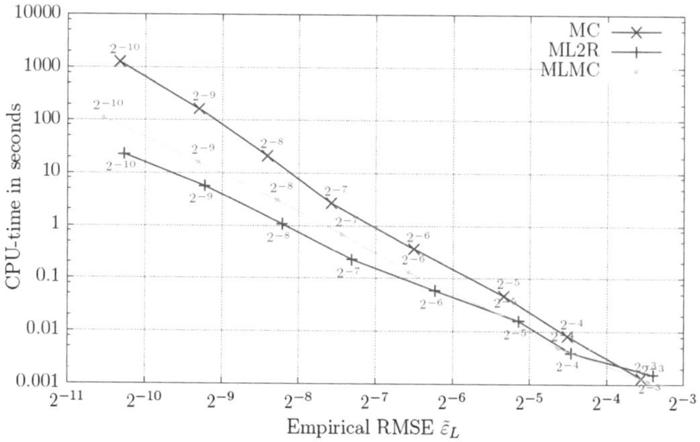

Fig. 9.6 DIGITAL OPTION ON A CALL: CPU-time (y–axis, log scale) as a function of the empirical RMSE$\tilde{\varepsilon}$ (x-axis, $\log_2$ scale) (with V. Lemaire)

We only tested the first two regular *MLMC* and *ML2R* estimators since antithetic pseudo-schemes are not efficient with non-smooth functions (here an indicator function).

Theorem 9.2(c) strongly suggests that weak error expansion holds with $\alpha = 1$ and Proposition 9.2(c) suggests that the strong error rate is satisfied with $\beta = 0.5^-$, so we adopt these values.

The multilevel estimators have been implemented for this simulation with the parameters from [198], which are slightly more conservative than those prescribed in this chapter, which explains why the crosses are slightly ahead of their optimal positions (the resulting empirical *RMSE* is lower than the prescribed one).

Note in Fig. 9.6 that the *ML2R* estimator is faster than the *MLMC* estimator as a function of the empirical *RMSE* $\tilde{\varepsilon}$ by a factor approximately equal to 5 within the range of our simulations. This is expected since we are in framework where $\beta < 1$.

Note that this toy example is close in spirit to the computation of Solvency Capital Requirement which consists by its very principle in computing a kind of quantile – or a value-at-risk – of a conditional expectation.

9.7.4 Multilevel Monte Carlo Research Worldwide

▷ The webpage below, created and maintained by Mike Giles, is an attempt to list the research groups working on multilevel Monte Carlo methods and their main contributions:

people.maths.ox.ac.uk/gilesm/mlmc_community.html

▷ For an interactive ride through the world of multilevel, have a look at the website simulations@lpsm at the url:

simulations.lpsm.paris/

9.8 Randomized Multilevel Monte Carlo (Unbiased Simulation)

We conclude this chapter with a section devoted to randomized multilevel estimators, which is a new name for a rather old idea going back to the very beginning of Monte Carlo simulation, brought back to light and developed more recently by McLeish [206], then Rhee and Glynn in [252].

We briefly expose in the next section a quite general abstract version. As a second step, we make the connection with our multilevel framework developed in this chapter.

9.8.1 *General Paradigm of Unbiased Simulation*

Let $(Z_n)_{n\geq 1}$ be a sequence of square integrable random variables defined on a probability space $(\Omega, \mathcal{A}, \mathbb{P})$ and let $\tau : (\Omega, \mathcal{A}, \mathbb{P}) \to \mathbb{N}^*$, independent of $(Z_n)_{n\geq 1}$, be such that

$$\pi_n = \mathbb{P}(\tau \geq n) > 0 \quad \text{for every } n \in \mathbb{N}^*.$$

We also set $p_n = \mathbb{P}(\tau = n) = \pi_n - \pi_{n+1}$ for every $n \in \mathbb{N}^*$.

We assume that $\sum_{n\geq 1} \|Z_n\|_2 < +\infty$ so that, classically ([6]),

$$Z_1 + \cdots + Z_n \overset{L^2\,\&\,a.s}{\longrightarrow} Y_0 = \sum_{n\geq 1} Z_n.$$

Our aim is to compute $I_0 = \mathbb{E}\, Y_0$ taking advantage of the assumption that the random variables Z_n are simulable (see further on) and to be more precise, to devise an *unbiased* estimator of I_0. To this end, we will use the random time τ (with the disadvantage of adding an exogenous variance to our estimator).

Let

$$I_\tau = \sum_{i=1}^{\tau} \frac{Z_i}{\pi_i} = \sum_{i\geq 1} \frac{Z_i}{\pi_i} \mathbf{1}_{\{i\leq \tau\}}$$

[6] As the L^1-norm is dominated by the L^2-norm, $\sum_{n\geq 1} \|Z_n\|_1 < +\infty$ so that $\mathbb{E}\big(\sum_{n\geq 1} |Z_n|\big) < +\infty$, which in turn implies that $\sum_{n\geq 1} |Z_n|$ is $\mathbb{P}$-$a.s.$ absolutely convergent.

and, for every $n \geq 1$,

$$I_{\tau\wedge n} = \sum_{i=1}^{n} \frac{Z_i}{\pi_i} \mathbf{1}_{\{i\leq\tau\}} = \sum_{i=1}^{\tau\wedge n} \frac{Z_i}{\pi_i}.$$

The following proposition provides a first answer to our problem.

Proposition 9.4 *(a) If the sequence $(Z_n)_{n\geq 1}$ satisfies*

$$\sum_{n\geq 1} \frac{\|Z_n\|_2}{\sqrt{\pi_n}} < +\infty, \tag{9.105}$$

then

$$I_{\tau\wedge n} \stackrel{L^2\,\&\,a.s}{\longrightarrow} I_\tau$$

(and $Z_1 + \cdots + Z_n \stackrel{L^2\,\&\,a.s}{\longrightarrow} Y_0$ as $n \to +\infty$). Then, the random variable I_τ is unbiased *with respect to Y_0 in the sense that*

$$\mathbb{E}\, I_\tau = \mathbb{E}\left[\sum_{i=1}^{\tau} \frac{Z_i}{\pi_i}\right] = \mathbb{E}\, Y_0$$

and

$$\mathrm{Var}\big(I_\tau\big) = \big\|I_\tau\big\|_2^2 - I_0^2 \quad \text{with} \quad \big\|I_\tau\big\|_2^2 \leq \left(\sum_{n\geq 1} \frac{\|Z_n\|_2}{\sqrt{\pi_n}}\right)^2. \tag{9.106}$$

(b) The L^2-norm of I_τ also satisfies

$$\big\|I_\tau\big\|_2^2 = \sum_{n\geq 1} p_n \left\|\sum_{i=1}^{n} \frac{Z_i}{\pi_i}\right\|_2^2 = \sum_{n\geq 1} \pi_n \left(\left\|\sum_{i=1}^{n} \frac{Z_i}{\pi_i}\right\|_2^2 - \left\|\sum_{i=1}^{n-1} \frac{Z_i}{\pi_i}\right\|_2^2\right).$$

(c) Assume the random variables $(Z_n)_{n\geq 1}$ are independent. If

$$\sum_{n\geq 1} \frac{\mathrm{Var}(Z_n)}{\pi_n} < +\infty \quad \text{and} \quad \sum_{n\geq 1} \frac{|\mathbb{E}\, Z_n|}{\sqrt{\pi_n}} < +\infty, \tag{9.107}$$

then $I_{\tau\wedge n} \stackrel{L^2\,\&\,a.s}{\longrightarrow} I_\tau$ as $n \to +\infty$ and

$$\big\|I_\tau\big\|_2 \leq \left(\sum_{n\geq 1} \frac{\mathrm{Var}(Z_n)}{\pi_n}\right)^{\frac{1}{2}} + \sum_{n\geq 1} \frac{|\mathbb{E}\, Z_n|}{\sqrt{\pi_n}}. \tag{9.108}$$

Remark. When the random variables Z_n are independent and $\mathbb{E}\, Z_n = o\big(\|Z_n\|_2\big)$, the assumption (9.107) in (c) is strictly less stringent than that in (a) (under the independence assumption). Furthermore, if one may neglect the additional term $\sum_{n\geq 1}\frac{|\mathbb{E}\, Z_n|}{\sqrt{\pi_n}}$ involving expectations in the upper-bound (9.108) of $\|I_\tau\|_2$, it is clear that the variance in the independent case is most likely lower than the general upper-bound established in (9.106)) since $a_n^2 = o(a_n)$ when $a_n \to 0$.

Proof of Proposition 9.4. (a) The space $L^2(\Omega, \mathcal{A}, \mathbb{P})$ endowed with the L^2-norm $\|\cdot\|_2$ is complete so it suffices to show that the series with general term $\frac{\|Z_n\|_2}{\sqrt{\pi_n}}$ is $\|\cdot\|_2$-normally convergent. Now, using that τ and the sequence $(Z_n)_{n\geq 1}$ are mutually independent, we have for every $n \geq 1$,

$$\left\|\frac{Z_n}{\sqrt{\pi_n}}\mathbf{1}_{\{n\leq\tau\}}\right\|_2 = \frac{\|Z_n\|_2}{\sqrt{\pi_n}}\left\|\mathbf{1}_{\{n\leq\tau\}}\right\|_2 = \frac{\|Z_n\|_2}{\pi_n}\sqrt{\pi_n} = \frac{\|Z_n\|_2}{\sqrt{\pi_n}}.$$

Then, the sequence $I_{\tau\wedge n}$ converges in $L^2(\mathbb{P})$, hence toward I_τ as it is clearly its $a.s.$ limit since $\tau < +\infty$.

The L^2-convergence of Y_n toward Y_∞ is straightforward under our assumptions since $1 \leq 1/\sqrt{\pi_i}$ by similar L^2-normal convergence arguments.

On the other hand $\mathbb{E}\left[\frac{Z_n}{\pi_n}\mathbf{1}_{\{n\leq\tau\}}\right] = \frac{\mathbb{E}\, Z_n}{\pi_n}\pi_n = \mathbb{E}\, Z_n$, $n \geq 1$. Now as a consequence of the L^2 (hence L^1)-convergence of $I_{\tau\wedge n}$ toward I_τ, their expectations converge, *i.e.* $\mathbb{E}\, I_\tau = \lim_n \mathbb{E}\, I_{\tau\wedge n} = \lim_n \sum_{i=1}^{n}\mathbb{E}\, Z_i = \mathbb{E}\, Y_0$. Finally,

$$\|I_\tau\|_2 \leq \sum_{n\geq 1}\frac{\|Z_n\|_2}{\sqrt{\pi_n}}.$$

(b) It is obvious after a preconditioning by τ that

$$\|I_\tau\|_2^2 = \mathbb{E}\left(\sum_{i=1}^{\tau}\frac{Z_i}{\pi_i}\right)^2 = \sum_{i\geq 1} p_i\, \mathbb{E}\left(\sum_{\ell=1}^{i}\frac{Z_\ell}{\pi_\ell}\right)^2.$$

Now, an Abel transform yields

$$\sum_{i=1}^{n} p_i \mathbb{E}\left(\sum_{\ell=1}^{i}\frac{Z_\ell}{\pi_\ell}\right)^2 = -\pi_{n+1}\mathbb{E}\left(\sum_{i=1}^{n}\frac{Z_i}{\pi_i}\right)^2 + \sum_{i=1}^{n}\pi_i\left[\mathbb{E}\left(\sum_{\ell=1}^{i}\frac{Z_\ell}{\pi_\ell}\right)^2 - \mathbb{E}\left(\sum_{\ell=1}^{i-1}\frac{Z_\ell}{\pi_\ell}\right)^2\right].$$

Now

$$\sqrt{\pi_{n+1}}\left\|\sum_{i=1}^{n}\frac{Z_i}{\pi_i}\right\|_2 \leq \sqrt{\pi_n}\sum_{i=1}^{n}\frac{\|Z_i\|_2}{\sqrt{\pi_i}\sqrt{\pi_i}} \to 0 \quad \text{as} \quad n \to +\infty$$

owing to Kronecker's Lemma (see Lemma 12.1 applied with $a_n = \frac{\|Z_i\|_2}{\sqrt{\pi_i}}$ and $b_n = 1/\sqrt{\pi_n}$).

(c) First we decompose $I_{\tau\wedge n}$ into

$$I_{\tau\wedge n} = \sum_{i=1}^{n} \frac{\widetilde{Z}_i}{\pi_i}\mathbf{1}_{\{i\le\tau\}} + \sum_{i=1}^{n} \frac{\mathbb{E}\,Z_i}{\pi_i}\mathbf{1}_{\{i\le\tau\}},$$

where the random variables $\widetilde{Z}_i = Z_i - \mathbb{E}\,Z_i$ are independent and centered.

Then, using that τ and $(Z_n)_{n\ge1}$ are independent, we obtain for every $n,\ m \ge 1$,

$$\mathbb{E}\left|\sum_{i=n+1}^{n+m}\frac{\widetilde{Z}_i}{\pi_i}\mathbf{1}_{\{i\le\tau\}}\right|^2 = \sum_{i=n+1}^{n+m}\frac{\mathbb{E}\,\widetilde{Z}_i^2}{\pi_i^2}\pi_i + 2\sum_{n+1\le i<j\le n+m}\frac{\mathbb{E}\,\widetilde{Z}_i\widetilde{Z}_j}{\pi_i\pi_j}\pi_{i\vee j}$$
$$= \sum_{i=n+1}^{n+m}\frac{\mathrm{Var}(Z_i)}{\pi_i}$$

since $\mathbb{E}\,\widetilde{Z}_i\widetilde{Z}_j = \mathbb{E}\,\widetilde{Z}_i\mathbb{E}\,\widetilde{Z}_j = 0$ if $i \ne j$. Hence, for all integers $n, m \ge 1$,

$$\left\|\sum_{i=n+1}^{n+m}\frac{\widetilde{Z}_i}{\pi_i}\mathbf{1}_{\{i\le\tau\}}\right\|_2 \le \left(\sum_{i=n+1}^{n+m}\frac{\mathrm{Var}(Z_i)}{\pi_i}\right)^{\frac{1}{2}}$$

which implies, under the assumption $\sum_{n\ge1}\frac{\mathrm{Var}(Z_n)}{\pi_n} < +\infty$, that the series $\displaystyle\sum_{i=1}^{\tau\wedge n}\frac{\widetilde{Z}_i}{\pi_i} = \sum_{i=1}^{n}\frac{\widetilde{Z}_i}{\pi_i}\mathbf{1}_{\{i\le\tau\}}$ is a Cauchy sequence hence convergent in the complete space $\big(L^2(\mathbb{P}), \|\,.\,\|_2\big)$. Its limit is necessarily $I_\tau = \displaystyle\sum_{i=1}^{\tau}\frac{\widetilde{Z}_i}{\pi_i}$ since that is its $a.s.$ limit.

On the other hand, as

$$\sum_{i\ge1}\left\|\frac{|\mathbb{E}\,Z_i|}{\pi_i}\mathbf{1}_{\{i\le\tau\}}\right\|_2 \le \sum_{i\ge1}\frac{|\mathbb{E}\,Z_i|}{\pi_i}\sqrt{\pi_i} = \sum_{i\ge1}\frac{|\mathbb{E}\,Z_i|}{\sqrt{\pi_i}} < +\infty,$$

one deduces that $\displaystyle\sum_{i=1}^{n}\frac{\mathbb{E}\,Z_i}{\pi_i}\mathbf{1}_{\{i\le\tau\}}$ is convergent in $L^2(\mathbb{P})$. Its limit is clearly $\displaystyle\sum_{i=1}^{\tau}\frac{|\mathbb{E}\,Z_i|}{\pi_i}$. Finally, $I_{\tau\wedge n} \to I_\tau$ in $L^2(\mathbb{P})$ (and $\mathbb{P}$-$a.s.$) and the estimate of $\mathrm{Var}(I_\tau)$ is a straightforward consequence of Minkowski's Inequality. ◇

Unbiased Multilevel estimator

The resulting unbiased multilevel estimator reads, for every integer $M \ge 1$,

$$\widehat{I}_M^{RML} = \frac{1}{M}\sum_{m=1}^{M} I_{\tau^{(m)}}^{(m)} = \frac{1}{M}\sum_{m=1}^{M}\sum_{i=1}^{\tau^{(m)}} \frac{Z_i^{(m)}}{\pi_i}, \tag{9.109}$$

where $(\tau^{(i)})_{i\geq 1}$ and $(Z_i^{(i)})_{i\geq 1}$, $i \geq 1$, are independent sequences, distributed as τ and $(Z_i)_{i\geq 1}$, respectively.

Definition 9.6 (Randomized Multilevel estimator) *The unbiased multilevel estimator* (9.109) *associated to the random variables* $(Z_i)_{i\in\mathbb{N}^*}$ *and the random time* τ *is called a* Randomized Multilevel estimator *and will be denoted in short by RML from now on.*

However, for a practical implementation, we need to specify the random time τ so that the estimator has both a finite mean complexity and a finite variance or, equivalently, a finite L^2-norm and such a choice is closely connected to the balance between the rate of decay of the variances of the random variables Z_i and the growth of their computational complexity. This is the purpose of the following section, where we will briefly investigate a geometric setting close to that of the multilevel framework with geometric refiners.

ℵ Practitioner's corner

In view of practical implementation, we make the following "geometric" assumptions on the complexity and variance (which are consistent with a multilevel approach, see later) and specify a distribution for τ accordingly. See the exercise in the next section for an analysis of an alternative non-geometric framework.

- *Distribution of* τ: we assume *a priori* that $\tau \stackrel{d}{=} G^*(p)$, $p\in(0,1)$, so that $p_n = p(1-p)^{n-1}$ and $\pi_n = (1-p)^{n-1}$, $n \geq 1$.
  This choice is motivated by the assumptions below.
- *Complexity*. The complexity of simulating Z_n is of the form

  $$\kappa(Z_n) = \kappa N^n, \ n \geq 1.$$

  Then, the mean complexity of the estimator I_τ is clearly

  $$\bar{\kappa}(I_\tau) = \kappa \sum_{n\geq 1} p_n N^n.$$

  As a consequence this mean complexity is finite if and only if

  $$(1-p)N < 1$$

  and $\bar{\kappa}(I_\tau) = \dfrac{\kappa p N}{1-(1-p)N}$.
- *Variance*. We make the assumption that there exists some $\beta > 0$ such that

$$\mathbb{E}\, Z_n^2 \le V_1 N^{-n\beta},\ n \ge 1,$$

for the same N as for the complexity. Then, the finiteness of the variance is satisfied if Assumption (9.105) holds true. With the above assumptions and choice, we have

$$\frac{\|Z_n\|_2}{\sqrt{\pi_n}} \le \sqrt{V_1}\frac{N^{-\frac{\beta}{2}n}}{(1-p)^{\frac{n-1}{2}}} = \sqrt{V_1(1-p)}\big(N^\beta(1-p)\big)^{-n/2}.$$

Consequently, the series in (9.105) converges if and only if $N^\beta(1-p) > 1$.

Now we are in a position to specify the parameters of the randomized multilevel estimator:

- *Choice of the geometric parameter p.* The two antagonist conditions on p induced by finiteness of the variance and the mean complexity read as

$$\frac{1}{N^\beta} < 1 - p < \frac{1}{N},$$

which may hold if and only if $\beta > 1$. A reasonable choice is then to set $1 - p$ to be the geometric mean of these bounds, namely

$$p = 1 - N^{\frac{1+\beta}{2}}. \tag{9.110}$$

- *Size of the simulation.* To specify our estimator for a given *RMSE*, we take advantage of its unbiased feature and note that

$$\big\|\widehat{I}_M^{RML} - I_0\big\|_2^2 = \mathrm{Var}\big(\widehat{I}_M^{RML}\big) = \frac{\mathrm{Var}\big(I_\tau\big)}{M}.$$

The prescription $\big\|\widehat{I}_M^{RML} - I_0\big\|_2 = \varepsilon$, reads like for any unbiased estimator,

$$M^*(\varepsilon) = \varepsilon^{-2}\mathrm{Var}\big(I_\tau\big).$$

In view of practical implementation a short pre-processing is necessary to roughly calibrate $\mathrm{Var}\big(I_\tau\big)$.

9.8.2 Connection with Former Multilevel Frameworks

The transcription to our multilevel framework (see Sect. 9.2) is straightforward: with our usual notations, we set for every $h \in \mathcal{H} = \big\{\frac{\mathbf{h}}{n},\ n \ge 1\big\}$, $\mathbf{h} \in (0, +\infty)$.

$$Z_1 = Y_h^{(1)}, \quad \text{and} \quad Z_i = Y_{\frac{h}{n_i}}^{(i)} - Y_{\frac{h}{n_{i-1}}}^{(i)},\ i \ge 2. \tag{9.111}$$

However, we insist upon the fact that this unbiased simulation method can only be implemented if the family $(Y_h)_{h\in\mathcal{H}}$ satisfies $(SE)_\beta$ with $\beta > 1$ (fast quadratic approximation). By contrast, at a first glance, *no assumptions on the weak error expansion* are needed. This observation should be tempered by the result stated in Claim (b) of the former proposition, where the Z_n are assumed independent. This setting is actually the closest to multilevel multilevel approach and the second series in (9.107) involves $\mathbb{E}\, Z_n$, which implies a control of the weak error to ensure its convergence.

► **Exercise** (*Linear refiners*). Set $n_i = i, i \geq 1$, in the multilevel framework. Assume $\mathcal{H} = \{\mathbf{k}/n,\ n \geq 1\}$ and the sequence $(Z_n)_{n\geq 1}$ is made of independent random variables (see Claim (c) of the above Proposition 9.4).

- Inverse linear complexity: there exists a $\kappa > 0$ such that $\kappa(Y_h) = \kappa/h, h \in \mathcal{H}$.
- $(SE)_\beta \;\equiv\; \text{Var}(Y_h - Y_{\frac{h}{i}}) \leq V_1 h^\beta \big(1 - 1/i\big)^\beta, h \in \mathcal{H}, i \geq 2.$
- $(WE)^1_\alpha \;\equiv\; \mathbb{E}\, Y_h + c_1 h + o(h)$ as $h \to 0$ in $\mathcal{H}$.

(a) Devise a new version of randomized Monte Carlo based on these assumptions (including a new distribution for τ).
(b) Show that, once again, it is unfortunately only *efficient* (by *efficient* we mean that it is possible to choose the distribution of τ so that the mean cost and the variance of the unbiased randomized multilevel estimate both remain finite) if $\beta > 1$.

Application to *weak exact* simulation of *SDE*s

In particular, in an *SDE* framework, a multilevel implementation of the Milstein scheme provides an efficient unbiased method of simulation of the expectation of a marginal function of a diffusion $\mathbb{E}\, f(X_T)$. This is always possible in one dimension (scalar Brownian motion, *i.e.* $q = 1$ with our usual notations) under appropriate smoothness assumptions on the drift and the diffusion coefficient of the *SDE*.

In higher dimensions ($q \geq 2$), an unbiased simulation method can be devised by considering the antithetic (meta-)scheme, see Sect. 9.6.1,

$$Z_1 = Y_h^{(1)}, \quad \text{and} \quad Z_i = \frac{1}{2}\big(Y^{[1],(i)}_{\frac{h}{n_i}} + Y^{[2],(i)}_{\frac{h}{n_i}}\big) - Y^{(i)}_{\frac{h}{n_{i-1}}},\ i \geq 2,$$

(with the notations of this section). Such an approach provides a rather simple *weak exact* simulation method of a diffusion among those mentioned in the introduction of this chapter.

9.8.3 Numerical Illustration

We tested the performances of the Randomized Multilevel estimator in the above with consistent and independent levels on a regular Black–Scholes Call in one dimension, discretized by a Milstein scheme in order to guarantee that $\beta > 1$. The levels have been designed following the multilevel framework described above. The parameters of the Call are

$$x_0 = 100, \quad K = 80, \quad \sigma = 0.4.$$

As for the simulation, we set $N = 6$ so that

$$p^* = 0.931959$$

owing to Formula (9.110) (further numerical experiments show that this choice is close to optimal).

First we compared the Randomized Multilevel estimator (see the former subsection) with consistent levels (the same Brownian underlying increments are used for the simulation at all levels) and the Randomized Multilevel estimator with independent levels (in the spirit of the multilevel paradigm described at the beginning of Sect. 9.5). Figure 9.7 depicts the respective performances of the two estimators. Together with the computations carried out in Proposition 9.4, it confirms the intuition that the Randomized Multilevel estimator with independent levels outperforms that with constant levels. To be more precise the variance of I_τ is approximately equal to 1617.05 in the consistent levels case and 1436.61 in the independent levels case.

As a second step, with the same toy benchmark, we tested the Randomized Multilevel estimator versus the weighted and regular Multilevel estimators described in the former sections. Figure 9.8 highlights that even for not so small prescribed *RMSE* ($\varepsilon = 2^{-k}$, $k = 1, \ldots, 5$ in this simulation), the Randomized Multilevel estimator (with independent levels) is outperformed by both weighted and regular multilevel

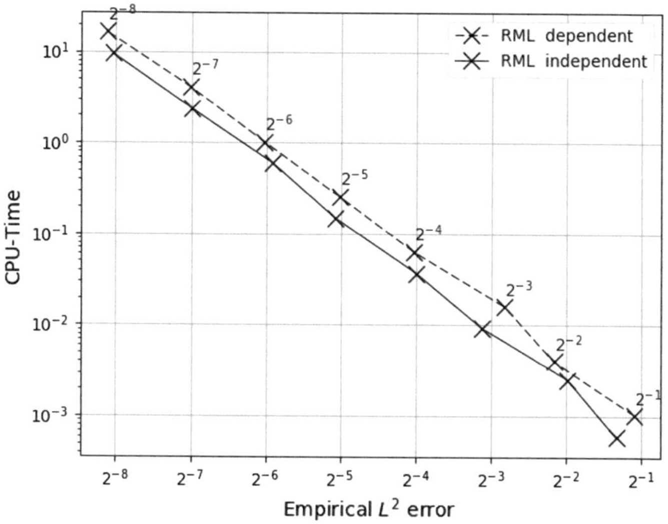

Fig. 9.7 *RML* ESTIMATORS: Constant vs independent levels (on a Black–Scholes Call). CPU-time (y–axis, log scale) as a function of the empirical RMSE $\tilde{\varepsilon}$ (x-axis, log scale) (with D. Giorgi)

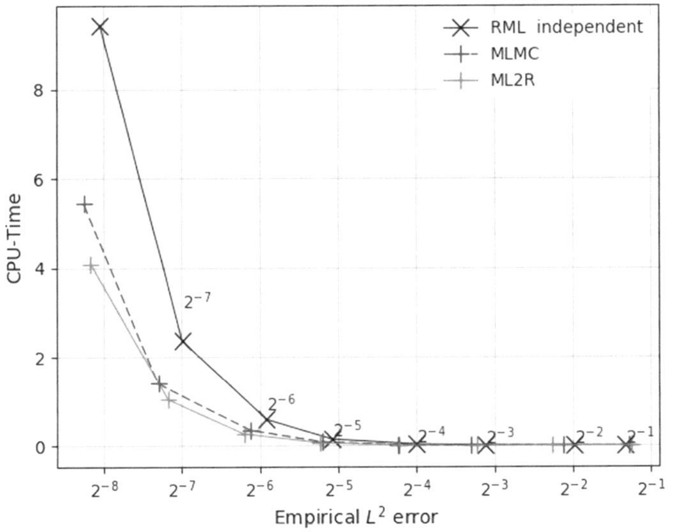

Fig. 9.8 *RML* ESTIMATOR VS ML2R AND MLMC ESTIMATORS: CPU-time as a function of the empirical RMSE $\tilde{\varepsilon}$ (x-axis, log scale) (with D. Giorgi)

estimators. This phenomenon can be attributed to exogenous variance induced by the random times τ_i.

As a conclusion of this action, numerical tests carried out with the Milstein scheme for vanilla options in a Black–Scholes model (see, for example, *e.g.* [113] for more details and further simulations) share the following features:

- The expected theoretical behavior as $\varepsilon \to 0$ is observed, corresponding to its counterpart $\beta > 1$ in the multilevel framework, as well as the absence of bias.
- The procedure shows better performances when implemented with independent levels than fully consistent levels.
- This *RML* estimator is always slower than the weighted and regular multilevel estimators introduced in former sections. This is undoubtedly due to the exogenous noise induced by the random times τ_i used to determine the number of levels on each path.

► **Exercise.** Compute by any means at your disposal the value of the constant

$$C = \mathbb{E}\,\|W\|_{L^2([0,1],dt)} \quad \text{with an accuracy of } \pm 3.10^{-7} \text{ at a confidence level } 95\%.$$

(This constant C was introduced in Sect. 5.3.1 – Exercise **4.** of Practitioner's corner to be precise – in order to produce optimal quantizers of the Lévy area.)

Chapter 10
Back to Sensitivity Computation

Let ([1]) $Z : (\Omega, \mathcal{A}, \mathbb{P}) \to (E, \mathcal{E})$ be an E-valued random variable where $(E, \mathcal{E})$ is an abstract measurable space, let I be a nonempty open interval of $\mathbb{R}$ and let $F : I \times E \to \mathbb{R}$ be a $\mathcal{B}or(I) \otimes \mathcal{E}$-measurable function such that, for every $x \in I$, $F(x, Z) \in L^2(\mathbb{P})$ ([2]). Then set

$$f(x) = \mathbb{E}\, F(x, Z).$$

Assume that the function f is regular, at least at some points. Our aim is to devise a method to compute by simulation $f'(x)$ or higher derivatives $f^{(k)}(x)$, $k \geq 1$, at such points x. If the functional $F(x, z)$ is differentiable in its first variable at x for $\mathbb{P}_Z$-almost every z, if a domination or uniform integrability property holds (like (10.1) or (10.5) below), if the partial derivative $\frac{\partial F}{\partial x}(x, z)$ can be computed and Z can be simulated both, at a reasonable computational cost, it is natural to compute $f'(x)$ using a Monte Carlo simulation based on the representation formula

$$f'(x) = \mathbb{E}\left(\frac{\partial F}{\partial x}(x, Z)\right).$$

[1]Although this chapter comes right after the presentation of multilevel methods, we decided to expose the main available approaches for simulation-based sensitivity computation on their own, in order not to be polluted by technicalities induced by a mix (or a merge). However, the multilevel paradigm may be applied efficiently to improve the performances of the Monte Carlo-based methods that we will present: in most cases we have at hand a strong error rate and a weak error expansion. This can be easily checked on the finite difference method, for example. For an application of the multilevel approach to Greek computation, we refer to [56].

[2]In fact, E can be replaced by the probability space Ω itself: Z becomes the canonical variable/process on this probability space (endowed with the distribution $\mathbb{P} = \mathbb{P}_Z$ of the process). In particular, Z can be the Brownian motion or any process at time T starting at x or its entire path, etc. The notation is essentially formal and could be replaced by the more general $F(x, \omega)$.

G. Pagès, *Numerical Probability*, Universitext,
https://doi.org/10.1007/978-3-319-90276-0_10

This approach has already been introduced in Chap. 2 and will be more deeply developed further on, in Sect. 10.2, mainly devoted to the tangent process method for diffusions.

Otherwise, when $\frac{\partial F}{\partial x}(x, z)$ does not exist or cannot be computed easily (whereas F can), a natural idea is to introduce a stochastic finite difference approach. Other methods based on the introduction of an appropriate weight will be introduced in the last two sections of this chapter.

10.1 Finite Difference Method(s)

The finite difference method is in some way the most elementary and natural method for computing sensitivity parameters, known as Greeks when dealing with financial derivatives, although it is an approximate method in its standard form. This is also known in financial Engineering as the "Bump Method" or "Shock Method". It can be described in a very general setting which corresponds to its wide field of application. Finite difference methods were been originally investigated in [117, 119, 194].

10.1.1 The Constant Step Approach

We consider the framework described in the introduction. We will distinguish two cases: in the first one – called the "regular setting" – the function $x \mapsto F(x, Z(\omega))$ is "not far" from being pathwise differentiable whereas in the second one – called the "singular setting" – f remains smooth but F becomes "singular".

The regular setting

Proposition 10.1 *Let $x \in \mathbb{R}$. Assume that F satisfies the following local mean quadratic Lipschitz continuous assumption at x*

$$\exists\, \varepsilon_0 > 0,\ \forall\, x' \in (x - \varepsilon_0, x + \varepsilon_0),\ \big\|F(x, Z) - F(x', Z)\big\|_2 \le C_{F,Z}|x - x'|. \quad (10.1)$$

Assume the function f is twice differentiable with a Lipschitz continuous second derivative on $(x - \varepsilon_0, x + \varepsilon_0)$. Let $(Z_k)_{k\ge1}$ be a sequence of i.i.d. random vectors with the same distribution as Z. Then for every $\varepsilon \in (0, \varepsilon_0)$, the mean quadratic error or Root Mean Square Error *(RMSE) satisfies*

$$
\begin{aligned}
&\left\| f'(x) - \frac{1}{M}\sum_{k=1}^{M} \frac{F(x+\varepsilon, Z_k) - F(x-\varepsilon, Z_k)}{2\varepsilon} \right\|_2 \\
&\quad \le \sqrt{\left([f'']_{\text{Lip}}\frac{\varepsilon^2}{2}\right)^2 + \frac{C_{F,Z}^2 - \left(f'(x) - \frac{\varepsilon^2}{2}[f'']_{\text{Lip}}\right)_+^2}{M}} \qquad (10.2)\\
&\quad \le \sqrt{\left([f'']_{\text{Lip}}\frac{\varepsilon^2}{2}\right)^2 + \frac{C_{F,Z}^2}{M}} \qquad (10.3)\\
&\quad \le [f'']_{\text{Lip}}\frac{\varepsilon^2}{2} + \frac{C_{F,Z}}{\sqrt{M}}.
\end{aligned}
$$

Furthermore, if f is three times differentiable on $(x-\varepsilon_0, x+\varepsilon_0)$ with a bounded third derivative, then one can replace $[f'']_{\text{Lip}}$ by $\frac{1}{3}\sup_{|\xi-x|\le\varepsilon_0}|f^{(3)}(\xi)|$.

Remark. In the above sum $[f'']_{\text{Lip}}\frac{\varepsilon^2}{2}$ represents the *bias* and $\frac{C_{F,Z}}{\sqrt{M}}$ is the *statistical* error.

Proof. Let $\varepsilon \in (0, \varepsilon_0)$. It follows from the Taylor formula applied to f between x and $x \pm \varepsilon$, respectively, that

$$
\left| f'(x) - \frac{f(x+\varepsilon) - f(x-\varepsilon)}{2\varepsilon} \right| \le [f'']_{\text{Lip}}\frac{\varepsilon^2}{2}. \qquad (10.4)
$$

On the other hand

$$
\mathbb{E}\left(\frac{F(x+\varepsilon, Z) - F(x-\varepsilon, Z)}{2\varepsilon}\right) = \frac{f(x+\varepsilon) - f(x-\varepsilon)}{2\varepsilon}
$$

and

$$
\begin{aligned}
&\text{Var}\left(\frac{F(x+\varepsilon, Z) - F(x-\varepsilon, Z)}{2\varepsilon}\right)\\
&\quad = \mathbb{E}\left(\left(\frac{F(x+\varepsilon, Z) - F(x-\varepsilon, Z)}{2\varepsilon}\right)^2\right) - \left(\frac{f(x+\varepsilon) - f(x-\varepsilon)}{2\varepsilon}\right)^2\\
&\quad = \frac{\mathbb{E}\,(F(x+\varepsilon, Z) - F(x-\varepsilon, Z))^2}{4\varepsilon^2} - \left(\frac{f(x+\varepsilon) - f(x-\varepsilon)}{2\varepsilon}\right)^2\\
&\quad \le C_{F,Z}^2 - \left(\frac{f(x+\varepsilon) - f(x-\varepsilon)}{2\varepsilon}\right)^2 \le C_{F,Z}^2. \qquad (10.5)
\end{aligned}
$$

Using the bias-variance decomposition (sometimes called Huygens' Theorem in Statistics ([3])), we derive the following upper-bound for the mean squared error

[3] which formally simply reads $\mathbb{E}\,|X-a|^2 = (a - \mathbb{E}\,X)^2 + \text{Var}(X)$.

$$\begin{aligned}
&\left\| f'(x) - \frac{1}{M}\sum_{k=1}^{M} \frac{F(x+\varepsilon, Z_k) - F(x-\varepsilon, Z_k)}{2\varepsilon} \right\|_2^2 \\
&\qquad = \left(f'(x) - \frac{f(x+\varepsilon) - f(x-\varepsilon)}{2\varepsilon} \right)^2 \\
&\qquad\quad + \mathrm{Var}\left(\frac{1}{M}\sum_{k=1}^{M} \frac{F(x+\varepsilon, Z_k) - F(x-\varepsilon, Z_k)}{2\varepsilon} \right) \\
&\qquad = \left(f'(x) - \frac{f(x+\varepsilon) - f(x-\varepsilon)}{2\varepsilon} \right)^2 \\
&\qquad\quad + \frac{1}{M}\mathrm{Var}\left(\frac{F(x+\varepsilon, Z) - F(x-\varepsilon, Z)}{2\varepsilon} \right) \\
&\qquad \le \left([f'']_{\mathrm{Lip}} \frac{\varepsilon^2}{2} \right)^2 + \frac{1}{M}\left(C_{F,Z}^2 - \left(\frac{f(x+\varepsilon) - f(x-\varepsilon)}{2\varepsilon} \right)^2 \right)
\end{aligned}$$

where we combined (10.4) and (10.5) to derive the last inequality, *i.e.* (10.3). To get the improved bound (10.2), we first derive from (10.4) that

$$\frac{f(x+\varepsilon) - f(x-\varepsilon)}{2\varepsilon} \ge f'(x) - \frac{\varepsilon^2}{2}[f'']_{\mathrm{Lip}}$$

so that

$$\begin{aligned}
\left| \mathbb{E}\left(\frac{F(x+\varepsilon, Z) - F(x-\varepsilon, Z)}{2\varepsilon} \right) \right| &= \left| \frac{f(x+\varepsilon) - f(x-\varepsilon)}{2\varepsilon} \right| \\
&\ge \left(\frac{f(x+\varepsilon) - f(x-\varepsilon)}{2\varepsilon} \right)_+ \\
&\ge \left(f'(x) - \frac{\varepsilon^2}{2}[f'']_{\mathrm{Lip}} \right)_+ .
\end{aligned}$$

Plugging this lower bound into the definition of the variance yields the announced inequality. ◇

ℵ **Practitioner's corner.** The above result suggests to choose $M = M(\varepsilon)$ (and ε) to equalize the variance and the statistical error term by setting

$$M(\varepsilon) = \left(\frac{2C_{F,Z}}{\varepsilon^2 [f'']_{\mathrm{Lip}}} \right)^2 = \left(\frac{2C_{F,Z}}{[f'']_{\mathrm{Lip}}} \right)^2 \varepsilon^{-4}.$$

However, this is not realistic since $C_{F,Z}$ is only an upper-bound of the variance term and $[f'']_{\mathrm{Lip}}$ is usually unknown. An alternative is to choose

$$M(\varepsilon) = o\left(\varepsilon^{-4}\right)$$

to be sure that the statistical error becomes smaller. However, it is of course useless to carry on the simulation too far since the bias error is not impacted. Note that such specification of the size M of the simulation breaks the recursive feature of the estimator. Another way to use such an error bound is to keep in mind that, in order to reduce the error by a factor of 2, we need to reduce ε and increase M as follows:

$$\varepsilon \rightsquigarrow \varepsilon/\sqrt{2} \quad \text{and} \quad M \rightsquigarrow 4\,M.$$

Warning (what should never be done)! Imagine that we are using two *independent* samples $(Z_k)_{k\geq 1}$ and $(\widetilde{Z}_k)_{k\geq 1}$ to simulate copies $F(x-\varepsilon, Z)$ and $F(x+\varepsilon, Z)$. Then,

$$\begin{aligned}\operatorname{Var}\left(\frac{1}{M}\sum_{k=1}^{M}\frac{F(x+\varepsilon, Z_k)-F(x-\varepsilon, \widetilde{Z}_k)}{2\varepsilon}\right) &= \frac{1}{4M\varepsilon^2}\Big(\operatorname{Var}\big(F(x+\varepsilon, Z)\big)+\operatorname{Var}\big(F(x-\varepsilon, Z)\big)\Big)\\ &\simeq \frac{\operatorname{Var}\big(F(x, Z)\big)}{2M\varepsilon^2}.\end{aligned}$$

Note that the asymptotic variance of the estimator of $\frac{f(x+\varepsilon)-f(x-\varepsilon)}{2\varepsilon}$ explodes as $\varepsilon \to 0$ and the resulting quadratic error reads approximately

$$[f'']_{\text{Lip}}\frac{\varepsilon^2}{2}+\frac{\sigma\big(F(x, Z)\big)}{\varepsilon\sqrt{2M}},$$

where $\sigma(F(x, Z)) = \sqrt{\operatorname{Var}(F(x, Z))}$ is the standard deviation of $F(x, Z)$. This leads to consider the unrealistic constraint $M(\varepsilon) \propto \varepsilon^{-6}$ to keep the balance between the bias term and the variance term; or equivalently to switch $\varepsilon \rightsquigarrow \varepsilon/\sqrt{2}$ and $M \rightsquigarrow 8\,M$ to reduce the error by a factor of 2.

⊳ **Examples (Greeks computation). 1.** *Sensitivity in a Black–Scholes model.* Vanilla payoffs viewed as functions of a normal distribution correspond to functions F of the form

$$F(x, z) = e^{-rt}h\left(x\,e^{(r-\frac{\sigma^2}{2})T+\sigma\sqrt{T}z}\right), \quad z\in\mathbb{R},\ x\in(0, +\infty),$$

where $h:\mathbb{R}_+\to\mathbb{R}$ is a Borel function with linear growth. If h is Lipschitz continuous, then

$$|F(x, Z)-F(x', Z)| \leq [h]_{\text{Lip}}|x-x'|e^{-\frac{\sigma^2}{2}T+\sigma\sqrt{T}Z}$$

so that elementary computations show, using that $Z \stackrel{d}{=} \mathcal{N}(0;1)$,

$$\|F(x, Z)-F(x', Z)\|_2 \leq [h]_{\text{Lip}}|x-x'|e^{\frac{\sigma^2}{2}T}.$$

The regularity of f follows from the following easy change of variable

$$f(x) = e^{-rt}\int_{\mathbb{R}} h\left(x\, e^{\mu T+\sigma\sqrt{T}z}\right)e^{-\frac{z^2}{2}}\frac{dz}{\sqrt{2\pi}} = e^{-rt}\int_0^{+\infty} h(y)e^{-\frac{(\log(x/y)+\mu T)^2}{2\sigma^2 T}}\frac{dy}{y\sigma\sqrt{2\pi T}}$$

where $\mu = r - \frac{\sigma^2}{2}$. This change of variable makes the integral appear as a "log"-convolution on $(0, +\infty)$ ([4]) with similar regularizing effects as the standard convolution on the whole real line. Under the appropriate growth assumption on the function h (say polynomial growth), one shows from the above identity that the function f is in fact infinitely differentiable over $(0, +\infty)$. In particular, it is twice differentiable with Lipschitz continuous second derivative over any compact interval included in $(0, +\infty)$.

2. *Diffusion model with Lipschitz continuous coefficients.* Let $X^x = (X_t^x)_{t\in[0,T]}$ denote the Brownian diffusion solution of the *SDE*

$$dX_t = b(X_t)dt + \vartheta(X_t)dW_t, \quad X_0 = x,$$

where b and ϑ are locally Lipschitz continuous functions (on the real line) with at most linear growth (which implies the existence and uniqueness of a strong solution $(X_t^x)_{t\in[0,T]}$ starting from $X_0^x = x$). In such a case, one should instead write

$$F(x, \omega) = h\big(X_T^x(\omega)\big).$$

The Lipschitz continuity of the flow of the above *SDE* (see Theorem 7.10) shows that

$$\|F(x, .) - F(x', .)\|_2 \le C_{b,\vartheta}[h]_{\mathrm{Lip}}|x - x'|e^{C_{b,\vartheta}T}$$

where $C_{b,\vartheta}$ is a positive constant only depending on the Lipschitz continuous coefficients of b and ϑ. In fact, this also holds for multi-dimensional diffusion processes and for path-dependent functionals.

The regularity of the function f is a less straightforward question. But the answer is positive in two situations: either h, b and σ are regular enough to apply results on the flow of the SDE which allows pathwise differentiation of $x \mapsto F(x, \omega)$ (see Theorem 10.1 further on in Sect. 10.2.2) or ϑ satisfies a uniform ellipticity assumption $\vartheta \ge \varepsilon_0 > 0$.

3. *Euler scheme of a Brownian diffusion model with Lipschitz continuous coefficients.* The same holds for the Euler scheme. Furthermore, Assumption (10.1) holds uniformly with respect to n if $\frac{T}{n}$ is the step size of the Euler scheme.

4. F can also be a functional of the whole path of a diffusion, provided F is Lipschitz continuous with respect to the sup-norm over $[0, T]$.

---

[4] The convolution on $(0, +\infty)$ is defined between two non-negative functions f and g on $(0, +\infty)$ by $f \odot g(x) = \displaystyle\int_0^{+\infty} f(x/y)g(y)dy$.

As emphasized in the section devoted to the tangent process below, the generic parameter x can be the maturity T (in practice the *residual maturity* $T-t$, also known as *seniority*), or any finite-dimensional parameter on which the diffusion coefficients depend since they can always be seen as an additional component or a starting value of the diffusion.

► **Exercises. 1.** Adapt the results of this section to the case where $f'(x)$ is estimated by its "forward" approximation

$$f'(x) \simeq \frac{f(x+\varepsilon)-f(x)}{\varepsilon}.$$

2. *Richardson–Romberg extrapolation.* Assume f is $\mathcal{C}^3$ with Lipschitz third derivative in the neighborhood of x. Starting from the expansion

$$\frac{f(x+\varepsilon)-f(x-\varepsilon)}{2\varepsilon} = f'(x) + f^{(3)}(x)\frac{\varepsilon^2}{6} + O(\varepsilon^3),$$

propose a Richardson–Romberg extrapolation method (see Sect. 9.4) based on the estimator $\widehat{f'(x)}_{\varepsilon,M} = \frac{1}{M}\sum_{1\le k\le M}\frac{F(x+\varepsilon,Z_k)-F(x-\varepsilon,Z_k)}{2\varepsilon}$ with ε and $\varepsilon/2$

3. *Multilevel estimators (see Sect.* 9.5). (a) Assume f is $\mathcal{C}^4$ and refine the above expansion given in the previous exercise. Give a condition on the random function $F(x,\omega)$ which makes it possible to design and implement a regular multilevel estimator for $f(x)$.
(b) Under which assumptions is it possible to design and implement a weighted Multilevel estimator (*ML2R*)?
(c) Which conclusions should be drawn from (a) and (b).

4. Apply the above method(s) to approximate the γ-parameter by considering that

$$f''(x) \simeq \frac{f(x+\varepsilon)+f(x-\varepsilon)-2f(x)}{\varepsilon^2}$$

under suitable assumptions on f and its derivatives.

The singular setting

In the above setting described in Proposition 10.1, we are close to a framework in which one can interchange derivation and expectation: the (local) Lipschitz continuous assumption on the random function $x'\mapsto F(x',Z)$ implies that $\left(\frac{F(x',Z)-F(x,Z)}{x'-x}\right)_{x'\in(x-\varepsilon_0,x+\varepsilon_0)\setminus\{x\}}$ is a uniformly integrable family. Hence, as soon as $x'\mapsto F(x',Z)$ is $\mathbb{P}$-$a.s.$ pathwise differentiable at x (or even simply in $L^2(\mathbb{P})$), one has $f'(x)=\mathbb{E}\,F'_x(x,Z)$.

Consequently, it is important to investigate the *singular setting* in which f is differentiable at x and $F(\,.\,,Z)$ is not Lipschitz continuous in L^2. This is the purpose of the next proposition (whose proof is quite similar to the Lipschitz continuous setting and is subsequently left to the reader as an exercise).

Proposition 10.2 *Let $x \in \mathbb{R}$. Assume that F satisfies in a x open neighborhood $(x-\varepsilon_0, x+\varepsilon_0)$, $\varepsilon_0 > 0$, of x the following local mean quadratic θ-Hölder assumption, $\theta \in (0,1]$, at x: there exists a positive real constant $C_{Hol,F,Z}$*

$$\forall\, x', x'' \in (x-\varepsilon_0, x+\varepsilon_0), \quad \big\|F(x',Z) - F(x'',Z)\big\|_2 \le C_{Hol,F,Z}|x'-x''|^{\theta}.$$

Assume the function f is twice differentiable with a Lipschitz continuous second derivative on $(x-\varepsilon_0, x+\varepsilon_0)$. Let $(Z_k)_{k\ge1}$ be a sequence of i.i.d. random vectors with the same distribution as Z. Then, for every $\varepsilon \in (0, \varepsilon_0)$, the RMSE satisfies

$$\begin{aligned}\left\| f'(x) - \frac{1}{M}\sum_{k=1}^{M}\frac{F(x+\varepsilon, Z_k) - F(x-\varepsilon, Z_k)}{2\varepsilon}\right\|_2 &\le \sqrt{\left([f'']_{\rm Lip}\frac{\varepsilon^2}{2}\right)^2 + \frac{C^2_{Hol,F,Z}}{(2\varepsilon)^{2(1-\theta)}M}}\\ &\le [f'']_{\rm Lip}\frac{\varepsilon^2}{2} + \frac{C_{Hol,F,Z}}{(2\varepsilon)^{1-\theta}\sqrt{M}}. \qquad (10.6)\end{aligned}$$

This variance of the finite difference estimator explodes as $\varepsilon \to 0$ as soon as $\theta < 1$. As a consequence, in such a framework, to divide the quadratic error by a factor of 2, we need to switch

$$\varepsilon \rightsquigarrow \varepsilon/\sqrt{2} \qquad \text{and} \qquad M \rightsquigarrow 2^{1-\theta} \times 4\,M.$$

A dual point of view in this singular case is to (roughly) optimize the parameter $\varepsilon = \varepsilon(M)$, given a simulation size of M in order to minimize the quadratic error, or at least its natural upper-bounds. Such an optimization performed on (10.6) yields

$$\varepsilon_{opt} = \left(\frac{2^{\theta}C_{Hol,F,Z}}{[f'']_{\rm Lip}\sqrt{M}}\right)^{\frac{1}{3-\theta}}$$

which of course depends on M so that it breaks the recursiveness of the estimator. Moreover, its sensitivity to $[f'']_{\rm Lip}$ (and to $C_{Hol,F,Z}$) makes its use rather unrealistic in practice.

The resulting rate of decay of the quadratic error is $O\left(M^{-\frac{2-\theta}{3-\theta}}\right)$. This rate shows that when $\theta \in (0,1)$, the lack of L^2-regularity of $x \mapsto F(x,Z)$ slows down the convergence of the finite difference method by contrast with the Lipschitz continuous case where the standard rate of convergence of the Monte Carlo method is preserved.

$\rhd$ **Example of the digital option.** A typical example of such a situation is the pricing of digital options (or equivalently the computation of the δ-hedge of a Call or Put options).

Let us consider, still in the standard risk neutral Black–Scholes model, a digital Call option with strike price $K > 0$ defined by its payoff

$$h(\xi) = \mathbf{1}_{\{\xi \ge K\}}$$

and set $F(x, z) = e^{-rt} h\left(x\, e^{(r-\frac{\sigma^2}{2})T+\sigma\sqrt{T}z}\right)$, $z \in \mathbb{R}$, $x \in (0, +\infty)$ (r denotes the constant interest rate as usual). We know that the premium of this option is given for every initial price $x > 0$ of the underlying risky asset by

$$f(x) = \mathbb{E}\, F(x, Z) \quad \text{with} \quad Z \stackrel{d}{=} \mathcal{N}(0; 1).$$

Set $\mu = r - \frac{\sigma^2}{2}$. It is clear since $Z \stackrel{d}{=} -Z$ that

$$\begin{aligned} f(x) &= e^{-rt}\mathbb{P}\Big(x\, e^{\mu T+\sigma\sqrt{T}Z} \geq K\Big) \\ &= e^{-rt}\mathbb{P}\Big(Z \geq -\frac{\log(x/K) + \mu T}{\sigma\sqrt{T}}\Big) \\ &= e^{-rt}\Phi_0\Big(\frac{\log(x/K) + \mu T}{\sigma\sqrt{T}}\Big), \end{aligned}$$

where Φ_0 denotes the c.d.f. of the $\mathcal{N}(0; 1)$ distribution. Hence the function f is infinitely differentiable on $(0, +\infty)$.

On the other hand, still using $Z \stackrel{d}{=} -Z$, for every $x, x' \in \mathbb{R}$,

$$\begin{aligned} &\big\|F(x, Z) - F(x', Z)\big\|_2^2 \\ &= e^{-2rT}\left\|\mathbf{1}_{\left\{Z \geq -\frac{\log(x/K)+\mu T}{\sigma\sqrt{T}}\right\}} - \mathbf{1}_{\left\{Z \geq -\frac{\log(x'/K)+\mu T}{\sigma\sqrt{T}}\right\}}\right\|_2^2 \\ &= e^{-2rT}\mathbb{E}\left|\mathbf{1}_{\left\{Z \leq \frac{\log(x/K)+\mu T}{\sigma\sqrt{T}}\right\}} - \mathbf{1}_{\left\{Z \leq \frac{\log(x'/K)+\mu T}{\sigma\sqrt{T}}\right\}}\right|^2 \\ &= e^{-2rT}\left(\Phi_0\Big(\frac{\log(\max(x, x')/K) + \mu T}{\sigma\sqrt{T}}\Big) - \Phi_0\Big(\frac{\log(\min(x, x')/K) + \mu T}{\sigma\sqrt{T}}\Big)\right). \end{aligned}$$

Using that Φ_0' is bounded by $\kappa_0 = \frac{1}{\sqrt{2\pi}}$, we derive that

$$\big\|F(x, Z) - F(x', Z)\big\|_2^2 \leq \frac{\kappa_0 e^{-2rT}}{\sigma\sqrt{T}}\big|\log x - \log x'\big|.$$

Consequently for every interval $I \subset (0, +\infty)$ *bounded away from* 0, there exists a real constant $C_{r,\sigma,T,I} > 0$ such that

$$\forall\, x,\ x' \in I, \quad \|F(x, Z) - F(x', Z)\|_2 \leq C_{r,\sigma,T,I}\sqrt{|x - x'|},$$

i.e. the functional F is $\frac{1}{2}$-Hölder in $L^2(\mathbb{P})$ and the above proposition applies.

► **Exercises. 1.** Prove the above Proposition 10.2.

2. *Digital option.* (a) Consider in the risk neutral Black–Scholes model a digital option defined by its payoff

$$h(\xi) = \mathbf{1}_{\{\xi \geq K\}}$$

and set $F(x, z) = e^{-rt} h\left(x\, e^{(r-\frac{\sigma^2}{2})T+\sigma\sqrt{T}z}\right)$, $z \in \mathbb{R}$, $x \in (0, +\infty)$ (r is a constant interest rate as usual). We still consider the computation of $f(x) = \mathbb{E}\, F(x, Z)$ where $Z \stackrel{d}{=} \mathcal{N}(0; 1)$.

Verify on a numerical simulation that the variance of the finite difference estimator introduced in Proposition 10.1 explodes as $\varepsilon \to 0$ at the rate expected from the preceding computations.

(b) Derive from the preceding a way to "synchronize" the step ε and the size M of the simulation.

10.1.2 A Recursive Approach: Finite Difference with Decreasing Step

In the former finite difference method with constant step, the bias never fades. Consequently, increasing the accuracy of the sensitivity computation, it has to be resumed from the beginning with a new ε. In fact, it is easy to propose a recursive version of the above finite difference procedure by considering some variable steps ε which go to 0. This can be seen as an application of the Kiefer–Wolfowitz principle originally developed for Stochastic Approximation purposes.

We will focus on the "regular setting" (F Lipschitz continuous in L^2) in this section, the singular setting is proposed as an exercise. Let $(\varepsilon_k)_{k\geq 1}$ be a sequence of positive real numbers decreasing to 0. With the notations and the assumptions of the former section, consider the estimator

$$\widehat{f'(x)}_M := \frac{1}{M}\sum_{k=1}^{M} \frac{F(x+\varepsilon_k, Z_k) - F(x-\varepsilon_k, Z_k)}{2\varepsilon_k}. \tag{10.7}$$

It can be computed in a recursive way since

$$\widehat{f'(x)}_{M+1} = \widehat{f'(x)}_M + \frac{1}{M+1}\left(\frac{F(x+\varepsilon_{M+1}, Z_{M+1}) - F(x-\varepsilon_{M+1}, Z_{M+1})}{2\varepsilon_{M+1}} - \widehat{f'(x)}_M\right).$$

Elementary computations show that the mean squared error satisfies

$$\left\| f'(x) - \widehat{f'(x)}_M \right\|_2^2 = \left(f'(x) - \frac{1}{M} \sum_{k=1}^{M} \frac{f(x+\varepsilon_k) - f(x-\varepsilon_k)}{2\varepsilon_k} \right)^2$$
$$+ \frac{1}{M^2} \sum_{k=1}^{M} \frac{\mathrm{Var}(F(x+\varepsilon_k, Z_k) - F(x-\varepsilon_k, Z_k))}{4\varepsilon_k^2}$$
$$\leq \frac{[f'']_{\mathrm{Lip}}^2}{4M^2} \left(\sum_{k=1}^{M} \varepsilon_k^2 \right)^2 + \frac{C_{F,Z}^2}{M} \tag{10.8}$$
$$= \frac{1}{M} \left(\frac{[f'']_{\mathrm{Lip}}^2}{4M} \left(\sum_{k=1}^{M} \varepsilon_k^2 \right)^2 + C_{F,Z}^2 \right)$$

where we again used (10.4) to get Inequality (10.8). As a consequence, the *RMSE* satisfies

$$\left\| f'(x) - \widehat{f'(x)}_M \right\|_2 \leq \frac{1}{\sqrt{M}} \sqrt{\frac{[f'']_{\mathrm{Lip}}^2}{4M} \left(\sum_{k=1}^{M} \varepsilon_k^2 \right)^2 + C_{F,Z}^2}. \tag{10.9}$$

L^2-rate of convergence (erasing the asymptotic bias)
An efficient way to prove an $\frac{1}{\sqrt{M}}$ L^2-rate like in a standard Monte Carlo simulation is to erase the bias by an appropriate choice of the sequence $(\varepsilon_k)_{k\geq 1}$. The bias term will fade as $M \to +\infty$ if

$$\left(\sum_{k=1}^{M} \varepsilon_k^2 \right)^2 = o(M).$$

This leads us to choose ε_k of the form

$$\varepsilon_k = o\big(k^{-\frac{1}{4}}\big) \quad \text{as} \quad k \to +\infty,$$

since, then, $\sum_{k=1}^{M} \varepsilon_k^2 = o\left(\sum_{k=1}^{M} \frac{1}{k^{\frac{1}{2}}} \right)$ as $M \to +\infty$ and

$$\sum_{k=1}^{M} \frac{1}{k^{\frac{1}{2}}} = \sqrt{M} \frac{1}{M} \sum_{k=1}^{M} \left(\frac{k}{M} \right)^{-\frac{1}{2}} \sim \sqrt{M} \int_0^1 \frac{dx}{\sqrt{x}} = 2\sqrt{M} \quad \text{as} \quad M \to +\infty.$$

However, the choice of too small steps ε_k may introduce numerical instability in the computations, so we recommend to choose the ε_k of the form

$$\varepsilon_k = o\big(k^{-(\frac{1}{4}+\delta)}\big) \quad \text{with} \quad \delta > 0 \text{ small enough.}$$

One can refine the bound obtained in (10.9): note that

$$\sum_{k=1}^{M} \frac{\mathrm{Var}(F(x+\varepsilon_k, Z_k) - F(x-\varepsilon_k, Z_k))}{4\varepsilon_k^2} = \sum_{k=1}^{M} \frac{\left\|F(x+\varepsilon_k, Z_k) - F(x-\varepsilon_k, Z_k)\right\|_2^2}{2\varepsilon_k} - \sum_{k=1}^{M}\left(\frac{f(x+\varepsilon_k) - f(x-\varepsilon_k)}{2\varepsilon_k}\right)^2.$$

Now, since $\dfrac{f(x+\varepsilon_k) - f(x-\varepsilon_k)}{2\varepsilon_k} \to f'(x)$ as $k \to +\infty$,

$$\frac{1}{M}\sum_{k=1}^{M}\left(\frac{f(x+\varepsilon_k) - f(x-\varepsilon_k)}{4\varepsilon_k^2}\right)^2 \to f'(x)^2 \quad \text{as} \quad M \to +\infty.$$

Plugging this into the above computations yields the refined asymptotic upper-bound

$$\overline{\lim_{M\to+\infty}} \sqrt{M}\left\| f'(x) - \widehat{f'(x)}_M \right\|_2 \le \sqrt{C_{F,Z}^2 - \left(f'(x)\right)^2}.$$

This approach has the same quadratic rate of convergence as a regular Monte Carlo simulation (*e.g.* a simulation carried out with $\frac{\partial F}{\partial x}(x, Z)$ if it exists).

Now, we show that the estimator $\widehat{f'(x)}_M$ is consistent, *i.e.* convergent toward $f'(x)$.

Proposition 10.3 *Under the assumptions of Proposition 10.1 and if ε_k goes to zero as k goes to infinity, the estimator $\widehat{f'(x)}_M$ $\mathbb{P}$-$a.s.$ converges to its target $f'(x)$.*

Proof. This amounts to showing that

$$\frac{1}{M}\sum_{k=1}^{M} \frac{F(x+\varepsilon_k, Z_k) - F(x-\varepsilon_k, Z_k)}{2\varepsilon_k} - \frac{f(x+\varepsilon_k) - f(x-\varepsilon_k)}{2\varepsilon_k} \xrightarrow{a.s.} 0 \tag{10.10}$$

as $M \to +\infty$. This is (again) a straightforward consequence of the $a.s.$ convergence of L^2-bounded martingales combined with the Kronecker Lemma (see Lemma 12.1): first define the martingale

$$L_M = \sum_{k=1}^{M} \frac{1}{k} \frac{F(x+\varepsilon_k, Z_k) - F(x-\varepsilon_k, Z_k) - (f(x+\varepsilon_k) - f(x-\varepsilon_k))}{2\varepsilon_k}, \quad M \ge 1.$$

One checks that

$$\begin{aligned}\mathbb{E}\,L_{_M}^2 &= \sum_{k=1}^{M}\mathbb{E}\,(\Delta L_k)^2 = \sum_{k=1}^{M}\frac{1}{4k^2\varepsilon_k^2}\mathrm{Var}(F(x+\varepsilon_k, Z_k) - F(x-\varepsilon_k, Z_k))\\ &\le \sum_{k=1}^{M}\frac{1}{4k^2\varepsilon_k^2}\mathbb{E}\,(F(x+\varepsilon_k, Z_k) - F(x-\varepsilon_k, Z_k))^2\\ &\le \sum_{k=1}^{M}\frac{1}{4k^2\varepsilon_k^2}C_{F,Z}^2\,4\,\varepsilon_k^2 = C_{F,Z}^2\sum_{k=1}^{M}\frac{1}{k^2}\end{aligned}$$

so that

$$\sup_M \mathbb{E}\,L_{_M}^2 < +\infty.$$

Consequently, $L_{_M}$ $a.s.$ converges to a square integrable (hence $a.s.$ finite) random variable $L_{_\infty}$ as $M \to +\infty$. The announced $a.s.$ convergence in (10.10) follows from the Kronecker Lemma (see Lemma 12.1). ◇

► **Exercises. 1.** *Central Limit Theorem.* Assume that $x \mapsto F(x, Z)$ is Lipschitz continuous from $\mathbb{R}$ to $L^{2+\eta}(\mathbb{P})$ for an $\eta > 0$. Show that the convergence of the finite difference estimator with decreasing step $\widehat{f'(x)}_{_M}$ defined in (10.7) satisfies the following property: from every subsequence (M') of (M) one may extract a subsequence (M'') such that

$$\sqrt{M''}\Big(\widehat{f'(x)}_{_{M''}} - f'(x)\Big) \stackrel{\mathcal{L}}{\longrightarrow} \mathcal{N}(0; v),\ \ v \in [0, \bar{v}] \text{ as } \ M \to +\infty$$

where $\bar{v} = C_{_{F,Z}}^2 - \big(f'(x)\big)^2$.

[Hint: Note that the sequence $\Big(\mathrm{Var}\big(\frac{F(x+\varepsilon_k, Z_k)-F(x-\varepsilon_k, Z_k)}{2\varepsilon_k}\big)\Big)$, $k \ge 1$, is bounded and use the following Central Limit Theorem: if $(Y_n)_{n\ge1}$ is a sequence of i.i.d. random variables such that there exists an $\eta > 0$ satisfying

$$\sup_n \mathbb{E}\,|Y_n|^{2+\eta} < +\infty \quad\text{and}\quad \exists\, N_n \to +\infty \text{ with } \frac{1}{N_n}\sum_{k=1}^{N_n}\mathrm{Var}(Y_k) \stackrel{n\to+\infty}{\longrightarrow} \sigma^2 > 0$$

then

$$\frac{1}{\sqrt{N_n}}\sum_{k=1}^{N_n} Y_k \stackrel{\mathcal{L}}{\longrightarrow} \mathcal{N}(0; \sigma^2) \quad\text{as } n \to +\infty.]$$

2. *Hölder framework.* Assume that $x \mapsto F(x, Z)$ is only θ-Hölder from $\mathbb{R}$ to $L^2(\mathbb{P})$ with $\theta \in (0, 1)$, like in Proposition 10.2.

(a) Show that a natural upper-bound for the quadratic error induced by the symmetric finite difference estimator with decreasing step $\widehat{f'(x)}_{_M}$ defined in (10.7) is given by

$$\frac{1}{M}\sqrt{\frac{[f'']^2_{\rm Lip}}{4}\left(\sum_{k=1}^{M}\varepsilon_k^2\right)^2+\frac{C^2_{F,Z}}{2^{2(1-\theta)}}\sum_{k=1}^{M}\frac{1}{\varepsilon_k^{2(1-\theta)}}}.$$

(*b*) Show that the resulting estimator $\widehat{f'(x)}_M$ *a.s.* converges to its target $f'(x)$ as soon as

$$\sum_{k\geq 1}\frac{1}{k^2\varepsilon_k^{2(1-\theta)}}<+\infty.$$

(*c*) Assume that $\varepsilon_k=\frac{c}{k^a}$, $k\geq 1$, where c is a positive real constant and $a\in(0,1)$. Show that the exponent a corresponds to an admissible step if and only if $a\in(0,\frac{1}{2(1-\theta)})$. Justify the choice of $a^*=\frac{1}{2(3-\theta)}$ for the exponent a and deduce that the resulting rate of decay of the (*RMSE*) is $O\Big(\sqrt{M}^{\,-\frac{2}{3-\theta}}\Big)$.

The above exercise shows that the (lack of) regularity of $x\mapsto F(x,Z)$ in $L^2(\mathbb{P})$ impacts the rate of convergence of the finite difference method.

10.2 Pathwise Differentiation Method

10.2.1 (Temporary) Abstract Point of View

We retain the notation of the former section. We assume that there exists a $p\in[1,+\infty)$ such that

$$\forall\,\xi\in(x-\varepsilon_0,x+\varepsilon_0),\quad F(\xi,Z)\in L^p(\mathbb{P}).$$

Definition 10.1 *The function $\xi\mapsto F(\xi,Z)$ from $(x-\varepsilon_0,x+\varepsilon_0)$ to $L^p(\mathbb{P})$ is L^p-differentiable at x if there exists a random variable denoted by $\partial_x F(x,\,.\,)\in L^p(\mathbb{P})$ such that*

$$\lim_{\substack{\xi\to x\\ \xi\neq x}}\left\|\frac{F(x,Z)-F(\xi,Z)}{x-\xi}-\partial_x F(x,\omega)\right\|_p=0.$$

Proposition 10.4 *If $\xi\mapsto F(\xi,Z)$ is L^p-differentiable at x, then the function f defined by $f(\xi)=\mathbb{E}\,F(\xi,Z)$ is differentiable at x and*

$$f'(x)=\mathbb{E}\,\big(\partial_x F(x,\omega)\big).$$

Proof. This is a straightforward consequence of the inequality $\big|\mathbb{E}\,Y\big|\leq\|Y\|_p$ (which holds for any $Y\in L^p(\mathbb{P})$) applied to $Y(\omega)=\frac{F(x,Z(\omega))-F(\xi,Z(\omega))}{x-\xi}-\partial_x F(x,\omega)$ by letting ξ converge to x. ◇

ℵ **Practitioner's corner.** • As soon as $\partial_x F(x,\omega)$ is simulable at a reasonable computational cost, it is usually preferable to compute $f'(x) = \mathbb{E}\,\partial_x F(x,\omega)$ by a direct Monte Carlo simulation rather than by a finite difference method. The performances are similar in terms of size of samples but the complexity of each path is lower. In a Brownian diffusion framework it depends on the smoothness of the functional of the diffusion and on the existence of the tangent process introduced in the next Sect. 10.2.2.

• *When the functional $F(x,\omega)$ is not pathwise differentiable* (or in L^p hereafter), whereas the function $f(x) = \mathbb{E}\,F(x,\omega)$ is differentiable, *the finite difference method is outperformed by* weighted estimators *whose variance appear significantly lower.* This phenomenon is illustrated in Sect. 10.4.4,

• In this section, we deal with functions $F(x,z)$ of a real variable x, but the extension to functionals depending on $x \in \mathbb{R}^d$ by replacing the notion of derivative by that of differential (or partial derivatives).

The usual criterion to establish L^p-differentiability, especially when the underlying source of randomness comes from a diffusion (see below), is to establish the pathwise differentiability of $\xi \mapsto F(\xi, Z(\omega))$ combined with an L^p- uniform integrability property of the ratio $\frac{F(x,Z)-F(\xi,Z)}{x-\xi}$ (see Theorem 12.4 and the Corollary that follows for a short background on uniform integrability).

Usually, this is applied with $p = 2$ since one needs $\partial_x F(x,\omega)$ to be in $L^2(\mathbb{P})$ to ensure that the Central Limit Theorem applies and rules the rate of convergence in the Monte Carlo simulation.

This can be summed up in the following proposition, whose proof is obvious.

Proposition 10.5 *Let $p \in [1,+\infty)$. If*

(i) there exists a random variable $\partial_x F(x,\,.\,)$ such that $\mathbb{P}(d\omega)$-a.s. $\xi \mapsto F(\xi, Z(\omega))$ is differentiable at x with derivative $\partial_x F(x,\omega)$,

(ii) there exists an $\varepsilon_0 > 0$ such that the family $\left(\frac{F(x,Z)-F(\xi,Z)}{x-\xi}\right)_{\xi \in (x-\varepsilon_0, x+\varepsilon_0)\setminus\{x\}}$ is L^p-uniformly integrable,

then $\partial_x F(x,\,.\,) \in L^p(\mathbb{P})$ and $\xi \mapsto F(\xi, Z)$ is L^p-differentiable at x with derivative $\partial_x F(x,.)$.

10.2.2 *The Tangent Process of a Diffusion and Application to Sensitivity Computation*

In a diffusion framework, it is important to keep in mind that, like for *ODE*s in a deterministic setting, the solution of an *SDE* is usually smooth when viewed as a (random) function of its starting value. This smoothness even holds in a pathwise sense with an (almost) explicit differential which appears as the solution to a linear *SDE*. The main result in this direction is due to Kunita (see [176], Theorem 3.3, p. 223).

This result must be understood as follows: when a sensitivity (the δ-hedge and the γ parameter but also other "greek parameters", as will be seen further on) related to the premium $\mathbb{E}\, h(X_T^x)$ of an option cannot be computed by "simply" interchanging differentiation and expectation, this lack of differentiability comes from the payoff function h. This also holds true for path-dependent options.

Let us come to a precise statement of Kunita's Theorem on the regularity of the flow of an *SDE*.

Theorem 10.1 (Smoothness of the flow (Kunita's Theorem))
(a) Let $b : \mathbb{R}_+ \times \mathbb{R}^d \to \mathbb{R}^d$, $\vartheta : \mathbb{R}_+ \times \mathbb{R}^d \to \mathcal{M}(d, q, \mathbb{R})$, with regularity C_b^1 i.e. with bounded α-Hölder partial derivatives for an $\alpha > 0$. Let $X^x = (X_t^x)_{t\geq 0}$ denote the unique $(\mathcal{F}_t^W)$-adapted strong solution of the SDE (on the whole non-negative real line)

$$dX_t = b(t, X_t)dt + \vartheta(t, X_t)dW_t, \quad X_0 = x \in \mathbb{R}^d, \tag{10.11}$$

where $W = (W^1, \ldots, W^q)$ is a q-dimensional Brownian motion defined on a probability space $(\Omega, \mathcal{A}, \mathbb{P})$. Then, at every $t \in \mathbb{R}_+$, the mapping $x \mapsto X_t^x$ is a.s. continuously differentiable and its Jacobian $Y_t(x) := \nabla_x X_t^x = \left[\frac{\partial (X_t^x)^i}{\partial x^j}\right]_{1\leq i,j\leq d}$ satisfies the linear stochastic differential system

$$\forall\, t \in \mathbb{R}_+, \quad Y_t^{ij}(x) = \delta_{ij} + \sum_{\ell=1}^{d} \int_0^t \frac{\partial b^i}{\partial y^\ell}(s, X_s^x) Y_s^{\ell j}(x) ds$$
$$+ \sum_{\ell=1}^{d} \sum_{k=1}^{q} \int_0^t \frac{\partial \vartheta_{ik}}{\partial y^\ell}(s, X_s^x) Y_s^{\ell j}(x) dW_s^k, \quad 1 \leq i, j \leq d,$$

where δ_{ij} denotes the Kronecker symbol.
(b) Furthermore, the tangent process *$Y(x)$ takes values in the set $GL(d, \mathbb{R})$ of invertible square matrices. As a consequence, for every $t \geq 0$, the mapping $x \mapsto X_t^x$ is a.s. a homeomorphism of $\mathbb{R}^d$.*

We will admit this theorem, which is beyond the scope of this monograph. We refer to Theorem 3.3, p. 223, from [176] for Claim (a) and Theorem 4.4 (and its proof), p. 227, for Claim (b).

Remarks. • If b and σ are $C_b^{n+\alpha}$ for some $n \in \mathbb{N}$, then, a.s. the flow $x \mapsto X_t^x$ is $C_b^{n+\beta}$, $0 < \eta < \alpha$ (see again Theorem 3.3 in [176]).
One easily derives from the above theorem the slightly more general result about the tangent process to the solution $(X_s^{t,x})_{s\in[t,T]}$ starting from x at time t. This process, denoted by $(Y(t, x)_s)_{s\geq t}$, can be deduced from $Y(x)$ by the following inverted transition formula

$$\forall\, s \geq t, \quad Y_s(t, x) = Y_s(x)[Y_t(x)]^{-1}.$$

This is a consequence of the uniqueness of the solution of a linear *SDE*.

• Higher-order differentiability properties hold true if b are ϑ are smoother. For a more precise statement, see Sect. 10.2.2 below.

> **Example.** If $d = q = 1$, the above *SDE* reads

$$dY_t(x) = Y_t(x)\Big(b'_x(t, X_t)dt + \vartheta'_x(t, X^x_t)dW_t\Big), \qquad Y_0(x) = 1$$

and elementary computations show that

$$Y_t(x) = \exp\left(\int_0^t \left(b'_x(s, X^x_s) - \frac{1}{2}(\vartheta'_x(s, X^x_s))^2\right) ds + \int_0^t \vartheta'_x(s, X^x_s)dW_s\right) > 0 \tag{10.12}$$

so that, in the Black–Scholes model ($b(t, x) = rx$, $\vartheta(t, x) = \vartheta x$), one retrieves that

$$\frac{d}{dx}X^x_t = Y_t(x) = \frac{X^x_t}{x}.$$

▶ **Exercise.** Let $d = q = 1$. Show that under the assumptions of Theorem 10.1, the tangent process at x, $Y_t(x) = \frac{d}{dx}X^x_t$, satisfies

$$\sup_{s,t\in[0,T]} \frac{Y_t(x)}{Y_s(x)} \in L^p(\mathbb{P}), \quad p \in (0, +\infty).$$

Applications to δ-hedging

The tangent process and the δ hedge are closely related. Assume for convenience that the interest rate is 0 and that a basket is made up of d risky assets whose price dynamics $(X^x_t)_{t\in[0,T]}$, $X^x_0 = x \in (0, +\infty)^d$, is a solution to (10.11).

Then the premium of the payoff $h(X^x_T)$ on the basket is given by

$$f(x) := \mathbb{E}\, h(X^x_T).$$

The δ-hedge vector of this option (at time 0 and) at $x = (x^1, \ldots, x^d) \in (0, +\infty)^d$ is given by $\nabla f(x)$.

We have the following proposition that establishes the existence and the representation of $f'(x)$ as an expectation (with in view its computation by a Monte Carlo simulation). It is a straightforward application of Theorem 2.2(b).

Proposition 10.6 *(δ-hedge of vanilla European options) If a Borel function h : $\mathbb{R}^d \to \mathbb{R}$ satisfies the following assumptions:*

(i) $A.s.$ differentiability: $\nabla h(y)$ *exists* $\mathbb{P}_{X^x_T}(dy)$-*a.s.*,

(ii) Uniform integrability condition ([5]): *there exists a neighborhood* $(x - \varepsilon_0, x + \varepsilon_0)$ $(\varepsilon_0 > 0)$ *of* x *such that,*

$$\left(\frac{|h(X_T^x) - h(X_T^{x'})|}{|x - x'|}\right)_{|x'-x|<\varepsilon_0, x'\neq x} \quad \textit{is uniformly integrable.}$$

Then f *is differentiable at* x *and its gradient* $\nabla f(x) = \left[\frac{\partial f}{\partial x^i}(x)\right]_{1\le i\le d}$ *admits the following representation as an expectation:*

$$\frac{\partial f}{\partial x^i}(x) = \mathbb{E}\left[\left(\nabla h(X_T^x) \,\Big|\, \frac{\partial X_T^x}{\partial x^i}\right)\right], \; i = 1, \ldots, d. \tag{10.13}$$

Remark. One can also consider the payoff of a *forward start option* $h(X_{T_1}^x, \ldots, X_{T_N}^x)$, $0 < T_1 < \cdots < T_N$, where $h : (\mathbb{R}^d)^N \to \mathbb{R}_+$. Then, under similar $a.s.$ differentiability assumptions on h (with respect to $\mathbb{P}_{(X_{T_1},\ldots,X_{T_N})}$ and uniform integrability, its premium $f(x)$ is differentiable and

$$\frac{\partial f}{\partial x^i}(x) = \mathbb{E}\left[\left(\nabla h(X_{T_1}^x, \ldots, X_{T_N}^x) \,\Bigg|\, \left[\frac{\partial X_{T_j}^x}{\partial x^i}\right]_{1\le j\le d}\right)\right].$$

Toward computation by Monte Carlo simulation
These formulas can be used to compute various sensitivities by Monte Carlo simulation since sensitivity parameters are functions of the pair made by the diffusion process X^x, the solution to the *SDE* starting at x, and its tangent process $\nabla_x X^x$ at x: it suffices to consider the *Euler scheme of this pair* $(X_t^x, \nabla_x X_t^x)$ over $[0, T]$ with step $\frac{T}{n}$.

Thus, if $d = 1$ (for notational convenience), and setting $Y_t = \nabla_x X_t^x$, we see that the pair (X^x, Y) satisfies the stochastic differential system

$$\begin{aligned} dX_t^x &= b(X_t^x)dt + \vartheta(X_t^x)dW_t, \quad X_0^x = x \in \mathbb{R} \\ dY_t &= Y_t\big(b'(X_t^x)dt + \vartheta'(X_t^x)dW_t\big), \quad Y_0 = 1. \end{aligned}$$

In one dimension, one can take advantage of the semi-closed formula (10.12) obtained in the above exercise for the tangent process.

Extension to an exogenous parameter θ
All theoretical results obtained for the δ, *i.e.* for the differentiation of the flow of an *SDE* with respect to its initial value, can be, at least first formally, extended to any parameter provided no ellipticity is required. This follows from the remark that if

[5]If h is Lipschitz continuous, or even locally Lipschitz with polynomial growth ($|h(x) - h(y)| \le C|x - y|(1 + |x|^r + |y|^r)$ $r > 0$) and X^x is a solution to an *SDE* with Lipschitz continuous coefficients b and ϑ in the sense of (7.2), this uniform integrability is always satisfied: it follows from Theorem 7.10 applied with $p > 1$.

the coefficient(s) b and/or ϑ of a diffusion depend(s) on a parameter θ, then the pair $(X_t, \theta_t)_{t\in[0,T]}$ is still a Brownian diffusion process, namely

$$\begin{aligned} dX_t &= b(\theta, X_t)dt + \vartheta(\theta, X_t)dW_t, \quad & X_0 = x,\\ d\theta_t &= 0, & \theta_0 = \theta. \end{aligned}$$

Set $\widetilde{x} = (x, \theta)$ and $\widetilde{X}_t^{\widetilde{x}} := (X_t^x, \theta_t)$. Thus, following Theorems 3.1 and 3.3 of Sect. 3 in [176], if $(\theta, x) \mapsto b(\theta, x)$ and $(\theta, x) \mapsto \vartheta(\theta, x)$ are $\mathcal{C}_b^{k+\alpha}$ $(0 < \alpha < 1)$ with respect to x and θ ([6]) then the solution of the *SDE* at a given time t will be $\mathcal{C}^{k+\beta}$ (for every $\beta \in (0, \alpha)$) as a function of (x and θ). A more specific approach would show that some regularity in the sole variable θ would be enough, but then this result does not follow for free from the general theorem of the differentiability of the flows.

Assume $b = b(\theta, \,.\,)$ and $\vartheta = \vartheta(\theta, \,.\,)$ and the initial value $x = x(\theta)$, $\theta \in \Theta$, $\Theta \subset \mathbb{R}^q$, open set. One can also differentiate an *SDE* with respect to this parameter θ. We can assume that $q = 1$ (by considering a partial derivative if necessary). Then, one gets

$$\begin{aligned} \frac{\partial X_t(\theta)}{\partial \theta} = \frac{\partial x(\theta)}{\partial \theta} &+ \int_0^t \left(\frac{\partial b}{\partial \theta}\big(\theta, X_s(\theta)\big) + \frac{\partial b}{\partial x}\big(\theta, X_s(\theta)\big)\Big(\frac{\partial X_s(\theta)}{\partial \theta}\Big)\right) ds\\ &+ \int_0^t \left(\frac{\partial \vartheta}{\partial \theta}\big(\theta, X_s(\theta)\big) + \frac{\partial \vartheta}{\partial x}\big(\theta, X_s(\theta)\big)\Big(\frac{\partial X_s(\theta)}{\partial \theta}\Big)\right) dW_s. \end{aligned}$$

ℵ Practitioner's corner

Once coupled with the original diffusion process X^x, this yields some expressions for the sensitivity with respect to the parameter θ, possibly closed, but usually computable by *a Monte Carlo simulation of the Euler scheme of the couple* $\left(X_t(\theta), \frac{\partial X_t(\theta)}{\partial \theta}\right)$.

As a conclusion, let us mention that this tangent process approach is close to the finite difference method applied to $F(x, \omega) = h(X_T^x(\omega))$ in the "regular setting": it appears as a limit case of the finite difference method. Their behaviors are similar, except that for a given size of Monte Carlo simulation, the tangent process method has a lower computational complexity. The only constraint is that it is more demanding in terms of "preliminary" human – hence expansive– calculations. This is no longer true since the recent emergence (in quantitative finance) of Automatic Differentiation (*AD*) as described by Griewank in [135–137], Naumann in [215] and Hascoet in [144], which, once included in the types of the variable spare much time prior to and during the simulation itself (at least for the adjoint version of *AD*).

---

[6] A function g has a $\mathcal{C}_b^{k+\alpha}$ regularity if g is $\mathcal{C}^k$ with k-th order partial derivatives globally α-Holder and if all its partial derivatives up to k-th order are bounded.

10.3 Sensitivity Computation for Non-smooth Payoffs: The Log-Likelihood Approach (II)

The approach based on the tangent process clearly requires smooth payoff functions, typically almost everywhere differentiable, as emphasized above (and in Sect. 2.2). Unfortunately, this assumption is not fulfilled by many of the usual payoffs functions like the digital payoff

$$h_T = h(X_T) \quad \text{with} \quad h(x) := \mathbf{1}_{\{x \geq K\}}$$

whose δ-hedge parameter cannot be computed by the tangent process method. In fact, $\frac{\partial}{\partial x}\mathbb{E}\, h(X_T^x)$ is the probability density of X_T^x evaluated at the strike K. For the same reason, this is also the case for the γ-sensitivity parameter of a vanilla Call option.

We also saw in Sect. 2.2 that in the Black–Scholes model, this problem can be overcome since integration by parts or differentiating the log-likelihood leads to some sensitivity formulas for non-smooth payoffs. Is it possible to extend this idea to more general models?

10.3.1 A General Abstract Result

In Sect. 2.2, we saw that a family of random vectors $X(\theta)$ indexed by a parameter $\theta \in \Theta$, Θ an open interval of $\mathbb{R}$, all having an *a.e.* positive probability density $p(\theta, y)$ with respect to a reference non-negative measure μ on $\mathbb{R}^d$. Assume this density is positive on a domain D of $\mathbb{R}^d$ for every $\theta \in \Theta$ where D does not depend on θ. Then one can derive the sensitivity with respect to θ of functions of the form

$$f(\theta) := \mathbb{E}\, \varphi(X(\theta)) = \int_D \varphi(y) p(\theta, y) \mu(dy),$$

where φ is a Borel function with appropriate integrability assumptions, provided the density function p is smooth enough as a function of θ, *regardless of the regularity of* φ. This idea has been briefly developed in a one-dimensional Black–Scholes framework in Sect. 2.2.3. The extension to a more general framework is actually straightforward and yields the following result.

Proposition 10.7 *If the parametrized probability density* $p(\theta, y) : \Theta \times D \to \mathbb{R}_+$ *satisfies*

(*i*) $\theta \longmapsto p(\theta, y)$ *is differentiable on* Θ $\mu(dy)$*-a.e.*

(*ii*) $\exists\, g : \mathbb{R}^d \overset{Borel}{\longrightarrow} \mathbb{R}$, *such that* $\begin{cases} \text{-}\ g\varphi \in L^1(\mu) \\ \text{-}\ \forall\, \theta \in \Theta,\ \Big(|\partial_\theta p(\theta, y)| \leq g(y)\ \mu(dy)\text{-}a.e.\Big) \end{cases}$, *then*

$$\forall\,\theta \in \Theta, \qquad f'(\theta) = \mathbb{E}\left[\varphi\big(X(\theta)\big)\frac{\partial \log p}{\partial \theta}\big(\theta, X(\theta)\big)\right].$$

10.3.2 The log-Likelihood Method for the Discrete Time Euler Scheme

As presented in an abstract framework, this approach looks attractive but unfortunately, in most situations, we have no explicitly computable form for the density $p(\theta, y)$ even when its existence is proved. However, if one considers diffusion approximation by some discretization schemes like the Euler scheme in a non-degenerate setting, the application of the log-likelihood method becomes much less unrealistic, for computing by simulation some proxies of the greek parameters by a Monte Carlo simulation.

As a matter of fact, under light ellipticity assumption, the (constant step) Euler scheme of a diffusion does have a probability density at each time $t > 0$ which can be made explicit in some way (see below). This is a straightforward consequence of the fact that it is a discrete time Markov process with conditionally Gaussian increments. The principle is the following: we consider an autonomous diffusion $(X_t^x(\theta))_{t\in[0,T]}$ depending on a parameter $\theta \in \Theta$, say

$$dX_t^x(\theta) = b\big(\theta, X_t^x(\theta)\big)\,dt + \vartheta\big(\theta, X_t^x(\theta)\big)dW_t, \quad X_0(\theta) = x. \tag{10.14}$$

We considered an autonomous *SDE* mainly for simplicity and to get nice formulas, but the extension to general *SDE* s is just a matter of notation.

Let $p_T(\theta, x, y)$ and $\bar{p}_T(\theta, x, y)$ denote the density of $X_T^x(\theta)$ and its discrete time Euler scheme $\big(\bar{X}_{t_k^n}^x(\theta)\big)_{0\le k\le n}$ with step size $\frac{T}{n}$ (the superscript ^{n} is dropped until the end of the section). Then one may naturally propose the following naive approximation

$$f'(\theta) = \mathbb{E}\left(\varphi(X_T^x(\theta))\frac{\partial \log p_T}{\partial \theta}(\theta, x, X_T^x(\theta))\right) \simeq \mathbb{E}\left(\varphi(\bar{X}_T^x(\theta))\frac{\partial \log \bar{p}_T}{\partial \theta}(\theta, x, \bar{X}_T^x(\theta))\right).$$

It consists in making the risky hypothesis that the derivative of an approximation is an approximation of the derivative (supported by various theoretical results about the convergence of the densities of the discrete time Euler scheme and their derivatives).

In fact, even at this stage, the story is not as straightforward as expected because only the density of the whole n-tuple $(\bar{X}_{t_1^n}^x, \ldots, \bar{X}_{t_k^n}^x, \ldots, \bar{X}_{t_n^n}^x)$ (with $t_n^n = T$) can be made explicit and tractable.

Proposition 10.8 (Density of the Euler scheme) *Let $q \ge d$. Assume $\vartheta\vartheta^*(\theta, x) \in GL(d, \mathbb{R})$ for every $x \in \mathbb{R}^d$ and $\theta \in \Theta$.*
(a) Then the distribution $\mathbb{P}_{\bar{X}^x_{\frac{T}{n}}}(dy)$ of $\bar{X}^x_{\frac{T}{n}}$ is absolutely continuous with respect to the Lebesgue measure with a probability density given by

$$\bar{p}_{\frac{T}{n}}(\theta, x, y) = \frac{1}{(2\pi\frac{T}{n})^{\frac{d}{2}}\sqrt{\det \vartheta\vartheta^*(\theta, x)}} e^{-\frac{n}{2T}\left(y-x-\frac{T}{n}b(\theta,x)\right)^*(\vartheta\vartheta^*(\theta,x))^{-1}\left(y-x-\frac{T}{n}b(\theta,x)\right)}.$$

(b) The distribution $\mathbb{P}_{(\bar{X}^x_{t^n_1},\ldots,\bar{X}^x_{t^n_k},\ldots,\bar{X}^x_{t^n_n})}(dy_1,\ldots,dy_n)$ *of the* n*-tuple* $(\bar{X}^x_{t^n_1},\ldots,\bar{X}^x_{t^n_k},\ldots,\bar{X}^x_{t^n_n})$ *has a probability density given by*

$$\bar{p}_{t^n_1,\ldots,t^n_n}(\theta, x, y_1,\ldots,y_n) = \prod_{k=1}^{n} \bar{p}_{\frac{T}{n}}(\theta, y_{k-1}, y_k) \qquad (\textit{convention: } y_0 = x).$$

Proof. (a) is a straightforward consequence of the first step of the definition (7.5) of the Euler scheme at time $\frac{T}{n}$ applied to the diffusion (10.14) and the standard formula for the density of a Gaussian vector since $\vartheta(\theta, x)\vartheta^*(\theta, x)$ is invertible. Claim (b) follows from an easy induction based on the Markov property satisfied by the Euler scheme. First, it is clear, again from the recursion satisfied by the discrete time Euler scheme of (10.14) (adapted from (7.5)) that

$$\mathcal{L}\Big(\bar{X}^x_{t^n_{k+1}}(\theta)\,|\,\mathcal{F}^W_{t^n_k}\Big) = \mathcal{L}\Big(\bar{X}^x_{t^n_{k+1}}(\theta)\,|\,\bar{X}_{t^n_k}\Big), \quad k = 0,\ldots,n-1$$

and that

$$\mathcal{L}\Big(\bar{X}^x_{t^n_{k+1}}(\theta)\,|\,\bar{X}^x_{t^n_k}(\theta) = y_k\Big) = \mathcal{L}\Big(\bar{X}_{t^n_1}(\theta)\,|\,\bar{X}_0(\theta) = y_k\Big) = \bar{p}_{\frac{T}{n}}(\theta, x, y_k)\lambda_d(dx).$$

One concludes by an easy induction that, for every bounded Borel function $g : (\mathbb{R}^d)^k \to \mathbb{R}_+$,

$$\mathbb{E}\, g\big(\bar{X}^x_{t^n_1},\ldots,\bar{X}^x_{t^n_k}\big) = \int_{(\mathbb{R}^d)^k} g(y_1,\ldots,y_k) \prod_{\ell=1}^{k} \bar{p}_{\frac{T}{n}}(\theta, y_{\ell-1}, y_\ell)$$

with the convention $y_0 = 0$, using the Markov property satisfied by the discrete time Euler scheme. ◇

The above proposition proves that every marginal $\bar{X}^x_{t^n_k}$ has a density which, unfortunately, cannot be made explicit. Moreover, to take advantage of the above closed form for the density of the n-tuple, we can write

$$\begin{aligned} f'(\theta) &\simeq \mathbb{E}\left(\varphi(\bar{X}^x_T(\theta))\frac{\partial \log \bar{p}_{t^n_1,\ldots,t^n_n}}{\partial\theta}(\theta, x, \bar{X}^x_{t^n_1}(\theta),\ldots,\bar{X}^x_{t^n_1}(\theta))\right) \\ &= \sum_{k=1}^{n} \mathbb{E}\left(\varphi\big(\bar{X}^x_T(\theta)\big)\frac{\partial \log \bar{p}_{\frac{T}{n}}}{\partial\theta}\big(\theta, \bar{X}^x_{t^n_{k-1}}(\theta), \bar{X}^x_{t^n_k}(\theta)\big)\right). \end{aligned}$$

At this stage it is clear that the method also works for path-dependent options, *i.e.* when considering $F(\theta) = \mathbb{E}\,\Phi\big((X_t(\theta))_{t\in[0,T]}\big)$ instead of $f(\theta) = \mathbb{E}\,\varphi\big(X_T(\theta)\big)$

(at least for specific functionals Φ involving time averaging, a finite number of instants, supremum, infimum, etc). This raises new difficulties in connection with the Brownian bridge method for diffusions, that need to be encompassed.

Finally, let us mention that evaluating the rate of convergence of these approximations from a theoretical point of view is quite a challenging problem since it involves not only the rate of convergence of the Euler scheme itself, but also that of the probability density functions of the scheme toward that of the diffusion (see [25] where this problem is addressed).

▶ **Exercise.** Apply the preceding to the case $\theta = x$ (starting value) when $d = 1$.

Comments. In fact, there is a way to get rid of non-smooth payoffs by regularizing them over one time step before maturity and then applying to tangent process method. Such an approach has been developed by M. Giles under the name of *Vibrato Monte Carlo* which appears as a kind of degenerate nested Monte Carlo combined with a pathwise differentiation based on the tangent process (see [108]).

10.4 Flavors of Stochastic Variational Calculus

A subtitle could be "from Bismut's formula to Malliavin calculus".

10.4.1 Bismut's Formula

In this section, for the sake of simplicity, we assume that $d = 1$ and $q = 1$ (scalar Brownian motion).

Theorem 10.2 (Bismut's formula) *Let $W = (W_t)_{t\in[0,T]}$ be a standard Brownian motion on a probability space $(\Omega, \mathcal{A}, \mathbb{P})$ and let $\mathcal{F}^W := (\mathcal{F}_t^W)_{t\in[0,T]}$ be its augmented natural filtration. Let $X^x = (X_t^x)_{t\in[0,T]}$ be a diffusion process, unique $(\mathcal{F}_t)_{t\in[0,T]}$-adapted solution to the autonomous SDE*

$$dX_t = b(X_t)dt + \vartheta(X_t)dW_t, \quad X_0 = x,$$

where b and ϑ are C_b^1 (hence Lipschitz continuous). Let $f : \mathbb{R} \to \mathbb{R}$ be a continuously differentiable function such that

$$\mathbb{E}\Big(f(X_T^x)^2 + \big(f'(X_T^x)\big)^2\Big) < +\infty.$$

Let $(H_t)_{t\in[0,T]}$ be an $\mathcal{F}^W$-progressively measurable ([7]) process lying in $L^2([0,T]\times\Omega, dt\otimes d\,\mathbb{P})$, i.e. satisfying $\mathbb{E}\int_0^T H_s^2 ds < +\infty$. Then

[7]This means that for every $t \in [0, T]$, $(H_s(\omega))_{(s,\omega)\in[0,t]\times\Omega}$ is $\mathcal{B}or([0,t]) \otimes \mathcal{F}_t$-measurable.

$$\mathbb{E}\left(f(X_T^x)\int_0^T H_s dW_s\right) = \mathbb{E}\left(f'(X_T^x)Y_T\int_0^T \frac{\vartheta(X_s^x)H_s}{Y_s}ds\right)$$

where $Y_t = \frac{dX_t^x}{dx}$ *is the tangent process of* X^x *at* x.

Proof (Sketch). To simplify the arguments, we will assume in the proof that $|H_t| \le C < +\infty$, where C is a real constant and that f and f' are bounded functions.

Let $\varepsilon \ge 0$. Set on the probability space $(\Omega, \mathcal{F}_T, \mathbb{P})$,

$$\mathbb{P}^{(\varepsilon)} = L_T^{(\varepsilon)}.\mathbb{P}$$

where

$$L_t^{(\varepsilon)} = \exp\left(-\varepsilon\int_0^t H_s dW_s - \frac{\varepsilon^2}{2}\int_0^t H_s^2 ds\right), \quad t \in [0, T],$$

is a $\mathbb{P}$-martingale since H is bounded. It follows from Girsanov's Theorem (see *e.g.* Sect. 3.5, Theorem 5.1, p. 191, in [162]) that

$$\widetilde{W}^\varepsilon := \left(W_t + \varepsilon\int_0^t H_s ds\right)_{t\in[0,T]} \quad \text{is a} \quad \left(\mathbb{P}^{(\varepsilon)}, (\mathcal{F}_t)_{t\in[0,T]}\right)\text{-Brownian motion.}$$

Now it follows from the definition of $L_T^{(\varepsilon)}$ and the differentiation Theorem 2.2 that

$$\mathbb{E}\left[f(X_T^x)\int_0^T H_s dW_s\right] = -\frac{\partial}{\partial\varepsilon}\left[\mathbb{E}\left(f(X_T^x)L_T^{(\varepsilon)}\right)\right]_{|\varepsilon=0}.$$

On the other hand

$$\frac{\partial}{\partial\varepsilon}\left[\mathbb{E}\left(f(X_T)L_T^{(\varepsilon)}\right)\right]_{|\varepsilon=0} = \frac{\partial}{\partial\varepsilon}\left[\mathbb{E}_{\mathbb{P}^{(\varepsilon)}}\left(f(X_T)\right)\right]_{|\varepsilon=0}.$$

Now we can rewrite the *SDE* satisfied by X as follows

$$\begin{aligned} dX_t^x &= b(X_t^x)dt + \vartheta(X_t^x)dW_t \\ &= \big(b(X_t^x) - \varepsilon H_t\vartheta(X_t^x)\big)dt + \vartheta(X_t^x)d\widetilde{W}_t^{(\varepsilon)}. \end{aligned}$$

Consequently (see [251], Theorem 1.11, p. 372), the process X^x has the same distribution under $\mathbb{P}^{(\varepsilon)}$ as $X^{(\varepsilon)}$, the solution to

$$dX_t^{(\varepsilon)} = \big(b(X_t^{(\varepsilon)}) - \varepsilon H_t\vartheta(X_t^{(\varepsilon)})\big)dt + \vartheta(X_t^{(\varepsilon)})dW_t, \quad X_0^{(\varepsilon)} = x.$$

Now we can write

$$\mathbb{E}\left[f(X_T)\int_0^T H_s\,dW_s\right] = -\frac{\partial}{\partial\varepsilon}\left[\mathbb{E}\left(f(X_T^{(\varepsilon)})\right)\right]_{|\varepsilon=0} = -\mathbb{E}\left[f'(X_T)\left(\frac{\partial X_T^{(\varepsilon)}}{\partial\varepsilon}\right)_{|\varepsilon=0}\right],$$

where we used once again Theorem 2.2 and the obvious fact that $X^{(0)} = X$.

Using the tangent process method with ε as an auxiliary variable, one derives that the process $U_t := \left(\frac{\partial X_t^{(\varepsilon)}}{\partial\varepsilon}\right)_{|\varepsilon=0}$ satisfies

$$dU_t = U_t\big(b'(X_t^x)dt + \vartheta'(X_t^x)dW_t\big) - H_t\vartheta(X_t^x)dt.$$

Plugging the regular tangent process Y into this equation yields

$$dU_t = \frac{U_t}{Y_t}dY_t - H_t\vartheta(X_t^x)dt. \tag{10.15}$$

We know that Y_t is never 0, hence (up to some localization if necessary) we can apply Itô's formula to the ratio $\frac{U_t}{Y_t}$: elementary computations of the partial derivatives of the function $(u, y) \mapsto \frac{u}{y}$ on $\mathbb{R}\times(0,+\infty)$ combined with Eq. (10.15) show that

$$\begin{aligned} d\left(\frac{U_t}{Y_t}\right) &= \frac{dU_t}{Y_t} - \frac{U_t dY_t}{Y_t^2} + \frac{1}{2}\left(-2\frac{d\langle U,Y\rangle_t}{Y_t^2} + \frac{2U_t d\langle Y\rangle_t}{Y_t^3}\right)\\ &= -\frac{H_t\vartheta(X_t^x)}{Y_t}dt + \frac{1}{2}\left(-2\frac{d\langle U,Y\rangle_t}{Y_t^2} + \frac{2U_t d\langle Y\rangle_t}{Y_t^3}\right). \end{aligned}$$

Then we derive from (10.15) that

$$d\langle U, Y\rangle_t = \frac{U_t}{Y_t}d\langle Y\rangle_t,$$

which yields

$$d\left(\frac{U_t}{Y_t}\right) = -\frac{\vartheta(X_t^x)H_t}{Y_t}\,dt.$$

Noting that $U_0 = \frac{dX_0^{(\varepsilon)}}{d\varepsilon} = \frac{dx}{d\varepsilon} = 0$ finally leads to

$$U_t = -Y_t\int_0^t \frac{\vartheta(X_s)H_s}{Y_s}ds, \quad t\in[0,T],$$

which completes this step of the proof.

The extension to more general processes H can be done by introducing for every $n \geq 1$

$$H_t^{(n)}(\omega) := H_t(\omega)\mathbf{1}_{\{|H_t(\omega)|\leq n\}}.$$

It is clear by the Lebesgue dominated convergence theorem that $H^{(n)}$ converges to H in $L^2([0,T]\times\Omega, dt\otimes d\,\mathbb{P})$. Then one checks that both sides of Bismut's identity are continuous with respect to this topology (using Hölder's Inequality).

The extension to unbounded functions and derivatives f follows by approximation of f by bounded $\mathcal{C}_b^1$ functions, *e.g.* by "smooth" truncations. ◇

Application to the computation of the δ-parameter

Assume b and ϑ are $\mathcal{C}_b^1$. If f is continuous with polynomial growth and satisfies

$$\mathbb{E}\left(f^2(X_T^x)+\int_0^T\left(\frac{Y_t}{\vartheta(X_t^x)}\right)^2dt\right)<+\infty,$$

then $\frac{\partial}{\partial x}\mathbb{E}\,f(X_T^x)$ appears as a *weighted* expectation of $f(X_T)$, namely

$$\frac{\partial}{\partial x}\mathbb{E}\,f(X_T^x)=\mathbb{E}\left(f(X_T^x)\underbrace{\frac{1}{T}\int_0^T\frac{Y_s}{\vartheta(X_s^x)}dW_s}_{\text{weight}}\right). \tag{10.16}$$

Proof. We proceed like we did with the Black–Scholes model in Sect. 2.2: we first assume that f is regular, namely bounded and differentiable with bounded derivative. Then, using the tangent process method approach

$$\frac{\partial}{\partial x}\mathbb{E}\,f(X_T^x)=\mathbb{E}\left(f'(X_T^x)Y_T\right),$$

still with $Y_t=\frac{dX_t^x}{dx}$. Then, we set

$$H_t=\frac{Y_t}{\vartheta(X_t^x)}.$$

Under the above assumption, we can apply Bismut's formula to get

$$T\,\mathbb{E}\left(f'(X_T^x)Y_T\right)=\mathbb{E}\left(f(X_T^x)\int_0^T\frac{Y_t}{\vartheta(X_t^x)}dW_t\right),$$

which yields the announced result. The extension to continuous functions with polynomial growth relies on an approximation argument (*e.g.* by convolution). ◇

Remarks. • One retrieves in the case of a Black–Scholes model the formula (2.8) for the δ, obtained in Sect. 2.2 by an elementary integration by parts, since $Y_t=\frac{X_t^x}{x}$ and $\vartheta(x)=\sigma x$.

• Note that the assumption

$$\int_0^T \left(\frac{Y_t}{\vartheta(X_t^x)}\right)^2 dt < +\infty$$

is essentially an ellipticity assumption. Thus, if $\vartheta^2(x) \geq \varepsilon_0 > 0$, one checks that the assumption is always satisfied.

► **Exercises. 1.** Apply the preceding to get a formula for the γ-parameter in a general diffusion model. [Hint: Apply the above "derivative free" formula to the δ-hedge formula obtained using the tangent process method.]

2. Show that if $b' - b\frac{\vartheta'}{\vartheta} - \frac{1}{2}\vartheta''\vartheta = c \in \mathbb{R}$, then

$$\frac{\partial}{\partial x}\mathbb{E}\, f(X_T^x) = \frac{e^{cT}}{\vartheta(x)}\mathbb{E}\left(f(X_T^x)W_T\right).$$

10.4.2 The Haussman–Clark–Occone Formula: Toward Malliavin Calculus

In this section we state an elementary version of the so-called Haussman–Clark–Occone formula, following the seminal paper by Haussman [147]. We still consider the standard *SDE*

$$dX_t = b(X_t)dt + \vartheta(X_t)dW_t, \quad X_0 = x, \quad t \in [0, T],$$

with Lipschitz continuous coefficients b and σ and $X^x = (X_t^x)_{t\in[0,T]}$ its unique $(\mathcal{F}_t^W)$-adapted solution starting at x, where $(\mathcal{F}_t)_{t\in[0,T]}$ is the (augmented) filtration of the Brownian motion W. We admit the following theorem, stated in a one-dimensional setting for (at least notational) convenience.

Theorem 10.3 (Haussman(–Clark–Occone) formula)
Let $F : \big(\mathcal{C}([0,T],\mathbb{R}), \|\cdot\|_{\sup}\big) \to \mathbb{R}$ be a differentiable functional with differential DF. Then

$$F(X^x) = \mathbb{E}\, F(X^x) + \int_0^T \mathbb{E}\left(DF(X)\cdot(\mathbf{1}_{[t,T]}Y_.^{(t)}) \mid \mathcal{F}_t\right)\vartheta(X_t^x)dW_t,$$

where $Y^{(t)}$ is the tangent process of X^x at time t, the solution to the SDE

$$dY_s^{(t)} = Y_s^{(t)}\left(b'(X_s^x)ds + \vartheta'(X_s^x)dW_s\right), \quad Y_t^{(t)} = 1, \quad s \in [t, T].$$

Note that $Y^{(t)}$ also reads

$$Y_s^{(t)} = \frac{Y_s}{Y_t}, \quad s \in [t, T] \text{ where } Y_t = Y_t^{(0)} \text{ is the tangent process of } X^x \text{ at the origin.}$$

Remarks. • The starting point to understand this formula is to regard it as a more explicit version of the classical representation formula of Brownian martingales (see Proposition 3.2, p. 191, in [251]). Thus, $M_t = \mathbb{E}(F(X) \mid \mathcal{F}_t)$, which admits a formal representation as a Brownian stochastic integral

$$M_t = M_0 + \int_0^T H_t \, dW_t,$$

where $(H_t)_{t\in[0,T]}$ is an $(\mathcal{F}_t^W)$-progressively measurable process satisfying $\int_0^T H_s^2 ds < +\infty$ $\mathbb{P}$-$a.s.$ So, the Clark-Occone-Haussman formula provides a kind of closed form for the process H.

• The differential $DF(\xi)$ of the functional F at an element $\xi \in \mathcal{C}([0,T],\mathbb{R})$ is a continuous linear form on $\mathcal{C}([0,T],\mathbb{R})$. Hence, following the Riesz representation Theorem (see *e.g.* [52]), it can be represented by a finite signed measure, say $\mu_{DF(\xi)}(ds)$, so that the term $DF(\xi)\cdot(\mathbf{1}_{[t,T]}Y_.^{(t)})$ reads

$$DF(\xi)\cdot(\mathbf{1}_{[t,T]}Y_.^{(t)}) = \int_0^T \mathbf{1}_{[t,T]}(s) Y_s^{(t)} \mu_{DF(\xi)}(ds) = \int_t^T Y_s^{(t)} \mu_{DF(\xi)}(ds).$$

Toward the Malliavin derivative

Assume $F(x) = f\big(x(t_0)\big)$, $x \in \mathcal{C}([0,T],\mathbb{R})$, $f : \mathbb{R} \to \mathbb{R}$ differentiable with derivative f'. Then $\mu_x(ds) = f'(x(t_0))\delta_{t_0}(ds)$, where $\delta_{t_0}(ds)$ denotes the Dirac mass at time t_0. Consequently

$$DF(X)\cdot(\mathbf{1}_{[t,T]}Y_.^{(t)}) = f'(X_{t_0}^x) Y_{t_0}^{(t)} \mathbf{1}_{[0,t_0]}(t),$$

whence one derives that

$$f(X_{t_0}^x) = \mathbb{E}\, f(X_{t_0}^x) + \int_0^{t_0} \mathbb{E}\left(f'(X_{t_0}) Y_{t_0}^{(t)} \mid \mathcal{F}_t\right) \vartheta(X_t^x)\, dW_t. \tag{10.17}$$

This leads us to introduce the *Malliavin derivative* $D_t F(X^x)$ of $F(X^x)$ at time t by

$$D_t F(X^x) := \begin{cases} DF(X^x)\cdot(\mathbf{1}_{[t,T]}Y_.^{(t)})\vartheta(X_t^x) & \text{if } t \le t_0 \\ 0 & \text{if } t > t_0 \end{cases}.$$

The simplest interpretation (and original motivation!) is that the Malliavin derivative is a derivative with respect to the Brownian path of W (viewed at time t). It can be easily understood owing to the following formal chain rule of differentiation

$$D_t F(X^x) = \frac{\partial F(X^x)}{\partial X_t^x} \times \frac{\partial X_t^x}{\partial W}.$$

If one notes that, for any $s \geq t$, $X_s^x = X_s^{X_t^x,t}$, the first term "$\frac{\partial F(X^x)}{\partial X_t^x}$" in the above product is clearly equal to $DF(X)\cdot(\mathbf{1}_{[t,T]}Y_.^{(t)})$ whereas the second term is the result of a formal differentiation of the *SDE* at time t with respect to W, namely $\vartheta(X_t^x)$.

An interesting feature of this derivative in practice is that it satisfies the usual chaining rules like

$$D_t F^2(X^x) = 2F(X^x) D_t F(X^x)$$

and more generally

$$D_t \Phi(F(X^x)) = D\Phi(F(X^x)) D_t F(X^x),$$

etc.

What is called *Malliavin calculus* is a way to extend this notion of differentiation to more general functionals using some functional analysis arguments (closure of operators, etc) using, for example, the domain of the operator $D_t F$ (see *e.g.* [16, 208]).

Using the Haussman–Clark–Occone formula to get Bismut's formula

As a first conclusion we will show that the Haussman–Clark–Occone formula contains the Bismut formula. Let X^x, H, f and T be as in Sect. 10.4.1. We consider the two true martingales

$$M_t = \int_0^t H_s dW_s \quad \text{and} \quad N_t = \mathbb{E}\, f(X_T^x) + \int_0^t \mathbb{E}\left(f'(X_T)Y_T^{(s)} \,|\, \mathcal{F}_s\right) dW_s, \quad t \in [0, T]$$

and perform a (stochastic) integration by parts. Owing to (10.17), we get, under appropriate integrality conditions,

$$\begin{aligned}
\mathbb{E}\left(f(X_T^x)\int_0^T H_s dW_s\right) &= 0 + \mathbb{E}\int_0^T [\ldots]\, dM_t + \mathbb{E}\int_0^T [\ldots]\, dN_t \\
&\quad + \mathbb{E}\left(\int_0^T \mathbb{E}\,(f'(X_T)Y_T^{(s)} \,|\, \mathcal{F}_s)\, \vartheta(X_s^x) H_s\, ds\right) \\
&= \int_0^T \mathbb{E}\left(\mathbb{E}\,(f'(X_T)Y_T^{(s)} \,|\, \mathcal{F}_s)\, \vartheta(X_s^x) H_s\right) ds
\end{aligned}$$

owing to Fubini's Theorem. Finally, using the characterization of conditional expectation to get rid of the conditioning, we obtain

$$\mathbb{E}\left(f(X_T^x)\int_0^T H_s dW_s\right) = \int_0^T \mathbb{E}\left(f'(X_T)Y_T^{(s)}\, \vartheta(X_s^x) H_s\right) ds.$$

Finally, a reverse application of Fubini's Theorem and the identity $Y_T^{(s)} = \frac{Y_T}{Y_s}$ leads to

$$\mathbb{E}\left(f(X_T^x)\int_0^T H_s dW_s\right) = \mathbb{E}\left(f'(X_T)Y_T \int_0^T \frac{\vartheta(X_s^x)}{Y_s} H_s\, ds\right),$$

which is simply Bismut's formula (10.2).

► **Exercise.** (*a*) Consider the functional $F : \xi \mapsto F(\xi) = f\left(\int_0^T \xi(s)ds\right)$, where $f : \mathbb{R} \to \mathbb{R}$ is a differentiable function. Show that

$$F(X^x) = \mathbb{E}\, F(X^x) + \int_0^T \mathbb{E}\left[f'\left(\int_0^T X_s^x ds\right)\int_t^T Y_s ds \mid \mathcal{F}_t\right]\frac{\vartheta(X_t^x)}{Y_t} dW_t.$$

(*b*) Derive, using the homogeneity of the Eq. (10.11), that

$$\mathbb{E}\left(f'\left(\int_0^T X_s^x ds\right)\int_t^T \frac{Y_s}{Y_t} ds \mid \mathcal{F}_t\right) = \left[\mathbb{E}\left(f'\left(\bar{x} + \int_0^{T-t} X_s^x ds\right)\int_0^{T-t} Y_s\, ds\right)\right]_{|x = X_t^x, \bar{x} = \int_0^t X_s^x ds}$$
$$=: \Phi\left(T - s, X_t^x, \int_0^t X_s^x ds\right).$$

10.4.3 *Toward Practical Implementation: The Paradigm of Localization*

For practical implementation, one should be aware that *the weighted estimators of sensitivities, obtained by* log*-likelihood methods (exact, see Sect.* 2.2.3, *or approximate, see* Sect. 10.3.1), *by Bismut's formula or more general Malliavin inspired methods, often suffer from high variance, especially for short maturities.*

This phenomenon can be observed in particular when such a formula coexists with a pathwise differentiated formula involving the tangent process for smooth enough payoff functions.

This can be easily understood on the formula for the δ-hedge when the maturity is small (see the toy-example in Sect. 3.1).

Consequently, weighted formulas for sensitivities need to be speeded up by efficient variance reduction methods. The usual approach, known as localization, is to isolate the singular part (where differentiation does not apply) from the smooth part.

Let us illustrate the principle of localization functions on a very simple toy-example ($d = 1$). Assume that, for every $\varepsilon > 0$,

$$|F(x, z) - F(x', z)| \le C_{F,\varepsilon}|x - x'|, \qquad x, x', z \in \mathbb{R}, |x - z|, |x' - z| \ge \varepsilon > 0$$

with $\underline{\lim}_{\varepsilon \to 0} C_{F,\varepsilon} = +\infty$. Assume furthermore that,

$$\forall\, x, z \in \mathbb{R}^d,\ x \ne z, \quad F'_x(x, z) \text{ exists}$$

(hence bounded by $C_{F,\varepsilon}$ if $|x - z| \ge \varepsilon$).

On the other hand, assume that *e.g.* $F(x, Z) = h(G(x, Z))$ where $G(x, Z)$ has a probability density which is regular in x whereas h is "highly" singular when in the neighborhood of $\{G(z, z),\ z \in \mathbb{R}\}$ (think of an indicator function). Then, the function $f(x) := \mathbb{E}\, F(x, Z)$ is differentiable.

Then, one considers a function $\varphi \in \mathcal{C}^\infty(\mathbb{R}, [0, 1])$ such that $\varphi \equiv 1$ on $[-\varepsilon, \varepsilon]$ and $\text{supp}(\varphi) \subset [-2\varepsilon, 2\varepsilon]$. Then, one may decompose

$$F(x, Z) = \big(1 - \varphi(x - Z)\big)F(x, Z) + \varphi(x - Z)F(x, Z) := F_1(x, Z) + F_2(x, Z).$$

Functions φ can be obtained as *mollifiers* in convolution theory but other choices are possible, like simply Lipschitz continuous functions (see the numerical illustration in Sect. 10.4.4). Set $f_i(x) = \mathbb{E}\, F_i(x, Z))$ so that $f(x) = f_1(x) + f_2(x)$. Then one may use a direct differentiation to compute

$$f_1'(x) = \mathbb{E}\left(\frac{\partial F_1(x, Z)}{\partial x}\right)$$

(or a finite difference method with constant or decreasing increments). As concerns $f_2'(x)$, since $F_2(x, Z)$ is singular, it is natural to look for a weighted estimator of the form

$$f_2'(x) = \mathbb{E}\left(F_2(x, Z)\Pi\right),$$

where Π is a random "weight" (possibly non-positive) obtained, for example, by the above described method if we are in a diffusion framework.

When working with a vanilla payoff in a Brownian diffusion framework at a fixed time T, the above Bismut formula (10.16) does the job, in its multi-dimensional version if necessary. When working with path-dependent payoff functionals (like barriers, etc) in local or stochastic volatility models, Malliavin calculus methods are often a powerful and elegant tools even if, in many settings, more elementary approaches can often be used to derive explicit weights. For an example of weight computation by means of Malliavin calculus in the case of Lookback or barriers options, we refer to [125]. These weights are not unique – even prior to any variance reduction –, see *e.g.* [121].

10.4.4 Numerical Illustration: What is Localization Useful for?

(written with V. Lemaire). Let us consider in a standard Black–Scholes model $(S_t)_{t\in[0,T]}$ with interest rate $r > 0$ and volatility $\sigma > 0$, two binary options:

– a *digital Call* with strike $K > 0$ and

– an *Asset-or-Nothing Call* with strike $K > 0$,

defined respectively by their payoff functions

$$h_1(\xi) = \mathbf{1}_{\{\xi \ge K\}} \quad \text{and} \quad h_s(\xi) = \xi\, \mathbf{1}_{\{\xi \ge K\}},$$

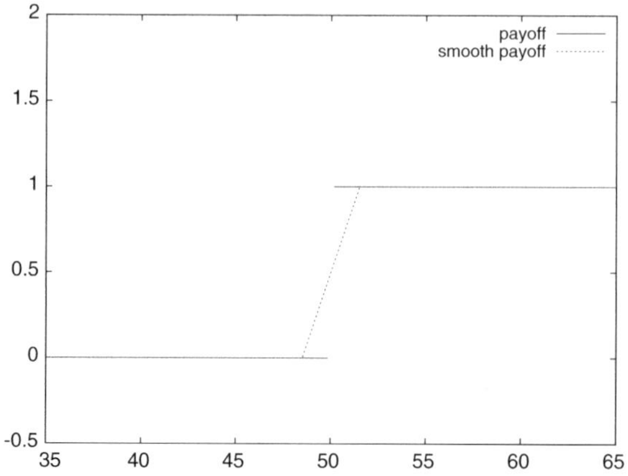

Fig. 10.1 Payoff h_1 of the digital Call with strike $K = 50$

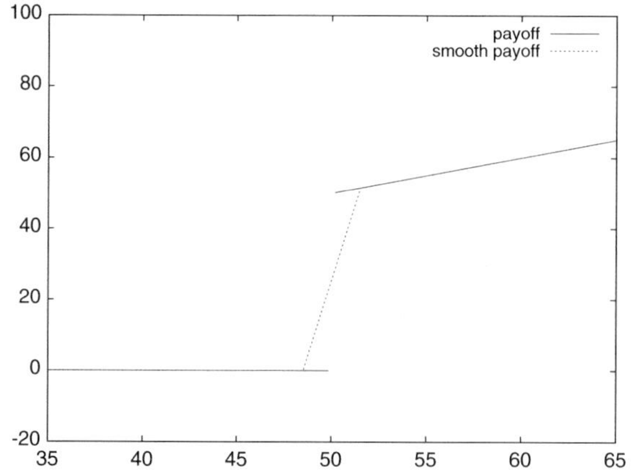

Fig. 10.2 Payoff h_S of the Asset-or-Nothing Call with strike $K = 50$)

reproduced in Figs. 10.1 and 10.2 (pay attention to the scales of the y-axis in each figure).

Let $F(x, z) = e^{-rt} h_1\big(x\, e^{(r-\frac{\sigma^2}{2})T+\sigma\sqrt{T}z}\big)$ in the digital Call case and let $F(x, z) = e^{-rt} h_S\big(x\, e^{(r-\frac{\sigma^2}{2})T+\sigma\sqrt{T}z}\big)$ in the Asset-or-Nothing Call. We define $f(x) = \mathbb{E}\, F(x, Z)$, where Z is a standard Gaussian variable. With both payoff functions, we are in the singular setting in which $F(\,.\,, Z)$ is not Lipschitz continuous but only $\frac{1}{2}$-Hölder in L^2 (see the singular setting in (10.1.1)) whereas f is $\mathcal{C}^\infty$ on $(0, +\infty)$. We are interested in computing the *delta* (δ-hedge) of the two options, *i.e.* $f'(x)$.

In such a singular case, the variance of the finite difference estimator explodes as $\varepsilon \to 0$ (see Proposition 10.2) and $\xi \mapsto F(\xi, Z)$ is not L^p-differentiable for $p \geq 1$ so that the tangent process approach cannot be used (see Sect. 10.2.2).

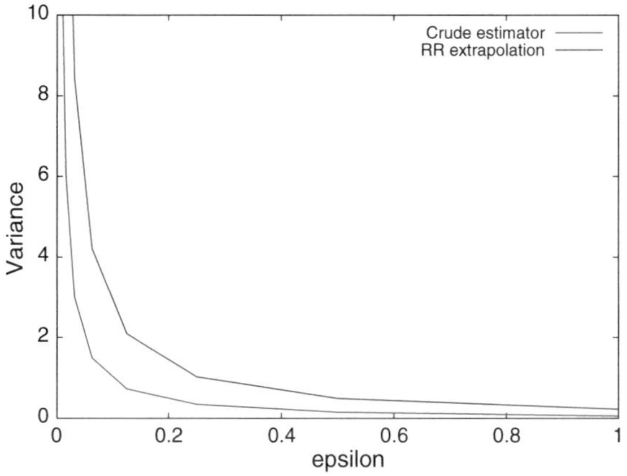

Fig. 10.3 Variance of the two estimators as a function of ε (digital Call)

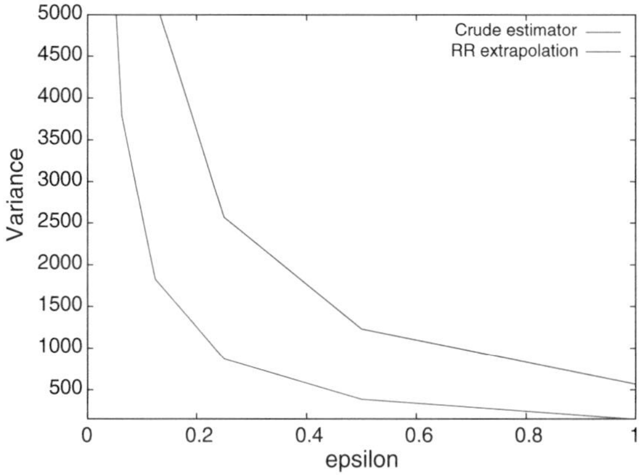

Fig. 10.4 Variance of the two estimators as a function of ε (Asset-or-Nothing Call)

We first illustrate the variance explosion in Figs. 10.3 and 10.4 where the parameters have been set to $r = 0.04$, $\sigma = 0.1$, $T = 1/12$ (one month), $x_0 = K = 50$ and $M = 10^6$.

To avoid the explosion of the variance one considers a smooth (namely Lipschitz continuous) approximation of both payoffs. Given a small parameter $\eta > 0$, one defines

$$h_{1,\eta}(\xi) = \begin{cases} h_1(\xi), & \text{if } |\xi - K| > \eta, \\ \frac{1}{2\eta}\xi + \frac{1}{2}(1 - \frac{K}{\eta}), & \text{if } |\xi - K| \leq \eta \end{cases}$$

and

$$h_{\eta,S}(\xi) = \begin{cases} h_s(\xi), & \text{if } |\xi - K| > \eta, \\ \frac{K+\eta}{2\eta}\xi + \frac{K+\eta}{2}(1 - \frac{K}{\eta}), & \text{if } |\xi - K| \leq \eta. \end{cases}$$

We define $F_\eta(x, Z)$ and $f_\eta(x)$ similarly as in the singular case.

In this numerical section, we introduce a Richardson–Romberg (*RR*) extrapolation of the finite difference estimators

$$\widehat{f'(x)}_{\varepsilon,M} = \frac{1}{M}\sum_{k=1}^{M} \frac{F(x+\varepsilon, Z_k) - F(x-\varepsilon, Z_k)}{2\varepsilon}$$

and

$$\widehat{f'_\eta(x)}_{\varepsilon,M} = \frac{1}{M}\sum_{k=1}^{M} \frac{F_\eta(x+\varepsilon, Z_k) - F_\eta(x-\varepsilon, Z_k)}{2\varepsilon}.$$

The extrapolation is done using a linear combination of the estimators with step ε and $\frac{\varepsilon}{2}$, respectively. As $\mathbb{E}\,\widehat{f'(x)}_{\varepsilon,M} = f'(x) + \frac{f^{(3)}(x)}{6}\varepsilon^2 + \frac{f^{(5)}(x)}{5!}\varepsilon^4 + O(\varepsilon^6)$, we easily check that the combination that kills the first term of this bias (in ε^2) is $\frac{4}{3}\widehat{f'(x)}_{\frac{\varepsilon}{2},M} - \frac{1}{3}\widehat{f'(x)}_{\varepsilon,M}$ (see Exercise **4.** in Sect. 10.1.1). The same holds for f_η.

Then, as in the proof of Propositions 10.1 (with $\theta = \frac{1}{2}$) and 10.2, we then prove that

$$\left\| f'(x) - \left(\frac{4}{3}\widehat{f'(x)}_{\frac{\varepsilon}{2},M} - \frac{1}{3}\widehat{f'(x)}_{\varepsilon,M}\right)\right\|_2 = O(\varepsilon^4) + O\left(\frac{1}{\sqrt{\varepsilon M}}\right), \tag{10.18}$$

and

$$\left\| f'_\eta(x) - \left(\frac{4}{3}\widehat{f'_\eta(x)}_{\frac{\varepsilon}{2},M} - \frac{1}{3}\widehat{f'_\eta(x)}_{\varepsilon,M}\right)\right\|_2 = O(\varepsilon^4) + O\left(\frac{1}{\sqrt{M}}\right). \tag{10.19}$$

► **Exercise.** Prove (10.18) and (10.19).

The control of the variance in the smooth case is illustrated in Figs. 10.5 and 10.6 when $\eta = 2$, and in Figs. 10.7 and 10.8 when $\eta = 0.5$. The variance increases when η decreases to 0 but does not explode as ε goes to 0.

For a given ε, note that the variance is usually higher using the standard *RR* extrapolation. However, in the Lipschitz continuous case the variance of the *RR*-estimator and that of the crude finite difference converge toward the same value when ε goes to 0. Moreover, from (10.18) we deduce the choice $\varepsilon = O(M^{-1/9})$ to keep the balance between the bias term of the *RR* estimator and the variance term.

As a consequence, for a given level of the L^2-error, we can choose a bigger ε with the *RR* estimator, which reduces the bias *without increasing the variance*.

The parameters of the model are the following: $x_0 = 50$, $K = 50$, $r = 0.04$, $\sigma = 0.3$ and $T = \frac{1}{52}$ (one week). The size of the Monte Carlo simulation is fixed to $M = 10^6$. We now compare the following estimators in the two cases with two different maturities $T = 1/12$ (one month) and $T = 1/52$ (one week):

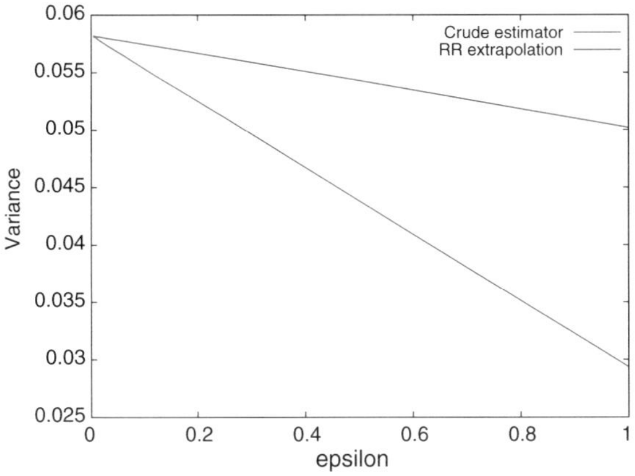

Fig. 10.5 DIGITAL CALL: Variance of the two estimators as a function of ε. $T = 1/52$. Smoothed payoff with $\eta = 1$

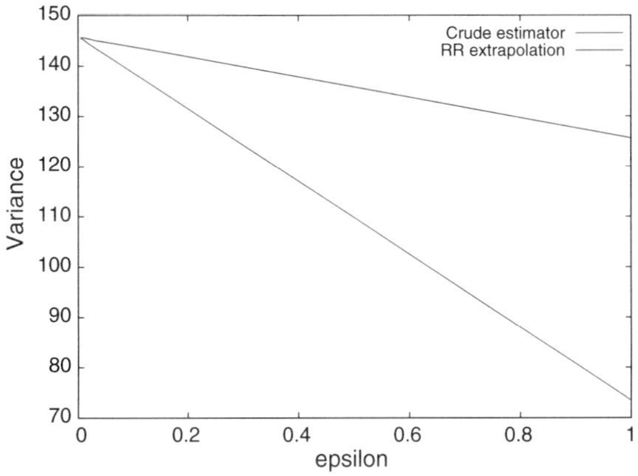

Fig. 10.6 ASSET- OR- NOTHING CALL: Variance of the two estimators as a function of ε, $T = 1/52$. Smoothed payoff with $\eta = 1$

- Finite difference estimator on the non-smooth payoffs h_1 and h_s with $\varepsilon = M^{-\frac{1}{5}} \simeq 0.0631$ since the *RMSE* has the form $f^{(3)}(x)\varepsilon^2 + O\big(\frac{1}{\sqrt{\varepsilon M}}\big)$ (see Proposition 10.1 with $\theta = \frac{1}{2}$).
- Finite difference estimator with *RR*-extrapolation on the non-smooth payoffs with $\varepsilon = M^{-\frac{1}{9}} \simeq 0.2154$ (based on (10.18)).
- Crude weighted estimator (with the standard Black–Scholes δ-weight 2.7) on the non-smooth payoffs h_1 and h_s.
- Localization (smoothing): Finite difference estimator on the smooth payoffs $h_{1,\eta}$ and $h_{S,\eta}$ with $\eta = 1$ and $\varepsilon = M^{-\frac{1}{4}}$ combined with the weighted estimator on the (non-smooth) differences $h_1 - h_{1,\eta}$ and $h_s - h_{S,\eta}$.

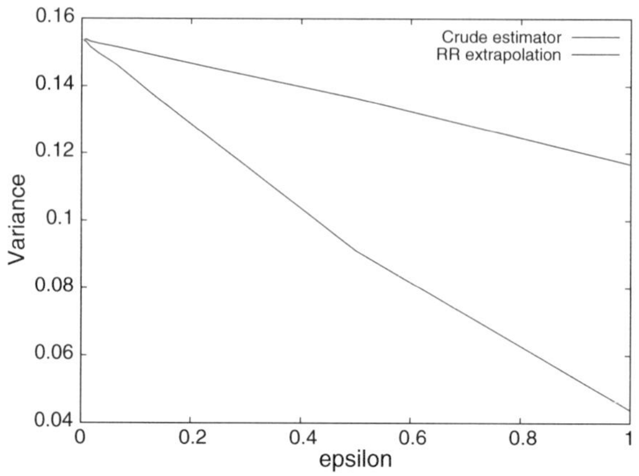

Fig. 10.7 DIGITAL CALL: Variance of the two estimators as a function of ε, $T = 1/52$. Smoothed payoff with $\eta = 0.5$

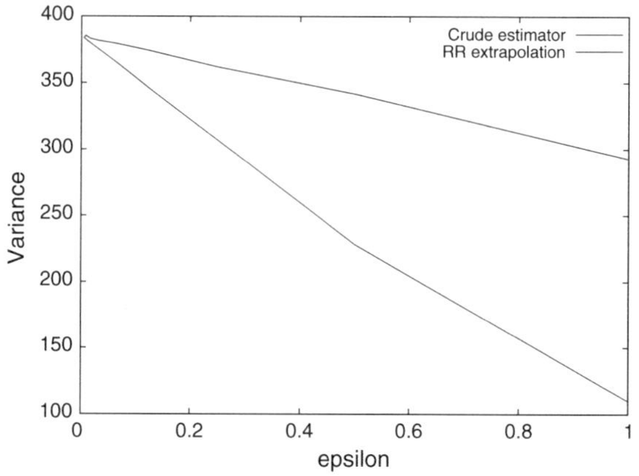

Fig. 10.8 ASSET- OR- NOTHING CALL: Variance of the two estimators as a function of ε, $T = 1/52$. Smoothed payoff with $\eta = 0.5$

- Localization (smoothing): Finite difference estimator with *RR*-extrapolation on the smooth payoffs $h_{1,\eta}$ and $h_{S,\eta}$ with $\eta = 1$ and $\varepsilon = M^{-\frac{1}{8}}$ combined with the weighted estimator on the (non-smooth) differences $h_1 - h_{1,\eta}$ and $h_s - h_{s,\eta}$ (based on (10.19)).

Table 10.1 DIGITAL CALL: *results with* $T = 1/52$ *(one week)*

Estimator	Mean	Variance	95% CI
Finite difference	0.19038	1.4735	[0.188; 0.19276]
FD + RR	0.19083	1.1967	[0.18869; 0.19297]
Weighted estimator	0.19196	0.07904	[0.19141; 0.19251]
Loc. FD + WE	0.1919	0.057459	[0.19141; 0.19239]
Loc. FD + WE + RR	0.19182	0.056925	[0.19133; 0.19231]

Table 10.2 ASSET- OR- NOTHING CALL: *results with* $T = 1/52$ *(one week)*

Estimator	Mean	Variance	95% CI
Finite difference	10.156	3736.2	[10.036; 10.276]
FD + RR	10.094	3001.1	[9.9865; 10.201]
Weighted estimator	10.101	227.98	[10.071; 10.131]
Loc. FD + WE	10.085	143.74	[10.061; 10.11]
Loc. FD + WE + RR	10.078	142.4	[10.054; 10.103]

The results are summarized in the following Tables 10.1 for the delta of the digital Call option and 10.2 for the delta of the Asset-or-Nothing Call option.

In practice, the variance of these estimators can be improved by appropriate control variates, not implemented in this numerical experiment.

Chapter 11
Optimal Stopping, Multi-asset American/Bermudan Options

11.1 Introduction

11.1.1 Optimal Stopping in a Brownian Diffusion Framework

This chapter is devoted to probabilistic numerical methods for solving optimal stopping problems, in particular the pricing of multi-asset American and Bermudan options. However, we do not propose an in-depth presentation of optimal stopping theory in discrete and continuous time, which would be far beyond the scope of this monograph. When necessary, we will refer to reference books or papers on this topic, and many important theoretical results will be admitted throughout this chapter. For an introduction to optimal stopping theory in discrete time we refer to [217] or more recently [183] and, for a comprehensive overview of the whole theory, including continuous time models and its applications to the pricing of American derivatives, we recommend Shiryaev's book [182, 262], which we will often refer to. On our side, we will focus on time and space discretization aspects in Markovian models, usually in connection with Brownian diffusion processes.

We will first explain how one can approximate a continuous time optimal stopping problem, namely for Brownian diffusions, by a discrete time optimal stopping problem for $\mathbb{R}^d$-valued Markov chains. The specificity of direct Markovian optimal stopping problems is that the quantity of interest (the Snell envelope) satisfies a backward Dynamical Programming Principle (*BDPP*) which is a major step toward an operating numerical scheme. Once this time discretization phase is achieved, we will present two different methods of space discretization which allow us to implement approximations of the *BDPP*: Least squares regression methods, also known as *American Monte Carlo* and optimal quantization methods (*quantization trees*).

As it does not add any specific difficulty, we will consider a general diffusion framework (7.1), namely

$$dX_t = b(t, X_t)dt + \sigma(t, X_t)dW_t, \quad t \in [0, T], \quad X_0 \in L^r_{\mathbb{R}^d}(\Omega, \mathcal{A}, \mathbb{P}) \tag{11.1}$$

G. Pagès, *Numerical Probability*, Universitext,
https://doi.org/10.1007/978-3-319-90276-0_11

($r \geq 1$), where $b : [0, T] \times \mathbb{R}^d \to \mathbb{R}^d$, $\sigma : [0, T] \times \mathbb{R}^d \to \mathcal{M}(d, q, \mathbb{R})$ are continuous functions, Lipschitz in x uniformly in $t \in [0, T]$ and W is a q-dimensional standard Brownian motion defined on $(\Omega, \mathcal{A}, \mathbb{P})$, independent of X_0. If we consider the augmented natural filtration of the Brownian motion $\mathcal{F}_t = \sigma(X_0, \mathcal{N}_{\mathbb{P}}, W_s, 0 \leq s \leq t)$, then this equation admits a unique $(\mathcal{F}_t)$-adapted solution defined on $(\Omega, \mathcal{A}, \mathbb{P})$. These assumptions are in force throughout the chapter.

Moreover, we will denote by $X^x = (X_t^x)_{t\in[0,T]}$ the unique solution starting at $X_0 = x \in \mathbb{R}^d$. More generally, under the above assumptions, for every $t \in [0, T]$ and every $x \in \mathbb{R}^d$, there exists a unique solution $(X_s^{t,x})_{s\in[t,T]}$ to (11.1) living between t and T and starting from x at time t, *i.e.* such that $X_t^{t,x} = x$ (note that $X_s^x = X_s^{0,x}$). The process $\big((X_s^{t,x})_{s\in[t,T]}\big)_{t\in[0,T],x\subset\mathbb{R}^d}$ is called the flow of (11.1).

An optimal stopping problem in that framework can be presented as follows. The process $(X_t^x)_{t\in[0,T]}$ is representative of a continuous time flow of information. Typically on a financial market, it can be the quotation of d traded risky assets (stocks, interest rates, currency exchange rates, quotation of various commodities, gas, oil, etc). It can also contain a lot of non-tradable information of an economical nature (inflation, industrial activity, unemployment rate,...) or a meteorological nature (temperature), etc. If X entirely consists of traded risky assets, all its components X_t^i are assumed to be positive and, in a risk neutral world, among other assumptions, the drift b is of the form $b(t, x) = r(t, x)x$, where $r : [0, T] \times \mathbb{R}_+^d \to \mathbb{R}_+$ is the instantaneous interest rate ([1]). Then, the above dynamics for the risky assets is known as a *local volatility model*, provided it satisfies some conditions on b and σ which preserve the non-negativity of the components X^i (see further on).

A stochastic game can be associated to the process X as follows: the player pays at time $t \in [0, T]$ to enter the game. She may leave the game whenever she wants at a $[t, T]$-valued random time τ. Then she receives $f(\tau, X_\tau)$ where the function $f : [0, T] \times \mathbb{R}^d \to \mathbb{R}_+$ is Borel. Not any τ is authorized, only "honest" stopping strategies. By "honest", we mean that the player made her decision in real time from her observations of the information flow $(X_t)_{t\in[0,T]}$, that is, in a non-anticipative way. That means that she "observes" the (augmented) filtration $(\mathcal{F}_t^X)_{t\in[0,T]}$ of the process $(X_t)_{t\in[0,T]}$ ([2]). If σ degenerates at some places, the filtrations $(\mathcal{F}_t)_{t\in[0,T]}$ and $(\mathcal{F}_t^X)_{t\in[0,T]}$ may differ, but of course $\mathcal{F}_t^X \subset \mathcal{F}_t$. However, as $(X_t)_{t\in[0,T]}$ is an $(\mathcal{F}_t)$-Markov process (see [256]), it turns out not to be a restriction to consider that the player observes the filtration $(\mathcal{F}_t)_{t\in[0,T]}$ to make her decisions. In fact, *a posteriori*, the decision only depends on the observed filtration $(\mathcal{F}_t^X)_{t\in[0,T]}$. In what follows, we will still denote by $\mathcal{F}_t$ any of these two filtrations when there is no necessity to distinguish

[1]This statement should be considered with caution since the traded asset may not correspond to what is quoted: thus in the case of Foreign Exchange, the quoted process is the exchange rate itself whereas the risky asset in arbitrage theory is the foreign bond quoted in domestic currency, *i.e.* the value of the foreign bond (of an appropriate maturity) multiplied by the exchange rate. Thus, if r_F denotes the foreign interest rate, r should be replaced by $r - r_F$ in the dynamics of the exchange rate. A similar situation occurs with a stock continuously distributing dividends at an exponential rate λ: r should be replaced by $r - \lambda$, or with a future contract where r should be replaced by 0 in the dynamics of the *quoted price*.

[2]$\mathcal{F}_t = \sigma(\mathcal{N}_{\mathbb{P}}, X_s, s \in [0, t])$.

them. In mathematical terms, this means that admissible stopping strategies τ are those which satisfy that the decision to stop between 0 and t is $\mathcal{F}_t$-measurable, *i.e.*

$$\forall t \in [0, T], \qquad \{\tau \le t\} \in \mathcal{F}_t.$$

In other words, τ is a $[0, T]$-valued $(\mathcal{F}_t)$-stopping time. Note that simply asking $\{\tau = t\}$ to lie in $\mathcal{F}_t$ (the player decides to stop exactly at time t) may seem more natural as an assumption, but, if it is the right notion in discrete time (equivalent to the above condition), in continuous time, this simpler condition turns out to be too loose to devise a consistent theory.

► **Exercise.** Show that for any $\mathcal{F}$-stopping time τ one has, for every $t \in [0, T]$, $\{\tau = t\} \in \mathcal{F}_t$ (the converse is not true in general in a continuous time setting).

For the sake of simplicity, we will assume that the instantaneous interest rate is constant and invariant, *i.e.* the interest rate curve is constant equal to r so that the discounting factor at time t is given by e^{-rt} (this also has consequences on the dynamics of risky assets under a risk neutral probability, see further on). At this stage, the supremum of the mean gain over all admissible stopping rules at each time is the best the player can hope for. Concerning measurability, it is not possible to consider such a supremum in a naive way. It should be replaced by a $\mathbb{P}$-essential supremum (see Sect. 12.9 in the Miscellany Chapter). This quantity, defined below, is known as the *Snell envelope*.

It is clear from Theorem 7.2 that, under the above assumptions made on b and σ, if $X_0 \in L^r(\mathbb{P})$ for some $r \ge 1$, then $\big\| \sup_{t\in[0,T]} |X_t| \big\|_r < +\infty$, so that as soon as the Borel function $f : [0, T] \times \mathbb{R}^d \to \mathbb{R}_+$ has r-polynomial growth in x uniformly in $t \in [0, T]$, that is, $0 \le f(t, x) \le C(1 + |x|^r)$, then $\sup_{t\in[0,T]} f(t, X_t) \in L^r(\mathbb{P})$. In particular, for any random variable $\tau : (\Omega, \mathcal{A}) \to [0, T]$, $0 \le f(\tau, X_\tau) \le \sup_{t\in[0,T]} f(t, X_t) \in L^r(\mathbb{P})$.

Definition 11.1 *(a) The $(\mathcal{F}_t)$-adapted process $f(t, X_t)$, $t \in [0, T]$ is called the* pay-off *process of the game (or the* American payoff*). Its discounted version, denoted by Z_t, reads* ([3])

$$Z_t = e^{-rt} f(t, X_t),\ t \in [0, T].$$

(b) Assume that $X_0 \in L^r(\mathbb{P})$ and f has r-polynomial growth. The $(\mathcal{F}, \mathbb{P})$-Snell envelope of the discounted payoff $(Z_t)_{t\in[0,T]}$, is defined for every $t \in [0, T]$ by

$$\begin{aligned} U_t &= \mathbb{P}\text{-esssup}\left\{\mathbb{E}\left(Z_\tau \,|\, \mathcal{F}_t\right),\ \tau \in \mathcal{T}^{\mathcal{F}}_{t,T}\right\} \\ &= \mathbb{P}\text{-esssup}\left\{\mathbb{E}\Big(e^{-r\tau} f(\tau, X_\tau) \,|\, \mathcal{F}_t\Big),\ \tau \in \mathcal{T}^{\mathcal{F}}_{t,T}\right\} < +\infty \ \ \mathbb{P}\text{-}a.s. \end{aligned} \tag{11.2}$$

where $\tau \in \mathcal{T}^{\mathcal{F}}_{t,T}$ denotes the set of $(\mathcal{F}_s)_{s\in[0,T]}$-stopping times having values in $[t, T]$.

[3] In this chapter, to preserve the internal consistency of the notations, we will not use $\tilde{}$ to denote discounted payoffs or prices.

In fact, since the Brownian diffusion process X is an $\mathcal{F}$-Markov process, one shows (see [263]) that the filtration $\mathcal{F}$ can be replaced by $\mathcal{F}^X$ in the above definition and characterizations of $F(t,x)$. This is clearly more in accordance with the usual models in Finance (and elsewhere) since X represents in most models the observable structure process.

Proposition 11.1 (Value function, [84, 262]) *Assume that $f : [0,T]\times\mathbb{R}^d \to \mathbb{R}_+$ is continuous with r-polynomial growth, $r \geq 1$.*
(a) There exists a continuous function $F : [0,T]\times\mathbb{R}^d \to \mathbb{R}_+$ defined by

$$F(t,x) = \sup_{\tau\in\mathcal{T}_{t,T}^{\mathcal{F}}} \mathbb{E}\left(e^{-r(\tau-t)} f(\tau, X_\tau^{t,x})\right),$$

where $\tau \in \mathcal{T}_{t,T}^{\mathcal{F}}$ denotes the set of $[t,T]$-valued $(\mathcal{F}_s)_{s\in[0,T]}$-stopping times. Then, if $X_0 = x$, the Snell envelope of $(Z_t)_{t\in[0,T]}$ satisfies

$$\forall\, t\in[0,T] \quad U_t = e^{-rt} F(t, X_t^x) \quad \mathbb{P}\ a.s. \tag{11.3}$$

This function F is called the value function. *If furthermore, f is Lipschitz continuous in $x\in\mathbb{R}^d$ uniformly in $t\in[0,T]$ then, so is the function F.*
(b) Autonomous case. *If $b(t,x) = b(x)$ and $\sigma(t,x) = \sigma(x)$, then*

$$F(t,x) = \sup_{\tau\in\mathcal{T}_{0,T-t}^{\mathcal{F}}} \mathbb{E}\left(e^{-r\tau} f(t+\tau, X_\tau^x)\right).$$

As a by-product of this intuitive, powerful (but quite demanding) result that we will admit, we derive that *the Snell envelope $(U_t)_{t\in[0,T]}$ has a continuous modification* given by (11.3) (at least when X_0 is deterministic, since the diffusion is $a.s.$ pathwise continuous).

The major results on the Snell envelope that we will admit in this continuous time framework (see [84, 182, 262]) are the following:

- $(U_t)_{t\in[0,T]}$ is a continuous $(\mathbb{P},\mathcal{F})$-super-martingale dominating the (discounted) payoff process $(Z_t)_{t\in[0,T]}$ (*i.e.* $\mathbb{P}$-$a.s.$ $(U_t \geq Z_t, \forall\, t\in[0,T])$) and
- If $(V_t)_{t\in[0,T]}$ is a continuous $(\mathbb{P},\mathcal{F})$-super-martingale dominating $(Z_t)_{t\in[0,T]}$, then $\mathbb{P}$-$a.s.$ $(V_t \geq U_t, \forall\, t\in[0,T])$.

Remark. • If the components X^i of X are positive such that $e^{-rt}X_t^i$ are $(\mathbb{P},\mathcal{F}_t)$-martingales, and if the representation theorem of $(\mathcal{F}_t^W)$-martingales as stochastic integrals with respect to W can be reformulated as a representation theorem with respect to the above (discounted) processes X_t^i, then the diffusion $(X_t)_{t\in[0,T]}$ is an abstract model of complete financial market where X^i are the quoted prices of the risky assets. The last representation assumption is essentially satisfied under uniform ellipticity of the diffusion coefficient of the *SDE* satisfied by the log-risky assets. In such a model, one shows by hedging arguments that $F(0,x)$ is the premium of an

American option with payoff $f(t, X_t^x)$, $t \in [0, T]$. We refer to [182] for a detailed presentation with a focus on a model of stock market distributing dividends (see also Sect. 11.1.2 for few more technical aspects).

▶ **Exercise.** Prove the "Lipschitz claim" of the theorem: if f is Lipschitz continuous in $x \in \mathbb{R}^d$ uniformly in $t \in [0, T]$, then so is the function F. [Hint: use that the flow of the SDE is Lipschitz continuous in $L^1(\mathbb{P})$, see Theorem 7.10.]

The Lipschitz continuity assumption on the payoff function f will be in force throughout the chapter, namely

$$(\text{Lip}) \equiv f \text{ is Lipschitz continuous in } x \in \mathbb{R}^d, \text{ uniformly in } t \in [0, T].$$

This assumption will sometimes be refined into a semi-convex condition

$(SC) \equiv f$ satisfies (Lip) and is semi-convex in the following sense: there exists $\delta_f : [0, T] \times \mathbb{R}^d \to \mathbb{R}^d$, Borel and bounded, and $\rho \in (0, +\infty)$ such that

$$\forall t \in [0, T], \forall x, y \in \mathbb{R}^d, f(t, y) - f(t, x) \geq \big(\delta_f(t, x) \,|\, y - x\big) - \rho|y - x|^2. \quad (11.4)$$

⊳ **Examples. 1.** If f is Lipschitz continuous and convex, then f is semi-convex with

$$\delta_f(t, x) = \left[\left(\frac{\partial f}{\partial x_i}\right)_r (t, x)\right]_{1 \leq i \leq d}$$

where $\left(\frac{\partial f}{\partial x_i}\right)_r (t, x)$ denotes the right derivative with respect to x_i at (t, x).

2. If $f(t, .) \in \mathcal{C}^1(\mathbb{R}^d, \mathbb{R}^d)$ and, for every $t \in [0, T]$, $\nabla_x f(t, .)$ is Lipschitz continuous in x, uniformly in $t \in [0, T]$, then

$$\begin{aligned} f(t, y) - f(t, x) &\geq (\nabla_x f(t, x)|y - x) - [\nabla_x f]_{\text{Lip}}|\xi - x||y - x|, \quad \xi \in [x, y] \\ &\geq \langle \nabla_x f(t, x)|y - x\rangle - [\nabla_x f]_{\text{Lip}}|y - x|^2, \end{aligned}$$

where $[\nabla_x f]_{\text{Lip}} = \sup_{t \in [0,T]} [\nabla_x f(t, .)]_{\text{Lip}}$.

11.1.2 Interpretation in Terms of American Options (Sketch)

As far as derivative pricing is concerned, the (non-negative) function f defined on $[0, T] \times \mathbb{R}_+^d$ is an American payoff function. This means that the holder of the contract exercises his/her option at time t and he/she will receive a monetary flow equal to $f(t, x)$ if the vector of market prices is equal to $x \in \mathbb{R}_+^d$ at time t. When modeling a financial market, one usually assumes that the quoted prices of the d risky assets X_t^i, $i = 1, \ldots, d$, are non-negative, which has important consequences on the possible choices for b and σ. As we assume that all components are observable, our framework corresponds to a multi-asset local volatility model. To ensure the positivity of the risky assets, a natural condition is to assume that

$$b(t,x) = \mathrm{Diag}(x^1, \ldots, x^d)\beta(t,x) \quad \text{and} \quad \sigma(t,x) = \mathrm{Diag}(x^1, \ldots, x^d)\vartheta(t,x),$$

where $\beta : [0,T] \times \mathbb{R}^d \to \mathbb{R}^d$ and $\vartheta : [0,T] \times \mathbb{R}^d \to \mathcal{M}(d,q,\mathbb{R})$ are bounded continuous functions, Lipschitz in x uniformly in $t \in [0,T]$ (other conditions are needed to ensure that b and σ themselves satisfy a Lipschitz continuity assumption in x uniformly in t). In practice, if one assumes that the interest rate curve is flat and stays invariant in time, the drift of the dynamics under a *risk-neutral probability* is simply of the form $b(t,x) = rx$ so ([4]) that, for every i $t \in \{1, \ldots, d\}$,

$$\forall\, t \in [0,T],\; X_t^i = x_0^i \exp\left(rt - \frac{\int_0^t |\vartheta_{i.}(s,X_s)|^2 ds}{2} + \int_0^t \big(\vartheta_{i.}(s,X_s)|dW_s\big)_q \right),\quad x_0^i > 0.$$

Note that, a necessary condition for such a market to be complete is that $q \geq d$ and a sufficient condition is that $\vartheta\vartheta^* \geq \varepsilon_0 I_d$ (uniform ellipticity). If this is the case, the risk neutral probability (equivalent to the historical probability) is unique and one shows (see [181] in a more general setting) that, for hedging purposes, one should price all the American option contracts under this *risk-neutral* probability measure: this premium is the minimal amount of cash that makes it possible to manage a self-financed portfolio to face any early – optimal or not – exercise of the American option contract during its entire life between 0 and T (see [182] or, in a discrete time models, [59]). Once again, one should be cautious in interpreting the preceding, the first task being to determine what the risky assets really are (since it may differ from the "quoted asset").

However, as far as numerical aspects are concerned, this kind of restriction is of little interest in the sense that it has no impact on the methods or on their performances. This is the reason why, in what follows, we will keep a general form for the drift b of our Brownian diffusion dynamics (and always denote by $\mathbb{P}$ the probability measure on $(\Omega, \mathcal{A})$).

[4] A risk neutral probability is a probability $\mathbb{P}^*$ on $(\Omega, \mathcal{A}, \mathbb{P})$, equivalent to $\mathbb{P}$, such that the discounted value of the d risky assets are $\mathbb{P}^*$-martingales. If $b(t,x) = rx$ and ϑ is bounded, $\mathbb{P}$ itself is a risk-neutral probability.

11.2 Optimal Stopping for Discrete Time $\mathbb{R}^d$-Valued Markov Chains

11.2.1 *General Theory, the Backward Dynamic Programming Principle*

Optimal stopping in discrete time can be developed in a more general framework, starting from a sequence $(Z_k)_{k=0,\dots,n}$ of non-negative integrable payoffs defined on a filtered probability space $(\Omega, \mathcal{A}, (\mathcal{F}_k)_{k=0,\dots,n}, \mathbb{P})$. For such a presentation we refer to [217] or [183] (on a finite probability space Ω of scenarii). In view of the applications we have in mind – the approximation of a continuous time Snell envelope where the underlying process is a Brownian diffusion – we chose to develop our presentation in a Markovian framework. However it remains very close to the abstract framework.

Let $(X_k)_{0\le k\le n}$ be an $\mathbb{R}^d$-valued $(\mathcal{F}_k)_{0\le k\le n}$-Markov chain defined on a filtered probability space $\big(\Omega, \mathcal{A}, (\mathcal{F}_k)_{k=0,\dots,n}, \mathbb{P}\big)$ with Markov transitions ([5])

$$P_k(x, dy) = \mathbb{P}\big(X_{k+1} \in dy \mid X_k = x\big),\ \ k = 0, \dots, n-1,$$

and let $Z = (Z_k)_{0\le k\le n}$ be an $(\mathcal{F}_k)_{0\le k\le n}$-adapted payoff sequence of non-negative *integrable* random variables of the form

$$Z_k = f_k(X_k) \in L^1_{\mathbb{R}_+}(\Omega, \mathcal{F}_k, \mathbb{P}),\ \ k = 0, \dots, n, \tag{11.5}$$

where $f_k : \mathbb{R}^d \to \mathbb{R}_+$ are non-negative Borel functions. We want to compute the $(\mathbb{P}, (\mathcal{F}_k)_{0\le k\le n})$-*Snell envelope* $U = (U_k)_{0\le k\le n}$ defined ([6]) by

$$U_k = \mathbb{P}\text{-esssup}\Big\{\mathbb{E}\left(f_\tau(X_\tau) \mid \mathcal{F}_k\right),\ \ \tau \in \mathcal{T}^{\mathcal{F}}_{k:n}\Big\}, \tag{11.6}$$

where $\mathcal{T}^{\mathcal{F}}_{k:n} = \big\{\tau : (\Omega, \mathcal{A}) \to \{k, \dots, n\}, (\mathcal{F}_\ell)_{0\le k\le n}\text{-stopping time}\big\}$ and its sequence of values $(\mathbb{E}\, U_k)_{0\le k\le n}$.

Note that $U_k \ge Z_k$ *a.s.* for every $k \in \{0, \dots, n\}$.

Proposition 11.2 (Backward Dynamic Programming Principle (*BDPP*)) (*a*) Pathwise Backward Dynamic Programming Principle. *The Snell envelope* $(U_k)_{0\le k\le n}$ *satisfies the following Backward Dynamic Programming Principle*

---

[5] Let us recall that this means that for every $k \in \{0, \dots, n-1\}$, for every bounded or non-negative Borel function $f : \mathbb{R}^d \to \mathbb{R}$,

$$\mathbb{E}\,(f(X_{k+1}) \mid \mathcal{F}_k) = \mathbb{E}\,(f(X_{k+1}) \mid X_k) = P_k(f)(X_k)\ \ \mathbb{P}\text{-}a.s.$$

[6] A random variable $\tau : (\Omega, \mathcal{A}) \to \{0, \dots, n\}$ is an $(\mathcal{F}_k)_{0\le k\le n}$-stopping time if $\{\tau \le k\} \in \mathcal{F}_k$ for every $k = 0, \dots, n$, or, equivalently, if $\{\tau = k\} \in \mathcal{F}_k$ for every $k = 0, \dots, n$.

$$U_n = Z_n \ \ and \ \ U_k = \max\Big(Z_k, \mathbb{E}\left(U_{k+1} \mid X_k\right)\Big),\ k = 0, \ldots, n-1. \tag{11.7}$$

(*b*) Optimal stopping times. *The sequence of stopping times* $(\tau_k)_{0\le k\le n}$ *defined by*

$$\tau_n = n \ \ and \ \ \tau_k = k\mathbf{1}_{\left\{Z_k \ge \mathbb{E}\left(Z_{\tau_{k+1}}|X_k\right)\right\}} + \tau_{k+1}\mathbf{1}_{\left\{Z_k < \mathbb{E}\left(Z_{\tau_{k+1}}|X_k\right)\right\}},\ k = 0, \ldots, n-1 \tag{11.8}$$

satisfies

$$\tau_k = \min\left\{\ell \in \{k, \ldots, n\},\ U_\ell = Z_\ell\right\} \quad and \quad U_k = \mathbb{E}\left(Z_{\tau_k}\,|X_k\right) a.s.,\ k = 0, \ldots, n. \tag{11.9}$$

In particular, this means that, for every $k \in \{0, \ldots, n\}$, *the* $\{k, \ldots, n\}$*-valued* τ_k *is an optimal stopping time for the optimal stopping problem starting at* k.

(*c*) Functional Backward Dynamic Programming Principle. *Furthermore, for every* $k \in \{0, \ldots, n\}$ *there exists a Borel function* $u_k : \mathbb{R}^d \to \mathbb{R}$ *such that*

$$U_k = u_k(X_k),\ k = 0, \ldots, n,$$

where

$$u_n = f_n \quad and \quad u_k = \max\left(f_k,\, P_k u_{k+1}\right),\ k = 0, \ldots, n-1.$$

Proof. Let us introduce (temporarily) the sequence

$$\theta_n = n \ \text{ and } \ \theta_k = k\mathbf{1}_{\left\{Z_k \ge \mathbb{E}\left(Z_{\theta_{k+1}}|\mathcal{F}_k\right)\right\}} + \theta_{k+1}\mathbf{1}_{\left\{Z_k < \mathbb{E}\left(Z_{\theta_{k+1}}|\mathcal{F}_k\right)\right\}},\ k = 0, \ldots, n-1.$$

The fact that θ_k is a stopping time is left as an exercise since $\{Z_k \ge \mathbb{E}\left(Z_{\theta_{k+1}}|\mathcal{F}_k\right)\}$, $\{Z_k < \mathbb{E}\left(Z_{\theta_{k+1}}|\mathcal{F}_k\right)\} = \{Z_k \ge \mathbb{E}\left(Z_{\theta_{k+1}}|\mathcal{F}_k\right)\}^c \in \mathcal{F}_k$ and θ_{k+1} is an $(\mathcal{F}_\ell)$-stopping time. As a first step we will prove (*a*) and (*b*) where X_k is replaced by $\mathcal{F}_k$ in the conditioning.

We proceed by a backward induction on the discrete instant k. If $k = n$, $\mathcal{T}^{\mathcal{F}}_{k:n}$ is reduced to $\{n\}$ so that $U_n = \mathbb{E}(Z_n|\mathcal{F}_n) = Z_n$. Now, let $k \in \{0, \ldots, n-1\}$ and let $\theta \in \mathcal{T}^{\mathcal{F}}_{k:n}$. Set $\widetilde{\theta} = \theta \vee (k+1)$ so that $\widetilde{\theta} = k\mathbf{1}_{\{\theta=k\}} + \theta\mathbf{1}_{\{\theta\ge k+1\}}$. The random variable $\widetilde{\theta}$ is a $\{k+1, \ldots, n\}$-valued $(\mathcal{F}_\ell)$-stopping time since $\{\theta \ge k+1\} = \{\theta = k\}^c \in \mathcal{F}_k$. Then, using the chain rule for conditional expectation in the third line and the definition of U_{k+1} in the fourth line, we obtain

$$\begin{aligned}
\mathbb{E}\left(f_\theta(X_\theta)\,|\mathcal{F}_k\right) &= f_k(X_k)\mathbf{1}_{\{\theta=k\}} + \mathbb{E}\left(f_{\widetilde{\theta}}(X_{\widetilde{\theta}})\mathbf{1}_{\{\theta\ge k+1\}}\,|\mathcal{F}_k\right)\\
&= f_k(X_k)\mathbf{1}_{\{\theta=k\}} + \mathbb{E}\left(f_{\widetilde{\theta}}(X_{\widetilde{\theta}})\,|\mathcal{F}_k\right)\mathbf{1}_{\{\theta\ge k+1\}}\\
&= f_k(X_k)\mathbf{1}_{\{\theta=k\}} + \mathbb{E}\Big[\mathbb{E}\left(f_{\widetilde{\theta}}(X_{\widetilde{\theta}})\,|\mathcal{F}_{k+1}\right)|\mathcal{F}_k\Big]\mathbf{1}_{\{\theta\ge k+1\}}\\
&\le f_k(X_k)\mathbf{1}_{\{\theta=k\}} + \mathbb{E}\left(U_{k+1}\,|\mathcal{F}_k\right))\mathbf{1}_{\{\theta\ge k+1\}} \quad a.s.\\
&\le \max\Big(f_k(X_k), \mathbb{E}\left(U_{k+1}\,|\mathcal{F}_k\right)\Big) \quad a.s.
\end{aligned}$$

By the very definition of the Snell envelope, $U_k \geq \mathbb{E}\,(Z_{\theta_{k+1}}\,|\mathcal{F}_k)\ a.s.$ so that

$$\begin{aligned}
U_k \geq \mathbb{E}\,(Z_{\theta_k}\,|\mathcal{F}_k) &\overset{a.s.}{=} Z_k \mathbf{1}_{\{\theta_k=k\}} + \left[\mathbb{E}\left(Z_{\theta_{k+1}}\,|\mathcal{F}_k\right)\right]\mathbf{1}_{\left\{\theta_k \geq k+1\right\}} \\
&= Z_k \mathbf{1}_{\left\{\theta_k=k\right\}} + \left[\mathbb{E}\Big(\mathbb{E}\left(Z_{\theta_{k+1}}\,|\mathcal{F}_{k+1}\right)|\mathcal{F}_k\Big)\right]\mathbf{1}_{\left\{\theta_k \geq k+1\right\}} \\
&= Z_k \mathbf{1}_{\left\{\theta_k=k\right\}} + \left[\mathbb{E}\left(U_{k+1}\,|\mathcal{F}_k\right)\right]\mathbf{1}_{\{\theta_k \geq k+1\}} \quad a.s.
\end{aligned}$$

owing to the induction assumption. Going back to the definition of θ_k we check that $\{\theta_k = k\} = \{Z_k \geq \mathbb{E}\,(Z_{\theta_{k+1}}|\mathcal{F}_k)\}$ and $\{\theta_k \geq k+1\} = \{Z_k < \mathbb{E}\,(Z_{\theta_{k+1}}\,|\mathcal{F}_k)\}$. Hence

$$U_k \geq \mathbb{E}\,(Z_{\theta_k}\,|\mathcal{F}_k) = \max\Big(Z_k, \mathbb{E}\left(U_{k+1}\,|\mathcal{F}_k\right)\Big).$$

This completes the proof of (11.7) except for the conditioning and, as a consequence, $\mathbb{E}\,(Z_{\theta_k}\,|\mathcal{F}_k) = \max\Big(Z_k, \mathbb{E}\left(U_{k+1}\,|\mathcal{F}_k\right)\Big) = U_k\ a.s.$

(c) The result is straightforward by a backward verification procedure: one notices that, for every $k \in \{0, \ldots, n-1\}$

$$\mathbb{E}\,(U_{k+1}\,|\,\mathcal{F}_k) = \mathbb{E}\,(u_{k+1}(X_{k+1})\,|\,\mathcal{F}_k) = \mathbb{E}\,(u_{k+1}(X_{k+1})\,|\,X_k) = P_k u_{k+1}(X_k)$$

owing to the Markov property, so that $U_k = \max\big(f_k(X_k), P_k u_{k+1}(X_k)\big) = u_k(X_k)$. Consequently, $\mathbb{E}\,(U_{k+1}\,|\,\mathcal{F}_k) = \mathbb{E}\,(U_{k+1}\,|\,X_k)$ which completes the proof of (11.7) and also shows that, for every $k \in \{0, \ldots, n\}$,

$$U_k = u_k(X_k) = \mathbb{E}\,(U_k\,|X_k) = \mathbb{E}\Big(\mathbb{E}\left(Z_{\theta_k}\,|\mathcal{F}_k\right)|X_k\Big) = \mathbb{E}\left(Z_{\theta_k}\,|X_k\right).$$

Plugging this equality into the definition of the stopping times θ_k shows, again by a backward induction, that $\theta_k = \tau_k\ a.s.$ since

$$\begin{aligned}
\mathbb{E}\,(Z_{\tau_{k+1}}\,|\mathcal{F}_k) &= \mathbb{E}\Big(\mathbb{E}\left(Z_{\tau_{k+1}}\,|\mathcal{F}_{k+1}\,\right)|\mathcal{F}_k\Big) \\
&= \mathbb{E}\left(U_{k+1}\,|\mathcal{F}_k\right) = \mathbb{E}\left(U_{k+1}\,|\,X_k\right) \\
&= \mathbb{E}\Big(\mathbb{E}\,(Z_{\tau_{k+1}}\,|\mathcal{F}_{k+1})\,|\,X_k\Big) = \mathbb{E}\,(Z_{\tau_{k+1}}\,|X_k)
\end{aligned}$$

$a.s.$ for every $k \in \{0, \ldots, n-1\}$. This completes the proof of (11.8).

Finally, we derive that

$$\begin{aligned}
\tau_k &= k\mathbf{1}_{\left\{Z_k \geq \mathbb{E}\,(U_{k+1}|\mathcal{F}_k)\right\}} + \tau_{k+1}\mathbf{1}_{\left\{Z_k < \mathbb{E}\,(U_{k+1}|\mathcal{F}_k)\right\}} \\
&= k\mathbf{1}_{\left\{U_k = Z_k\right\}} + \tau_{k+1}\mathbf{1}_{\left\{U_k > Z_k\right\}}, \quad k = 0, \ldots, n,
\end{aligned}$$

which implies, again by a backward induction, that

$$\tau_k = \min\left\{\ell \in \{k, \ldots, n\},\ U_\ell = Z_\ell\right\}$$

since $U_n = Z_n$. ◇

Remarks. • The optimal stopping time (starting at time k) is in general not unique (see [182] or [59], Problem 21, p. 253 for an example dealing with the American Put in a binomial model) but

$$\tau_k = \min\left\{\ell \in \{k, \ldots, n\},\ U_\ell = Z_\ell\right\}$$

is the lowest such optimal stopping time.

• It immediately follows from (11.7) that $(U_k)_{k=0,\ldots,n}$ is a $\mathbb{P}$-super-martingale dominating the payoff process $(Z_k)_{k=0,\ldots,n}$ (*i.e.* $U_k \geq Z_k$ *a.s.*).

► **Exercise.** Show that any $(\mathcal{F}, \mathbb{P})$-super-martingale $(V_k)_{0\leq k\leq n}$ which dominates $(Z_k)_{0\leq k\leq n}$, dominates $(U_k)_{0\leq k\leq n}$ as well.

This proposition shows that two backward dynamic programming principles coexist, a first one based on the Snell envelope, which can be considered as the "regular" *BDPP* formula, and a second one, (11.8), based on a recursion on optimal stopping times depending on the effective date k of entry into the game, which can be seen as a dual form of the *BDPP*. It is on this second *BDDP* that the original paper [202] on least squares regression methods is based (see the next section).

11.2.2 Time Discretization for Snell Envelopes Based on a Diffusion Dynamics

We aim at discretizing the optimal stopping problem defined by (11.2). This discretization is two-fold: first we discretize the instants at which the payoffs can be "exercised" and, as a second step, we will discretize the diffusion itself, if necessary, in order to have at hand a simulable underlying structure process. In this first phase, the Markov chain of interest is $(X_{t_k^n})_{0\leq k\leq n}$ and the payoff functions are $f_k = f(t_k^n, .)$, $k = 0, \ldots, n$.

First we note that if $t_k^n = \frac{kT}{n}$, $k = 0, \ldots, n$, then $(X_{t_k^n})_{0\leq k\leq n}$ is an $(\mathcal{F}_{t_k^n})_{0\leq k\leq n}$-Markov chain with transitions

$$P_k(x, dy) = \mathbb{P}(X_{t_{k+1}^n} \in dy \mid X_{t_k^n} = x),\ k = 0, \ldots, n-1.$$

Proposition 11.3 *Let $n \in \mathbb{N}^*$. Set*

$$\mathcal{T}_k^n = \left\{\tau : (\Omega, \mathcal{A}, \mathbb{P}) \to \{t_\ell^n,\ \ell = k, \ldots, n\},\ (\mathcal{F}_{t_\ell^n})\text{-stopping time}\right\}$$

and, for every $x \in \mathbb{R}^d$, the value function (for discrete time observations or exercise):

$$F_n(t_k^n, x) = \sup_{\tau \in \mathcal{T}_k^n} \mathbb{E}\Big(e^{-r\tau} f\big(\tau, X_\tau^{t_k^n, x}\big)\Big), \tag{11.10}$$

where $(X_s^{t_k^n,x})_{s\in[t_k^n,T]}$ is the unique solution to the (SDE) starting from x at time t_k^n.

(a) *For every $x \in \mathbb{R}^d$, $\Big(e^{-rt_k^n} F_n(t_k^n, X_{t_k^n}^x)\Big)_{k=0,\dots,n} = \big(\mathbb{P}, (\mathcal{F}_{t_k^n})\big)\text{-Snell}\Big(e^{-rt_k^n} f(t_k^n, X_{t_k^n}^x)\Big)$.*

(b) *For every $x \in \mathbb{R}^d$, $\Big(F_n(t_k^n, X_{t_k^n}^x)\Big)_{k=0,\dots,n}$ satisfies the "pathwise" BDPP*

$$F_n(T, X_T^x) = f(T, X_T^x)$$

and

$$F_n(t_k^n, X_{t_k^n}^x) = \max\Big(f(t_k^n, X_{t_k^n}^x), e^{-r\frac{T}{n}}\mathbb{E}\big(F_n(t_k^n, X_{t_{k+1}^n}^x) \mid \mathcal{F}_{t_k^n}\big)\Big),\ k = 0, \dots, n-1. \tag{11.11}$$

(c) *The functions $F_n(t_k^n, .)$ satisfy the "functional" BDPP*

$$F_n(T, x) = f(T, x) \ \text{ and } \ F_n(t_k^n, x) = \max\Big(f(t_k^n, x), e_k^{-r\frac{T_n}{P}}\big(F_n(t_k^n, .)\big)(x)\Big),$$
$$k = 0, \dots, n-1.$$

(d) *If the function f satisfies the Lipschitz assumption (Lip), then the functions $F_n(t_k^n, .)$ are all Lipschitz continuous, uniformly in $k \in \{0, \dots, n\}$, $n \geq 1$.*

Proof. We proceed by a backward induction, starting from claim (c). We consider the functions $F_n(t_k^n, .)$, $k = 0, \dots, n$.

$(c) \Rightarrow (b)$. The result follows from the Markov property, which implies for every $k = 0, \dots, n-1$,

$$\mathbb{E}\Big(F_n(t_{k+1}^n, X_{t_{k+1}^n}) \mid \mathcal{F}_{t_k^n}\Big) = P_k\big(F_n(t_k^n, .)\big)(x).$$

$(b) \Rightarrow (a)$. This is a trivial consequence of the $(BDPP)$ since $\Big(e^{-rt_k^n} F_n(t_k^n, X_{t_k^n})\Big)_{k=0,\dots,n}$ is the $(\mathbb{P}, (\mathcal{F}_{t_k^n})_{0\leq k\leq n})$-Snell envelope associated to the obstacle sequence $\Big(e^{-rt_k^n} f(t_k^n, X_{t_k^n})\Big)_{k=0,\dots,n}$.

Applying the preceding to the case $X_0 = x$, we derive from the general theory on optimal stopping that

$$F_n(0, x) = \sup\Big\{\mathbb{E}\big(e^{-r\tau} f(\tau, X_\tau)\big),\ \tau \in \mathcal{T}_0^n\Big\}.$$

The extension to times t_k^n, $k = 1, \dots, n$, follows likewise from the same reasoning carried out with $\Big(X_{t_\ell^n}^{t_k^n, x}\Big)_{\ell=k,\dots,n}$.

(d) A slight adaptation of Theorem 7.10 shows that there exists a real constant $C_{[b]_{\mathrm{Lip}},[\sigma]_{\mathrm{Lip}},T}$ such that

$$\Big\| \sup_{0\le t\le s\le T} |X_s^{t,x} - X_s^{t,y}| \Big\|_p \le C_{[b]_{\mathrm{Lip}},[\sigma]_{\mathrm{Lip}},T} |x-y|.$$

Using the Lipchitz continuity of f, the conclusion follows from the definition (11.10) of the functions $F_n(t_k^n, \,.\,)$. ◇

The following rates of convergence of both the Snell envelope related to the optimal stopping problem with payoff $\big(f(t_k^n, X_{t_k^n})\big)_{k=0,\dots,n}$ and its value function toward the continuous time Snell envelope and its value function at the discretization times were established in [21]. Note that this rate can be significantly improved when f is semi-convex.

Theorem 11.1 (Diffusion at discrete times) (a) Discretization of the stopping rules for the structure process X: *If f satisfies (Lip), i.e. is Lipschitz continuous in x uniformly in $t\in[0,T]$, then so are the value functions $F_n(t_k^n, \,.\,)$ and $F(t_k^n, \,.\,)$, uniformly with respect to t_k^n, $k=0,\dots,n$, $n\ge 1$.*

Furthermore, $F(t_k^n, .) \ge F_n(t_k^n, \,.\,)$ and

$$\Big\| \max_{0\le k\le n} \Big(F(t_k^n, X_{t_k^n}) - F_n(t_k^n, X_{t_k^n})\Big)\Big\|_p \le \frac{C_{b,\sigma,f,T}}{\sqrt{n}}$$

and, for every compact set $K\subset\mathbb{R}^d$,

$$0 \le \sup_{x\in K}\Big(\max_{0\le k\le n}\Big(F(t_k^n, x) - F_n(t_k^n, x)\Big)\Big) \le \frac{C_{b,\sigma,f,T,K}}{\sqrt{n}}.$$

(b) Semi-convex payoffs. *If f is semi-convex, then there exist real constants $C_{b,\sigma,f,T}$ and $C_{b,\sigma,f,T,K} > 0$ such that*

$$\Big\| \max_{0\le k\le n} \Big(F_n(t_k^n, X_{t_k^n}) - F(t_k^n, X_{t_k^n})\Big)\Big\|_p \le \frac{C_{b,\sigma,f,T}}{n}$$

and, for every compact set $K\subset\mathbb{R}^d$,

$$0 \le \sup_{x\in K}\Big(\max_{0\le k\le n}\Big(F(t_k^n, x) - F_n(t_k^n, x)\Big)\Big) \le \frac{C_{b,\sigma,f,T,K}}{n}.$$

If the diffusion process is simulable at the instants t_k^n, as may happen if $X_{t_k^n} = \varphi(X_0, t_k^n, W_{t_k^n})$, where φ has an explicit (computable) form, the time discretization can be stopped here. Otherwise, we need to perform a second time discretization of the underlying diffusion process in order to be able to simulate the dynamics.

Now we pass to the second phase: we approximate the (usually not simulable) Markov chain $(X_{t_k^n})_{0\le k\le n}$ by the discrete time Euler scheme $(\bar{X}^n_{t_k^n})_{0\le k\le n}$ as defined by

Eq. (7.5) in Chap. 7. We recall its definition for convenience (to emphasize a change of notation concerning the Gaussian noise):

$$\bar{X}^n_{t^n_{k+1}} = \bar{X}^n_{t^n_k} + \frac{T}{n} b(t^n_k, \bar{X}^n_{t^n_k}) + \sigma(t^n_k, \bar{X}^n_{t^n_k})\sqrt{\frac{T}{n}}\xi^n_{k+1}, \quad \bar{X}^n_0 = X_0, \quad k = 0, \dots, n-1, \tag{11.12}$$

where $(\xi^n_k)_{1\le k\le n}$ denotes a sequence of i.i.d. $\mathcal{N}(0; I_q)$-distributed random vectors given by

$$\xi^n_k := \sqrt{\frac{n}{T}}\left(W_{t^n_k} - W_{t^n_{k-1}}\right), \quad k = 1, \dots, n.$$

In the absence of ambiguity, we will drop the superscript ^{n} in $\bar{X}^n$ and ξ^n_k to write $\bar{X}$ and ξ_k. Also note that we temporarily gave up the notation Z^n_k for the Gaussian innovations in favor of ξ^n_k since Z is often representative of the reward process in this chapter.

Proposition 11.4 (Euler scheme) *The above proposition remains true when replacing the sequence $(X_{t^n_k})_{0\le k\le n}$ by its Euler scheme $(\bar{X}^n_{t^n_k})_{0\le k\le n}$ with step $\frac{T}{n}$, still with the filtration $\mathcal{F}_{t^n_k} = \sigma(X_0, \mathcal{F}^W_{t^n_k})$, $k = 0, \dots, n$. In both cases one just has to replace the transitions $P_k(x, dy)$ of the original process by that of its Euler scheme with step $\frac{T}{n}$, namely $\bar{P}^n_k(x, dy)$ defined for every $k \in \{0, \dots, n-1\}$ by*

$$\bar{P}^n_k f(x) = \mathbb{E}\left(f\left(x + \frac{T}{n}b(t^n_k, x) + \sqrt{\frac{T}{n}}\sigma(t^n_k, x)\,\xi\right)\right), \quad \xi \overset{d}{=} \mathcal{N}(0; I_q),$$

and F_n by $\bar{F}_n$.

Remark. In fact, the result also holds true with the (smaller) innovation filtration $\mathcal{F}^{X_0,Z}_k = \sigma(X_0, Z_1, \dots, Z_k)$, $k = 0, \dots, n$, since the Euler scheme is still a Markov chain with the same transitions with respect to this filtration.

The rate of convergence between the Snell envelopes and their value functions when switching *mutatis mutandis* from the diffusion "sampled" at discrete times $(X_{t^n_k})_{k=0,\dots,n}$ to the Euler scheme $(\bar{X}_{t^n_k})_{k=0,\dots,n}$ is also established in [21].

Theorem 11.2 (Euler scheme approximation) *Under the assumptions of the former Theorem 11.1(a), there exists real constants $C_{b,\sigma,f,T}$ and $C_{b,\sigma,f,T,K} > 0$ such that*

$$\left\| \max_{0\le k\le n} |F_n(t^n_k, X_{t^n_k}) - \bar{F}_n(t^n_k, \bar{X}^n_{t^n_k})| \right\|_p \le \frac{C_{b,\sigma,f,T}}{\sqrt{n}}$$

and, for every compact set $K \subset \mathbb{R}^d$,

$$0 \le \sup_{x\in K}\left(\max_{0\le k\le n}\left(F(t^n_k, x) - \bar{F}_n(t^n_k, x)\right)\right) \le \frac{C_{b,\sigma,f,T,K}}{\sqrt{n}}.$$

ℵ Practitioner's corner

▷ If the diffusion process X is simulable at times t_k^n (in an exact way), there is no need to introduce the Euler scheme and it will be possible to take advantage of the semi-convexity of the payoff/obstacle function f to get a time discretization error of the value function F by F_n at a $O(1/n)$-rate.

A typical example of this situation is provided by the multi-dimensional Black–Scholes model (and its avatars for Foreign Exchange (Garman–Kohlhagen) or future contracts (Black)) and, more generally, by models where the process $(X_t^x)_{t\in[0,T]}$ can be written at each time $t \in [0, T]$ as an explicit function of W_t, namely

$$\forall t \in [0, T], \quad X_t = \varphi(t, W_t),$$

where $\varphi(t, x)$ can be computed at a very low cost. When $d = 1$ (although of smaller interest for application in view of the available analytical methods based on variational inequalities), one can also rely on the exact simulation method of one-dimensional diffusions (see [43]).

▷ In the general case, we will rely on these two successive steps of discretization: one for the optimal stopping rules and one for the underlying process to make it simulable. However, in both cases, as far as numerics are concerned, we will rely on the *BDPP*, which itself requires the computation of conditional expectations.

In both cases we now have access to a simulable Markov chain (either the Euler scheme or the process itself at times t_k^n). This task requires a second discretization phase, making possible the computation of the discrete time Snell envelope (and its value function).

11.3 Numerical Methods

11.3.1 The Regression Methods

The Longstaff–Schwartz approach

Regression methods are often generically known as Longstaff–Schwartz method(s) in reference to the paper [202] (see also [60]). However, they present in this paper a particular and original approach relying on the dual *BDPP* based on running optimal stopping times.

Assume that all the random variables $Z_k = f_k(X_k)$, $k = 0, \dots, n$, as defined in (11.5), are square integrable, then so are the payoffs Z_{τ_k}.

The idea is to replace the conditional expectation operator $\mathbb{E}(\cdot|X_k)$ by a linear regression on the first N elements of a Hilbert basis of $(L^2(\Omega, \sigma(X_k), \mathbb{P}), \langle \cdot, \cdot \rangle_{L^2(\mathbb{P})})$. This is a very natural idea to approximate conditional expectation (see *e.g.* [48], Chap. 2.D. for an introduction in a general framework).

Mostly for convenience, we will consider a sequence $e_i : \mathbb{R}^d \to \mathbb{R}$, $i \in \mathbb{N}^*$, of Borel functions such that, for every $k \in \{0, \ldots, n\}$, $\big(e_i(X_k)\big)_{i\in\mathbb{N}^*}$ is a complete system of $L^2(\Omega, \sigma(X_k), \mathbb{P})$, *i.e.*

$$\big\{e_\ell(X_k),\ \ell \in \mathbb{N}^*\big\}^\perp = \{0\}, \quad k = 0, \ldots, n.$$

In practice, one may choose different functions at every time k, *i.e.* families $e_{i,k}$ so that $\big(e_{i,k}(X_k)\big)_{i\geq 1}$ makes up a Hilbert basis of $L^2_{\mathbb{R}}(\Omega, \sigma(X_k), \mathbb{P})$.

⊳ **Example.** If $X_k = W_{t_k^n}$, or equivalently after normalization, if $X_0 = 0$ and $X_k = \sqrt{\frac{n}{kT}}\, W_{t_k^n}$, $k = 1, \ldots, n$, the family of Hermite polynomials provides an orthonormal basis of $L^2_{\mathbb{R}}\big(\Omega, \sigma(X_k), \mathbb{P}\big)$ at each time step t_k^n, by setting $e_\ell = H_{\ell-1}$, $\ell \geq 1$ (see *e.g.* [162], Chap. 3, p. 167).

META- SCRIPT OF LONGSTAFF–SCHWARTZ' REGRESSION PROCEDURE.

- **Approximation 1:** Dimension Truncation

▷ At every time $k \in \{0, \ldots, n\}$, truncate at level N_k

$$e^{[N_k]}(X_k) := \big(e_1(X_k), e_2(X_k), \ldots, e_{N_k}(X_k)\big)$$

and set

▷ $\tau_n^{[N_n]} := n,$

▷ $\tau_k^{[N_k]} := k\mathbf{1}_{\left\{Z_k > \left(\alpha_k^{[N_k]} | e^{[N_k]}(X_k)\right)\right\}} + \tau_{k+1}^{[N_{k+1}]}\mathbf{1}_{\left\{Z_k \leq \left(\alpha_k^{[N_k]} | e^{[N_k]}(X_k)\right)\right\}},$

where

$$\alpha_k^{[N_k]} := \operatorname{argmin}\left\{\mathbb{E}\Big(Z_{\tau_{k+1}^{[N_k]}} - \big(\alpha | e^{[N_k]}(X_k)\big)\Big)^2, \quad \alpha \in \mathbb{R}^{N_k}\right\}, k = 0, \ldots, n-1.$$

(Keep in mind that, at each step k, $(\,\cdot\,|\,\cdot\,)$ denotes the canonical inner product on $\mathbb{R}^{N_k}$.)

In fact, this finite-dimensional optimization problem has a well-known solution given by

$$\alpha_k^{[N_k]} = \operatorname{Gram}\big(e^{[N_k]}(X_k)\big)^{-1}\left[\big\langle Z_{\tau_{k+1}^{[N_{k+1}]}} \,|\, e_\ell^{[N_k]}(X_k)\big\rangle_{L^2(\mathbb{P})}\right]_{1\leq \ell\leq N_k},$$

where $\operatorname{Gram}(e^{[N_k]}(X_k))$ is the Gram matrix of $e^{[N_k]}(X_k)$ defined by

$$\operatorname{Gram}\big(e^{[N_k]}(X_k)\big) = \left[\big\langle e_\ell^{[N_k]}(X_k) \,|\, e_{\ell'}^{[N_k]}(X_k)\big\rangle_{L^2(\mathbb{P})}\right]_{1\leq \ell, \ell'\leq N_k}.$$

- **Approximation 2:** Monte Carlo approximation

  This second approximation phase can itself be decomposed into two successive phases:

  – a forward MC simulation of the underlying "structure" Markov process $(X_k)_{0\le k\le n}$, followed by

  – a backward approximation of $\tau_k^{[N]}$, $k=0,\dots,n$. In a more formal way, the idea is to replace the true distribution of the Markov chain $(X_k)_{0\le k\le n}$ by the empirical measure of a simulated sample of size M of the chain.

  ▷ *Forward Monte Carlo simulation phase:* Simulate (and store) M independent copies $X^{(1)},\dots,X^{(m)},\dots,X^{(M)}$ of $X=(X_k)_{0\le k\le n}$ in order to have access to the empirical measure

$$\frac{1}{M}\sum_{m=1}^{M}\delta_{X^{(m)}}.$$

  ▷ *Backward phase:*

  – At time n: For every $m\in\{1,\dots,M\}$,

$$\tau_n^{[N_n],m,M}:=n.$$

  – for $k=n-1$ down to 0:
  Compute

$$\alpha_k^{[N_k],M}:=\operatorname{argmin}_{\alpha\in\mathbb{R}^{N_k}}\left(\frac{1}{M}\sum_{m=1}^{M}Z^{(m)}_{\tau_{k+1}^{[N_{k+1}],m,M}}-\left(\alpha|e^{[N_k]}(X_k^{(m)})\right)\right)^2$$

  using the closed form formula

$$\alpha_k^{[N_k],M}=\left[\frac{1}{M}\sum_{m=1}^{M}e_\ell^{[N_k]}(X_k^{(m)})e_{\ell'}^{[N_k]}(X_k^{(m)})\right]^{-1}_{1\le\ell,\ell'\le N_k}\left(\frac{1}{M}\sum_{m=1}^{M}Z^{(m)}_{\tau_{k+1}^{[N_{k+1}],m,M}}e_\ell^{[N_k]}(X_k^{(m)})\right)_{1\le\ell\le N_k}.$$

  – For every $m\in\{1,\dots,M\}$, set

$$\tau_k^{[N_k],m,M}:=k\mathbf{1}_{\left\{Z_k^{(m)}>\left(\alpha_k^{[N_k],M}\,\middle|\,e^{[N_k]}(X_k)\right)\right\}}+\tau_{k+1}^{[N_{k+1}],m,M}\mathbf{1}_{\left\{Z_k^{(m)}\le\left(\alpha_k^{[N_k],M}\,\middle|\,e^{[N_k]}(X_k)\right)\right\}}.$$

Finally, the resulting approximation of the mean value at the origin of the Snell envelope reads

$$\mathbb{E}\,U_0=\mathbb{E}\,(Z_{\tau_0})\simeq\mathbb{E}\,(Z_{\tau_0^{[N_0]}})\simeq\frac{1}{M}\sum_{m=1}^{M}Z^{(m)}_{\tau_0^{[N_0],m,M}}\quad\text{as}\quad M\to+\infty.$$

Note that when $\mathcal{F}_0=\{\varnothing,\Omega\}$, (so that $X_0=x_0\in\mathbb{R}^d$), $U_0=\mathbb{E}\,U_0=u_0(x_0)$.

Remarks. • One may formally rewrite the second approximation phase by simply replacing the distribution of the chain $(X_k)_{0\le k\le n}$ by the empirical measure

$$\frac{1}{M}\sum_{m=1}^{M}\delta_{X^{(m)}}$$

where $X^{(m)}$ are i.i.d. copies of X.

• The Gram matrices $\left[\mathbb{E}\left(e_i^{[N_k]}(X_k)e_j^{[N_k]}(X_k)\right)\right]_{1\le i,j\le N_k}$ can be computed *off line* in the sense that they are not payoff dependent by contrast with the inner product term $\left(\langle Z_{\tau_{k+1}^{[N_k]}}\,|\,e_\ell^{[N_k]}(X_k)\rangle_{L^2(\mathbb{P})}\right)_{1\le\ell\le N_k}$.

In various situations, it is even possible for this Gram matrix to have a closed form, *e.g.* when $e_i(X_k)$, $i\ge 1$, happens to be an orthonormal basis of $L^2(\Omega,\sigma(X_k),\mathbb{P})$. In that case the Gram matrix is reduced to the identity matrix. This is is the case, for example, when $X_k=\sqrt{\frac{n}{kT}}W_{\frac{kT}{n}}$, $k=0,\dots,n$ and $e_i=H_{i+1}$, where $(H_i)_{i\ge 0}$ is the orthonormal basis of Hermite polynomials given by

$$H_i(x)=(-1)^i e^{\frac{x^2}{2}}\frac{d^i}{dx^i}e^{-\frac{x^2}{2}},\ i\ge 0.$$

• The algorithmic analysis of the above described procedure shows that its implementation requires

– a *forward* simulation of M paths of the Markov chain,

– a *backward* non-linear optimization phase in which all the (stored) paths have to interact through the computation at every time k of $\alpha_k^{[N],M}$, which depends on all the simulated values $X_k^{(m)}$, $m=1,\dots,M$.

However, still in very specific situations, the forward phase can be skipped if a backward simulation method for the Markov chain $(X_k)_{0\le k\le n}$ is available. Such is the case for the Brownian motion at times $\frac{k}{n}T$, using a backward recursive simulation method, or, equivalently, the Brownian bridge method (see Chap. 8).

The rate of convergence of the Monte Carlo phase of the procedure is ruled by a Central Limit Theorem due to Clément–Lamberton–Protter in [64], stated below.

Theorem 11.3 (CLT, see [64] (2003)) *The Monte Carlo approximation satisfies a Central Limit Theorem, namely*

$$\sqrt{M}\left(\frac{1}{M}\sum_{m=1}^{M}Z^{(m)}_{\tau_k^{[N_k],m,M}}-\mathbb{E}\,Z_{\tau_k^{[N_k]}},\ \alpha_k^{[N_k],M},\alpha_k^{[N_k]}\right)_{0\le k\le n-1}\xrightarrow{\mathcal{L}}\mathcal{N}(0;\Sigma)\quad as\quad M\to+\infty,$$

where Σ is a non-zero covariance matrix.

Regression on the continuation function

Another approach to introducing regression methods is to directly apply them to the *continuation function* $\mathbb{E}\left(U_{k+1}\,|\mathcal{F}_k\right) = \mathbb{E}\left(u_{k+1}(X_{k+1})\,|X_k\right)$. The – maybe more natural or straightforward – idea is again to project this function onto the finite-dimensional vector space $\text{span}\big(e^{[N_k]}(X_k)\big) := \big\langle e_1(X_k), e_2(X_k), \ldots, e_{N_k}(X_k)\big\rangle$ as a first step and then to plug this projection into the regular *BDPP* (11.7). This second approach was proposed and developed by Tsitsiklis and van Roy in [269]. We briefly describe it with the notation used for Longstaff–Schawartz' approach.

META- SCRIPT OF TSITSIKLIS–VAN ROY'S REGRESSION PROCEDURE.

– At time n:

$$U_n^{(m)} = Z_n^{(m)} = f_n\big(X_n^{(m)}\big), \quad m \in \{1, \ldots, M\}.$$

– For every $k = n-1, \ldots, 0$, one defines in a backward way

$$\begin{aligned}\alpha_k^{[N_k],M} &= \arg\min_{\alpha\in\mathbb{R}^{N_k}} \sum_{m=1}^{M} \Big(U_{k+1}^{(m)} - \big(\alpha|e^{[N_k]}(X_k^{(m)})\big)\Big)^2 \\ &= \left[\frac{1}{M}\sum_{m=1}^{M} e_\ell^{[N_k]}(X_k^{(m)})e_{\ell'}^{[N_k]}(X_k^{(m)})\right]^{-1}_{1\le \ell,\ell'\le N_k} \left(\frac{1}{M}\sum_{m=1}^{M} U_{k+1}^{(m)} e_\ell^{[N_k]}(X_k^{(m)})\right)_{1\le \ell\le N_k}\end{aligned}$$

and

$$U_k^{(m)} = \max\Big(f_k(X_k^{(m)}), \big(\alpha_k^{[N_k],M}|e^{[N_k]}(X^{(m)})\big)\Big), \; m = 1, \ldots, M.$$

– In particular, when $k = 0$, one obtains

$$\mathbb{E}\,U_0 \simeq \frac{1}{M}\sum_{m=1}^{M} U^{(m)}(0) = \frac{1}{M}\sum_{m=1}^{M} \max\Big(f_0(X_0^{(m)}), \big(\alpha_0^{[N_0],M}|e^{[N_0]}(X_k^{(m)})\big)\Big). \tag{11.13}$$

▶ **Exercise.** Justify the above algorithm starting from the *BDPP*.

Pros and Cons of the regression method(s) (ℵ Practitioner's corner)

PROS

• The method is "natural": it involves the approximation of a conditional expectation by an affine regression operator performed on a truncated generating system of $L^2(\sigma(X_k), \mathbb{P})$.

• The method appears to be "flexible": there is an opportunity to change or adapt the (truncated) basis of $L^2(\sigma(X_k), \mathbb{P})$ at each step of the procedure, *e.g.* by including the payoff function in the (truncated) basis, at each step at least during the first backward iterations of the induction.

CONS

• From a purely theoretical point of view the regression approach does not provide error bounds or a rate of approximation for the convergence of $\mathbb{E}\, Z_{\tau_0^{[N_0]}}$ toward $\mathbb{E}\, Z_{\tau_0} = \mathbb{E}\, U_0$, which is mostly ruled by the rate at which the family $e^{[N_k]}(X_k)$ "fills" $L^2(\Omega, \sigma(X_k), \mathbb{P})$ when N_k goes to infinity. This point is discussed, for example, in [118] in connection with the size of the Monte Carlo sample. However, in practice this information may be difficult to exploit, especially in higher dimensions, the possible choices for the size N_k are limited by the storing capacity of the computing device.

• Most computations are performed *on-line* since they are payoff dependent. However, note that the Gram matrix of $\big(e_\ell^{[N_k]}(X_k)\big)_{1\le \ell\le N_k}$ can be computed off-line since it only depends on the structure process.

• Due to the sub-optimality of the stopping times involved in the regression procedure, the method tends to produce lower bounds for the $u_0(x_0) = \mathbb{E}\, U_0$ since it has a methodological negative bias.

• The **choice of the functions** $\big(e_\ell(x)\big)_{\ell\in\mathbb{N}^*}$, is crucial and needs much care (and intuition). In practical implementations, it may vary at each time step (our choice of a unique family is mostly motivated by notational simplicity). Furthermore, it may have a biased effect for options deep in- or out-of-the-money since the coordinates of the functions $e_i(X_k)$ are computed "locally" from simulated data which of course lie most of the time where things happen (around the mean). This has an impact on the prices of options with long maturity and/or deep-in- or out-in the money since the calibration, mainly performed "at-the-money" of the coordinates of $e^{[N_k]}(X_k)$, induce their behavior at the aisles. On the other hand, this choice of the functions, if they are smooth, has a smoothing effect which can be interesting to users (provided it does not induce any hidden arbitrage...). To overcome the first problem one may choose local functions like indicator functions of a Voronoi diagram (see the next Sect. 11.3.2 devoted to quantization tree methods or Chap. 5) with the counterpart that a smoothing effect can no longer be expected.

When there is a family of distributions "related" to the underlying Markov structure process, a natural idea can be to consider an orthonormal basis of $L^2(\mu_0)$, where μ_0 is a normalized distribution of the family. A typical example is the sequence of Hermite polynomials for the normal distribution $\mathcal{N}(0; 1)$.

When no simple solution is available, considering the simple basis $(t^\ell)_{\ell\ge 0}$ remains a quite natural and efficient choice in one dimension.

In higher dimensions (in fact the only case of interest in practice since one-dimensional setting is usually solved through the associated variational inequality by specific numerical schemes, see [2, 158]), this choice becomes more and more influenced by the payoff itself.

• A huge RAM capacity is needed to store all the paths of the simulated Markov chain (forward phase) except when a backward simulation procedure is available. This induces a stringent limitation on the size M of the simulation, even with recent devices, to prevent a swapping effect which would dramatically slow down the procedure. By a swapping effect we mean that, when the quantity of data to be stored

becomes too large, the computer uses its hard disk to store it but the access to this ROM memory is incredibly slow compared to the access to RAM memory.
• Regression methods are strongly *payoff*-dependent in the sense that a significant part of the procedure (the product of the inverted Gram matrix by the projection of the payoff at every time k) has to be done for each payoff.

► **Exercise.** Write a regression algorithm based on the "primal" BDPP.

11.3.2 Quantization Methods II: Non-linear Problems (Quantization Tree)

Approximation by a quantization tree

In this section, we continue to deal with the simple discrete time Markovian optimal stopping problem introduced in the former section. The underlying idea of the quantization tree method is to approximate the whole Markovian dynamics of the chain $(X_k)_{0\le k\le n}$ using a *skeleton* of the distribution supported by a tree. In some sense, the underlying idea is to design some optimized tree methods in higher dimensions with a procedure optimally fitted to the underlying marginal distributions of the chain in order to prevent an explosion of their size (number of nodes per level).

For every $k \in \{0, \dots, n\}$, we replace the marginal X_k by a function $\widehat{X}_k$ of X_k taking values in a *grid* Γ_k, namely $\widehat{X}_k = \pi_k(X_k)$, where $\pi_k : \mathbb{R}^d \to \Gamma_k$ is a Borel function. The grid $\Gamma_k = \pi_k(\mathbb{R}^d)$ (also known as a *codebook* in Signal processing or Information Theory) will always be assumed finite in practice, with size $|\Gamma_k| = N_k \in \mathbb{N}^*$.

However, note that *all the error bounds established below still hold if the grids are infinite provided* π_k *is sub-linear, i.e.*

$$|\pi_k(x)| \le C\big(1 + |x|\big),$$

so that $\widehat{X}_k$ has at least as many finite moments as $\widehat{X}_k$. In such a situation, all in all integrability assumptions X_k and $\widehat{X}_k$ should be coupled, *i.e.* $X_k, \widehat{X}_k \in L^r(\mathbb{P})$ instead of $X_k \in L^r(\mathbb{P})$.

We saw in Chap. 5 an optimal way to specify the function π_k (including Γ_k) when trying to minimize the induced L^p-mean quadratic error $\|X_k - \widehat{X}_k\|_p$. This is the purpose of optimal quantization theory. We will return to these aspects later.

The starting point, being aware that the sequence $(\widehat{X}_k)_{0\le k\le n}$ has no reason to share a Markov property, is to force this Markov property in the *BDPP*. This means defining by induction a quantized pseudo-Snell envelope of $\big(f_k(X_k)\big)_{0\le k\le n}$ (assumed to lie at least in L^1), namely

$$\widehat{U}_n = f_n(\widehat{X}_n), \qquad \widehat{U}_k = \max\Big(f_k(\widehat{X}_k), \mathbb{E}\left(\widehat{U}_{k+1} \,|\, \widehat{X}_k\right)\Big). \tag{11.14}$$

The forced Markov property results from the conditioning by $\widehat{X}_k$ rather than by the σ-field $\widehat{\mathcal{F}}_k := \sigma\big(\widehat{X}_\ell,\, 0 \le \ell \le k\big)$.

It is straightforward by induction that, for every $k \in \{0, \dots, n\}$,

$$\widehat{U}_k = \widehat{u}_k(\widehat{X}_k) \quad \widehat{u}_k : \mathbb{R}^d \to \mathbb{R}_+, \text{ Borel function.}$$

See Subsection "Implementation of a quantization tree descent" for the detailed implementation.

Error bounds

The following theorem establishes the control on the approximation of the true Snell envelope $(U_k)_{0 \le k \le n}$ by the quantized pseudo-Snell envelope $(\widehat{U}_k)_{0 \le k \le n}$ using the L^p-mean approximation errors $\|X_k - \widehat{X}_k\|_p$.

Theorem 11.4 (see [20] (2001), [235] (2011)) *Assume that all functions $f_k : \mathbb{R}^d \to \mathbb{R}_+$ are Lipschitz continuous and that all the transitions $P_k(x, dy) = \mathbb{P}(X_{k+1} \in dy \mid X_k = x)$ are Lipschitz continuous in the following sense*

$$[P_k]_{\rm Lip} = \sup_{[g]_{\rm Lip} \le 1} [P_k g]_{\rm Lip} < +\infty, \quad k = 0, \dots, (n-1).$$

Set $[P]_{\rm Lip} = \max_{0 \le k \le n-1} [P_k]_{\rm Lip}$ and $[f]_{\rm Lip} = \max_{0 \le k \le n} [f_k]_{\rm Lip}$.

Let $p \in [1, +\infty)$. We assume that $\sum_{k=0}^{n} \|X_k\|_p + \|\widehat{X}_k\|_p < +\infty$.

(a) For every $k \in \{0, \dots, n\}$,

$$\|U_k - \widehat{U}_k\|_p \le 2[f]_{\rm Lip} \sum_{\ell=k}^{n} \big([P]_{\rm Lip} \vee 1\big)^{n-\ell} \|X_\ell - \widehat{X}_\ell\|_p.$$

(b) If $p = 2$, for every $k \in \{0, \dots, n\}$,

$$\|U_k - \widehat{U}_k\|_2 \le \sqrt{2}\,[f]_{\rm Lip} \left(\sum_{\ell=k}^{n} \big([P]_{\rm Lip} \vee 1\big)^{2(n-\ell)} \|X_\ell - \widehat{X}_\ell\|_2^2 \right)^{\frac{1}{2}}.$$

Proof. Step 1. First, we control the Lipschitz continuous constants of the functions u_k. It follows from the classical inequality

$$\forall\, a_i,\, b_i \in \mathbb{R}, \quad \Big| \sup_{i \in I} a_i - \sup_{i \in I} b_i \Big| \le \sup_{i \in I} |a_i - b_i|$$

that

$$\begin{aligned}[u_k]_{\rm Lip} &\le \max\big([f_k]_{\rm Lip}, [P_k u_{k+1}]_{\rm Lip}\big)\\ &\le \max\big([f]_{\rm Lip}, [P_k]_{\rm Lip}[u_{k+1}]_{\rm Lip}\big)\end{aligned}$$

with the convention $[u_{n+1}]_{\rm Lip} = 0$. An easy backward induction yields $[u_k]_{\rm Lip} \le [f]_{\rm Lip}\big([P]_{\rm Lip} \vee 1\big)^{n-k}$.

STEP 2. We focus on claim (b) when $p = 2$. It follows from both Backward Dynamic Programming formulas (original and quantized) and the above elementary inequality that

$$|U_k - \widehat{U}_k| \le \max\Big(|f_k(X_k) - f_k(\widehat{X}_k)|, \big|\mathbb{E}\big(U_{k+1}|X_k\big) - \mathbb{E}\big(\widehat{U}_{k+1}|\widehat{X}_k\big)\big|\Big)$$

so that

$$|U_k - \widehat{U}_k|^2 \le |f_k(X_k) - f_k(\widehat{X}_k)|^2 + \big|\mathbb{E}\big(U_{k+1}|X_k\big) - \mathbb{E}\big(\widehat{U}_{k+1}|\widehat{X}_k\big)\big|^2. \tag{11.15}$$

Keeping in mind that $U_{k+1} = u_{k+1}(X_{k+1}) =$ and $\widehat{U}_{k+1} = \widehat{u}_{k+1}(\widehat{X}_{k+1})$, we derive from the quantized approximation bounds for conditional expectation (5.16) applied with $F = u_{k+1}$, $G = \widehat{u}_{k+1}$, $X = X_{k+1}$ and $Y = X_k$ that

$$\begin{aligned}\Big\|\mathbb{E}\big(U_{k+1}|X_k\big) - \mathbb{E}\big(\widehat{U}_{k+1}|\widehat{X}_k\big)\Big\|_2^2 &= \big\|P_k u_{k+1}(X_k) - \mathbb{E}\big(\widehat{u}_{k+1}(\widehat{X}_{k+1})\,|\,\widehat{X}_k\big)\big\|_2^2\\ &= [P_k u_{k+1}]_{\rm Lip}^2\big\|X_k - \widehat{X}_k\big\|_2^2 + \big\|u_k(X_{k+1}) - \widehat{u}_{k+1}(\widehat{X}_{k+1})\big\|_2^2.\end{aligned} \tag{11.16}$$

Plugging this inequality into (11.15) and taking the expectation yields for every $k \in \{0, \ldots, n\}$,

$$\big\|U_k - \widehat{U}_k\big\|_2^2 \le \Big([f]_{\rm Lip}^2 + [P]_{\rm Lip}^2[u_{k+1}]_{\rm Lip}^2\Big)\big\|X_k - \widehat{X}_k\big\|_2^2 + \big\|U_{k+1} - \widehat{U}_{k+1}\big\|_2^2$$

still with the conventions $[u_{n+1}]_{\rm Lip} = 0$ and $\widehat{U}_{n+1} = U_{n+1} = 0$. Now,

$$\begin{aligned}[f]_{\rm Lip}^2 + [P]_{\rm Lip}^2[u_{k+1}]_{\rm Lip}^2 &\le [f]_{\rm Lip}^2 + [P]_{\rm Lip}^2\big(1 \vee [P_k]_{\rm Lip}\big)^{2(n-(k+1))}\\ &\le 2[f]_{\rm Lip}^2\big(1 \vee [P_k]_{\rm Lip}\big)^{2(n-k)}.\end{aligned}$$

Consequently,

$$\begin{aligned}\big\|U_k - \widehat{U}_k\big\|_2^2 &\le \sum_{\ell=k}^{n-1} 2[f]_{\rm Lip}^2\big(1 \vee [P_k]_{\rm Lip}\big)^{2(n-k)}\big\|X_\ell - \widehat{X}_\ell\big\|_2^2 + [f]_{\rm Lip}^2\big\|X_n - \widehat{X}_n\big\|_2^2\\ &\le 2[f]_{\rm Lip}^2 \sum_{\ell=k}^{n}\big(1 \vee [P_k]_{\rm Lip}\big)^{2(n-k)}\big\|X_\ell - \widehat{X}_\ell\big\|_2^2,\end{aligned}$$

which completes the proof. ◇

Remark. The above control emphasizes the interest of minimizing the L^p-mean quantization error $\|X_k - \widehat{X}_k\|_p$ at each time step of the Markov chain to reduce the final resulting error.

► **Exercise.** Prove claim (a) starting from the L^p-error bound (5.18).

⊳ **Example of application: the Euler scheme.** Let $(\bar{X}^n_{t^n_k})_{0\le k\le n}$ be the Euler scheme with step $\frac{T}{n}$ of the d-dimensional diffusion solution to the *SDE* (11.1). It defines a homogenous Markov chain with transition

$$\bar{P}^n_k g(x) = \mathbb{E}\, g\left(x + \frac{T}{n} b(t^n_k, \bar{X}^n_{t^n_k}) + \sigma(t^n_k, \bar{X}^n_{t^n_k})\sqrt{\frac{T}{n}} Z\right), \quad Z \stackrel{d}{=} \mathcal{N}(0; I_q).$$

If f is Lipschitz continuous, then

$$\begin{aligned}\left|\bar{P}^n_k g(x) - \bar{P}^n_k g(x')\right|^2 &\le [g]^2_{\rm Lip}\mathbb{E}\left|x - x' + \frac{T}{n}\big(b(t^n_k, x) - b(t^n_k, x')\big) + \sqrt{\frac{T}{n}}\big(\sigma(t^n_k, x) - \sigma(t^n_k, x')\big)Z\right|^2 \\ &\le [g]^2_{\rm Lip}\left(\left|x - x' + \frac{T}{n}\big(b(t^n_k, x) - b(t^n_k, x')\big)\right|^2 + \frac{T}{N}\left|\sigma(t^n_k, x) - \sigma(t^n_k, x')\right|^2\right) \\ &\le [g]^2_{\rm Lip}|x - x'|^2\left(1 + \frac{T}{n}[\sigma]^2_{\rm Lip} + \frac{2T}{n}[b]_{\rm Lip} + \frac{T^2}{n^2}[b]^2_{\rm Lip}\right).\end{aligned}$$

As a consequence, setting $C_{b,\sigma} = [b]_{\rm Lip} + \dfrac{[\sigma]^2_{\rm Lip}}{2}$ yields

$$[\bar{P}^n_k g]_{\rm Lip} \le \left(1 + \frac{C_{b,\sigma,T}}{n}\right)[g]_{\rm Lip}, \; k = 0, \ldots, n-1,$$

i.e.

$$[\bar{P}^n]_{\rm Lip} \le \left(1 + \frac{C_{b,\sigma,T}}{n}\right).$$

Applying the control established in claim (b) of the above theorem yields with obvious notations

$$\begin{aligned}\left\|U_k - \widehat{U}_k\right\|_2 &\le \sqrt{2}[f]_{\rm Lip}\left(\sum_{\ell=k}^{n}\left(1 + \frac{C_{b,\sigma,T}}{n}\right)^{2(n-\ell)}\left\|X_\ell - \widehat{X}_\ell\right\|_2^2\right)^{\frac{1}{2}} \\ &\le \sqrt{2}e^{C_{b,\sigma,T}}[f]_{\rm Lip}\left(\sum_{\ell=k}^{n} e^{2C_{b,\sigma,T}t^n_\ell}\left\|X_\ell - \widehat{X}_\ell\right\|_2^2\right)^{\frac{1}{2}}.\end{aligned}$$

► **Exercise.** Derive a result in the case $p \neq 2$ based on Claim (a) of theorem.

Background on optimal quantization

For some background on optimal quantization, we refer to Chap. 5.

Implementation of a quantization tree descent

▷ **Quantization tree.** The pathwise Quantized Backward Dynamic Programming Principle (11.14) can be rewritten in distribution as follows. Let, for every $k \in \{0, \ldots, n\}$,

$$\Gamma_k = \left\{x_1^k, \ldots, x_{N_k}^k\right\}.$$

Keeping in mind that $\widehat{U}_k = \widehat{u}_k(\widehat{X}_k)$, $k = 0, \ldots, n$, we first get

$$\begin{cases} \widehat{u}_n = f_n \text{ on} \Gamma_n, \\ \widehat{u}_k(x_i^k) = \max\left(f_k(x_i^k), \mathbb{E}\left(\widehat{u}_{k+1}(\widehat{X}_{k+1})|\widehat{X}_k = x_i^k\right)\right), \\ \qquad\qquad i = 1, \ldots, N_k, k = 0, \ldots, n-1, \end{cases} \tag{11.17}$$

which finally leads to

$$\begin{cases} \widehat{u}_n(x_i^n) = f_n(x_i^n), \; i = 1, \ldots, N_n, \\ \widehat{u}_k(x_i^k) = \max\left(f_k(x_i^k), \sum_{j=1}^{N_{k+1}} p_{ij}^k \widehat{u}_{k+1}(x_j^{k+1})\right), \\ \qquad\qquad i = 1, \ldots, N_k, k = 0, \ldots, n-1, \end{cases} \tag{11.18}$$

where the transition weight “super-matrix” $[p_{ij}^k]$ is defined by

$$p_{ij}^k = \mathbb{P}\big(\widehat{X}_{k+1} = x_j^{k+1} | \widehat{X}_x = x_i^k\big), \;\; 1 \le i \le N_k, \; 1 \le j \le N_{k+1}, \; k = 0, \ldots, n-1. \tag{11.19}$$

Although the above super-matrix defines a family of Markov transitions, the sequence $(\widehat{X}_k)_{0 \le k \le n}$ is definitely not a Markov Chain since there is no reason why $\mathbb{P}\big(\widehat{X}_{k+1} = x_j^{k+1} | \widehat{X}_k = x_i^k\big)$ and $\mathbb{P}\big(\widehat{X}_{k+1} = x_j^{k+1} | \widehat{X}_k = x_i^k, \widehat{X}_\ell = x_{i_\ell}^\ell, \ell = 0, \ldots, i-1\big)$ should be equal.

In fact, one should rather view the quantized transitions

$$\widehat{P}_k(x_i^k, dy) = \sum_{j=1}^{N_{k+1}} p_{ij}^k \delta_{x_j^{k+1}}, \;\; x_i^k \in \Gamma_k, \; k = 0 = \ldots, n-1,$$

as spatial discretizations of the original transitions $P_k(x, du)$ of the original Markov chain.

Definition 11.2 *The family of grids* (Γ_k), $0 \le k \le n$, *and the transition super-matrix* $[p_{ij}^k]$ *defined by (11.19) defines a* quantization tree *of size* $N = N_0 + \cdots + N_n$.

Remark. A quantization tree in the sense of the above definition does not characterize the distribution of the sequence $(\widehat{X}_k)_{0\le k\le n}$.

▷ **Implementation of the quantization tree descent.** The implementation of the whole quantization tree method relies on the computation of this transition super-matrix. Once the grids (optimal or not) have been specified and the weights of the super-matrix have been computed or, to be more precise, have been estimated, the computation of the approximate value function $\mathbb{E}\,\widehat{U}_0$ at time 0 amounts to an almost instantaneous "backward descent" of the quantization tree based on (11.18).

If we can simulate M independent copies of the Markov chain $(X_k)_{0\le k\le n}$ denoted by $X^{(1)},\ldots,X^{(M)}$, then the weights p^k_{ij} can be estimated by a standard Monte Carlo estimator

$$p^{(M),k}_{ij}=\frac{\big|\big\{m\in\{1,\ldots,M\},\ \widehat{X}^{(m)}_{k+1}=x^{k+1}_j\ \&\ \widehat{X}^{(m)}_k=x^k_i\big\}\big|}{\big|\big\{m\in\{1,\ldots,M\},\ \widehat{X}^{(m)}_k=x^k_i\big\}\big|}\overset{M\to+\infty}{\longrightarrow}p^k_{ij}$$

as $M\to+\infty$.

Remark. By contrast with the regression methods for which theoretical results are mostly focused on the rate of convergence of the Monte Carlo phase, here we will analyze the part of the procedure for which we refer to [17], where the transition super-matrix is computed and its impact on the quantization tree is deeply analyzed.

Application. We can apply the preceding, still within the framework of an Euler scheme, to our original optimal stopping problem. We assume that all random vectors X_k lie in $L^{p'}(\mathbb{P})$, for a real exponent $p'>2$, and that they have been optimally quantized (in L^2) by grids of size N_k, $k=0,\ldots,n$, then, relying on the non-asymptotic Zador Theorem (claim (*b*) from Theorem 5.1.2), we get with obvious notations,

$$\|\bar{U}^n_0-\widehat{\bar{U}^n_0}\|_2\le\sqrt{2}e^{C_{b,\sigma,T}}[f]_{\rm Lip}C'_{p',d}\left(\sum_{k=0}^{n}e^{2C_{b,\sigma,T}}\sigma_{p'}(\bar{X}^n_{t^n_k})^2N_k^{-\frac{2}{d}}\right)^{\frac12}.$$

▷ **Optimizing the quantization tree.** At this stage, one can process an optimization of the quantization tree. To be precise, one can optimize the size of the grids Γ_k subject to a "budget" (or total allocation) constraint, typically

$$\min\left\{\sum_{k=0}^{n}e^{2C_{b,\sigma,T}}\sigma_{p'}(\bar{X}^n_{t^n_k})^2N_k^{-\frac{2}{d}},\ N_k\ge1,\ N_0+\cdots+N_n=N\right\}.$$

► **Exercise.** (*a*) Solve (more) rigorously the constrained optimization problem

$$\min\left\{\sum_{k=0}^{n}e^{2C_{b,\sigma,T}}\sigma_{p'}(\bar{X}^n_{t^n_k})^2x_k^{-\frac{2}{d}},\ x_k\in\mathbb{R}_+,\ x_0+\cdots+x_n=N\right\}.$$

(*b*) Derive an asymptotically optimal choice for the grid size allocation (as $N \to +\infty$).

This optimization turns out to have a significant numerical impact, even if, in terms of rate, the uniform choice $N_k = \bar{N} = \frac{N-1}{n}$ (doing so we implicitly assume that $X_0 = x_0 \in \mathbb{R}^d$, so that $\widehat{X}_0 = X_0 = x_0$, $N_0 = 1$ and $\widehat{U}_0 = \widehat{u}_0(x_0)$) leads to a quantization error of the form

$$\begin{aligned}|\bar{u}_0^n(x_0) - \widehat{\bar{u}}_0^n(x_0)| &\le \big\|\bar{U}_0^n - \widehat{\bar{U}}_0^n\big\|_2\\ &\le \kappa_{b,\sigma,T}[f]_{\rm Lip} \max_{0\le k\le n} \sigma_{p'}(\bar{X}^n_{t_k^n}) \times \frac{\sqrt{n}}{\bar{N}^{\frac{1}{d}}}\\ &\le [f]_{\rm Lip}\kappa'_{b,\sigma,T,p'}\frac{\sqrt{n}}{\bar{N}^{\frac{1}{d}}}\end{aligned}$$

since we know (see Proposition 7.2 in Chap. 7) that $\sup_n \mathbb{E}\big(\max_{0\le k\le n} |\bar{X}^n_{t_k^n}|^{p'}\big) < +\infty$. If we plug that into the global estimate obtained in Theorem 11.2, we obtain the typical error bound

$$|u_0(x_0) - \widehat{\bar{u}}_0^n(x_0)| \le C_{b,\sigma,T,f,d}\left(\frac{1}{n^{\frac{1}{2}}} + \frac{n^{\frac{1}{2}}}{\bar{N}^{\frac{1}{d}}}\right). \tag{11.20}$$

Remark. If we can directly simulate the sampled $(X_{t_k^n})_{0\le k\le n}$ of the diffusion instead of its Euler scheme and if the obstacle/payoff function is semi-convex in the sense of Condition (SC), then we get as a typical error bound

$$\big|u_0(x_0) - \widehat{\bar{u}}_0^n(x_0)\big| \le C_{b,\sigma,T,f,d}\left(\frac{1}{n} + \frac{n^{\frac{1}{2}}}{\bar{N}^{\frac{1}{d}}}\right).$$

Remark. • The rate of decay $\bar{N}^{-\frac{1}{d}}$ becomes obviously bad as the spatial dimension d of the underlying Markov process increases, but this phenomenon cannot be overcome by such tree methods. This is a consequence of Zador's Theorem. This rate degradation is known as the *curse of dimensionality*.

• These rates can be significantly improved by introducing a Romberg-like extrapolation method and/or some martingale corrections to the quantization tree (see [235]).

ℵ Practitioner's corner

▷ ***Tree optimization*** *vs* $BDDP$ ***complexity***

If, n being fixed, we set

$$N_k = \left\lfloor \frac{(\sigma_{2+\delta}(X_k))^{\frac{2d}{d+2}}}{\sum_{0\le \ell\le n}(\sigma_{2+\delta}(X_k))^{\frac{2d}{d+2}}} N \right\rfloor \vee 1,\ k = 0,\ldots,n$$

(see exercise, in the previous section) we (asymptotically) minimize the resulting quantization error induced by the $BDDP$ descent.

$\triangleright$ **Examples: 1.** *Brownian motion* $X_k = W_{t_k}$. Then $\widehat{W}_0 = 0$ and

$$\|W_{t_k}\|_{2+\delta} = C_\delta \sqrt{t_k}, \quad k = 0, \ldots, n.$$

Hence $N_0 = 1$,

$$N_k \simeq \frac{2(d+1)}{d+2} \left(\frac{k}{n}\right)^{\frac{d}{2(d+1)}} N, \quad k = 1, \ldots, n,$$

and

$$|\mathcal{V}_0 - \widehat{v}_0(0)| \le C_{W,\delta} \left(\frac{2(d+1)}{d+2}\right)^{1-\frac{1}{d}} \frac{n^{1+\frac{1}{d}}}{N^{\frac{1}{d}}} = O\Big(\frac{n}{\bar{N}^{\frac{1}{d}}}\Big)$$

with $\bar{N} = \frac{N}{n}$.

Theoretically this choice may not look crucial since it has no impact on the convergence rate, but in practice, it does influence the numerical performances.

2. *Stationary processes.* The process $(X_k)_{0\le k\le n}$ is stationary and $X_0 \in L^{2+\delta}$ for some $\delta > 0$. A typical example in the Gaussian world, is, as expected, the stationary Ornstein–Uhlenbeck process

$$dX_t = -BX_t dt + \Sigma dW_t, \;\; B,\; \Sigma \in \mathcal{M}(d, \mathbb{R}),$$

where all eigenvalues of B have positive real parts and W is a d-dimensional Brownian motion defined on $(\Omega, \mathcal{A}, \mathbb{P})$ and X_0 is independent of W. Then, it is a classical result that, if X_0 is distributed according to the distribution

$$\nu_0 = \mathcal{N}\left(0; \int_0^{+\infty} e^{-tB} \Sigma\Sigma^* e^{-tB^*} dt\right),$$

then the process $(X_t)_{t\ge 0}$ is stationary, *i.e.* $(X_{t+s})_{s\ge 0}$ and $(X_s)_{s\ge 0}$ have the same distribution as processes, with common marginal distribution ν_0. If we "sample" (or observe) this process at times $t_k^n = \frac{kT}{n}$, $k = 0, \ldots, n$, between 0 and T, it takes the form of a Gaussian Autoregressive process of order 1

$$X_{t_{k+1}^n} = X_{t_k^n}\Big(I_d - \frac{T}{2n}B\Big) + \Sigma\sqrt{\frac{T}{n}} Z_{k+1}^n, \quad (Z_k^n)_{1\le k\le n} \text{ i.i.d., } \mathcal{N}(0; I_d) \text{ distributed.}$$

The key feature of such a setting is that the quantization tree only relies on one optimal $\bar{N}$-grid $\Gamma = \Gamma^{0,(\bar{N})} = \{x_1^0, \ldots, x_{\bar{N}}^0\}$ (say L^2 optimal for the distribution of X_0 at level $\bar{N}$) and one quantized transition matrix $\big[\mathbb{P}(\widehat{X}_1^\Gamma = x_j^0 \mid \widehat{X}_0^\Gamma = x_i^0)\big]_{1\le i,j\le n}$.

For every $k \in \{0,\dots,n\}$, $\|X_k\|_{2+\delta} = \|X_0\|_{2+\delta}$, hence $N_k = \left\lceil \frac{N}{n+1} \right\rceil$, $k = 0,\dots,n$,

$$\|\mathcal{V}_0 - \widehat{v}_0(\widehat{X}_0)\|_2 \le C_{X,\delta} \frac{n^{\frac{1}{2}+\frac{1}{d}}}{N^{\frac{1}{d}}} \le C_{X,\delta} \frac{n}{\bar{N}^{\frac{1}{d}}} \quad \text{with } \bar{N} = \frac{N}{n+1}.$$

Numerical optimization of the grids: Gaussian and non-Gaussian vectors

We refer to Sect. 5.3 in Chap. 5 devoted to optimal quantization and Sect. 6.3.5 in Chap. 6 devoted to Stochastic approximation and optimization.

► ***Richardson–Romberg extrapolation(s).*** Richardson–Romberg extrapolation in this framework is based on a *heuristic guess*: there exists a "sharp rate" of convergence of the quantization tree method as the total budget N goes to infinity and this rate of convergence is given by (11.20) when replacing $\bar{N}$ by N/n.

On can perform a Romberg extrapolation in N for fixed n or even a full Romberg extrapolation involving both the time discretization step n and the size N of the quantization tree.

► **Exercise (temporary).** Make the assumption that the error in a quantization scheme admits a first-order expansion of the form

$$\mathcal{E}rr(n, N) = \frac{c_1}{n^\alpha} + \frac{c_2 n^{\frac{1}{2}+\frac{1}{d}}}{N^{\frac{1}{d}}}.$$

Devise a Richardson–Romberg extrapolation based on two quantization trees with sizes $N^{(1)}$ and $N^{(2)}$, respectively.

► ***Martingale correction.*** When $(X_k)_{0\le k\le n}$ is a martingale, one can force this martingale property on the quantized chain by freezing the transition weight super-matrix and by moving in a backward way the grids Γ_k so that the resulting grids $\widetilde{\Gamma}_k$ satisfy

$$\mathbb{E}\left(\widehat{X}_k^{\widetilde{\Gamma}_k} \mid \widehat{X}_{k-1}^{\widetilde{\Gamma}_{k-1}}\right) = \widehat{X}_{k-1}^{\widetilde{\Gamma}_{k-1}}.$$

In fact, this identity defines by a backward induction new grids $\widetilde{\Gamma}_k$. As a final step, one translates these new grids so that Γ_0 and $\widetilde{\Gamma}_0$ have the same mean. Of course, such a procedure is entirely heuristic.

Pros and Cons of quantization method(s) (ℵ Practitioner's corner)

PROS

• The quantization tree, once optimized, appears as a skeleton of the distribution of the underlying Markov chain. This optimization phase can be performed *off-line* whereas the payoff dependent part, the *Quantized BDPP*, is almost instantaneous.
• In many situations, including the multi-dimensional Black–Scholes model, the quantization tree can be designed on the Brownian motion itself, for which pre-

computed grids are available (website `quantize.maths-fi.com`). It only remains to commute the transitions.

• Several natural and easy-to-implement methods are available to improve its crude performances when the dimension d of the state space of the underlying Markov chain increases: use the premium of a European option with payoff the terminal value of an American payoff as a control variate, Richardson–Romberg extrapolation, martingale correction as described above, etc.

• A new approach has been recently developed to dramatically speed up the contraction of performing quantization trees, so far in one dimension (see [234]) but a higher-dimensional extension is in progress (see [91]). This new quantization method preserves the Markov property, but it is slightly less accurate in its approximation of the distribution.

CONS

• The method clearly suffers from the curse of dimensionality as emphasized by its *a priori error bounds*. Indeed, *all methods suffer from this curse*: the regression method through the "filling" rate of $L^2\big(\sigma(X_k), \mathbb{P}\big)$ by the basis $e^{[N]}(X_k)$ as N goes to infinity, the Malliavin-Monte Carlo method (not developed here), *e.g.* through the variance of their estimator of conditional expectations (usually exploding like (timestep)$^{-d}$ in d dimensions).

• Quantization-based numerical schemes are local by construction: they rely on the indicator functions of Voronoi cells and, owing to this feature, they do not propagate regularity properties.

• The method may appear to be less flexible than its competitors, a counterpart of the off-line calibration phase. In particular, it may seem inappropriate to perform calibration of financial models where at each iteration of the procedure the model parameters, and subsequently the resulting quantization tree, are modified. This point may be discussed with respect to competing methods and current performance of modern computers. However, to take this objection into account, an alternative *fast quantization* method has been proposed and analyzed in [234] to design performing (though sub-optimal) quantization trees. They have been successfully tested on calibration problems in [57, 58] in local and stochastic volatility problems.

• The method is limited to medium dimensions, say up to $d = 10$ or 12, but this is also the case for other methods. For higher dimensions the complete paradigm should be modified.

11.4 Dual Form of the Snell Envelope (Discrete Time)

We rely in this brief section on the notations introduced in Sect. 11.2.1. So far we have seen that the $(\mathbb{P}, (\mathcal{F}_k)_{k=0,\dots,n})$-Snell envelope $(U_k)_{k=0,\dots,n}$ of a sequence $(Z_k)_{k=0,\dots,n}$ of an $(\mathcal{F}_k)_{0\le k\le n}$-adapted sequence of integrable non-negative random variables defined on a filtered probability space $\big(\Omega, \mathcal{A}, \mathbb{P}, (\mathcal{F}_k)_{k=0,\dots,n})\big)$ is defined as a $\mathbb{P}$-essential supremum (see Sect. 12.9) and satisfies a Backward Dynamic Programming Principle established in Proposition 11.7. The numerical methods (regression and optimal quantization) described in the former section relied on this *BDDP* (quantization) or its backward running optimal stopping times-based variant (regression). This second method is supposed to provide lower bounds for the Snell envelope (up to the Monte Carlo error). In the early 2000s a dual representation of the Snell envelope has been established by Rogers in [255] (see also [146]) where the Snell envelope is represented as an essential infimum (see Proposition 12.4 and Eq. (12.7) in Sect. 12.9) with the hope that numerical methods based on such a representation will provide some upper-bounds.

This dual representation reads as follows (we will not use the Markov property).

Theorem 11.5 (Dual form of the Snell envelope (Rogers)) *Let $(Z_k)_{k=0,\dots,n}$ be as above and let $(U_k)_{k=0,\dots,n}$ be its $(\mathbb{P}, (\mathcal{F}_k)_{0\le k\le n})$-Snell envelope. Let*

$$\mathcal{M}_k = \big\{M = (M_\ell)_{\ell=0,\dots,n}\ (\mathbb{P}, \underline{\mathcal{F}})\text{-martingale, } M_k = 0\big\}.$$

Then, for every $k \in \{0, \dots, n\}$,

$$U_k = \mathbb{P}\text{-essinf}\left\{\mathbb{E}\left(\sup_{\ell\in\{k,\dots,T\}}(Z_\ell - M_\ell)\,\Big|\,\mathcal{F}_k\right),\ M \in \mathcal{M}_k\right\}.$$

The proof relies on Doob's decomposition of a super-martingale:

Proposition 11.5 (Doob's decomposition) *Let $(S_k)_{k=0,\dots,n}$ be a $(\mathbb{P}, (\mathcal{F}_k)_{0\le k\le n})$-super-martingale, i.e. a sequence of integrable $(\mathcal{F}_k)_{0\le k\le n}$-adapted random variables satisfying*

$$\forall\, k \in \{0, \dots, n-1\},\quad \mathbb{E}\,(S_{k+1}\,|\,\mathcal{F}_k) \le S_k.$$

Then there exists a pair (M, A), unique up to $\mathbb{P}$-a.s. equality, where $M = (M_k)_{k=0,\dots,n}$ is a $(\mathbb{P}, (\mathcal{F}_k)_{0\le k\le n})$-martingale null at 0 and $A = (A_k)_{k=0,\dots,n}$ is a non-decreasing sequence of random variables such that $A_0 = 0$ and A_k is $\mathcal{F}_{k-1}$-measurable for every $k \in \{1, \dots, n\}$ such that

$$\forall\, k \in \{0, \dots, n-1\},\quad S_k = S_0 + M_k - A_k.$$

► **Exercise.** Prove this proposition. [Hint: First establish uniqueness by showing that $A_k = -\sum_{\ell=1}^{k}\mathbb{E}\big(\Delta S_\ell\,|\,\mathcal{F}_{\ell-1}\big)$, $k = 1, \dots, n$.]

Proof of Theorem 11.5 STEP 1 ($k = 0$). Let $M \in \mathcal{M}_0$ and let τ be an $(\mathcal{F}_k)_{0\le k\le n}$-stopping time. Then, by the optional stopping theorem (see [217]), $\mathbb{E}\,M_\tau = 0$ so that

$$\mathbb{E}\,\big(Z_\tau\,|\,\mathcal{F}_0\big) = \mathbb{E}\,\big(Z_\tau - M_\tau\,|\,\mathcal{F}_0\big) \le \mathbb{E}\left(\sup_{k\in\{0,\dots,n\}}(Z_k - M_k)\,|\,\mathcal{F}_0\right)\quad \mathbb{P}\text{-}a.s.$$

Hence, by the very definition of a $\mathbb{P}$-essinf,

$$\mathbb{E}\big(Z_\tau \,|\, \mathcal{F}_0\big) \le \mathbb{P}\text{-essinf}\left\{\mathbb{E}\left(\sup_{k\in\{0,\dots,n\}}(Z_k - M_k) \,|\, \mathcal{F}_0\right),\ M \in \mathcal{M}_0\right\} \quad \mathbb{P}\text{-}a.s.$$

Then by the definition (11.6) of a Snell envelope and $\mathbb{P}$-esssup

$$U_0 \le \mathbb{P}\text{-essinf}\left\{\mathbb{E}\left(\sup_{k\in\{0,\dots,n\}}(Z_k - M_k) \,|\, \mathcal{F}_0\right),\ M \in \mathcal{M}_0\right\} \quad \mathbb{P}\text{-}a.s.$$

Conversely, it is clear from the *BDPP* (11.7) satisfied by the Snell envelope that $(U_k)_{k=0,\dots,n}$ is a $(\mathbb{P}, (\mathcal{F}_k)_{0\le k\le n})$-super-martingale which dominates $(Z_k)_{k=0,\dots,n}$ (*i.e.* $U_k \ge Z_k$ $\mathbb{P}$-$a.s.$ for every $k = 0, \dots, n$). Hence, it admits a Doob decomposition (M^*, A^*) such that $U_k = U_0 + M_k^* - A_k^*$, $k = 0, \dots, n$, with M^* and A^* as in the above Proposition 11.5. Consequently, for every $k = 0, \dots, n$,

$$Z_k - M_k^* = \underbrace{Z_k - U_k}_{\le 0} - A_k^* + U_0 \le -A_k^* + U_k \le U_0,$$

which in turn implies $\mathbb{E}\left(\sup_{0\le k\le n} Z_k - M_k^* \,|\, \mathcal{F}_0\right) \le \mathbb{E}\left(U_0 \,|\, \mathcal{F}_0\right) = U_0$.

STEP 2 (*Generic k*). For a generic $k \in \{0, \dots, n\}$, one adapts the proof of the above step by considering for τ a $\{k, \dots, n\}$-valued stopping time and, in the second part of the proof, by considering the martingale $M_\ell^{*,k} = M_{\ell\vee k}^* - M_k^*$, $\ell = 0, \dots, n$. Conditioning by $\mathcal{F}_k$ completes the proof. ◇

Various papers take advantage of this representation to devise alternative American style option pricing algorithms, starting with [146, 255]. Let us also cite, among other references, [8, 36, 38] with extensions to multiple stopping time problems and to Backward Stochastic Differential Equations.

Chapter 12
Miscellany

12.1 More on the Normal Distribution

12.1.1 Characteristic Function

Proposition 12.1. *If $Z \stackrel{d}{=} \mathcal{N}(0;1)$, then its characteristic function χ_Z, defined for every $u \in \mathbb{R}$ by $\chi_Z(u) = \mathbb{E}\, e^{\tilde{\imath} u Z}$ is given by*

$$\forall\, u \in \mathbb{R},\ \ \chi_Z(u) := \int_{-\infty}^{+\infty} e^{\tilde{\imath} u x} e^{-\frac{x^2}{2}} \frac{dx}{\sqrt{2\pi}} = e^{-\frac{u^2}{2}}.$$

Proof. Differentiating under the integral sign yields

$$\chi_Z'(u) = i \int_{-\infty}^{+\infty} e^{\tilde{\imath} u x} e^{-\frac{x^2}{2}} x \frac{dx}{\sqrt{2\pi}}.$$

Now the integration by parts $\begin{cases} e^{iux} \stackrel{'}{\longrightarrow} \tilde{\imath} u e^{\tilde{\imath} u x} \\ x e^{-\frac{x^2}{2}} \stackrel{\int}{\longrightarrow} -e^{-\frac{x^2}{2}} \end{cases}$ yields

$$\chi_Z'(u) = \tilde{\imath}^2 u \int_{-\infty}^{+\infty} e^{\tilde{\imath} u x} e^{-\frac{x^2}{2}} x \frac{dx}{\sqrt{2\pi}} = -u \Phi_Z(u)$$

so that

$$\chi_Z(u) = \Phi_Z(0) e^{-\frac{u^2}{2}} = e^{-\frac{u^2}{2}}.$$

◇

Corollary 12.1 *If $Z = (Z^1, \ldots, Z^d) \stackrel{d}{=} \mathcal{N}(0; Id)$ is a multivariate normal distribution, then its characteristic function $\chi_Z(u) = \mathbb{E}\, e^{\tilde{\imath}(u\,|\,Z)}$, $u \in \mathbb{R}^d$, is given by*

$$\chi_Z(u) = e^{-\frac{|u|^2}{2}}.$$

G. Pagès, *Numerical Probability*, Universitext,
https://doi.org/10.1007/978-3-319-90276-0_12

Proof. For every $u \in \mathbb{R}^d$, $(u \mid Z) = \sum_{i=1}^d u^i Z^i$ has a Gaussian distribution with variance $\sum_{i=1}^d (u^i)^2 = |u|^2$ since the components Z^i are independent and $\mathcal{N}(0; 1)$-distributed. Consequently,

$$(u \mid Z) \stackrel{d}{=} |u|\,\zeta, \;\; \zeta \stackrel{d}{=} \mathcal{N}(0; 1),$$

so that the characteristic function χ_Z of Z defined by $\chi_Z(u) = \mathbb{E}\, e^{\tilde{\imath}(u|Z)}$ is given by

$$\forall\, u \in \mathbb{R}^d, \quad \chi_Z(u) = e^{-\frac{|u|^2}{2}}.$$

◇

Remark. An alternative argument is that the $\mathbb{C}$-valued random variables $e^{\tilde{\imath} u^i Z^i}$, $i = 1, \ldots, d$, are independent, so that

$$\chi_Z(u) = \mathbb{E} \prod_{i=1}^d e^{\tilde{\imath} u^i Z^i} = \prod_{i=1}^d \mathbb{E}\, e^{\tilde{\imath} u^i Z^i} = \prod_{i=1}^d e^{-\frac{(u^i)^2}{2}} = e^{-\frac{|u|^2}{2}}.$$

12.1.2 Numerical Approximation of the Cumulative Distribution Function Φ_0

To compute the c.d.f. (cumulative distribution function) of the normal distribution, one usually relies on the fast approximation formula obtained by continuous fraction expansion techniques (see [1]):

$$\forall\, x \in \mathbb{R}_+, \quad \Phi_0(x) = 1 - \frac{e^{-\frac{x^2}{2}}}{\sqrt{2\pi}} \left((a_1 t + a_2 t^2 + a_3 t^3 + a_4 t^4 + a_5 t^5 + O\left(e^{-\frac{x^2}{2}} t^6\right) \right),$$

where $t := \dfrac{1}{1 + px}$, $p := 0.231\,6419$ and

$$a_1 := 0.319\,381\,530, \quad a_2 := -0.356\,563\,782, \quad a_3 := 1.781\,477\,937,$$
$$a_4 := -1.821\,255\,978, \quad a_5 := 1.330\,274\,429.$$

inducing a maximal error of the form $O\left(e^{-\frac{x^2}{2}} t^6\right) \leq 7.5\,10^{-8}$.

12.1.3 Table of the Distribution Function of the Normal Distribution

The distribution function of the $\mathcal{N}(0; 1)$ distribution is given for every real number t by

$$\Phi_0(t) := \frac{1}{\sqrt{2\pi}} \int_{-\infty}^{t} e^{-\frac{x^2}{2}}\, dx.$$

Since the probability density is even, one easily checks that

$$\Phi_0(t) - \Phi_0(-t) = 2\,\Phi_0(t) - 1.$$

The following tables give the values of $\Phi_0(t)$ for $t = x_0, x_1 x_2$ where $x_0 \in \{0, 1, 2\}$, $x_1 \in \{0, 1, 2, 3, 4, 5, 6, 7, 8, 9\}$ and $x_2 \in \{0, 1, 2, 3, 4, 5, 6, 7, 8, 9\}$.

For example, if $t = 1.23$ (*i.e.* row 1.2 and column 0.03) one has $\Phi_0(t) \simeq 0.8907$.

t	0.00	0.01	0.02	0.03	0.04	0.05	0.06	0.07	0.08	0.09
0.0	0.5000	0.5040	0.5080	0.5120	0.5160	0.5199	0.5239	0.5279	0.5319	0.5359
0.1	0.5398	0.5438	0.5478	0.5517	0.5557	0.5596	0.5636	0.5675	0.5714	0.5753
0.2	0.5793	0.5832	0.5871	0.5910	0.5948	0.5987	0.6026	0.6064	0.6103	0.6141
0.3	0.6179	0.6217	0.6255	0.6293	0.6331	0.6368	0.6406	0.6443	0.6480	0.6517
0.4	0.6554	0.6591	0.6628	0.6661	0.6700	0.6736	0.6772	0.6808	0.6844	0.6879
0.5	0.6915	0.6950	0.6985	0.7019	0.7054	0.7088	0.7123	0.7157	0.7190	0.7224
0.6	0.7257	0.7290	0.7324	0.7357	0.7389	0.7422	0.7454	0.7486	0.7517	0.7549
0.7	0.7580	0.7611	0.7642	0.7673	0.7704	0.7734	0.7764	0.7794	0.7823	0.7852
0.8	0.7881	0.7910	0.7939	0.7967	0.7995	0.8023	0.8051	0.8078	0.8106	0.8133
0.9	0.8159	0.8186	0.8212	0.8238	0.8264	0.8289	0.8315	0.8340	0.8365	0.8389
1,0	0.8413	0.8438	0.8461	0.8485	0.8508	0.8531	0.8554	0.8577	0.8599	0.8621
1,1	0.8643	0.8665	0.8686	0.8708	0.8729	0.8749	0.8770	0.8790	0.8810	0.8830
1,2	0.8849	0.8869	0.8888	0.8907	0.8925	0.8944	0.8962	0.8980	0.8997	0.9015
1,3	0.9032	0.9049	0.9066	0.9082	0.9099	0.9115	0.9131	0.9147	0.9162	0.9177
1,4	0.9192	0.9207	0.9222	0.9236	0.9251	0.9265	0.9279	0.9292	0.9306	0.9319
1,5	0.9332	0.9345	0.9357	0.9370	0.9382	0.9394	0.9406	0.9418	0.9429	0.9441
1,6	0.9452	0.9463	0.9474	0.9484	0.9495	0.9505	0.9515	0.9525	0.9535	0.9545
1,7	0.9554	0.9564	0.9573	0.9582	0.9591	0.9599	0.9608	0.9616	0.9625	0.9633
1,8	0.9641	0.9649	0.9656	0.9664	0.9671	0.9678	0.9686	0.9693	0.9699	0.9706
1,9	0.9713	0.9719	0.9726	0.9732	0.9738	0.9744	0.9750	0.9756	0.9761	0.9767
2,0	0.9772	0.9779	0.9783	0.9788	0.9793	0.9798	0.9803	0.9808	0.9812	0.9817
2,1	0.9821	0.9826	0.9830	0.9834	0.9838	0.9842	0.9846	0.9850	0.9854	0.9857
2,2	0.9861	0.9864	0.9868	0.9871	0.9875	0.9878	0.9881	0.9884	0.9887	0.9890
2,3	0.9893	0.9896	0.9898	0.9901	0.9904	0.9906	0.9909	0.9911	0.9913	0.9916
2,4	0.9918	0.9920	0.9922	0.9925	0.9927	0.9929	0.9931	0.9932	0.9934	0.9936
2,5	0.9938	0.9940	0.9941	0.9943	0.9945	0.9946	0.9948	0.9949	0.9951	0.9952
2,6	0.9953	0.9955	0.9956	0.9957	0.9959	0.9960	0.9961	0.9962	0.9963	0.9964
2,7	0.9965	0.9966	0.9967	0.9968	0.9969	0.9970	0.9971	0.9972	0.9973	0.9974
2,8	0.9974	0.9975	0.9976	0.9977	0.9977	0.9978	0.9979	0.9979	0.9980	0.9981
2,9	0.9981	0.9982	0.9982	0.9983	0.9984	0.9984	0.9985	0.9985	0.9986	0.9986

One notes that $\Phi_0(t) = 0.9986$ for $t = 2.99$. This comes from the fact that the mass of the normal distribution is mainly concentrated on the interval $[-3, 3]$ as emphasized by the table of the "large" values hereafter (for instance, we observe that $\mathbb{P}(\{|X| \leq 4.5\}) \geq 0.99999$!).

t	3,0	3,1	3,2	3,3	3,4	3,5	3,6	3,8	4,0	4,5
$\Phi_0(t)$	.99865	.99904	.99931	.99952	.99966	.99976	.999841	.999928	.999968	.999997

12.2 Black–Scholes Formula(s) (To Compute Reference Prices)

In a risk-neutral Black–Scholes model, the quoted price of a risky asset is a solution to the *SDE* $dX_t = X_t(r dt + \sigma dW_t)$, $X_0 = x_0 > 0$, where r is the interest rate and $\sigma > 0$ is the volatility and W is a standard Brownian motion. Itô's formula (see Sect. 12.8) yields that

$$X_t^{x_0} = x_0\, e^{(r-\frac{\sigma^2}{2})t+\sigma W_t}, \quad W_t \stackrel{d}{=} \mathcal{N}(0;1).$$

A vanilla (European) payoff of maturity $T > 0$ is of the form $h_T = \varphi(X_T)$. A European option contract written on the payoff h_T is the right to receive h_T at the maturity T. Its price – or premium – at time $t = 0$ is given by $e^{-rt}\mathbb{E}\,\varphi(X_T^{x_0})$ and, more generally at time $t \in [0, T]$, it is given by $e^{-r(T-t)}\mathbb{E}\left(\varphi(X_T^{x_0}) \mid X_t^{x_0}\right) = e^{-r(T-t)}\mathbb{E}\,\varphi(X_{T-t}^{x_0})$. In the case where $\varphi(x) = (x - K)_+$ (*call* with *strike price* K) this premium at time t has a closed form given by

$$\text{Call}_t(x_0, K, R, \sigma, T) = \text{Call}_0(x_0, K, R, \sigma, T - t),$$

where

$$\text{Call}_0(x_0, K, r, \sigma, \tau) = x_0\Phi_0(d_1) - e^{-r\tau}K\,\Phi_0(d_2), \quad \tau > 0, \tag{12.1}$$

with

$$d_1 = \frac{\log\left(\frac{x_0}{K}\right) + (r + \frac{\sigma^2}{2})\tau}{\sigma\sqrt{\tau}}, \quad d_2 = d_1 - \sigma\sqrt{\tau}. \tag{12.2}$$

As for the *put* option written on the payoff $h_T = (K - X_T^{x_0})_+$, the premium is

$$\text{Put}_t(x_0, K, r, \sigma, T) = \text{Put}_0(x_0, K, r, \sigma, T - t),$$

where

$$\text{Put}_0(x_0, K, r, \sigma, \tau) = e^{-r\tau}K\,\Phi_0(-d_2) - x_0\Phi_0(-d_1). \tag{12.3}$$

The avatars of the regular Black–Scholes formulas can be obtained as follows:

- *Stock without dividend (Black–Scholes):* the risky asset is X.
- *Stock with continuous yield $\lambda > 0$ of dividends:* the risky asset is $e^{\lambda t}X_t$ and one has to replace x_0 by $e^{-\lambda\tau}x_0$ in the right-hand sides of (12.1), (12.2) and (12.3).

- *Foreign exchange (Garman–Kohlhagen):* though $X_t^{x_0}$ is quoted, the risky asset is $e^{r_F t}X_t$ where r_F denotes the foreign interest rate and one has to replace x_0 by $e^{-r_F T}x_0$ in the right-hand sides of (12.1), (12.2) and (12.3).
- *Future contract (Black):* the risky asset is the underlying asset of the future contract (with maturity $L > T$), *i.e.* it is $e^{r(L-t)}X_t$ and one has to replace x_0 by $e^{rT}x_0$ in the right-hand sides of (12.1), (12.2) and (12.3).

12.3 Measure Theory

Theorem 12.1 **(Baire σ-field Theorem)** *Let (S, d_S) be a metric space. Then*

$$\mathcal{B}or(S, d_S) = \sigma\big(\mathcal{C}(S, R)\big),$$

where $\mathcal{C}(S, R)$ denotes the set of continuous functions from (S, d_S) to $\mathbb{R}$. When S is σ-compact (i.e. is a countable union of compact sets), one may replace the space $\mathcal{C}(S, R)$ by the space $\mathcal{C}_K(S, R)$ of continuous functions with compact support.

Theorem 12.2 **(Functional monotone class Theorem)** *Let $(S, \mathcal{S})$ be a measurable space. Let V be a vector space of real-valued bounded measurable functions defined on $(S, \mathcal{S})$. Let C be a subset of V, stable under the product of two functions. Assume furthermore that V satisfies*

(i) $\mathbf{1} \in V$,
(ii) *V is closed under uniform convergence,*
(iii) *V is closed under "bounded non-decreasing convergence": if $\varphi_n \in V$, $n \geq 1$, $\varphi_n \leq \varphi_{n+1}$, $|\varphi_n| \leq K$ (real constant) and $\varphi_n(x) \to \varphi(x)$ for every $x \in S$, then $\varphi \in V$.*

Then H contains the vector subspace of all $\sigma(C)$-measurable bounded functions.

We refer to [216] for a proof of this result.

12.4 Uniform Integrability as a Domination Property

In this section, we present a brief background on uniform integrability for random variables taking values in $\mathbb{R}^d$. Mostly for notational convenience, all these random variables are defined on the same probability space $(\Omega, \mathcal{A}, \mathbb{P})$, although this is absolutely not mandatory. We leave it as an exercise to check in what follows that each random variable X_i can be defined on its own probability space $(\Omega_i, \mathcal{A}_i, \mathbb{P}_i)$ with a straightforward adaptation of the statements.

Theorem 12.3 **(Equivalent definitions of uniform integrability I)** *A family* $(X_i)_{i\in I}$ *of* $\mathbb{R}^d$*-valued random vectors, defined on a probability space* $(\Omega, \mathcal{A}, \mathbb{P})$*, is said to be* uniformly integrable *if it satisfies one of the following equivalent properties,*

(i) $\displaystyle\lim_{R\to+\infty}\sup_{i\in I}\mathbb{E}\Big(|X_i|\mathbf{1}_{\{|X_i|\ge R\}}\Big)=0.$

(ii) $\begin{cases}(\alpha)\ \sup_{i\in I}\mathbb{E}\,|X_i|<+\infty,\\ \\ (\beta)\ \forall\,\varepsilon>0,\ \exists\,\eta=\eta(\varepsilon)>0\ \textit{such that},\\ \ \forall\,A\in\mathcal{A},\ \mathbb{P}(A)\le\eta\Longrightarrow\sup\limits_{i\in I}\int_A|X_i|d\,\mathbb{P}\le\varepsilon.\end{cases}$

Remarks • All norms being strongly equivalent on $\mathbb{R}^d$, claims (i) and (ii) do not depend on the selected norm on $\mathbb{R}^d$.

• L^1-Uniform integrability of a family of probability distribution $(\mu_i)_{i\in I}$ defined on $(\mathbb{R}^d, \mathcal{B}or(\mathbb{R}^d))$ can be defined accordingly by

$$\lim_{R\to+\infty}\sup_{i\in I}\int_{\{|x|\ge R\}}|x|\,\mu_i(dx)=0.$$

All of the following can be straightforwardly "translated" in terms of probability distributions. Thus, more generally, let $f:\mathbb{R}^d\to\mathbb{R}_+$ be a non-zero Borel function such that $\underline{\lim}_{|x|\to+\infty} f(x)=0$. A family of probability distribution $(\mu_i)_{i\in I}$ defined on $(\mathbb{R}^d, \mathcal{B}or(\mathbb{R}^d))$ is f-uniformly integrable if

$$\lim_{R\to+\infty}\sup_{i\in I}\int_{\{f(x)\ge R\}}f(x)\,\mu_i(dx)=0.$$

Proof of Theorem 12.3. Assume first that (i) holds. It is clear that

$$\sup_{i\in I}\mathbb{E}\,|X_i|\le R+\sup_{i\in I}\mathbb{E}\Big(|X_i|\mathbf{1}_{\{|X_i|\ge R\}}\Big)<+\infty$$

at least for large enough $R\in(0,+\infty)$. Now, for every $i\in I$ and every $A\in\mathcal{A}$,

$$\int_A|X_i|\,d\,\mathbb{P}\le R\,\mathbb{P}(A)+\int_{\{|X_i|\ge R\}}|X_i|\,d\,\mathbb{P}.$$

Owing to (i) there exists a real number $R=R(\varepsilon)>0$ such that $\sup\limits_{i\in I}\displaystyle\int_{\{|X_i|\ge R\}}|X_i|\,d\,\mathbb{P}\le\frac{\varepsilon}{2}$. Then setting $\eta=\eta(\varepsilon)=\frac{\varepsilon}{2R}$ yields (ii).

Conversely, for every real number $R>0$, the Markov Inequality implies

$$\sup_{i\in I}\mathbb{P}(|X_i|\ge R)\le\frac{\sup_{i\in I}\mathbb{E}\,|X_i|}{R}.$$

Let $\eta=\eta(\varepsilon)$ be given by $(ii)(\beta)$. As soon as $R>\frac{\sup_{i\in I}\mathbb{E}\,|X_i|}{\eta}$, $\sup\limits_{i\in I}\mathbb{P}\big(\{|X_i|\ge R\}\big)\le\eta$ and $(ii)(\beta)$ implies that

$$\sup_{i\in I}\mathbb{E}\Big(|X_i|\mathbf{1}_{|X_i|\geq R}\Big)\leq\varepsilon$$

which completes the proof. ◇

As a consequence, one easily derives that

P0. $(X_i)_{i\in I}$ is uniformly integrable if and only if $(|X_i|)_{i\in I}$ is.

P1. If $X\in L^1(\mathbb{P})$ then the family (X) is uniformly integrable.

P2. If $(X_i)_{i\in I}$ and $(Y_i)_{i\in I}$ are two families of uniformly integrable $\mathbb{R}^d$-valued random vectors, then $(X_i+Y_i)_{i\in I}$ is uniformly integrable.

P3. If $(X_i)_{i\in I}$ is a family of $\mathbb{R}^d$-valued random vectors dominated by a uniformly integrable family $(Y_i)_{i\in I}$ of random variables in the sense that

$$\forall\, i\in I,\ |X_i|\leq Y_i\quad \mathbb{P}\text{-}a.s.$$

then $(X_i)_{i\in I}$ is uniformly integrable.

The four properties follow from characterization (i). To be precise, **P1** is a consequence of the Lebesgue Dominated Convergence Theorem, whereas **P2** follows from the obvious

$$\mathbb{E}\Big(|X_i+Y_i|\mathbf{1}_{\{|X_i+Y_i|\geq R\}}\Big)\leq\mathbb{E}\Big(|X_i|\mathbf{1}_{\{|X_i|\geq R\}}\Big)+\mathbb{E}\Big(|Y_i|\mathbf{1}_{\{|Y_i|\geq R\}}\Big).$$

Now let us pass to a simple criterion of uniform integrability.

Corollary 12.2 (**de La Vallée Poussin criterion**) *Let $(X_i)_{i\in I}$ be a family of $\mathbb{R}^d$-valued random vectors defined on a probability space $(\Omega,\mathcal{A},\mathbb{P})$ and let $\Phi:\mathbb{R}^d\to\mathbb{R}_+$ satisfying $\lim_{|x|\to+\infty}\dfrac{\Phi(x)}{|x|}=+\infty$. If*

$$\sup_{i\in I}\mathbb{E}\,\Phi(X_i)<+\infty,$$

then the family $(X_i)_{i\in I}$ is uniformly integrable.

Theorem 12.4 (**Uniform integrability II**) *Let $(X_n)_{n\geq 1}$ be a sequence of $\mathbb{R}^d$-valued random vectors defined on a probability space $(\Omega,\mathcal{A},\mathbb{P})$ and let X be an $\mathbb{R}^d$-valued random vector defined on the same probability space. If*

(i) $(X_n)_{n\geq 1}$ is uniformly integrable,

(ii) $X_n\xrightarrow{\mathbb{P}}X$,

then

$$\mathbb{E}\,|X_n-X|\longrightarrow 0\quad as\quad n\to+\infty.$$

In particular, $\mathbb{E}\,X_n\to\mathbb{E}\,X$. Moreover the converse is true: $L^1(\mathbb{P})$-convergence implies (i) and (ii).

Proof. One derives from (ii) the existence of a subsequence $(X_{n'})_{n\geq 1}$ such that $X_{n'} \to X$ $a.s$. Hence, by Fatou's Lemma,

$$\mathbb{E}\,|X| \leq \underline{\lim}_n \mathbb{E}\,|X_{n'}| \leq \sup_n \mathbb{E}\,|X_n| < +\infty.$$

Consequently, $X \in L^1(\mathbb{P})$ and, owing to **P1** and **P2**, $(X_n - X)_{n\geq 1}$ is a uniformly integrable sequence. Now, for every integer $n \geq 1$ and every $R > 0$,

$$\mathbb{E}\,|X_n - X| \leq \mathbb{E}\big(|X_n - X| \wedge M\big) + \mathbb{E}\big(|X_n - X|\mathbf{1}_{\{|X_n - X|\geq M\}}\big).$$

The Lebesgue Dominated Convergence Theorem implies $\lim_n \mathbb{E}(|X_n - X| \wedge M) = 0$ so that

$$\overline{\lim_n}\,\mathbb{E}\,|X_n - X| \leq \lim_{R\to+\infty} \sup_n \mathbb{E}\Big(|X_n - X|\mathbf{1}_{\{|X_n - X|\geq M\}}\Big) = 0.$$

◇

Corollary 12.3 (**L^p-uniform integrability**) *Let $p \in [1, +\infty)$. Let $(X_n)_{n\geq 1}$ be a sequence of $\mathbb{R}^d$-valued random vectors defined on a probability space $(\Omega, \mathcal{A}, \mathbb{P})$ and let X be an $\mathbb{R}^d$-valued random vector defined on the same probability space. If*

(i) $(X_n)_{n\geq 1}$ is L^p-uniformly integrable (i.e. $(|X_n|^p)_{n\geq 1}$ is uniformly integrable),

(ii) $X_n \xrightarrow{\mathbb{P}} X$,

then

$$\|X_n - X\|_p \longrightarrow 0 \quad \textit{as} \quad n \to +\infty.$$

(In particular, $\mathbb{E}\,X_n \to \mathbb{E}\,X$.) Moreover the converse is true: $L^p(\mathbb{P})$-convergence implies (i) and (ii).

Proof. By the same argument as above $X \in L^p(\mathbb{P})$ so that $(|X_n - X|^p)_{n\geq 1}$ is uniformly integrable by **P2** and **P3**. The result follows by the above theorem since $|X_n - X|^p \to 0$ in probability as $n \to +\infty$. ◇

12.5 Interchanging...

Theorem 12.5 (**Interchanging continuity and expectation**) *(see e.g. [52], Chap. 8) (a) Let $(\Omega, \mathcal{A}, \mathbb{P})$ be a probability space, let I be a nontrivial interval of $\mathbb{R}$ and let $\Psi : I \times \Omega \to \mathbb{R}$ be a $\mathcal{B}or(I) \otimes \mathcal{A}$-measurable function. Let $x_0 \in I$. If the function Ψ satisfies:*

(i) for every $x \in I$, the random variable $\Psi(x, .) \in L^1_{\mathbb{R}}(\mathbb{P})$,

(ii) $\mathbb{P}(d\omega)$-a.s., $x \mapsto \Psi(x, \omega)$ is continuous at x_0,

(iii) there exists $Y \in L^1_{\mathbb{R}_+}(\mathbb{P})$ such that for every $x \in I$,

$$\mathbb{P}(d\omega)\text{-}a.s. \;\; |\Psi(x, \omega)| \leq Y(\omega),$$

then the function $\psi(x) := \mathbb{E}\,\Psi(x\,,\,.)$ *is defined at every* $x \in I$ *and is continuous at* x_0.

(*b*) *The domination property* (*iii*) *in the above theorem can be replaced* mutatis mutandis *by a uniform integrability assumption on the family* $\big(\Psi(x\,,\,.)\big)_{x\in I}$.

The same extension as Claim (b), based on uniform integrability, holds true for the differentiation Theorem 2.2 (see the exercise following the theorem).

12.6 Weak Convergence of Probability Measures on a Polish Space

The main reference for this topic is [45]. See also [239].

The basic result of weak convergence theory is the so-called Portmanteau Theorem stated below (the definition and notation of weak convergence of probability measures are recalled in Sect. 4.1).

Theorem 12.6 *Let* $(\mu_n)_{n\ge1}$ *be a sequence of probability measures on a Polish (metric) space* (S, δ) *equipped with its Borel* σ*-field* $\mathcal{S}$ *and let* μ *be a probability measure on the same space. The following properties are equivalent:*

(*i*) $\mu_n \stackrel{\mathcal{S}}{\Longrightarrow} \mu$ *as* $n \to +\infty$,

(*ii*) *For every open set* O *of* (S, δ).

$$\mu(O) \le \underline{\lim}_n\, \mu_n(O).$$

(*ii*) *For every bounded Lipschitz continuous function,*

$$\lim_n \int_S f\,d\mu_n = \int_S f\,d\mu.$$

(*iii*) *For every closed set* F *of* (S, δ),

$$\mu(F) \ge \overline{\lim}_n\, \mu_n(F).$$

(*iv*) *For every Borel set* $A \in \mathcal{S}$ *such that* $\mu(\partial A) = 0$ (*where* $\partial A = \overline{A} \setminus \mathring{A}$ *is the boundary of* A),

$$\lim_n \mu_n(A) = \mu(A).$$

(*v*) *Weak Fatou Lemma: For every non-negative lower semi-continuous function* $f : S \to \mathbb{R}_+$,

$$0 \le \int_S f\,d\mu \le \underline{\lim}_n \int_S f\,d\mu_n.$$

(vi) For every bounded Borel function $f : S \to \mathbb{R}$ *such that* $\mu\big(\mathrm{Disc}(f)\big) = 0$,

$$\lim_n \int_S f\,d\mu_n = \int_S f\,d\mu.$$

where $\mathrm{Disc}(f) = \big\{x \in S : f \text{ is not continuous at } x\big\}$.

For a proof, we refer to [45], Chap. 1.

When dealing with unbounded functions, there is a kind of weak Lebesgue dominated convergence theorem.

Proposition 12.2 *Let* $(\mu_n)_{n\geq 1}$ *be a sequence of probability measures on a Polish (metric) space* (S, δ) *weakly converging to* μ.
(*a*) *Let* $g : S \to \mathbb{R}_+$ *be a (non-negative)* μ*-integrable continuous function and let* $f : S \to \mathbb{R}$ *be a* μ*-a.s. continuous Borel function. If*

$$0 \leq |f| \leq g \quad \text{and} \quad \int_S g\,d\mu_n \longrightarrow \int_S g\,d\mu \ \text{ as } n \to +\infty$$

then $f \in L^1(\mu)$ *and* $\displaystyle\int_S f\,d\mu_n \longrightarrow \int_S f\,d\mu$ *as* $n \to +\infty$.
(*b*) *The conclusion still holds if* f *is* $(\mu_n)_{n\geq 1}$*-uniformly integrable, i.e.*

$$\lim_{R\to+\infty} \sup_{n\geq 1} \int_{|f|\geq R} |f|\,d\mu_n = 0.$$

Proof. Let $R > 0$ such that $\mu(f = R) = \mu(g = R) = 0$. Set $f_{R} = f\mathbf{1}_{|f|\leq R}$ and $g_{R} = g\mathbf{1}_{g\leq R}$. The functions f_{R} and g_{R} are μ-*a.s.* continuous and bounded. Starting from

$$\left|\int f\,d\mu_n - \int f\,d\mu\right| \leq \left|\int f_R\,d\mu_n - \int f_R\,d\mu\right| + \int g\,d\mu_n - \int g_R\,d\mu_n + \int g\,d\mu - \int g_R\,d\mu,$$

we derive from the above proposition and the assumption made on g that, for every such R,

$$\overline{\lim_n}\left|\int f\,d\mu_n - \int f\,d\mu\right| \leq 2\int (g - g_{R})\,d\mu = 2\int g\mathbf{1}_{\{g>R\}}\,d\mu.$$

As there are at most countably many R which are μ-atoms for g and f, we may let R go to infinity, so that $\displaystyle\int g\mathbf{1}_{\{g\geq R\}}\,d\mu \to 0$ as $R \to +\infty$. This completes the proof. ◇

If $S = \mathbb{R}^d$, weak convergence is also characterized by the Fourier transform $\widehat{\mu}$ defined on $\mathbb{R}^d$ by

$$\widehat{\mu}(u) = \int_{\mathbb{R}^d} e^{\tilde{\imath}(u|x)}\mu(dx), \ \ u \in \mathbb{R}^d.$$

Proposition 12.3 *Let $(\mu_n)_{n\geq 1}$ be a sequence of probability measures on $(\mathbb{R}^d, \mathcal{B}or(\mathbb{R}^d))$ and let μ be a probability measure on the same space. Then*

$$\Big(\mu_n \stackrel{\mathcal{S}}{\Longrightarrow} \mu\Big) \Longleftrightarrow \Big(\forall\, u \in \mathbb{R}^d,\ \widehat{\mu}_n(u) \longrightarrow \widehat{\mu}(u)\Big).$$

For a proof we refer, for example, to [156], or to any textbook presenting a first course in Probability Theory.

Remark. The convergence in distribution of a sequence $(X_n)_{n\geq 1}$ defined on probability spaces $(\Omega^n, \mathcal{A}^n, \mathbb{P}^n)$ of random variables taking values in a Polish space is defined as *the weak convergence of their distributions* $\mu_n = \mathbb{P}^n_{X_n} = \mathbb{P}^n \circ X_n^{-1}$ on $(S, \mathcal{S})$.

12.7 Martingale Theory

Theorem 12.7 *Let $(M_n)_{n\geq 0}$ be a square integrable discrete time $(\mathcal{F}_n)$-martingale defined on a filtered probability space $(\Omega, \mathcal{A}, \mathbb{P}, (\mathcal{F}_n)_{n\geq 0})$.*
(a) Then there exists a unique non-decreasing $\mathcal{F}_n$-predictable process null at 0 denoted by $(\langle M\rangle_n)_{n\geq 0}$ such that

$$(M_n - M_0)^2 - \langle M\rangle_n \quad \textit{is an } \mathcal{F}_n\textit{-martingale.}$$

This process reads

$$\langle M\rangle_n = \sum_{k=1}^{n} \mathbb{E}\big((M_k - M_{k-1})^2 \,|\, \mathcal{F}_{k-1}\big).$$

(b) Set $\langle M\rangle_\infty := \lim_n \langle M\rangle_n$. Then

$$M_n \stackrel{n\to+\infty}{\longrightarrow} M_\infty \quad \textit{on the event } \big\{\langle M\rangle_\infty < +\infty\big\},$$

where M_∞ is a finite random variable. If, furthermore, $\mathbb{E}\,\langle M\rangle_\infty < +\infty$, then $M_\infty \in L^2(\mathbb{P})$.

We refer to [217] for a proof of this result.

Lemma 12.1 **(Kronecker Lemma)** *Let $(a_n)_{n\geq 1}$ be a sequence of real numbers and let $(b_n)_{n\geq 1}$ be a non-decreasing sequence of positive real numbers with $\lim_n b_n = +\infty$. Then*

$$\left(\sum_{n\geq 1} \frac{a_n}{b_n} \textit{ converges in } \mathbb{R} \textit{ as a series}\right) \Longrightarrow \left(\frac{1}{b_n}\sum_{k=1}^{n} a_k \longrightarrow 0 \ \textit{as}\ n \to +\infty\right).$$

Proof. Set $C_n = \sum_{k=1}^{n} \frac{a_k}{b_k}$, $n \geq 1$, $C_0 = 0$. The assumption says that $C_n \to C_\infty = \sum_{k\geq 1} \frac{a_k}{b_k} \in \mathbb{R}$ as $n \to +\infty$. Now, for every $n \geq 1$, $b_n > 0$ and

$$\begin{aligned}\frac{1}{b_n}\sum_{k=1}^{n} a_k &= \frac{1}{b_n}\sum_{k=1}^{n} b_k\, \Delta C_k \\ &= \frac{1}{b_n}\left(b_n C_n - \sum_{k=1}^{n} C_{k-1}\Delta b_k\right) \\ &= C_n - \frac{1}{b_n}\sum_{k=1}^{n} \Delta b_k\, C_{k-1},\end{aligned}$$

where we used an Abel Transform for series. The result follows by the extended Césaro Theorem since $\Delta b_n \geq 0$ and $\lim_n b_n = +\infty$. ◇

To establish Central Limit Theorems outside the "i.i.d.ßetting, we will rely on Lindeberg's Central limit theorem for arrays of martingale increments (see Theorem 3.2 and its Corollary 3.1, p. 58 in [142]), stated below in a simple form involving only one square integrable martingale.

Theorem 12.8 (Lindeberg's Central Limit Theorem for martingale increments, see [142]) *Let $(M_n)_{n\geq 1}$ be a square integrable martingale with respect to a filtration $(\mathcal{F}_n)_{n\geq 1}$ and let $(a_n)_{n\geq 1}$ be a non-decreasingsequence of positive real numbers going to infinity as n goes to infinity.*

If the following two conditions hold:

(i) $\frac{1}{a_n}\sum_{k=1}^{n} \mathbb{E}\Big((\Delta M_k)^2 \,|\, \mathcal{F}_{k-1}\Big) \longrightarrow \sigma^2 \in [0, +\infty)$ *in probability,*

(ii) $\forall\, \varepsilon > 0,\ \frac{1}{a_n}\sum_{k=1}^{n} \mathbb{E}\Big((\Delta M_k)^2 \mathbf{1}_{\{|\Delta M_k| \geq \varepsilon\sqrt{a_n}\}} \,|\, \mathcal{F}_{k-1}\Big) \longrightarrow 0$ *in probability,*

then

$$\frac{M_n}{\sqrt{a_n}} \xrightarrow{\mathcal{L}} \mathcal{N}\big(0; \sigma^2\big).$$

► **Exercise.** Derive from this result a d-dimensional theorem. [Hint: Consider a linear combination of the d-dimensional martingale under consideration.]

12.8 Itô Formula for Itô Processes

12.8.1 Itô Processes

An $\mathbb{R}^d$-valued stochastic process $(X_t)_{t\in[0,T]}$ defined on a filtered probability space $(\Omega, \mathcal{A}, \mathbb{P}, (\mathcal{F}_t)_{t\geq 0})$ of the form, *i.e.*

$$X_t = X_0 + \int_0^t K_s ds + \int_0^s H_s\, dW_s, \quad t \in [0, T], \tag{12.4}$$

where

- X_0 is $\mathcal{F}_0$-measurable,
- $(H_t)_{t\geq 0} = \big([H_t^{ij}]_{i=1:d,j=1:q}\big)_{t\geq 0}$ and $(K_t)_{t\geq 0} = \big([K_t^i]_{i=1:d}\big)_{t\geq 0}$ are $(\mathcal{F}_t)$-progressively measurable processes ([1]) having values in $\mathbb{R}^d$ and $\mathcal{M}(d, q, \mathbb{R})$, respectively,
- $\displaystyle\int_0^T |K_s| ds < +\infty$ $\mathbb{P}$-$a.s.$,
- $\displaystyle\int_0^T \|H_s\|^2 ds < +\infty$ $\mathbb{P}$-$a.s.$,
- W is a q-dimensional $(\mathcal{F}_t)$-standard Brownian motion, ([2])

is called an *Itô process* ([3]).

Note that the processes K and H in (12.4) are $\mathbb{P}$-$a.s.$ essentially unique (since a continuous (local) martingale null at zero with finite variation is indistinguishable from the null process).

In particular, an Itô process is a local martingale ([4]) if and only if, $\mathbb{P}$-$a.s$, $K_t = 0$ for every $t \in [0, T]$.

If $\mathbb{E}\displaystyle\int_0^t \|H_s\|^2 ds = \int_0^t \mathbb{E}\|H_s\|^2 ds < +\infty$ then the process $\left(\displaystyle\int_0^t H_s\, dW_s\right)_{t\geq 0}$ is an $\mathbb{R}^d$-valued square integrable martingale with a bracket process

$$\left[\left\langle\left(\int_0^t H_s\, dW_s\right)^i, \left(\int_0^t H_s\, dW_s\right)^j\right\rangle\right]_{1\leq i,j\leq d} = \left[\int_0^t \big[(HH^*)_s^{ij}\big]^2 ds\right]_{1\leq i,j\leq d}$$

(H^* denotes here the transpose of H) and

---

[1]A stochastic process $(Y_t)_{\geq 0}$ defined on $(\Omega, \mathcal{A}, \mathbb{P}$ is $(\mathcal{F}_t)$-progressively measurable if for every $t \in \mathbb{R}_+$, the mapping $(s,\omega) \mapsto Y_s(\omega)$ defined on $[0,t] \times \Omega$ is $\mathcal{B}or([0,t]) \otimes \mathcal{F}_t$-measurable.

[2]This means that W is $(\mathcal{F}_t)$-adapted and, for every $s,\ t \geq 0,\ s \leq t$, $W_t - W_s$ is independent of $\mathcal{F}_s$.

[3]The stochastic integral is defined by $\int_0^s H_s\, dW_s = \left[\sum_{j=1}^d H_s^{ij}\, dW_s^j\right]_{1\leq i\leq d}$.

[4]An $\mathcal{F}_t$-adapted continuous process is a local martingale if there exists an increasing sequence $(\tau_n)_{n\geq 1}$ of $(\mathcal{F}_t)$-stopping times, increasing to $+\infty$, such that $(M_{t\wedge\tau_n} - M_0)_{t\geq 0}$ is an $(\mathcal{F}_t)$-martingale for every $n \geq 1$.

$$\left(\int_0^t H_s\,dW_s\right)^i \cdot \left(\int_0^t H_s\,dW_s\right)^j - \left\langle \left(\int_0^t H_s\,dW_s\right)^i, \left(\int_0^t H_s\,dW_s\right)^j \right\rangle$$

are $(\mathbb{P}, \mathcal{F}_t)$-martingales for every $i, j \in \{1, \ldots, d\}$. Otherwise, these are only local martingales.

12.8.2 The Itô Formula

The Itô formula, also known as Itô's Lemma, applies to this family of processes: let $f \in \mathcal{C}^{1,2}(\mathbb{R}_+, \mathbb{R}^d)$. Then the process $\big(f(t, X_t)\big)_{t\geq 0}$ is still an Itô process reading

$$f(t, X_t) = f(0, X_0) + \int_0^t \big(\partial_t f(s, X_s) + Lf(s, X_s)\big)ds$$
$$+ \frac{1}{2}\int_0^t \big(\nabla_x f(s, X_s) \,|\, \sigma(s, X_s)dW_s\big), \tag{12.5}$$

where

$$Lf(t, X_t) = \big(b(t, x) \,|\, \nabla_x f(t, x)\big) + \frac{1}{2}\mathrm{Tr}\big(\sigma D^2 f \sigma^*\big)(t, x)$$

denotes the infinitesimal generator of the diffusion process.

For a comprehensive exposition of the theory of continuous martingales theory and stochastic calculus, we refer to [162, 163, 249, 251, 256] among many other references. For a more synthetic introduction with a view to applications to Finance, we refer to [183].

12.9 Essential Supremum (and Infimum)

The essential supremum is an object playing for $a.s.$ comparisons a similar role for random variables defined on a probability space $(\Omega, \mathcal{A}, \mathbb{P})$ to that played by the regular supremum for real numbers on the real line.

Let us understand on a simple example why the regular supremum is not the right notion in a framework where equalities and inequalities hold in an $a.s.$ sense. First, it is usually not measurable when taken over a non-countable infinite set of indices: indeed, let $(\Omega, \mathcal{A}, \mathbb{P}) = ([0, 1], \mathcal{B}or([0, 1]), \lambda)$, where λ denotes the Lebesgue measure on $[0, 1]$, let $I \subset [0, 1]$ and $X_i = \mathbf{1}_{\{i\}}$, $i \in I$. Then, $\sup_{i\in I} X_i = \mathbf{1}_I$ is measurable if and only if I is a Borel subset of the interval $[0, 1]$.

Furthermore, even when $I = [0, 1]$, $\sup_{i\in I} X_i = 1$ is certainly measurable but is therefore constantly 1, whereas $X_i = 0$ $\mathbb{P}$-$a.s.$ for every $i \in I$. We will see, however, that the notion of essential supremum defined below does satisfy $\operatorname{esssup}_{i\in I} X_i = 0$ $\mathbb{P}$-$a.s.$

Proposition 12.4 (Essential supremum) *(a) Let $(\Omega, \mathcal{A}, \mathbb{P})$ be a probability space and $(X_i)_{i\in I}$ a family of random variables having values in $\overline{\mathbb{R}} = [-\infty, +\infty]$ defined on $(\Omega, \mathcal{A})$. There exists a unique random variable on Ω, up to a $\mathbb{P}$-a.s. equality, denoted by $\mathbb{P}\text{-}\operatorname{esssup}_{i\in I} X_i$ (or simply $\operatorname{esssup}_{i\in I} X_i$ when there is no ambiguity on the underlying probability $\mathbb{P}$), satisfying:*

(i) $\forall\, i \in I,\ X_i \le \mathbb{P}\text{-}\operatorname{esssup}_{i\in I} X_i$ $\mathbb{P}$-a.s.,

(ii) for every random variable $Z : (\Omega, \mathcal{A}) \to \overline{\mathbb{R}}$, if $X_i \le Z$ $\mathbb{P}$-a.s. for every $i \in I$, then $\mathbb{P}\text{-}\operatorname{esssup}_{i\in I} X_i \le Z$ $\mathbb{P}$-a.s.

(b) A family $(X_i)_{i\in I}$ of real-valued random variables defined on $(\Omega, \mathcal{A})$ is stable under finite supremum (up to $\mathbb{P}$-a.s. equality), if

$$\forall i, j \in I,\ \exists\, k \in I \ \text{ such that } \ X_k \ge \max(X_i, X_j) \quad \mathbb{P}\text{-}a.s. \tag{12.6}$$

If $(X_i)_{i\in I}$ is stable under finite supremum, then there exists a subsequence $(X_{i_n})_{n\in\mathbb{N}}$ $\mathbb{P}$-a.s. non-decreasing such that

$$\mathbb{P}\text{-}\operatorname{esssup}_{i\in I} X_i = \sup_{n\in\mathbb{N}} X_{i_n} = \lim_{n\to+\infty} X_{i_n} \ \mathbb{P}\text{-}a.s.$$

(c) Let $(\Omega, \mathcal{P}(\Omega), \mathbb{P})$ be a finite probability space such that $\mathbb{P}(\{\omega\}) > 0$ for every $\omega \in \Omega$. Then $\sup_{i\in I} X_i$ is the unique essential supremum of the X_i.

We adopt the simplified notation $\operatorname{esssup}_{i\in I}$ in the proof below to alleviate notation.

Proof. (a) Let us begin by $\mathbb{P}$-$a.s.$ uniqueness. Assume there exists a random variable Z satisfying (i) and (ii). Following (i) for Z and (ii) for $\operatorname{esssup}_{i\in I} X_i$, we derive that $\operatorname{esssup}_{i\in I} X_i \le Z$ $\mathbb{P}$-$a.s.$ Finally, (i) for $\operatorname{esssup}_{i\in I} X_i$ and (ii) for Z give the reverse inequality.

As for the existence, first assume that the variables X_i are $[0, 1]$-valued. Set

$$M_{\scriptscriptstyle X} := \sup\left\{\mathbb{E}\Big(\sup_{j\in J} X_j\Big),\ J \subset I,\ J \text{ countable}\right\} \in [0, 1].$$

This definition is consistent since, J being countable, $\sup_{j\in J} X_j$ is measurable and $[0, 1]$-valued. As $M_{\scriptscriptstyle X} \in [0, 1]$ is a finite real number, there exists a sequence $(J_n)_{n\ge 1}$ of countable subset of I satisfying:

$$\mathbb{E}\Big(\sup_{j\in J_n} X_j\Big) \ge M_{\scriptscriptstyle X} - \frac{1}{n}.$$

Set

$$\operatorname*{esssup}_{i \in I} X_i := \sup_{j \in J} X_j \quad \text{where} \quad J = \cup_{n \geq 1} J_n.$$

This defines a random variable since J is itself countable as the countable union of countable sets. This yields

$$M_X \geq \mathbb{E}\Big(\operatorname*{esssup}_{i \in I} X_i\Big) \geq \lim_n \mathbb{E}\Big(\sup_{j \in J_n} X_j\Big) \geq M_X.$$

whence $M_X = \mathbb{E}\Big(\operatorname*{esssup}_{i \in I} X_i\Big)$. Then let $i \in I$ be fixed. The subset $J \cup \{i\}$ of I is countable, hence

$$\mathbb{E}\Big(\sup_{j \in J \cup \{i\}} X_j\Big) \leq M_X = \mathbb{E}\Big(\sup_{j \in J} X_j\Big).$$

Consequently,

$$\mathbb{E}\Big(\underbrace{\sup_{j \in J \cup \{i\}} X_j - \sup_{j \in J} X_j}_{\geq 0}\Big) \leq 0,$$

which implies $\sup_{j \in J \cup \{i\}} X_j - \sup_{j \in J} X_j = 0$ $\mathbb{P}$-$a.s.$, *i.e.*

$$X_i \leq \operatorname*{esssup}_{i \in I} X_i = \sup_{j \in J} X_j \quad \mathbb{P}\text{-}a.s.$$

Moreover, if $X_i \leq Z$ $\mathbb{P}$-$a.s.$ for every $i \in I$, in particular this holds true for $i \in J$ so that

$$\operatorname*{esssup}_{i \in I} X_i = \sup_{j \in J} X_j \leq Z \quad \mathbb{P}\text{-}a.s.$$

When the X_i have values in $[-\infty, +\infty]$, one introduces an increasing bijection $\Phi : \overline{\mathbb{R}} \to [0, 1]$ and sets

$$\operatorname*{esssup}_{i \in I} X_i := \Phi^{-1}\Big(\operatorname*{esssup}_{i \in I} \Phi(X_i)\Big).$$

It is easy to check that, thus defined, $\operatorname*{esssup}_{i \in I} X_i$ satisfies (i) and (ii). By uniqueness $\operatorname*{esssup}_{i \in I} X_i$ does not depend $\mathbb{P}$-$a.s.$ on the selected function Φ.

(b) Following (a), there exists a sequence $(j_n)_{n \in \mathbb{N}}$ such that $J = \big\{j_n,\ n \geq 0\big\}$ satisfies

$$\operatorname*{esssup}_{i \in I} X_i = \sup_{n \in \mathbb{N}} X_{j_n} = \lim_n X_{j_n} \quad \mathbb{P}\text{-}a.s.$$

Now, owing to the stability property of the supremum (12.6), one may build a sequence $(i_n)_{n \in \mathbb{N}}$ such that $X_{i_0} = X_{j_0}$ and, for every $n \geq 1$,

$$X_{i_n} \geq \max(X_{j_n}, X_{i_{n-1}}), \quad \mathbb{P}\text{-}a.s.$$

It is clear by induction that $X_{i_n} \geq \max(X_{j_0}, \ldots, X_{j_n})$ increases toward $\operatorname{esssup}_{i\in I} X_i$, which completes the proof of this item.

(c) When the probability space is endowed with the coarse σ-field $\mathcal{P}(\Omega)$, the mapping $\sup_{i\in I} X_i$ is measurable. Now, $\sup_{i\in I} X_i$ obviously satisfies (i) and (ii) and is subsequently $\mathbb{P}$-$a.s.$ equal to $\operatorname{esssup}_{i\in I} X_i$. As the empty set is the only $\mathbb{P}$-negligible set, the former equality is true on the whole Ω. ◇

We will not detail all the properties of the $\mathbb{P}$-essential supremum which are quite similar to those of regular supremum (up to an $a.s.$ equality). Typically, one has with obvious notations and $\mathbb{P}$-$a.s.$,

$$\operatorname{esssup}_{i\in I}(X_i + Y_i) \leq \operatorname{esssup}_{i\in I} X_i + \operatorname{esssup}_{i\in I} Y_i$$

and,

$$\forall \lambda \geq 0, \ \operatorname{esssup}_{i\in I}(\lambda.X_i) = \lambda.\operatorname{esssup}_{i\in I} X_i,$$

etc.

As a conclusion, we may define the essential infimum by setting, with the notation of the above theorem,

$$\mathbb{P}\text{-essinf}_{i\in I} X_i = -\mathbb{P}\text{-}\operatorname{esssup}_{i\in I}(-X_i). \tag{12.7}$$

12.10 Halton Sequence Discrepancy (Proof of an Upper-Bound)

This section is devoted to the proof of Theorem 4.2. We focus on dimension 2, assuming that the one dimensional result for the Vdc(p) sequence holds. This one dimensional case can be established directly following the lines of the proof below: it is easier and it does not require the Chinese Remainder Theorem. In fact, the general bound can be proved by induction on d by adapting the proof of the 2-dimensional setting hereafter.

Assume p_1 and p_2 are mutually prime.

STEP 1. We deal with the two-dimensional case for simplicity: let p_1, p_2, two mutually prime numbers and let $\xi_n^i = \Phi_{p_i}(n)$, $i = 1, 2$, $n \geq 1$, where Φ_{p_i} is the radical inverse function associated to p_i. Note that $\xi_n^i \in (0, 1)$ for every $n \geq 1$ and every $i = 1, 2$. We set, for every integer $n \geq 1$,

$$\Phi_n(x, y) = \sum_{k=1}^{n} \mathbf{1}_{\{\xi_k^1 \le x\} \cap \{\xi_k^2 \le y\}}, \ (x, y) \in [0, 1]^2$$

and

$$\widetilde{\Phi}_n(x, y) = \Phi_n(x, y) - nxy.$$

Let $n \ge 1$ be a fixed integer. Let $r_i = \left\lfloor \frac{\log n}{\log p_i} \right\rfloor$, $i = 1, 2$. We consider the increasing reordered n-tuples of both components, denoted by $(\xi_k^{i,(n)})_{k=1,\dots,n}$, $i = 1, 2$ with the two additional terms $\xi_0^{i,(n)} = 0$ and $\xi_{n+1}^{i,(n)} = 1$. Then, set

$$\mathcal{X}_{1,2}^n = \left\{ \left(\xi_{k_1}^{1,(n)}, \xi_{k_2}^{2,(n)}\right),\ 0 \le k_1, k_2 \le n \right\}.$$

Note that, for any pair $(x, y) \in \mathcal{X}_{1,2}^n$, x reads

$$x = \frac{x_1}{p_1} + \cdots + \frac{x_{r_1}}{p_1^{r_1}}, \quad y = \frac{y_1}{p_2} + \cdots + \frac{y_{r_2}}{p_2^{r_2}}.$$

As Φ_n is constant over rectangles of the form $\left(\xi_k^{1,(n)}, \xi_{k+1}^{1,(n)}\right] \times \left(\xi_\ell^{1,(n)}, \xi_{\ell+1}^{1,(n)}\right]$, $0 \le k, \ell \le n$, and $\widetilde{\Phi}_n$ is decreasing with respect to the componentwise partial order on $[0, 1]^2$ (the order induced by inclusions of boxes $[\![0, x]\!]$). Finally, $\widetilde{\Phi}_n$ is "right-continuous" in the sense that $\widetilde{\Phi}_n(x, y) = \lim_{(u,v)\to(x,y), u>x, v>y} \widetilde{\Phi}_n(u, v)$. Moreover, as both sequences are $(0, 1)$-valued, $\widetilde{\Phi}_n$ is continuous at $(1, 1)$, $\widetilde{\Phi}_n(1, 1) = 0$ and, for $x, y \in [0, 1)$,

$$\widetilde{\Phi}_n(x, 1) = \lim_{(u,v)\to(x,1), u>x, v<1} \widetilde{\Phi}_n(u, v),\ \widetilde{\Phi}_n(1, y) = \lim_{(u,v)\to(x,y), u<1, v>y} \widetilde{\Phi}_n(u, v).$$

This shows that

$$\begin{aligned}\sup_{(x,y)\in[0,1]^2} \widetilde{\Phi}_n(x, y) &= \sup_{(x,y)\in[0,1)^2} \widetilde{\Phi}_n(x, y)\\ &= \max_{0\le k,\ell\le n} \widetilde{\Phi}_n\left(\xi_k^{1,(n)}, \xi_\ell^{2,(n)}\right) = \max_{(x,y)\in\mathcal{X}_{1,2}^n} \widetilde{\Phi}_n(x, y) \ge 0.\end{aligned}$$

Likewise, setting $\Phi_n(x_-, y_-) = \lim_{(u,v)\to(x,y), u<x, v<y} \Phi_n(u, v)$, one shows that

$$\inf_{(x,y)\in[0,1]^2} \widetilde{\Phi}_n(x, y) = \min_{1\le k,\ell\le n+1} \widetilde{\Phi}_n\left(\xi_{k-}^1, \xi_{\ell-}^2\right).$$

Note that, as $\xi_k^2 \in (0, 1)$, $k = 1, \dots, n$,

$$\widetilde{\Phi}_n(1_-, y_-) = \lim_{v\to y, v<y} \widetilde{\Phi}_n(1, y) \ge -nD_n^*(\xi^2) \text{ and } \widetilde{\Phi}_n(x_-, 1_-) \ge -nD_n^*(\xi^1)$$

so that

$$\inf_{(x,y)\in[0,1]^2} \widetilde{\Phi}_n(x, y) \geq \min\left(\min_{(x,y)\in\mathcal{X}^n_{1,2}} \widetilde{\Phi}_n(x, y), -nD^*_n(\xi^1), -nD^*_n(\xi^2)\right). \quad (12.8)$$

Consequently,

$$\sup_{(x,y)\in[0,1]^2} \left|\widetilde{\Phi}_n(x, y)\right| \leq \max\left(nD^*_n(\xi^1), nD^*_n(\xi^2), \max_{(x,y)\in\mathcal{X}^n_{1,2}} \left|\widetilde{\Phi}_n(x, y)\right|\right). \quad (12.9)$$

STEP 2. Let $k \in \{0, \dots, n\}$ and let $(x, y) \in \mathcal{X}^n_{1,2}$. Write $k = k_0 + k_1 p_1 + \cdots + k_{r_1} p_1^{r_1}$, $k_i \in \{0, \dots, p_1 - 1\}$, $k_{r_1} \neq 0$ in base p_1. Let us focus first on the inequality, $\xi^1_k < x$. It reads

$$\xi^1_k = \frac{k_0}{p} + \cdots \frac{k_{r_1}}{p^{r_1+1}} < \frac{x_1}{p} + \cdots + \frac{x_r}{p^{r_1+1}},$$

which is equivalent to

$$\exists\, r \in \{0, \dots, r_1\} \text{ s.t. } k_s = x_{s+1},\ s = 0, \dots, r-1 \text{ and } k_r < x_{r+1}.$$

These conditions can be rewritten as

$$\begin{aligned}&\exists\, r \in \{0, \dots, r_1\},\ \exists\, u_r \in \{0, \dots, x_{r+1}\} \text{ s.t.}\\ &\qquad\qquad k \equiv x_1 + \cdots x_r\, p_1^{r-1} + u_r p_1^r \mod p_1^{r+1}.\end{aligned}$$

Consequently, the joint conditions $\xi^1_k < x$ and $\xi^2_k < y$ read: there exists $r \in \{0, \dots, r_1\}$, $s \in \{0, \dots, r_2\}$, $u_r \in \{0, \dots, x_{r+1} - 1\}$, $v_s \in \{0, \dots, y_{s+1} - 1\}$ such that

$$S_{r_1,r_2}(u_r, v_s) \equiv \begin{cases} k \equiv x_1 + \cdots + x_r\, p_1^{r-1} + u_r p_1^r \mod p_1^{r+1} \\ \\ k \equiv y_1 + \cdots + y_s\, p_2^{s-1} + v_s p_2^s \mod p_2^{s+1}. \end{cases}$$

Since p_1^{r+1} and p_2^{s+1} are mutually prime, we know by the Chinese Remainder Theorem that the system $S_{r_1,r_2}(u_r, v_r)$ has a unique solution in $\{0, \dots, p_1^{r+1} p_2^{r+1} - 1\}$, hence $\left\lfloor \frac{n}{p^{r+1} q^{r+1}} \right\rfloor + \eta_{r,s,u_r,v_s}$ solutions lying in $\{0, \dots, n\}$ with $\eta_{r,s,u_r,v_r} \in \{0, 1\}$. The solution $k = 0$ to the system $S_{0,0}(0, 0)$ should be removed since it is not admissible. Consequently, as $(x, y) \in \mathcal{X}^n_{1,2}$,

$$\begin{aligned}\Phi_n(x,y) &= 1+\sum_{k=1}^{n}\mathbf{1}_{\{\xi_k^1<x\}}\mathbf{1}_{\{\xi_k^2<y\}}\\ &= 1-1+\sum_{r=0}^{r_1}\sum_{s=0}^{r_2}\sum_{u_r=0}^{x_{r+1}-1}\sum_{v_s=0}^{y_{s+1}-1}\left\lfloor\frac{n}{p_1^{r+1}p_2^{s+1}}\right\rfloor+\eta_{r,s,u_r,v_s}\\ &= \sum_{r=0}^{r_1}\sum_{s=0}^{r_2}x_{r+1}y_{s+1}\left(\left\lfloor\frac{n}{p_1^{r+1}p_2^{s+1}}\right\rfloor+\eta_{r,s}\right),\end{aligned}$$

where $\eta_{r,s}\in[0,1]$. Owing to (12.8), (12.9) and the obvious fact that $nxy = \sum_{0\le r\le r_1}\sum_{0\le s\le r_2}\frac{n\,x_{r+1}y_{s+1}}{p^{r+1}q^{s+1}}$, we derive that for every $(x,y)\in\mathcal{X}_{1,2}^n$,

$$\begin{aligned}\left|\widetilde{\Phi}_n(x,y)\right| &\le \sum_{r=0}^{r-1}\sum_{s=0}^{r_2}\sum_{u_r=0}^{x_{r+1}-1}\sum_{v_s=0}^{y_{s+1}-1}x_{r+1}y_{s+1}\left|\underbrace{\left\lfloor\frac{n}{p_1^{r+1}p_2^{s+1}}\right\rfloor-\frac{n}{p_1^{r+1}p_2^{s+1}}}_{\in[-1,0]}+\underbrace{\eta_{r,s}}_{\in[0,1]}\right|\\ &\le (r_1+1)(r_2+1)(p_1-1)(p_2-1).\end{aligned}$$

We conclude by noting that, on the one hand, $r_i+1=\left\lfloor\frac{\log p_i n}{\log p_i}\right\rfloor$, $i=1,2$ and, on the other hand,

$$n\,D_n^*(\xi)\le\max\left(\sup_{(x,y)\in[0,1]^2}\left|\widetilde{\Phi}_n(x,y)\right|,\,n\,D_n^*(\xi^1),\,n\,D_n^*(\xi^2)\right).$$

Finally, following the above lines – in a simpler way – one shows that

$$n\,D_n^*(\xi^i)\le(r_i+1)(p_i-1),\, i=1,2.$$

This completes the proof since $(r_i+1)(p_i-1)\ge 1$, $i=1,2$, and $\max(a,b)\le ab$ for every $a,b\in\mathbb{N}^*$. ◇

12.11 A Pitman–Yor Identity as a Benchmark

We aim at computing $\mathbb{E}\cos(X_t^2)$, where X^2 denotes the second component of a Clark–Cameron oscillator defined by (9.104), namely

$$X_t^2=\sigma\int_0^t(W_s^1+\mu s)dW_s^2,\quad t\in[0,T].$$

Conditional on the process $(W_s^1)_{0\le s\le t}$, X_t^2 has a centered Gaussian distribution with stochastic variance

$$\sigma^2 \int_0^t (\mu s + W_s^1)^2 ds.$$

Let us compute $\mathbb{E}\, e^{\tilde{\imath} X_t^2}$. As W^1 and W^2 are independent,

$$\begin{aligned}\mathbb{E}\, e^{\tilde{\imath} X_t^2} &= \mathbb{E}\, e^{\tilde{\imath}\sigma \int_0^t (\mu s + W_s^1) dW_s^2} = \mathbb{E}\left[\mathbb{E}\left[e^{\tilde{\imath}\sigma \int_0^t (\mu s + w_s) dW_s^2} \,\middle|\, W^1 = w\right]\right] \\ &= \mathbb{E}\, e^{-\frac{\sigma^2}{2}\int_0^t (\mu s + W_s^1)^2 ds}.\end{aligned}$$

We apply Girsanov's Theorem and we make a change of probability $\mathbb{P}^* = e^{-\mu W_t^1 - \frac{\mu^2}{2} t}.\,\mathbb{P}$ such that $B_s = \mu s + W_s^1$, $s \in [0, t]$, is a $\mathbb{P}^*$-standard Brownian motion. Hence, $\dfrac{d\mathbb{P}}{d\mathbb{P}^*} = e^{\mu B_t - \frac{\mu^2}{2} t}$, which yields

$$\begin{aligned}\mathbb{E}\, e^{\tilde{\imath} X_t^2} &= \mathbb{E}^*\left[e^{\mu B_t - \frac{1}{2}\mu^2 t} e^{-\frac{\sigma^2}{2}\int_0^t (B_s)^2 ds}\right] && (12.10)\\ &= \mathbb{E}^*\left[e^{\mu B_t - \frac{1}{2}\mu^2 t}\mathbb{E}^*\left[e^{-\frac{\sigma^2}{2}\int_0^t (B_s)^2 ds} \,\middle|\, B_t\right]\right] \\ &= e^{-\frac{1}{2}\mu^2 t}\mathbb{E}^*\left[e^{\mu B_t}\mathbb{E}^*\left[e^{-\frac{\sigma^2}{2}\int_0^t (B_s)^2 ds} \,\middle|\, (B_t)^2\right]\right] && (12.11)\end{aligned}$$

where we used in the third line that, as $B \stackrel{d}{=} -B$ and $\int_0^t (B_s)^2 ds = \int_0^t (-B_s)^2 ds$,

$$\mathbb{E}^*\left[e^{-\frac{\sigma^2}{2}\int_0^t (B_s)^2 ds} \,\middle|\, B_t\right] = \mathbb{E}^*\left[e^{-\frac{\sigma^2}{2}\int_0^t (B_s)^2 ds} \,\middle|\, (B_t)^2\right]. \qquad (12.12)$$

We recall now a formula established in [244] (p. 427):

$$\mathbb{E}_x\left[e^{-\frac{\sigma^2}{2}\int_0^t B_s^2 ds} \,\middle|\, (B_t)^2 = y\right] = \left(\frac{\sigma t}{\sinh \sigma t} e^{\frac{y}{2t}(1 - \sigma t \coth \sigma t)}\right) \frac{I_{-\frac{1}{2}}\left(\frac{\sigma z}{\sinh \sigma t}\right)}{I_{-\frac{1}{2}}\left(\frac{z}{t}\right)}$$

with $z = \sqrt{xy}$ and where I_ν is the Bessel function which, for $\nu = -1/2$, reads $I_{-\frac{1}{2}}(z) = \sqrt{\frac{2}{\pi}} \frac{\cosh z}{\sqrt{z}}$. Here $x = 0$, hence

$$\mathbb{E}^*\left[e^{-\frac{\sigma^2}{2}\int_0^t (B_s)^2 ds} \,\middle|\, (B_t)^2 = y\right] = \sqrt{\frac{\sigma t}{\sinh \sigma t}} e^{\frac{y}{2t}(1 - \sigma t \coth \sigma t)}.$$

Plugging this into Equation (12.10) yields

$$\mathbb{E}\, e^{\tilde{\imath} X_t^2} = \sqrt{\frac{\sigma t}{\sinh \sigma t}} e^{-\frac{1}{2}\mu^2 t}\mathbb{E}^*\left[e^{\mu B_t + \frac{B_t^2}{2t}(1 - \sigma t \coth \sigma t)}\right].$$

As $B_t \stackrel{d}{=} \mathcal{N}(0; t)$,

$$\mathbb{E}^*\left[e^{\mu B_t+\frac{B_t^2}{2t}(1-\sigma t\coth\sigma t)}\right]=\int_{-\infty}^{+\infty}e^{\mu x+\frac{x^2}{2t}(1-\sigma t\coth\sigma t)}e^{-\frac{x^2}{2t}}\frac{dx}{\sqrt{2\pi}}$$
$$=\frac{1}{\sqrt{t}}\int_{-\infty}^{+\infty}\exp\Big(-\frac{1}{2}\big(x^2\sigma\coth\sigma t-2\mu x\big)\Big)\frac{dx}{\sqrt{2\pi}}.$$

We set $a=\sqrt{\sigma\coth\sigma t}$ and $b=\mu/a$ and we get

$$\mathbb{E}^*\left[e^{\mu B_t+\frac{B_t^2}{2t}(1-\sigma t\coth(\sigma t))}\right]=\frac{1}{\sqrt{t}}e^{\frac{b^2}{2}}\int_{-\infty}^{+\infty}\exp\Big(-\frac{(ax-b)^2}{2}\Big)\frac{dx}{\sqrt{2\pi}}=\frac{e^{\frac{b^2}{2}}}{a\sqrt{t}}.$$

Hencc,

$$\mathbb{E}\,e^{\tilde{\imath}X_t^2}=\sqrt{\frac{\sigma t}{\sinh\sigma t}}\frac{1}{\sqrt{t}}\frac{1}{\sqrt{\sigma\coth\sigma t}}e^{-\frac{1}{2}\mu^2t+\frac{b^2}{2}}=\frac{1}{\sqrt{\cosh\sigma t}}e^{-\frac{\mu^2t}{2}(1-\frac{\tanh\sigma t}{\sigma t})}.$$

Since the right term of the above equality has no imaginary part, we obtained the announced formula (9.105)

$$\mathbb{E}\,\cos(\sigma X_t^2)=\frac{e^{-\frac{\mu^2t}{2}(1-\frac{\tanh\sigma t}{\sigma t})}}{\sqrt{\cosh\sigma t}}.$$

Remark. Note that these computations are shortened when there is no drift term, *i.e.*

$$X_t^1=W_t^1,\quad X_t^2=\sigma\int_0^t X_s^1dW_s^1.$$

Indeed, owing to the Cameron–Martin formula (see [251] p. 445),

$$\mathbb{E}\,e^{-\sigma\int_0^1(W_s^1)^2ds}=\Big(\cosh\sqrt{2\sigma}\Big)^{-\frac{1}{2}}$$

and the scaling property of the Brownian motion yields

$$\mathbb{E}\,e^{\tilde{\imath}X_t^2}=\mathbb{E}\,e^{-\frac{\sigma^2}{2}\int_0^t(W_s^2)^2ds}=\mathbb{E}\,e^{-\frac{\sigma^2t^2}{2}\int_0^1(W_s^2)^2ds}=(\cosh\sigma t)^{-\frac{1}{2}}.$$

► **Exercise.** Prove in detail the identity (12.12).

Bibliography

1. M. Abramovicz, I.A. Stegun, *Handbook of Mathematical Functions* (National Bureau of Standards, Washington, 1964), pp. xiv+1046
2. Y. Achdou, O. Pironneau, *Computational Methods for Option Pricing, Collection Frontiers in Applied Mathematics*. Society for Industrial and Applied Mathematics, vol. 30 (SIAM, Philadelphia, 2005), pp. xviii+297. ISBN: 0-89871-573-3
3. A. Alfonsi, High order discretization schemes for the CIR process: application to affine term structure and Heston models. Math. Comput. **79**(269), 209–237 (2010)
4. A. Alfonsi, *Affine Diffusions and Related Processes: Simulation, Theory and Applications*. Bocconi and Springer Series, vol. 6 (Springer, Cham; Bocconi University Press, Milan, 2015), pp. xiv+252
5. A. Alfonsi, B. Jourdain, A. Kohatsu-Higa, Pathwise optimal transport bounds between a one-dimensional diffusion and its Euler scheme. Ann. Appl. Probab. **24**(3), 1049–1080 (2014)
6. A. Al Gerbi, B. Jourdain, E. Clément, Ninomiya-Victoir scheme: strong convergence, antithetic version and application to multilevel estimators. Monte Carlo Methods Appl. **22**(3), 197–228 (2016)
7. L. Andersen, Simple and efficient simulation of the Heston stochastic volatility model. J. Comput. Financ. **11**(3), 1–42 (2015)
8. L. Andersen, M. Broadie, Primal-Dual simulation algorithm for pricing multi-dimensional American options. Manag. Sci. **50**(9), 1222–1234 (2004)
9. P. Andersson, A. Kohatsu-Higa, Exact simulation of stochastic differential equations using parametrix expansions. Bernoulli **23**(3), 2028–2057 (2017)
10. I.A. Antonov, V.M. Saleev, An economic method of computing LP_τ-sequences. Zh. vȳ chisl. Mat. mat. Fiz. **19**, 243–245 (1979); English translation. U.S.S.R. Comput. Maths. Math. Phys. **19**, 252–256 (1979)
11. D.G. Aronson, Bounds for the fundamental solution of a parabolic equation. Bull. Am. Math. Soc. **73**, 890–903 (1967)
12. B. Arouna, Adaptive Monte Carlo method, a variance reduction technique. Monte Carlo Methods Appl. **10**(1), 1–24 (2004)
13. M.D. Ašić, D.D. Adamović, Limit points of sequences in metric spaces. Am. Math. Monthly **77**(6), 613–616 (1970)
14. S. Asmussen, J. Rosinski, Approximations of small jumps of Lévy processes with a view towards simulation. J. Appl. Probab. **38**, 482–493 (2001)
15. P. Baldi, Exact asymptotics for the probability of exit from a domain and applications to simulation. Ann. Probab. **23**(4), 1644–1670 (1995)

G. Pagès, *Numerical Probability*, Universitext,
https://doi.org/10.1007/978-3-319-90276-0

16. V. Bally, *An elementary introduction to Malliavin calculus*, technical report (Inria) (2003), https://hal.inria.fr/inria-00071868/document
17. V. Bally, The central limit theorem for a non-linear algorithm based on quantization. Proc. R. Soc. **460**, 221–241 (2004)
18. V. Bally, A. Kohatsu-Higa, A probabilistic interpretation of the parametrix method. Ann. Appl. Probab. **25**(6), 3095–3138 (2015)
19. V. Bally, G. Pagès, J. Printems, A stochastic quantization method for non-linear problems. Monte Carlo Methods Appl. **7**(1), 21–34 (2001)
20. V. Bally, G. Pagès, A quantization algorithm for solving discrete time multi-dimensional optimal stopping problems. Bernoulli **9**(6), 1003–1049 (2003)
21. V. Bally, G. Pagès, Error analysis of the quantization algorithm for obstacle problems. Stoch. Process. Appl. **106**(1), 1–40 (2003)
22. V. Bally, G. Pagès, J. Printems, First order schemes in the numerical quantization method. Math. Financ. **13**(1), 1–16 (2003)
23. V. Bally, G. Pagès, J. Printems, A quantization tree method for pricing and hedging multi-dimensional American options. Math. Financ. **15**(1), 119–168 (2005)
24. V. Bally, D. Talay, The distribution of the Euler scheme for stochastic differential equations: I. Convergence rate of the distribution function. Probab. Theory Relat. Fields **104**(1), 43–60 (1996)
25. V. Bally, D. Talay, The law of the Euler scheme for stochastic differential equations. II. Convergence rate of the density. Monte Carlo Methods Appl. **2**(2), 93–128 (1996)
26. C. Barrera-Esteve, F. Bergeret, C. Dossal, E. Gobet, A. Meziou, R. Munos, D. Reboul-Salze, Numerical methods for the pricing of Swing options: a stochastic control approach. Methodol. Comput. Appl. Probab. **8**(4), 517–540 (2006). https://doi.org/10.1007/s11009-006-0427-8
27. O. Bardou, S. Bouthemy, G. Pagès, Pricing swing options using optimal quantization. Appl. Math. Financ. **16**(2), 183–217 (2009)
28. O. Bardou, S. Bouthemy, G. Pagès, When are Swing options bang-bang? Int. J. Theor. Appl. Financ. **13**(6), 867–899 (2010)
29. O. Bardou, N. Frikha, G. Pagès, Computing VaR and CVaR using stochastic approximation and adaptive unconstrained importance sampling. Monte Carlo Appl. J. **15**(3), 173–210 (2009)
30. J. Barraquand, D. Martineau, Numerical valuation of high dimensional multivariate American securities. J. Financ. Quant. Anal. **30**, 383–405 (1995)
31. D. Bauer, A. Reuss, D. Singer, On the calculation of the solvency capital requirement based on nested simulations. Astin Bull. **42**(2), 453–499 (2012)
32. D. Belomestny, T. Nagapetyan, Multilevel path simulation for weak approximation schemes with application to Lévy-driven SDEs. Bernoulli **23**(2), 927–950 (2017)
33. M. Benaïm, Dynamics of stochastic approximation algorithms, in *Séminaire de Probabilités XXXIII*, ed. by J. Azéma, M. Émery, M. Ledoux, M. Yor. LNM, vol. 1709 (1999), pp. 1–68
34. M. Ben Alaya, A. Kebaier, Central limit theorem for the multilevel Monte Carlo Euler method. Ann. Appl. Probab. **25**(1), 211–234 (2015)
35. M. Ben Alaya, A. Kebaier, Multilevel Monte Carlo for Asian options and limit theorems. Monte Carlo Methods Appl. **20**(3), 181–194 (2014)
36. C. Bender, Dual pricing of multi-exercise options under volume constraints. Financ. Stoch. **15**(1), 1–26 (2011)
37. C. Bender, C. Gärtner, N. Schweizer, Iterative improvement of lower and upper bounds for backward SDEs. SIAM J. Sci. Comput. **39**(2), B442–B466 (2017)
38. C. Bender, J. Schoenmakers, J. Zhang, Dual representations for general multiple stopping problems. Math. Financ. **25**(2), 339–370 (2015)
39. A. Benveniste, M. Métivier, P. Priouret, *Algorithmes Adaptatifs et Approximations Stochastiques* (Masson, Paris, 1987), pp. 367 (*Adaptive Algorithms and Stochastic Approximations* (Springer, Berlin, English version, 1993))
40. J. Bertoin, *Lévy Processes*. Cambridge tracts in Mathematics, vol. 121 (Cambridge University Press, Cambridge, 1996), pp. 262

41. A. Berkaoui, M. Bossy, A. Diop, Euler scheme for SDEs with non-Lipschitz diffusion coefficient: strong convergence. ESAIM Probab. Stat. **12**, 1–11 (2008)
42. A. Beskos, O. Papaspiliopoulos, G.O. Roberts, Retrospective exact simulation of diffusion sample paths with applications. Bernoull **12**(6), 1077–1098 (2006)
43. A. Beskos, G.O. Roberts, Exact simulation of diffusions. Ann. Appl. Probab. **15**(4), 2422–2444 (2005)
44. P. Billingsley, *Probability and Measure* (First edition, 1979), 3rd edn. Wiley Series in Probability and Mathematical Statistics (A Wiley–Interscience Publication, Wiley, New York, 1995), pp. xiv+593
45. P. Billingsley, *Convergence of Probability Measures*, 1st edn. (Second edition 1999, 277pp.) (Wiley, New York, 1968), pp. 253
46. J.-P. Borel, G. Pagès, Y.J. XIao, Suites à discrépance faible and intégration numérique, in *Probabilités Numériques*, ed. by N. Bouleau, D. Talay (Coll. didactique, INRIA, 1992). ISBN-10: 2726107087
47. B. Bouchard, I. Ekeland, N. Touzi, On the Malliavin approach to Monte Carlo approximation of conditional expectations. Financ. Stoch. **8**(1), 45–71 (2004)
48. N. Bouleau, D. Lépingle, *Numerical Methods for Stochastic Processes*. Wiley Series in Probability and Mathematical Statistics: Applied Probability and Statistics (A Wiley-Interscience Publication; Wiley, New York, 1994), pp. 359
49. C. Bouton, Approximation gaussienne d'algorithmes stochastiques à dynamique markovienne. Annales de l'I.H.P. Probabilités et Statistiques **24**(1), 131–155 (1998)
50. P.P. Boyle, Options: a Monte Carlo approach. J. Financ. Econ. **4**(3), 323–338 (1977)
51. O. Brandière, M. Duflo, Les algorithmes stochastiques contournent-ils les pièges? (French) [Do stochastic algorithms go around traps?]. Ann. Inst. H. Poincaré Probab. Stat. **32**(3), 395–427 (1996)
52. M. Briane, G. Pagès, *Théorie de l'intégration, convolution and transformée de Fourier, in Cours and Exercices*, 6th edn. (Vuibert, Paris, 2015), pp. 400
53. M. Broadie, Y. Du, C.C. Moallemei, Efficient risk estimation via nested sequential simulation. Manag. Sci. **57**(6), 1172–1194 (2011)
54. R. Buche, H.J. Kushner, Rate of convergence for constrained stochastic approximation algorithm. SIAM J. Control Optim. **40**(4), 1001–1041 (2002)
55. K. Bujok, B. Hambly, C. Reisinger, Multilevel simulation of functionals of Bernoulli random variables with application to basket credit derivatives. Methodol. Comput. Appl. Probab. **17**(3), 579–604 (2015)
56. S. Burgos, M. Giles, *Computing Greeks using multilevel path simulation, in Monte Carlo and Quasi-Monte Carlo Methods 2010* (Springer, Berlin, 2012), pp. 281–296
57. G. Callegaro, L. Fiorin, M. Grasselli, Quantized calibration in local volatility, in *Risk (Cutting Hedge: Derivatives Pricing)*, (2015), pp. 56–67, https://www.risk.net/risk-magazine/technical-paper/2402156/quantized-calibration-in-local-volatility
58. G. Callegaro, L. Fiorin, M. Grasselli, Pricing via quantization in stochastic volatility models. Quant. Financ. **17**(6), 855–872 (2017)
59. L. Carassus, G. Pagès, *Finance de marché, Modèles mathématiques à temps discret* (Vuibert, Paris, 2016), pp. xii+ 385. ISBN 978-2-311-40136-3
60. J.F. Carrière, Valuation of the early-exercise price for options using simulations and nonparametric regression. Insur. Math. Econ. **19**, 19–30 (1996)
61. N. Chen, Y. Liu, Estimating expectations of functionals of conditional expected via multilevel nested simulation, Presentation at *MCQMC*'12 Conference, Sydney (2010)
62. D.-Y. Cheng, A. Gersho, B. Ramamilrthi, Y. Shoham, Fast search algorithms for vector quantization and pattern matching. Proc. IEEE Int. Conf. Acoust. Speech Signal Process **1**, 9.11.1–9.11.4 (1984)
63. K.L. Chung, An estimate concerning the Kolmogorov limit distribution. Trans. Am. Math. Soc. **67**, 36–50 (1949)
64. É. Clément, A. Kohatsu-Higa, D. Lamberton, A duality approach for the weak approximation of stochastic differential equations. Ann. Appl. Probab. **16**(3), 1124–1154 (2006)

65. É. Clément, D. Lamberton, P. Protter, An analysis of a least squares regression method for American option pricing. Financ. Stoch. **6**(2), 449–471 (2002)
66. S. Cohen, M.M. Meerschaert, J. Rosinski, Modeling and simulation with operator scaling. Stoch. Process. Appl. **120**(12), 2390–2411 (2010)
67. P. Cohort, Sur quelques problèmes de quantification, thèse de l'Université Paris 6, Paris (2000), pp. 187
68. P. Cohort, Limit theorems for random normalized distortion. Ann. Appl. Probab. **14**(1), 118–143 (2004)
69. L. Comtet, *Advanced Combinatorics. The Art of Finite and Infinite Expansions*. Revised and enlarged edition. D. Reidel Publishing Co., Dordrecht, 1974, pp. xi+343. ISBN: 90-277-0441-4 05-02
70. S. Corlay, G. Pagès, Functional quantization-based stratified sampling methods. Monte Carlo Methods Appl. J. **21**(1), 1–32 (2015). https://doi.org/10.1515/mcma-2014-0010
71. R. Cranley, T.N.L. Patterson, Randomization of number theoretic methods for multiple integration. SIAM J. Numer. Anal. **13**(6), 904–914 (1976)
72. D. Dacunha-Castelle, M. Duflo, *Probabilités et Statistique II : Problèmes à temps mobile* (Masson, Paris, 1986). English version: Probability and Statistics II (Translated by D. M, McHale) (Springer, New York, 1986), pp. 290
73. R.B. Davis, D.S. Harte, Tests for Hurst effect. Biometrika **74**, 95–101 (1987)
74. S. Dereich, F. Heidenreich, A multilevel Monte Carlo algorithm for Lévy-driven stochastic differential equations. Stoch. Process. Appl. **121**, 1565–1587 (2011)
75. S. Dereich, S. Li, Multilevel Monte Carlo for Lévy-driven SDEs: central limit theorems for adaptive Euler schemes. Ann. Appl. Probab. **26**(1), 136–185 (2016)
76. L. Devineau, S. Loisel, Construction d'un algorithme d'accélération de la méthode des simulations dans les simulation pour le calcul du capital économique Solvabilité II. Bull. Français d'Actuariat, Institut des Actuaires **10**(17), 188–221 (2009)
77. L. Devroye, *Non Uniform Random Variate Generation* (Springer, New York, 1986), 843pp. First edition available at http://www.eirene.de/Devroye.pdf or http://www.nrbook.com/devroye/
78. J. Dick, Higher order scrambled digital nets achieve the optimal rate of the root mean square error for smooth integrands. Ann. Stat. **39**(3), 1372–1398 (2011)
79. Q. Du, V. Faber, M. Gunzburger, Centroidal Voronoi tessellations: applications and algorithms. SIAM Rev. **41**, 637–676 (1999)
80. Q. Du, M. Emelianenko, L. Ju, Convergence of the Lloyd algorithm for computing centroidal Voronoi tessellations. SIAM J. Numer. Anal. **44**, 102–119 (2006)
81. M. Duflo, *Algorithmes stochastiques, coll, in Mathématiques and Applications*, vol. 23 (Springer, Berlin, 1996), pp. 319
82. M. Duflo, *Random Iterative Models*, translated from the 1990 French original by Stephen S. Wilson and revised by the author. Applications of Mathematics (New York), vol. 34 (Springer, Berlin, 1996), pp. 385
83. D. Egloff, M. Leippold, Quantile estimation with adaptive importance sampling. Ann. Stat. **38**(2), 1244–1278 (2010)
84. N. El Karoui, J.P. Lepeltier, A. Millet, A probabilistic approach of the réduite in optimal stopping. Probab. Math. Stat. **13**(1), 97–121 (1988)
85. M. Emelianenko, L. Ju, A. Rand, Nondegeneracy and weak global convergence of the Lloyd algorithm in $\mathbb{R}^d$. SIAM J. Numer. Anal. **46**(3), 1423–1441 (2008)
86. P. Étoré, B. Jourdain, Adaptive optimal allocation in stratified sampling methods. Methodol. Comput. Appl. Probab. **12**(3), 335–360 (2010)
87. M. Fathi, N. Frikha, Transport-entropy inequalities and deviation estimates for stochastic approximations schemes. Electron. J. Probab. **18**(67), 36 (2013)
88. H. Faure, Suite à discrépance faible dans T^s, Technical report, University Limoges (France, 1986)
89. H. Faure, Discrépance associée à un système de numération (en dimension s). Acta Arith. **41**, 337–361 (1982)

90. O. Faure, *Simulation du mouvement brownien des diffusions, Thèse de l'ENPC* (France, Marne-la-Vallée, 1992), pp. 133
91. L. Fiorin, G. Pagès, A. Sagna, Markovian and product quantization of an $\mathbb{R}^d$-valued Euler scheme of a diffusion process with applications to finance, to appear in Methodology and Computing in Applied Probability (2017), arXiv:1511.01758v3
92. H. Föllmer, A. Schied, in *Stochastic Finance, An Introduction in Discrete Time* (2nd edn. in 2004). De Gruyter Studies in Mathematics, vol. 27 (De Gruyter, Berlin, 2002), pp. 422
93. G. Fort, É. Moulines, A. Schreck, M. Vihola, Convergence of Markovian stochastic approximation with discontinuous dynamics. SIAM J. Control Optim. **54**(2), 866–893 (2016)
94. J.C. Fort, G. Pagès, Convergence of stochastic algorithms: from the Kushner & Clark Theorem to the Lyapunov functional. Adv. Appl. Probab. **28**, 1072–1094 (1996)
95. J.C. Fort, G. Pagès, Decreasing step stochastic algorithms: $a.s.$ behavior of weighted empirical measures. Monte Carlo Methods Appl. **8**(3), 237–270 (2002)
96. J.C. Fort, G. Pagès, Asymptotics of optimal quantizers for some scalar distributions. J. Comput. Appl. Math. **146**(2), 253–275 (2002)
97. N. Fournier, A. Guillin, On the rate of convergence in Wasserstein distance of the empirical measure. Probab. Theory Relat. Fields **162**(3–4), 707–738 (2015)
98. A. Friedman, *Stochastic Differential Equations and Applications Volume 1–2. Probability and Mathematical Statistics*, vol. 28 (Academic Press [Harcourt Brace Jovanovich, Publishers], New York, 1975), pp. 528
99. J.H. Friedman, J.L. Bentley, R.A. Finkel, An algorithm for finding best matches in logarithmic expected time. ACM Trans. Math. Softw. **3**(3), 209–226 (1977)
100. N. Frikha, S. Menozzi, Concentration bounds for stochastic approximations. Electron. Commun. Probab., **183 17**(47), 1–15 (2012)
101. S. Gadat, F. Panloup, Optimal non-asymptotic bound of the Ruppert-Polyak averaging without strong convexity, pre-print, (2017), arXiv:1709.03342v1
102. J.G. Gaines, T. Lyons, Random generation of stochastic area integrals. SIAM J. Appl. Math. **54**(4), 1132–1146 (1994)
103. S.B. Gelfand, S.K. Mitter, Recursive stochastic algorithms for global optimization in $\mathbb{R}^d$. SIAM J. Control Optim. **29**, 999–1018 (1991)
104. S.B. Gelfand, S.K. Mitter, Metropolis-type annealing algorithms for global optimization in $\mathbb{R}^d$. SIAM J. Control Optim. **31**, 111–131 (1993)
105. A. Gersho, R.M. Gray, Special issue on quantization, **I-II** (A. Gersho and R.M. Gray eds.) IEEE Trans. Inf. Theory **28**, 139–148 (1988)
106. A. Gersho, R.M. Gray, *Vector Quantization and Signal Compression* (Kluwer, Boston, 1992), pp. 732
107. M.B. Giles, Multilevel Monte Carlo path simulation. Oper. Res. **56**(3), 607–617 (2008)
108. M.B. Giles, *Vibrato Monte Carlo Sensitivities, Monte Carlo and quasi-Monte Carlo Methods 2008* (Springer, Berlin, 2009), pp. 369–382
109. M.B. Giles, Multilevel Monte Carlo methods. Acta Numer. **24**, 259–328 (2015)
110. M.B. Giles, L. Szpruch, Antithetic multilevel Monte Carlo estimation for multi-dimensional SDEs, in *Monte Carlo and quasi-Monte Carlo methods 2012*. Springer Proceedings in Mathematics and Statistics, vol. 65, (Springer, Heidelberg, 2013), pp. 367–384
111. M.B. Giles, L. Szpruch, *Multilevel Monte Carlo methods for applications in finance, in Recent Developments in Computational Finance, Interdisciplinary Mathematics Sciences*, vol. 14 (World Scientific Publishing, Hackensack, 2013), pp. 3–47
112. D. Giorgi, V. Lemaire, G. Pagès, Limit theorems for weighted and regular Multilevel estimators. Monte Carlo Methods Appl. **23**(1), 43–70 (2017)
113. D. Giorgi, Théorèmes limites pour estimateurs Multilevel avec et sans poids. Comparaisons et applications, PhD thesis (2017), pp. 123 (cf 68. cohort). (UPMC)
114. D. Giorgi, V. Lemaire, G. Pagès, Weak error for nested Multilevel Monte Carlo, (2018), arXiv:1806.07627.
115. P. Glasserman, *Monte Carlo Methods in Financial Engineering* (Springer, New York, 2003), pp. 596

116. P. Glasserman, P. Heidelberger, P. Shahabuddin, Asymptotically optimal importance sampling and stratification for pricing path-dependent options. Math. Financ. **9**(2), 117–152 (1999)
117. P. Glasserman, D.D. Yao, Some guidelines and guarantees for common random numbers. Manag. Sci. **36**(6), 884–908 (1992)
118. P. Glasserman, B. Yu, Number of paths versus number of basis functions in American option pricing. Ann. Appl. Probab. **14**(4), 2090–2119 (2004)
119. P. Glynn, Optimization of stochastic systems via simulation, in *Proceedings of the 1989 Winter Simulation Conference, Society for Computer Simulation, San Diego* (1989), pp. 90–105
120. E. Gobet, Weak approximation of killed diffusion using Euler schemes. Stoch. Proc. Appl. **87**, 167–197 (2000)
121. E. Gobet, Revisiting the Greeks for European and American options, in *Stochastic Processes and Applications to Math. Finance* (Proceedings of the Ritsumeikan International Symposium, Kusatsu, Shiga, Japan 5–9 March 2003), ed. by S. Watanabe, J. Akahori, S. Ogawa (2003), pp. 53–71. ISBN-13: 978-9812387783
122. E. Gobet, *Advanced Monte Carlo methods for barrier and related exotic options, in Mathematical Modeling and Numerical Methods in Finance, Handbook of Numerical Analysis*, vol. XV (North-Holland, Special Volume, Elsevier, Netherlands, 2009), pp. 497–528
123. E. Gobet, S. Menozzi, Exact approximation rate of killed hypo-elliptic diffusions using the discrete Euler scheme. Stoch. Process. Appl. **112**(2), 201–223 (2004)
124. E. Gobet, S. Menozzi, Stopped diffusion processes: boundary corrections and overshoot. Stoch. Process. Appl. **120**(2), 30–162 (2010)
125. E. Gobet, A. Kohatsu-Higa, Computation of Greeks for barrier and look-back options using Malliavin calculus. Electron. Commun. Probab. **8**, 51–62 (2003)
126. E. Gobet, R. Munos, Sensitivity analysis using Itô-Malliavin calculus and martingales. Application to optimal control. SIAM J. Control Optim. **43**(5), 1676–1713 (2005)
127. E. Gobet, G. Pagès, H. Pham, J. Printems, Discretization and simulation for a class of SPDE's with applications to Zakai and McKean-Vlasov equations. SIAM J. Numer. Anal. **44**(6), 2505–2538 (2007)
128. M.B. Gordy, S. Juneja, Nested simulation in portfolio risk measurement. Manag. Sci. **56**(10), 1833–1848 (2010)
129. S. Graf, H. Luschgy, *Foundations of Quantization for Probability Distributions*. LNM, vol. 1730 (Springer, Berlin, 2000), pp. 230
130. S. Graf, H. Luschgy, G. Pagès, Optimal quantizers for Radon random vectors in a Banach space. J. Approx. **144**, 27–53 (2007)
131. S. Graf, H. Luschgy, G. Pagès, Distortion mismatch in the quantization of probability measures. ESAIM P&S **12**, 127–154 (2008)
132. S. Graf, H. Luschgy, G. Pagès, The local quantization behavior of absolutely continuous probabilities. Ann. Probab. **40**(4), 1795–1828 (2012)
133. C. Graham, D. Talay (2013). *Stochastic Simulation and Monte Carlo Methods. Mathematical Foundations of Stochastic Simulation*. Stochastic Modelling and Applied Probability, vol. 68 (Springer, Heidelberg), pp. xvi+260
134. C. Graham, D. Talay, *Analysis and Simulation of Hypoelliptioc, Stopped, Reflected and Ergodic Diffusions, Mathematical Foundations of Stochastic Simulation*. Stochastic Modelling and Applied Probability (Springer, Berlin, 2017) (to appear)
135. A. Griewank, On automatic differentiation, in *Mathematical Programming: Recent Developments and Applications*, ed. by M. Iri, K. Tanabe (Kluwer Academic Publishers, Dordrecht, 1989), pp. 83–108
136. A. Griewank, A. Walther, *Evaluating Derivatives: Principles and Techniques of Algorithmic Differentiation*. Frontiers in Applied Mathematics (SIAM, Philadelphia, 2008), pp. xxi+426. ISBN 978-0-89871-659-7
137. A. Griewank, A. Walther, *ADOL-C: A Package for the Automatic Differentiation of Algorithm Written in C/C++* (University of Paderborn, Germany, 2010)
138. J. Guyon, Euler scheme and tempered distributions. Stoch. Process. Appl. **116**(6), 877–904 (2006)

139. M. Hairer, On Malliavin's proof of Hörmander's theorem, Notes of a mini-course, Warwick University (2010), pp. 19, arXiv:1103.1998v1
140. A.L. Haji-Ali, Pedestrian Flow in the Mean-Field Limit, MSc. Thesis, 2012, KAUST (2012)
141. A.L. Haji-Ali, F. Nobile, E. von Schwerin, R. Tempone, Optimization of mesh hierarchies in multilevel Monte Carlo samplers. Stoch. Partial Differ. Equ. Anal. Comput. **4**(1), 76–112 (2016)
142. P. Hall, C.C. Heyde, *Martingale Limit Theory and Its Applications* (Academic Press, New York, 1980), pp. 308
143. G.H. Hardy, J.E. Littlewood, G. Pùlya, *Inequalities*. Reprint of the 1952 edition. Cambridge Mathematical Library (Cambridge University Press, Cambridge, 1952, Reprinted on 1988), pp. xii+324
144. L. Hascoë, V. Pascual, The Tapenade automatic differentiation tool: principles, model, and specification. ACM Trans. Math. Softw. **39**(3), 20:1–20:43 (2013)
145. R.Z. Has'minskiĭ, *Stochastic Stability of Differential Equations* (1980). Translated from the Russian by D. Louvish. Monographs and Textbooks on Mechanics of Solids and Fluids: Mechanics and Analysis, 7. Sijthoff & Noordhoff, Alphen aan den Rijn-Germantown, Md., 1980, pp. xvi+344
146. M.B. Haugh, L. Kogan, Pricing American options: a duality approach. Oper. Res. **52**(2), 258–270 (2004)
147. U.G. Haussmann, On the integral representation of functionals of Itô Processes. Stochastics **3**(1–4), 17–27 (1980)
148. S. Heinrich, *Multilevel Monte Carlo methods, in In Large-Scale Scientific Computing* (Springer, Berlin, 2001), pp. 58–67
149. P. Henry-Labordère, X. Tan, N. Touzi, Exact simulation of multi-dimensional stochastic differential equations (2015), arXiv:1504.06107
150. S.L. Heston, A closed-form solution for options with stochastic volatility with applications to bond and currency options. Rev. Financ. Stud. **6**(2), 327–343 (1993)
151. W. Hoeffding, Probability inequalities for sums of bounded random variables. J. Am. Stat. Assoc. **58**(301), 13–30 (1963)
152. M. Hutzenthaler, A. Jentzen, Numerical approximations of stochastic differential equations with non-globally Lipschitz continuous coefficients. Mem. Am. Math. Soc. **236**, 1112 (2015), pp. v+99
153. M. Hutzenthaler, A. Jentzen, P.E. Kloeden, Divergence of the multilevel Monte Carlo Euler method for non-linear stochastic differential equations. Ann. Appl. Probab. **23**(5), 1913–1966 (2013)
154. J. Jacod, A.N. Shiryaev, *Limit Theorems for Stochastic Processes*, 2nd edn. (2003) (first edition 1987). Grundlehren der Mathematischen Wissenschaften [Fundamental Principles of Mathematical Sciences], vol. 288 (Springer, Berlin, 2003), pp. xx+661
155. J. Jacod, P. Protter, Asymptotic error distributions for the Euler method for stochastic differential equations. Ann. Probab. **26**(1), 267–307 (1998)
156. J. Jacod, P. Protter, *Probability Essentials*, 2nd edn. (Universitext) (Springer, Berlin, 2003), pp. x+254
157. J. Jacod, T. Kurtz, S. Méléard, P. Protter, The approximate Euler method for Lévy driven stochastic differential equations. Ann. Inst. H. Poincaré Probab. Stat. **41**(3), 523–558 (2005)
158. P. Jaillet, D. Lamberton, B. Lapeyre, Variational inequalities and the pricing of American options. Acta Appl. Math. **21**, 263–289 (1990)
159. H. Johnson, Call on the maximum or minimum of several assets. J. Financ. Quant. Anal. **22**(3), 277–283 (1987)
160. B. Jourdain, A. Kohatsu-Higa, A review of recent results on approximation of solutions of stochastic differential equations. Prog. Probab. **65**, 141–164 (2011) (Springer Basel AG)
161. B. Jourdain, J. Lelong, Robust adaptive importance sampling for normal random vectors. Ann. Appl. Probab. **19**(5), 1687–1718 (2009)
162. I. Karatzas, S.E. Shreve, *Brownian Motion and Stochastic Calculus*. Graduate Texts in Mathematics (Springer, New York, 1998), pp. xxix+470

163. I. Karatzas, S.E. Shreve, *Methods of Math. Finance*. Applications of Mathematics, vol. 39 (Springer, New York, 1998), xvi+407pp
164. O. Kavian, Introduction à la théorie des points critiques and applications aux problèmes elliptiques (1993) (French) [Introduction to critical point theory and applications to elliptic problems]. coll. *Mathématiques and Applications* [Mathematics and Applications], vol. 13 (Springer, Paris [Berlin]), pp. 325. ISBN: 2-287-00410-6
165. H.L. Keng, W. Yuan, *Application of Number Theory to Numerical Analysis* (Springer and Science Press, Beijing, 1981), pp. 241
166. A.G. Kemna, A.C. Vorst, A pricing method for options based on average asset value. J. Bank. Financ. **14**, 113–129 (1990)
167. A. Kebaier, Statistical Romberg extrapolation: a new variance reduction method and applications to option pricing. Ann. Appl. Probab. **15**(4), 2681–2705 (2005)
168. J.C. Kieffer, Exponential rate of convergence for Lloyd's method I. IEEE Trans. Inf. Theory (Special issue on quantization) **28**(2), 205–210 (1982)
169. J. Kiefer, J. Wolfowitz, Stochastic estimation of the maximum of a regression function. Ann. Math. Stat. **23**, 462–466 (1952)
170. P.E. Kloeden, E. Platten, *Numerical Solution of Stochastic Differential Equations. Applications of Mathematics*, vol. 23 (Springer, Berlin (New York), 1992), pp. 632
171. A. Kohatsu-Higa, R. Pettersson, Variance reduction methods for simulation of densities on Wiener space. SIAM J. Numer. Anal. **40**(2), 431–450 (2002)
172. V. Konakov, E. Mammen, Edgeworth type expansions for Euler schemes for stochastic differential equations. Monte Carlo Methods Appl. **8**(3), 271–285 (2002)
173. V. Konakov, S. Menozzi, S. Molchanov, Explicit parametrix and local limit theorems for some degenerate diffusion processes. Ann. Inst. H. Poincaré Probab. Stat. (série B) **46**(4), 908–923 (2010)
174. U. Krengel, *Ergodic Theorems*. De Gruyter Studies in Mathematics, vol. 6 (Springer, Berlin, 1987), pp. 357
175. L. Kuipers, H. Niederreiter, *Uniform Distribution of Sequences* (Wiley, New York, 1974), pp. 390
176. H. Kunita, *Stochastic differential equations and stochastic flows of diffeomorphisms, in Cours d'école d'été de Saint-Flour XII–1982, LN-1097* (Springer, Berlin, 1984), pp. 143–303
177. H. Kunita, *Stochastic flows and stochastic differential equations*. Reprint of the 1990 original. Cambridge Studies in Advanced Mathematics, vol. 24 (Cambridge University Press, Cambridge, 1997), pp. xiv+346
178. T.G. Kurtz, P. Protter, Weak limit theorems for stochastic integrals and stochastic differential equations. Ann. Probab. **19**, 1035–1070 (1991)
179. H.J. Kushner, H. Huang, Rates of convergence for stochastic approximation type algorithms. SIAM J. Control **17**(5), 607–617 (1979)
180. H.J. Kushner, G.G. Yin, *Stochastic Approximation and Recursive Algorithms and Applications*, 2nd edn. Applications of Mathematics, Stochastic Modeling and Applied Probability, vol. 35 (Springer, New York, 2003)
181. J.P. Lambert, Quasi-Monte Carlo, low-discrepancy, and ergodic transformations. J. Comput. Appl. Math. **12–13**, 419–423 (1985)
182. D. Lamberton, Optimal stopping and American options, a course at the Ljubljana Summer School on Financial Mathematics (2009), https://www.fmf.uni-lj.si/finmath09/ShortCourseAmericanOptions.pdf
183. D. Lamberton, B. Lapeyre, *Introduction to Stochastic Calculus Applied to Finance* (Chapman & Hall, London, 1996), pp. 185
184. D. Lamberton, G. Pagès, P. Tarrès, When can the two-armed bandit algorithm be trusted? Ann. Appl. Probab. **14**(3), 1424–1454 (2004)
185. B. Lapeyre, G. Pagès, Familles de suites à discrépance faible obtenues par itération d'une transformation de [0, 1]. Comptes rendus de l'Académie des Sciences de Paris, Série **I**(308), 507–509 (1989)

186. B. Lapeyre, G. Pagès, K. Sab, Statistics. Sequences with low discrepancy. Generalization and application to Robbins-Monro algorithm **21**(2), 251–272 (1990)
187. B. Lapeyre, E. Temam, Competitive Monte Carlo methods for the pricing of Asian options. J. Comput. Financ. **5**(1), 39–59 (2001)
188. S. Laruelle, C.-A. Lehalle, G. Pagès, Optimal split of orders across liquidity pools: a stochastic algorithm approach. SIAM J. Financ. Math. **2**, 1042–1076 (2011)
189. S. Laruelle, C.-A. Lehalle, G. Pagès, Optimal posting price of limit orders: learning by trading. Math. Financ. Econ. **7**(3), 359–403 (2013)
190. S. Laruelle, G. Pagès, Stochastic Approximation with averaging innovation. Monte Carlo Methods Appl. J. **18**, 1–51 (2012)
191. V.A. Lazarev, Convergence of stochastic approximation procedures in the case of a regression equation with several roots (transl. from). Problemy Pederachi Informatsii **28**(1), 66–78 (1992)
192. A. Lejay, V. Reutenauer, A variance reduction technique using a quantized Brownian motion as a control variate. J. Comput. Financ. **16**(2), 61–84 (2012)
193. M. Ledoux, M. Talagrand, *Probability in Banach Spaces. Isoperimetry and Processes*. Reprint of the 1991 edition. Classics in Mathematics (Springer, Berlin, 2011), pp. xii+480
194. P. L'Ecuyer, G. Perron, On the convergence rates of IPA and FDC derivatives estimators. Oper. Res. **42**, 643–656 (1994)
195. M. Ledoux, Concentration of measure and logarithmic Sobolev inequalities, technical report (1997), http://www.math.univ-toulouse.fr/~ledoux/Berlin.pdf
196. J. Lelong, Almost sure convergence of randomly truncated stochastic algorithms under verifiable conditions. Stat. Probab. Lett. **78**(16), 2632–2636 (2008)
197. J. Lelong, Asymptotic normality of randomly truncated stochastic algorithms. ESAIM. Probab. Stat. **17**, 105–119 (2013)
198. V. Lemaire, G. Pagès, Multilevel Richardson-Romberg extrapolation. Bernoulli **23**(4A), 2643–2692 (2017)
199. V. Lemaire, G. Pagès, Unconstrained recursive importance sampling. Ann. Appl. Probab. **20**(3), 1029–1067 (2010)
200. L. Ljung, Analysis of recursive stochastic algorithms. IEEE Trans. Autom. Control **22**(4), 551–575 (1977)
201. S.P. Lloyd, Least squares quantization in PCM. IEEE Trans. Inf. Theory **28**, 129–137 (1982)
202. F.A. Longstaff, E.S. Schwarz, Valuing American options by simulation: a simple least-squares approach. Rev. Financ. Stud. **14**, 113–148 (2001)
203. H. Luschgy, *Martingale in diskreter Zeit : Theory und Anwendungen* (Springer, Berlin, 2013), pp. 452
204. H. Luschgy, G. Pagès, Functional quantization of Gaussian processes. J. Funct. Anal. **196**(2), 486–531 (2001)
205. H. Luschgy, G. Pagès, Functional quantization rate and mean regularity of processes with an application to Lévy processes. Ann. Appl. Probab. **18**(2), 427–469 (2008)
206. D. McLeish, A general method for debiasing a Monte Carlo estimator. Monte Carlo Methods Appl. **17**(4), 301–315 (2011)
207. J. Mac Names, A fast nearest-neighbor algorithm based on principal axis search tree. IEEE T. Pattern. Anal. **23**(9), 964–976 (2001)
208. P. Malliavin, A. Thalmaier *Stochastic Calculus of Variations in Mathematical Finance*. Springer Finance. (Springer, Berlin, 2006), pp. xii+142
209. S. Manaster, G. Koehler, The calculation of implied variance from the Black-Scholes model: a note. J. Financ. **37**(1), 227–230 (1982)
210. G. Marsaglia, T.A. Bray, A convenient method for generating normal variables. SIAM Rev. **6**, 260–264 (1964)
211. M. Matsumata, T. Nishimura, Mersenne twister: a 623-dimensionally equidistributed uniform pseudorandom number generator. ACM Trans. Model. Comput. Simul. **8**(1), 3–30 (1998)
212. G. Marsaglia, W.W. Tsang, The ziggurat method for generating random variables. J. Stat. Softw. **5**(8), 363–372 (2000)

213. M. Métivier, P. Priouret, Théorèmes de convergence presque sûre pour une classe d'algorithmes stochastiques à pas décroissant. (French. English summary) [Almost sure convergence theorems for a class of decreasing-step stochastic algorithms]. Probab. Theory Relat. Fields **74**(3), 403–428 (1987)
214. G.N. Milstein, A method of second order accuracy for stochastic differential equations. Theor. Probab. Appl. (USSR) **23**, 396–401 (1976)
215. U. Naumann, *The Art of Differentiating Computer Programs: An Introduction to Algorithmic Differentiation. Software, Environments and Tools* (SIAM, RWTH Aachen University, Aachen, Germany, 2012), pp. xviii+333
216. J. Neveu, *Bases mathématiques du calcul des Probabilités, Masson, Paris, 213 pp; English translation: Mathematical Foundations of the Calculus of Probability (1965)* (Holden Day, San Francisco, 1964)
217. J. Neveu, *Martingales à temps discret* (Masson, 1972, 218pp. English translation: *Discrete-Parameter Martingales* (North-Holland, New York, 1975), 236pp
218. D.J. Newman, The Hexagon Theorem. IEEE Trans. Inf. Theory (A. Gersho and R.M. Gray eds.) **28**, 137–138 (1982)
219. H. Niederreiter, *Random Number Generation and Quasi-Monte Carlo Methods CBMS-NSF regional conference series in Applied mathematics* (SIAM, Philadelphia, 1992)
220. Numerical Recipes: Textbook available at http://www.ma.utexas.edu/documentation/nr/bookcpdf.html. See also [247]
221. A. Owen, Local antithetic sampling with scrambled nets. Ann. Stat. **36**(5), 2319–2343 (2008)
222. G. Pagès, *Sur quelques problèmes de convergence*, thèse University Paris 6, Paris (1987) pp. 144
223. G. Pagès, Van der Corput sequences, Kakutani transform and one-dimensional numerical integration. J. Comput. Appl. Math. **44**, 21–39 (1992)
224. G. Pagès, A space vector quantization method for numerical integration. J. Comput. Appl. Math. **89**, 1–38 (1998) (Extended version of "Voronoi Tessellation, space quantization algorithms and numerical integration, ed. by M. Verleysen, Proceedings of the ESANN' 93, Bruxelles, Quorum Editions, (1993), pp. 221–228
225. G. Pagès, Multistep Richardson-Romberg extrapolation: controlling variance and complexity. Monte Carlo Methods Appl. **13**(1), 37–70 (2007)
226. G. Pagès, Quadratic optimal functional quantization methods and numerical applications, in *Proceedings of MCQMC, Ulm'06* (Springer, Berlin, 2007), pp. 101–142
227. G. Pagès, *Functional co-monotony of processes with an application to peacocks and barrier options, in Séminaire de Probabilités XXVI, LNM*, vol. 2078 (Springer, Cham, 2013), pp. 365–400
228. G. Pagès, Introduction to optimal quantization for numerics. ESAIM Proc. Surv. **48**, 29–79 (2015)
229. G. Pagès, H. Pham, J. Printems, Optimal quantization methods and applications to numerical problems in finance, in *Handbook on Numerical Methods in Finance*, ed. by S. Rachev (Birkhauser, Boston, 2004), pp. 253–298
230. G. Pagès, H. Pham, J. Printems, An optimal Markovian Quantization algorithm for multi-dimensional stochastic control problems. Stoch. Dyn. **4**(4), 501–545 (2004)
231. G. Pagès, J. Printems, Optimal quadratic quantization for numerics: the Gaussian case. Monte Carlo Methods Appl. **9**(2), 135–165 (2003)
232. G. Pagès, J. Printems (2005), http://www.quantize.maths-fi.com, website devoted to optimal quantization
233. G. Pagès, J. Printems, Optimal quantization for finance: from random vectors to stochastic processes. *Mathematical Modeling and Numerical Methods in Finance* (special volume) ed. by A. Bensoussan, Q. Zhang guest éds.), coll. *Handbook of Numerical Analysis*, ed. by P.G. Ciarlet (North Holland, 2009), pp. 595–649
234. G. Pagès, A. Sagna, Recursive marginal quantization of the Euler scheme of a diffusion process. Appl. Math. Financ. **22**(5), 463–98 (2015)

235. G. Pagès, B. Wilbertz, Optimal Delaunay and Voronoi quantization methods for pricing American options, in *Numerical Methods in Finance*, ed. by R. Carmona, P. Hu, P. Del Moral, N. Oudjane (Springer, New York, 2012), pp. 171–217
236. G. Pagès, B. Wilbertz, *Sharp rate for the dual quanization problem, forthcoming, in Séminaire de Probabilités XLIX* (Springer, Berlin, 2018)
237. G. Pagès, Y.-J. Xiao, Sequences with low discrepancy and pseudo-random numbers: theoretical results and numerical tests. J. Stat. Comput. Simul. **56**, 163–188 (1988)
238. G. Pagès, J. Yu, Pointwise convergence of the Lloyd I algorithm in higher dimension. SIAM J. Control Optim. **54**(5), 2354–2382 (2016)
239. K.R. Parthasarathy, Probability measures on metric spaces, in *Probability and Mathematical Statistics*, vol. 3 (Academic Press, Inc., New York-London, 1967), pp. xi+276
240. L. Paulot, Unbiased Monte Carlo Simulation of Diffusion Processes (2016), arXiv:1605.01998
241. M. Pelletier, Weak convergence rates for stochastic approximation with application to multiple targets and simulated annealing. Ann. Appl. Probab. **8**(1), 10–44 (1998)
242. R. Pemantle, Non-convergence to unstable points in urn models and stochastic approximations. Ann. Probab. **18**(2), 698–712 (1990)
243. H. Pham, Some applications and methods of large deviations in finance and insurance, in *Paris-Princeton Lectures on Mathematical Finance 2004*. LNM, vol. 1919 (Springer, New York, 2007), pp. 191–244
244. J. Pitman, M. Yor, A decomposition of Bessel bridges. Z. Wahrsch. Verw. Gebiete **59**(4), 425–457 (1982)
245. B.T. Polyak, A new method of stochastic approximation type, (Russian) Avtomat. i Telemekh. **7**, 98–107 (1991); translation in Automation Remote Control, **51**, part 2, 937–946
246. B.T. Polyak, A.B. Juditsky, Acceleration of stochastic approximation by averaging. SIAM J. Control Optim. **30**(4), 838 (1992). https://doi.org/10.1137/0330046
247. W.H. Press, S.A. Teukolsky, W.T. Vetterling, B.P. Flannery, *Numerical Recipes in* C++. *The Art of Scientific Computing*, 2nd edn. updated for C++ (Cambridge University Press, Cambridge, 2002), pp. 1002
248. P.D. Proïnov, Discrepancy and integration of continuous functions. J. Approx. Theory **52**, 121–131 (1988)
249. P.E. Protter, *Stochastic integration and differential equations*, 2nd edn. Version 2.1. Corrected Third Printing. Stochastic Modelling and Applied Probability, vol. 21 (Springer, Berlin, 2004), pp. xiv+419
250. P. Protter, D. Talay, The Euler scheme for Lévy driven stochastic differential equations. Ann. Probab. **25**(1), 393–423 (1997)
251. D. Revuz, M. Yor, *Continuous martingales and Brownian motion*, 3rd edn. Grundlehren der Mathematischen Wissenschaften [Fundamental Principles of Mathematical Sciences], vol. 293 (Springer, Berlin, 1999), pp. 602
252. C. Rhee, P.W. Glynn, Unbiased estimation with square root convergence for SDE models. Oper. Res. **63**(5), 1026–1043 (2012)
253. H. Robbins, S. Monro, A stochastic approximation method. Ann. Math. Stat. **22**, 400–407 (1951)
254. R.T. Rockafellar, S. Uryasev, Optimization of conditional value-at-risk. J. Risk **2**(3), 21–41 (2000)
255. L.C.G. Rogers, Monte Carlo valuation of American options. Math. Financ. **12**(3), 271–286 (2002)
256. L.C.G. Rogers, D. Williams, in *Diffusions, Markov Processes, and Martingales*, vol. 2 (Itô calculus) (2000). Reprint of the second (1994) edition. Cambridge Mathematical Library (Cambridge University Press, Cambridge 2000), pp. xiv+480
257. K.F. Roth, On irregularities of distributions. Mathematika **1**, 73–79 (1954)
258. W. Rudin, *Real and Complex Analysis* (McGraw-Hill, New York, 1966), pp. xiii+416
259. D. Ruppert, Efficient Estimations from a Slowly Convergent Robbins-Monro Process, Cornell University Operations Research and Industrial Engineering, Technical Report 781 Ithaca, New York, 1988, pp. 29

260. A. Sagna, Pricing of barrier options by marginal functional quantization. Monte Carlo Methods Appl. **17**(4), 371–398 (2011)
261. K.-I. Sato, *Lévy processes and Infinitely Divisible Distributions* (Cambridge University Press, Cambridge, 1999)
262. A.N. Shyriaev, Optimal stopping rules (2008). Translated from the 1976 Russian second edition by A.B. Aries. Reprint of the, translation *Stochastic Modelling and Applied Probability*, vol. 8 (Springer, Berlin, 1978), pp. 217
263. A.N. Shiryaev, *Probability*. Graduate Texts in Mathematics, 2nd edn. (Springer, New York, 1995), pp. 621; Original Russian edition, 1980, original Russian second edition, 1989
264. P. Seumen-Tonou, *Méthodes numériques probabilistes pour la résolution d'équations du transport and pour l'évaluation d'options exotiques, Thèse de l'Université de Provence* (France, Marseille, 1997), pp. 116
265. I.M. Sobol', Distribution of points in a cube and approximate evaluation of integrals, Zh. Vych. Mat. Mat. Fiz., **7**, 784–802 (1967) (in Russian); U.S.S.R Comput. Math. Math. Phys. **7**, 86–112 (1967) (in English)
266. I.M. Sobol', Y.L. Levitan, The production of points uniformly distributed in a multidimensional cube. Technical Report 40, Institute of Applied Mathematics, USSR Academy of Sciences (1976) (in Russian)
267. C. Soize *The Fokker–Planck Equation for Stochastic Dynamical Systems and Its Explicit Steady State Solutions*. Series on Advances in Mathematics for Applied Sciences, vol. 17 (World Scientific Publishing Co., Inc., River Edge, 1994), pp. xvi+321
268. M. Sussman, W. Crutchfield, M. Papakipos, Pseudo-random numbers Generation on GPU, in *Graphic Hardware 2006, Proceeding of the Eurographics Symposium Vienna, Austria, September 3–4, 2006*, ed. by M. Olano, P. Slusallek (A K Peters Ltd., 2006), pp. 86–94
269. J.N. Tsitsiklis, B. Van Roy, Regression methods for pricing complex American-style options. IEEE Trans. Neural Netw. **12**(4), 694–703 (2001)
270. D. Talay, L. Tubaro, Expansion of the global error for numerical schemes solving stochastic differential equations. Stoch. Anal. Appl. **8**, 94–120 (1990)
271. B. Tuffin, Randomization of Quasi-Monte Carlo Methods for error estimation: survey and normal approximation. Monte Carlo Methods Appl. **10**(3–4), 617–628 (2004)
272. C. Villani, *Topics in Optimal Transportatio*. Graduate Studies in Mathematics, vol. 58 (American Mathematical Society, Providence, 2003), pp. xvi+370
273. S. Villeneuve, A. Zanette, Parabolic A.D.I. methods for pricing American option on two stocks. Math. Oper. Res. **27**(1), 121–149 (2002)
274. H.F. Walker, P. Ni, Anderson acceleration for fixed-point iterations. SIAM J. Numer. Anal. **49**(4), 1715–1735 (2011)
275. D.S. Watkins, *Fundamentals of Matrix Computations*. Pure and Applied Mathematics (Hoboken), 3rd edn. (Wiley, Hoboken, 2010), pp. xvi+644
276. M. Winiarski, Quasi-Monte Carlo Derivative valuation and Reduction of simulation bias, Master thesis (Royal Institute of Technology (KTH), Stockholm (Sweden), 2006)
277. A. Wood, G. Chan, Simulation of stationary Gaussian processes in $[0, 1]^d$. J. Comput. Gr. Stat. **3**(4), 409–432 (1994)
278. Y.-J. Xiao, *Contributions aux méthodes arithmétiques pour la simulation accélérée* (Thèse de l'ENPC, Paris, 1990), pp. 110
279. Y.-J. Xiao, Suites équiréparties associées aux automorphismes du tore. C.R. Acad. Sci. Paris (Série I) **317**, 579–582 (1990)
280. P.L. Zador, Development and evaluation of procedures for quantizing multivariate distributions. Ph.D. dissertation, Stanford University (1963), pp. 111
281. P.L. Zador, Asymptotic quantization error of continuous signals and the quantization dimension. IEEE Trans. Inf. Theory **28**(2), 139–149 (1982)

Index

G. Pagès, *Numerical Probability*, Universitext,
https://doi.org/10.1007/978-3-319-90276-0

D

E

F

G

H

Printed in the United States
By Bookmasters